Pharmaceutical Physical Chemistry

Theory and Practices

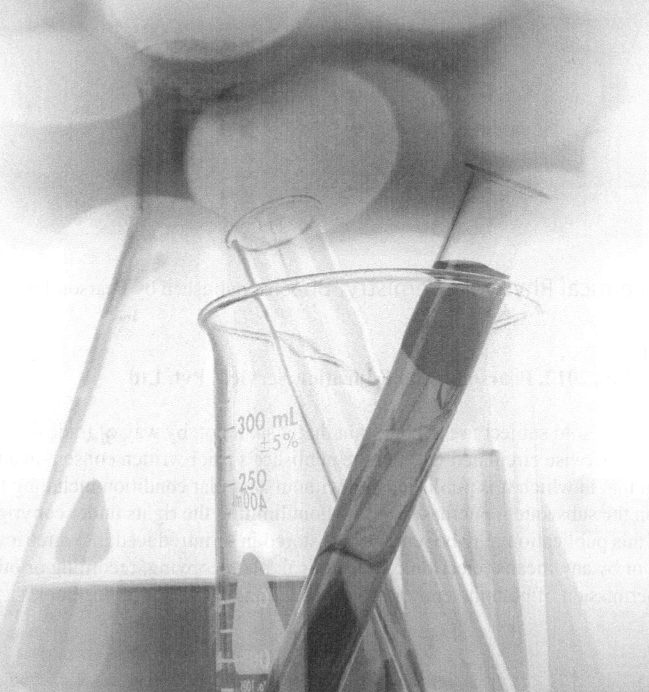

S K Bhasin

Pearson

ISBN 9788131765272
First Impression, 2012
Third Impression, 2024
Fourth Impression, 2025

Published by Pearson India Education Services Pvt. Ltd, CIN: U72200TN2005PTC057128.

Head Office: 1st Floor, Berger Tower, Plot No. C-001A/2, Sector 16B, Noida - 201 301, Uttar Pradesh, India.
Registered Office: Featherlite, 'The Address' 5th Floor, Survey No 203/10B, 200 Ft MMRD Road, Zamin Pallavaram, Chennai – 600 044.
Website: in.pearson.com; Email: companysecretary.india@pearson.com

Printed in India at: Manipal Technologies Limited, Manipal

Dedicated to
All Those Who
Toiled in Shaping Me
into What I Am Today

Contents

Preface xxi

About the Author xxii

1. Behaviour of Gases 1

1.1 Introduction 2

1.2 Gas Laws 2

1.2.1 Boyle's Law 2

1.2.2 Charles Law 3

1.2.3 Avogadro's Law 4

1.2.4 The Combined Gas Law Equation or the Gas Equation 4

1.2.5 Graham's Law of Diffusion 6

1.2.6 Dalton's Law of Partial Pressure 6

1.3 Kinetic Theory of Gases 6

1.3.1 Postulates (Assumptions) of Kinetic Theory 6

1.4 Derivation of Kinetic Gas Equation 7

1.5 Derivation of Gas Laws from Kinetic Equation 9

1.5.1 Some Useful Deductions from Kinetic Theory of Gases 12

1.6 Ideal and Real Gases 17

1.6.1 Ideal Gases 17

1.6.2 Real Gas 17

1.7 Deviations of Real Gases from Gas Laws 18

1.7.1 Deviations from Boyle's Law 18

1.8 Causes of the Derivations from Ideal Behaviour 20

1.9 van der Waals' Equation (Reduced Equation of State) (Equation of State for Real Gases) 20

1.9.1 Units of van der Waals' Constants 23

1.9.2 Significance of van der Waals' Constant 24

1.10 Explanation of Behaviour of Real Gases on the Basis of van der Waals' Equation 24

1.11 Isotherms of Carbon Dioxide—Critical Phenomenon 28

1.12 Principle of Continuity of States 30

1.13 Critical Constants 31

1.13.1 Relations Between van der Waals' Constants and Critical Constants 31

1.13.2 Derivation of $P_cV_c = \frac{3}{8} RT_c$ from van der Waals' Equation 33

1.13.3 Calculation of van der Waals' Constants in terms of T_c and P_c 34
1.14 Law of Corresponding States 34
1.14.1 Significance of Law of Corresponding States 35
1.15 Limitations of van der Waals' Equation 36
Revision Questions 36
Multiple Choice Questions 39
Answers 42

2. The Liquid State **43**

2.1 Introduction 44
2.2 General Characteristics of Liquids 44
2.3 Classification of Physical Properties of Liquids 46
2.4 Surface Tension 46
2.4.1 Some Important Results 47
2.4.2 Effect of Temperature on Surface Tension 48
2.4.3 Measurement of Surface Tension 48
2.4.4 Surface Tension in Everyday Life 51
2.4.5 Surface Tension and Chemical Constitution (Parachor) 52
2.5 Viscosity 57
2.5.1 Coefficient of Viscosity 58
2.5.2 Measurement of Viscosity 59
2.5.3 Effect of Temperature on Viscosity 60
2.5.4 Factors Affecting Viscosity 61
2.5.5 Viscosity and Chemical Constitution 62
2.6 Refractive Index 64
2.6.1 Measurement of Refractive Index 65
2.6.2 Refractive Index and Chemical Constitution 66
2.7 Optical Activity 69
2.7.1 Optical Activity 70
2.7.2 Specific Rotation 70
2.7.3 Optical Activity and Chemical Constitution 71
2.8 Polarity of Bonds 74
2.8.1 Polar Character of Covalent Bond 75
2.9 Dipole Moment 75
2.9.1 Unit of Dipole Moment 75
2.9.2 Dipole Moment and Molecular Structure 76
2.9.3 Application of Dipole Moments 76
Revision Questions 80
Multiple Choice Questions 81
Answers 83

3. Solution **85**

3.1 Introduction 86
3.2 Modes of Expressing Concentration of Solutions 87
3.3 Raoult's Law 89
3.3.1 For a Solution of Volatile Liquids 89
3.3.2 For a Solution of Non-volatile Solute 90
3.4 Ideal Solution 90
3.4.1 Non-ideal Solution 91
3.4.2 Solutions Showing Positive Deviations 91
3.4.3 Solutions Showing Negative Deviations 91
3.4.4 Factors Responsible for Deviations 92
3.4.5 Distinction Between Ideal and Non-ideal Solutions 93
3.4.6 Difference Between Solutions of Positive and Negative Deviations 93
3.5 Colligative Properties of Dilute Solution 95
3.6 Lowering of Vapour Pressure 95
3.6.1 Determination of Molecular Masses of Non-volatile Solute 96
3.7 Elevation in Boiling Point 99
3.7.1 Expression for the Elevation in Boiling Point 99
3.7.2 Calculation of Molecular Masses 101
3.8 Depression of Freezing Point 102
3.8.1 Expression for the Depression in Freezing Point 103
3.8.2 Calculation of Molecular Masses 104
3.9 Osmotic Pressure 105
3.9.1 Difference Between Osmosis and Diffusion 105
3.9.2 Osmotic Pressure 105
3.9.3 Determination of Osmotic Pressure Berkley and Hertley's Method 106
3.9.4 Osmotic Pressure is a Colligative Property 107
3.9.5 Isotonic Solutions 107
3.9.6 Calculation of Molecular Masses from Osmotic Pressure 108
3.10 Abnormal Molecular Masses 112
3.10.1 Modified Equation for Colligative Properties in Case of Abnormal Molecular Masses 114

Revision Question 115
Multiple Choice Questions 117
Answers 118

4. Thermodynamics **119**

4.1 Introduction 119
4.1.1 Objective of Thermodynamics 120
4.1.2 Limitation of Thermodynamics 120

4.2 Some Common Thermodynamics Terms 121
4.2.1 Thermodynamic Equilibrium 122
4.2.2 Thermodynamic Processes 122
4.2.3 Reversible and Irreversible Processes 123
4.2.4 Thermodynamic Properties 124
4.3 Zeroth Law of Thermodynamics 126
4.3.1 Absolute Scale of Temperature 126
4.4 Work, Heat and Energy Changes 127
4.4.1 Work 127
4.4.2 Heat 129
4.4.3 Equivalence Between Mechanical Work and Heat 130
4.4.4 Internal Energy 130
4.5 First Law of Thermodynamics 131
4.5.1 Mathematical Formulation of First Law of Thermodynamics 132
4.5.2 Some Special Forms of First Law of Thermodynamics 132
4.5.3 Limitations of the First Law of Thermodynamics 133
4.6 The Heat Content or Enthalpy of a System 135
4.7 Heat Capacities at Constant Pressure and at Constant Volume 136
4.7.1 Heat Capacity at Constant Volume 137
4.7.2 Heat Capacity at Constant Pressure 137
4.7.3 Relationship Between C_p and C_v 138
4.8 Joule-Thomson Effect 138
4.9 Reversible-Isothermal Expansion of an Ideal Gas 140
4.9.1 Maximum Work 141
4.10 Second Law of Thermodynamics 142
4.10.1 Spontaneous Processes and Reactions (Basis of Second Law) 143
4.10.2 Spontaneous Reactions 143
4.11 Entropy 145
4.11.1 Mathematical Explanation of Entropy 145
4.11.2 Entropy Change in Chemical Reaction 147
4.11.3 Units of Entropy 147
4.11.4 Physical Significance of Entropy 147
4.11.5 Entropy Change Accompanying Change of Phase 147
4.11.6 Entropy Changes in Reversible Processes 148
4.11.7 Entropy Changes in Irreversible Processes 149
4.11.8 Entropy as Criterion of Spontaneity 150
4.11.9 Entropy Changes for an Ideal Gas 150
Revision Questions 154
Multiple Choice Questions 156
Answers 157

5. Adsorption and Catalysis 159

5.1 Adsorption 160
5.2 Types of Adsorption 161
5.3 Factors Affecting Adsorption of Gases on Solids 162
5.4 Adsorption Isobar (Effect of Temperature on Adsorption) 163
5.5 Adsorption Isotherm (Effect of Pressure) 163
5.5.1 Explanation of Type I Isotherm 164
5.5.2 Freundlich Adsorption Isotherm 164
5.5.3 The Langmuir Adsorption Isotherm 165
5.5.4 Verification 166
5.5.5 Explanation of Type II and III Isotherms 167
5.5.6 Explanation of Type IV and V Isotherms 167
5.6 Theory of Adsorption 168
5.7 Gibbs' Adsorption Equation 169
5.8 Applications of Gibbs' Adsorption Equation 173
5.9 Equation for Multi-Layer Adsorption (B.E.T. Equation) 176
5.9.1 Determination of Surface Area of the Adsorbent 178
5.10 Catalysis 179
5.10.1 Positive and Negative Catalyses 179
5.11 Homogeneous and Heterogeneous Catalyses 179
5.12 How Does a Catalyst Work? 180
5.12.1 Characteristics of Catalytic Reactions 181
5.12.2 Acid–Base Catalysis 182
5.12.3 Enzyme Catalysis 183
5.13 Mechanism of Homogeneous and Heterogeneous Catalyses 184
5.13.1 Significant Characteristics of Heterogeneous Catalysis 186
5.13.2 Facts Explained by Adsorption Theory 187
Revision Questions 188
Multiple Choice Questions 190
Answers 192

6. Photochemistry 193

6.1 Introduction 193
6.2 Thermochemical and Photochemical Reactions 194
6.3 Laws Governing Light Absorption — Lambert's Law and Beer's Law 195
6.3.1 Limitations of Lambert–Beer's Law 198
6.4 Laws of Photochemistry 201
6.4.1 Grotthus–Drapper Principle of Photochemical Activation: (First Law of Photochemistry) 201
6.4.2 Stark–Einstein's Law of Photochemical Equivalence—The Second Law of Photochemistry 201

6.5 Quantum Efficiency 204
6.5.1 Explanation of the Unexpected Behaviour 204
6.5.2 Classification of Photochemical Reactions (Based on their Quantum Efficiencies) 205
6.6 Study of Some Photochemical Reactions 208
6.7 Fluorescence and Phosphorescence 211
6.7.1 Fluorescence 211
6.7.2 Phosphorescence 212
6.7.3 Photophysical Process—Consequence of Light Absorption (Jablonski Diagram) 213
6.7.4 Mechanism of Fluorescence and Phosphorescence 215
6.7.5 Difference between Fluorescence and Phosphorescence 215
Revision Questions 216
Multiple Choice Questions 217
Answers 219

7. Chemical Kinetics 221

7.1 Introduction 222
7.2 Rate of a Reaction 222
7.2.1 Measurement of Rate of a Reaction 223
7.2.2 Expressing the Rate of a Reaction 224
7.2.3 Factors Influencing Rate of a Reaction 225
7.3 Rate Constant and Rate Equation 225
7.3.1 Differences Between Rate of a Reaction and Rate Constant 226
7.4 Order of a Reaction 227
7.4.1 Units for Rate Constant or Specific Reaction Rate 228
7.5 Molecularity of a Reaction 229
7.5.1 Differences Between Order and Molecularity of a Reaction 230
7.6 Zero-order Reactions 231
7.6.1 Characteristics of a Zero-order Reaction 232
7.7 Intergrated Rate Law Equation for First-order Reactions 234
7.7.1 Characteristics or Significance of First-order Reaction 235
7.7.2 Examples of the Reactions of First Order 236
7.7.3 Pseudo First-order Reaction 239
7.8 Second-order Reactions 242
7.8.1 Characteristics of Second-order Reaction 243
7.8.2 Example of the Second-order Reaction 245
7.9 Third-order Reaction 247
7.9.1 Characteristics of a Third-order Reaction 248
7.9.2 Example of Third-order Reactions 250

7.10 Reactions of Higher Order 251

7.11 Determination of Rate Law, Rate Constant and Order of Reaction 251

7.12 Some Complications in Determination of Order of a Reaction 256

7.12.1 Consecutive Reactions 256

7.13 Temperature Dependence of Reaction Rates 258

7.13.1 Explanation of Effect of Temperature 258

7.13.2 Arrhenius Equation 259

7.14 Mechanism of a Reaction (Concept of Molecularity and Order of a Reaction) 261

7.15 Theories of Reaction Rates 264

Revision Questions 269

Multiple Choice Questions 270

Answers 272

8. Quantum Mechanics 273

8.1 Introduction 274

8.2 Classical Mechanics and Its Limitations 274

8.2.1 Limitations 274

8.3 Origin of Quantum Mechanics 275

8.3.1 Classical Mechanics versus Quantum (or wave) Mechanics 275

8.4 Black Body Radiations 276

8.5 Kirchoff's Law 277

8.5.1 Spectral Distribution of Black Body Radiation 278

8.6 Stefan-Boltzmann Fourth Power Law 279

8.7 Wien's Displacement Law 279

8.8 Planck's Radiation Law 281

8.9 Postulates of Quantum Mechanics 283

8.10 Operators in Quantum Mechanics 285

8.10.1 Types of Operators 286

8.11 Schrödinger Wave Equation 288

8.11.1 Derivation of Schrödinger Wave Equation 288

8.12 Eigenvalues and Eigenfunctions (or Wave Functions) 290

8.12.1 Physical Significance of the Wave Function 290

8.13 Normalized and Orthogonal Eigenfunctions 290

8.14 Concept of Atomic Orbital 291

8.15 Probability Distribution Curves 292

8.16 Radial Probability Distribution Curves 292

8.16.1 Radial Probability Distribution Curve for 1s Orbital 293

8.16.2 Radial Probability Distribution Curves for other s Orbitals 293

8.16.3 Comparison of Radial Probability, Distribution Curves for 1s with Other s Atomic Orbitals 294

8.16.4 Radial Probability Distribution Curves for p Orbitals 295

8.16.5 Comparison of Radial Probability Distribution Curves for 2s and 2p Orbitals 295

8.16.6 Comparison of Radial Probability Distribution Curves for 3s, 3p and 3d Orbitals 295

Revision Questions 296

Multiple Choice Questions 298

Answers 299

9. Ionic Equilibria **301**

9.1 Introduction 301

9.2 Arrhenius Theory of Ionization 302

9.2.1 Degree of Dissociation or Ionization 303

9.3 Ionisation of Weak Electrolytes—Ostwald's Dilution Law 304

9.3.1 Verification of Ostwald's Dilution Law 305

9.4 Arrhenius Concept of Acids and Bases 306

9.4.1 Limitation of Arrhenius Theory 306

9.5 Ionisation Constant of Weak Acids and Bases (Arrhenius Concept) 307

9.6 Bronsted–Lowry Concept of Acids and Bases 308

9.6.1 Conjugate Acid Base Pairs 309

9.6.2 Relative Strength of Acids and Bases 310

9.6.3 Limitation of Bronsted—Lowry Theory 311

9.7 Lewis Concept of Acids and Bases 311

9.7.1 Limitations of Lewis Concept 312

9.8 Ionic Product of Water 313

9.8.1 Concentrations of H_3O^+ and OH^- ions in Aqueous Solutions of Acids and Bases 314

9.9 pH Scale 315

9.9.1 The pOH Scale 316

9.10 Buffer Solution 321

9.10.1 Buffer Action of Buffer Solutions 322

9.10.2 Applications of Buffer Solutions 323

Revision Questions 324

Multiple Choice Questions 325

Answers 326

10. Distribution Law **327**

10.1 Introduction 327

10.2 Conditions for the Validity of the Distribution Law 328

10.3 Effect of Temperature on Distribution Coefficient 329

10.4 Thermodynamic Derivation of Distribution Law 329

10.4.1 Principle 329

10.5 Distribution Law and Molecular State of Solute 331

10.5.1 Case I: When the Solute Undergoes Association in One of the Solvents 331

10.5.2 Case II: When the Solute Undergoes Dissociation in One of the Solvents 334

10.5.3 Case III: When the Solute Enters into Chemical Combination with One of the Solvents 335

10.6 Applications of Distribution Law 336

10.6.1 Determination of Solubility of a Solute in a Solvent 337

10.6.2 Determination of Molecular State of Solute in Different Solvents 337

10.6.3 Determination of Distribution Indicators 337

10.6.4 Study of Complex Ions 338

10.6.5 In the Process of Extraction 339

10.6.6 Application of Principle of Extraction To Desilverization of Lead 342

10.6.7 Determination of Degree of Hydrolysis 343

Revision Questions 344

Multiple Choice Questions 346

Answers 347

11. Electrochemistry 349

11.1 Introduction 349

11.2 Electrolysis 350

11.2.1 Faraday's First Law of Electrolysis 351

11.2.2 Faraday's Second Law of Electrolysis 351

11.2.3 Application of Electrolysis 352

11.3 Electrolytic Conduction 354

11.3.1 Differences Between Metallic Conductor and Electrolytic Conductor 355

11.3.2 Factors Affecting Electrolytic Conduction 355

11.3.3 Electrical Conductance 356

11.3.4 Specific Conductance 357

11.3.5 Equivalent Conductance and Molecular Conductance 358

11.3.6 Relation Between Specific Conductance and Equivalent Conductance 358

11.3.7 Experimental Measurement of Conductance 359

11.3.8 Effect of Dilution on Conductance 363

11.4 Kohlrausch Law 364

11.4.1 Applications of Kohlrausch's Law 365

11.5 Migration of Ions 367

11.5.1 Migration Velocity of Ions and Change in Concentration—Hittorf Theoretical Device 367

11.6 Transport Number 369

11.6.1 Important Relations Concerning Transport Number 369

11.6.2 Factors Controlling Transport Number 370

11.6.3 Determination of Transport Numbers 371

11.7 Limitations of Arrhenius Theory 375

11.8 Modern Theory of Strong Electrolytes 376

Revision Questions 379

Multiple Choice Questions 380

Answers 382

12. Electromotive Force and Oxidation–Reduction System 383

12.1 Introduction 383

12.2 Single Electrode Potential 384

12.2.1 Definition 385

12.3 Standard Electrode Potential 387

12.4 Measurement of Single Electrode Potential 388

12.4.1 Sign Conventions 388

12.5 Reference Electrodes 389

12.5.1 Primary Reference Electrodes 389

12.5.2 Secondary Reference Electrodes 390

12.5.3 Advantages of Glass Electrode 392

12.5.4 Limitations of Glass Electrode 393

12.6 Electrochemical Series 393

12.6.1 Applications of Electrochemical Series 394

12.7 Cell Potential or EMF of a Cell 396

12.7.1 Calculation of EMF of a Cell 396

12.8 Derivation of Nernst Equation (Concentration Dependence of Electrode Potential) 399

12.8.1 Application of Nernst Equation 400

Revision Questions 406

Multiple Choice Questions 407

Answers 410

13. Solid State (Crystalline State) 411

13.1 Introduction 411

13.2 Crystalline and Amorphous Solids 412

13.3 Some Terms Used in Crystal Structure 415

13.4 Crystal Lattice and Unit Cell 416

13.5 Elements of Symmetry 419

13.5.1 Plane of Symmetry and Reflections 419
13.5.2 Axis of Symmetry or Axis of Rotation 419
13.5.3 Centre of Symmetry or Inversion Centre 421
13.5.4 Improper Axis or Rotation Reflector Axis and Improper Rotation 421
13.5.5 Axis of Rotation Inversion 421
13.5.6 Total Elements of Symmetry 421
13.6 Crystallographic Designations 423
13.6.1 Weiss Indices (Parameter System of Weiss) 423
13.6.2 Index System of Miller (Miller Indices) 423
13.7 Laws of Crystallography 424
13.7.1 The Law of Constancy of Interfacial Angles 424
13.7.2 The Law of Rationality of Indices 424
13.7.3 The Law of Symmetry 425
13.8 Crystal Systems 427
13.9 Types of Unit Cells in Crystal System (Bravais Lattice) 428
13.10 Bragg's Method of Crystal Analysis 430
13.10.1 Principle 430
13.10.2 Derivation of Bragg's Equation 430
13.10.3 Bragg's Method for Determining Crystal Structure 431
13.10.4 Applications of Bragg's Equation 432
13.11 Types of Crystalline Solids 433
13.11.1 Ionic Solids 433
13.11.2 Metallic Solids 434
13.11.3 Covalent Solids 435
13.11.4 Molecular Solids 435
Revision Questions 437
Multiple Choice Questions 438
Answers 439

14. Chemical Bonding 441

14.1 Introduction 441
14.2 Valence Bond (VB) Theory 442
14.3 Molecular Orbital (MO) Theory 443
14.3.1 Molecular Orbitals 443
14.3.2 Conditions for Atomic Orbitals to Form Molecular Orbitals 444
14.3.3 Difference Between Atomic and Molecular Orbitals 445
14.3.4 Formation of Bonding and Anti-bonding Molecular Orbitals (LCAO Method) 445
14.3.5 Bonding and Anti-bonding Molecular Orbitals in Terms of Wave Functions 447

14.3.6 Characteristics of Bonding and Anti-bonding Molecular Orbitals 448
14.3.7 Combination of Atomic Orbitals—Sigma (σ) and Pi (π) Molecular Orbitals 448
14.4 Relative Energies of Molecular Orbitals and Filling of Electrons 451
14.5 Stability of Molecules 453
14.5.1 Stability of Molecules in Terms of Bonding and Anti-bonding Electrons 453
14.5.2 Stability of Molecules in Terms of Bond Order 453
14.6 Molecular Orbital Configurations 454
14.6.1 Bonding in Some Homonuclear Diatomic Molecules and Ions – Electronic Configurations 454
14.6.2 Helium Ion, He_2^+ 457
14.6.3 Nitrogen Molecule, N_2 459
14.6.4 Oxygen Molecule, O_2 460
14.6.5 The Fluorine Molecule, F_2 462
14.6.6 Hypothetical Neon Molecule, Ne_2 462
14.6.7 Molecular Orbital Electronic Configuration of Some Common Heteronuclear Molecules 462
14.7 Comparison of Valence Bond (VB) Theory and Molecular Orbital (MO) Theory 463
14.7.1 Points of Similarly 463
14.7.2 Points of Difference 463
Revision Questions 465
Multiple Choice Questions 466
Answers 467

15. Phase Equilibria 469

15.1 Introduction 469
15.2 Explanation of the Terms Involved 470
15.2.1 True and Metastable Equilibrium 470
15.2.2 Phase 470
15.2.3 Components 471
15.2.4 Degrees of Freedom or Variance 474
15.3 Mathematical Statement of Phase Rule 474
15.4 Phase Diagrams 476
15.5 One-component Systems 477
15.6 The Water System 478
15.7 Sulphur System 481
15.8 Application of Phase Rule To Two-component Systems (Liquid–Solid Phase Diagram) 484
15.9 Type A—Simple Eutectic System 485

15.9.1 Characteristics of Eutectic Point 485
15.9.2 Use of Eutectic Systems 486
15.9.3 Lead–Silver System 486
15.9.4 Pattinson's Process for Desilverization of Lead 488
15.10 Type B—System in Which Two Components form a Stable Compound (Zinc–Magnesium Alloy System) 488
15.10.1 Eutectic Points and Congruent Melting Point 490
15.11 Type C—The Two-component Form: A Compound With Incongruent Melting Point 490
15.11.1 Sodium – Potassium System 490
15.12 Thermal Analysis (Cooling Curve) 492
Revision Questions 493
Multiple Choice Questions 495
Answers 496

Experiments

Refractometry 501
Refractive Index 501, Measurement of Refractive Index 502

Experiment 1 **503**
Object 503, Apparatus 503, Theory 503, Procedure 503, Observations 503, Result 503, Precautions 504, Viva-voce 504

Polarimetry 506
Optical Activity 506, Specific Rotation 507

Experiment 2 **508**
Object 508, Apparatus 508, Theory 508, Procedure 508, Observations and Results 509, Viva-voce 509

Experiment 3 **512**
Object 512, Apparatus 512, Theory 512, Procedure 512, Calculation 513, Result 513, Viva-voce 513

Experiment 4 **514**
Object 514, Theory 514, Apparatus/Reagents Required 515, Procedure 515, Observations 515, Calculations 516, Result 516, Precautions 516, Graphical Method 516, Viva-voce 517

Experiment 5 **520**
Object 520, Theory 520, Cell Constant 520, Measurement of Cell Constant 520, Apparatus Required 521, Materials Required 521, Procedure 521, Determination and Calculations 522, Precautions 522, Viva-voce 523

Experiment 6 **524**

Object 524, Apparatus and Reagents 524, Theory 524, Procedure 525, Observation 525, Calculation 526, Result 526, Precautions 526, Viva-voce 526

Experiment 7 **528**

Object 528, Theory 528, Requirements 529, Procedure 529, Important Note 529, Calibration of the Instrument 529, Titration of HCl vs. NaOH Solution 530, Observation and Calculations 530, Result 531, Note 531, Precautions 531, Viva-voce 531

Experiment 8 **535**

Object 535, Theory 535, Materials 535, Description of the Apparatus 535, Procedure 536, Observations and Calculations 536, Calculations 537, Result 537, Precautions 537, Viva-voce 538

Experiment 9 **540**

Object 540, Theory 540, Apparatus/Reagents Required 540, Procedure 540, Observations 541, Calculations 541, Calculate Values of K for Each Set 542, Results 542, Precautions 542, Viva-voce 542

Experiment 10 **543**

Object 543, Theory 543, Apparatus Reagents Required 543, Method 543, Observations 544, Burette Readings 544, Calculations 544, Result 545, Precautions 545, Viva-voce 545

Experiment 11 **548**

Object 548, Apparatus 548, Description 548, Theory 549, Procedure 549, Observations 549, Calculations 550, Result 550, Precautions 550, Viva-voce 550

Experiment 12 **552**

Object 552, Apparatus 552, Theory 552, Indicators 553, Procedure 554, Observations 554, Viva-voce 554

Experiment 13 **555**

Object 555, Apparatus 555, Theory 555, Procedure and Observations 555, Result 556, Viva-voce 556

Experiment 14 **558**

Object 558, Apparatus 558, Theory 558, Procedure 558, Calculations 559, Result 559, Precautions 559, Viva-voce 560

Experiment 15 **562**

Object 562, Theory 562, Apparatus and Materials 562, Procedure 562, Observations 563, Calculations 564, Precautions 564, Viva-voce 564

Experiment 16 **566**

Object 566, Theory 566, Apparatus and Materials Required 566, Procedure 567, Observations 568, Calculations 568, Result 569, Precautions 569, Viva-voce 570

Experiment 17 **573**

Object 573, Theory 573, Materials Required 573, Procedure 574, Precautions 575, Viva-voce 575

Experiment 18 **577**

Object 577, Theory 577, Materials Required 577, Procedure 578, Observations 578, Result 579, Viva-voce 579

Experiment 19 **580**

Object 580, Theory 580, Apparatus and other Materials Required 580, Procedure 581, Observations and Calculations 581, Viva-voce 582

Experiment 20 **584**

Object 584, Theory 584, Apparatus and Other Materials Required 584, Procedure 585, Observation and Calculations 585, Viva-voce 586

Index *587*

Preface

It gives me immense pleasure to place before a large community of pharmacy students, my humble work on *Pharmaceutical Physical Chemistry*, written in accordance with the recent syllabus prescribed for the B.Pharma. (II Semester) students of all Indian universities. My aim in writing this book is to present the fundamental principles of physical chemistry for the pharmacy students on modern lines. No single book on pharmacy that covers the revised syllabus exclusively is available in the market. Keeping in view the requirement of the students and the teachers, this book has been written to cover all the topics with the desired limits of the prescribed syllabus. I hope the book will be useful and meets the requirements of students at large.

Based on my vast teaching experience, I have prepared the text in a simple, lucid and comprehensive style, keeping in view the difficulties of the students; in addition, the manuscript has been presented as if the teacher is talking to the students in the class. Throughout the text, special care has been taken to add 'Review Problems for Tutorials' at relevant stages for the students to assess their grasp of the topic covered. Students are advised to solve the problems themselves and look for their solution only afterwards.

An added feature of this book is that it contains the 'Laboratory Manual', which contains 20 experiments covering the syllabus in the practical of all Indian universities along with the 'Viva-Voce' at the end of each experiment. In this section, we have described the theory of the experiment in details before giving the procedural details.

A large number of solved and unsolved 'Numerical Problems' have also been included wherever required. Students should solve the unsolved numerical problems in the tutorial classes under the guidance of their learned teachers.

Here is a book that resolves all your queries and doubts. This book discusses the fundamentals of pharmaceutical physical chemistry required for a pharmacist. The book will not only help you to score better in the examination but would also develop your skills to apply your knowledge of pharmaceutical physical chemistry in solving the problems that you may face during your pharmaceutical career.

I sincerely express my thanks to the authors and the publishers whose works I have consulted in producing this book. I am grateful to the editors of Pearson Education for the sustained interest shown by them during the publication of this book.

Although care has been taken while preparing and typing the manuscript, yet errors and misprints might have crept in; I shall be grateful to the students and teachers who would be kind enough to send their suggestions for the further improvement of the book.

I shall consider my efforts amply rewarded if all those for whom the book is intended are benefited by it.

Dr. S. K. Bhasin

About the Author

S.K. Bhasin is presently Director and Professor in Chemistry in the Global Research Institute of Management and Technology, Radaur. A matured academician, experienced teacher, established author and devoted researcher, Dr Bhasin has a teaching experience of over 50 years; 35 years of which relates to teaching of undergraduate and postgraduate students in chemistry and 15 years of teaching in professional institutes.

He is a 'life fellow' of four professional bodies, namely, the International Congress of Chemistry and Environment (FICCE), Fellow of Indian Council of Chemists (FICC), Fellow of Indian Association of Environmental Management (FIAEM) and Fellow of *Indian Journal of Environmental Protection* (FIJEP). He is a member on the Editorial Board of *International Journal of Environmental Research.* He has published 64 research papers in reputed international and national journals and has published 14 articles in national-level magazines and newspapers. He has also convened a number of national-level conferences and workshops and has acted as resource person in many conferences.

He has presented his research papers in international conferences abroad—two research papers in Kuwait in 2007 and one paper in Thailand in 2009. Recently, he has presented his research paper in an international conference held at Malaysia, May 2011 organized by the Indian Congress of Chemistry and Environment (ICCE) and yet another international conference organized by the Indian Council of Chemists (ICC), June 2011, at Bangkok (Thailand).

He has acted as Chairman of ICC chapter at M.L.N. College Yamuna Nagar for consecutive three years. He has been honoured with appreciation awards twice at international conferences organized by the ICCE in the year 2009 and in Malaysia in 2011. Has also been decorated with Dronacharya Award in recognition of his services in the field of education and research. He has over 24 graduate-level books in chemistry, engineering and pharmacy to his credit.

Behaviour of Gases

1

CHAPTER OBJECTIVES

1.1 Introduction
1.2 Gas Laws
1.3 Kinetic Theory of Gases
1.4 Derivation of Kinetic Gas Equation
1.5 Derivation of Gas Laws from Kinetic Equation
1.6 Ideal and Real Gases
1.7 Deviations of Real Gases from Gas Laws
1.8 Causes of the Derivations from Ideal Behaviour
1.9 van der Waals' Equation (Reduced Equation of State) (Equation of State for Real Gases)
1.10 Explanation of Behaviour of Real Gases on the Basis of van der Waals' Equation
1.11 Isotherms of Carbon Dioxide-Critical Phenomenon
1.12 Principle of Continuity of States
1.13 Critical Constants
1.14 Law of Corresponding States
1.15 Limitations of van der Waals' Equation

A vacuum is nothing and what is nothing cannot exist

Thomas Hobbes

As the pressure increased, the volume of the gas decreased.

Robert Boyle

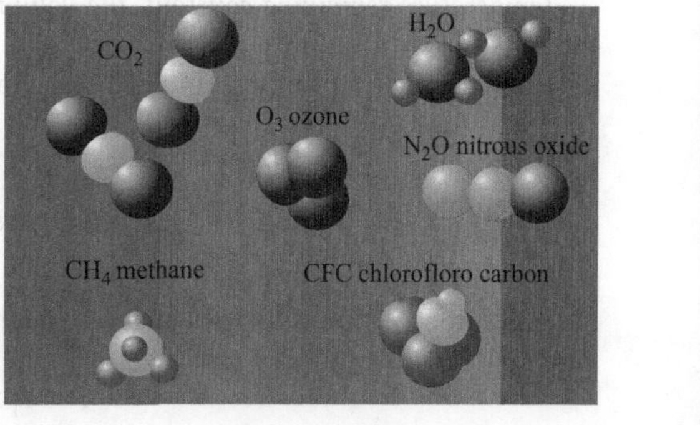

1.1 INTRODUCTION

All matter exists in one of the three states of aggregation: solids, liquids or gases. A solid has definite shape and volume due to strong intermolecular attractive forces. A substance will be a solid if it melts at a temperature higher than room temperature under atmospheric pressure.

In the liquid state, the attractive forces are relatively weak and it has a definite volume but no definite shape. A substance will be a liquid if its freezing point (i.e., melting point) is below the room temperature under atmospheric pressure.

In the gaseous state, the molecular forces are much weaker. Hence, due to possible random motion in all directions, gases have no bounding surface and lend to fill completely any available space. The gases thus have neither definite shape nor definite volume. A substance will be a gas if its boiling point is below room temperature under atmospheric pressure.

Out of the three states of matter, i.e., solid, liquid and gases in which the different substances exist, the gases show the most uniform behaviour irrespective of the nature of the gas. Some common properties of gases are given below:

(i) They *expand indefinitely* and fill up the whole vessel in which they are placed.
(ii) They *can be compressed* by the application of pressure.
(iii) They *can be liquefied* by cooling and applying pressure.
(iv) They *possess low densities* under ordinary conditions of temperature and pressure.
(v) They *intermix spontaneously*, i.e., they show the property of *diffusion.*
(vi) They *exert pressure on the walls of the vessel in which they are contained.*

In addition to the above general properties of gases, another important feature of all gases is that they obey certain gas laws such as Boyle's law and Charles' law which are briefly outlines below. *A gas which obeys the gas laws under all conditions of temperature and pressure is known as an ideal gas.* However, gases deviate from behaviour especially at low temperature and high pressure. The concept of ideal gases is only hypothetical. Gases obey gas laws at high temperature and low pressure usually and are called real gases. In this chapter, we will discuss the deal gas behaviour in terms of kinetic theory of gases and also the deviation of real gases from ideal behaviour.

1.2 GAS LAWS

1.2.1 Boyle's Law

This law was proposed by **Robert Boyle** in 1662 based on his experimental study of the variation of volume of a gas with change of pressure at constant temperature. The law states as follow:

Temperature remaining constant, the volume of a given mass of a gas is inversely proportional to its pressure.

Mathematically, if V is the volume of a gas at pressure P, then

$$V \propto \frac{1}{P}, \text{ if temperature is kept constant.}$$

or

$$PV = \text{constant, if temperature is kept constant.} \tag{1.1}$$

In other words, if at constant temperature, V_1 is the volume of a gas at pressure P_1 and on changing the pressure to P_2, the volume changes to V_2, then

$$P_1 \times V_1 = P_2 \times V_2$$

1.2.2 Charles Law

Charles in 1787 observed that at constant pressure, the volume of every gas increased by the same amount when heated from 0 °C to a particular temperature. Gay-Lussac in 1802 found the exact increase in volume for every degree rise of temperature. Based on their study, they put forward the following generalization.

Pressure remaining constant, the volume of a given mass of a gas increases or decreases by 1/273 of its volume at 0 °C for one degree rise or fall of temperature.

Thus, if V_0 is the volume of a gas at 0 °C and V_t is the volume at t °C, then mathematically, the law may be proposed as

$$V_1 = V_0 + \frac{V_0}{273} \times t = V_0\left(1 + \frac{t}{273}\right) = V_0\left(\frac{273 + t}{273}\right) \qquad (1.2)$$

Thus, at $t = -273\ °C$

$$V_t = V_0\left(\frac{273 - 273}{273}\right) = 0$$

This means that at –273 °C, the volume of the gas is reduced to zero i.e., the gas ceases to exist; in fact, the gas gets liquefied. This temperature (–273 °C) at which the gas ceases to exist is called **absolute zero.** It is represented by 0 °A or 0 K (K stands for kelvin. Kelvin is the SI unit of temperature. According to SI unit, it should be written without putting the symbol of degree i.e., we should write 0 K; 273 K etc.). Thus,

$$-273\ °C = 0\ K \quad \text{or} \quad 0\ °C = 273\ K$$

This means that to convert t °C into K, we should add 273

$$T°\ C = (t + 273)\ K = T\ K$$

Substituting 273 + t °C = T K in Eq. (1.1), we get

$$V_t = V_0 \frac{t}{273}$$

This implies that $V_t \propto T$ if pressure remains constant or simply $V \propto T$ if pressure constant pressure.

Thus, the Charles Gay-Lussac's law may also be defined as follows:

Pressure remaining constant, the volume of a given mass of a gas is directly proportional to its absolute temperature

Mathematically, the law may also written as: $V \propto T$. (at constant pressure)

$$\frac{V}{T} = \text{constant, if pressure is kept constant}$$

or

$$\frac{V_1}{T_1} = \frac{V_2}{T_2} \quad \text{under constant pressure} \qquad (1.3)$$

where V_1, is the volume of the gas at pressure P_1 and V_2 is the volume at pressure P_2.

Absolute scale/Kelvin scale of temperature: Taking –273 °C as 0 K, the scale of temperature obtained is called absolute scale or Kelvin scale. On this scale, the temperature in degrees kelvin is obtained by adding 273 to the temperature in degrees centigrade.

1.2.3 Avogadro's Law

Avogadro, in 1811, put forward the following hypothesis.

Under similar conditions of temperature and pressure, equal volumes of all gases contain the same number of molecules.

In other words, if temperature and pressure are kept constant, the volume (V) of a gas is directly proportional to the number of molecules (N) present in the gas. Mathematically,

$$V \propto N, \qquad \text{if } T \text{ and } P \text{ are kept constant} \tag{1.4}$$

However, the number of molecules present in 1 mole of the substance (solid, liquid or gas) has been determined experimentally. It has been found that one mole of every gas (i.e., 22.4 litre at N.T.P.) contains 6.023×10^{23} molecules. This number is known as **Avogadro's number** and is usually represented by N_0.

Thus, what was originally known as hypothesis is now known as Avogadro's law. Obviously, the number of moles (n) of a gas may be calculated as follows:

$$n = \frac{N}{N_0}$$

As N_0 is a constant quantity, we can say that

$$n \propto N$$

Hence, the equation $V \propto N$ may also be written as

$$V \propto n$$

where n is the number of moles of the gas.

1.2.4 The Combined Gas Law Equation or the Gas Equation

The above three gas laws may be combined to give a general equation, called the gas equation. This may be done as follows:

According to *Boyle's law*

$$V \propto \frac{1}{P} \quad \text{if } T \text{ is kept constant} \tag{1.5}$$

According to *Charles law*

$$V \propto T \quad \text{if } P \text{ is kept constant} \tag{1.6}$$

According to *Avogadro's law*

$$V \propto n \quad \text{if } T \text{ and } P \text{ are kept constant} \tag{1.7}$$

Combining Eqs (1.5), (1.6) and (1.7), we get

$$V \propto \frac{1}{P} \times T \times n$$

or
$$V = R \times \frac{1}{P} \times T \times n$$

where R is constant of proportionality and is called gas constant.

or
$$PV = nRT \text{ (for } n \text{ moles of the gas.)} \quad (1.8)$$

This equation is known as *gas equation* (or more correctly, ***ideal gas equation***), which will be explained later.

Standard Temperature and Pressure or Normal Temperature and Pressure (N.T.P.)

As discussed above, volume of a given mass of gas changes with temperature and pressure. Hence, to compare the volume of different gases, they have to be taken or converted to same condition of temperature and pressure using gas equation. The conditions most commonly chosen are $P = 1$ atm and $T = 0\,°C$ or 273 K.

Nature/significance of gas constant, R: From the gas equation, $PV = nRT$, we have

$$R = \frac{P \times V}{n \times T} = \frac{\text{Pressure} \times \text{Volume}}{\text{Moles} \times \text{temperature K}}$$

$$R = \frac{\frac{\text{Force}}{\text{Area}} \times (\text{Length})^3}{\text{Moles} \times \text{Degrees}} = \frac{\frac{\text{Force}}{(\text{Length})^2} \times (\text{Length})^3}{\text{Moles} \times \text{Degrees}}$$

$$R = \frac{\text{Force} \times \text{Length}}{\text{Moles} \times \text{Degrees}} = \frac{\text{Work}}{\text{Moles} \times \text{Degrees}}$$

$$R = \text{Work done per degree per mole.}$$

Numerical Values of the Gas Constant, R

(i) At N.T.P. conditions, for 1 mole of the gas

$$P = 1 \text{ atm}, V = 22.4 \text{ l}$$

$$T = 273 \text{ K}, n = 1 \text{ mole}$$

$$\therefore \quad R = \frac{P \times V}{n \times T} = \frac{1 \times 22.4}{1 \times 273} = \mathbf{82.1\ ml\ atm\ K^{-1} mol^{-1}}$$

(ii) In C.G.S. units, 1 mole of gas at N.T.P.

$$P = 76 \text{ cm} = 76 \times 13.6 \times 981 \text{ dyns/sq.cm}$$

$$V = 22400 \text{ cm}^3, T = 273 \text{ K}, n = 1 \text{ mole}$$

$$\therefore \quad R = \frac{(76 \times 13.6 \times 981) \times 22400}{1 \times 273} = \mathbf{8.314 \times 10^7\ erg\ K^{-1} mol^{-1}}$$

(iii) For expressing in SI units, put 10^7 erg = 1 J

$$\therefore \qquad R = \frac{8.314 \times 10^7}{10^7} \text{joules deg}^{-1}\text{mol}^{-1} = \mathbf{8.314\,JK^{-1}mol^{-1}}$$

(iv) Further putting 4.184 J = 1 calorie, we have

$$\therefore \qquad R = \frac{8.314}{4.184} = \mathbf{1.987 \simeq 2\ cal\,K^{-1}mol^{-1}}$$

1.2.5 Graham's Law of Diffusion

It states that: *Under similar conditions of temperature and pressure, the rates of diffusion of different gases are inversely proportional to the square root of their densities. Mathematically,*

$$\frac{r_1}{r_2} = \sqrt{\frac{d_2}{d_1}}$$

As molecular weight = 2 vapour density, weight can also be written as

$$\frac{r_1}{r_2} = \sqrt{\frac{M_2}{M_1}} \qquad (1.9)$$

1.2.6 Dalton's Law of Partial Pressure

If states that: *If a number of gases which do not react chemically with one another are enclosed together in the same vessel, the total pressure exerted by the mixture of gases is the sum of their partial pressures,* i.e., the pressures which each gas would exert if present alone in the same vessel. Mathematically,

$$P = p_1 + p_2 + p_3 + \ldots\ldots \qquad (1.10)$$

where P is the total pressure and p_1, p_2, p_3 etc. are partial pressures.

1.3 KINETIC THEORY OF GASES

It was observed that the gas laws were based on experimental observations. The theoretical foundation or mathematical representation was missing. However, several workers studied the properties of gases and found that the gases are essentially composed of freely moving molecules. To explain the behaviour of gases, they put forward a theory called *kinetic theory of gases*. This theory succeeded to attain a rigid mathematical form. The *assumptions* made in this theory are given below.

1.3.1 Postulates (Assumptions) of Kinetic Theory

(i) A gas consists of a large number of very small spherical tiny particles called molecules.
(ii) The volume occupied by the molecules is negligible in comparison to the total volume of the gas.

NOTEWORTHY POINTS

1. Air is dense at sea level, because it is compressed by the mass of air above it. However, the density and pressure decrease with altitude. The atmospheric pressure at Mount Everest is only 0.5 atm.
2. Pressure of a pure gas is measured by manometer while that of a mixture of gases by a barometer.
3. 1 Bar = 0.9862 atm = 1.0 atm.
   1 atm = 1.03125×10^5 Pa
4. At 4.58 mm Hg pressure and at 0.0098 °C, ice (solid), water (liquid) and vapour (gas) may be present simultaneously and all are stable.
   Ice/water/vapour
   The corresponding temperature at which all the three states coexist is called triple point.
5. Centigrade scale and Fahrenheit scale of temperature were related as $C = 5/9$ (F–32)
6. Charle's law is not applicable to liquids.
7. Gas constant per molecule is known as Boltzmann constant (K).
   K = R/N = 1.38×10^{-10} erg/K/molecule or 1.38×10^{-23} JK^{-1} mol^{-1}.
8. Numerical value of R = 0.0821 litre atm K^{-1} mol^{-1} = 8.314 JK^{-1} mol^{-1}
   = 1.987 cal K^{-1} mol^{-1}

(iii) The molecules are in a state of constant rapid motion which is completely random. During their motion, they collide into one another and with the walls of the vessel.
(iv) The pressure of a gas is due to the collisions of the molecules with the walls of the vessel.
(v) The molecules are perfectly elastic, i.e., there is no loss of energy when they collide with one another or with the walls of the vessel.
(vi) There is no force of attraction or repulsion amongst the molecules.
(vii) The average kinetic energy of a molecule is directly proportional to the absolute temperature of gas.
(viii) The laws of the mechanics in particular Newton's second law of motion are applicable to the molecules in motion.

1.4 DERIVATION OF KINETIC GAS EQUATION

Based on the above postulates of the kinetic theory, we can calculate the pressure of a gas in terms of molecular quantities. Suppose a volume of gas is enclosed in a cubical vessel (Fig. 1.1) of side of l centimetres. Suppose the number of molecules present in it be n and the mass of each molecule be m.

The molecules move with different velocities in different directions colliding with each other and with the walls of the containing vessel. Suppose u is the velocity of a molecule at any instant. This velocity may be resolved in three rectangular components, u_x, u_y and u_z, along the three axes x, y and z of the cube. Then, we have

$$u^2 = u_x^2 + u_y^2 + u_z^2 \qquad (1.11)$$

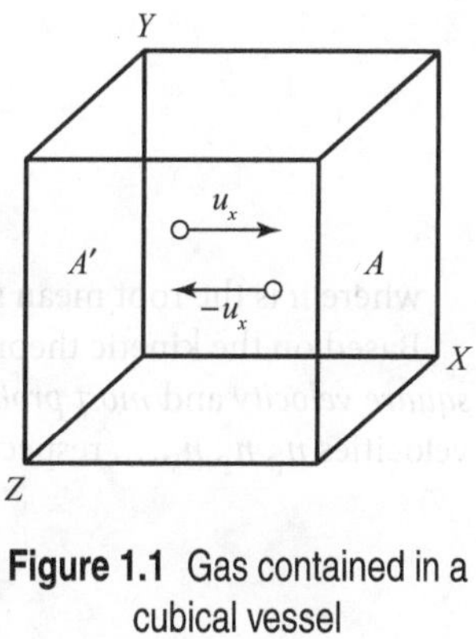

Figure 1.1 Gas contained in a cubical vessel

Now, consider the molecule moving with a velocity ux along x-axis striking the walls A and A'. Since the collisions are perfectly elastic, the molecule after colliding with the wall A would rebound with exactly the same velocity but in opposite direction.

The momentum of the molecule before it strikes

$$A = mu_x$$

The momentum of the molecule after it strikes

$$= m(-u_x) = -mu_x$$

Therefore, the change of momentum due to one impact

$$= mu_x - (-mu_x) = 2mu_x$$

After colliding with the wall A, it will travel a distance of l centimetres to strike the face A' (i.e., opposite to A) and then returns to face A again. Hence, after traversing a distance of l every time, it will make one impact on the wall. The total number of impacts the molecule makes on the two walls A and A' per unit time would thus be u_x/l.

Therefore, the change of momentum per molecule along x-axis per second

$$= 2mu_x \times \frac{u_x}{l} = \frac{2mu_x^2}{l}$$

Similarly, the change of momentum per molecule along y- and z-axes per second would be $2mu_y^2/l$ and $2mu_z^2/l$, respectively.

Therefore, the total change of momentum along all the three axes per second per molecule

$$= \frac{2mu_x^2}{l} + \frac{2mu_y^2}{l} + \frac{2mu_z^2}{l}$$

$$= \frac{2m}{l}\left(u_x^2 + u_y^2 + u_z^2\right) = \frac{2mu^2}{l}$$

Now, we consider all the molecules having different velocities u_1, u_2, u_3,... etc. The change of momentum per second of all the molecules on all the walls of the cube:

$$= \frac{2mnu_1^2}{l} + \frac{2mnu_2^2}{l} + \frac{2mnu_3^2}{l} + \ldots\ldots + \frac{2mnu_n^2}{l}$$

$$= \frac{2mn}{l}\left(u_1^2 + u_2^2 + u_3^2 + \ldots\ldots + u_n^2\right) = \frac{2mnu^2}{l}$$

where u is the root mean square velocity.

Based on the kinetic theory of gases, we have three types of velocities, viz. *average velocity*, *root mean square velocity* and *most probable velocity*. Suppose in a gas, there are n_1, n_2, n_3,... molecules possessing velocities n_1, n_2, n_3,..., respectively. The average velocity (v) is then given by,

$$v = \frac{n_1u_1 + n_2u_2 + n_3u_3 + \ldots\ldots}{n_1 + n_2 + n_3 + \ldots\ldots}$$

The root mean square velocity (u) is defined as the square root of the mean of squares of all the velocities and is given by

$$u = \sqrt{\left(\frac{n_1u_1^2 + n_2u_1^2 + n_3u_3^2 + ...}{n_1 + n_2 + n_3 + ...}\right)}$$

Most probable velocity ($\overline{u}$) is defined as the velocity possessed by the maximum number of molecules of the gas. We have

$$v = 09213 \times u \qquad \text{and} \qquad \overline{u} = \sqrt{\left(\frac{2u}{3}\right)}$$

According to Newton's second law of motion, the change of momentum per unit time is the force (F) exerted by the molecules on the walls of the cube.

$$\therefore \qquad F = \frac{2mnu^2}{l}$$

Since, pressure is the force per unit area (A), we have

$$P = \frac{F}{A} = \frac{2mnu^2}{lA} = \frac{2mnu^2}{l\,6l^2} \qquad \left[\text{As } A = \text{Area of cube} = 6l^2\right]$$

$$P = \frac{1}{3}\frac{mnu^2}{V} \qquad \left[\text{As } l^3 = \text{Volume of cube} = V\right]$$

or

$$PV = \frac{1}{3}mnu^2 \tag{1.12}$$

If d be the density of the gas, then from Eq. (1.12), we have

$$P = \frac{1}{3}\frac{mn}{V}u^2 = \frac{1}{3}du^2 \tag{1.13}$$

For 1 mole of a gas, n can be replaced by N (Avogadro number). Therefore, kinetic equation (1.12) then becomes,

$$PV = \frac{1}{3}mNu^2 \tag{1.14}$$

Expressions (1.12), (1.13) and (1.14) are different forms of kinetic equation of gases.

1.5 DERIVATION OF GAS LAWS FROM KINETIC EQUATION

(i) Boyle's law: According to it, at constant temperature, the volume of a given mass of a gas is inversely proportional to pressure, i.e.,

$$V \propto \frac{1}{P}, \quad \text{at constant } T$$

or $$PV = \text{constant}, \quad \text{at constant } T$$

From kinetic equation, we have

$$PV = \frac{1}{3}mnu^2 = \frac{2}{3} \times \frac{1}{2}mnu^2$$

The kinetic energy of the gas

$$\therefore \quad PV = \frac{2}{3}\text{E} \tag{1.15}$$

At constant temperature, the kinetic energy (E) of the gas is constant. Therefore, at constant temperature

$$PV = \text{constant}$$ [This is Boyle's law.]

(ii) Charles' law: According to it, at constant pressure, the volume of a given mass of a gas is directly proportional to its absolute temperature, i.e.,

$$V \propto T, \quad \text{at constant } P$$

As already derived, from Eq. (1.15), we have

$$PV = \frac{2}{3}\text{E} \quad \text{or} \quad V = \frac{2}{3}\frac{\text{E}}{P}$$

At constant pressure,

$$V = \text{constant} \times \text{E} \quad \text{or} \quad V \propto \text{E}$$

However, we know that $\text{E} \propto T$, where T is absolute temperature.

$$V \propto T$$ [This is Charles' law.]

(iii) Avogadro's hypothesis: According to it, equal volumes of all gases under similar conditions of temperature and pressure contain equal number of molecules. For any two gases, the kinetic equation can be written as:

$$P_1V_1 = \frac{1}{3}m_1n_1u_1^2 = \frac{2}{3} \times \frac{1}{2}m_1n_1u_1^2$$

$$P_2V_2 = \frac{1}{3}m_2n_2u_2^2 = \frac{2}{3} \times \frac{1}{2}m_2n_2u_2^2$$

When pressures and volumes of the gases are the same, i.e., $P_1 = P_2$ and $V_1 = V_2$, it follows that

$$\frac{1}{2}m_1n_1u_1^2 = \frac{1}{2}m_2n_2u_2^2 \tag{1.16}$$

When the temperatures of the gases are the same, their mean kinetic energy will also be the same, i.e.,

$$\frac{1}{2}m_1u_1^2 = \frac{1}{2}m_2u_2^2 \tag{1.17}$$

Dividing Eq. (1.16) by Eq. (1.17), we get, $n_1 = n_2$ [This is Avogadro's hypothesis.]

(iv) Graham's law of diffusion: According to it, the rate of diffusion (r) of a gas is inversely proportional to the square root of the density (d) of the gas, at constant pressure, i.e.,

$$r \propto \frac{1}{\sqrt{d}}, \text{ at constant pressure}$$

From kinetic equation, we have

$$PV = \frac{1}{3}mnu^2 \quad \text{or} \quad u = \sqrt{\left(\frac{3PV}{mn}\right)} = \sqrt{\left(\frac{3P}{d}\right)}$$

$$\left[\text{As } \frac{mn}{V} = \frac{\text{Total mass of the gas}}{\text{Volume}} = \text{Density of the gas, } d\right]$$

Therefore, at constant pressure, $u \propto \frac{1}{\sqrt{d}}$

The rate of diffusion (r) of the gas is directly proportional to the mean velocity (u) of the molecules, i.e., $u \propto r$

$$\therefore \quad r \propto \frac{1}{\sqrt{d}}$$ [This is Graham's law of diffusion.]

(v) Dalton's law of partial pressures: Suppose n_1 molecules, each of mass m_1, of a gas, A, are contained in a vessel of volume V. Then, according to the kinetic theory, the pressure, p_A, of the gas A will be given by:

$$p_A = \frac{1}{3}\frac{m_1n_1u_1^2}{V}$$

Now, suppose n_2 molecules, each of mass m_2, of another gas, B, are contained in the same vessel at the same temperature and there is no other gas present at that time. The pressure, p_B, of the gas B is then given by,

$$p_B = \frac{1}{3}\frac{m_2n_2u_2^2}{V}$$

If both gases are present in the same vessel at the same time, the total pressure, P, is given by,

$$P = \frac{1}{3}\frac{m_1n_1u_1^2}{V} + \frac{1}{3}\frac{m_2n_2u_2^2}{V} = p_A + p_B$$

Similarly, if three, four or more gases are present in the same vessel, the total pressure is given by,

$$P = p_A + p_B + p_C + p_D + \ldots$$ [This is Dalton's law of partial pressures.]

(vi) Derivation of ideal gas equation:

$$PV = \frac{1}{3} mnu^2$$

$$= \frac{2}{3}\frac{1}{2} mnu^2 = \frac{2}{3}(\text{K.E.})$$

$$= \frac{2}{3}kT \qquad [\because \text{K.E.} \propto \text{T}, \therefore \text{K.E.} = kT, \text{ where } k \text{ is constant}]$$

$$= k'T \qquad \left[k' = \frac{2}{3}k \text{ is a new constant}\right]$$

$$\therefore \qquad PV = k'T$$

It is combined gas equation.

If $k' = \text{R}$, (gas constant),

then

$$PV = RT \quad \text{(for one mole of gas)}$$

$$PV = n\,RT \quad \text{(for } n \text{ moles of gas)}$$

which is the required ideal gas equation.

1.5.1 Some Useful Deductions from Kinetic Theory of Gases

1.5.1.1 Calculation of Molecular Velocities

According to kinetic gas equation

$$PV = \frac{1}{3} mnu^2$$

$$u^2 = \frac{3PV}{mn} = \frac{3PV}{M} \qquad [\because mn = M \text{ is the mass of a gas}]$$

If 1 mole of the gas is taken, M will be equal to the molecular weight of the gas.

$$u = \sqrt{\frac{3PV}{M}} \tag{1.18}$$

Further, for one mole of the gas (assuming it is ideal)

$$PV = RT$$

$$\therefore \qquad u = \sqrt{\frac{3RT}{M}} \qquad (1.19)$$

Knowing that $\frac{M}{V} = d$ is the density of the gas, equation $u = \sqrt{\frac{3PV}{M}}$ can also be written as

$$\therefore \qquad u = \sqrt{\frac{3P}{d}} \qquad (1.20)$$

Depending upon the conditions given, any one of the Eqs (1.18), (1.19) and (1.20) can be used for the calculation of the root mean square velocity.

Numerical Problems

Example 1 **Calculate the root mean square velocity of CO_2 molecule at 1,000 °C.**

Solution: We know that $u = \sqrt{\frac{3PV}{M}} = \sqrt{\frac{3RT}{M}}$ $\left[\because PV = RT \text{ for 1 mole}\right]$

$$R = 8.314 \times 10^7 \text{ erg deg}^{-1} \text{ mol}^{-1}$$

$$= 0.88314 \times 10^8 \text{ erg deg}^{-1} \text{ mol}^{-1}$$

$$u = \sqrt{\frac{3 \times 0.8314 \times 10^8 \times T}{M}}$$

$$= 1.58 \times 10^4 \times \sqrt{\frac{T}{M}} \text{ cm sec}^{-1}$$

We have $$u = 1.58 \times 10^4 \times \sqrt{\frac{1273}{44}}$$

$$u = 84{,}985 \text{ cm sec}^{-1} \quad \text{or} \quad \mathbf{849.85 \text{ m sec}^{-1}}$$

Example 2 **Calculate the RMS velocity of chlorine molecules at 12 ϒC and 78 cm pressure.**

Solution: At given condition: At STP

$V_2 = ?$ $\qquad V_1 = 22{,}400$ ml

$T_2 = 12 + 273 = 285$ K $\qquad T_1 = 273$ K

$P_2 = 78$ cm $\qquad P_1 = 76$ cm

Applying $$\frac{P_1 V_1}{T_1} = \frac{P_2 V_2}{T_2}$$

We have $$V_2 = \frac{P_1 V_1 T_2}{T_1 P_2} = \frac{76 \times 22400 \times 283}{273 \times 78} = 22{,}785 \text{ ml}$$

$$u = \sqrt{\frac{3PV}{M}}$$

$$P = \text{hdg} = 78 \times 13.6 \times 981 \text{ dynes cm}^{-2}$$

$$V = 22{,}785 \text{ ml};\ M = 71$$

$$u = \sqrt{\frac{3 \times 78 \times 13.6 \times 981 \times 22785}{71}}$$

$$u = 31{,}652 \text{ cm sec}^{-1} \quad \text{or} \quad \mathbf{316.52\ m\ sec^{-1}}$$

Example 3 **Calculate the average velocity of nitrogen molecule at STP.**

Solution: Substituting the given values in this equation

$$u = \sqrt{\frac{3PV}{M}}$$

$$= \sqrt{\frac{3 \times 76 \times 13.6 \times 981 \times 22400}{28}}$$

$$= 49.330 \text{ cm sec}^{-1}$$

$\therefore$ Average velocity $= 0.9213 \times 49{,}330$ cm sec^{-1} = **45,447 cm sec^{-1}**

Example 4 **Oxygen at 1 atmosphere pressure and 0 ϒC has density of 1.4290 grams per litre. Find the RMS velocity of oxygen molecules.**

Solution: We have $P = 1 \text{ atom} = 76 \times 13.6 \times 981 \text{ dynes cm}^{-2}$

$$D = 1.4290 \text{ gl}^{-1} = \frac{1.4290}{1000} \text{ gml}^{-1} = 0.001429 \text{ gml}^{-1}$$

Applying $$u = \sqrt{\frac{3P}{D}},$$

We get $$u = \sqrt{\frac{3 \times 76 \times 13.6 \times 981}{0.001429}} = \mathbf{46{,}138\ cm\ sec^{-1}}$$

1.5.1.2 Calculation of the Kinetic Energy of the Molecules

For 1 mole of the gas containing *n* molecules, each of the mass *m*, the kinetic energy will be

$$\text{K.E.} = \frac{1}{2} mnu^2$$

According to kinetic gas equation

$$PV = \frac{1}{3} mnu^2 \quad \text{or} \quad mnu^2 = 3PV$$

$$\therefore \qquad \frac{1}{2} mnu^2 = \frac{3}{2} PV$$

$$\therefore \qquad \text{K.E.} = \frac{3}{2} PV$$

Further, for 1 mole of the ideal gas,

$$PV = RT$$

$$\therefore \qquad \text{K.E.} = \frac{3}{2} RT$$

From the equation, we observe that the kinetic energy of an ideal gas is independent of the nature or pressure of the gas and it depends only on the absolute temperature, being directly proportional to it. The average kinetic energy per molecule can be obtained from the above equation by dividing with the Avogadro's number *N*. Thus, average K.E. per molecule

$$= \frac{\text{K.E./mol}}{N} = \frac{3}{2}\frac{R}{N}T = \frac{3}{2}kT$$

where $k = \dfrac{R}{N}$ is called Boltzmann constant.

Example 5 **Calculate the total and average kinetic energy of 32 g methane molecules at 27ϒC ($R = 8.314$ JK^{-1} mol^{-1}).**

Solution: Total K.E.$= \dfrac{3}{2} nRT = \dfrac{3}{2} \times \dfrac{32}{16}$ mol $\times$ 8.314 JK^{-1} mol^{-1} $\times$ 300 K = 7482.6 J

(molar mass of CH_4 = 16 g mol–1)

Average K.E., $\quad s = \dfrac{3}{2} kT = \dfrac{3}{2} \times \dfrac{8.314 \text{ JK}^{-1}\text{mol}^{-1}}{6.02 \times 10^{23} \text{ molecules mol}^{-1}} \times 300 \text{ K} = 6.21 \times 10^{-21} \text{ J mol}^{-1}$

or $\quad = \dfrac{3}{2} RT \text{ mol}^{-1} = \dfrac{3}{2} \times 8.314 \text{ JK}^{-1} \text{ mol}^{-1} \times 300 \text{ K} = \mathbf{3741.3 \text{ J mol}^{-1}}$

Example 6 Calculate the root mean square speed of methane molecules at 27 °C.

Solution:

$$u_{r.m.s} = \sqrt{\frac{3RT}{M}}$$

Using C.G.S. units, put $R = 8.314 \times 10^7$ erg K^{-1} mol^{-1}, $T = 27 + 273 = 300$ K,

M (for CH_4) = 16 g mol^{-1}, we get $u_{r.m.s.} = \sqrt{\frac{3 \times 8.314 \times 10^7 \times 300}{16}} = 6.839 \times 10^4$ cm s^{-1} = 683.9 m s^{-1}

Using S.I. units, put $R = 8.314$ Jk^{-1} mol^{-1}, $T = 300$ K, $M = 16 \times 10^{-3}$ kg mol^{-1}, we get

$$u_{r.m.s.} = \sqrt{\frac{3 \times 8.314 \times 300}{16 \times 10^{-1}}} = \mathbf{683.9\ m\ s^{-1}}$$

Example 7 The average kinetic energy of a gas molecule at 0 ϒC is 5.621 × 10^{-21} J. Calculate Boltzmann constant. In addition, calculate the number of molecules present in one mole of the gas.

Solution: Average Kinetic energy = $\frac{3}{2}$ kTs

$$\therefore \qquad K = \frac{2}{3} \times \frac{(K.E.)}{T} = \frac{2}{3} \times \frac{5.621 \times 10^{-21}\ J\ molecule^{-1}}{273K} = 1.373 \times 10^{-23}\ JK^{-1}\ mol^{-1}$$

No. of molecules in 1 mole of the gas (Avogadro's No.) $= \frac{R}{K} = \frac{8.314 K^{-1} mol^{-1}}{1.373 \times 10^{-21} JK^{-1} molecule^{-1}}$

$= \mathbf{6.055\ \ 10^{23}\ molecules\ mol^{-1}}$

1.5.1.3 Concept of Absolute Zero

We know that

$$K.E. = \frac{1}{2} mnu^2$$

$$\therefore \qquad K.E. \propto u^2 \qquad (1.21)$$

Further as derived above,

$$K.E. = \frac{3}{2} RT$$

$$\therefore \qquad K.E. \propto T \qquad (1.22)$$

Combining Eqs (1.21) and (1.22), we get

$$u^2 \propto T \quad \text{or} \quad u \propto \sqrt{T}$$

Thus, the molecular velocity of any gas is directly proportional to the square root of its absolute temperature. Hence, when $T = 0$, u also becomes zero. This temperature is called **absolute zero**. Hence, absolute zero is that temperature at which the molecular motion creases.

Remember $\qquad 0\,°A \quad \text{or} \quad 0\,K = -273\,°C$

1.5.1.4 Slow Diffusion of Gases

In spite of the fact that the gas molecules move with high speeds, yet their motion is not so fast. For example, it takes quite some time for a smell to spread from one corner of a room to the other. This is obviously due to the fact the molecules of a gas undergo an extremely large number of collisions, per second. In every collision, the path of the molecule is deflected, so much so that sometimes it is even deflected back. Hence, the net linear movement of the molecules in a particular direction is quite slow.

1.6 IDEAL AND REAL GASES

1.6.1 Ideal Gases

Gases which obey gas equation $PV = RT$ rigidly for all values of temperature T and pressure P are called ***ideal or perfect gases****. Thus, an ideal or perfect gas is one which rigidly obeys Boyle's and Charles law for all values of temperature and pressure.*

1.6.1.1 Characteristics of an Ideal Gas

(i) The product of pressure and volume of a given mass of an ideal gas at constant temperature should be constant, i.e., if product *PV* is plotted against pressure *P* at constant temperature, the curve obtained should be a straight line parallel to the pressure axis (Boyle's law).

(ii) The compressibility factor $\left(Z = \dfrac{PV}{RT}\right)$ of an ideal gas should be unity, i.e., $PV = n\ RT$ (Gas equation).

(iii) If, at a constant pressure, the temperature is reduced, the volume should reduce uniformly and become zero at −273 °C or 0 K (Charles law).

(iv) If a gas is allowed to expand without doing any external work, it should show no thermal effect (no force of attraction between the molecules).

(v) At constant temperature, if pressure is doubled, the volume should reduce to half (Boyle's law).

(vi) According to Avogadro's law, one mole of every gas should occupy 22.4 l at N.T.P. (0 °C and 1 atm). Experiments show that this is true only approximately because the actual value for different gases does not come out to be the same (Avogadro's law).

If we study, however, the actual behaviour of the gases, we notice that no gas is ideal or perfect in this sense although some gases approach perfection as the temperature rises above their boiling points.

Thus, an ideal or perfect gas is only *hypothetical.* It has been observed that at high pressures and low temperatures, gases do not obey the gas laws. However, at low pressures and high temperatures, they obey the gas laws to a fair degree of approximation. Such gases are, therefore, known as real gases.

1.6.2 Real Gas

Real gas is one which obeys the gas laws fairly well under low pressure and high temperature. It means all gases are real gases, but they show deviation from ideal behaviour. The deviations are more pronounced, as the pressure is increased or the temperature is decreased. Further, in case of real gases, it is not possible to reduce uniformly the volume of a gas to zero at −273 °C, as all real gases liquefy when sufficiently cooled. Real gases such as O_2, H_2 and N_2 which are difficult to be liquefied differ from an ideal gas to a greater extent than the gases such as NH_3, SO_2 and CO_2 which can be easily liquefied. Helium shows an ideal gas behaviour to some extent.

Since we know that no gas is ideal or perfect, the gas equation is obeyed as an approximation by a real gas at low pressure and high temperature, while at high pressure and low temperature, there is much deviation from this ideal behaviour.

1.7 DEVIATIONS OF REAL GASES FROM GAS LAWS

1.7.1 Deviations from Boyle's Law

1.7.1.1 Effect of Pressure

The plots of PV versus P for various gases at 0 °C are depicted in Fig. 1.2. For an ideal gas, the plot should be a straight line parallel to pressure axis. Such a plot is indicated by a dotted line. It is clear from the various plots that all gases deviate considerably from the ideal gas behaviour and the product PV does not remain constant with increasing pressure. Only in the limit of zero pressure, all gases are behaving ideally.

A close examination of Fig. 1.2 reveals that all the curves can be classified into two types:

Type 1: Here, the curves start ascending at once from the value of PV and the value of PV increases continuously with increasing pressure. This is true in case of gases having smaller molecules, e.g., H_2, He etc.

Type 2: Here, the curves show a decrease in PV in the beginning passes through a minimum and then begins to increase with increasing pressure. This is true in case of gases having larger molecules e.g., O_2, N_2, CO_2 etc.

Above 400 atm, the product of PV increases with the increase in pressure for all gases.

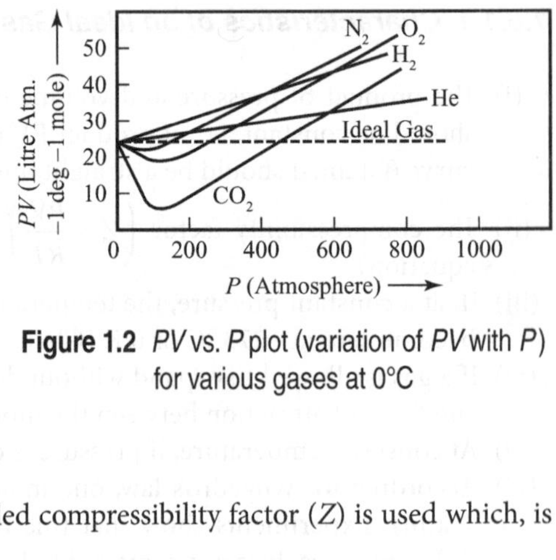

Figure 1.2 *PV* vs. *P* plot (variation of *PV* with *P*) for various gases at 0°C

1.7.1.2 Compressibility Factor

To show both the deviations more clearly, a quantity, called compressibility factor (Z) is used which, is defined as:

$$Z = \frac{PV}{RT}$$

where V is the volume occupied by one mole of the gas at a given temperature T and pressure P.

For an ideal gas, $Z = 1$ and is independent of pressure and temperature.

For a real gas, Z is a function of both temperature and pressure and its value varies.

The amount by which the actual factor differs from a value of unity gives measure of deviation from the ideal behaviour.

Since $$(PV)_{\text{ideal}} = RT$$

$$\therefore \quad Z = \frac{PV}{RT} = \frac{PV}{(PV)_{\text{ideal}}}$$

So compressibility factor is a measure of deviation from ideality.

Figure 1.3 shows a plot of Z versus P for some gases at 0 °C. For H2, the value of Z continuously increases but for N2, CH4 and CO2, Z first decreases and then increases quite rapidly with increasing pressure.

It has been found that for those gases which can be easily liquefied, Z decreases sharply below the value of unity at low pressure. Since carbon dioxide is a gas which can be more easily liquefied than nitrogen and methane, therefore, the dip in value of Z is much more pronounced in carbon dioxide and, persists over a range of pressures. However, as hydrogen gas cannot be easily liquefied, its curve does not show any dip in the value of Z with increasing pressure.

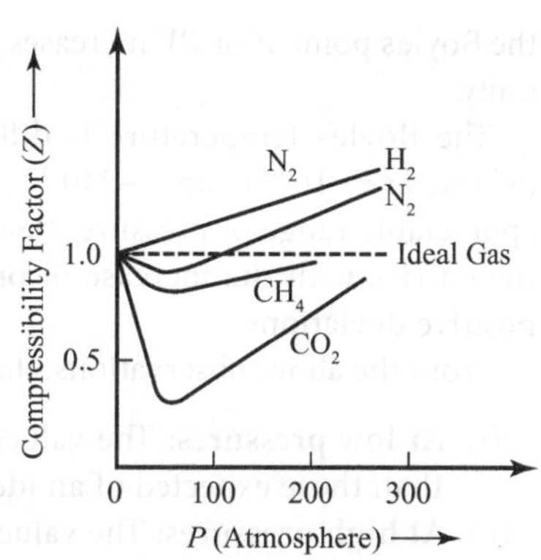

Figure 1.3 *Z* vs. *P* curves for various gases at 0°C

1.7.1.3 Effect of Temperature

If a plot of Z (compressibility) or relative PV versus P is constructed for a gas at several temperatures, it is fund that as the temperature decreases, the dip in the curve is more pronounced.

Figure 1.4 shows various curves of Z (or relative PV) versus P for nitrogen at various temperatures. All the curves start with $Z = 1$ at $P = 0$ and value of Z varies from unity as pressure increases in a manner dependent on temperature. As the temperature increases, the minimum in the curve (dip) becomes smaller and shifts towards lower pressure.

However, a temperature is reached at which Z remains close to unity over an appreciable range of pressure.

It is seen that in case of nitrogen when PV values are plotted against pressure at different temperatures, the shape of PV curve changes with temperature. In general, the deviations are greater at higher pressures and near the temperature of liquefaction. *Further, it is also observed that nitrogen gas behaves like hydrogen gas at higher temperatures; whereas at lower temperatures, it behaves like CO_2.*

Examination of the Fig. 1.4 reveals that at 50 °C the value of PV remains constant and is close to the value of the ideal gas for an appreciable range of pressure. In fact up to 100 atmosphere pressure, the compressibility factor (Z) changes from 1.0 to 1.02. Beyond 100 atm., the value of Z increases rapidly and attains values much greater than 1.

The temperature at which a real gas obeys the ideal gas law over an appreciable range of pressure is known as the **Boyle's temperature or the Boyle point**. It is so named because at this temperature, Boyle's law is obeyed over a range of pressures. Obviously about the Boyle's temperature, a gas shows positive deviation only.

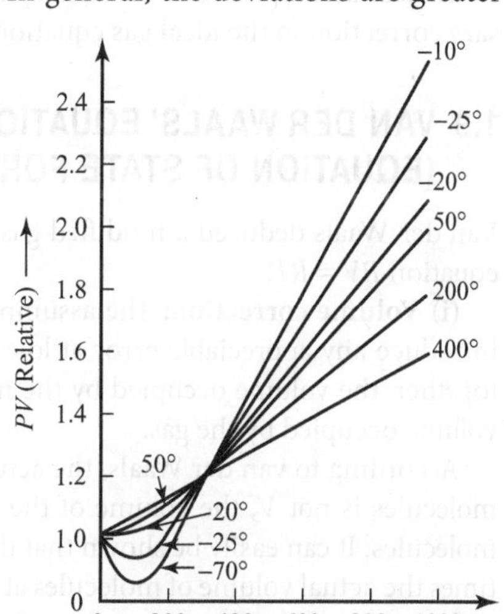

Figure 1.4 Variation of *Z* (relative *PV*) with *P* for nitrogen at various temperatures

For every gas when the temperature is below the Boyle's temperature, the value of Z or PV will first decrease with increasing pressure, pass through a minimum and then increase continuously. However, if the temperature is above the Boyle's point, Z or PV increases sharply, with increasing pressure and the values of Z are higher than unity.

The Boyle's temperature is different for different gases. Boyle's temperature for hydrogen and helium are −165 °C and −240 °C, respectively. Hydrogen gas obeys Boyle's law at −165 °C for an appreciable range of pressure. Below this temperature, plot of PV versus P will first show a fall and then a rise with the increase of pressure. Obviously, above the Boyle's temperature, all gases show positive deviations.

From the above observations, the following three points are evident:

(i) **At low pressures:** The values of PV for all gases except hydrogen, helium and neon are lower than those expected of an ideal gas.

(ii) **At high pressures:** The value of PV for all gases including H_2, He and Ne are higher than those expected of an ideal gas.

(iii) **Effect of temperature**: At a particular pressure, the deviation of a gas from ideal behaviour is more pronounced at low temperature. The lower the temperature is, the greater are the deviations.

1.8 CAUSES OF THE DERIVATIONS FROM IDEAL BEHAVIOUR

The failure of the ideal gas equation $PV = RT$ over an extensive range of temperatures and pressures has been explained by *van der Waals* to be due to the following two faulty assumptions made in the kinetic theory:

(i) *The actual volume occupied by the gas molecules is negligible as compared to the total volume enclosing the gas.*

(ii) *The molecules exert no inter-molecular attraction among themselves.*

Both these assumptions are not true, particularly *at low temperature and high pressure.* Therefore, necessary correction to the ideal gas equation should be made.

1.9 VAN DER WAALS' EQUATION (REDUCED EQUATION OF STATE) (EQUATION OF STATE FOR REAL GASES)

Van der Waals deduced a modified gas equation by making following necessary corrections to the gas equation $PV = RT$.

(i) Volume correction: The assumption that actual volume of the molecules is negligible, does not introduce any appreciable error at low pressures. At high pressure when the molecules are quite close together, the volume occupied by the molecules themselves becomes considerable fraction of the total volume occupied by the gas.

According to van der Waals, the actual free space available inside the vessel for the movement of the molecules is not V, the volume of the container, but $(V - b)$, where 'b' is the **effective volume** of the molecules. It can easily be shown that the effective volume of molecules in motion is equal to about four times the actual volume of molecules at rest. The effective volume (b) is also called *co-volume* or *excluded volume.*

Let us consider two molecules approaching each other. Let δ and r represent the diameter and the radius of each molecule. The molecules cannot approach more closely than the distance as shown in Fig. 1.5. It is because the molecules are impenetrable. It is clear from the figure that within a sphere of radius $\delta\,(= 2r)$, no volume will be occupied by this pair of molecules.

TRY YOURSELF

1. What type of graph will you get when *PV* is plotted against *P* at constant temperature?
2. What is the relationship between average kinetic energy and the temperature of a gas?
3. Why in case of hydrogen and helium, the compressibility factor is always greater than 1 and increases with increase of pressure?
4. What is the difference between total kinetic energy and translational kinetic energy? For what type molecules, the two are equal?
5. What would have happened to the gas if the molecular collisions were not elastic?
6. CO_2 is heavier than O_2 and N_2 gases present in the air but it does not form the lower layer of the atmosphere. Why?

Answers

1. A straight line parallel to pressure axis.
2. K.E. $= \frac{3}{2} kT$ where k is Boltzmann constant $= \frac{R}{N_A}$.
3. Refer to 'Explanation of exceptional behaviour of H_2 and He'.
4. Total kinetic energy is the sum of translational, vibrational and rotational kinetic energies. The total kinetic energy is equal to the translational kinetic energy for monoatomic gases (He, Ne etc.), as they do not posses vibrational and rotational motions but have only translational motion.
5. On every collision, there would have been loss of energy. As a result, the molecules would have slowed down and ultimately settle down in the vessel. Moreover, the pressure would have gradually reduced to zero.
6. Gases possess the property of diffusion which is independent of the force of gravitation. Due to diffusion, the gases mix into each other and remain almost uniformly distributed in the atmosphere.

The excluded volume per pair of molecules

$$= \frac{4}{3}\pi(2r)^3 = 8\left(\frac{4}{3}\pi r^3\right)$$

Therefore, excluded volume for a single molecule

$$= \frac{8}{2}\left(\frac{4}{3}\pi r^3\right) = 4\,V_m$$

where V_m is the volume occupied by a single molecule.

Thus, the excluded volume (b) for a mole of a gas is given as

$$b = 4\left(\frac{4}{3}\pi r^3\right) \times N$$

$$= 4\,V_m N$$

where N is the Avogadro's number.

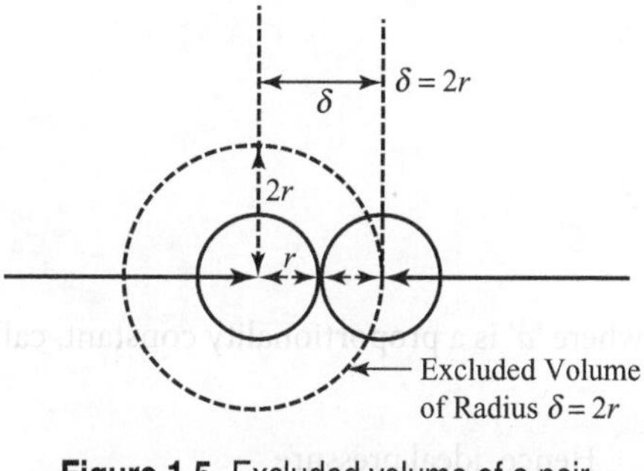

Figure 1.5 Excluded volume of a pair of molecule

Hence, the corrected volume = $(V - b)$ where b is the volume to be excluded which is four times the actual volume of the molecules and it is a characteristic constant for each gas.

(ii) Attraction correction – Pressure correction: The assumption that molecules exert no appreciable intermolecular attraction is not justified especially at high pressure when *the molecules are quite close together.*

Consider particles A and B in the gas Fig. 1.6; the net force of intermolecular attraction on particle A situated in the interior of the gas is nil, because it is surrounded by molecules which are equally distributed in opposite directions, thereby the attractions on all sides are mutually cancelled.

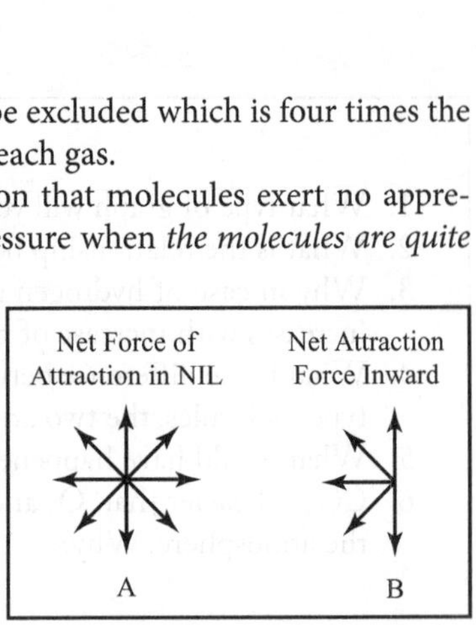

Figure 1.6 Molecular attraction

However, molecule B just near the boundary, which is about to strike the wall, is subjected to a net *inward force of attraction or pull* due to the interior molecules, Therefore, when a molecule is about to strike the wall, this inward pull act, so as to reduce the force exerted by the impact and consequently *the measured pressure P is less than the ideal pressure,* i.e., when no attraction were there. The observed pressure is thus less than that would be expected from the kinetic theory. This will tend to make PV less than RT and so will tend to cause a negative or under perfect deviation. Hence, a correction term p_a should be added to P to get ideal pressure.

Ideal pressure = $P + p_a$. The inward force, p_a exerted on molecules about to strike the walls of the vessel is proportional to:

(i) *The number of molecules striking the wall at any given instant, which is proportional to the density of the gas.*

(ii) *The number of molecules in the interior which are attracting the striking molecules, which is also proportion to the density of the gas.*

$$\therefore \quad p_a \propto d^2$$

However,
$$\text{density} \propto \frac{1}{\text{volume}}$$

$$\therefore \quad p_a \propto \frac{1}{V^2}$$

$$\therefore \quad p_a = \frac{a}{V^2}$$

where 'a' is a proportionality constant, called *co-efficient of attraction.*

Hence, ideal pressure
$$= \left(p + \frac{a}{V^2}\right)$$

Incorporating the volume correction and attraction correction (or pressure correction) in the gas equation, we get

$$(V-b)\left(p+\frac{a}{V^2}\right)=RT \tag{1.23}$$

This modified gas equation is called the van der Waals' equation of state, as it was first established by J.D. van der Waals – a Dutch chemist in 1873. It represents the behaviour of real gases over a wide range of temperature and pressure for one mole of gases.

If there are n moles of the gas occupying volume V, then the effective volume of the molecules is given by nb and the actual free space available inside the vessel for the movement of the molecules is then $V - nb$.

The pressure correction $\quad p_a \propto n^2 d^2$

$$\propto n^2 \frac{1}{V^2}$$

$$p_a = \frac{n^2 a}{V^2}$$

Therefore, the van der Waals' equation for n moles of a gas will be

$$\left(p+\frac{n^2 a}{V^2}\right)(V-nb)=nRT \tag{1.24}$$

1.9.1 Units of van der Waals' Constants

The units for van der Waals' constants a and b depend on the units in which P and V are expressed.

From equation $\quad p_a = \frac{an^2}{V^2}$

$$\therefore \quad a=\frac{p_a V^2}{n^2} \quad \text{or} \quad a=\frac{\text{Pressure} \times (\text{Volume})^2}{(\text{moles})^2}$$

If pressure is expressed in atmospheres and volume in litres, 'a' will be expressed in atm. litre^2 mole^{-2}.

In SI Units: Now, volume correction for n moles of a gas is nb

$$\text{Volume correction} = nb$$

$$\therefore \quad b=\frac{\text{Volume}}{n}$$

In SI Units: If volume is expressed in litres, the units of b are litre mol^{-1}.

1.9.2 Significance of van der Waals' Constant

The values of van der Waals' constants for some important gases are given in Table 1.1.

Table 1.1 Van der Waals' constants for some common gases.

Gas	(a) atm. L^2 mol^{-2}	(b) atm. L mol^{-1}
H_2	0.024	0.0266
He	0.034	0.237
Ne	0.21	0.0171
O_2	1.36	0.0318
N_2	1.39	0.0391
CO_2	3.59	0.0427
HCl	3.67	0.0408
NH_3	4.17	0.0371
SO_2	6.71	0.0564

The gases such as CO_2, NH_3 and HCl, which can be easily liquefied, have high values of van der Waals' constant *a* and *b* and show maximum departure from the ideal gas equation.

The van der Waals' constant '*a*' is a measure of strength of the van der Waals' force between the molecules of the gas. Greater the magnitude of '*a*' is, the stronger are the van der Waals' forces. The greater the magnitude of these forces, i.e., a, is, the greater is the ease with which gas can be liquefied.

The van der Waals' constant '*b*' represents effective volume of gas molecules. It has been observed that its value remains constant over a wide range of temperature and pressure for a gas. This shows that gas molecules are incompressible.

1.10 EXPLANATION OF BEHAVIOUR OF REAL GASES ON THE BASIS OF VAN DER WAALS' EQUATION

$$\left(P+\frac{a}{V^2}\right)(V-b)=RT \quad \text{van der Waals' equation of state.}$$

or

$$PV+\frac{a}{V}-Pb-\frac{ab}{V^2}=RT$$

Neglecting $\frac{ab}{V^2}$, a product of two very small quantities, we get

or

$$PV-Pb+\frac{a}{V}=RT=P'V' \tag{1.25}$$

where P' and V' are the ideal pressure and ideal volume of the gas, respectively.

(i) At low pressure: At lower pressure, the gas has a large volume and volume taken up by the molecules is relatively unimportant. Small addition of pressure due to mutual attraction plays an important part while b is negligible as compared to the large total volume V. The factor a/V, therefore, overweight Pb i.e.,

$$\frac{a}{V} > Pb$$

Let $$\frac{a}{V} - Pb = Z, \text{ positive quantity}$$

From Eq. (1.23), we get

$$PV + Z = RT = P'V'$$

$$PV = RT - Z = P'V' - Z$$

i.e., the observed PV is less than $P'V'$.

The observed product PV at low pressure is less than $P'V'$, product of pressure and volume if the gases were ideal. Thus, the intermolecular attractive forces cause a *negative or under perfect deviation.*

At low pressures, therefore, PV decreases with the increase in pressure in case of N_2, O_2, CO_2 etc.

(ii) At high pressure: When the pressure is high, the volume is small and small volume occupied by molecules themselves cannot be neglected, whereas in comparison with the high pressure, small addition of pressure due to mutual attraction is negligible. Hence, Pb over weights $\frac{a}{V}$, i.e., $Pb > \frac{a}{V}$

i.e., $$\frac{a}{V} - Pb = -y, \text{ a negative quantity}$$

$$PV + (-y) = P'V'$$

$$PV - y = P'V'$$

$$PV = P'V' + y$$

Therefore, the observed PV is greater than the value expected from an ideal gas. The volume occupied by the gas molecules cause a *positive or over perfect deviation.* This explains the fact that PV, after reaching a minimum, increases with further increase in pressure.

(iii) At extremely low pressures and high temperatures: When P is very small, V will be very large. In this case, both the correction terms a/V^2 and b, in the van der Waals' equation are negligibly small in comparison to P and V and the equation reduces to $PV = RT$, the equation for an ideal gas. Similarly, when the temperature is very high, V will be very large (since volume $\propto$ temperature) or P will be very small and again $PV = RT$ is obtained. Thus, at extremely low pressures and high temperatures, all real gases obey the ideal gas laws.

(iv) At ordinary (medium) pressure: We have seen that at ordinary temperature, the factor a/V^2 predominates at low pressure, whereas the factor Pb predominates at high pressures. Therefore, these two factors will balance each other at some intermediate range of pressures in which the gas will show the ideal behaviour.

Thus, at medium pressure, $$Pb = \frac{a}{V}$$

The Eq. (1.25) is reduced to

$$PV = RT = P'V'$$

and the gas behaves like an ideal gas.

(v) At high temperature: At high temperatures, V is very large and hence both Pb and a/V will be negligibly small, so that we have

$$PV = RT = P'V'$$

i.e., at high temperatures, the behaviour of a real gas nearly agrees with that expected for ideal gas.

(vi) At low temperature: Both P and V are small, hence, both pressure (attraction) and volume corrections are appreciable, with the result that the deviations are more pronounced.

(vii) Exceptional behaviour of hydrogen and helium: The molecules of these gases are of very small mass and consequently the force of attraction between the molecules is always negligible.

Thus, neglecting the term a/V, the Eq. (1.25) is reduced to

$$PV - Pb = RT = P'V'$$

$$PV = P'V' + Pb$$

i.e., the product PV is greater than $P'V'$ even at low pressures.

Numerical Problems

Example 8 Using van der Walls' equation, calculate the pressure exerted by one mole of carbon dioxide. When it occupies a volume of 0.05 l at 100 ϒC, given that $a = 3.592$, $b = 0.0426$ and $R = 0.0821$ l atm/degree mole.

Solution: It is given that, n = one mole, $V = 0.05$ l

$$a = 3.592 \text{ atm l}^2/\text{mole}^{-2}, \qquad b = 0.0426 \text{ l/mole}$$

$$R = 0.0821 \text{ l atm/degree/mole}$$

$$T = 100\,°\text{C} + 273 = 373\,°\text{A}$$

van der Waals' equation, for one mole

$$\left(P + \frac{a}{V^2}\right)(V - b) = RT$$

$$P + \frac{a}{V^2} = \frac{RT}{V - b}$$

$$P = \frac{RT}{V - b} - \frac{a}{V^2}$$

Substituting the values of a, b, V and T in the van der Waals' equation,

$$P = \frac{0.0821 \times 373}{0.05 - 0.0426} - \frac{3.592}{0.0025} = \mathbf{2701.5 \text{ atm}}$$

TRY YOURSELF

1. Show that van der Waals' equation for one mole of a gas can be written the form

$$\frac{PV}{RT}=1+\frac{bP}{RT}-\frac{a}{RTV}+\frac{ab}{RTV^2}$$

2. Comment on the nature of the gas whose equation of state is

$$P(\mathrm{V}-\mathrm{b})=RT$$

3. Calculate the pressure exerted by 10.2 g NH_3 in a 3.0 dm^3 vessel at 25°C (a) using ideal gas equation (b) = 0.0371 $dm^3\ mol^{-1}$.
4. 180 g of steam is contained in a vessel of 25 L capacity under a pressure of 50 atm. Calculate the temperature of the steam. Given that for water vapour, $a = 5.46$ bar $L^2\ mol^{-2}$ and $b = 0.031$ $L\ mol^{-1}$.
5. Out of N_2 and NH_3, which one will have greater value for van der Waals' constant '*a*' and which one will have greater value for van der Waals' constant '*b*'?
6. The van der Waals' constants for two gases are as follows:

Gas	a (atm $L^2\ mol^{-2}$)	b ($L\ mol^{-1}$)
X	1.39	0.0391
Y	3.59	0.0427

Which of them is more easily liquefiable and which has greater molecular size?

Answers

1. Divide van der Waals' equation by RT and rearrange. **2.** This equation is applicable at high pressure because pressure correction can be neglected at high pressure. The volume at high pressure is low, so the volume correction can not be neglected. The factor Pb overweight the factor a/V i.e Pb>a/v. **3.** (a) 4.9 atm (b) 4.8 atm. **4.** 1531.7 K. **5.** (i) As NH_3 is more easily liquefiable (due to hydrogen bonding), intermolecular forces of attraction are stronger than in N_2. Hence, NH_3 will have greater value for '*a*'. (ii) As NH_3 molecule is larger in size than N_2. Hence, NH_3 will have greater value for '*b*'. (For NH_3, $a = 4.17$ L^2 atm mol^{-2}, $b = 0.0371$ $L\ mol^{-1}$. For N_2, $a = 1.39$ L^2 atm mol^{-2}, $b = 0.0319$ $L\ mol^{-1}$). **6.** Greater the value of '*a*' is, the more easy it is liquefiable. Similarly, greater the value of '*b*', greater is the molecular size. Hence, gas *Y* will be more easily liquefiable and will have greater molecular size.

Example 9 Two moles of ammonia are enclosed in a five-litre flask at 27ΥC. Calculate the pressure exerted by the gas assuming that

(i) the gas behaves like an ideal gas

(ii) the gas behaves like a real gas.

Solution: It is given that, $V = 5$ l, $a = 4.17$ atm l^{-2} mol^{-2}

$b = 0.037$ l mol^{-1}, $R = 0.082$ l atm. $degree^{-1}$ mol^{-1}

$T = 27 + 273 = 300$ K,

Number of moles $(n) = ?$

(i) When the gas behaves like an ideal gas, we have

$$PV = n\,RT$$

$$P = \frac{nRT}{V} = \frac{2 \times 0.082 \times 300}{5} = \mathbf{9.84\ atm}$$

(ii) When the gas behaves like a real gas, we apply van der Waals' equation

$$\left(P + a\frac{n^2}{V^2}\right)(V - nb) = nRT \text{ (for } n \text{ moles)}$$

i.e.,
$$P = \frac{nRT}{V - nb} - \frac{an^2}{V^2}$$

Substituting the values of *a*, *b*, *R*, *V* and *T* in above equation

$$P = \frac{2 \times 0.082 \times 300}{5 - 2 \times 0.037} - \frac{4.17 \times 2 \times 2}{5 \times 5}$$

$$P = 9.99 - 0.67 = \mathbf{9.32\ atm}$$

Example 10 **A vessel of 50 l capacity contains 10 moles of steam under 50 atmospheres of pressure. Calculate the temperature of the steam using van der Waals' equation for water $a = 5.46$ atm l² mole^{-2} and $b = 0.031$ l mole.**

Solution: We known that $\left(P + \frac{an^2}{V^2}\right)(V - nb) = nRT$

i.e.,
$$T = \left(P + \frac{an^2}{V^2}\right)\left(\frac{V - nb}{nR}\right)$$

Here, $P = 50$ atm, $n = 10$ moles

$V = 50$ l, $a = 5.46$ atm l² mole^{-2}

and $b = 0.031$ l/mole, $R = 0.082$ l atm/degree/mole

Substituting the values, we get

$$T = \left(50 + \frac{(10)^2 \times 5.46}{(50)^2}\right)\left(\frac{50 - 0.031 \times 10}{10 \times 0.082}\right)$$

$$T = \mathbf{306.2\ K}$$

1.11 ISOTHERMS OF CARBON DIOXIDE—CRITICAL PHENOMENON

The curves representing the variation of volume and pressure at constant temperature are called **Isotherms** (Greek; *iso* = 'equal'). In the case of an idea gas, the isotherms would be rectangular hyperbolas.

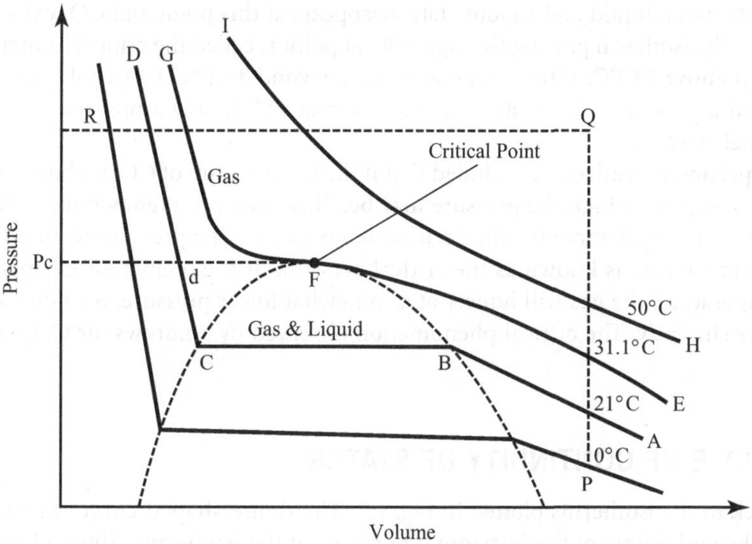

Figure 1.7 Andrews isotherms of CO_2 at different temperatures

In 1869, T. Andrews studied the effect of pressure on volume of CO_2 at different temperatures. When the volume readings were plotted against the pressure, a set of curves of the type shown in Fig. 1.7 was obtained. Each curve corresponds to one particular temperature and is, therefore, called isotherm. The curves should be rectangular, hyperbolas like that expected for an ideal gas. It is quite clear that CO_2 does not give rectangular hyperbolas although with increasing temperature the curves approximate to one. Let us consider the various isotherms separately.

(i) Isotherm at 0 ΥC: The isotherm shows that carbon dioxide is entirely gaseous at low pressures and the volume decreases with increasing pressure in the normal fashion.

(ii) Isotherm below 31 ΥC: The isotherm below 31 °C is not continuous. For example, in the isotherm of 21 °C, the point A corresponds to gaseous CO_2 occupying a certain pressure. At low pressures, if the pressure is increased, the volume decrease as indicated by the curve AB.

At B liquefaction commences; vapour is converted into liquid of very much greater density and the decreases suddenly as the volume occupied by a liquid is much less than that of BC is approximately parallel to the volume axis indicating the sudden decrease in volume taking place at the same pressure.

Liquefaction is complete at C and CD shows the small effect of pressure on a liquid. The steep line CD which is almost vertical represents the compressibility of a liquid and shows that the liquids are much less compressible, i.e., increase in pressure produces a very small decrease in volume.

To summarize, along curve AB there is only vapour, along CD only liquid and along BC liquid and vapour co-exists. Since BC is parallel to the volume axis, there can be only one pressure at which liquid and vapour co-exits. This pressure is called vapour pressure of the gas.

(iii) Isotherm at 31 ΥC: As the temperature is raised, this horizontal portion becomes less and less till at 31.1 °C; it reduces almost to a point F. Above this temperature, there is no trace of horizontal portion, i.e., beyond 31.1 °C, there is no indication of liquefaction at all. Point F is the critical point for CO_2.

The distinction between liquid and vapour state disappears at this point and CO_2 exits in a state called the critical state. The isotherm passing through critical point is called the critical isotherm.

(iv) Isotherm above 31 ϒC: If the temperature was beyond 31.1 °C, CO_2 could not be liquefied. Yet Andrew's applied a pressure of 300–400 atm. However, at 21 °C, only a pressure of 75 atm was quite sufficient to liquefy CO_2.

From the experiment, Andrews concluded that if the temperature of CO_2 is above 31.1 °C, it cannot be liquefied no matter how high the pressure may be. This maximum temperature (31.1 °C) at which liquefaction can be brought about is called critical temperature. The pressure required to liquefy a gas at the critical temperature is known as the critical pressure. If the temperature happens to be below the critical temperature, the gas will liquefy at a somewhat lower pressure. All other gases have been found to behave similarly. The critical phenomenon observed by Andrews for CO_2 may be observed for any gas.

1.12 PRINCIPLE OF CONTINUITY OF STATES

Let us refer **back** to the isotherms plotted in Fig. 1.7. The dome-shaped curve shown by the dashed line connects the end points of the horizontal portions of the isotherms. Since a horizontal portion represents two phases in equilibrium, all points inside the dome correspond to equilibrium between liquid and vapour. For such points, it is always possible to draw any sharp distinction between liquid and vapour, since there is a definite boundary surface between the two phases. For points lying outside the dome, only one phase (liquid or vapour) is present, and here it is not possible to draw any sharp distinction between liquid and vapour.

Actually there is no dividing line between the liquid and the gaseous states and there is a gradual transition from gas to liquid and vice versa without any abrupt change in state. The fact that it is not possible to make distinction between a gas and a liquid is known as the **principle of continuity of state.** The distinction between gas and liquid is possible only within the dome area when both are present in equilibrium and there is boundary surface. In the absence of this boundary, no fundamental distinction between gas and liquid is possible.

When a gas is compressed at a temperature below its critical temperature, its volume is decreased, i.e., the free space between the molecules is decreased and the inter-molecular forces are increased. Consequently, the gaseous state changes to the liquid state. Therefore, **the liquid state is only a continuation of the gaseous state into a region of low volume and high inters molecular forces of attraction**. Below the critical temperature whenever gaseous state is changed to liquid state, we have to pass through a stage when both the liquid and the gaseous states change into liquid state and vice versa without passing through a stage in which both the phases are present together. **In this transition, the continuity of state is said to be maintained.** This can be done as follows.

Consider a certain volume of CO_2 gas represented by a first point P on the isotherm. Let us increase the pressure of the gas at constant volume from point P to point Q by increasing the temperature. After arriving at point Q, the volume is decreased at constant pressure along the line QR by decreasing the temperature. At R, the temperature is again 13.1 °C, but CO_2 exists as liquid; there has never been more than one phase present at any time. The volume of the gas decreases gradually, as the temperature is decreased from Q to R. Ultimately, the molecules are so close that the inter-molecular forces of attraction become strong enough to cause their condensation into the liquid state. Thus, it is not always possible to distinguish a gas from a liquid and the conversion of gas to liquid or vice-versa is continuous. Therefore, the liquid state may be taken as highly compressed gaseous state.

1.13 CRITICAL CONSTANTS

(i) **Critical temperature of a gas (T_c)** is the minimum temperature which must be reached before the gas can be liquefied by the application of pressure. A gas cannot be liquefied by pressure above its critical temperature, no matter how high the applied pressure may be. Each gas has a definite critical temperature, e.g., T_c of CO_2 is 31.1 °C or 304.1 K.

(ii) **Critical pressure of gas (P_c)** is the minimum pressure necessary for the liquefaction of a gas at critical temperature, e.g., P_e for CO_2 is 72.9 atm.

(iii) **Critical volume of a gas (V_c)** is the volume occupied by one mole of the gas or liquid under critical pressure and under critical temperature e.g., V_e for CO_2 is 95.65 ml.

From Andrews experiment, very useful conditions for the liquefaction of a gas may be stated as

(i) The temperature must be at or below the critical temperature.

(ii) A suitable pressure must be applied equal to the critical pressure if the temperature is at the critical temperature but becoming less if the temperature is reduced below this value.

1.13.1 Relations Between van der Waals' Constants and Critical Constants

A very good confirmation of the general character of van der Waals' equation is furnished by its application to critical phenomenon.

The van der Waals' equation is

$$\left(P+\frac{a}{V^2}\right)(V-b)=RT \quad \text{(putting } n=1\text{)}.$$

Alternatively,

$$PV+\frac{a}{V}-Pb-\frac{ab}{V^2}=RT$$

Multiplying throughout by V^2, we get

$$PV^3+aV-PbV^2-ab=RTV^2$$

$$PV^3+aV-PbV^2-ab-RTV^2=0$$

Arranging in descending powers of V.

$$PV^3-PbV^2-RTV^2+aV-ab=0$$

Dividing by P,

$$V^3-bV^2-\frac{RTV^2}{P}+\frac{aV}{P}-\frac{ab}{P}=0$$

$$V^3-V^2\left(b+\frac{RT}{P}\right)+\frac{aV}{P}-\frac{ab}{P}=0 \quad (1.26)$$

This is a third-degree equation in terms of V (cubic equation in V) and below a certain value of temperature T, the equation can have three real roots. With the rise of temperature, these roots approach one another and become identical ultimately. This can be easily understood if we plot a theoretical $P-V$

graph from the van der Waals' equation. The curves obtained are of the type ABLMNCD as shown in the Fig. 1.8.

The three real unequal roots correspond to point B, M and C. The points B and C represent saturated vapour and liquid, respectively, but M has no physical significance. As the temperature rises, the S-shaped dotted portion of the curve gets reduced to a point. The values for the volume of liquid and gas approach each other until at this point, the three values of V become identical. This point corresponds to the critical temperature and the volume V_c. At the critical point, then the three roots of the van der Waals' equation become identical and equal to V_c. Since this is the value of the three roots of the van der Waals' equation at the critical point,

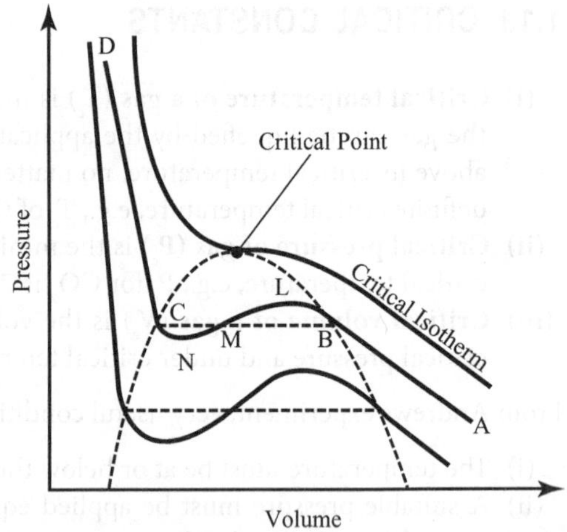

Figure 1.8 Isotherms of CO_2 from van der Waals' equation

$$(V - V_c)^3 = 0$$

or $$V^3 - 3V_cV^2 + 3V_c^2V - V_c^3 = 0 \tag{1.27}$$

Under critical conditions, $T = T$ and $P = P_c$.
Substituting these values in Eq. (1.26),

$$V^3 - V^2\left(b + \frac{RT_c}{P_c}\right) + \frac{V_a}{P_c} - \frac{ab}{P_c} = 0 \tag{1.28}$$

As both the (1.27) and (1.28) are identical, then, the coefficients of the equal powers of V in the two equations may be equated as

$$3V_c = b + \frac{RT_c}{P_c} \tag{1.29}$$

$$3V_c^2 = \frac{a}{P_c} \tag{1.30}$$

$$V_c^3 = \frac{ab}{P_c} \tag{1.31}$$

Dividing Eq. (1.31) by Eq. (1.30), we get

$$\frac{V_c^3}{3V_c^2} = \frac{ab}{P_c} \times \frac{P_c}{a} \quad \text{or} \quad \frac{V_c}{3} = b \quad \text{or} \quad V_c = 3b \tag{1.32}$$

Substituting the value of V_c from Eq. (1.26) in Eq. (1.24), we get

or $$\frac{a}{P_c} = 3 \times (3b)^2 = 27b^2$$

or $$P_c = \frac{a}{27b^2} \tag{1.33}$$

Substituting the values of V_c and P_c in Eq. (1.23)

$$3 \times 3b = b + RT_c \times \frac{27b^2}{a}$$

$$9b = b + RT_c \frac{27b^2}{a}$$

$$8b = RT_c \times \frac{27b^2}{a}$$

$$T_c = 8b \times \frac{a}{27b^2} \times \frac{1}{R}$$

Alternatively,

$$T_c = \frac{8a}{27bR} \tag{1.34}$$

1.13.2 Derivation of $P_c V_c = \frac{3}{8} RT_c$ from van der Waals' Equation

From van der Waals' equation, we have

$$V^3 - V^2\left(b + \frac{RT_c}{P_c}\right) + V\frac{a}{P_c} - \frac{ab}{P_c} = 0 \tag{1.35}$$

In addition,

$$V^3 - 3V_cV^2 + 3V_c^2V - V_c^3 = 0 \tag{1.36}$$

Equating the coefficients of Eqs (1.35) and (1.36), we get

$$3V_c = \frac{RT_c}{P_c} + b \tag{1.37}$$

$$3V_c^2 = \frac{a}{P_c} \tag{1.38}$$

$$V_c^3 = \frac{ab}{P_c} \tag{1.39}$$

Dividing Eq. (1.39) by Eq. (1.38), we get

$$V_c = 3b \quad \text{or} \quad b = \frac{1}{3}V_c \tag{1.40}$$

Substituting the value of b in Eq. (1.37), we get

$$3V_c = \frac{RT_c}{P_c} + \frac{V_c}{3}$$

or

$$3V_c - \frac{V_c}{3} = \frac{RT_c}{P_c}$$

or

$$\frac{8}{3}V_c = \frac{RT_c}{P_c}$$

$$P_cV_c = \frac{3}{8}RT_c \tag{1.41}$$

1.13.3 Calculation of van der Waals' Constants in Terms of T_c and P_c

Rearranging Eq. (1.41), we get

$$V_c = \frac{3}{8} \times \frac{RT_c}{P_c}$$

Squaring we get

$$V_c^2 = \frac{9}{64} \times \frac{R^2 V_c^2}{P_c} \tag{1.42}$$

Substituting the value of $V_c^2 = \frac{a}{3P_c}$ from Eq. (1.38) in Eq. (1.42), we get

$$\frac{a}{3P_c} = \frac{9}{64} \times \frac{R^2 T_c^2}{P_c^2}$$

$$a = \frac{27 R^2 T_c^2}{64 P_c} \tag{1.43}$$

Substituting the value of $V_c = 3b$ in Eq. (1.37), we get

$$9b = \frac{RT_c}{P_c} + b \quad \text{or} \quad 8b = \frac{RT_c}{P_c}$$

or

$$b = \frac{RT_c}{8P_c} \tag{1.44}$$

1.14 LAW OF CORRESPONDING STATES

According to van der Waals' equation,

$$\left(P + \frac{a}{V^2}\right) \quad (V - \text{b}) = RT$$

In this equation, we have constants 'a' and 'b' which cannot be measured directly. Therefore, this equation is not applicable to gases and in order to make it so, these constants must be removed.

It has been observed experimentally that near the critical point, all gases give the same type of isotherm as CO_2, but the actual position of the critical isotherm varies with the critical constants of the gas studied. The similarity is great enough to suggest that if for each different gas, a suitable scale of pressure, volume and temperature is chosen; it will be possible to superimpose all the critical isotherms on one another. If this were done for the critical isotherms, it is probable that all other isotherms would coincide. Under such circumstances, we should have one isothermal diagram and one fundamental equation for all gases. Such a scale will be one in which the pressure, volume and temperature will be expressed a fraction of the critical pressure, volume and temperature.

In order to get such an equation, the values of P, V and T must be expressed as corresponding fractions of P_e, V_c and T_e, i.e., the critical pressure, volume and temperature.

Putting

$$\pi = P/P_e \quad \text{or} \quad P = \pi P_e$$

$$\phi = V/V_c \quad \text{or} \quad V = \phi V_e$$

$$\theta = T/T_e \quad \text{or} \quad T = \theta T_e$$

where π, ϕ and θ are as the reduced pressure the reduced volume and the reduced temperature, respectively.

Substituting the values of P, V and T in the van der Waals' equation:

$$\left(\pi P_e + \frac{a}{\phi^2 V_c^2}\right)(\phi V_e - b) = R\,\theta\,T_c$$

We have already seen that

$$P_e = \frac{a}{27\,b^2},\ V_e = 3b \text{ and } T_c = \frac{8a}{27R\,b}$$

Substituting these values in above equation, we get

$$\left(\pi\frac{a}{27b^2} + \frac{a}{9\phi^2 b^2}\right)(3\,\phi - b - b) = R\theta\left(\frac{8a}{27\,Rb}\right)$$

or,

$$\frac{a}{27b^2}\left(\pi + \frac{3}{\phi^2}\right)\times b(3\,\phi - 1) = \frac{a8\theta}{27b}$$

or,

$$\frac{a}{27b}\left(\pi + \frac{3}{\phi^2}\right)\times b\,(3\,\phi - 1) = \frac{a}{27b}\,8\theta$$

$$\left(\pi + \frac{3}{\phi^2}\right)(3\,\phi - 1) = 8\theta$$

This equation involving neither of the characteristic constants a and b and T_e, P_e, and V_e is known as **van der Waals' reduced equation of state.** This equation is applicable to all substances (liquid or gaseous) irrespective of their nature, and is thus a perfectly general equation.

Two or more substances are said to be in the corresponding states, when they have the same reduced temperature, pressure and volume.

It is evidently clear from the above equation that when two substances are, at the same, reduced temperature (θ) and pressure (π), they must have the same reduced volume (ϕ). This statement is called the law of corresponding states or principle of corresponding states.

1.14.1 Significance of Law of Corresponding States

It has a great significance in the study of the relationship between physical properties and chemical constitution of various liquids. As we know, pressure has practically no effect on liquid; therefore, while studying the relation between physical properties and chemical constitution, the determination of these properties should be carried out at the same reduced temperature.

It has been found that the boiling point of a liquid at atmospheric pressure is approximately two-third of its critical temperature (T_c) when expressed in absolute scale. It means that liquids are in corresponding states at their boiling points (in degrees absolute) approximately. Consequently, boiling point is regarded as the corresponding temperature. Therefore, various physical properties of liquid, such as molecular volumes of various liquids, should be determined at their respective boiling points.

1.15 LIMITATIONS OF VAN DER WAALS' EQUATION

Van der Waals' equation has been very successful in expressing the behaviour of real gases qualitatively over wide ranges of temperature and pressure, but fails in quantitative agreement. The limitations of the equation may be enumerated as:

(i) At very high pressures and low temperatures near the critical points, the equation becomes inaccurate.
(ii) The values of '*a*' and '*b*', which are constants in the equation, are not independence of temperature, thus '*a*' and '*b*' are no longer regarded as absolute constants.
(iii) The critical constants as calculated by van der Waals' equation do not agree well with experimental values.
(iv) Law of corresponding state is followed only approximately.
(v) Further, van der Waals' equation shows that the coefficient $\frac{RT_c}{P_c V_c}$ is equal to 2.67 for all gases but experimental results are generally higher and vary from gas to gas.
(vi) According to van der Waals' equation, the isotherm should be continuous at all temperature but experimental isotherms are found to be discontinuous below the critical temperature.

SOME NOTEWORTHY POINTS

1. Compressibility factor is $Z = \frac{PV}{RT}$. Its value is 1 for ideal gases.
2. Real gases behave in nearly ideal manner at high temperature and low pressure.
3. Easily liquefiable gases, e.g., CO_2, SO_2 and NH_3 deviate more from ideal behaviour.
4. H_2 gas shows positive deviation at all pressures.
5. The van der Waals' constant '*a*' is a measure of inter molecular forces of attraction and increases with the ease of liquification of gas.
   It has value in the order $SO_2 > NH_3 > H_2O > CO$. The constant '*b*' is effective volume of gas molecules. Larger the value of '*b*' is, larger will be the molecular volume and diameter.

REVISION QUESTIONS

1. List the general properties of gases.
2. State and explain the following laws:
   (a) Boyle's law
   (b) Charle's law
   (c) Avogardo's law
3. What is an ideal gas equation? How will you derive it?
4. Write the physical significance of gas constant R. Calculate its numerical value in terms of different units.

5. What are the postulates of kinetic theory of gases?
6. Derive expression for the pressure exerted by an ideal gas or derive kinetic gas equation.
7. Derive kinetic gas equation keeping in view the assumptions of kinetic molecular theory of gases.
8. Derive Boyle's and Charle's law from the kinetic gas equation.
9. What is the significance of the assumption that
   (a) Molecular collisions are elastic?
   (b) Gravity has no effect on the motion of gas molecules?
10. Derive Charle's law, Graham's law of diffusion and Dalton's law of partial pressure from the kinetic gas equation.
11. State Avogadro's law and derive it from kinetic gas equation.
12. (a) Distinguish an ideal gas and a real gas.
    (b) Define compressibility factor and Boyle's temperature.
    (c) Define ideal gas and explain how it is different from real gases.
13. Explain the deviations of real gases from ideal behaviour.
14. Do real gases obey the equation $PV = RT$? Give reasons for you answer.
15. Explain why gases fail to obey the general gas equation of high pressure and low temperature.
16. What are the limitations of the equation $PV = RT$? What improvements have been suggested by van der Waals?
17. Derive van der Waals' equation of state for n moles of a gas.
18. Show that van der Waals' equation for one mole of a gas can be written in the form

$$\frac{PV}{RT} = 1 + \frac{bP}{RT} - \frac{a}{RTV} + \frac{ab}{RTV^2}$$

19. What is excluded volume? Show that the excluded volume is four times the actual volume of a gas.
20. How does van der Waals' equation explain the behaviour of real gases when
    (a) pressure is low,
    (b) pressure is high and
    (c) temperature is high
21. (a) How does van der Waals' equation explain the exceptional behaviour of hydrogen and helium?
    (b) What are the limitations of van der Waals' equation?
22. (a) What is the S.I. units of van der Waals' constants 'a' and 'b'?
    (b) Explain the significance of the van der Waals' constants 'a' and 'b'?
23. What is van der Waals' equation? Discuss the equation when
    (a) pressure is not too high
    (b) temperature is high.
24. (a) What are the assumptions of the kinetic theory of gases? Which of the assumptions led to the development of van der Waals' equation?
    (b) Using kinetic gas equation, show that kinetic energy of gas molecules depend only on absolute temperature.
    (c) Write gas equation for one mole of each of ideal and real gases.

25. Suggest explanation for the fact that
    (a) Ideal volume of a gas is less than the observed volume.
    (b) Ideal pressure of a gas is more than the observed pressure.
    (c) Gases approach ideal behaviour at low pressure and high temperature.
26. Explain that
    (a) The ideal volume of gas is different for its observed volume.
    (b) The ideal pressure of the gas is different from its observed pressure.
    (c) No real gas obeys Boyle's law strictly.
    (d) Gases deviate from ideal behaviour at high pressure and low temperature.
27. 'The liquid state is a continuation of the gaseous and not a sharp and discontinuous transition from one phase to the other to the other.' Prove this statement giving behaviour of CO_2 at different temperatures.
28. Explain what is meant by critical constants of a gas? Derive their relationship with van der Waals' constant.
29. Derive the following:

$$P_c V_c = \frac{3}{8} RT_c$$

30. Define law of corresponding states and give its significance.
31. State Joule–Thomson effect. What is the significance of inversion temperature in this regard?
32. (a) Distinguish real and ideal gases.
    (b) Define:
        (i) critical temperature
        (ii) law of corresponding states
    (c) What do you understand by isotherms of CO_2?
    (d) What are the units of van der Waals' constants 'a' and 'b'?
    (e) Under what conditions liquefaction of gases can take place?
33. (a) Why do real gases show deviations from behaviour?
    (b) Show that $T_c = \dfrac{8a}{27Rb}$
    (c) What are van der Waals' constants 'a' and 'b'?
34. (a) What is an ideal gas? Why do real gas deviates from ideal behaviour? Derive van der Waals' equation of state.
    (b) Define the following terms:
        (i) critical temperature
        (ii) critical pressure
        (iii) critical volume
35. From Kinetic gas equation, derive
    (a) Boyle's law
    (b) Avogadro's law
    (c) Graham's law of diffusion.
36. (a) Define Charles law. Derive this law from kinetic gas equation.
    (b) What is ideal gas? Explain the causes of deviation of gas from ideal behaviour.

37. (a) Write Van der Waals' equation. Discuss the deviations of real gases from ideal behaviour at
(i) moderate pressure
(ii) very high pressure
(b) What is meant by Boyle's temperature and Triple points?
(c) Derive an expression for the pressure of ideal gas by means of Kinetic theory of gases.
38. Give reasons for the following:
(a) The tyre of an automobile in inflated to lesser pressure in summer than in winter. why?
(b) Liquid amonia bottle is cooled before opening the seal. Why?
(c) Gases can be liquified by increasing pressure alone, discuse.

MULTIPLE CHOICE QUESTIONS

1. The value of gas constant per degree per mole is approximately
(a) 1 cals
(b) 2 cals
(c) 3 cals
(d) 4 cals
2. In the gas equation PV = nRT, the value of universal gas constant would depend only on
(a) the nature of the gas
(b) the pressure of the gas
(c) the temperature of the gas
(d) the units of measurement
3. Which set of conditions represents easiest way to liquify a gas
(a) Low temperature & high pressure
(b) High temperature & low pressure
(c) Low temperature & low pressure
(d) High temperature & high pressure
4. Gases deviate from ideal behaviour because their molecules
(a) posses negligible volume
(b) have forces of attraction between them
(c) are polyatomic
(d) are not attached to one another
5. At relatively high pressure, van der Waals' equation reduces to
(a) PV = RT0..
(b) PV = RT-a/V
(c) PV = RT + Pb
(d) PV = RT-a/V2
6. The density of gas A is four times that of gas B. If the molecular weight of A is M, that of B is
(a) 2M
(b) 4M
(c) M/4
(d) M/2
7. The pressure of real gases is less that that of ideal gas because of
(a) increase in the number of collisions
(b) finite size of particles
(c) intermolecular attraction
(d) increase in kinetic energy of the molecules
8. At lower temperatures, all gases show
(a) negative deviation
(b) positive deviation
(c) positive and negative deviation
(d) none
9. The molecular velocity of a gas is
(a) independent of temperature
(b) proportional to absolute temperature
(c) proportional to square root of absolute temperature
(d) proportional to square of absolute temperature
10. According to the ideal gas laws, the molar volume of a gas is given by
(a) 2.24L
(b) RT/P
(c) 8RT/PV
(d) RT/PV
11. When the universal gas constant (R) is divided by Avogadro's number (N) their ratio is called

(a) Planck's constant
(b) Rydberg constant
(c) Boltzman constant
(d) Van dar Waals' constant

12. A gas can be liquefied by pressure alone when its temperature is
(a) higher than the critical temperature
(b) lower than its critical temperature
(c) equal to its critical temperature
(d) none

13. The kinetic energy of gas molecules is equal to
(a) 3/2 RT
(b) 2/3 RT
(c) RT/2
(d) 2R/3

14. The kinetic theory of gases predicts that total kinetic energy of a gaseous assembly depends on
(a) pressure of the gas
(b) temperature of the gas
(c) volume of the gas
(d) pressure, volume and temperature of the gas

15. The temperature at which real gases obey the ideal gas laws over a wide range of pressure is called
(a) Critical temperature
(b) Boyle temperature
(c) Inversion temperature
(d) Reduced temperature

16. For a given mass of a gas, if pressure is reduced to half and temperature is increased two times, then the volume would become
(a) V/4
(b) $2V^2$
(c) 6 V
(d) 4V

17. Gases deviate from ideal behavior at high pressure. Which of the following is correct for non-ideality
(a) At high pressure, the collision between the gas molecules becomes enormous
(b) At high pressure, the gas molecules move only in one direction
(c) At high pressure, the volume of gas becomes insignificant
(d) At high pressure the intermolecular interaction become significant

18. Among the following gases which one has the lowest root mean square velocity at 25°C
(a) SO_2
(b) N_2
(c) O_2
(d) Cl_2

19. If a gas expands at constant temperature:
(a) The pressure increases
(b) K.E. of the molecules decreases
(c) K.E. of the molecules remains constant
(d) The number of gaseous molecules increases

20. Which of the following is not correct in case of kinetic theory of gases ?
(a) Gases are made up of small particles called molecules
(b) The molecules are in constant motion
(c) When molecules collide, they lose energy
(d) When the gas is heated, the molecules move faster

21. When do you expect a real gas to behave like an ideal gas ?
(a) When temperature and pressure are low
(b) When temperature and pressure are high
(c) When temperature is low and pressure is high
(d) When temperature is high and pressure is low

22. When an ideal gas undergoes unrestrained expansion, no cooling occurs because the molecules
(a) are above inversion temperature
(b) exert no attractive forces on each other
(c) do work equal to loss in K.E.
(d) collide with loss of energy

23. Two substances with the same reduced pressure and the same reduced temperature will have the same reduced volume. This is called
(a) Gay-Lussac law
(b) Law of corresponding states
(c) Law of reciprocal proportions
(d) Continuity of state

24. The compressibility factor of gas is defined as Z = PV/RT. The compressibility factor of an ideal gas is
 (a) 0
 (b) Infinite
 (c) 1
 (d) –1
25. The Joule Thomson coefficient is zero at
 (a) inversion temperature
 (b) critical temperature
 (c) absolute temperature
 (d) below 0°C
26. Most probable velocity, average velocity and R.M.S. velocity are related as
 (a) 1 : 1.128 : 1.224
 (b) 1 : 1.128 : 1.424
 (c) 1 : 2.128 : 1.224
 (d) 1 : 1.428 : 1.442
27. The average K.E. of an ideal gas per mole is approximately equal to
 (a) three times the absolute temperature
 (b) absolute temperature
 (c) two times the absolute temperature
 (d) 1.5 times the absolute temperature
28. In deriving the kinetic gas equation, use is made of root mean square velocity of the molecules because it is
 (a) the average velocity of the molecules
 (b) the most probable velocity of molecules
 (c) the square root of the average square velocity of molecules
 (d) the most accurate form in which velocity can be used in these calculations
29. A gas deviates from ideal bevaiour when it is subject to
 (a) high temp & low pressure
 (b) high temp. & high pressure
 (c) low temp. & high pressure
 (d) low temp. & low pressure
30. The critical temperature of a substance is
 (a) the temperature above which the substance undergoes decomposition
 (b) the temperature above which a substance can exist only as a gas
 (c) B.P. of the substance
 (d) All are wrong
31. Which of the following are not the basic assumptions of the kinetic theory ?
 (a) Molecules of the gas are perfectly elastic hard spherical particles and are all alike
 (b) The molecules of the gas are always in the state of ceaseless haphazard motion
 (c) The molecules exert appreciable attraction on each other. Therefore energy is lost due to collision or mutual attraction
 (d) The molecules move in a straight line with uniform velocity between two collisions
32. According to kinetic theory of gases,
 (a) the pressure exerted by a gas is proportional to mean square velocity of the molecules
 (b) the pressure exerted by the gas is proportional to the root mean square velocity of the molecules
 (c) the root mean square velocity is inversely proportional to the temperature
 (d) the mean translational K.E. of the molecule is directly proportional to the absolute temperature
33. At the same temperature and pressure, which of the following gases will have the highest K.E. per mole?
 (a) Hydrogen
 (b) Oxygen
 (c) Methane
 (d) All the same
34. Energy in an ideal gas is
 (a) completely kinetic
 (b) completely potential
 (c) KE + PE
 (d) All the above
35. An ideal gas is
 (a) an imaginary gas which obeys gas laws strictly
 (b) a real gas at high pressure and low pressure
 (c) a real gas at high temperature and low pressure
 (d) Helium gas

36. Hydrogen and helium are kept in two identical bottles at room temperature and atmospheric pressure
    (a) Both contain the same number of atoms
    (b) Number of atoms of hydrogen is one-half that of helium
    (c) Number of atoms of hydrogen is double that of helium
    (d) None of the above statements is true
37. At absolute zero, which of the following statements is true?
    (a) The volume of an ideal gas is zero
    (b) All molecular motion ceases
    (c) The average kinetic energy of translation of molecules is zero
    (d) All are correct
38. At very low pressures
    (a) a/V^2 approaches zero
    (b) b is negligible compared to V
    (c) Vander Waals' equation reduces to PV = RT
    (d) All of the above
39. For a given mass of gas, if pressure is reduced to half and temperature is doubled, then volume V will becomes
    (a) 4V
    (b) $2V^2$
    (c) V/4
    (d) 8V
40. An ideal gas obeying kinetic gas equation can be liquefied if
    (a) its temperature is more than critical temperature
    (b) it cannot be liquefied at any value of P and T
    (c) its pressure is more than critical pressure
    (d) its pressure is more than critical pressure but temperature is less than critical temperature
41. The root mean square velocity of a gas is doubled when the temperature is
    (a) reduced to half
    (b) reduced to one fourth
    (c) increased four times
    (d) increased two times
42. Attractive forces and size effects in a gas can be neglected at
    (a) low temperature
    (b) low pressure
    (c) low pressure and high temperature
    (d) independent of the temperature
43. The kinetic energy of particles of an ideal gas is a measure of
    (a) density
    (b) temperature
    (c) viscosity
    (d) refractive index

ANSWERS

1. (b)	12. (b)	23. (b)	34. (a)
2. (d)	13. (a)	24. (c)	35. (a)
3. (a)	14. (b)	25. (a)	36. (c)
4. (b)	15. (b)	26. (a)	37. (d)
5. (c)	16. (d)	27. (d)	38. (d)
6. (c)	17. (d)	28. (d)	39. (a)
7. (c)	18. (d)	29. (c)	40. (b)
8. (a)	19. (c)	30. (b)	41. (c)
9. (c)	20. (c)	31. (c)	42. (c)
10. (b)	21. (d)	32. (b)	43. (b)
11. (c)	22. (b)	33. (d)	

The Liquid State 2

CHAPTER OBJECTIVES

- 2.1 Introduction
- 2.2 General Characteristics of Liquids
- 2.3 Classification of Physical Properties of Liquids
- 2.4 Surface Tension
- 2.5 Viscosity
- 2.6 Refractive Index
- 2.7 Optical Activity
- 2.8 Polarity of Bonds
- 2.9 Dipole Moment

If I were in the world of molecules, I would rather be in liquid state with some freedom of gases and some order of solids.

Anon

Water existed on earth long before life began and is fundamental to most life processes. Although simple in structure, the ubiquitous water molecule is one of the most complex substances known; a complete understanding of its intermolecular behaviour still eludes us.

Felex Franks

2.1 INTRODUCTION

On the one hand, we have gases which are characterized by complete randomness on the molecular scale and on the other hand, we have ideal crystals which have the most orderly arrangement. Between the two extremes, we have liquids which represent a compromise between order and disorder. From the viewpoint of kinetic theory, a liquid can be thought of as continuation of the gaseous phase into the region of small volumes and very high cohesive forces. The molecular attractions in a liquid are much stronger than in gas even at high pressures and that is why the liquid molecules are confined to a definite volume. However, the molecules of a liquid are not as rigid as molecules of a solid; rather, they have some freedom of motion. This is expressed by saying that *liquid state is intermediate between gaseous state and solid state.*

The difference in the behaviour of the three states of matter is mainly due to the intermolecular distances of separation. In a solid, the constituent particles are closely packed (almost touching each other) and may be treated as hard spheres. In a liquid, the molecules are separated by small distances; while in a gas, the intermolecular distances are so large that the molecules can move randomly within the vessel. These intermolecular distances suggest that the intermolecular forces of attraction in a solid are very strong; whereas in case of gases, they are very weak and in case liquids, they are in between i.e. they are fairly strong to keep them in the condensed state but not strong enough to hold them in fixed positions (as in a solid). In fact, all the characteristic properties of liquids such as vapour pressure, surface tension, viscosity etc. arise due to the nature of the intermolecular forces existing in the liquids. Some of these properties are described in the following sections.

2.2 GENERAL CHARACTERISTICS OF LIQUIDS

The important general characteristics of liquids are as follows:

(i) **Shape:** Like gases, molecules in a liquid are in constant random motion. Therefore, *a liquid has no shape of its own but assumes the shape of the container in which it is placed.*

(ii) **Volume:** Unlike gases, liquids have a definite volume. This is because the molecules of a liquid are sufficiently close together and the intermolecular forces are quite strong. As a result, the molecules are not free to occupy any space. Therefore, *the volume of a given amount of a liquid remains unchanged irrespective of the shape or size of the vessel in which it is present.*

(iii) **Density:** *The densities of the substances in the liquid state are about a thousand times greater than the densities of the corresponding gaseous state under similar conditions.* For example, the density of liquid water at l00 °C and 1 atmospheric pressure is 0.958 g cm^{-3} as compared with that of water vapour which is 0.000588 g cm^{-3} under the same conditions. The higher densities of liquids are due to the closer packing of the molecules in the liquid state.

(iv) **Compressibility:** Liquids are practically incompressible. This is due to the very little free space available in the liquids.

(v) **Diffusion:** *Molecules in the liquid state diffuse in the same manner as those in the gaseous state.* However, their rate of diffusion is much slower. This is due to the fact that molecules in the liquid are much closer to one another as compared to those in the gaseous state. The number of collisions per second per unit volume between molecules, therefore, is much larger, so that the mean distance travelled between two successive collisions is very small. As such diffusion takes place very slowly.

(vi) **Evaporation:** Evaporation is the process of escaping of molecules from liquid to vapour (or gaseous) state. As in gaseous state, all the molecules in liquid state do not possess the same velocity or kinetic energy. Some of the molecules, because of mutual collisions and exchange of energy, possess higher kinetic energy than others. These molecules tend to escape into the space above liquid surface and form vapour. *This process of escaping of molecules from liquid to vapour (or gaseous) state is called evaporation.*

As the molecules with high kinetic energies keep on leaving the liquid phase by evaporation, the average kinetic energy of the remaining molecules continues to fall. This is because each escaping molecule takes away with it more than the average amount of energy. *Since the molecules left behind have lower average kinetic energy, the temperature of liquid falls* ($\because$ K.E. $\propto$ T). *This is how evaporation causes cooling.*

(vii) **Vapour pressure:** When evaporation takes place in a closed and limited space in a closed vessel at constant temperature, evaporation proceeds for some time and then seems to stop.

This is due to the fact that an equilibrium is established in which the rate of evaporation becomes equal to the rate of condensation. Thus, a dynamic equilibrium known as **phase equilibrium** is set up. At this stage, the system does not undergo any further change although the two opposing proceed on with equal rates.

$$H_2O_{(l)} \underset{\text{Condensation}}{\overset{\text{Evaporation}}{\rightleftharpoons}} H_2O_{(g)}$$

The molecules of the liquid which escape into the vapour phase exert a pressure on the liquid phase. Once the equilibrium has been set up at a particular temperature, the concentration of the vapour and the pressure that it exerts also becomes constant. *The pressure exerted by the vapour in equilibrium with its liquid at a given temperature is known as the vapour pressure of the liquid at that temperature.*

Factors affecting vapour pressure: The vapour pressure of a liquid depends upon the following factors.

(1) **Nature of liquid:** It is well known that each liquid has a characteristic vapour pressure at a given temperature. This is because each liquid has different magnitudes of intermolecular attractive forces and, therefore, different tendencies to evaporate.

(2) **Temperature:** It has been observed that the vapour pressure of every liquid increases, as the temperature of the liquid increases. As the temperature of the liquids is increased, the average kinetic energy of the molecules of the liquid also goes up with this the number of energetic molecules capable of escaping into the vapour phase also becomes large. As a result, the rate of evaporation and the concentration of the molecules in the vapour phase increase.

(viii) **Boiling point:** When a liquid is heated in an open container, its vapour pressure keeps on increasing with the rise in temperature till it becomes equal to the atmospheric pressure. At this point, molecules of the liquid are readily converted to the vapour phase and the liquid is said to boil. *The temperature at which vapour pressure of the liquid becomes equal to the atmospheric pressure is known as the boiling point of the liquid.*

(ix) **Heat of vaporization:** *The amount of heat necessary to vaporize one mole of the liquid at constant temperature is known as the molar heat of vaporization.* This quantity gives a measure of the intermolecular attractive forces characteristic of the liquid. For example, the heat of vaporization of water at 100 °C is 40.6 kJ per mole while that of benzene at 80 °C is 31 kJ per mole. This means attractive forces between water molecules are stronger than between benzene molecules.

2.3 CLASSIFICATION OF PHYSICAL PROPERTIES OF LIQUIDS

The physical properties of liquids are classified into the following four types:

(i) **Additive property:** *It is defined as the property which is the sum of the corresponding properties of the atoms constituting the molecule.* For example, molecular mass of a molecule can be obtained by adding up the atomic masses of the constituent atoms. It is, therefore, additive property.

(ii) **Constitutive property:** *It is defined as the property which depends upon the arrangement of the atoms within the molecule, i.e., upon the constitution of the molecules.* For example, optical activity of a molecule depends upon the spacial arrangement of the constituent atoms or groups in molecule. It is, therefore, constitutive property.

(iii) **Additive and constitutive property:** *It is defied as the property which is obtained by adding the corresponding properties of the constituent atoms as well as depends upon the arrangement of atoms within the molecule.* For example, parachor, molar viscosity, molar refraction etc.

(iv) **Colligative property:** *It is the property which depends upon the number of molecules, atoms and ions present in the solution, i.e. their concentration and not upon their nature.* For example, osmotic pressure, lowering of vapour pressure, elevation in boiling point and depression in freezing point etc.

2.4 SURFACE TENSION

A molecule in the interior of a liquid is completely surrounded by other molecules and thus it is attracted equally in all directions. However, a molecule in the surface is attracted inwards, since the number of the molecules is greater in the bulk of the liquid than in the vapour. As a result of this inward pull, the surface of a liquid tries to attain the smallest possible area and is always stretched. In order to extend the area of an interface, i.e. to bring molecules from the bulk into the surface, work must be done against the cohesive forces in the liquid. Thus, the surface portion of the liquid has a higher energy than the bulk liquid. The extra surface energy is best described by saying that there is surface tension acting parallel to the surface, which opposes any attempt to extend the surface. A tension is also called negative pressure and pressure is force per unit length. Thus, *surface pressure is a force per unit length.* If the surface tension is γ, the work done on the surface in extending its area by an amount dA is $dW = \gamma dA$.

Definition of surface tension: *Surface* tension may be defined *as the force (in newtons or dynes) acting per unit length along the surface of liquid at right angle to any line taken in the surface of liquid in any direction.* The S.I. unit of surface tension is newton per metre (Nm^{-1}) and in C.G.S., it is dyne cm^{-1}.

Surface energy: To increase the area of the surface of a liquid, some work has to be done against the inward pull. *The work in ergs required to be done to increase the surface area by 1 sq. cm is called surface energy.*

$$\text{Surface energy} = \text{work per sq. cm} = (\text{force} \times \text{length}) \text{ per sq. cm}$$

$$= \frac{\text{Dynes}}{\text{cm}^2} \times \text{cm} = \text{ dynes cm}^{-1}$$

This unit is the same as that of surface tension. The surface energy is the same thing as surface tension. Hence, the definition of surface energy may be regarded as the second definition of surface tension.

2.4.1 Some Important Results

Some important results related to the property of the surface tension are briefly described as follows:

(i) **Spherical shape of drops:** Surface tension tries to decrease the surface area of the liquid to the minimum. The drops of a liquid (or bubbles of a gas) are spherical because for given volume, a sphere has minimum surface area.

(ii) **Rise of a liquid in a capillary tube:** Suppose one end of a capillary tube is put into liquid that wets glass (Fig. 2.1). It is found that the liquid rises into the capillary tube to a certain height. This rise is obviously due to the inward pull of surface tension acting on the surface which pushes the liquid into the capillary tube. It is because of the same reason that oil rises into the wick of an oil lamp or water below the surface of the earth rises to the plants through the roots or ink rises in a blotting paper.

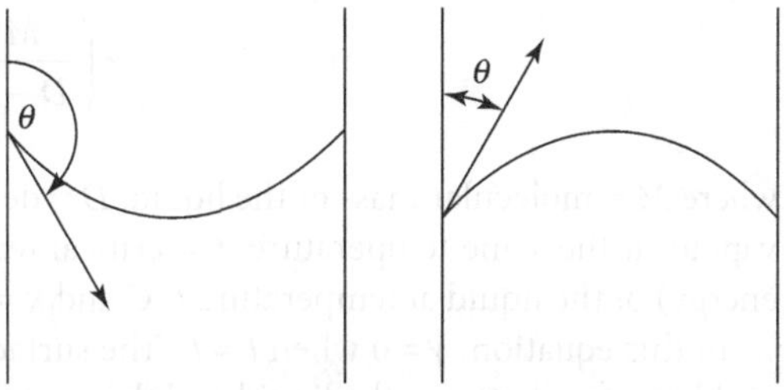

Figure 2.1 Contact angle of meniscus

It may be mentioned here that in case of liquids which do not wet glass, e.g. mercury, the level inside the capillary falls below the level outside. Further, whereas the upper surface of a liquid that wets glass is concave; however, that of mercury is convex. Such a surface is, therefore, *called meniscus. The angle which the curved surface (i.e. meniscus) makes with the wall of the tube is called contact angle and is found by drawing a tangent at that point* (Fig. 2.1).

(iii) **Interfacial tension:** If two immiscible or partially miscible liquids are taken in a vessel, then the surface tension acting along their surface of separation, i.e. along the *interface*, is called the interfacial tension. Its value is generally intermediate between the surface tension of the two liquids, but some times it is less than both. This is due to the fact that at the interface the molecules of one liquid are attracted by the molecules of the other.

In fact, the surface tension of a liquid is also an interfacial tension; the two phases involved being the liquid and the vapour above it.

(iv) **Surface active agents:** There are certain substances such as soaps and certain sulphonic acids and certain other organic compounds such as methyl alcohol, ethyl alcohol and acetone which when added to water even in small amount decrease the surface tension of water to a considerable extent. *Such substances which lower the surface tension of water are called surface active agents.* It is for this reason that the soap acts as a detergent. For example, if grease is sticking on the surface of a cloth, water does not wet the grease and hence it is not removed by water alone. However, if soap is added to water, it lowers the interfacial tension between water and grease. As a result, the grease mixes into the water (soap solution) and hence is removed from the surface of the cloth.

2.4.2 Effect of Temperature on Surface Tension

The surface tension of liquids generally decreases with the increase of temperature and becomes zero at the critical temperature. Critical temperature is defined as the temperature at which the meniscus between the liquid and the vapour disappears. The decrease in surface tension with the increase of temperature is obviously due to the fact that with the increase of temperature, the kinetic energy of the molecules increases. In other words, the speeds of molecules increase and, therefore, intermolecular attraction decreases.

An important relationship between temperature and surface tension, as deduced by Eotvas (in 1886) and modified by Ramsay and Shields (in 1893), is as follows:

$$\gamma\left(\frac{M}{D-d}\right)^{2/3} = k(t_c - t)$$

where M = molecular mass of the liquid, D = density of the liquid at temperature t °C, d = density of the vapour at the same temperature, t_c = critical temperature of the liquid, γ = surface tension (or surface energy) of the liquid at temperature t °C and k = a constant.

In this equation, $\gamma = 0$ when $t = t_c$. The surface tension vanishes at the critical temperature when the visible surface between the liquid and the vapour disappears.

2.4.3 Measurement of Surface Tension

The surface tension of a liquid can be obtained by various methods; out of these, the three most commonly used methods are:

(i) *The stalagmometer or drop pipette method*
(ii) *Capillary rise method*
(iii) *Du Nouy tensiometer method*

2.4.3.1 The Stalagmometer Method

This method is carried out in two different ways as follows:

(a) **Drop-weight method:** This method is based upon the principle that the weight of drop falling slowly out of a capillary held vertically is directly proportional to its surface tension. Thus, for two liquids, the drops of which fall almost at the same rate out of the same capillary,

$$\frac{\gamma_1}{\gamma_2} = \frac{m_1}{m_2} \tag{2.1}$$

where γ_1 and m_1 are respectively the surface tension and weight of each drop of the first liquid and γ_2 and m_2 are the corresponding values for the second liquid. Thus, knowing the surface tension of one of the liquids, that of the other can be determined.

The apparatus used is called a *stalagmometer* and is shown in Fig. 2.2. It resembles a pipette with a bulb in the middle and a mark M on the tube above the bulb. However, the lower portion is made up of a capillary with a flattened tip.

Procedure: The various steps are:

(i) The stalagmometer is first cleaned with chromic acid to remove any greasy matter sticking on it and then thoroughly washed with water.

(ii) It is clamped vertically in a clamp stand and a rubber tube with a pinch cock is attached to the upper portion.

(iii) Distilled water is sucked into the apparatus and then the pressure is adjusted with the help of pinch cock, so that the rate of fall of drops is about 15–20 drops per minute.

(iv) A definite number of drops (say 20) of water are allowed to fall into a clean weighing bottle which has already been weighed when empty. The weight of the weighing bottle along with the drops is taken.

(v) The apparatus is now rinsed with acetone and then allowed to dry.

(vi) Now, the experimental liquid is sucked into the stalagmometer; the rate of fall of drops adjusted as before and 20 drops of the liquid taken into the empty weighing bottle and the bottle weighed.

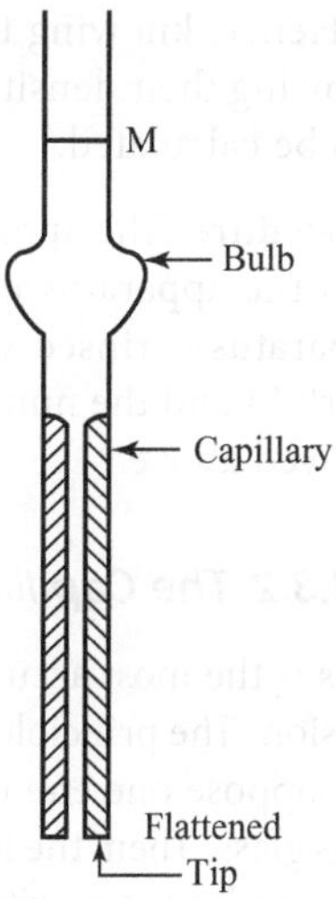

Figure 2.2 A Stalagmometer

Calculations:

Suppose weight of empty bottle = w_1g
Weight of bottle + 20 drops of water = w_2 g
Weight of bottle + 20 drops of liquid = w_3 g
Weight of 20 drops of water = $(w_2 - w_1)$ g
Weight of 20 drops of liquid = $(w_3 - w_1)$ g

If γ_w is the surface tension of water then the surface tension of the experimental liquid, γ_l will be given by

$$\frac{\gamma_l}{\gamma_w} = \frac{w_3 - w_1}{w_4 - w_2} \quad \text{or} \quad \gamma_l = \frac{w_3 - w_1}{w_4 - w_2} \times \gamma_w$$

Knowing γ_w, γ_l can be calculated.

(b) **Drop-number method:** It is more convenient to count the number of drops formed from a certain volume of the liquid than to take the weight of a drop of a liquid. Therefore, the above method may be carried out in an alternative manner as follows:

Suppose the number of drops formed from v millilitre of a liquid = n. Then, the weight of one drop of the liquid can be calculated as

$$\frac{v \times d}{n},$$

where d is the density of the liquid.

Thus, if from the same volume (say v millilitre) of the two liquids, the number of drops formed are n_1, n_2 and d_1, d_2 are their respective densities, then

$$\frac{\gamma_1}{\gamma_2} = \frac{m_1}{m_2} = \frac{v \times d_1}{n_1} \div \frac{v \times d_2}{n_2} = \frac{n_2 \times d_1}{n_1 \times d_2}$$

or

$$\frac{\gamma_1}{\gamma_2} = \frac{n_2 \times d_1}{n_1 \times d_2} \tag{2.2}$$

Hence, knowing the number of drops (n_1 and n_2) formed from the same volume of two liquids, knowing their densities (d_1 and d_2) and knowing the surface tension of one of them, that of the other can be calculated.

Procedure: The apparatus is cleaned and clamped vertically as before. Distilled after is first sucked into the apparatus up to mark M. The number of drops falling out of the capillary is noted. The apparatus is rinsed with acetone and dried. The experimental liquid is known sucked in up to the mark M and the number of drops formed is counted. The result is then calculated using the formula derived above.

2.4.3.2 The Capillary Rise Method

This is the most accurate method for the measurement of surface tension. The principle of the method is as follows.

Suppose one end of a capillary tube is put into a liquid which wets glass. Then, the force of surface tension pushes the liquid into the capillary to a certain height, as shown in Fig. 2.3, till the force of surface tension acting upwards balances the weight of the liquid column acting downwards.

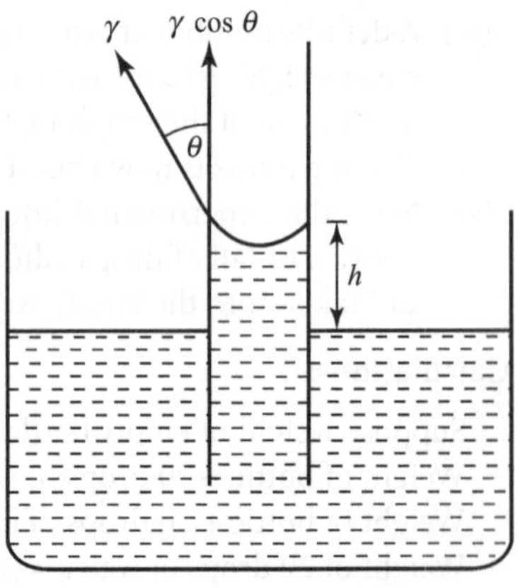

Figure 2.3 Rise of liquid in a capillar tube; surface tension (γ) acts along tangent to the meniscus and its vertical component is $\gamma \cos \theta$

Suppose, height to which liquid rises in the capillary = h cm, density of the liquid = d g/cm^3, inner radius of the capillary = r cm, surface tension of the liquid = γ dynes/cm and contact angle = θ. Then, the component of surface tension acting in the vertical direction = $\gamma \cos\theta$.

Inner circumference of the capillary = $2\pi r$. Force of surface tension acting upwards all along the circumference = $2\pi r \gamma \cos\theta$.

Weight of the liquid acting downwards = $\pi r^2 hdg$, where g is acceleration due to gravity and is the volume of the liquid column.

Hence, at equilibrium

$$2\pi r\gamma \cos\theta = \pi r^2 hdg \quad \text{or} \quad \gamma = \frac{rhdg}{2\cos\theta}$$

For most of the liquids which wet glass,

$$\theta = 0, \quad \text{so that} \quad \cos\theta = 1$$

Hence,

$$\gamma = \frac{rhdg}{2} \tag{2.3}$$

2.4.3.3 The Tensiometer or The Tension Balance Method

It is commercial method used for the rapid measurement of surface tension of a liquid. The method is based upon the principle that the force required to take out a horizontal platinum ring dipping just below the surface of a liquid is directly proportional to the surface tension of the liquid (Table 2.1).

Table 2.1 Surface tension of some common liquids at 20 °C in dynes cm^{-1}.

Liquid	Surface Tension	Liquid	Surface Tension
Water	72.8	Chloroform	27.1
Nitrobenzene	41.8	Carbon tetrachloride	26.8
Carbon disulphide	33.5	Acetone	23.7
Benzene	28.9	Methyl alcohol	22.6
Toluene	28.4	Ethyl alcohol	22.3
Acetic acid	27.6	Diethyl ether	17.0

2.4.4 Surface Tension in Everyday Life

1. Cleansing action of soap and detergents is due to their property of lowering the interfacial tension between water and grease. Soap solution can penetrate into the cloth fibre due to its lower surface tension to form an emulsion with grease on which the dirt sticks.
2. Surface tension is related to life because: (i) plants rely on surface tension to transport water during the biological process known as transpiration and (ii) surface tension allows organisms such as water insects to walk on water.
3. Certain substances which have the ability to lower surface tension are added to tooth pastes, mouthwashes and nasal jellies. Due to lowering in surface tension, these preparations spread uniformly on the surface on which these are applied. Thus, the efficiency of their antiseptic action is increased.

MEMORY FOCUS

1. **Surface tension** is the energy required to stretch a unit change of the surface area. Its units are newton per metre (Nm^{-1}).
2. The surface tension is due to the unbalanced forces experienced by molecules at the surface of a liquid.
3. As a result of surface tension, a drop of liquid tends to form a sphere because a sphere offers the smallest area for a definite volume.
4. Substances with low surface tension have a tendency to form films. When detergent is added to water, it lowers the surface tension.
5. Blowing soap water with a straw form bubbles, due to the low surface.
6. **Cohesion** is the intermolecular attraction between like molecules and **adhesion** is that between unlike molecules. Liquids with high surface tension have strong cohesion forces and they are poor wetting liquids due to low adhesion forces.
7. A detergent or wetting agent is a substance that increases the adhesion force between two different materials. Molecules of detergent usually have a polar and a non-polar portion. When added, the wetting agent increases the wetting action of water with the non-polar material. By this action, dirt is removed when washed with water.

(*Continued*)

MEMORY FOCUS (*Continued*)

8. When a small tube is dipped into a liquid, the level in the tube is usually higher or lower than that of the bulk liquid. If adhesion force between the tube material and the liquid is stronger than the cohesion force, the level is higher as in the case of water. Otherwise, the level is lower as in the case of mercury or coloured solutions. Such phenomena are called **capillary action.**
9. Capillary action is one of the factors responsible for the transport of liquid and nutrients in plants and sometimes in animals.
10. Surface tension decreases with the rise in temperature. It is because the kinetic energy of the molecules increases with the rise in temperature, thereby decreasing the intermolecular forces and hence the surface tension.

2.4.5 Surface Tension and Chemical Constitution (Parachor)

In 1923, Macleod gave a relation between surface tension and density which is given as

$$\frac{\gamma^{1/4}}{D-d} = \text{constant}$$

and the multiplication of this equation with molecular weight (M) gives

$$\frac{M\gamma^{1/4}}{D-d} = M \times \text{constant} = |P|$$

where γ = surface tension of the liquid, D = density of liquid, d = density of vapour at the same temperature and $|P|$ is called parachor.

Usually $d <<< D$ at ordinary temperatures; therefore, if we ignore d in comparison to D then, we get

$$\frac{M\gamma^{1/4}}{D} = |P| \tag{2.4}$$

i.e., $\quad$ Molar volume $\times \gamma^{1/4} = |P|$

When $\quad \gamma = 1$ unit $\quad$ [Nm^{-1} or dyne cm^{-1}],

Molar volume = $|P|$

Therefore, when surface tension is unity, parachor represents the value of moral volume. Hence, the parachor of a liquid is defined as *the molar volume of the liquid at a temperature at which its surface tension is unity.*

If M_1 and M_2 represent the molecular weights of two liquids 1 and 2 and D_1 and D_2 are their densities and γ_1 and γ_2 be their surface tensions, we have

$$[P]_1 = \frac{M_1}{D_1}\gamma_1^{1/4} = V_1\gamma_1^{1/4}$$

$$[P]_2 = \frac{M_2}{D_2}\gamma_2^{1/4} = V_2\gamma_2^{1/4}$$

$$\therefore \quad \frac{[P]_1}{[P]_2} = \frac{V_1\gamma_1^{1/4}}{V_2\gamma_2^{1/4}}$$

where V_1 and V_2 are the molecular volumes of the two liquids.

If $$\gamma_1 = \gamma_2, \text{ then } \frac{[P]_1}{[P]_2} = \frac{V_1}{V_2} \tag{2.5}$$

Therefore, *a comparison of parachors of different liquids gives the ratio of their molecular volumes under such conditions that their surface tensions are equal.*

It was observed that parachor is largely an additive property; however, it is also partly constitutive in nature. It has been found that:

(a) *The parachors of isomeric substances of the same family are identical,* e.g., substances of the same series having the same molecular formula $C_6H_{12}O_2$ will have the same parachors.

Ester	Parachor	Ester	Parachor
Methyl valerate	292.4	Iso-amyl connate	293.1
Ethyl isobutyrate	292.1	Iso-butyl acetate	293.5
Ethyl butyrate	292.9	*n*-Propyl propionate	294.5

(b) *The difference in parachor values of successive members in a homologous series is found to be constant, irrespective of the nature of the series.* It is true as shown in the following table.

Hydrocarbons			Esters		
Formula	**[*P*]**	**Δ[*P*]**	**Formula**	**[*P*]**	**Δ[*P*]**
C_2H_6	110.5		$C_2H_4O_2$	138.1	
C_3H_8	150.8	40.3	$C_3H_6O_2$	177.3	39.2
C_4H_{10}	190.0	39.2	$C_4H_8O_2$	216.1	38.8
C_6H_{14}	270.1	40.0 × 2	$C_6H_{12}O_2$	293.8	38.8 × 2
C_7H_{16}	309.3	39.2	$C_7H_{14}O_2$	332.3	38.5
C_8H_{18}	345.0	35.7	—	—	—
$C_{10}H_{22}$	424.2	39.6 × 2	—	—	—

(i) **Atomic parachors:** From the above conclusions, we can say that parachors are additive functions of the chemical composition. The contribution of a $-CH_2$ group to the parachor value is about 39. The molar parachor would be the sum of the atomic parachors. Based on this assumption, the parachor value of hydrogen, P_H, can be calculated as follows:

The parachor value of *n*-decane ($C_{10}H_{22}$) is experimentally found to be 424.2. Thus, following the additivity rule,

$$P_{C_{10}H_{22}} = 10 \times P_{CH_2} + 2 \times P_H$$

$$424.2 = 10 \times 39 + 2 \times P_H$$

or $$P_H = \frac{424.2 - 390}{2}$$

or $$P_H = 17.1$$

Table 2.2 Atomic and structural parachors.

Atomic Parachors		Structural Parachors	
Atom	**Parachor**	**Structure**	**Parachor**
H	17.1	Single bond	0
O	20.0	Double bond	23.2
N	12.5	Triple bond	46.6
P	37.7	Co-ordinate bond	–1.6
S	48.5	3-Membered ring	16.7
C	4.8	4-Membered ring	11.6
Cl	53.8	5-Membered ring	8.5
Br	68.0	6-Membered ring	6.1

Similarly, the atomic parachor of a carbon atom is calculated as follows:

$$P_H = P_{CH_2} - 2P_H = 39 - 2\times 1.1 = 4.8$$

In the above way, we can calculate the atomic parachors of other elements also.

(ii) **Structure parachors:** A comparison of the parachor values calculated from atomic parachor—equivalents with those experimentally observed—showed a divergence in the cases of unsaturated and ring compounds. In other words, we can say that parachors are not only additive but also constitutive. For example, the parachor of ethylene (C_2H_4), which should be = 2 × 4.8 + 4 × 17.1 = 78.0, is in fact 99.5. Thus, the double bond present in ethylene makes a contribution of 21.5. From a study of different olefinic compounds, the contribution of a double bound towards parachor has been taken to be 23.2. The parachor value of a single bond has been taken as zero. The values of some atomic and structural parachors are given Table 2.2.

From what has been discussed above, it may be concluded that the total parachor of a molecule is the sum of atomic and structural parachors. The application of parachor in deciding chemical constitution of a liquid may be illustrated with the help of the following examples:

1. **Structure of quinone:** The following two structural formulae had been proposed for quinone.

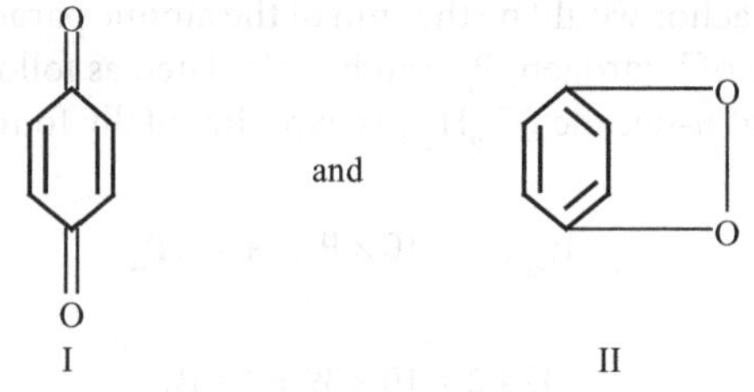

Molecular formula of quinone is $C_6H_4O_2$.

Parachor of quinone according to structure I

$$= 6[P]_C + 4[P]_H + 2[P]_O + 4[P]_{\text{Double bond}} + [P]_{\text{6-ring}}$$

$$= (6\times 4.8) + (4\times 17.1) + (2\times 20.0) + (4\times 23.2) + 6.1 = \mathbf{236.1}$$

Parachor of quinone according to structure II

$$= 6[P]_C + 4[P]_H + 2[P]_O + 3[P]_{\text{Double bond}} + 2[P]_{\text{6-ring}}$$

(Because, there are two 6-membered rings in structure II.)

$$= (6\times 4.8) + (4\times 17.1) + (2\times 20.0) + (3\times 23.2) + (2\times 6.1) = \mathbf{219.0}$$

The observed value of parachor for quinone (from experiment) = 236.8.
This value is close to the value calculated for structure I.
Hence, structure I must be the correct structure for quinone.

2. **Structure of paraldehyde:** Acetaldehyde (CH_3CHO) forms a polymer called paraldehyde, having formula (CH_3 CHO). The two possible structures for this compound are

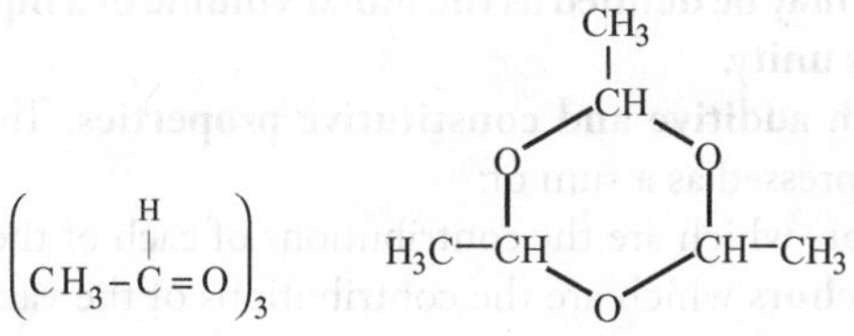

I II

Calculated parachor for structure I = 363.6
Calculated parachor for structure II = 300.1
The experimental value of parachor is found to be 298.7 which is close to that calculated for structure II. Hence, structure II must be the correct structure for paraldehyde.

3. **Structure of nitro group:** The nitro group may be assumed to have the following structures:

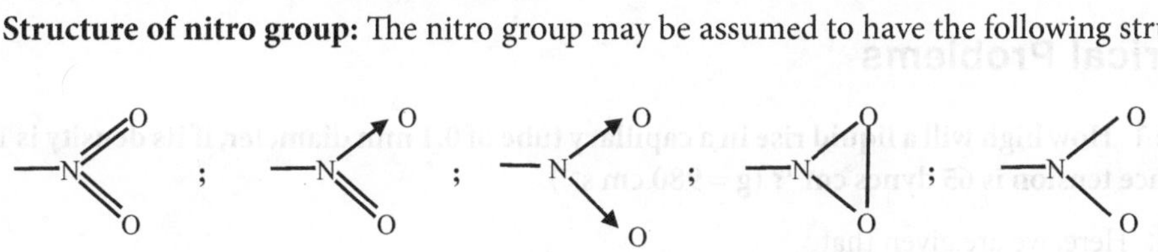

I II III IV V

The parachor values are as follows:

I : $1\times 12.5 + 2\times 20.0 + 2\times 23.2 = 98.9$

II : $1\times 12.5 + 2\times 20.0 + 1\times 23.2 + 1\times(-1.6) = 74.1$

III : $1\times 12.5 + 2\times 20.0 + 2\times(-1.6) = 49.3$

IV : $1\times 12.5 + 2\times 20.0 + 1\times 1.67 = 69.2$

V : $1\times 12.5 + 2\times 20.0 = 52.5$

From the parachors of a large number of organic compounds, we observe that the contribution of $-NO_2$ group on an average, corresponds to 73.6. Thus, the constitution of a nitro group should be given by structure (II).

MEMORY FOCUS

1. According to Sugden

$$\frac{M\gamma^{1/4}}{D-d}=[P]$$

The quality $[P]$, which is constant for a liquid, is called **parachor.** As d is negligible compared to D, the equation reduces to

$$\frac{M}{D}\gamma^{1/4}=[P] \quad \text{or} \quad V_m\gamma^{1/4}=[P] \text{ where } V_m \text{ is the molar volume of the liquid.}$$

If surface tension (γ) is unity, from equation, we may write $[P] = V_m$

The parachor [*P*] may be defined as the molar volume of a liquid at a temperature, so that its surface tension is unity.

2. **The parachor is both additive and constitutive properties.** The parachor of an individual compound can be expressed as a sum of:
   (i) **Atomic parachors** which are the contributions of each of the atoms present in molecule.
   (ii) **Structural parachors** which are the contributions of the various bonds and rings present in the molecule.
3. Parachor helps us in confirming the structure of any compound in liquid form. The theoretical value calculated based on atomic and structural parachor values is compared with the experimentally determined value. In case these two values are approximately the same, the assigned structure is taken to be correct.

Numerical Problems

Example 1 **How high will a liquid rise in a capillary tube of 0.1 mm diameter, if its density is 1.3 g cm^{-3} and surface tension is 65 dynes cm^{-1}? (g = 980 cm s^{-1})**

Solution: Here, we are given that

Diameter of the capillary $(2r) = 0.1$ mm $= 0.0 1$ cm

$$r = 0.005 \text{ cm}$$

Density of the liquid (d) = 1.3 g cm^{-3}

Surface tension of the liquid (γ) = 65 dynes cm

The formula for the rise of liquid in the capillary is

$$\gamma=\frac{rhdg}{2} \quad \text{or} \quad h=\frac{2\gamma}{rdg}$$

$$=\frac{2\times 65}{0.005\times 1.3\times 980} \qquad \left[g = 980 \text{ cm s}^{-2}\right]$$

$$= \mathbf{20.41 \text{ cm}}$$

Example 2 **The density of acetone at 20 °C is 0.7910 g/ml. Calculate the surface tension of acetone, given that the parachor equivalents of C, H, O and double bond are 7.2, 16.2, 20.0 and 23.2, respectively.**

Solution: Here, we are given that

Density of acetone at 20 °C (D) = 0.791 g/ml

$$\therefore \quad \text{Parachor for acetone} \left(CH_3 - \overset{\overset{\displaystyle O}{\|}}{C} - CH_3 \right)$$

$$= 3[P]_C + 6[P]_H + [P]_O + [P]_{\text{Double bond}}$$

$$= (3\times 7.2) + (6\times 16.2) + +20.0 + 23.2 = \mathbf{162}$$

Molecular weight of acetone (M) = 58

Surface tension is related to parachor according to the formula

$$\frac{M\gamma^{1/4}}{D} = [P]$$

$$\therefore \quad \gamma^{1/4} = \frac{[P]D}{M}$$

$$\text{or} \quad \gamma = \left[\frac{[P]D}{M}\right]^4 = \left[\frac{162\times 0.791}{58}\right]^4 = \mathbf{23.83\ dynes/cm}$$

Example 3 **The density and surface tension of benzene (C_6H_6) at 20 °C are 0.378 g/ml and 29.3 dynes/cm respectively. Prove that the Kekule structure of benzene is correct. Given that Parachor value of C, H, double bond and six membered ring of carbon atoms is 4.8, 17.1, 23.2 and 6.1 respectively.**

Solution: The calculated value of parachor for benzene will be:

$$= 6[P]_C + 4[P]_H + 3[P]_{\text{Double bond}} + [P]_{6\,\text{membered ring}}$$

$$= (6\times 4.8) + (4\times 17.1) + (3\times 23.2) + (6.1) = \mathbf{207.1}$$

Density of benzene at 20 °C, (D) = 0.378 g/ml

Surface tension of benzene at 20 °C, (γ) = 29.3 dynes/cm

Molecular weight of benzene (C_6H_6) = 78

Therefore, experimental value of parachor, (P)

$$= \frac{M\gamma^{1/7}}{D} = \frac{78\times(29.3)^{1/4}}{0.878} = \mathbf{206.7}$$

Thus, the experimental value is in close agreement with the calculated value. Hence, Kekule's structure must be correct.

2.5 VISCOSITY

Viscosity of a liquid is the resistance offered by any liquid towards the flow of the liquid. It is reverse of *fluidity.* The greater the viscosity of a liquid is, the smaller would be its fluidity and vice versa. For example, honey or glycerine is more viscous than water, but the fluidity of water is more when compared to honey or glycerine. *The internal friction which opposes the relative motion of two adjacent layers of a liquid is known as viscosity.* Like gases, in case of liquids also there will be exchange of molecules between two layers due to the motion. As a result, some part of momentum will pass over the other layer and this will result in the fall of rate of movement of one layer relative to the other. Therefore, to maintain a uniform velocity gradient, a force will have to be applied and this applied force is a measure of the viscosity of the liquid. The viscosities of liquids are measured in terms of coefficient of viscosity.

2.5.1 Coefficient of Viscosity

The coefficient of viscosity of a liquid at a given temperature is the steady force required to maintain a unit velocity difference between two parallel layers unit distance apart and having unit area of contact.

The steady force (f) applied to the liquid column in the direction of flow is

(i) directly proportional to the area (A) of contact between two layers and
(ii) directly proportional to the velocity gradient dv/dx

$$\therefore \quad f \propto A\frac{dv}{dx}$$

or

$$f = \eta A\frac{dv}{dx} \tag{2.6}$$

η, the proportionality constant is called coefficient of viscosity. When A = 1 m^2 and When $f = \eta$.

The S.I. unit of the coefficient of viscosity of a liquid is Nm^{-2}s (newton metre^{-2} second) or Pa S (pascal–second) and is defined as *the force required to maintain a unit velocity gradient of one metre per metre between two parallel layers one metre apart and having one metre area of contact.*

The C.G.S. unit of coefficient of viscosity is poise and is defined as the force in dynes required to maintain a unit velocity gradient of one centimetre per second per centimetre between two parallel layers one centimetre apart and having one centimetre area of contact.

$$1 \text{ poise} = 1 \text{ g cm}^{-1} \text{ s}^{-1} = 0.1 \text{ kg m}^{-1} \text{ s}^{-1} = 0.1 \text{ Nm}^{-2} \text{ S} = 0.1 \text{ PaA}$$

The coefficient of viscosity of some liquids is given in Table 2.3.

Table 2.3 Viscosity of liquids at different temperatures (poise × 10^3).

Liquid	273 K	293 K	313 K	333 K	353 K
Water	17.92	10.05	6.56	4.69	3.57
Mercury	16.85	15.54	14.50	13.67	12.98
Ethyl alcohol	17.73	12.00	8.34	5.92	—
Carbon tetrachloride	13.29	9.69	7.39	5.85	4.68
Benzene	9.12	6.52	5.03	3.92	3.29

The viscosity of the liquid depends upon the nature of the liquid. Branch-chained molecule offer more resistance to flow whereas straight-chained molecules offer relatively less resistance. Spherical molecules have low viscosity when compared to rod-like or plate-like molecule, as they offer less resistance to flow and hence are associated with high viscosity.

The intermolecular attractions in liquids are opposed by the thermal agitation of the molecules and the resistance that liquids offer to flow one molecule over the other is due to unbalanced part of intermolecular interactions. Increased intermolecular attractions cause a corresponding larger resistance to flow. Therefore, polar substances such as water and alcohols which have larger intermolecular interactions than hydrocarbons (which involve relatively less metre molecular attraction) have larger coefficient of viscosity.

Learning Plus

How to increase viscosity of a liquid?

A standard way to increase viscosity is to add long-chain molecules (polymers) that are soluble in that liquid. For water, polyethlyteneoxide or polyvinyl alcohol may be used. These are industrially used and are effective in quite small concentrations.

For glycols, the above two or polyethylene is employed. Organic compounds such as glycerine/glycol are good viscosity modifiers. Viscosity modifiers are added to reduce splashing of water from fountains, for thickening ink etc.

Learning Plus

Is viscosity of liquid related to its density?

Viscosity and density are not related. Liquids with similar densities may have very different viscosities. Density remains essentially the same regardless of the temperature of a liquid, but viscosity generally changes quite dramatically with temperature. If you look at different substances/liquids, there is no simple relationship, i.e. liquids of very similar density can have widely varying viscosities. Density is an almost static property of a liquid and in most cases, the density is similar to that of the solid form of the same substance over the whole liquid range (i.e. from melting to boiling), whereas viscosity is a transport property and can be very sensitive to temperature changes.

2.5.2 Measurement of Viscosity

Most methods employed for the measurement of viscosity of liquids are based on either Poiseuille equation or stokes equation. One of the most used methods is Ostwald's viscometer. This method is based on Poiseuille equation. **The Poiseuille equation** for the coefficient of viscosity of a fluid is

$$\eta = \frac{\pi P r^4 t}{8VL} \tag{2.7}$$

where η = coefficient of viscosity of liquid, V = volume of the liquid, r = radius of the capillary, L = length and P = pressure.

To obtain the viscosity of a liquid by this equation, it is not always necessary to measure all the quantities indicated when once the viscosity of some reference liquid, usually water, is known with accuracy the time of flow of same volume of two different liquids through the same capillary, then according to the Poiseuille equation, the ratio of the viscosity coefficient of two liquids is given by

$$\frac{\eta_1}{\eta_2} = \frac{\pi P_1 r^4 t_1}{8LV} \frac{8LV}{\pi P_2 r^4 t_2} = \frac{P_1 t_1}{P_2 t_2} \tag{2.8}$$

Now, $P_1 = hd_1g$ and $P_2 = hd_2g$

$$\frac{\eta_1}{\eta_2} = \frac{hd_1\, g\, t_1}{4d_2\, g t_2} = \frac{d_1 t_1}{d_2 t_2} \tag{2.9}$$

Thus, if d_1, d_2 and are known, then by determining the time of flow t_1 and t_2, the viscosity of the liquid can be calculated.

Ostwald's Viscometer The quantities t_1 and t_2 can most conveniently be determined using an Ostwald's viscometer (Fig. 2.4). A known quantity of liquid is introduced into the viscometer immersed in a thermostat and is taken drawn up by suction into the bulb X and the liquid level is above the mark *x*. The liquid is allowed to drain and time for the liquid level to fall from *x* and *y* is measured with the help of an electronic timer or stop watch. The viscometer is cleaned and the liquid of known viscosity is introduced and the entire procedure is repeated again and the time of flow (t_2) between the same points *x* and *y* is obtained. The densities d_1 and d_2 can be obtained with the help of specific gravity bottles or pyknometer. Knowing d_1, d_2, t_1, t_2 and η_1, η_2 can be calculated. The viscosity of some common liquids is given in Table 2.4.

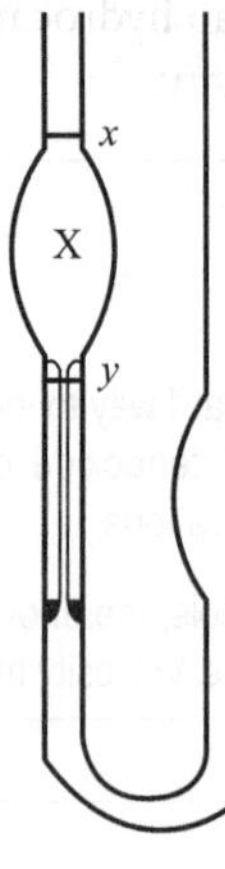

Figure 2.4 Ostwald's viscometer

Table 2.4 Coefficients of viscosity of some common liquids (in millipoises) at 20 °C.

Liquid	Viscosity (in millipoises)	Liquid	Viscosity (in millipoises)
Ethyl ether	2.33	Chlorobenzene	8.0
Acetone	3.29	Carbon tetrachloride	9.68
Carbon disulphide	3.68	Ethyl alcohol	12.0
Chloroform	5.63	Acetic acid	12.2
Methyl alcohol	5.93	Nitrobenzene	20.13
Benzene	6.47	Glycerol	10^4

2.5.3 Effect of Temperature on Viscosity

With the increase of temperature, the kinetic energy of the molecules increases. Hence, the liquid starts flowing faster. In other words, the viscosity of a liquid decreases with the increase of temperature. It has been found that the decrease is about 2% per degree rise of temperature.

A number of equations have been proposed connecting viscosity and temperature. The best out of these is the one given by Arrhenius (in 1912) for pure liquids. The equation is

$$\eta = Ae^{E/RT} \tag{2.10}$$

where A and E are constants for the liquid and *T* is the absolute temperature. E is called activation energy for viscous flow.

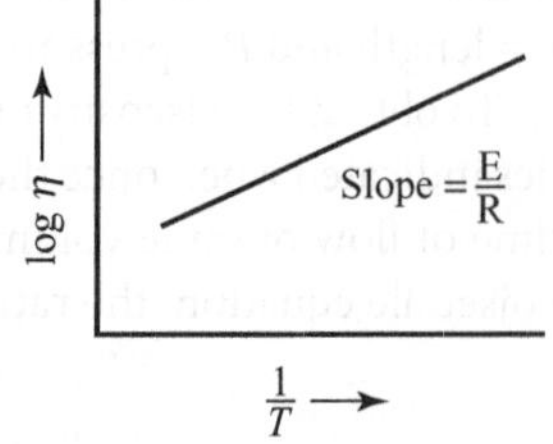

Figure 2.5 Plot of $\log \eta$ vs $\frac{1}{T}$

The above equation may be written as

$$\log \eta = \log A + \frac{E}{RT}$$

Thus, the plot of log $\frac{1}{T}$ is a straight line with slop = $\frac{E}{R}$ as shown in Fig. 2.5.

It has been found that for most of the liquids, viscosity decrease by about 2% for every degree rise of temperature.

Learning Plus

The Hole Theory

The viscosity of a liquid generally decreases with the rise in temperature. This has been explained in terms of 'hole theory' of liquids. According to this theory, there are holes and vacancies in a liquid. The liquid molecules keep on moving continuously into these vacancies and the vacancies also keep on moving around as other wise the liquid will not be able to flow. This process requires energy. Increases in temperature provide the activation energy to the molecules to move into a hole. Therefore, the liquid can flow more easily at higher temperature.

2.5.4 Factors Affecting Viscosity

(i) **Size and weight of molecules:** Viscosity increases with the increase of molecular weight for compounds belonging to the same class.

(ii) **Presence of other substances:** The addition of cane sugar increases the viscosity of water, while the addition of highly ionized salt decreases it. The presence of suspended matter and colloids increases the viscosity of water.

(iii) **Intermolecular forces:** The viscosity of water is high due to hydrogen bonding, similarly the viscosity of glycerol with three OH groups is also high due to the same reason.

(iv) **Shape of molecules:** With an increase in the branching in the chain, the viscosity increases.

(v) **Temperature:** It has been observed that the viscosity of liquids decreases with the rise in temperature.

MEMORY FOCUS

1. Viscosity of a liquid is a measure of its inability to flow. It is a reverse of fluidity. It is a measure of the frictional resistance between the molecules of liquid.
2. In C.G.S. units, viscosity is expresses in dyne cm^{-2} s. This unit is known as poise. Lower value of viscosity is expressed in centipoise $\left(\frac{1}{100}\text{poise}\right)$ and millipoise $\left(\frac{1}{100}\text{poise}\right)$.
3. Coefficient of viscosity is the force of resistance per unit area which will maintain unit viscosity difference between two layers of a liquid at a unit distance from each other. The force in dynes per square centimetre required to maintain a difference of velocity of 1 cm per second between two parallel layers of liquid held at a distance of 1 cm from one other.
4. Both castor oil and glycerol have very high viscosity; the values being 986 cP and 1490 cP, respectively. Being pure compounds, they have definite value viscosity of a particular temperature.

(Continued)

MEMORY FOCUS (*Continued*)

5. Honey also has very high viscosity. However, its viscosity cannot be given any value, because it is not a pure compound and its viscosity varies with composition and temperature too.
6. Viscosity can be correlated with molecular structures of liquids. In general, liquids of large molecules have high viscosity.
7. There is no direct correlation between viscosity and surface tension. These two properties are independent of each other.
8. Viscosity of any liquid decreases with increases in temperature.

2.5.5 Viscosity and Chemical Constitution

No signification relationship between viscosity and chemical constitution has been put forth so far. However, some general results have been observed which are summarized as follows:

(i) In a homologous series, viscosity increases as we go up a series and the increase per CH_2 group is almost constant.

(ii) In compounds involving chain isomerism, the *n*-isomers have generally greater viscosity than the branched chain isomer.

(iii) In compounds involving geometrical isomerism, *trans*-isomers have higher value of viscosity than the corresponding *cis*-isomers.

(iv) To predict whether a liquid is associated or not, Dunstan proposed a rule which is called *Dunstan's rule.* According to this rule, for normal liquid, i.e. non-associated liquids:

$$\frac{d}{M}\eta \times 10^6 = 40 \text{ to } 80 \tag{2.11}$$

where d = density of the liquid, η = coefficient of viscosity and M = Molecular mass. For associated liquid, the value is much higher than 80.

(v) **Rheochor:** Just as parachor, rheochor is an additive and constitutive property related with surface tension; efforts were made to derive a similar quantity related with viscosity which should be additive and constitutive in nature. One such quantity found has been named as rheochor [R], which is related to viscosity by the following expression

$$[R] = \frac{M}{D}(\eta)^{1/8} \tag{2.12}$$

where M, D and η are the molecular mass, density and coefficient of viscosity of the liquid. Thus, if

$$\eta = 1,$$

Then,

$$[R] = \frac{M}{D} = V, \text{ molar volume}$$

Hence, rheochor may be defined as the molar volume of the liquid at a temperature at which its coefficient of viscosity is unity.

However, the use of rheochor as an additive and constitutive property was found to be very limited. Hence, the term could not gain any significance.

(vi) **Molecular viscosity:** Molecular viscosity is defined as the product of molecular surface and viscosity,

i.e. Molecular viscosity = Molecular surface × Viscosity = $(M/d)^{2/3} \times \eta$

The molecular surface is equal to $(M/d)^{2/3}$ and is the surface area of one mole of the liquid. Thorpe and Rodger showed molecular viscosity to be all additive and constitutive properties. They prepared a table of values of molecular viscosity contributions by different atoms or structural factors as shown in the following table:

Atom	C	H	O (in –OH gp)	O (in > C = 0)	S
Molecular viscosity	–98	80	196	248	155

From these constants, the molecular viscosities of substances of known constitution could be calculated. If the calculated and experimental values coincide, it is taken as a confirmation of the constitution of the substance. For example, the structure of propionic acid is written as

$$CH_3-CH_2-\overset{\overset{\displaystyle O}{\|}}{C}-OH$$

Its molecular viscosity can be calculated as follows:

$$\begin{aligned} 6H = 6 \times 80 &= 480 \\ O\ (\text{in OR}) = 1 \times 196 &= 196 \\ O\ (\text{in} > C = 0) = 1 \times 248 &= 248 \\ 3C = 3\ (-98) &= \underline{-294} \\ \text{Total} &= \underline{630} \end{aligned}$$

The experimental molecular viscosity of propionic acid is found to be 631.8. If the experimental and calculated values being the same, the structural formula assigned to propionic acid is correct.

MEMORY FOCUS

1. Dunstan showed that viscosity coefficient (η) and molecule volume (d/M) were related as:

$$\frac{d}{M} \times \eta \times 10^6 = 40 \text{ to } 60$$

This expression holds only for normal (unassociated) liquids.

2. The product of molar surface and viscosity is termed **molar viscosity**.

$$\text{Molar viscosity} = \text{Molar surface} \times \text{Viscosity} = \left(\frac{M}{d}\right)^{2/3} \times \eta$$

Molar viscosity is an additive property at the boiling point.

3. The quantity $[R]$ is termed **Rheochor**,

$$\frac{M}{d} \times \eta = [R]$$

The Rheochor may be defined as the molar volume of the liquid at the temperature at which its viscosity is unity. Like parachor, rheochor is both additive and constitutive.

Numerical Problems

Example 4 **The water flow time for an Ostwald viscometer is 59.2 s at 25 °C. If 46.2 s are required for the same volume of ethyl benzene (density = 0.867 g cm³) to flow through the capillary, calculate its absolute viscosity at 25 °C, that of water being 0.00895 poise at the same temperature.**

Solution: Here, we are given that

For water: $t_1 = 59.2\text{s}$, $1 = 0.00895$ poise, $d_1 = 1.0$ g/cc
For ethyl benzene: $t_2 = 46.2\text{s}$, $2 = ?$, $d_2 = 0.867$ g/cc

Substituting the values in the formula $\frac{\eta_1}{\eta_2} = \frac{d_1 t_1}{d_2 t_2}$, we get

$$\eta_2 = \frac{\eta_1 \times d_2 t_2}{d_1 t_1} = \frac{0.00895 \times 0.687 \times 46.2}{1 \times 59.2} = \mathbf{0.00606 \ poise}$$

Example 5 **At 20 °C, the density of water is 0.9983 g cm⁻³ and its viscosity is 0.010087 poise. Explain how this figure shows that water is an associated liquid.**

Solution: Here, we are given that

$$d = 0.9983 \text{ g cm}^{-3}, \quad \eta = 0.010087 \text{ poise}$$

For H_2O, $M = 18$

$$\therefore \quad \text{Factor } \frac{d}{M}\eta \times 10^6 = \frac{0.9983}{18} \times 0.010087 \times 10^6 = \mathbf{559.5}$$

which is greater than 80. Hence, water is an associated liquid.

2.6 REFRACTIVE INDEX

When a ray of light travels from a less denser medium (i.e., air) to a more denser medium (i.e., liquid), the ray is bent towards the normal and the ratio of sine of the angle of incidence (sin *i*) to sin *r* of the angle of refraction (sin *r*) is constant and is called refractive index of the medium. If *n* is the refractive index then,

$$n = \frac{\sin i}{\sin r} \tag{2.13}$$

Refractive index above is defined as

$$n = \frac{\text{velocity of light in vacuum}}{\text{velocity of light in the medium}}$$

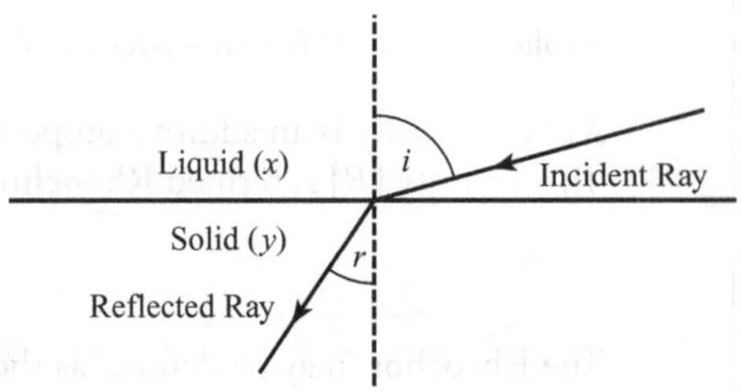

In addition, if *x* and *y* are refractive indices of less dense (liquid) and more dense media, respectively,

$$\frac{\sin i}{\sin r} = \frac{y}{x}$$

2.6.1 Measurement of Refractive Index

Refractive index depends upon the temperature as well as the wavelength of light used. Generally, the D-line of sodium is used for standard measurements. Refractive indices of liquids can be measured directly by means of an instrument known as refractometer. One such refractometer is Abbe's refractometer Fig. 2.6.

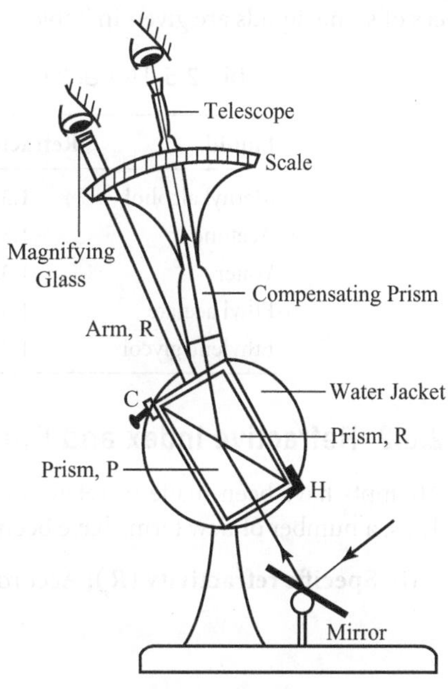

Figure 2.6 Abbe's refractometer

The principle of Abbe's refractometer can be discussed with the help of Fig. 2.7. A beam of light from a suitable source is reflected by a mirror M, passes through a lower prism P and illuminates its upper surface AB. The surface of the prism is ground and thus it serves as a diffusing screen providing rays in every direction. The small space between the lower prism P and the upper prism R contains a very thin layer of experimental liquid (the refractive index of the liquid should be smaller than that of the prism P, so as to allow the **critical angle phenomenon**).

The rays after passing through the diffused surface AB enter into the liquid medium at different angles of incidence. A particular ray going along the grazing incidence, i.e., at an angle only slightly less that 90 °, will pass through the prism R at an angle *e* which is equal to the critical angle. According to critical angle phenomenon,

$$\sin e = \frac{n}{N} \tag{2.14}$$

where N is the refractive index of the prism R and *n* that of the liquid (Fig. 2.7).

The critical ray coming out from the face AC at an angle α is observed through a telescope T. It can be seen that the refractive index of the liquid and the angle α are related by the equation:

$$n = \sin \widehat{ABC} \sqrt{(n^2 - \sin^2 \alpha)} - \cos \widehat{BAC} \sin \alpha \tag{2.15}$$

All rays entering the liquid at angles less than the grazing incidence will pass through the prism R at angles less than the critical angle. All other rays will suffer the phenomenon of total internal reflection; therefore, they will not pass through the prism R at all. Thus, the band of light viewed through the telescope finishes sharply at the point where the critical angle emerges out of the face AC. Therefore, the angle a can be accurately measured.

In experimental practice, the telescope is fixed and the prism box is rotated such that the critical ray coincides with the cross wire of the eye piece. The setting of the prism at this position corresponds to a certain critical angle and, therefore, to a definite value of refractive index. This value can be read directly from the scale made on the instrument.

Within the apparatus, there is an arrangement to remove chromatic aberration when white light is used. The temperature of the liquid can be varied by

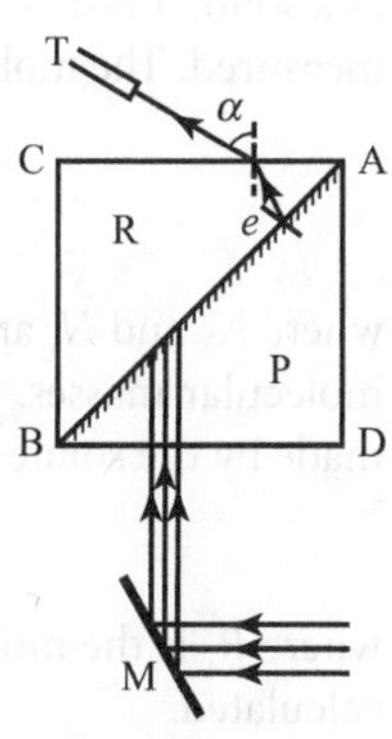

Figure 2.7 Principle of Abbe's refractometer

flowing water at the required temperature through a jacket surrounding the prism box. Refractive indices of some liquids are given in Table 2.5.

Table 2.5 Refractive indices of some liquids at 293 K for D-sodium line.

Liquid	Refractive Index	Liquid	Refractive Index
Methyl alcohol	1.3289	Benzene	1.5005
Acetone	1.3585	Toluene	1.4950
Water	1.3330	Aniline	1.5859
Ethyl acetate	1.3725	Ethyl alcohol	1.3611
Ethylene glycol	1.4314	Cyclohexanol	1.4663

2.6.2 Refractive Index and Chemical Constitution

Attempts have been made to relate the chemical constitution of a substance with its refractive index. Thus, a number of new terms have been put forward as explained in the following list:

(i) **Specific refractivity (R_s):** According to Lorentz and Lorentz,

$$R_s = \frac{n^2 - 1}{n^2 + 2}\frac{1}{d} \tag{2.16}$$

where R_s = specific refraction of the substance, n = the index of refraction and d = the density of the liquid at the same temperature.
For any given substance, specific refraction is constant and independent of temperature.

(ii) **Molar refractivity:** Multiplying the specific refractivity by the molecular mass M of the substance, we get what is known as molar refractivity or molar refraction R. Thus,

$$R_m = r \times M = \left(\frac{n^2 - 1}{n^2 + 2}\right)\frac{M}{d} \tag{2.17}$$

The above formula is applicable to gases, liquids as well as to solid. To determine the molar refractivity of a solid, it is dissolved in a suitable solvent and the refractive index n and density d of the solution are measured. The molar refractivity, R, of the solution is then given by the formula

$$R_s = \left(\frac{n^2 - 1}{n^2 + 2}\right)\left(\frac{N_1M_1 + N_2M_2}{d}\right)$$

where N_1 and N_2 are the mole fractions of the solvent and solute, respectively, and M_1 and M_2 are their molecular masses. However, the total molar refractivity of the solution is the sum of the contributions made by the solute (solid) and the solvent. Hence,

$$R_s = N_1R_1 + N_2R_2$$

where R_1 is the molar refractivity of the solvent and R_2 is that of the solid. Thus, knowing R_1, R_2 can be calculated.

Like parachor, the molar refractivity has been found to be additive as well as constitutive. The additive nature is supported again by the facts that (i) the difference in the molar refractivity of the successive members of a homologous series is nearly constant and (ii) the isomeric compounds with the similar structures (e.g. isopropyl alcohol and *n*-propyl alcohol) are found to have nearly same value of the molar refractivity. Similarly, the constitutive nature is supported by the fact that in case of compounds containing double bond, triple bond or a closed ring etc., the observed value of molar refractivity is higher than the value obtained by simply adding the atomic refractivities.

The values of atomic refractivities have been calculated in a manner similar to that done for parachor. Tables are available for consulting these values. However, while calculating the molar refractivity, the following two points must be kept in mind:

1. The atomic refractivity of oxygen atom is different in alcohols, ethers and ketones.
2. When a compound contains more than one double or triple bond, the molar refractivity depends not only on the number of double or triple bonds but also on their relative position in the molecule. When the double bonds or triple bonds are present in conjugated position, e.g. –CH = CH–CH = CH–, the molar refractivity is higher than the calculated value. This is known as **optical exaltation**.

Thus for 1, 2, 3 hexatriene,

$$CH_2 = CH–CH = CH–CH = CH_2$$

Calculated value of $[R] = 28.52$
Experimental value of $[R] = 30.58$
Optical exaltation $= 30.58 - 28.52 = 2.06$

The application of molar refractivity in deciding chemical constitution of a substance may be illustrated with the help of example given below.

However, even this method has found only limited application in deciding chemical constitution.

Numerical Problems

Example 6 **A substance having the formula C_3H_6O might be either acetone** **or allyl alcohol CH_2=CH–CH_2–OH). Determine which substance it actually is if its molar refractivity is found to be 16.974 by experiment. Given that refractivity for C = 2.418, H = 1.100, O (carbonyl) = 2.211, O (hydroxyl) = 1.525 and C=C double bond = 1.733.**

Solution:

Acetone		**Allyl Alcohol**	
		3 Carbon	= 7.254
3 Carbon	$= 3 \times 2.418 = 7.254$	6 Hydrogen	= 6.600
6 Hydrogen	$= 6 \times 1.100 = 6.600$	1 Double bond (C=C)	= 1.733
1 Carbonyl oxygen	= 2.211	1 Hydroxyl oxygen	= 1.525
	16.065		17.112

The value for double bond in the carbonyl group (>C = O) is included in the atomic refractivity of oxygen.

Example 7 **Calculate the molar refraction of acetic acid (CH_3COOH) at temperature at which its density is 1.046 g cm^{-3}. The experimentally observed value of refractive index at this temperature is 1.3715.**

Solution: Here, we are given

$$d = 1.046 \text{ g cm}^{-3}; \qquad n = 1.3715$$

Molecular mass of CH_3COOH is $M = 60$

Substituting this value in the formula $R_m = \left(\frac{n^2-1}{n^2+2}\right)\frac{M}{d}$, we get

$$R_m = \frac{(1.3715)^2 - 1}{(1.3715)^2 + 2} \cdot \frac{60}{1.046} = \mathbf{13.021\ cm^3\ mol^{-1}}$$

Example 8 **Density of acetic acid at 25 °C is 1.046 g/ml. Calculate the refractive index of acetic acid. Value of R_m (molecular refractivity) for C = 2.42, H = 1.10, O in OH = 1.50, O in CO = 2.21.**

Solution:

First step: To calculate the value of R_m for CH_3COOH

Here, we are given $R_C = 2.42$, $R_H = 1.10$

R_O in –OH = 1.50 $\qquad$ R_O in > CO = 2.21

$$R_m \text{ for } CH_3-\overset{\overset{\displaystyle O}{\|}}{C}-OH = 2R_C + 4R_H + R_O \text{ in} -OH + R_O \text{ in} > CO$$

$$= 2 \times 2.42 + 4 \times 1.10 + 1.50 + 2.21$$

$$= \mathbf{12.95\ cm^3\ mol^{-1}}$$

Second step: To calculate the value of n

In addition, we are given $d = 1.046$ g/ml

Molecular mass of CH_3COOH (M) = 60

Substituting this value in the formula $R_m = \left(\frac{n^2-1}{n^2+2}\right)\frac{M}{d}$, we get

$$12.95 = \left(\frac{n^2-1}{n^2+2}\right)\frac{60}{1.046}$$

or $\quad \frac{n^2-1}{n^2+2} = 0.2258 \quad$ or $\quad n^2 - 1 = 0.2258\, n^2 + 0.4516$

or $\quad 0.7742\, n^2 = 1.4516$

or $\quad n^2 = \frac{1.4516}{0.7742} = 1.8753$

$\therefore \quad \mathbf{n = 1.369}$

MEMORY FOCUS

1. The refractive Index (n) of a substance is defined as the ratio of the velocity of light in vacuum or air to that in the substance:

$$n = \frac{\text{Velocity of light in substance}}{\text{Velocity of light in air}}$$

2. The refractive index of the liquid with respect to air is given by **Snelle's law.** According to it,

$$n = \frac{\sin i}{\sin r}$$

3. **Molar refraction** is defined as the product of specific refraction and molecular mass.

$$R_M = \frac{n^2 - 1}{n^2 + 2} \cdot \frac{M}{d}$$

4. The value of molar refraction is characteristic of a substance and is temperature-independent.
5. The molar refraction (R_M) is an additive and constitutive property. The molar refraction of a molecule is thus a sum of the contributions of the atoms (atomic refractions) and bonds (bond refractions) present. On comparing the calculated and experimentally observed value of R_M, the structure of compounds can be confirmed.
6. **Optical exaltation.** A compound containing conjugated double bonds (C = C – C = C) has a higher observed R_M than that calculated from atomic and bond refractions. The molar refraction is thus said to be exalted (raised) by the presence of a conjugated double bond and the phenomenon is called optical exaltation. For example, for hexatriene,

$$CH_3 = CH - CH = CH - CH = CH_2$$

The observed value of R_M is 30.58 cm^{3}mol^{-1} as against the calculated value 28.28 cm^{3}mol^{-1}. If present in a closed structure as benzene, the conjugated double bonds do not cause exaltation.

2.7 OPTICAL ACTIVITY

Plane polarized light: According to wave theory of light, ordinary light consists of waves vibrating in all plane perpendicular to its path of propagation. When, by a special device, the waves in all plane except one are cut off, the light radiations vibrating only plane are called plane polarized light. The device which is capable of producing plane polarized light is named Nicol prism or *polarizer.*

Nicol prism is made by cutting a rhombohedron piece of calcite diagonally and cementing them along the diagonal by means of Canada balsam. When light falls on one side of the prism, all other waves are reflected from the interface to one side and the waves vibrating in one plane pass through it.

Even monochromatic light, such as light emitted by a sodium lamp, is composed of waves which are vibrating in a number of different planes. If such a beam is passed through a Nicol prism, it is converted into plane polarized light (Fig. 2.8).

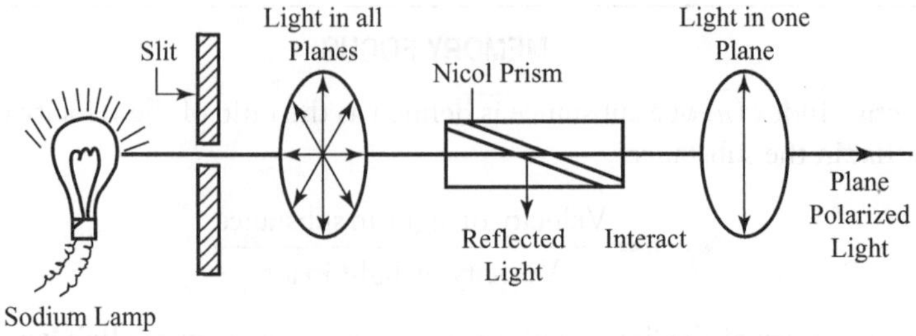

Figure 2.8 How to obtain plane polarized light

2.7.1 Optical Activity

There are certain substances, such as quartz, turpentine oil or sugar solution, which when placed in the path of plane polarized light, rotate the plane of polarized light towards left or right. *Such substances which rotate the plane of polarized light towards right (clockwise) or towards left (anticlockwise) are called* ***optically active substances*** *and the property is called* ***optical activity.***

A substance which rotates the plane of polarized light to right (clockwise) is called ***dextrorotatory*** (Latin: *dextro* meaning right) and the one capable of rotating the plane of polarized light to left (anticlockwise) is termed as ***laevorotatory*** (Latin: *laevous* meaning left). The *dextro-* and *laevorotatory* compounds are also represented as *d-* and *l-* (or + and –) forms respectively.

The phenomenon of the rotation of plane of polarized light can be observed experimentally as explained below.

Allow the plane polarized light after coming out of the polarizer to pass through another Nicol prism with its axis parallel to that of the first (polarizer). The light in this case will be visible. Now, rotate the second prism so that its axis is at right angle to the first. In this case, no light will get through and the field of view will be dark. If a glass cell containing some turpentine oil or a solution of cane sugar is placed between two Nicol prisms. The field gets lighted up and the second prism has to be rotated through a certain angle for 'extinction' of light. This clearly indicates that after passing through the optically active substance, the plane of polarized light gets rotated.

2.7.2 Specific Rotation

The angle in degrees through which the plane of polarized light gets rotated on passing through the solution of an optically active compound is known as **angle of rotation.**

The angle of rotation produced by an optically active compound depends upon:

(i) the nature of the optically active compound,
(ii) the concentration of the solution (in g/ml),
(iii) the length of the liquid column through which the light passes,
(iv) the nature of the solvent,
(v) the temperature of the solution, and
(vi) the wavelength of light used.

If *g* grams of an optically active substance are dissolved in *v* ml of the solution and the solution taken in a glass tube of length *l* cm, then optical rotation produced will be found proportional to the concentration of the solution (g/ml) and length of the tube (in decimetres), i.e.,

$$\alpha \propto \frac{m}{v} \qquad \text{(the concentration of solution)}$$

$$\propto l \qquad \text{(the length of the tube in decimetres)}$$

or

$$\alpha \propto \frac{m}{v} l$$

or

$$\alpha = [\alpha]_{\lambda}^{t} \frac{m \times l}{v} \tag{2.18}$$

where $[\alpha]_{\lambda}^{t}$ is a constant characteristic of the substance under conditions of temperature (1 °C) and wavelength of light used (λ). Since *d* light of sodium flame at 25 °C is generally used. It is generally expressed as $[\alpha]_{D}^{25°}$ and in known specific rotation of the substance.

From Eq. (2.18), we get

$$[\alpha]_{\lambda}^{t} = \frac{v}{l} \times \frac{\alpha}{m} \tag{2.19}$$

If $m = 1$ g, $v = 1$ ml and $l = 1$ dm,

then $[\alpha]_{\lambda}^{t} = \alpha$

Thus, **specific rotation** of a substance may be defined as the angle of rotation produced when the plane polarized light passes through the decimetre length of a solution containing one gram per millilitre of the optically substance.

If *c* grams of substance are dissolved in 100 ml of the solution, then Eq. (2.19) reduces to

$$[\alpha]_{\lambda}^{t} = \frac{100 \times \alpha}{l \times c}$$

In case of pure solids or liquids, the above expression may be written as

$$[\alpha]_{\lambda}^{t} = \frac{\alpha}{l \times d} \tag{2.20}$$

where *d* is the density of pure substance at temperature *t* °C.

2.7.3 Optical Activity and Chemical Constitution

Optical activity is an important property for determining the structure of different substances even in these days of highly developed electronic techniques. It is purely a constitutive property which depends upon the arrangements of atoms within the molecule. It is found, in general, that optically active organic compound consist of molecules which have at least *one asymmetric carbon atom*, i.e., a carbon atom attached to four different atoms or groups. Lactic acid represents a very simple case of optically active compounds. It contains an *asymmetric carbon atom* (marked with asterisks) attached to four different atoms or groups.

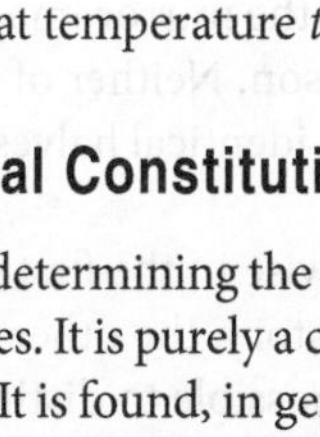

It is observed that lactic acid exists in two forms; one of which rotates the plane of polarized light towards right and is called *dextrorotatory* (*d*-form) and the other rotates the plane of polarized light towards left and is called *laevorotatory* (*l*-form). Both the forms have the same constitution, i.e., contain an asymmetric carbon atom attached to four different atoms or groups but differ in their arrangements of atoms or groups in space. Such forms, which otherwise, have the same molecular formula but differ only in their arrangement of atoms or groups in space are called *optical isomers*.

Vant' Hoff and Le bel explained the relationship between optical activity and molecular asymmetry. They pointed out that in all such compounds, the asymmetric carbon atom is present at the centre of a regular tetrahedron and four different atoms or groups attached to it are present at the corners of a tetrahedron. In such a case, it can be shown by models that there are two ways of arranging the groups about the central carbon atom as shown in Fig. 2.9.

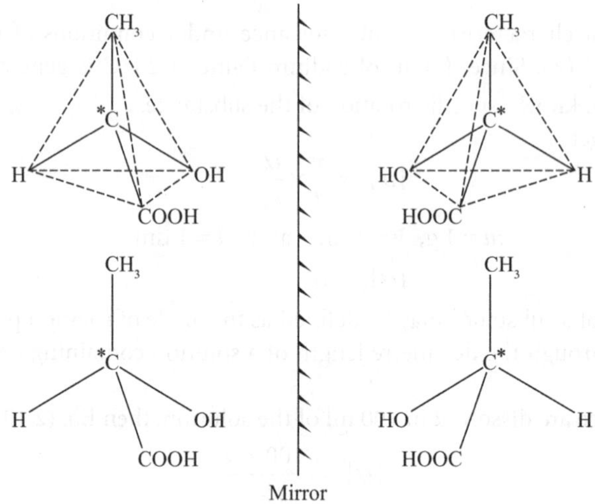

Figure 2.9 Two space models of latic acid

Evidently, these two arrangements, although having similar structure or constitutions, cannot be superimpose on each other. One structure is the mirror image of other and they are related to each other just as the right hand is to left hand of a person. Neither of the two structures has a plane of symmetry, i.e. the structures cannot be divided into two identical halves by a single plane. In a simple way, the above two forms can be represented as follows.

The arrangement from H to OH is clockwise in the first form, while it is anticlockwise in the second form. Therefore, if one of them is dextrorotatory, the other must be laevorotatory (Fig. 2.10). However, simply by seeing its configuration, it is not possible to distinguish as to which one of them is dextro- or laevorotatory. These two forms of lactic acid are also known as *enantiomorphs*.

If *d*- and *l*- forms of a compound are mixed with one another in equimolar quantities, the resulting product is optically inactive. This is due to mutual or external compensation of the two constituents, i.e., the rotation produced by one form completely nullifies the equal rotation in the opposite direction produced by the second form. The optically inactive products thus obtained are called **racemic mixtures** and are represented as *dl* form or ± form.

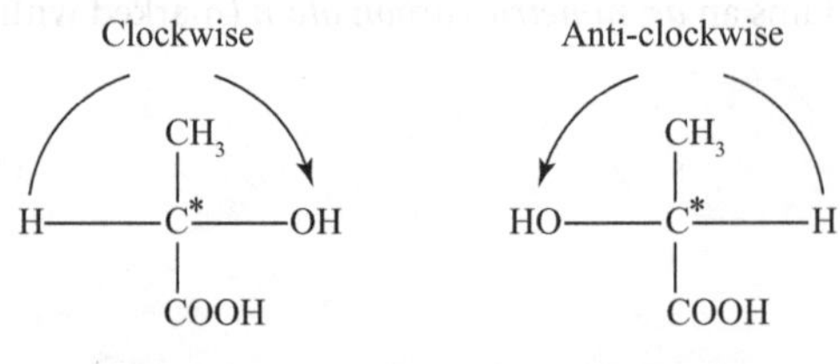

Figure 2.10 Two forms of lactic acid

It is interesting to note that whenever any optically active compound is synthesized in the laboratory, the product obtained in most of the cases is a racemic mixture. This is because that there are equal chances of the formation of two enantiomers. *The racemic mixtures obtained above can be separated into d- and l-forms by suitable physical and chemical methods. This process of separation of racemic mixture into d- and l- forms is called* ***resolution****.*

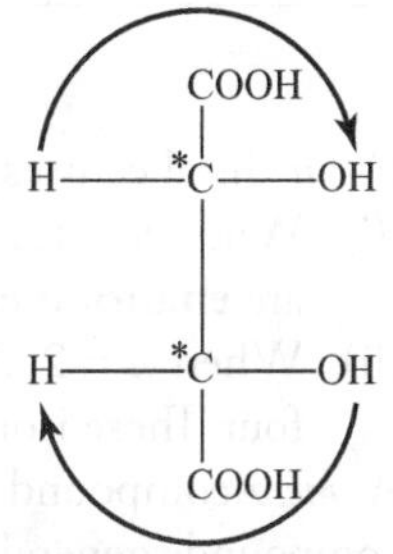

Figure 2.11 Meso form of tartaric acid

In case of certain compounds, which contain more than one asymmetric carbon atoms, another form is found to exist. In this form, the rotation caused by one half of the molecule is exactly cancelled by equal and opposite rotation caused by other half of the molecule (Fig. 2.11) and the molecule on the whole becomes optically inactive. Such a form which becomes optically inactive due to internal composition is called *mesoform.*

From above discussion, it appears as if only the organic compounds containing an asymmetric carbon atom show optical isomerism. However, there are certain inorganic compounds, which contain other asymmetric atoms such as cobalt, silicon and sulphur and show optical isomerism. Lastly, it may appear from above that the presence of an asymmetric carbon atom is the essential condition for a compound to be optically active. However, this is not so. 2,3-pentadiene, for example, does not contain an asymmetric carbon atom, yet it is optically active. This is because, the molecule, as a whole, is asymmetric and does not possess any plane of symmetry.

We may conclude from above discussion that:

(a) Optical activity can be used to prove a structure but cannot be used to disprove any structure.
(b) In general, optical activity can be used to test the presence of asymmetry in the molecules.

MEMORY FOCUS

1. Some substances possess a special property of rotating the plane of polarized light. These substances are said to be **optically active substances** and this property is called **optical activity**.
2. Ordinary light has vibrations in all the planes perpendicular to the path of light. When the vibrations are restricted in a single plane, it is said to be plane **polarized light**.
3. If the plane polarized light is rotated towards right, the compound is called **dextrorotatory** (*d* or + form).
4. If the plane polarized light is rotated towards left, the compound is called **laevorotatory** (*l* or – form).
5. **Specific rotation** is defined as, 'the angle of rotation of the plane of polarized light by a liquid, which in a volume of 1 ml contains 1 gram of active substance when the length of the column through which the light passes is one decimetre'.
6. **With one tetrahedral chiral centre, a molecule is always chiral.** CHBrClF is a chiral molecule containing one chiral centre.
7. **With two or more chiral centres, a molecule may or may not be chiral.**
8. To distinguish chiral and achiral compounds, keep in mind the followings:
   (i) A plane of symmetry is a mirror plane that cuts a molecule in half, so that one half of the molecule is a reflection of the other.
   (ii) Achiral molecules usually contain a plane of symmetry, i.e., they can be cut into two identical halves by a simple plane, but chiral molecules do not.

(*Continued*)

MEMORY FOCUS (*Continued*)

9. For n chiral centres, the maximum number of stereoisomers is 2^n.
   (i) When $n = 1$, $2^1 = 2$. With one chiral centre, there are always two stereoisomers and they are **enantiomers**.
   (ii) When $n = 2$, $2^2 = 4$. With two chiral centres, the maximum number of stereoisomers is four. These isomers include enantiomers, diastereomers and *meso*-compounds.
10. A *meso*-compound is an achiral compound that contains tetrahedral chiral centres. *Meso*-compounds generally have a plane of symmetry; therefore, they possess two identical halves.
11. Because two enantiomers have identical physical properties, they cannot be separated by common physical techniques such as distillation.
12. Diastereomers and constitutional isomers have different physical properties, and therefore they can be separated by common physical techniques.

2.8 POLARITY OF BONDS

Covalent bonds are of two types:
(i) Non-polar covalent bond. (ii) Polar covalent bond.

(i) **Non-polar covalent bond**: In the non-polar covalent bonds, the two atoms have equal electronegativity, i.e., equally attract the pair of shared electrons.

Thus, the electron pair is shared equally between two atoms, e.g., the covalent bond in H_2 and Cl_2 is non-polar. As a result, the molecule is neutral or non-polar.

$$\underset{\text{Hydrogen molecule}}{H:H} \qquad \underset{\text{Chlorine molecule}}{:\ddot{\underset{..}{Cl}}:\ddot{\underset{..}{Cl}}:}$$

(ii) **Polar covalent bond:** In the case of polar covalent bonds, the two atoms have unequal electronegativities or the electron pair is shared unequally. For example, in HCl, the electron pair is more attracted or displaced towards chlorine, as it is more electronegative than hydrogen. Due to this, chlorine end of the molecule appears negative and the hydrogen end appears positive. Such molecules having the oppositely charged poles are called **polar molecules** and the bond is said to be **a polar covalent bond.**
Displaced to Cl atom

$$H \;\; :\ddot{\underset{..}{Cl}}: \longrightarrow \overset{\delta \rightarrow \delta-}{H - - Cl} \text{ or } \overset{\delta+}{H} \;\; \overset{\delta-}{:\ddot{\underset{..}{Cl}}:}$$

Similarly, BrCl molecule is a polar molecule, as chlorine is more electronegative than bromine. The shared electron pair is more attracted towards chlorine than towards bromine. As a result, the chlorine end of the molecule is negative and the bromine end is positive.

If there is a single bond in a molecule, the molecular dipole moment is the same as that of the individual bond, e.g., molecular dipole moment of HCl is the same as that of single HCl bond, i.e., 1.03 D. In case of the molecule having more than one polar bond, the molecular dipole moment is not measured by the values of individual bonds but by the arrangement of the polar bonds in space.

$$:\ddot{\underset{.}{Br}} \;{}_{x}\overset{xx}{\underset{xx}{Cl}}{}_{x}^{x} \longrightarrow \overset{\delta \rightarrow \delta-}{Br - - Cl}$$

2.8.1 Polar Character of Covalent Bond

When a covalent bond is formed between two dissimilar atoms, one of which has larger value of electronegativity and the bonding pair of electrons is displaced towards the more electronegative atom. In other words, electron cloud containing the bonding electrons gets distorted and the charge density gets concentrated around the more electronegative atom and it acquires a partial negative charge (indicated by δ^-), where as less electronegative atom acquires a partial positive charge (indicated as δ^+). Such a bond is called a polar covalent bond and it develops a partial ionic character. It is represented by an arrow pointing towards the more electronegative atom as shown in the case of hydrochloric acid

$$\overset{\delta^+ \longrightarrow \delta^-}{\text{H} \text{———} \text{Cl}}$$

Whereas the covalent bond between two similar atoms is called *non-polar covalent bond*, as the shared pair of electrons lies midway between the nuclei of two atoms (Fig. 2.12).

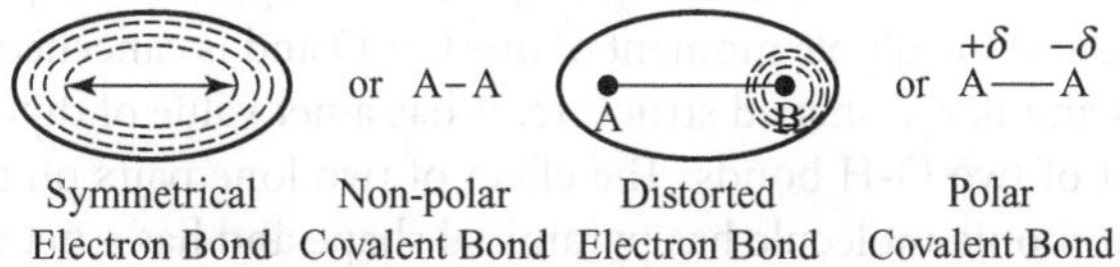

Figure 2.12 Polarity of covalent bond

The extent of ionic character in a covalent bond depends upon the difference of electronegativities of the atoms forming a covalent bond. A bond is considered to be ionic if it has more than 50% ionic character. If the difference in electronegative values of the two bonded atoms is more than 2, it is primarily ionic. The difference of about 1.17 gives 50% ionic character.

2.9 DIPOLE MOMENT

The degree of polarity developed in polar molecule is expressed in terms of dipole moment.

Dipole moment may be defined as the product of the magnitude of charge on either atom and the distance between the centres of the nuclei of boding atoms forming a polar covalent bond.

Dipole moment is represented by Greek letter μ and is expressed as:

$$\text{Dipole moment, } = e \times d \qquad (2.21)$$

where e is the charge on either atom and d is the distance between them.

2.9.1 Unit of Dipole Moment

Since charge is of the order of 10^{-10} esu and distance is of the order of 10^{-8} cm, dipole moment is of the order of 10^{18} esu-cm and this unit is known as Debye unit (D), i.e.

$$1\text{ D} = 10^{-18}\text{ esu-cm}$$

In S.I. system, the unit of dipole moment is coulomb metre, C-m.

$$1\text{ D} = 3.336 \times 10^{-30}\text{ C–m.}$$

Dipole moment is a vector quantity and can be determined experimentally. It can be represented by an arrow pointing towards negative charge with a small tail at the positive charge, e.g.,

$$\overset{\delta+}{\text{H}}\text{—}\overset{\delta-}{\text{F}} \quad \text{or} \quad \text{H—F}$$

2.9.2 Dipole Moment and Molecular Structure

(i) **Diatomic molecules:** Dipole moment is a vector quantity, i.e., it has magnitude as well as direction. As a polar diatomic molecule has only one polar bond, the dipole moment of that molecule is equal to the dipole moment of the polar bond. Thus, the greater the electronegativity difference between the bonded atoms is, the greater is the dipole moment. For example, the dipole moments of hydrogen halides are in the order:

H–F	>	H–Cl	>	H–Br	H–I
(1.91 D)		(1.03 D)		(0.78 D)	(0.38 D)

(2) **Polyatomic molecules:** As polyatomic molecule has more than one polar bond, the dipole moment of the molecule is equal to the resultant dipole moment of all the individual bonds (called *bond moments*). The dipole moments of lone pairs also make their contribution to the resultant dipole moment. The magnitude of the resultant dipole moment not only depends upon the values of individual moments but also on their arrangements in space i.e shape. For example, CO_2 has linear structure because the dipole moment of one C = O and is cancelled by that of the other C = O bond. Water has angular V-shaped structure. It has a net value of dipole moments (= 1.85 D), which the resultant of two O-H bonds. The effect of two lone pairs on the oxygen atom is cancelled mutually. Ammonia molecule has pyramidal shape and has a net resultant value of dipole moment (= 1.46 D) due to unsymmetrical structure.

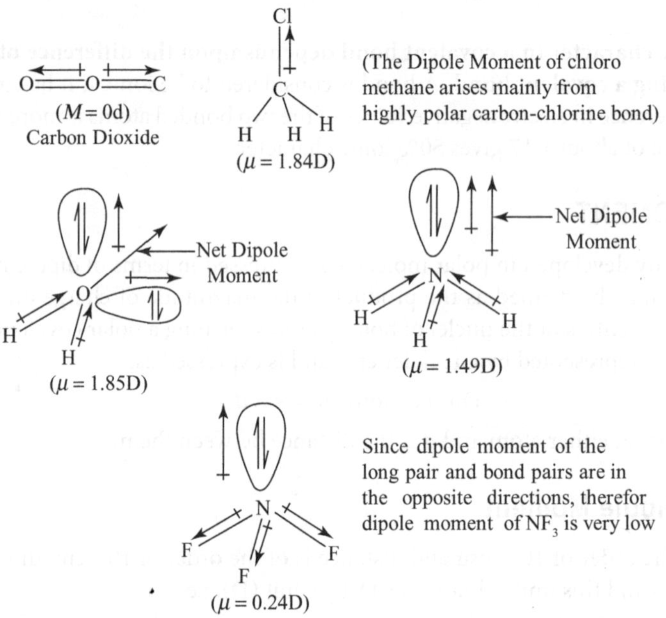

2.9.3 Application of Dipole Moments

(i) **To distinguish polar and non-polar molecules:** The molecules having zero dipole moment are non-polar, e.g., O_2, CO_2, CCl_4, CH_4, CS_2, SiF_4, $SnCl_4$ and BF_3. The bonds of such molecules may be polar or non-polar. The molecules having dipole moments are polar, e.g., NH_3, H_2O and HF.

(ii) **To predict degree of polarity:** Dipole moments help in determining the degree of polarity developed in molecules. More the dipole moment is, the more will be the degree of polarity in molecules. For example, HF has more dipole moment than HCl. Thus, HF is more polar than HCl.

(iii) **To predict the shape or the symmetry of molecules:** Dipole moment helps to determine the shape of the molecules. Tri and polyatomic molecules having zero double moment have **symmetrical structure.**

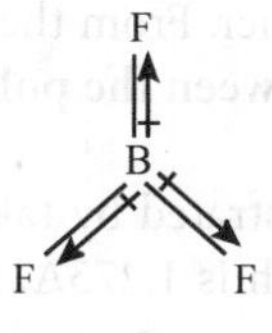

Boron Trifluoride
$\mu = 0.0D$
Plane Triangular Structure

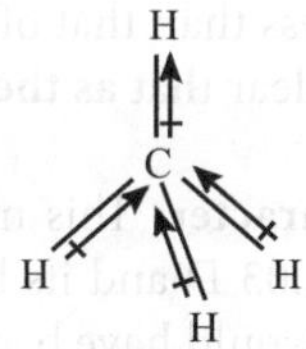

Methane
$\mu = 0.0$
Symmetrical
Tetrahedral Structure

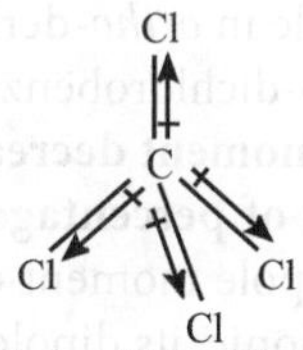

Carbon Tetrachloride
$\mu = 0.0D$
Symmetrical
Tetrahedral Structure

The bonds in such molecules (being polar in nature) have some dipole moment. However, the resultant dipole moment becomes zero due to symmetrical structure.

(iv) **To distinguish *cis*- and *trans*-isomers:** *Trans*-isomer has almost zero dipole moment, whereas *cis*-isomer has significant dipole moment, e.g.,

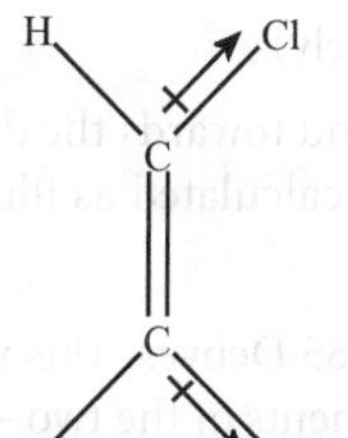

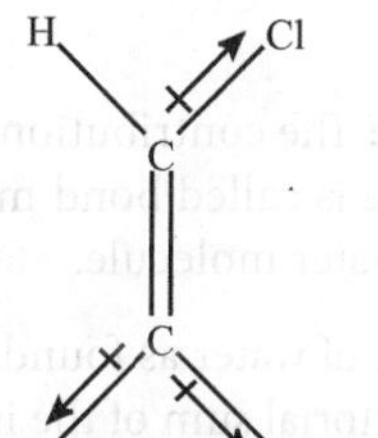

(v) **To distinguish *ortho*-, *meta*- and *para*-isomer:** The position of the two electronegative atoms/groups in the benzene ring, i.e., *o*-, *m*- and *p*-isomers.

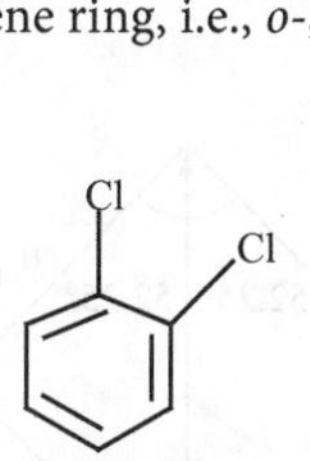

$\mu = 2.54$ D

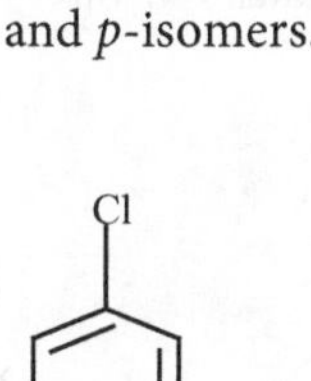

$\mu = 1.48$ D

$\mu = 0$ D

The role of bond angle in the dipole moment is explained as

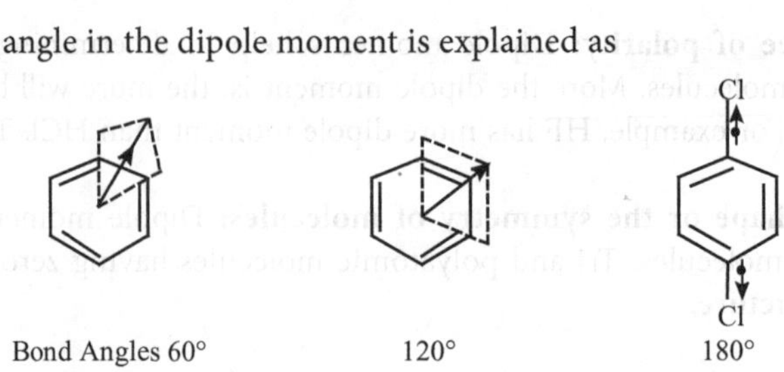

Bond angle in *ortho*-derivative is less than that of *meta*-isomer. From the above examples of *o*-, *m*- and *p*-dichlorobenzenes, it is clear that **as the angle between the polar bonds increase, the dipole moment decreases.**

(vi) **Calculation of percentage ionic character:** This may be illustrated by taking an example of HCl. The dipole moment of HCl is 1.03 D and its bond length is 1.275Å. If the bond in HCl were 100% ionic, its dipole moment would have been

$$= (4.8 \times 10^{-10}) \times (1.275 \times 10^{-8}) \text{ esu-cm}$$

where 4.8×10^{-10} esu is one unit charge which would have been present on end in case the molecule was 100% ionic.

$$= 6.11 \times 10^{-18} \text{ esu-cm} = 6.11 \text{ D}$$

Actual dipole moment of HCl = 1.03 D

$$\text{Hence, percentage ionic character} = \frac{1.03}{6.11} \times 100$$

$$= 17\% \text{ (approximately)}$$

(vii) **Bond moments:** The contribution of an individual bond towards the dipole moment of a polyatomic molecule is called bond moment. This can be calculated as illustrated below by taking an example of water molecule.

The dipole moment of water as found experimentally is 1.85 Debyes. This value of dipole moment is supposed to be the vectorial sum of the individual bond moments of the two –OH bonds directed at an angle of 104.5° (see Fig. 2.13).

Thus, the dipole moment of water is given by,

$$\mu_{\text{observed}} = 2\mu_{\text{OH}} \cos\frac{104.5}{2}$$

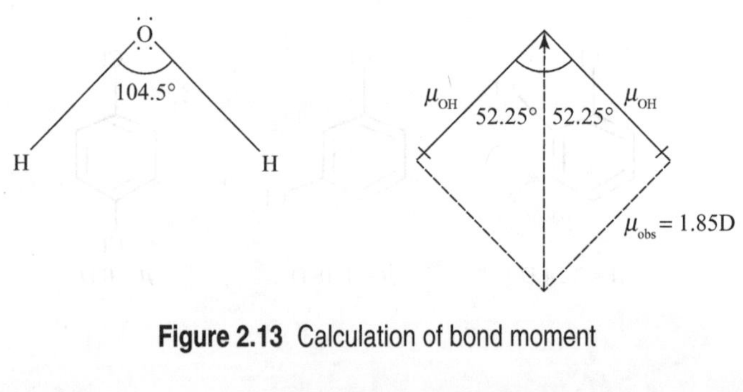

Figure 2.13 Calculation of bond moment

$$\mu_{\text{observed}} = 2\mu_{\text{OH}} \cos 52.25°$$

$$\mu_{\text{observed}} = 2\mu_{\text{OH}} \times 0.6129$$

However, μ_{observed} in case of water = 1.85 D

$\therefore$
$$1.85 = 2\mu_{\text{OH}} \times 0.6129$$

or
$$\mu_{\text{OH}} = \frac{1.85}{2 \times 0.6129} = 1.51 \text{ Debyes}$$

The values of calculated bond moments of some bonds are given in Table 2.6.

Table 2.6 Bond moments of some common bonds.

Bond	Dipole Moment (Debye)	Bond	Dipole Moment
H–F	1.9	F–CI	0.9
H–CI	1.0	F–Br	1.3
H–Br	0.8	Br–CI	0.6
H–I	0.4	C–H	0.4
H–O	1.5	C–F	1.4
H–N	1.3	C–CI	1.5
H–P	0.4	C–Br	1.4
P–CI	0.8	C–I	1.2
P–Br	0.4	C–O	0.7
As–F	2.0	C–N	0.2
As–CI	1.6	C–O	2.3
AS–Br	1.3	C–NO_2	3.54

MEMORY FOCUS

1. The dipole moment of a polar molecule is given by the product of the charge at one end and the distance between the opposite charges, i.e., $\mu = q \times r$.
2. The dipole moment (μ) is a vector quality. It is represented by an arrow with a crossed tail.
3. The C.G.S. unit for dipole moment is the Debye, symbolized by D,

   $\text{ID} = 1 \times 10^{-18}$ esu-cm and $\text{ID} = 3.336 \times 10^{-30}$ cm.
4. The net dipole moment of the molecule is the vector resultant of all the individual bond moments.
5. Dipole moment can provide important information about the geometry of molecular. For example:
   (i) H_2O has a bent structure because of its resulted dipole moments of the two O–H bonds.
   (ii) CO_2 has a linear structure because its dipole moment of two C–O bonds cancels each other; the resulted dipole is zero.

(Continued)

MEMORY FOCUS (*Continued*)

(iii) SO_2 has a bent structure because dipole moment is 1.63 D and the dipole moment of two S–O bonds does not cancel.
(iv) BF_3 has a planar and NH_3 a pyramid structure.
(v) CH_4 has tetrahedral structure.
(vi) It helps in the identification of *cis*- and *trans*-isomers.
(vii) It helps in the identification of *ortho*-, *meta*- and *para*-isomers of aromatics.
(viii) The magnitude of the dipole moment of a diatomic molecule determines its ionic character.

REVISION QUESTIONS

1. Discuss briefly the main characteristics of liquids.
2. How would you classify the physical properties of liquids? Give examples of each class.
3. Define surface tension. What is its unit? Discuss any one method of its determination.
4. What is stalagmometer? Explain the drop-number as well as drop-weight methods of determining the surface tension of a liquid.
5. How will you determine the surface tensions of liquids?
6. Give reasons for the followings:
   (a) Liquid drops are spherical in shape.
   (b) Water rises in capillary.
   (c) Surface tension of mercury is much larger that of water at the same temperature.
   (d) Mercury forms a different type of meniscus as compared to water.
7. What is parachor? Give two examples to explain the importance of parachor in the elucidation of molecular structure.
8. What do you mean by atomic and structural parachor values? Illustrate with an example the use of parachor in confirming molecular formula of a compound.
9. Define viscosity. How will you measure the coefficient of viscosity in the laboratory? What is the effect of temperature on viscosity?
10. What is the effect of temperature on viscosity? What light viscosity throws on molecular structure?
11. Describe Ostwald's method for determining viscosity of a liquid.
12. What are additive and constitutive properties? Explain any two such properties.
13. Write a short note on the followings:
    (a) Parachor (b) Rheochor (c) Molecular viscosity
14. What is meant by molar refraction? How can this be experimentally determined? Explain the role of molar refraction in determining the chemical constitution of a compound.
15. Explain the term refractive index. How it is determined experimentally?
16. Explain how refractivity measurements can be used to explain the molecular structure of a compound.
17. Define specific and molar refraction. Discuss the principle, constructive and working of Abbe's refractometer.
18. How will you measure optical activity of a system? Give factors affecting optical activity of a compound.
19. What is meant by optical rotation? How is specific rotation measured? Give the principle of the instrument used.

20. What is the cause of optical activity. How is the measurement of optical rotation helpful in elucidation of chemical constitution of a substance?
21. 'Optical activity is a constitutive property'. Explain with examples.
22. What are polar and non-polar molecules? Give two examples of each.
23. Define dipole moment. How it is measured? Illustrate by giving any two methods.
24. Illustrate the application of dipole moment in the determination of (i) percentage of ionic character of a polar bond and (ii) molecular structure.
25. State and explain dipole moment. How is it helpful in elucidating the molecular structure of a compound?
26. (a) Define relative and absolute viscosity.
    (b) Write a note on dipole moment.
    (c) Explain the terms: Specific refraction and molar refraction.
27. Discuss any two methods for determining dipole moment.
28. (a) Define the terms: Refractive index and optical rotations
    (b) Write short notes on the following:
    (i) Rheochor (ii) Viscosity
    (iii) Optical activity (iv) Parachor
29. (a) Define surface tension. How does surface tension of a liquid vary with temperature?
    (b) Give one experimental method for the determination of surface tension.
30. (a) Explain the followings:
    (i) Drops of a liquid assumed a spherical shape. Why?
    (ii) Which one of glycerine and kerosene possess high viscosity? Why?
    (b) Discuss one method of measuring surface tension of any liquid.
    (c) Define the following terms:
    (i) optical rotation
    (ii) dipole moment
31. (a) What is coefficient of viscosity? Describe one method to measure it.
    (b) What is molar refraction? How is it related to chemical constitution?
32. (a) Explain the terms Surface tension and Surface energy. How are they related to each other?
    (b) Define the following:
    (i) specific refraction (ii) Viscosity
    (iii) Specific rotation (iv) Dipole moment

MULTIPLE CHOICE QUESTIONS

1. The boiling point of a liquid is the temperature at which
   (a) the vapour pressure of the liquid is equal to the atmospheric pressure.
   (b) the vapour pressure of the liquid is less than the atmospheric pressure.
   (c) the vapour pressure of the liquid is greater than the atmospheric pressure.
   (d) the vapour pressure of the liquid is equal to the square root of the atmospheric pressure.
2. With the rise in temperature, the surface tension of a liquid
   (a) increases
   (b) decreases
   (c) remains the same
   (d) none of the above

3. The unit in which surface tension is measured is
   (a) dyne cm
   (b) dyne cm^{-1}
   (c) $dyne^{-1}$ cm
   (d) $dyne^{-1}$ cm^{-1}
4. The formula used for the determination of surface tension by capillary rise method is
   (a) $2y = hrdg$
   (b) $2y = hr^2dg$
   (c) $2y = \pi r \cos\theta$
   (d) $2y = \pi hr^2dg$
5. In drop-weight method for the determination of surface tension of liquids, the formula used is
   (a) $\frac{y_1}{y_2} = \frac{m_1}{m_2}$ (b) $\frac{y_1}{y_2} = \frac{m_2}{m_1}$
   (c) $\frac{y_1}{y_2} = \frac{m_2^2}{m_1^2}$ (d) $\frac{y_1}{y_2} = \frac{m_1^2}{m_2^2}$
6. The formula used in drop-number method for the determination of surface tension of liquids is
   (a) $\frac{y_1}{y_2} = \frac{n_2 d_1}{n_1 d_2}$ (b) $\frac{y_1}{y_2} = \frac{n_1 d_1}{n_2 d_2}$
   (c) $\frac{y_1}{y_2} = \frac{n_1 d_2}{n_2 d_1}$ (d) $\frac{y_1}{y_2} = \frac{n_2 d_2}{n_1 d_1}$
7. Viscosity of a liquid is a measure of
   (a) repulsive forces between the liquid molecules
   (b) frictional resistance
   (c) intermolecular forces between the molecules
   (d) none of the above
8. The reciprocal of viscosity called
   (a) surface tension
   (b) frictional resistance
   (c) fluidity
   (d) none of these
9. The S.I. unit of coefficient of viscosity is
   (a) kg m^2s
   (b) kg m s^{-1}
   (c) kg $m^{-1}s$
   (d) kg $m^{-1}s^{-1}$
10. The formula used for the determination of relative viscosity by Ostwald's method is
    (a) $\frac{n_1}{n_2} = \frac{d_1 t_1}{d_2 t_2}$ (b) $\frac{n_1}{n_2} = \frac{d_1 t_2}{d_2 t_1}$
    (c) $\frac{n_1}{n_2} = \frac{d_2 t_1}{d_1 t_2}$ (d) $\frac{n_1}{n_2} = \frac{d_2 t_2}{d_1 t_1}$
11. In general, the viscosity with temperature
    (a) decreases
    (b) increases
    (c) remains the same
    (d) none of these
12. The specific refraction is given by the formula
    (a) $R_s = \frac{n^2 - 1}{n^2 + 1} \times \frac{1}{d}$ (b) $R_s = \frac{n^2 + 1}{n^2 - 1} \times \frac{1}{d}$
    (c) $R_s = \frac{n^2 - 1}{n^2 - 2} \times \frac{1}{d}$ (d) $R_s = \frac{n^2 + 1}{n^2 + 2} \times \frac{1}{d}$
13. The molar refraction is given by which of the formula given below:
    (a) $R_m = \frac{n^2 + 1}{n^2 + 2} \times \frac{M}{d}$
    (b) $R_m = \frac{n^2 - 1}{n^2 + 2} \times \frac{M}{d}$
    (c) $R_m = \frac{n^2 + 1}{n^2 - 2} \times \frac{M}{d}$
    (d) $R_m = \frac{n^2 - 1}{n^2 - 2} \times \frac{M}{d}$
14. The units in which molar refraction is expressed are
    (a) cm^3mol^1 (b) $cm^{-3}mol$
    (c) $cm^{-3}mol^{-1}$ (d) cm^3mol^{-1}
15. The specific rotation, a characteristic of an optically active substance, is expressed as
    (a) $[\alpha] = \frac{\alpha}{l \times c}$ (b) $[\alpha] = \frac{\alpha}{l^2 \times c}$
    (c) $[\alpha] = \frac{\alpha}{l \times c}$ (d) $[\alpha] = \frac{\alpha^2}{l \times c}$
16. A liquid is in equilibrium with its vapours at its boiling point. On the average, the molecules in the two phases have equal

(a) potential energy
(b) kinetic energy
(c) van der Waal's forces
(d) total energy

17. Small droplets are spherical in shape. It is due to
(a) high viscosity
(b) their tendency to acquire minimum surface area
(c) their tendency to acquire maximum surface area
(d) less viscosity

18. Insects can walk on the surface of water due to
(a) viscosity
(b) surface tension
(c) refractivity
(d) optical activity

19. The works in ergs required to be done to increase the surface area by 1 sq. cm is called
(a) surface tension
(b) internal friction
(c) fluidity
(d) surface energy

20. The vapour pressure of a liquid at a given temperature is given by
(a) $\frac{n}{V}RT$ (b) $\frac{n}{V}RT^2$
(c) $\frac{n}{2v}RT$ (d) $\frac{n}{V}RT^3$

21. The boiling point of a liquid ________ with increase in its external pressure.
(a) changes
(b) always increases
(c) always decreases
(d) remains the same

22. The liquids with high intermolecular attractive forces have __________ viscosity.
(a) low
(b) intermediate
(c) high
(d) none of these

ANSWERS

1. (a)	5. (a)	9. (d)	13. (b)	17. (b)	21. (a)
2. (b)	6. (a)	10. (a)	14. (d)	18. (b)	22. (c)
3. (b)	7. (b)	11. (a)	15. (c)	19. (d)	
4. (a)	8. (c)	12. (a)	16. (b)	20. (a)	

Solution 3

CHAPTER OBJECTIVES

3.1 Introduction

3.2 Modes of Expressing Concentration of Solutions

3.3 Raoult's Law

3.4 Ideal Solution

3.5 Colligative Properties of Dilute Solution

3.6 Lowering of Vapour Pressure

3.7 Elevation in Boiling Point

3.8 Depressing of Freezing Point

3.9 Osmotic Pressure

3.10 Abnormal Molecular Masses

Some of the liquids we drink are solutions of caffeine, sugar and so on in water. The coins we exchange are solid solutions of metals. We inhale a gaseous solution of a particular composition and exhale a solution of another composition. Solution chemistry is indeed very pervasive, and many interesting chemical processes are studied in the solution phase.

Almost all the chemical processes occuring in nature, whether in animal or vegetable organisms or in the non-living surfaces of the earth, take place between reacting substances in solution

W. Ostwald

3.1 INTRODUCTION

A **solution** is a *single-phase homogeneous system,* containing two or more substances, whose compositions can be varied within wide limits. The component of a solution forming the larger proportion is referred to as *solvent,* while the other component which is present in minor proportion is called *solute.* Examples of solutions are common salt in water, alcohol in water, sugar in water, etc. A solution in water is called aqueous solution (e.g., cane-sugar dissolved in water); while a solution in a solvent other than water is called *non-aqueous solution* (e.g., iodine in carbon disulphide).

In chemistry, the term 'solution' has wide meaning. It stands for homogeneous mixture of two or more substances which may exist in any of the three states of matter. Air, which is a mixture of many gases, is an example of *gaseous solution.* Petrol, which is a mixture of various liquid hydrocarbons, is an example of *liquid solution.* An alloy, which is mixture of different metals, is an example of *Solid Solution.* Depending upon the physical states of solute and solvent, the solution is of nine types (Table 3.1).

Table 3.1 Types of solutions.

Solvent	Solute	Examples(s)
Gas	Gas	Air (mixture of N_2, O_2, Ar, etc.)
Gas	Liquid	Water vapour in air, mist.
Gas	Solid	Smoke, NH_4CI vapour in air.
Liquid	Gas	Aerated drink (CO_2 dissolved in water).
Liquid	Liquid	Alcohol in water, benzene in toluene.
Liquid	Solid	Sugar in water, common salt in water.
Solid	Gas	Dissolved gases in minerals.
Solid	Liquid	Amalgam, e.g., gold in mercury.
Solid	Solid	Homogeneous alloys (e.g., brass, steel).

Out of the nine types of solutions, the most important are with *liquid solvent,* i.e., solutions of solids, liquids or gases in liquids.

In this chapter, we shall deal with ideal and non-ideal solutions. For a proper understanding of this chapter, we must understand what vapour is? We should also learn Raoult's law, which relates to vapour pressure of liquids and their solutions.

We are familiar with the process of evaporation. When evaporation takes place in a closed vessel, a stage is ultimately reached when the rate of evaporation becomes equal to that of condensation and a phase equilibrium is said to be established. Once the equilibrium is attained at a particular temperature, there is no further change in the number of molecules in either the liquid or the vapour state. *The pressure of the vapour above the liquid at equilibrium refers to as the vapour pressure of the liquid. It will have a characteristic value at a particular temperature and will increase with temperature. The vapour pressure of a liquid at a given temperature may thus be defined as the pressure of the vapour in equilibrium with the liquid at that temperature in an enclosed space.* The magnitude of the equilibrium vapour pressure depends on (1) the nature of the liquid and (2) temperature.

3.2 MODES OF EXPRESSING CONCENTRATION OF SOLUTIONS

(1) Percentage of a solution can be expressed as:

(i) *Mass percentage* is the number of parts by mass of the solute per hundred parts by mass of solution. If w g of solute is dissolved in W g of solvent, then mass percentage of solute is given as follows:

$$\text{Mass percentage} = \frac{w}{w + W'} \times 100$$

(ii) *Volume percentage* is the number of parts by volume of solute per hundred parts by volume of solution. If V_A and V_B be the respective volumes of component A and B of a solution, then volume percentage of A is given as follows:

$$\text{Volume percentage of A} = \frac{V_A}{V_A + V_B} \times 100$$

(2) *Normality of a solution* is the number of gram equivalents of the solute dissolved per litre of the solution. Thus, normality (N)

$$= \frac{\text{Gram-equivalents of solute present in solution}}{\text{Volume of solution in litres}}$$

$$= \frac{\text{Mass of solute in g}}{\text{Gram eq. mass of solute}} \times \frac{1}{\text{Volume of solution in litres}}$$

$$= \frac{\text{W in g}}{\text{E}_{solute} \times \text{V in L}} = \frac{\text{W} \times 1{,}000}{\text{E}_{solute} \times \text{V in mL}}$$

Thus, xN solution means one which contains x gram equivalents of solute per litre. The terms *semi-*, *deci-* and *centi-*normal are used to express N/2, N/10 and N/100 solutions, respectively.

Equal volumes of acid and alkali (or oxidizing agent and reducing agent) of identical normality neutralize or react just completely with each other, since they contain identical number of gram equivalent. Thus,

$$N_1V_1 = N_2V_2$$

where N_1 and V_1 are the normality and volume of one solution; while N_2 and V_2 are the normality and volume of the second solution.

(3) *Molarity* is defined as the number of moles (molecular weight expressed in grams) of the solute present in 1 l of the solution. Thus,

$$\text{Molarity (M)} = \frac{\text{Number of moles of solute present}}{\text{Volume of solution in litres}}$$

$$= \frac{\text{No. of g of solut present/Mol.wt. of solute}}{\text{Volume of solution in litres}}$$

Thus, if w g of solute (mol. Mass = m) is present in V cm³ of a solution, then,

$$\text{Molarity (M)} = \frac{w}{m} \times \frac{1{,}000}{V}$$

(4) *Molality* is defined as the number of moles of the solute dissolved in 1,000 g (or 1 kg) of the *solvent* (*not solution*). Thus,

$$\text{Molarity}\,(m) = \frac{\text{Number of moles of solute dissolved}}{\text{Number of kg of solvent}}$$

$$= \frac{\text{Number of g of solute/Mol.wt.of solute}}{\text{Number of kg of solvent}}$$

Thus, if w g of solute (mol. Mass = m) is dissolved in W g of the solvent, then,

$$\text{Molality}\,(m) = \frac{w}{m} \times \frac{1{,}000}{W}$$

Effect of temperature on molarity and molality: (a) Molarity of a solution is the number of moles present per litre of solution. When temperature is *raised* or *lowered, the* volume of the solution *increases* or *decreases*, respectively. Consequently, the number of moles present per litre in the changed volume also *alters*. Hence, molarity of a solution *decreases*, when temperature of solution is *raised; increases*, when temperature of solution is *lowered*. (b) Molality of a solution remains *unchanged* with the change in temperature, because *temperature has no effect on weights* (of both solute as well as solvent).

Relation between molarity and normality: These are inter-related as

$$\frac{\text{Molarity (M)}}{\text{Normality (N)}} = \frac{\text{Molecular mass of the solute (m)}}{\text{Equivalent mass of the solute (E)}}$$

For acids: $\text{Normality} = \text{Molarity} \times \text{Basicity of acid}$

where *basicity* is the number of H^+ ions furnished by each molecule of the acid on complete dissociation in aqueous solution.

For bases: $\text{Normality} = \text{Molarity} \times \text{Acidity}$

where *acidity* is the number of OH^- ions furnished by each molecule of the base on complete dissociation in aqueous solution.

(5) *Mole fraction* ***(X)*** of any component in a solution is the ratio of the number of moles of that component to the total number of moles of solute plus solvent in the solution. Thus,

$$X_{solute} = \frac{\text{No. of moles of solute}}{\text{No. of moles of solute} + \text{No. of moles of solvent}}$$

$$= \frac{n}{n+N} = \frac{w/n}{(w/m)+(W/M)}$$

and mole fraction of solvent.

$$X_{solute} = \frac{\text{No. of moles of solvent}}{\text{No. of moles of solute} + \text{No. of moles of solvent}}$$

$$= \frac{N}{n+N} = \frac{W/M}{(w/m)+(W/M)}$$

where n = number of moles of solute; N = number of moles of solvent; w = weight of solute; W = weight of solvent; m = molar mass of solute; M = molar mass of solvent.

From the above, it is evident that:

$$X_{solute} + X_{solvent} = 1 \text{ or } X_{solute} = 1 - X_{solvent}$$

(6) *Mass fraction* of any component in a solution is the ratio of the mass of that component to the total mass of both (or all) components of the solution. Thus,

$$\text{Mass fraction of solute} = \frac{\text{Mass of solute}}{\text{Mass of (solute + Solvent)}} = \frac{w}{(w+W)}$$

$$\text{Mass fraction of solvent} = \frac{\text{Mass of solvent}}{\text{Mass of (solute + Solvent)}} = \frac{w}{(w+W)}$$

(7) *Parts per million (ppm)* is the mass of solute present per 10^6 (million) parts by mass of a very dilute solution. Thus, ppm of solute

$$= \frac{\text{Mass of solute}}{\text{Mass of solution}} \times 10^6$$

3.3 RAOULT'S LAW

This law can be stated in two different ways

3.3.1 For a Solution of Volatile Liquids

Raoult studied the vapour pressure of a number of binary solutions of volatile liquids and presented a generalization called Raoult's law. It states that,

For a solution of volatile liquids, the partial vapour pressure of any volatile component at a given temperature is equal to product of the vapour pressure of the component in pure state and mole fraction of that component in solution.

For a solution of two volatile liquids A and B, let X_A and X_B be their mole fractions and P_A and P_B be their partial pressures, respectively, then,

$$P_A = X_A P_A^0$$

and $$P_B = X_B P_B^0 \qquad (3.1)$$

where P_A^0 and P_B^0 are the vapour pressure of volatile liquids A and B in their pure states.

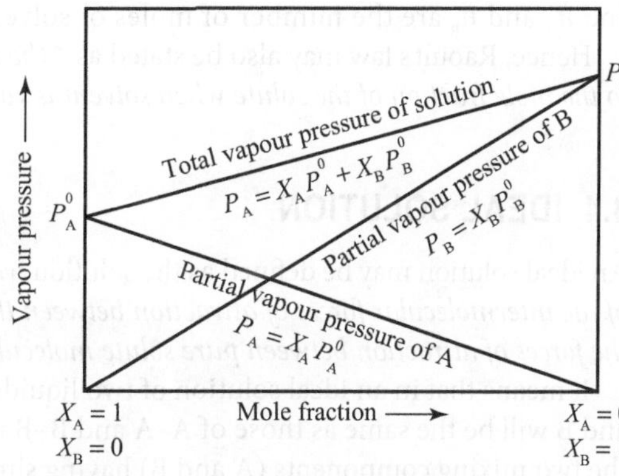

Figure 3.1 Total and partial vapour pressure of an ideal solution

For an ideal solution, a graph plot between P_A or P_B against X_A or X_B is always a straight line (Fig. 3.1). Suppose the component B is more volatile than component A and P_A^0 and P_B^0 represent their vapour pressures in the pure state.

From Fig. 3.1, it is clear that vapour pressure of the solution P at any composition is equal to the sum of vapour pressure of each component. The vapours above the solution must obey Dalton's law of partial pressures, so that

$$P = P_A + P_B$$

$$P = X_A P_A^0 + X_B P_B^0 \qquad (3.2)$$

3.3.2 For a Solution of Non-volatile Solute

In case of solution of a non-volatile solute in a volatile solvent, there is hardly any contribution of solute to the total vapour pressure of the solution. Therefore, the vapour pressure of the solution is in fact due to that of solvent only. In such a case, the preceding equation becomes

$$P = P_A = X_A\, P_A^0$$

For a binary solution, we know

$$X_A + X_B = 1$$

$$X_A = 1 - X_B$$

Putting this value, we get

$$P = (1 - X_B)\, P_A^0$$

$$P = P_A^0 - P_A^0\, X_B$$

$$X_B\, P_A^0 = P_A^0 - P$$

Dividing by P_A^0 on both sides, we get

or

$$\frac{P_A^0 - P}{P_A^0} = X_B = \frac{n_B}{n_A + n_B} \tag{3.3}$$

where $P_A^0 - P$ is the lowering of vapour pressure, $\frac{P_A^0 - P}{P_A^0}$ is the relative lowering of vapour pressure and n_A and n_B are the number of moles of solvent and solute, respectively.

Hence, Raoult's law may also be stated as, '*The negative lowering of vapour pressure of solution is equal to the mole fraction of the solute when solvent is volatile*'.

3.4 IDEAL SOLUTION

An ideal solution may be defined as the solution *which obeys Raoult's law and in which the average strength of the intermolecular forces of attraction between the solute and solvent is equal to the average strenvgth of the forces of attraction between pure solute molecules and between pure solvent molecules.*

It means that in an ideal solution of two liquids A and B, the force of attraction between molecules A and B will be the same as those of A–A and B–B molecular attractions. This is because the molecules of the two mixing components (A and B) having similarity in shape, size and polarity.

A solution is said to be ideal if it fulfils the following conditions:

(i) *It obeys Raoult's law over the entire range of concentrations* ($P_A = X_A\, P_A^0$).
(ii) *There should be no volume change on mixing the components, i.e.,* ΔV_{mix} = *zero.*
(iii) *There should be no heat change on mixing the components, i.e.,* ΔH_{mix} = *zero.*
(iv) *There should be no heat change on diluting the solution, i.e.,* $\Delta H_{dil} = 0$.

Thus, an ideal solution may also be defined as that solution in which no change in volume takes place on mixing a solute and solvent in any proportion and no heat change is observed on mixing or diluting it further.

We have only a concept of ideal solutions but no solution is exactly ideal. There are many solutions, which are nearly ideal, i.e., they tend to approach ideal behaviour. For example, a mixture of *n*-hexane and *n*-heptane, ethyl bromide and ethyl iodide, benzene and toluene, and chlorobenzene and bromobenzene are almost ideal solutions. It may be noted that the concentration of the solute approaches zero, all solutions tend to behave ideally. In other words, *all dilute solutions are almost ideal.*

3.4.1 Non-ideal Solution

A solution is said to be non-ideal if it does not fulfil the required conditions for an ideal solution and does not obey Raoult's law. Most of the solutions are non-ideal solutions and they show considerable deviations from ideal behaviour. Non-ideal solutions are of two types, which are discussed in the following sections.

3.4.2 Solutions Showing Positive Deviations

In these solutions, the partial vapour pressure of a component, A, in the solution obtained by mixing B and A is more than what it should be according to Raoult's law. Similarly, the partial vapour pressure of B in the solution obtained by mixing A with B is more than that calculated from Raoult's law. The total vapour pressure of the solution is more than that of an ideal solution of the same composition. In Fig. 3.2, the dotted lines represent ideal behaviour, whereas full-line curves represent non-ideal behaviour.

The molecular interaction between A and B is weaker than A–A and B–B interactions. In such a case, both A and B find easier to escape from the solution. Thus, their partial pressure and their total pressure are higher than expected. In case of solution showing positive deviation, ΔV_{mix} and ΔH_{mix} are always positive. Ethyl alcohol–chloroform, propyl alcohol and water are some the examples of this type of solution.

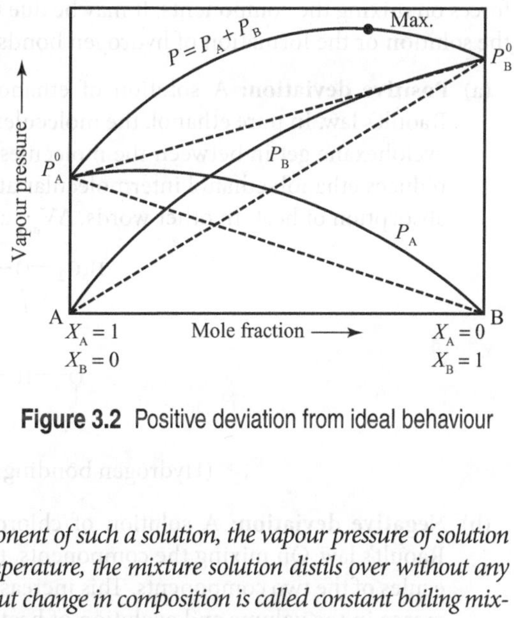

Figure 3.2 Positive deviation from ideal behaviour

Azeotropic Mixture: *At some composition of the component of such a solution, the vapour pressure of solution is the highest and its boiling point is the lowest. At this temperature, the mixture solution distils over without any change in composition. A liquid mixture which distils without change in composition is called constant boiling mixture or azeotropic mixture or azeotrope. Such solutions are minimum boiling point azeotropes.*

3.4.3 Solutions Showing Negative Deviations

In some cases, the vapour pressure of component A in a solution obtained by mixing B with it is less than that of the one calculated from Raoult's law. Similarly, the vapour pressure of a component B and also that of the solution is to be less than the calculated value (Fig. 3.3).

The molecular interaction between A and B is stronger than A–A and B–R interactions. In such a case, the escaping tendency (vapour pressure) of A and B becomes less and total pressure is lower than expected. ΔV_{mix} and ΔH_{mix} for such solution will have negative value. Water–sulphuric acid and acetone–chloroform are some common examples showing negative deviation.

Azeotropic Mixture: *At some composition of such a solution, the vapour pressure of solution is the lowest and its boiling point is the highest. Solutions of this composition which boil at a constant temperature and distil over without any change in composition are called azeotropic mixtures or azeotropes. Such solutions are called maximum boiling point azeotropes.*

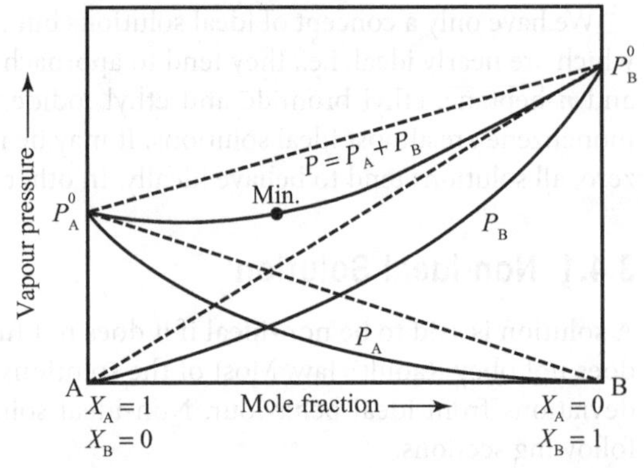

Figure 3.3 Negative deviation from ideal behaviour

3.4.4 Factors Responsible for Deviations

The deviations from ideal behaviour in solutions are due to change in the magnitude of intermolecular forces on mixing the components. It may be due to the breaking of hydrogen bonds in a component of the solution or the formation of hydrogen bonds between the two components of the liquid mixture.

(a) Positive deviation: A solution of ethanol and cyclohexane shows positive deviation from Raoult's law. In pure ethanol, the molecules are hydrogen bonded. On mixing, the molecules of cyclohexane get in between the molecules of ethanol. Thus, hydrogen bonds are broken. This reduces ethanol–ethanol intermolecular attractions. It leads to a slight increase in volume and absorption of heat. In other words, ΔV_{mix} and ΔH_{mix} are positive.

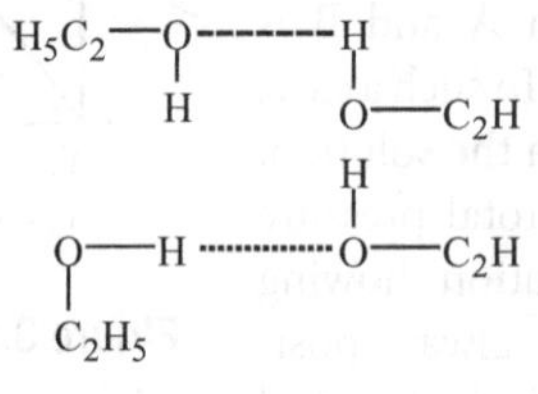

(Hydrogen bonding between ethanol molecules)

(b) Negative deviation: A solution of chloroform and acetone shows negative deviation from Raoult's law. On mixing the components, the hydrogen bonds are developed between the molecules of the two components. This increases the intermolecular forces which leads to slight decrease in the volume and evolution of heat. In other words, both ΔV_{mix} and ΔH_{mix} are negative.

Besides hydrogen bonding, there may be some other factors which are responsible for changing the magnitude of intermolecular attractions.

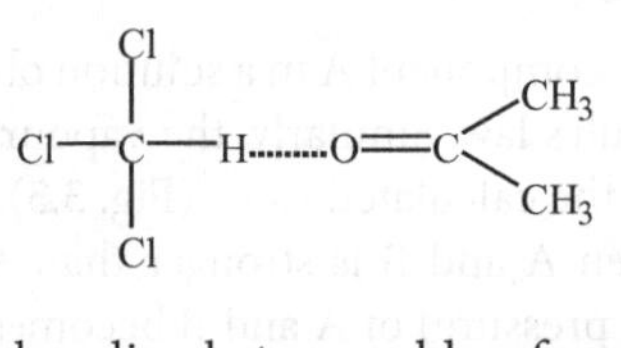

(Hydrogen bonding between chloroform and acetone)

Pairs of miscible liquids forming Ideal and non-ideal solutions.

Nearly Ideal Solution	Mixture Showing Positive Deviations	Mixtures Showing Negative Deviations
Toluene + benzene	Acetone + CS_2	$CH_3COOH + C_5H_5N$ (pyridine)
n-hexane + *n*-heptane	Acetone + C_2H_5OH	$CHCl_3$ + acetone
	Acetone + C_6H_6	$CHO_3 + C_6H_6$
	$CCl_4 + C_6H_6$	CHO_3 + ether
$C_2H_5Cl + C_2H_5Br$	CCl_4 + toluene	$H_2O + HCl$
$C_6H_sCl + C_6H_6Br$	$H_2O + CH_3OH$	$H_2O + HNO_3$
	$H_2O + C_2H_5OH$	Acetone + aniline

Azeotrope: *An azeotrope is a solution of definite composition which distils over at a constant temperature, minimum or maximum, without any change in its composition. It means that the vapours formed have the same composition of their components as in the solution. As a result, an azeotrope cannot be separated into its constituents by fractional distillation. For example, water and ethyl alcohol form an azeotrope of composition 4.43 and 95.57% respectively boiling at 351.15 K. The two components cannot be separated from this mixture.*

3.4.5 Distinction Between Ideal and Non-ideal Solutions

The differences between ideal and non-ideal solutions are summarized in Table 3.2.

Table 3.2 Distinction between ideal and non-ideal solutions.

Ideal Solutions	Non-ideal Solutions
1. The molecular interactions are similar to those in the pure components.	The molecular interactions are different from the pure components.
2. There are no enthalpy (H) change and volume change (V).	There are enthalpy change and volume change.
3. These solutions obey Raoult's law.	These solutions do not obey Raoult's law.

3.4.6 Difference Between Solutions of Positive and Negative Deviations

The differences between solutions of positive and negative deviations are summarized in Table 3.3.

Table 3.3 Different between positive and negative deviations.

Solutions of +ve Deviation	Solutions of −ve Deviation
1. The forces between A and B are less than A–A and B–B forces. (A and B are components of solution.)	The forces between A and B are more than A–A and B–B forces.
2. ΔH_{mix} is +ve.	ΔH_{mix} is −ve.
3. ΔV_{mix} = +ve or increases.	ΔV_{mix} = −ve or decreases.
4. Heating increases solubility.	Heating decreases solubility.

MEMORY FOCUS

1. A solution of two or more constituents is said to be ideal if it obeys **Raoult's law exactly at all concentrations and all temperatures.**
2. Criteria for idea solutions are:
   (a) $\Delta V_{mix} = 0$ *(there should not be any change in volume),*
   (b) $\Delta H_{mix} = 0$ *(there should not be any change in heat) and*
   (c) solutions should be dilute.
3. All solutions tend to be ideal when dilute.
4. If the vapour pressure of the liquid mixture is more than that calculated from Raoult's law, it is said to be showing positive deviation.
5. Binary liquid mixtures which show negative deviation from Raoult's law boil at a higher temperature than the components.
6. Binary liquid mixtures which show positive deviation from Raoult's law boil at a lower temperature than the components.
7. Examples for pairs of liquids which show negative deviation are:
   (a) acetone + chloroform,
   (b) water + sulphuric acid and
   (c) water + nitric acid
8. Examples for pairs of liquids which show positive deviation are:
   (a) acetaldehyde + carbon disulphide,
   (b) water + propyl alcohol and
   (c) ethyl alcohol + chloroform.
9. Negative deviation from Raoult' law occurs:
   (a) when $\Delta H_{mix} = -ve$,
   (b) when $\Delta V_{mix} = -ve$ and
   (c) when the interaction of unlike molecules is stronger than like molecules. (A–B interactions are stronger than A–A or B–B.)
10. Positive deviation from Raoult's law occurs:
    (a) when $\Delta H_{mix} = +ve$,
    (b) when $\Delta V_{mix} = +ve$ and
    (c) when the interaction between unlike molecules is weaker than like molecules. (A–B interactions are weaker than A–A or A–B.)
11. *Solutions which distil without change in composition or boiling point are called azeotropes or constant boiling liquids.*
12. Water forms azeotropic mixtures with ethyl alcohol, **HCI, HNO_3, HBr, HI,** formic acid etc.
13. Azeotropes cannot be separated easily by simple or fractional distillations.

3.5 COLLIGATIVE PROPERTIES OF DILUTE SOLUTION

There are certain properties of solutions which depend upon the mole fraction of solute in the solutions but are independent of nature of solute. Such properties of solutions which depend only on the relative concentration of the solute and the solvent regardless of the nature of the solute are called *colligative properties* or *democratic properties*. Some of these properties are:

(i) Relative lowering of vapour pressure
(ii) Elevation of boiling point
(iii) Depression of freezing point
(iv) Osmotic pressure

It may be noted that these properties are independent of the nature of the solute, but are considerably influenced by the nature of solvent. For a particular solvent, they depend upon the concentration of solute and not upon the nature of solute.

3.6 LOWERING OF VAPOUR PRESSURE

We know that every liquid has a definite vapour pressure over its surface at a given temperature. When any non-volatile substance is dissolved in liquid, the vapour pressure over the surface of the solution is lowered. This is called the lowering of vapour pressure. The lowering of vapour pressure can be explained based on the escaping tendency of the energetic molecules from the surface.

Each volatile component in a solution has certain vapour pressure which is less than the vapour pressure of the pure substance at the same temperature. This is due to the fact that the other component occupies a part of the surface and thus decreases the chances of the molecule of the first component to escape in the vapour state and exert vapour pressure. When a non-volatile solute is dissolved in a volatile liquid solvent, the number of molecules of the solvent on the surface of solution becomes less than that of the pure solvent. The chances of the escape of solvent molecules in the vapour phase are reduced in the case of solution and the vapour pressure is lowered (Fig. 3.4).

Suppose P_A^0 is the vapour pressure of pure liquid component and P is the vapour pressure of the solution when some non-volatile solutes are dissolved in it ($P_A^0 - P$) is called the lowering of vapour pressure. This factor has a different value at different temperature for a particular solution. However, the factor $\left(\dfrac{P_A^0 - P}{P_A^0}\right)$, which is called the **relative lowering of vapour pressure**, is independent of temperature for a particular solution. The relative lowering of vapour pressure is a colligative property.

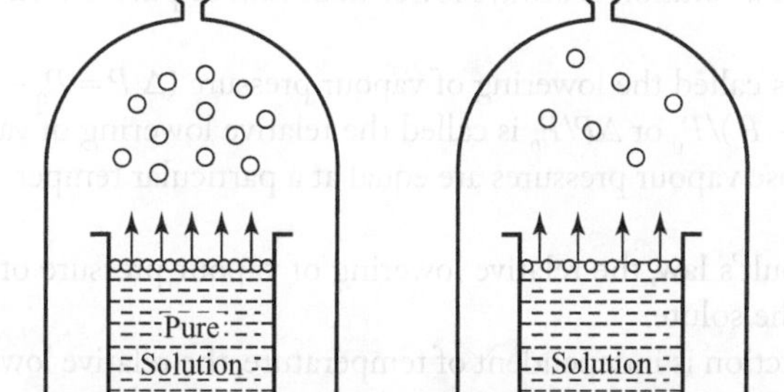

Figure 3.4 Cause of lowering of vapour pressure

According to Raoult's law, the relative lowering of vapour pressure is equal to the mole fraction of the solute.

$$\frac{P_A^0 - P}{P_A^0} = X_B = \frac{n_B}{n_A + n_B} \tag{3.4}$$

where n_A and n_B are the number of moles of solvent and solute, respectively.

For a very dilute solution: $n_A >> n_B$; n_B may be neglected from denominator.

$$\therefore \quad \frac{P_A^0 - P}{P_A^0} = \frac{n_B}{n_A}$$

For a given volume of solvent, n_A is constant.

$$\therefore \quad \frac{P_A^0 - P}{P_A^0} \propto n_B$$

Thus, the relative lowering of vapour pressure depends upon the number of moles of solute. *Hence the relative lowering of vapour pressure is a colligative property.*

3.6.1 Determination of Molecular Masses of Non-volatile Solute

According to Raoult's law, the relative lowering of vapour pressure of a solution is equal to the mole fraction of the solute.

$$\frac{P_A^0 - P}{P_A^0} = X_B = \frac{n_B}{n_A + n_B}$$

MEMORY FOCUS

1. Properties which depend on the number of particles of the solute but not on the nature of the solute are called colligative properties.
2. All the colligative properties can be used to determine the molecular mass of the non-volatile solute. However, the best method is osmotic pressure.
3. Vapour pressure of a pure solvent at constant temperature is constant (P_0).
4. When a non-volatile solute is dissolved in a solvent, its vapour pressure is lowered (P_s).
5. Vapour pressure of a solution is always lower than that of pure solvent, i.e. $P_0 > P_s$ at the same temperature.
6. The term $P_0 - P_s$ is called the lowering of vapour pressure ($\Delta P = P_0 - P_s$).
7. The fractions $(P_0 - P_s)/P_0$ or $\Delta P/P_0$ is called the relative lowering of vapour pressures.
8. The solutions whose vapour pressures are equal at a particular temperature are called isopiestic solution.
9. According to Raoult's law, the relative lowering of vapour pressure of solution is equal to the mole fraction of the solute.
10. Since the mole fraction is independent of temperature, the relative lowering of vapour pressure of a liquid is also independent of temperature.

where n_A and n_B represent the number of moles of solvent A and the solute B, respectively.

Suppose W_B g of the solute B of a molecular M_B is dissolved in W_A g of the solvent having molecular mass, M_A.

$$\therefore \quad \frac{P_A^0 - P}{P_A^0} = \frac{n_B}{n_A + n_B} = \frac{\frac{W_B}{M_B}}{\frac{W_B}{M_B} + \frac{W_A}{M_A}}$$

For a dilute solution, the number of moles of the solute $\frac{W_B}{M_B}$ may be neglected in the denominator.

$$\therefore \quad \frac{P_A^0 - P}{P_A^0} = \frac{\frac{W_B}{M_B}}{\frac{W_A}{M_A}} = \frac{W_B \times M_A}{W_B \times M_A} \qquad (3.5)$$

Knowing the lowering of vapour pressure (P), the vapour pressure of the pure solvent (p_A^0) and the molecular weight of the solvent (M_A), we can calculate the molecular weight of solute.

Numerical Problems

Example 1 At 298 K, the vapour pressure of a solution containing 9 g of substance in 90 g of water is 3049.02 Nm^{-2}. Calculate the molecular mass of substance taking molecular mass of water as 18 and vapour pressure of pure water at 298 K as 3139.68 Nm^{-2}.

Solution: Vapour pressure of solvent (water), $P_A^0 = 3139.68$ Nm^{-1}

Vapour pressure of solution, $P = 3049.02$ Nm^{-1}

Lowering of vapour pressure, $P_A^0 - P = 3139.68 - 3049.02 = 90.66$ Nm^{-2}

$$\text{Relative lowering of vapour pressure} = \frac{P_A^0 - P}{P_A^0} = \frac{90.66}{3139.68}$$

Weight of solute (W_B) = 9 g

Weight of solvent (W_A) = 90 g

Molecular mass of water (M_A) = 18

Molecular mass of water (M_B) = ?

According to Raoult's law,

$$\frac{P_A^0 - P}{P_A^0} = \frac{W_B \times M_A}{M_A \times W_B} = \frac{90.66}{3139.68} = \frac{9 \times 18}{M_B \times 90}$$

i.e.

$$M_B = \frac{9 \times 18}{90} \times \frac{3139.68}{90.66} = 62.33$$

Hence, the molecular mass of substance = **62.33.**

Example 2 The vapour pressure of water at 293 K is 226.7 Nm^{-2}. Calculate the vapour pressure of solution containing 3 g of urea in 50 g of water. (Molecular mass of urea = 60 a.m.u.).

Solution: Weight of solute (urea) $W_B = 3$ g
Molecular mass of solute $M_B = 60$
Weight of solvent (water) $W_A = 50$ g
Molecular mass of solvent $M_A = 18$ g
Vapour pressure of solvent = 2267 Nm^{-2}
According to Raoult's law,

$$\frac{P_A^0 - P}{P_A^0} = \frac{W_B \times M_A}{M_A \times W_B}$$

$$\therefore \quad \frac{2266.7 - P}{2266.7} = \frac{3 \times 18}{60 \times 50}$$

$$\therefore \quad 2266.7 - P = \frac{3 \times 18 \times 2266.7}{60 \times 50} = 40.80$$

$$P = 2266.7 - 40.8 = 225.9 \text{ Nm}^{-2}$$

Hence, the vapour pressure of solution = **2225.9 Mm^{-2}**.

Example 3 **The vapour pressure of a dilute aqueous solution of glucose ($C_6H_{12}O_6$) is 750 mm Hg at 373 K. Calculate (i) molality and (ii) mole fraction of the solute.**

Solution: Vapour pressure of solvent (water), $P_A^0 = 760$ mm
Vapour pressure of solution, $P = 750$ mm

From Raoult's law, $$\frac{P_A^0 - P}{P_A^0} = X_B$$

where X_B is the mole fraction of the solute.
Putting the values, we get

$$X_B = \frac{760 - 750}{760} = \frac{10}{760} = 0.01316$$

However, the mole fraction of the solute in dilute solution can be expressed as

$$X_B = \frac{n_B}{n_A} = \frac{W_B M_A}{W_A M_B} \quad \text{(i)}$$

and molality $$m = \frac{n_B}{W_A} \times 1000 = \frac{W_B \times 1000}{M_B \times W_A} \quad \text{(ii)}$$

Dividing Eq. (i) by Eq. (ii), we get

$$\frac{m}{X_B} = \frac{1000}{M_A} \quad \text{or} \quad m = \frac{1000 \times X_B}{M_A}$$

$$= \frac{1000 \times 0.01316}{18} = \mathbf{0.731}$$

Example 4 **The vapour pressure of water at room temperature is 23.8 mm Hg. What will be the vapour pressure of aqueous solution of sucrose with mole fraction equal to 0.1?**

Solution: $P_A^0 = 23.8$ mm, $P_S = ?$

$$\frac{P_0 - P_S}{P_0} = 0.1 \quad \text{(mole fraction of sucrose)}$$

$$\frac{23.8 - P_S}{23.8} = 0.1$$

or $$23.8 - P_S = 2.38 \quad \text{or} \quad \mathbf{P_S = 21.42\ mm.}$$

3.7 ELEVATION IN BOILING POINT

The boiling point of a liquid is the temperature at which the vapour pressure of the liquid becomes equal to the atmosphere pressure. It has been observed that the boiling point of a solution is always higher than that of the pure solvent. Since the vapour pressure over the surface of solution is lower than that of pure solvent, the solution shall have to be heated to a higher temperature, so that the vapour pressure over its surface becomes equal to the atmosphere pressure. The difference between the boiling point of a solution and that of a pure solvent is called **elevation in boiling point.** It is denoted by ΔT_b and is given by

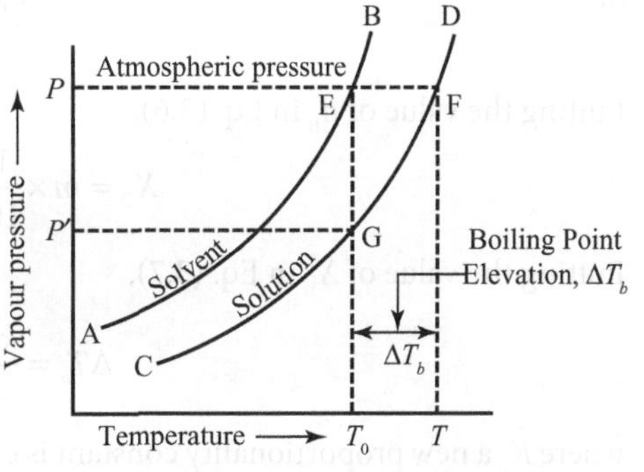

Figure 3.5 Plot of vapour pressure against temperature showing rise in boiling point

$$\Delta T_b = T_b - T_b^0$$

where ΔT_b is the boiling point of solution and is the boiling point of pure solvent.

The effect of a non-volatile solute on the boiling point of a pure solvent is illustrated in Fig. 3.5. The vapour pressure of a pure solvent and a solution has been plotted against temperature, represented by the curves AD and CD, respectively.

It is clear that the vapour pressure of solution is less than that of pure solvent at any temperature. The vapour pressure of pure solvent becomes equal to the atmospheric pressure at temperature T_0, which is the boiling point of the solvent. Similarly, T represents the boiling points of the solution. The elevation in the boiling points for solution is $\Delta T_b = T_b - T_b^0$.

3.7.1 Expression for the Elevation in Boiling Point

It also clear that the elevation of boiling point is proportional to the lowering of vapour pressure, which in turn is proportional to the mole fraction of the solute.

$$\Delta T_b \propto \Delta P$$

$$\Delta P \propto X_B$$

or
$$\Delta T_b \propto X_B$$

$\therefore$
$$\Delta T_b = kX_B \tag{3.6}$$

where k is proportionally constant and X_B is the mole fraction of solute.

For dilute solutions,

$$X_B = \frac{n_B}{n_A} = \frac{n_B}{\frac{W_A}{M_A}} = n_B \times \frac{M_A}{W_A} \tag{3.7}$$

Molality (m) of a solute is written as

$$m = \frac{n_B}{W_A} \times 1000$$

or
$$n_B = \frac{m \times W_A}{1000}$$

Putting the value of n_B in Eq. (3.6),

$$X_B = m \times \frac{W_A}{1000} \times \frac{M_A}{W_A} = \frac{m \times M_A}{1000}$$

Putting the value of X_B in Eq. (3.7),

$$\Delta T_b = \frac{k \times m \times M_A}{1000} = K_b m \tag{3.8}$$

where K_b a new proportionality constant is called the **molal elevation constant** or **ebulisscopic constant** or **the solvent**.

$$\text{If } m = 1, \Delta T_b = K_b$$

Thus, molal elevation constant (K_b) may be defined as the elevation in boiling point when molality of the solution is unity, i.e., when one mole of the solute is dissolved in 1000 g of the solvent. The value of K_b is different for different solvents, e.g., K_b values for water, ethyl alcohol and benzene are 0.53, 1.20 and 2.53, respectively. Thus, ΔT_b depends upon the number of moles of solute and not upon the nature

MEMORY FOCUS

1. When a non-volatile solute is dissolved in a solvent, its vapour pressure is lowered; consequently, the boiling point of the solutions becomes higher.
2. Sea water boils at a higher temperature than distilled water.
3. The phenomenon of the increase in the boiling points of solvent when a solute is dissolved in it is known as **elevation of boiling point**.
4. The study of elevation in the boiling points of a liquid in which a non-volatile solute is dissolved is called as **ebullioscopy**.
5. The elevation of the boiling point of a dilute solution is directly proportional to the number of moles (particles) of solute and inversely proportional to the molecular mass of the solute.

of the solute. It also depends upon K_b, which depends upon the nature of the solvent. Therefore, elevation or boiling point is colligative property.

3.7.2 Calculation of Molecular Masses

Suppose W_B g of solute of molecular mass M_B is dissolved in W_A g of solvent. The number of moles of the solute present in the solution is W_B equal to W_B/M_B moles. The molality (m) of the solution, i.e., the number of moles of the solute dissolved in 1000 g of solvent $\frac{W_B}{M_B} \times \frac{1000}{W_A}$.

Substituting the value of molality in the formula,

$$\Delta T_b = K_b m$$

$$\Delta T_b = K_b \times \frac{W_B}{M_B} \times \frac{1000}{W_A}$$

or

$$M_B = \frac{K_b \times W_B \times 1000}{\Delta T_b W_A} \quad (3.9)$$

By knowing elevation in b.p., it is possible to determine the molecular mass of the dissolved solute. Here, M_B is the molecular weight of solute and can be calculated.

Numerical Problems

Example 5 The boiling point of water (100 ΥC) becomes 100.52 ΥC if 3 g of a non-volatile solute is dissolved in 20 ml of it. Calculate the molecular mass of the solute (K_b for water is 0.52 K/m).

Solution: Elevation of boiling point $\Delta T_b = T_b - T_b^0$

$$= 100.52 - 100 = 0.52\,°\text{C}$$

Mass of solute $W_B = 3$ g

Mass of solvent $W_A = 20$ ml $= 20$ g

Substituting the values in the expression:

$$M_B = \frac{1000 K_b W_B}{\Delta T_b W_A} = \frac{1000 \times 0.52 \times 3}{20 \times 0.52} = \mathbf{150\ g\ mol^{-1}}$$

Example 6 A solution containing 15.5 g of a non-electrolyte substance in 175 g of water gave the elevation of the boiling point of 0.70 K. Calculate the molar mass of the substance.

Solution: The molecular mass, M_B of the unknown substance can be calculated from the relation.

$$M_B = \frac{1000 K_b W_B}{\Delta T_b W_A}$$

It is given that $W_B = 12.5$ g, $W_A = 175$ g

$\Delta T_b = 0.070\ K$, $K_b = 0.52$ K kg mol^{-1}

$$\therefore \quad M_B = \frac{1000 \times 0.52 \times 12.5}{0.70 \times 175} = \mathbf{53.06}$$

Example 7 **What is the boiling point of a solution containing 18 g of glucose in 100 g of H_2O. M is the elevation constant of water, which is 0.52 K kg mol^{-1}.**

Solution: The elevation in boiling, ΔT_b can calculated from the relation

$$\Delta T_b = \frac{1000 K_b W_B}{M_B W_A}$$

Putting the given values, we get $\Delta T_b = \frac{1000 \times 0.52 \times 18}{18 \times 1000} = 0.52 K$

We know that the boiling point of water $= 100\,°C = 3$

Therefore, the boiling point of the solution = **373.52 K.**

3.8 DEPRESSION OF FREEZING POINT

Freezing point of any liquid is that the temperature at which the vapour pressure of the liquid and the solid become equal. In other words, it is the temperature at which the solid and the liquid forms of the substance co-exist.

We know that the vapour pressure of a solution of a non-volatile substance is lower than that of the pure solvent at any given temperature. It means that the vapour pressure of the solution would become equal to the vapour pressure of the solvent at a lower temperature as compared to a pure solvent. Obviously, the freezing point of a solution is always lower than that of the purely vent. The difference between the freezing point of a pure solvent and the freezing point of a solution is called the depression in freezing point. It is denoted by ΔT_f and is given by

$$\Delta T_f = T_f^0 - T_f$$

where ΔT_f^0 and ΔT_f are the freezing points of pure solvent and solution, respectively.

The effect of a non-volatile solute on the freezing point of a pure solvent is illustrated in Fig. 3.6. The vapour pressure of a pure solvent and a solution has been plotted against temperature in the figure.

It is clear from the figure that the vapour pressure of the pure substance in the liquid state becomes equal to that of the solid state at the point A, which represents the freezing point of the pure solvent (T_0). At the point A′, the vapour pressure curve of the liquid and solid phases meets. Similarly, T_1 represents the freezing points of the solution.

The depression in freezing point in case of solution is $\Delta T_f = T_0 - T_1$.

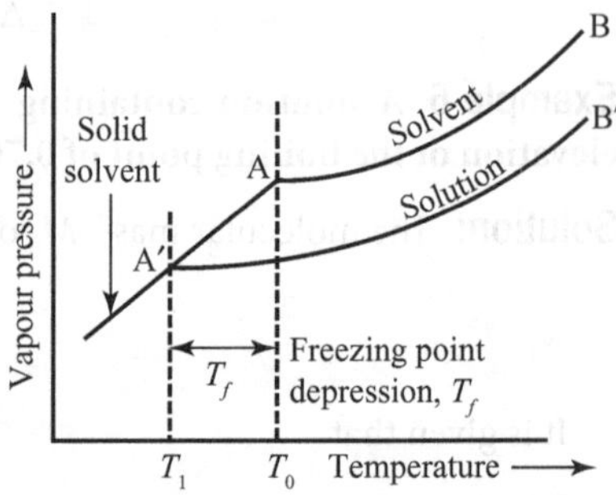

Figure 3.6 Plot of vapour pressure against temperature showing depression in freezing point

> **CAUTION**
>
> Molal elevation constant (K_b) and molal freezing point depression constant (K_f) have characteristic values for different solvents. These are expressed as degree/molal i.e. K/m for water K_f=1.86 km^{-1}. It means that for 1 molal solution of a substance in water, the freezing point will be lowered by 1.86°. However, sometimes, K_f and K_b are also expressed as mole per 100 g i.e. K/mol per 100 g. These values are 10 times the values expressed as km^{-1}. For example, K_f for water =1.86 Km or 18.6 K/mol per 100 g. One should be cautious while using the value in solving numerical problems.

3.8.1 Expression for the Depression in Freezing Point

It is also clear that the depressions in freezing point are proportional to the lowering of vapour pressure which in turn is proportional to the mole fraction of the solute.

$$\Delta T_f \propto \Delta P \quad \text{and} \quad \Delta P \propto X_B$$

$$\therefore \qquad \Delta T_f = kX_B \qquad (3.10)$$

where k is the proportionality constant and X_B is the mole fraction of the solute.

For dilute solutions,

$$X_B = n_B \times \frac{M_A}{W_A} \text{ (by definition)}$$

$$n_B = \frac{m}{1000} \times W_A$$

where m is the molality of the solute and M_A is the molecular mass of the solvent. Putting the value of X_B in Eq. (3.10),

$$\Delta T_f = k \times m \times \frac{M_A}{1000} = k_f \times m \qquad (3.11)$$

where K_f is called the molal depression constant of the solvent. If $m = 1$ and $T_f = K_f$. Thus, the molal depression constant may be defined as the depression in freezing point when the molality of the solution is unity, i.e., when one mole of the solute is dissolved in 1000 g of the solvent. The value of K_f is different for different solvents, e.g., K_f values for water, ethyl alcohol and benzene are 1.86, 1.99 and 5.1, respectively. Thus, T_f depends upon the number of moles of the solute and not upon the nature of the solute. It also depends upon K_f, which depends upon the nature of the solvent. Thus, depression in freezing point is a colligative property.

MEMORY FOCUS

1. The temperature at which solid and liquid states of a substance have the same vapour pressure is called its **freezing point**.
2. Sea water freezes at a temperature that is lower than pure water.
3. The phenomenon of the decrease in freezing point of a solvent when a solute is dissolved in it is known as depression in freezing point.
4. The study of the depression in freezing point of a liquid in which non-volatile solute is dissolved is called as **cryoscopy**.
5. The depression in freezing point of a solvent in a dilute solution is directly proportional to the number of moles (particles) of solute and inversely proportional to the molecular weight of solute.
6. A sensitive and rapid method to determine the molecular weight of a solute using the depression in freezing point is **Rast method**.
7. In Rast method, the solvent used is camphor (solid).

3.8.2 Calculation of Molecular Masses

If W_B g of solute of molecular mass M_B is dissolved in W_A g of solvent, the number of moles of the solute present in the solution is equal to W_B/M_B.

The molality (m) of the solution, i.e., the number of moles of the solution dissolved is 1000 g of solvent.

$$= \frac{W_B}{M_B} \times \frac{1000}{W_B}$$

Substituting the value of molality in the formula, $\Delta T = K_f\, m$

$$\Delta T_f = K_f \frac{W_B}{M_B} \frac{1000}{W_A}$$

$$\therefore \qquad \overline{M}_B = \frac{K_f \times W_B \times 1000}{\Delta T_f \times W_A} \qquad (3.12)$$

Thus, knowing the depression in freezing point, it is possible to determine the molecular mass of the dissolved solute.

Numerical Problems

Example 8 **The freezing point of pure nitrobenzene is 278.8 K. When 2.5 g of an unknown substance is dissolved in 100 g of nitrobenzene, the freezing point of the solution is found to be 276.8 K. If the freezing point depression constant of nitrobenzene is 8.0 K kg mol^{-1}, what is the molar mass of the unknown substance?**

Solution: Depression in freezing point of nitrobenzene (ΔT_f) = 278.8–276.8 =2.0 K

Applying the relation, $$M_B = \frac{1000 K_f W_B}{\Delta T_f W_A}$$

$$M_B \text{ (molar mass of unknown substance)} = \frac{1000 \times 8.0 \times 2.5}{2.0 \times 100} = \mathbf{100\ amu.}$$

Example 9 **A solution containing 25.6 g of sulphur dissolved in 1000 g of naphthalene, whose melting point is 80.1 ϒC gave the freezing point lowering of 0.680 ϒC. Calculate the formula of sulphur (K_f, for naphthalene = 6.8 K/m and atomic mass of sulphur = 32).**

Solution: Molecular mass of sulphur can be calculated from the formula

$$M_B = \frac{1000 K_f W_B}{\Delta T_f W_A}$$

Putting the given values, we get $M_B = \dfrac{1000 \times 6.8 \times 25.6}{0.68 \times 1000} = 256$

Atomic mass of $S = 32$

The number of atoms of sulphur in one molecule $= \dfrac{256}{32} = 8$

Therefore, the formula of sulphur = $\mathbf{S_8}$.

3.9 OSMOTIC PRESSURE

Semipermeable membrane *is a membrane which allows the passage of solvent molecules through it but not that of solute when a solution is separated from the solvent by it.* These membranes are of two types:

(i) **Natural semipermeable membranes:** Vegetable membranes and animal membranes are the examples of the natural semipermeable membranes, which are just found under the outer skin of plants and animals. The pig's bladder is the most common animal membrane.

(ii) **Artificial semipermeable membranes:** Some of the examples of artificial semipermeable membrane are parchment paper, cellophane, collodion and certain freshly precipitated inorganic substance, e.g., copper ferrocyanide, silicate of iron, cobalt, nickel etc.

If the two solutions of different concentrations are kept in contact, the molecules of the solute and solvent both move from the region of higher concentration to the region of lower concentration till a uniform concentration is obtained. *The spontaneous mixing of the particles of solute and solvent to form a homogeneous mixture is called* ***diffusion in solutions.***

A different phenomenon has been observed whenever a solution and a pure solvent or two solutions of different concentrations are kept in contact through a semipermeable membrane. In this case, only the solvent molecules penetrate through the membrane while the solute particles are held back. The solvent molecules pass through the semipermeable membrane from the solvent to the solution side or from the less concentrated solution to the more concentrated solution till an equilibrium is attained. *This process of flow of the solvent molecules from the solvent to the solution or from a less concentrated to a more concentrated solution through a semipermeable membrane is called* ***osmosis.***

3.9.1 Difference Between Osmosis and Diffusion

The main differences between osmosis and diffusion are:

(i) Osmosis can be carried out only with the help of semipermeable membrane, but no such external agency is required for diffusion to take place.

(ii) In diffusion, the movement of the molecules is from both sides. The solute molecules move in one direction, whereas the solvent molecules in the other direction. However, in osmosis, the movement of molecules is one sided.

(iii) In diffusion, the molecules of solute move from solution of higher concentration to the lower concentration giving a homogeneous mixture, whereas in osmosis, the solvent molecules move from the solution of lower concentration to that of the higher concentration making it dilute.

(iv) Osmosis occurs in biological systems.

(v) Osmosis can be stopped by applying pressure to the solution; however, diffusion cannot be stopped.

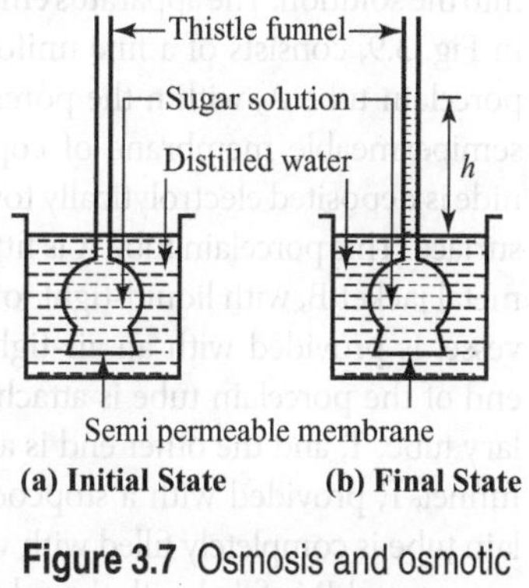

Figure 3.7 Osmosis and osmotic pressure

3.9.2 Osmotic Pressure

The phenomenon of osmosis can be illustrated by a simple experiment. Let a solution of sugar in water be kept in contact with water through a semipermeable membrane in a thistle funnel as shown in Fig. 3.7. Here, water is separated from sugar solution with the help of a semipermeable membrane. It is observed that water enters the given solution through the semipermeable membrane and the level of water in the funnel rises gradually. Thus, experiment illustrates osmosis. A stage is ultimately reached when the level of the solution becomes constant. At this stage, the hydrostatic pressure developed is just sufficient to prevent the inflow of the solvent into the solution. *The equilibrium hydrostatic pressure developed on the solution side due to the net flow of the solvent molecules through a semipermeable membrane is called the* ***osmotic pressure.***

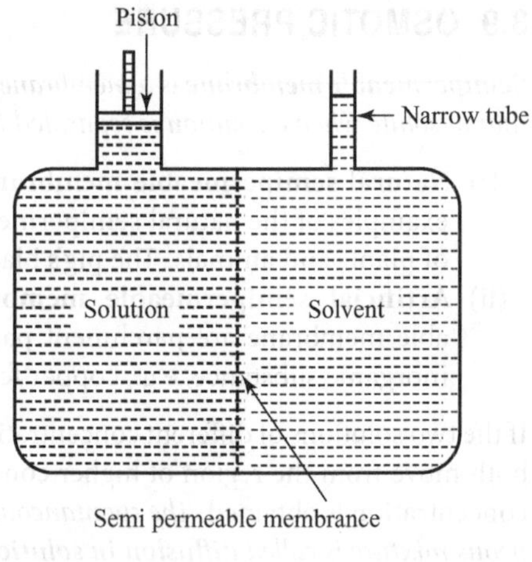

Figure 3.8 Concept of osmotic pressure

The term osmotic pressure of a solution is, however, not strictly correct. A solution does not, in itself, have any osmotic pressure. Osmotic pressure is the hydrostatic pressure developed by osmosis under the conditions given above. Thus, we see that, it is osmosis that develops osmotic pressure and not the osmotic pressure, which develops osmosis. Osmotic pressure is a measure of the force, which produces osmosis.

This flow of solvent into the solution may also be stopped by the application of sufficient pressure with the help of a piston as illustrated in Fig. 3.8. This extra pressure which must be applied to a solution in order to stop the flow of the solvent into it through a semipermeable membrane separating the two is a measure of osmotic pressure of the solution.

3.9.3 Determination of Osmotic Pressure Berkley and Hertley's Method

The measurement of osmotic pressure involves essentially the determination of the excess pressure that has to be applied upon the solvent to produce a state of equilibrium in the system (i.e., to stop osmosis).

The method commonly employed for the determination of osmotic pressure is due to Berkley and Hartley (1904). In this method, the osmotic pressure is balanced by a counter pressure, which prevents the passage of the solvent into the solution. The apparatus employed, shown in Fig. 3.9, consists of a fine uniformly textured porcelain tube A, within the pores of which the semipermeable membrane of copper ferrocyanide is deposited electrolytically towards its outer surface. This porcelain tube A is fitted into a gunmetal jacket B, with liquid-tight joints. The outer vessel is provided with an air-tight piston. One end of the porcelain tube is attached with capillary tube, T, and the other end is attached with a funnel, F, provided with a stopcock. The porcelain tube is completely filled with water while the outer vessel B is filled with the solution. External

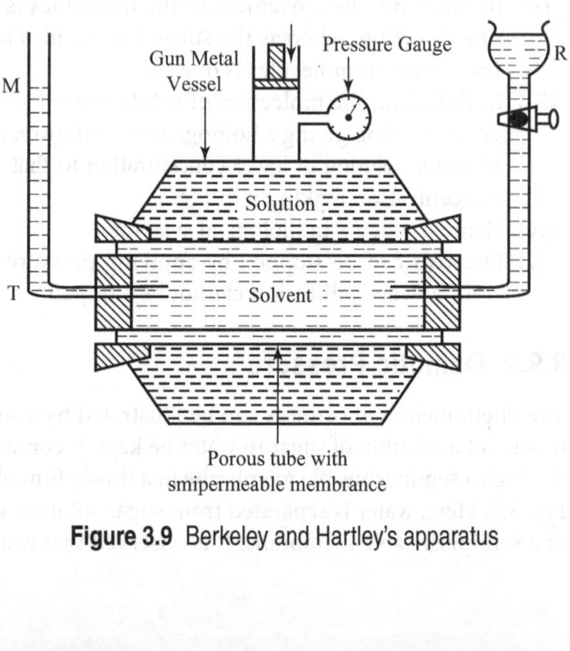

Figure 3.9 Berkeley and Hartley's apparatus

pressure, which is measured by a pressure gauge, is applied on the solution by working the piston C so as to prevent the entrance of water into the solution, which is indicated when the level of water in the capillary tube T remains stationary. This pressure measures the osmotic pressure of the solution.

So long as the hydraulic pressure is less than the osmotic pressure, the level of the solvent in the capillary tube will fall. When the hydraulic pressure exceeds the osmotic pressure, the liquid in the capillary tube remains constant, which indicates that the two opposing pressures balance each other. This method has following advantages:

(i) The equilibrium is established very quickly and so it gives the results in very short time.
(ii) Since the external pressure applied balances the osmotic pressure, there is a little strain on the semipermeable membrane.
(iii) The concentration of the solution does not change during the determination of osmotic pressure.

Learning Plus

Silica gardens. When small crystals of copper sulphate, nickel sulphate etc. are added to 5% sodium silicate solution, the so-called silicate gardens or chemical gardens are produced because of the diffusion of metal ions from crystals to form precipitate of metal silicate. These precipitates act as semipermeable membrane around the crystal. The osmosis takes place and water flows from dilute sodium silicate solution to stronger salt solution. The membrane bursts at its weaker points so as to liberate more metal ions and these ions further give rise to new membrane. As a result, the growth of metal ions looks like gardens around the crystals.

3.9.4 Osmotic Pressure is a Colligative Property

It has been observed that behaviour of non-electrolyte substances in dilute solutions is very much similar to the behaviour of gases. Vant Hoff pointed out that substances in dilute solutions obey certain laws, which are very much similar to the gas laws.

Just like gas equation, we have a Vant Hoff equation for dilute solutions $\pi V = n\,RT$, where π is the osmotic pressure of the solution at temperature T containing n moles of the solute dissolved in V litres of the solution R is a constant called solution constant. The value of the constant R is the same as for the gas constant. The value is taken as 0.0821 litre atm per degree per mole or 8.3141 K^{-1} $mole^{-1}$.

From the preceding equations, we have $$\pi = \frac{n}{V}RT$$

means the number of moles per litre of the solution, which may be taken as the molar concentration (C) of the solution.

$$\therefore \quad \pi = CRT \tag{3.13}$$

It is clear that osmotic pressure (π) is proportional to the concentration of the solute and not upon the nature of the solute. *Therefore, osmotic pressure is a colligative property.*

3.9.5 Isotonic Solutions

It is evident from the equation = CRT that, if two solutions have the same value of C and T, they must have the same value of π. It means that the solutions of equimolar concentrations at the same temperature must have same osmotic pressure. Such solutions, which have the same osmotic pressure at the same temperature, are called *isotonic solutions.*

3.9.6 Calculation of Molecular Masses from Osmotic Pressure

According to Vant Hoff equation for dilute solutions, we have $\pi V = n\,RT$. If W_B of the solute is dissolved in V litres of the solution and M_B is the molecular mass of the solute, then

$$n = \frac{W_B}{M_B}$$

Substitute the value of n in the preceding equation, we get

$$\pi V = \frac{W_B}{M_B} RT$$

$$M_B = \frac{W_B RT}{\pi V} \tag{3.14}$$

Thus, measuring the osmotic pressure (π) of the solution of known concentration at a given temperature (T), the molecular mass (M_B) of the solute can be calculated. The value of R is taken as 0.0821 litre atm/degree/mole or 8.314 JK^{-1} mol^{-1}.

It may be noted that osmotic pressure method for determining the molar masses of bio-molecules such as proteins and polymers such as polythene and PVC are preferred to other colligative properties because of the following reasons:

(i) The magnitude of other colligative concentrations may be too low to be measured accurately due to very big size of the particles.
(ii) The experimental determination may require high temperature at which the big molecules get decomposed, e.g., in case of elevation in boiling point.

MEMORY FOCUS

1. The spontaneous flow of solvent, through a semipermeable membrane, from a solution of low concentration or solvent to a solution of high concentration is called as '**Osmosis**'.
2. The excess pressure that must be applied to a solution to prevent the flow of the solvent into it through a semipermeable membrane is known as **osmotic pressure.**
3. The equilibrium hydrostatic pressure from the solution to the solvent side, when the two are separated by a semipermeable membrane is called **osmotic pressure.**
4. The negative pressure which must be applied to the solvent in order to just stop the osmosis is equal to the **osmotic pressure** of that solution.
5. Solutions having same osmotic pressure are said to be isotonic with each other.
6. Isotonic solutions must have same molar concentrations.
7. When isotonic solutions are separated by a semipermeable membrane, no transfer of solvent takes place.
8. If the given solution has lower osmotic pressure than the one under consideration, then the given solution is said to be **hypotonic solution.**
9. If the given solution has higher osmotic pressure than the one under consideration, then the given solution is said to be **hypertonic solution.**
10. Solutions having same osmotic pressure with reference to a perfect semipermeable are said to **isosmotic.** Only for a perfect semipermeable membrane, isotonic solutions are isosmotic.

(*Continued*)

MEMORY FOCUS (*Continued*)

11. Osmotic pressure of a solution is proportional to:
    (a) concentration of solute at constant temperature and
    (b) the absolute temperature at constant concentration.
12. The osmotic pressure of a dilute solution is equal to the pressure in which the solute would exert if it existed as a gas at the same temperature and occupied the same volume as the solution. This statement is called as Vant Hoff's theory of dilute solutions.
13. Swelling of resins (dried grapes) when put in water for sometime is another example of osmosis in everyday life.

Numerical Problems

Example 10 **A solution of sucrose (molecular mass = 342) is prepared by dissolving 68.4 g of it per litre of solution. What is its osmotic pressure at 300 K?**

Solution: We know $\pi = \frac{nRT}{V}$

Here, $n = \frac{68.4\ \text{g}}{342\ \text{g mol}^{-1}} = 0.2\ \text{mol}$

$R = 0.0821$ litre atom K^{-1} mol^{-1} and $V = 1$ litre.

Putting the values, we get

$$\pi = \frac{0.2 \times 0.0821 \times 300}{1} = 4.92\ \text{atm}$$

where R is expressed as 8.314 KP_a dm^2 K^{-1} $mole^{-1}$

$$\pi = \frac{0.2 \times 8.314 \times 330}{1} = \mathbf{448.84\ KP_a}$$

Example 11 **Osmotic pressure of a solution containing 6 g of dissolved protein per 100 mol of the solution is 0.0392 atm at 310 K. What is the molecular mass of the protein? The value of *R* is 0.8 litre deg^{-1} mol^{-1}.**

Solution: We know that $\pi V = n\ RT$

$\pi = 0.0329$ atm

$n_B = \frac{6}{M_B}$, $V = 100$ mol $= 0.1$ litre

$R = 0.0820$, $T = 310$ K

Putting the value in the expression,

$$0.0329 \times 0.1 = \frac{6}{M_B} \times 0.0820 \times 310$$

or

$$M_B = \frac{6 \times 0.0820 \times 310}{0.1 \times 0.0329} = \mathbf{46.387}$$

Example 12 **A 6% sucrose ($C_{12}H_{22}O_{11}$) is isotonic with a 3% solution of an unknown organic substance. Calculate the molecular mass of the unknown substance.**

Solution: Since the two solutions are isotonic, therefore, they are equimolar, i.e., they have same mol concentration.

Now, molecular mass of sucrose = 342

$\therefore$ Molar concentration of sucrose $= \dfrac{6}{100} \times \dfrac{1000}{342} = \dfrac{60}{342}$

Let M_B be the molecular mass of the unknown substance.

Then, molar concentration of unknown substance $= \dfrac{3 \times 1000}{100 \times M_B} = \dfrac{30}{M_B}$

Thus, we have
$$\frac{60}{342} = \frac{30}{M_B}$$

or
$$M_B = \frac{30 \times 342}{60} = \mathbf{171}$$

Hence, the molecular mass of unknown organic substance = **171 amu**

Example 13 **An aqueous solution freezes at 272 K, while pure water freezes at 273.0 K. Determine (i) the molality of the solution, (ii) the boiling point of the solution and (iii) the lowering of vapour pressure of water at 298 K.**

[Given K_f = 1.86 K kg mol^{-1}, K_b = 0.512 K kg mol^{-1} and vapour pressure of water at 298 K = 23.75 mm Hg].

Solution:

(i) From the given values,
$$\Delta T_f = (273\text{ K} = 272.4\text{K}) = 0.6\text{K}$$

We know
$$\Delta T_f = K_f m \quad \text{or} \quad m = \frac{\Delta T_f}{k_f}$$

$$\text{Molality}\,(m) = \frac{0.06\text{ K mol}^{-1}}{1.86\text{ K kg mol}^{-1}} = 0.3225\text{ mol kg}^{-1}$$

(ii)
$$\Delta T_b = k_b \times m$$
$$= 0.512\text{ K kg mol}^{-1} \times 0.3225\text{ mol kg}^{-1}$$
$$= 0.1612\text{ K}$$

$\therefore$
$$T_b = 373\text{K} + 0.1612\text{K} = 373.1612\text{K}$$

(iii) we know that
$$\frac{P_A^0 - P_A}{P_A^0} = x_B = \frac{n_B}{n_B + n_A}$$

$\therefore$
$$\frac{P_A^0 - P_A}{27.756} = \frac{0.3225}{0.32 + 55.55} = \frac{0.3225}{55.8725}$$

or
$$P_A^0 - P_A = \frac{0.3225 \times 23.756}{55.8725} = \mathbf{0.137\ mm\ Hg}$$

Example 14 **A solution or sucrose (molar mass = 342 g) has been prepared by dissolving 68.4 g of sucrose in one kg of water. Calculate the following:**

(i) the vapour pressure of the solution at 298 K,

(ii) the osmotic pressure of the solution at 298 K and

(iii) the freezing point of the solution.

[Given: the vapour pressure of water at 298 K = 0.024 atm, K_f for water = 1.86 K kg mol^{-1}, R = 0.082 L atm mol^{-1} K^{-1}]

Solution:

(i) The number of moles of sucrose $= \frac{68.4}{342} = 0.2$

The number of moles of water $= \frac{1000}{18} = 55.55$

We know that $\frac{P_A^0 - P_A}{P_A^0} = x_B$

$$\frac{0.024 - P_A}{0.024} = \frac{n_B}{n_A + n_B} = \frac{0.2}{55.55 + 0.2}$$

$$0.024 \times 55.75 - P_A \times 55.75 = 0.2 \times 0.024$$

$$1.338 - 55.75 \times P_A = 0.0048$$

$$-55.75P_A = -1.338 + 0.0048$$

$$-55.75P_A = -1.3332$$

or $$P_A = \frac{1.3332}{55.75} = 0.0239 \text{ atm}$$

(ii) We know $$\pi = \frac{W_B RT}{M_B V} = \frac{68.4 \times 0.082 \times 298}{342 \times 1}$$

$$= \frac{1671.4224}{342} \text{ atm} = 4.8872 \text{ atm}$$

(iii) We know that $\Delta T_f = K_f m = 1.86 \times 0.2 = 0.372 \text{K}$

$\therefore$ $T_f = 273 - 0.372 = \mathbf{272.628\ K}$

Example 15 **1% solution of KCl is dissociated to the extent of 80%. What would be osmotic pressure at 27 ϒC (R = 0.0821 L atm K^{-1} mol^{-1}?)**

Solution:
$$\underset{1-\alpha}{\overset{1}{\text{KCl}}} \rightarrow \underset{\alpha}{\overset{0}{\text{K}^+}} + \underset{\alpha}{\overset{0}{\text{Cl}^-}}$$

$$i = 1 - \alpha + \alpha + \alpha = 1 + \alpha = 1 + 0.80 = 1.80$$

$$C = \frac{1}{74.5} \times \frac{1000}{100} = 0.34$$

$$T = 273 + 27 = 300 \text{ K}$$

Now $$\pi = iCRT = 1.80 \times 0.134 \times 0.0821 \times 300 = \mathbf{5.98\ atm}$$

NOTEWORTHY POINTS

1. Osmotic pressure is the best colligative property to determine the molecular mass of a non-volatile substance.
2. Substances having high vapour pressure (e.g. petrol) evaporate more quickly than the substances of low vapour pressure (e.g. motor oil).
3. K_b and K_f are also called ebullioscopic constant and cryoscopic constant, respectively.
4. Ethylene glycol is commonly added to car radiators to depress the freezing point of water. It is known as **antifreeze**.
5. NaCl or $CaCl_2$ (anhydrous) is used to clear snow on roads. It depresses the freezing point of water and reduces the temperature at which ice is expected to be formed.
6. **Raoult's law** is applicable for only dilute solutions.
7. **Konowaloff's rule:** In case of non-ideal solutions, the vapour is relatively richer in the component whose addition to the required mixture results in an increase of total vapour pressure.
8. **Babo's law:** It states that the vapour pressure of a liquid gets decreased when a non-volatile solute is added, the amount of decrease being proportional to the amount of solute dissolved.
9. Relative lowering of vapour pressure is measured by Ostwald and Walker method. Elevation in boiling point by Landsberger method, depression in freezing point by Beckmann method and osmotic pressure by a Berkeley and Hertley's method.

3.10 ABNORMAL MOLECULAR MASSES

We know that colligative properties depend upon the molar concentration of the solute in solution. These properties depend upon the number and not upon the nature of the solute particles in solution. Therefore, in many cases where the solute associate or dissociate in solution, certain abnormal values of colligative properties are obtained. The values of molecular masses calculated based on colligative properties in such cases will also be abnormal.

(i) Association: If a solute associates in the solution, the number of particles present in the solution will be less than the actual number dissolved and hence the value of osmotic pressure and other colligative properties will be smaller and consequently the molecular mass indicated will be higher than the true molecular mass. For example, carboxylic acids such as acetic acid and benzoic acid associate in benzene to form dimers due to hydrogen bonding.

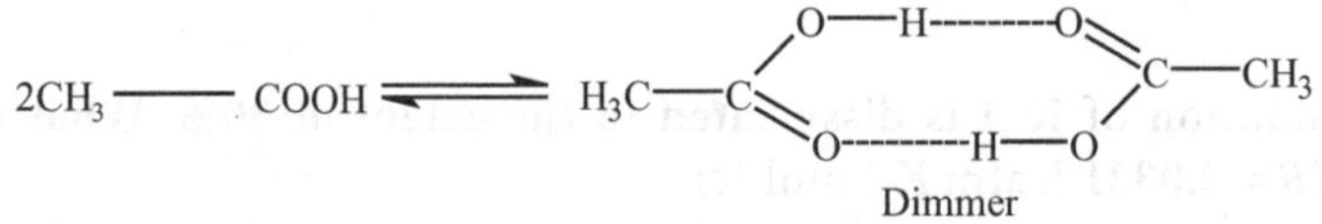

Dimmer

(ii) Dissociation: If a solute dissociates in the solution, the number of particles in the solution will be more than that actually dissolved and the value of colligative properties will be higher and hence the molecular masses observed will be smaller than the true molecular mass. For example, the electrolyte such as NaCl dissociates in solution and exists as Na^+ and Cl^- ions.

$$NaCl_{(a)} \xrightarrow{\text{Water}} Na^+_{(aq)} + Cl^+_{(aq)}$$

Vant Hoff's factor (*i*)

To account for the abnormal results, Vant Hoff introduced a factor (*i*) known after his name. It is given by the expression:

$$i = \frac{\text{Observed value of the colligative property}}{\text{Calculated or normal value assuming no dissociation or association}}$$

For example, the osmotic pressure observed in the case of electrolytes is much greater than what could come out from the equation $\pi V = n\,RT$. Vant Hoff explained this abnormally by modifying the equation by putting factor (*i*), thus

$$\pi_{obs} V = i n\, RT \tag{3.15}$$

Now, the equation in case of calculated osmotic pressure is

$$\pi_{cal} V = n\, RT \tag{3.16}$$

Dividing Eq. (3.15) by Eq. (3.16), $\dfrac{\pi_{obs}}{\pi_{cal}} = i$

Osmotic pressure of the solution is proportional to the vapour pressure (ΔP), elevation in boiling point (ΔT_b) and depression in freezing point (ΔT_f). Hence, Vant Hoff's factor can be expressed as

$$i = \frac{\pi_{obs}}{\pi_{cal}} = \frac{\Delta P_{obs}}{\Delta P_{cal}} = \frac{\Delta T_{b\,obs}}{\Delta T_{b\,cal}} = \frac{\Delta T_{f\,obs}}{\Delta T_{f\,cal}}$$

Each colligative property is inversely proportional to the molecular mass of the solute. Therefore,

$$i = \frac{\text{Observed colligative property}}{\text{Normal or calculated colligative property}}$$

$$i = \frac{\text{Normal Molecular mass}}{\text{Observed Molecular mass}} \tag{3.17}$$

Since a colligative property is directly proportional to the number of moles and inversely proportional to the molar mass of the solute, Vant Hoff's factor (*i*) may also be expressed as

$$i = \frac{\text{Number of particles after dissoiaton or association}}{\text{Normal number of particles}} \tag{3.18}$$

In case of ***association****, since the observed molecular mass is more than the normal molecular mass, the factor i has a value less than 1.*

In case of ***dissociation****, since the observed molecular mass is less than the normal molecular mass, the factor i has a value greater than 1.*

In case there is ***no association*** *or dissociation, the value of i is equal to 1.*

Degree of Dissociation

Equation (3.14) can be used to determine the degree of dissociation of solutes in different solutions.

(i) Dissociation: Let a molecule of an electrolyte when dissolved in a solvent gives *n* ions and α is the degree of dissociation.

$$\underset{(1-\alpha)\text{ mole}}{\overset{1\text{ mole}}{AB}} \xrightleftharpoons{\text{Water}} \underset{\alpha}{\overset{0}{A^+(aq)}} + \underset{\alpha}{\overset{0}{B^-(aq)}} \quad \begin{matrix}\text{....Before dissociation}\\ \text{....After dissociation}\end{matrix}$$

The number of moles before dissociation = 1 mole

The number of moles after dissociation $= 1-\alpha+n\alpha = 1(n-1)\alpha$

Therefore, $$i = \frac{1(n-1)\alpha}{1} \quad \text{or} \quad \frac{i-1}{n-1} = \alpha$$

Thus, α can be calculated.

(ii) Association: Let n molecules of solution 'A' associate to form one molecule and α is the degree of association.

$$\underset{1-\alpha}{\overset{0}{nA}} \longrightarrow \underset{\alpha/n}{\overset{0}{(A)_n}} \quad \begin{matrix}\text{....Before dissociation}\\ \text{....After dissociation}\end{matrix}$$

when n = factor of association.
Total number of moles after association

$$= 1-\alpha+\frac{\alpha}{n}$$

Therefore, $$i = \frac{1-\alpha+\frac{\alpha}{n}}{1}$$

or $$\alpha = (1-i)\frac{n}{n-1}$$

3.10.1 Modified Equation for Colligative Properties in Case of Abnormal Molecular Masses

(i) Elevation of boiling point, $\Delta T_b = iK_b m$

(ii) Depression in freezing point, $\Delta T_f = iK_f m$

(iii) Osmotic pressure, $\pi V = i\,n\,RT$

or $$\pi = i\frac{n}{V}RT = iCRT$$

MEMORY FOCUS

1. When the solute undergoes either dissociation or association, abnormalities are observed in the colligative properties.
2. Colligative properties are inversely proportional to the molecular mass of the solute and directly proportional to the number of solute particles.
3. If the solute undergoes association,
   (a) molecular mass increases
   (b) the number of particles decreases
   (c) observed colligative property is less than the theoretical value.

(*Continued*)

MEMORY FOCUS (*Continued*)

4. If the solute under goes dissociation,
   (a) molecular mass decreases
   (b) the number of particles increases
   (c) observed colligative property is higher than the theoretical value.
5. To explain the abnormal values of colligative properties and consequently molecular masses, a correction factor known as Vant Hoff's factor (i) is employed.

$$\text{Vant Hoff's factor } (i) = \frac{\text{Observed colligative property}}{\text{Calculated colligative property}}$$

6. Vant Hoff's factor for binary electrolytes such as NaCl and $CuSO_4$ is close to 2, whereas of ternary electrolytes such as $CaCl_2$ and Na_2SO_4 is close to 3.
7. Vant Hoff's factor (i) for acetic acid or benzoic acid in benzene is close to 0.5.
8. Vant Hoff's factor is useful to calculate the degree of dissociation or association and the number of particles (ions) produced by complex salts in solution.

NOTEWORTHY POINTS

1. **Plasmolysis:** When a plant cell is placed in a hypertonic solution, the fluid from the plant cell comes out and the cell shrinks. This phenomenon is called plasmolysis and is due to osmosis.
2. **Reverse Osmosis:** When the external pressure applied on the solution is more than the osmotic pressure, the solvent will start following from the solution to the pure solvent. It is called reverse osmosis. It is used in the desalination of sea water to obtain pure water.
3. Bursting of red blood cells when placed in water is due to osmosis.
4. A 0.91% solution of pure NaCl is isotonic with human RBC's.
5. Association generally occurs in the non-aqueous solvents because in the aqueous solution, the high dielectric constant of water helps in the dissociation of the associated molecules.
6. When outflow of water occurs from a cell, it is called exo-**osmosis** and when inward flow of water into a cell takes place, it is called **endo-osmosis**.
7. Gelatinous $Cu_2[Fe(CN)_6]$ and gelatinous $Ca_3(PO_4)_2$ are artificial semipermeable membranes.
8. Semipermeable membrane of $Cu_2[Fe(CN)_6]$ does not work in non-aqueous solutions because it gets dissolved in non-aqueous solvents.
9. For solutes showing dissociation, $i > 1$ and in case of association, $i < 1$.
10. The phenomenon of osmosis was first observed by Abbe Nollet.

REVISION QUESTION

1. Define or explain the following terms:
   (a) Colligative properties   (b) Raoult's law
   (c) Molecular mass   (d) Boiling point elevation
   (e) Boiling point constant   (f) Molal elevation constant
   (g) Freezing point depression   (h) Electrolytes

2. Explain the following:
   (a) Vapour pressure of a liquid does not depend upon the size of the container.
   (b) Boiling point of a liquid increase on adding non-volatile solute in it.
   (c) Vapour pressure of a liquid varies with temperature.
3. (a) What are isotonic solutions? Explain.
   (b) Write a note on Vant Hoff's factor.
4. Differentiate ideal and real solutions.
5. State Raoult's law and write about the solubility of gases in liquids.
6. Write briefly a short note on Raoult's law and azeotropes.
7. What do you understand by colligative properties of solutions? List important colligative properties.
8. Define osmosis and osmotic pressure. How does it differ from diffusion?
9. What do you understand by relative lowering of vapour pressure and mole fraction of solute? Describe an experimental method for determining the molecular weight of solute by measuring the lowering of vapour pressure.
10. Derive the expression for determining the elevation of boiling point.
11. Describe the method used for determining the molecular weight of a non-volatile solute by the boiling point method.
12. Derive a relationship between the depression of freezing point and the molecular weight of a non-volatile solute.
13. Define depression of freezing point and describe any one method to determine it.
14. Explain what is meant by molal depression constant of solvent? How it is related to latent heat of fusion? Describe a method for determining the molecular weight of a substance by depression in freezing point.
15. Write short notes on the following:
    (a) Abnormal molecular weights or abnormal solutes.
    (b) Vant Hoff's factor.
    (c) Isotonic solutions.
16. (a) Explain the term lowering of vapour pressure and relative lowering of vapour pressure.
    (b) What are the colligative properties? Explain. Why electrolytes have abnormally high value of colligative properties?
17. Explain the following by giving reasons:
    (a) Addition of non-volatile solute lowers the freezing point and elevates the boiling point of a solvent.
    (b) Equimolar solutions of sucrose and sodium chloride in water are not isotonic.
18. State and explain Van't Hoff theory of dilute solutions
19. (a) State and explain Henry's Law
    (b) Write a note on ideal solution
    (c) Write a short note on depression in freezing point
20. (a) Explain the following terms
        (i) Molarity (ii) Molality
        (iii) Normality (iv) Mole fraction
    (b) What is fractional distillation? Explain it for binary liquid solutions
    (c) Write a short note on reverse osmosis and depression in freezing point

MULTIPLE CHOICE QUESTIONS

1. Very dilute solutions which show deviations (positive or negative) from Raoult's law are called
   (a) Ideal solution (c) Non-ideal solutions
   (b) True solution (d) Colloidal solutions
2. Colligative properties of a solution depend upon
   (a) Nature of solute
   (b) Nature of solvent
   (c) The relative number of solute and solvent particles
   (d) None of these
3. If 5.85 g of NaCl is dissolved in 90 g of water, the mole fraction of NaCl is
   (a) 0.1 (c) 0.2
   (b) 0.01 (d) 0.0196
4. The freezing point of a solution containing 18 g of a non-volatile solute dissolved in 200 g of water is 0.93 °C. Find the molecular weight of the solute. K for water is 1.86° C/mole.
   (a) 160 (c) 210
   (b) 320 (d) 180
5. In an equimolar solution of (A) electrolyte and (B) non-electrolyte,
   (a) Freezing point of A is greater than that of B.
   (b) Freezing point of B is greater than that of A.
   (c) Osmotic pressure of B is greater than that of A.
   (d) Boiling point of B is greater than that of A.
6. How many grams of CH_3OH would have to be added to water to prepare 150 ml of a solution that is 2.0 MCH_3OH?
   (a) 9.6 (c) 9.6×10^4
   (b) 2.4 (d) 4.3×10^2
7. At 25 °C, the highest osmotic pressure is exhibited by 0.1 M solution of
   (a) $CaCl_2$ (c) Glucose
   (b) KCl (d) Urea
8. 10 ml of conc. H_2SO4 (18 molar) is diluted to 1 litre. The approximate strength of dilute acid could be
   (a) 0.18 N (c) 0.36 N
   (b) 0.09 N (d) 1800 N
9. Equal volumes of 0.1 M $AgNO_3$ and 0.2 M NaCl are mixed. The conc. of NO_3 ions in the mixture will be
   (a) 0.1 M (c) 0.2 M
   (b) 0.05 M (d) 0.15 M
10. Addition of common salt to a sample of water will
    (a) increase its freezing point and increase the boiling point.
    (b) decrease its freezing point and increase the boiling point.
    (c) increase both the boiling and freezing points.
    (d) decrease both the boiling and freezing points.
11. Partial vapour pressure of a solution component is directly proportional to its mole fraction. This statement is known as
    (a) Henry's law
    (b) Raoult's law
    (c) Ostwald dilution law
    (d) Distribution law
12. The law which states that the mass of a gas dissolved in a given mass of a solvent at any temperature is directly proportional to the pressure of the gas above the solution is called
    (a) Joule's law (c) Charle's law
    (b) Boyle's law (d) Henry's law
13. Vant Hoff's factor for 0.1 M ideal solution is
    (a) 0.1 (c) 0.01
    (b) 1 (d) none of the three
14. The depression in freezing point of 0.01 M aqueous solutions containing urea, sodium chloride and sodium sulphate is in the ratio of
    (a) 1:1:1 (c) 1:2:4
    (b) 1:2:3 (d) 2:2:3
15. The solution of sugar in water contains
    (a) free atoms
    (b) free molecules
    (c) free ions
    (d) free atoms and free molecules
16. Vant Hoff's factor more than unity indicates that the solute has

(a) dissociated
(b) associated
(c) both
(d) cannot say anything

17. Isotonic solutions have same
(a) molar concentration
(b) molality
(c) normality
(d) none of these

18. Unit of molarity is
(a) g/litre (c) kg/litre
(b) mole/litre (d) none of these

19. When mango is placed in very dilute aqueous solution of hydrochloric acid, it
(a) shrinks (c) bursts
(b) swells (d) nothing happens

20. 4.0 g of caustic soda is dissolved in 100 cc of solution; the normality of solution is
(a) 1 (c) 0.5
(b) 0.1 (d) 4.0

21. Molecular weight of urea is 60. A solution of urea containing 6 g of urea in one litre is a
(a) 1 molar (c) 0.1 molar
(b) 1.5 molar (d) 0.01 molar

22. Which of the following solution in water possesses the lowest vapour pressure?
(a) 0.1 (M) NaCl (c) 0.1 (M) KCl
(b) 0.1 (M) $AlCl_3$ (d) None of these

23. 2N–HCl will have the same molar concentration as
(a) 0.5 N–H_2SO_4 (c) 2N–H_2SO_4
(b) 1.0N–H_2SO_4 (d) 4N–H_2SO_4

24. The relationship between osmotic pressure at 273 K when 10 g of glucose (P_1) 10 g of urea (P_2) and 10 g of sucrose (P_3) are dissolved, respectively, in 250 ml of water is
(a) $P_1 > P_2 > P_3$ (c) $P_2 > P_1 > P_3$
(b) $P_3 > P_1 > P_2$ (d) $P_2 > P_3 > P_1$

25. Lowering of vapour pressure is the highest for
(a) Urea (c) 0.1 M $MgSO_4$
(b) 0.1 M glucose (d) 0.1 M $BaCl_2$

26. Which of the following 0.10 M aqueous solution will have the lowest freezing point?
(a) $Al_2(SO_4)_3$ (c) KI
(b) $C_2H_{10}O_5$ (d) $C_{12}H_{22}O_{11}$

27. Which is not affected by temperature?
(a) Normality (c) Molarity
(b) Formality (d) Molality

28. A 5% solution of cane sugar (molecular weight= 342) is isotonic with 1% solution of a substance *X*. The molecular weight of *X* is
(a) 34.2 (c) 68.4
(b) 171.2 (d) 136.8

29. If the increase in the boiling point of sucrose solution is 0.1 K, then what is the increase in boiling point of the same concentration of NaCl solution?
(a) 0.1 K (c) 0.4 K
(b) 0.2 K (d) 0.58 K

30. An ideal solution is formed, when its constituent components
(a) have only zero heat of mixing.
(b) have only zero volume change.
(c) are converted into ideal gases.
(d) both 'a' and 'b'.

ANSWERS

1. (c)	6. (a)	11. (b)	16. (a)	21. (c)	26. (a)
2. (c)	7. (a)	12. (d)	17. (a)	22. (b)	27. (d)
3. (d)	8. (c)	13. (b)	18. (b)	23. (d)	28. (c)
4. (d)	9. (b)	14. (b)	19. (b)	24. (c)	29. (d)
5. (b)	10. (b)	15. (b)	20. (a)	25. (d)	30. (d)

Thermodynamics 4

CHAPTER OBJECTIVES

4.1 Introduction
4.2 Some Common Thermodynamics Terms
4.3 Zeroth Law of Thermodynamics
4.4 Work, Heat and Energy Changes
4.5 First Law of Thermodynamics
4.6 The Heat Content or Enthalpy of a System
4.7 Heat Capacities at Constant Pressure and at Constant Volume
4.8 Joule–Thomson Effect
4.9 Reversible-Isothermal Expansion of an Ideal Gas
4.10 Second Law of Thermodynamics
4.11 Entropy

First Law of Thermodynamics

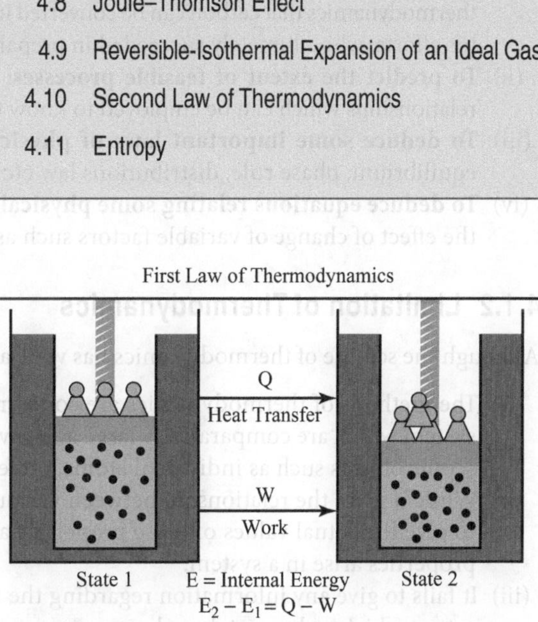

Any thermodynamic system in an equilibrium state possesses a state variable called the internal energy (E). Between any two equilibrium states, the change in internal energy is equal to the difference of the heat transfer into the system and work done by the system

4.1 INTRODUCTION

Thermodynamics deals with the energy changes accompanying physical and chemical transformations. Energy may be manifested in several forms; heat, work, light, mechanical and electrical energies etc. These forms of energies are interconvertible. Thus, ***'the branch of science which deals with the study of the quantitative relationships between heat and other form of energy' is known as thermodynamics.*** When the study is confined to chemical changes only, the branch is known as **chemical thermodynamics.**

Thermodynamics is an exact mathematical science and is based upon four laws namely, zeroth, first, second and third laws, which form the basis of thermodynamics. These laws have proved of fundamental importance in the domains of physics, chemistry and engineering. The first law deals with the conservation of energy, the second law provides the explanation for the condition, extent and direction of the energy changes, whereas the third law helps in the evaluation of thermodynamics parameters. The zeroth law provides us the definition of temperature. Applying these laws, we derive certain relations in terms of thermodynamic functions such as internal energy (E), heat content or enthalpy (H), entropy (S), work function (A) and Gibb's free energy (G) which relate reaction parameters such as temperature, pressure and concentration with other properties. These relations have become the basic principles of physical chemistry. In this chapter, we will take up the study of first and second laws of thermodynamics and some relations derived from them.

4.1.1 Objective of Thermodynamics

The main objectives of chemical thermodynamics are listed below:

(i) **To predict the feasibility of a process:** The primary objective of chemical thermodynamics is to predict the feasibility of a given transformation which may be physical or chemical. Thermodynamic predictions are very certain, because thermodynamics is a mathematical science. Normally, we rely more on practical results rather than theory, but predictions from thermodynamics are sure and dependable just like information from practical experiments. For instance, it was predicted by thermodynamics that carbon can be converted into diamond. This led Moissan, a chemist, to continue his efforts, who ultimately succeeded in preparing artificial diamond starting from carbon.

(ii) **To predict the extent of feasible processes:** Thermodynamics has given certain mathematical relationships which can be employed to know the extent to which a transformation can take place.

(iii) **To deduce some important laws of physical chemistry:** For example, the law of chemical equilibrium, phase rule, distributions law etc.

(iv) **To deduce equations relating some physical properties:** From these equations, we can predict the effect of change of variable factors such as temperature and pressure on different processes.

4.1.2 Limitation of Thermodynamics

Although the science of thermodynamics has wide applications, yet it has the following limitations.

(i) The methods of thermodynamics are concerned with matter on the macroscopic scale, i.e., the system which are comparatively large and involve many molecules. It does not deal with microscopic objects such as individual atoms, molecules or ions.

(ii) While it gives the relationship between various properties experimentally observed, it is unable to give the actual values of these properties and does not offer any explanation as to why these properties arise in a system.

(iii) It fails to give any information regarding the time taken for the completion of a change and the rate at which a change takes place.

(iv) It cannot explain completely the behaviour of a system away from equilibrium. It deals with substances in stationary state or equilibrium state.

(v) It fails to tell us the way the system undergoes change and is also silent about the path that the change follows. It deals only with the initial and final states of the system and not with the path by which the change has been brought about.

4.2 SOME COMMON THERMODYNAMICS TERMS

(1) **Thermodynamic system and its surroundings:** In performing any experiment, we select a part of the universe of interest to us, and attempt to isolate it from any uncontrolled disturbances. This object, whose properties we wish to study, is called the system. All other parts of the universe, whose properties are not of immediate interest, are called the surroundings.

System: *Any portion of universe or any portion of matter chosen to study the effect of various variables such as temperature, pressure or volume is called system.*

Surroundings: The remaining portion of the universe excluded from the system, which may be capable of exchanging energy and matter with it, is called its surroundings. This separated from its surroundings by real or imaginary boundaries.

For example, a mixture of compounds A and B taken in a beaker; the mixture is called the system, while the beaker in which mixture is kept and the room where the beaker is kept are surroundings.

(2) **Types of system:** Depending on the interactions between system and its surroundings, three kinds of systems are distinguishable; they are isolated system, closed system and open system.

(i) **Isolated system:** A system which neither exchanges energy nor matters with its surroundings is called an isolated system. For example, a liquid in contact with its vapour in an insulated closed vessel is an isolated system.

(ii) **Closed system:** A system which may exchange energy but not matter with its surroundings is called a closed system. If in the above example the vessel containing liquid in contact with its vapour is closed one but is not insulated, it is a closed system.

(iii) **Open system:** A system which may exchange both energy and matter with its surroundings is called an open system. One such example is the evaporation of water in an open beaker. Animals and vegetables can be considered open systems from thermodynamics viewpoint. These are depicted in Fig. 4.1 (a), (b), and (c).

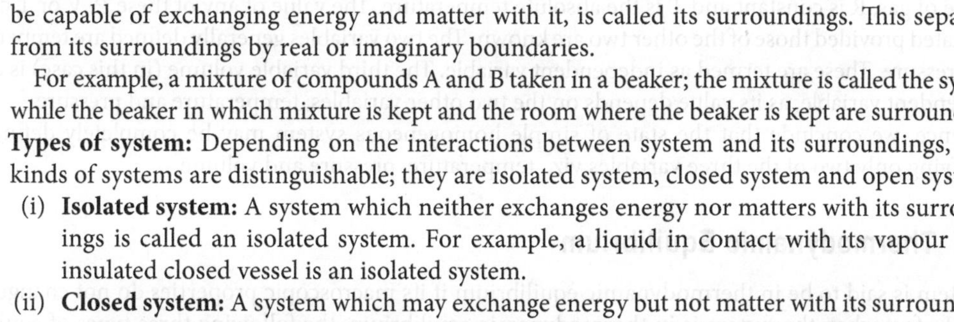

Figure 4.1 (a) Isolated system (b) Closed system (c) Open system

(3) **State of a system and state variable:**

(i) **State of a system:** The system is said to have a definite state when it is in such a condition where all of its macroscopic properties have definite values. The term macroscopic signifies a large number of particles, atoms or ions.

It follows that under any other condition where the value of anyone of its macroscopic properties change, the system is said to have a different state. Thus, the state of a system is defined when its macroscopic properties are specified, i.e., the state of a system is fixed by its macroscopic properties.

(ii) **State variables:** The macroscopic properties of a system are often called **state variables,** since the state of a system changes with the change in any of the macroscopic properties.

Of all the macroscopic properties (state variables), pressure, volume temperature and composition assume particular importance in thermodynamics, since they can be controlled more easily. If the system is homogeneous and consists of only one phase, the composition is automatically fixed. The state of this system can, therefore, be defined by pressure, volume and temperature alone. If these three properties are specified, all other properties (macroscopic) such as mass, density and refractive index are also fixed.

For example, for an ideal gas, $PV = RT$ is the equation of state where P is the pressure, V is the volume of 1 mole of gas, R is constant and T is the absolute temperature, The value of any of these P, V or T can be calculated provided those of the other two are known. The two variables generally defined are temperature and pressure. These are termed as independent variable. The third variable volume (in this case) is called a dependent variable, as its value depends on the two other variables; temperature and pressure.

Hence, we conclude that the state of simple homogeneous system may be completely defined by specifying only two of the three variables viz., temperature, pressure and volume.

4.2.1 Thermodynamic Equilibrium

A system is said to be in thermodynamic equilibrium if its macroscopic properties do not change with time. In fact when the system is in thermodynamic equilibrium, the following three types of equilibria exist simultaneously.

(i) **Thermal equilibrium:** Thermal equilibrium demands that there is no flow of 'heat' from one macroscopic portion of the system to another. This is possible if the temperature in all parts of the system remains same throughout (constancy of temperature).

(ii) **Mechanical equilibrium:** Mechanical equilibrium implies that there is no 'work' done by one macroscopic portion of the system over another. This is possible if the pressure in all parts of the system remains same throughout (constancy of pressure). In case of mechanical equilibrium, there is no macroscopic movement of constituents, within the system or the system as a whole with respect to surroundings.

(iii) **Chemical equilibrium:** Chemical equilibrium demands that the composition (if the system consists of more than one phase) of various phases in the system remains same throughout (constancy of composition).

4.2.2 Thermodynamic Processes

A thermodynamic process (physical or chemical) may be defined as a path of change of a system from one equilibrium state to another which is usually accompanied by the change in energy and or mass.

The different types of processes commonly met within the study of chemical thermodynamics are:

(i) **Isothermal process** is that in which the temperature of the system remains constant during each step of the change in the state of the system. The chemical reaction may be exothermic or endothermic. For an exothermic reaction, the heat evolved is instantaneously given to the surroundings and the temperature of the system does not rise during the change. While for an endothermic reaction, the required amount of heat is instantaneously taken from the surroundings and the temperature of the system does not fall during the change.

Thus, during the isothermal processes (or change), the system exchanges heat with surroundings and the temperature of the system remains constant.

(ii) **Adiabatic process** is that in which no heat enters or no heat leaves the system during each step of the change in the state of system. In other words, the system is completely insulated (thermally) from its surroundings and it does not exchange heat with its surroundings. In such a process for an exothermic reaction, the heat evolved remains within the system and thus the temperature of the system increases. For an endothermic reaction, heat is supplied by the system and thus the temperature of the system falls.

Thus, in an adiabatic process, the temperature of the system may change according to the conditions. The difference between isothermal and adiabatic processes may be summarized in the following table.

Isothermal Process	Adiabatic Process
(i) The temperature of the system remains constant during the process.	(i) The temperature of the system may rise or fall.
(ii) The system exchanges heat with the surroundings.	(ii) The system is completely insulated from the surroundings and thus it does not exchange heat with the surroundings.

(iii) **Isobaric process** is that in which the pressure of the system remains constant during each step of the change in the state of the system.

In an isobaric process (i.e., a process occurring at constant pressure), change in volume may take place.

(iv) **Isochoric process** is that in which the volume of the system remains constant during each step of the change in the state of the system.

In an isochoric process, i.e., a process occurring at constant volume, it may be followed by the change of pressure.

In case of a reaction, $H_2\,(g) + Cl_2\,(g) \rightarrow 2\,HCl\,(g)$

which is not accompanied by any change in volume and there will be no change in pressure of the reacting system.

4.2.3 Reversible and Irreversible Processes

A reversible process may be defined as the process which can be performed in the opposite direction in such a way that all changes occurring in the direct process are exactly reversed in the corresponding stages of reverse process; the process being carried out infinitesimally slowly, so that the system remains almost in a state of equilibrium at all the times.

A strictly reversible process is thus purely a concept; nevertheless, it is extremely useful in the analysis of our problems of energy changes.

For a clear understanding of this concept, imagine a gas confined within a cylinder provided with a weightless and frictionless piston, upon which is piled some grains of sand (Fig. 4.2). At the start of the experiment, the piston is motionless, indicating that equilibrium exists between the pressure of the gas inside the cylinder and the combined pressure of the sand and the atmosphere outside. Now, remove one grain of sand; the gas will expand very slightly but equilibrium will be restored almost immediately. Thus, during the process of expansion, the system is not far from equilibrium. Such a change

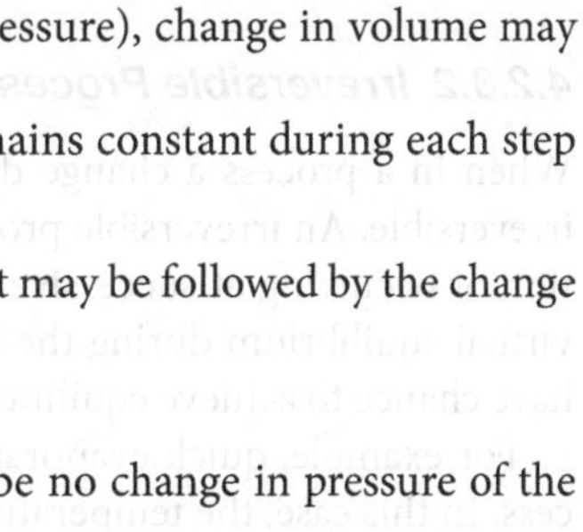

Figure 4.2 Reversible process

approximates an infinitesimal change. If the grain of sand is replaced, the gas will return to its original volume. By the continued removal of grains of sand, the gas can be allowed to undergo a finite expansion, but each step in this expansion is an infinitesimal one and one can be reversed by an infinitesimal change in the external conditions. At all times, the system is only an infinitesimal distance away from equilibrium. A process carried out in this way is called **reversible process.**

In the above example, if the combined pressure of the sand and atmosphere outside, i.e., P_{ext} (external pressure), is equal to the pressure of the gas inside the cylinder, i.e., P, the gas will neither expand nor contract. If P_{ext} is slightly smaller than P, the gas will expand; whereas if P_{ext} is slightly more than P, the gas will be compressed (or contract).

4.2.3.1 Characteristics of a Reversible Process

(i) At any time during the reversible process, the driving force is only infinitesimally greater than the opposing force.
(ii) Such a process can be reversed by an infinitesimal increase in the opposing force.
(iii) At all times, the system is in virtual equilibrium with the surroundings; therefore, at anytime, the initial state can be regained by infinitesimal small changes in the state variables.
(iv) Reversible changes are very slow.
(v) Work obtainable in a reversible change is maximum.

4.2.3.2 Irreversible Process

When in a process a change does not follow the conditions of reversibility, the process is said to be irreversible. An irreversible process is one which occurs suddenly or spontaneously without the restriction of occurring at successive stages of infinitesimal quantities. The system would also not remain in virtual equilibrium during the transition. When the process is carried out rapidly, the system does not have chance to achieve equilibrium during the process.

For example, quick evaporation by sudden decrease in the external pressure is an irreversible process. In this case, the temperature and pressure within the system change and disturb the equilibrium.

It is of interest to note that most of the spontaneous processes are irreversible in nature, e.g., water flowing down a hill, gases expanding from the regions of high pressure to the regions of low pressure, heat flowing of its own accord from hot bodies to cooler bodies, etc.

4.2.4 Thermodynamic Properties

The properties of a system may be divided into two types, namely, extensive and intensive properties.

(i) Extensive Properties

Extensive property of a system is any property whose magnitude depends upon the amount of the substance present (the quantity of the matter). Examples of such properties are total volume, mass and energy.

(ii) Intensive Properties

Intensive properties are those properties whose value is independent of the total amount of substance present, i.e., their values do not depend upon the quantity of matter present in the system. Examples of intensive properties are pressure, temperature, density, refractive index, surface tension, mass per unit mole, volume per unit mole or energy per mole.

MEMORY FOCUS

1. Thermodynamics deals with energy changes accompanying all types of physical and chemical changes.
2. Most of the important generalizations in physical chemistry can be deduced from the laws of thermodynamics.
3. The laws of thermodynamics are not concerned with the atomic or molecular structure of matter.
4. Thermodynamics helps us to predict the favourable conditions for the feasibility or spontaneity of a process (physical or chemical).
5. Laws of thermodynamics apply to vast number of molecules (matter in bulk) but not to individual atoms or molecules.
6. A specific portion of matter under study which is isolated from the rest of the universe with a bounding surface is called a **system.**
7. The rest of the universe which might be in a position to exchange energy and matter with the system is called the **surroundings.**
8. A system which can exchange neither energy nor matter with its surroundings is called an **isolated system.**
9. A system which can exchange energy but not matter with its surroundings is called a **closed system.**
10. A system which can exchange matter as well as energy with its surroundings is called **open system.**
11. A property which depends upon the amount of the substance(s) present in the system is called **extensive property.**
12. A property which is independent of the amount of substance(s) present in the system is called an **intensive property.**
13. The operation by which a system changes from one state to the other is called a process or (**change of state**).
14. A process that take place at constant temperature in each stage is called **isothermal process** ($\Delta T = 0$).
15. A process that takes place such that no heat enters or leaves the system in any step of the process is called an **adiabatic process** ($\Delta q = 0$).
16. In case of expansion of a gas, maximum work is done if the process is isothermal and reversible.
17. In adiabatic expansion of a gas, it gets cooled because of the decrease in its internal energy.
18. A process in which pressure of the system remains constant through all the steps is called an **isobaric process** ($\Delta P = 0$).
19. A process that takes place in such a manner so that there is no change in volume in any step is called **isochoric process** ($\Delta V = 0$).
20. Change of state is an isothermal process. During melting, freezing, evaporation or condensation, the temperature remains constant.
21. A reaction conducted in an isolated system is an example for **adiabatic process.**
22. The expansion of a gas in open system is an example for **isobaric process.**
23. The change taking place in a closed system is an example for **isochoric process.**
24. In any phenomenon if the change in a property depends only on the initial and final states of the system, that property is called as a **state function.**
25. **State functions** do not depend on path of process. Internal energy and enthalpy are state functions. Work and heat energy are not state functions.

It is interesting to note that extensive property may become intensive property by specific unit amount of the substance concerned. Thus, mass and volume are extensive; however, density (mass per unit volume) and specific volume (volume per unit, mass), respectively, are intensive properties of the substance or system. Similarly, heat capacity is an extensive property, but specific heat is an intensive property.

4.3 ZEROTH LAW OF THERMODYNAMICS

The primary importance of this law was stated when the first law of thermodynamics had already been established. It states that, *'When two bodies A and B are separately in thermal equilibrium with a third body C, then they must also be in mutual thermal equilibrium if placed together'*. In other words, *'If body A is in thermal equilibrium with body C and body B is also in thermal equilibrium with body C, then bodies A and B are in thermal equilibrium with each other.'* These three bodies, A, B and C, are said to be equally hot or cold. In other words, they are said to be at the same temperature. Thus the Zeroth law introduced the concept of temperature and also provided the means for measuring it.

Consider a small system is placed in contact with a large system, and a thermal equilibrium is established between them without any significant loss of energy. Some suitable property of the small system which changes with degree of hotness can be made the basis of temperature measurement. The magnitude of the chosen property at two different standard conditions (say temperatures) is measured and the difference between the values is divided into a convenient number of scale divisions. The small system may be considered as a *thermometer* and the large system which loses only an insignificant thermal energy is called a *thermostat* for the small system.

The suitable properties chosen for temperature measurement are volume, length, electrical resistance and potential difference. Based upon these properties, the corresponding thermometers in use are *air thermometer, mercury in glass capillary thermometer, platinum resistance thermometer and thermo couple thermometer, respectively. All these thermometers agree at the standard or fixed point, but are liable to differ slightly at the remaining scale points.* ***'If body A is in equilibrium with body C and body B is also in equilibrium with body C, then bodies A and B are in equilibrium with each other.'***

Energy manifests itself not only in the form of mechanical work, but also in the form of heat energy, electrical energy and chemical energy. Energy is composed of two factors, namely, intensity factor and capacity factor. The product of these two factors gives the energy. Heat energy is measured by the product of temperature (intensity factor) and heat capacity (capacity factor) of the system. The product gives the energy of the system. If a substance of mass *m* kg and of specific heat *s* kJ/kg is heated through temperature *t*, then the heat energy involved is given by *mst* kJ.

For any thermodynamical system in equilibrium, there exists a certain single-valued function *f* of the state variables, *P* and *V*, which have the same values for all the system. Mathematically, it can be expressed as,

$$f_1(P_1, V_1) = f_2(P_2, V_2) = f_3(P_3, V_3) = \ldots = \mathrm{T}$$

where the subscripts 1, 2, 3, refer to different systems. The above equation may be considered to define empirical temperature and been found possible to express the last equation in finite or closed form for most substances. Its existence for most substances will, however, be taken for granted.

4.3.1 Absolute Scale of Temperature

Let us define the true zero or natural zero of the temperature scale. Several temperature scales are known, but they are all arbitrarily based on boiling and freezing points of water or on some other transition temperature. A true scale is based on the observation that *PV—P* graphs of all real gases when extrapolated to zero pressure give the same value of PV at a given temperature. This value is represented by $(PV)_0$, in general and by $(PV_m)_0$, if one mole of gas is taken.

The values of $(PV_m)_0$ change linearly with temperature on the centigrade scale, as shown in Fig. 4.3. The change is expressed by the equation,

$$(PV_m)_0 = a + bt \qquad (4.1)$$

where t is the centigrade temperature measured by a mercury thermometer and a and b are constants. The values of $(PV_m)_0$ at 0°C and 100°C have been found to be 22.4136 and 30.6192 dm³ atm, respectively. Substituting these values in equation (4.1), we get

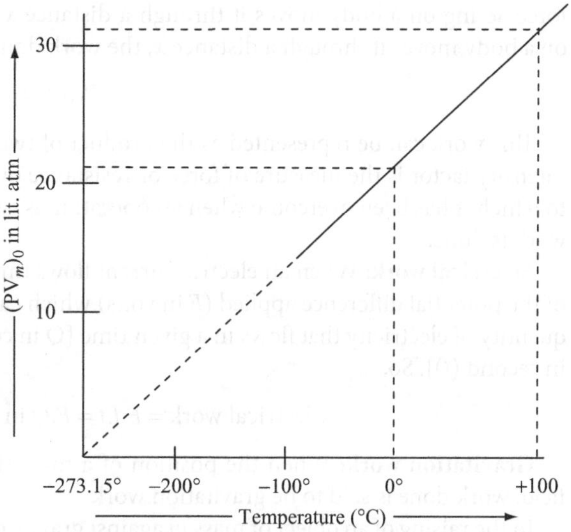

Figure 4.3 Absolute scale of temperature

$$22.4136 \text{ dm}^3 \text{ atm} = a + 0$$

$$30.6192 \text{ dm}^3 \text{ atm} = a + 100b$$

Therefore, $\quad b = 0.082056 \text{ dm}^3\text{atm}$

On substituting the values of 'a' and 'b' in equation (4.1), we can calculate the temperature at which $(PV_m)_0 = 0$. So,

$$0 = 22.4136 \text{ dm}^3 \text{ atm} + (0.082056 \text{ dm}^3\text{atm})^t$$

or $\quad t = -273.15°\text{C}.$

The temperature −273.15°C is the natural or true zero. It is defined as the temperature at which the limiting value of PV_m, i.e., the product PV_m at pressures approaching zero, is zero itself. This zero defines an *absolute temperature scale*, and as it is based on the behaviour of real gases at zero pressure when they behave ideally, it is called the *ideal gas temperature scale.* However, the size of the degree on the absolute scale and centigrade scale is the same.

The t temperature on the absolute scale is represented by T. So, the freezing temperature of water on this scale is given by $T = 273.15 + 0 = 273.15$. For practical purpose, the absolute zero temperature is taken as equal to −273°C.

Therefore, $\quad T = t°\text{C} + 273$

4.4 WORK, HEAT AND ENERGY CHANGES

Whenever a change in the state of a system takes place, there always occurs a change in energy. The energy change accompanying a physical or chemical process is usually in the form of (i) work, (ii) heat and (iii) radiations. Let us understand the nature of these terms used in thermodynamics.

4.4.1 Work

It is a mode of transference of energy between the system and the surroundings. In thermodynamics, *work is defined as any quantity of energy that flows across the boundary of a system during a change in its state and is completely convertible into the lifting of a weight in the surroundings.* In general, when a force acting on a body moves it through a distance x, the work is set to be done. If a force (F) acting on a body moves it through a distance x, the work done is expressed as

$$\text{Work} = F \times x$$

This work can be represented as the product of two factors, intensity factor and capacity factor. The intensity factor is the measure of force or resistance overcome and capacity factor represents the extent to which it has been overcome when an operation is carried out. The following are some ways by which work is done.

Electrical work: When an electric current flows through a circuit, work done is equal to the product of the potential difference applied (E in volts) which causes the flow of current (intensity factor) and the quantity of electricity that flows in a given time [Q in coulombs = current strength in amperes (I) × time in second (t)]. So,

$$\text{Electrical work} = E.I.t = E.Q \text{ in volts Coulombs or Joules.}$$

Gravitation work: When the position of a mass (body) is changed with respect to the gravitation field, work done is said to be gravitation work.

In the raising of an object of mass m against gravity g, through a height h, the force exerted in opposition to gravity is mg *(intensity factor)* and the Capacity factor is the height h. The energy expanded or the work done becomes, therefore, the product mgh.

$$\text{Gravitation work} = mgh$$

Pressure–volume work *(mechanical work):* In most processes, the interest of a chemist in thermodynamics is pressure–volume or mechanical work. This type of work is associated with increase or decrease of the volume of the system.

Consider an ideal gas contained in a cylinder fitted with a frictionless piston (Fig. 4.4).

If P_{ext} is the pressure (of surrounding atmosphere) acting on a piston of cross-sectional area A, then the force F_{ext} acting on the piston is given by

$$F_{ext} = P_{ext} A$$

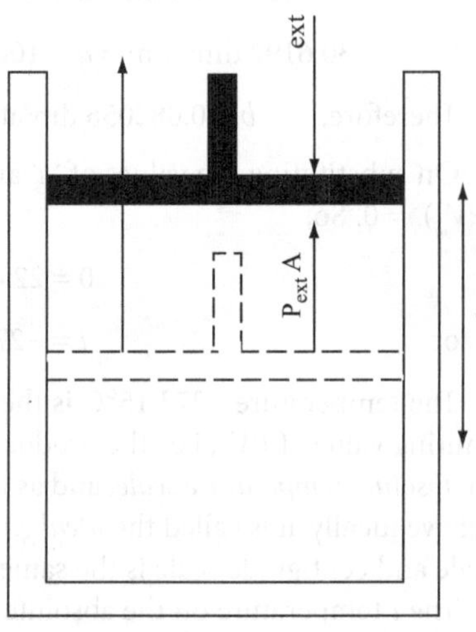

Figure 4.4 Pressure–volume work or mechanical work in the expansion of an ideal gas

The force is balanced by an equal force exerted by the gas on the piston. Suppose the gas in the cylinder expands by an infinitesimally small amount and as a result, the piston is pushed out through an infinitesimally small distance dx (displacement dx by force **F** from the point of its application in the direction of the force). The small amount of work done (dw) by the system, i.e., a gas on the surrounding is given by,

$$dw = F_{ext} \times dx = P_{ext} \times A \times dx = P_{ext} \times dV \tag{4.2}$$

The quantity $A \times dx$ represents infinitesimally small increase in volume in the expansion process and is represented by dV. It may be noted here that P_{ext} (the force per unit area F_{ext}/A) is always the external pressure or the pressure of the surroundings, against which the gas expands.

If this expansion is brought about infinitesimally slowly in a reversible manner in a number of stages from an initial volume V_1 to the final volume V_2, i.e., the volume of the system changes by a finite quantity V_2-V_1, the total work (w) done on the surroundings by the system can be obtained by the integration of $P_{ext}\, dV$,

$$w = \int_{V_1}^{V_2} P_{ext} dv \qquad \text{[When the volume inceases from } V_1 \text{ to } V_2\text{]}$$

If the contraction is carried out infinitesimally slowly in reversible manner in a number of stages, the work done (w) on the system by the surroundings is given by the same relation.

If the pressure P_{ext} remains constant during the process, the total work done (w) is given as follows:

$$w = P_{ext} \int_{V_1}^{V_2} dv = P_{ext}(V_2 - V_1) = P_{ext}\,\Delta V \tag{4.3}$$

where V_1 is the volume of the system in the initial state and V_2 is the volume of the system in the final state.

Sign of w. If ΔV is positive, it means the gas expands and the work done (w) will be positive and the work is done by the surroundings on the surroundings.

If ΔV is negative, it means the gas undergoes contraction and the work done (w) will be negative and the work is done by the surroundings on the system (gas).

Units of w. The units of work are the same as the units of energy, i.e., erg or Joule. An erg is equal to the *maximum work* done when a force of one dyne acts through 1 cm.

$$10^7 \text{ erg} = 1 \text{ J}$$

In the above example, if the pressure applied on the piston, P_{ext}, is very small as compared with the pressure of gas contained in the cylinder, then the gas will expand rapidly (not in small stages), i.e., *irreversibly*, and the work done by the system will be very small as the opposing force is very small. If the gas is allowed to expand against vacuum when P_{ext}, the opposing pressure, is zero, then the work done will be zero.

i.e., $$P_{ext} \times \Delta V = 0$$

Thus for free expansion of gas against vacuum, no work is done because $P_{ext} = 0$.

It may be noted here that the magnitude of the opposing (external) pressure determines the amount of work done by the system in expansion. The smaller the difference between the opposing (external pressure) and the pressure of the gas (system) contained in the cylinder, the greater will be the work done by it in expansion. Thus if the difference between the magnitudes of two pressures is infinitesimally small, which is a condition for ideal reversible process, then the work done is maximum.

4.4.2 Heat

Heat is another mode of energy exchange between the system and the surroundings which results from the difference in the temperature of the system and the surroundings. If a temperature difference exists between a system and its surroundings, then energy may be transferred by heat flow (radiation or conduction).

Just like work, heat is also a method by which system exchanges energy. The proof of this was given by James Joule who showed that the same change in state (i.e., a certain rise in temperature) can be accomplished either by doing work on a body, or by heating it. To distinguish between them, we need only to say the work is transferred by virtue of a mechanical link between systems, and that heat is the energy transferred due to a temperature difference.

Units of heat. The heat change involved in any process is represented by the letter 'q' and measured in calories. A calorie is the quantity of heat required to raise the temperature of 1 g of water through 1°C.

Sign of q. Regarding the convention of sign, a positive value of q, $(+q)$, means that the system has absorbed heat from its surroundings and a negative value of q, $(-q)$, means that the system has given heat to the surroundings.

4.4.3 Equivalence Between Mechanical Work and Heat

All other forms of energy tend to change ultimately into heat. Thus, mechanical work gets transformed into heat by friction, chemical energy through chemical reactions, and electrical energy by its passage through a resistance. It is now well-known that heat, which is a form of energy, cannot be completely converted into work, but work can be converted into equivalent of heat energy. James Joule showed that the amount of mechanical work (W) is proportional to the quantity of heat produced (H).

$$W \propto H \text{ or } W = JH$$

where J is Joule's mechanical equivalent of heat. When H=1, W=J.

Joule mechanical equivalent of heat may, therefore, be defined as the amount of work done to produce one calorie of heat and is numerically equal to 4.184×10^7 erg.

$$1 \text{ calorie} = 4.184 \text{ J} = 4.184 \times 10^7 \text{ erg.}$$

4.4.4 Internal Energy

Energy may be defined as the capacity to do work. It may be classified into two types: *kinetic* and *potential energy*. In chemical thermodynamics, we are not concerned with such two forms of energy. We are concerned with an energy which a chemical substance may possess because of the arrangement of atoms or electrons within its molecule or which a system possesses because of its temperature, volume, pressure and chemical composition. This is known as the *internal energy* or *intrinsic energy* and is designated by the letter 'E'.

Every system consisting of one or more than one substance is associated with some *inherent energy*. If it were not so, it would not have been possible to obtain heat or some other form of energy during chemical reaction. When petrol burns in air, heat energy is obtained. *The inherent energy which is possessed by a substance or by a system by virtue of its existence is called its internal energy. It is, infact, the sum of different forms of molecular energies such as the kinetic energy of translation, vibration and rotation, the potential energy due to intermolecular forces, the chemical bond energy, and the electronic and nuclear energies of the constituent atoms in the molecule of the substance.* The internal energy of a substance depends upon various factors such as quantity and chemical nature of the substance, temperature and pressure.

It may be noted that each system can be imagined to have an *absolute value* of the internal energy. But it is neither necessary nor possible to ascertain this absolute amount of energy. This is due to the fact that it is not feasible to determine most of the quantities (mentioned above) which contribute towards internal energy. However, the change in the internal energy of a system, accompanying the changes in temperature, volume, pressure and composition, is a measurable quantity and can be determined experimentally. Thus if E_1 and E_2 are the internal energies of a system in the initial state and the final state respectively, then the change in the internal energy (ΔE) is given by,

$$\Delta E = E_2 - E_1$$

4.5 FIRST LAW OF THERMODYNAMICS

The first law of thermodynamics is not derived from any assumptions; rather, it summarizes our experience with material processes. It is a law of human experience and has no theoretical proof. This law has not been contradicted by any experimental result so far. It is simply the law of conservation of energy which states that, 'Energy can neither be created nor be destroyed although it may be converted from one form to the other.' The law may also be stated in the following ways:

(i) *The total energy of an isolated system remains constant although it can undergo a change from one form to the other.*
(ii) *Whenever a quantity of some form of energy disappears, an exactly equivalent quantity of some other form of energy makes it appearance.*
(iii) *It is impossible to construct a perpetual motion machine which would produce work without consuming energy.*
(iv) *For a system in contact with its surroundings, the sum of the energies of the system and the surroundings remains constant; however, differently, it may be shared between the two.*
(v) *In modern times, mass is also considered as a form of energy. According to Einstein, mass can be converted into energy and these two forms of energy are related to each other by the expressions:*

$$E = mc^2$$

where E is the energy obtained by the conversion of mass m and c is the velocity of light and is equal to 3×10^{10} ms^{-1}. The value of c being very high conversion of small amount of mass will give rise to enormous amount of energy.

Therefore, the first law has been modified and stated as: '*The total mass and energy isolated system remains constant at all tile times*'.

Validity of the law:
The following examples illustrate the validity of the law:

(i) One calorie of heat is produced for every 4.184 J of work done irrespective of the manner in which the work is done. It has also been observe that the same amount of heat is produced by the same amount of work irrespective of the manner how the work is done.
(ii) When one mole of hydrogen gas is treated with half a mole of oxygen gas at room temperature (298 K), one mole of liquid water is formed with the evolution of heat.

$$H_2(g)+\frac{1}{2}O_2(g)\rightarrow H_2O(l); \qquad \Delta H = -285.8 \text{ kJ}$$

Conversely, electrical energy equivalent to 285.8 kJ is required for the electrolysis of one mole of liquid water at room temperature (298 K),

$$H_2O(l)\rightarrow H_2(g)+\frac{1}{2}O_2(g); \qquad \Delta H = +285.8 \text{ kJ}$$

The potential energy of water in a dam is changed to kinetic energy and is finally changed to electrical energy.

4.5.1 Mathematical Formulation of First Law of Thermodynamics

The internal energy of a system can be changed in the following two ways:

(i) By allowing heat to flow into the system or out of the system.
(ii) By doing work on the system or allowing to system to do work.

Let the initial internal energy of the system be E_1. Suppose a certain amount of heat q is supplied to the system so that the internal energy of the system increases and becomes $E_1 + q$. Now, if work, w, is done on the system, its internal energy further increases and becomes $E_1 + q + w$. It is obvious that final energy (E_2) of the system is equal to this factor:

$$E_2 = E_1 + q + w$$

or
$$E_2 - E_1 = q + w$$

or
$$\Delta E = q + w \tag{4.4}$$

Here, work is done by the system on the surrounding, so w is taken as positive. If, however, work is done by the system on the surrounding, w is taken as negative;

$$\Delta E = q - w \tag{4.5}$$

The above expression is the mathematical formulation of the first law of thermodynamics. w, here, includes all kinds of work such as pressure volume work and electrical work.

If the work done in the above process is only pressure volume work, $w = -P\Delta V$. Therefore, the above expression can be written as:

$$\Delta E = q - P\Delta V \tag{4.6}$$

4.5.2 Some Special Forms of First Law of Thermodynamics

Under special conditions, Eq. 4.4 for the law gets modified e.g.,

Case 1 For a *cyclic process* returning to its original state, $\Delta E = 0$ and $q = -w$, i.e., the work done by the system is exactly equal to the heat absorbed.

Case 2 For an *isothermal change*, $\Delta E = 0$ and again $q = -w$, i.e., the heat absorbed by the system is equal to work done by the system.

Case 3 For a process *at constant volume*, no work of expansion is done so that $w = 0$ and $\Delta E = q_v$, i.e., the increase in the internal energy is equal to the heat absorbed at constant volume.

Case 4 For *an adiabatic process*, i.e., when the system is so well insulated that no heat is gained or lost to the surrounding, $q = 0$ so that $\Delta E = w$, i.e. the change in internal energy is exactly equal to the work done.

Aid to Memory

1. For an **isothermal reversible process**

$\Delta E = 0, \Delta H = 0$ (it is because the energy is function of temperature)

$q = -W$

Since $W = -nRT\ln\dfrac{v_2}{v_1} = -nRT\ln\dfrac{p_1}{p_2}$

(*Continued*)

Aid to Memory (*Continued*)

or $$q = -nRT \ln \frac{v_2}{v_1} = -nRT \ln \frac{p_1}{p_2}$$

2. For an **isothermal isobaric expansion.** At constant T and P

$$q = -W, \Delta E = 0, \Delta H = 0$$

3. For an **isochoric process,** $\Delta V = 0$: $q = \Delta E$
4. For an **adiabatic process,** $q = 0$: $\Delta E = W$
5. For a **cyclic process,** $\Delta E = 0$: $q = -W$
6. q and W are not state functions since their values depend upon the path by which the change is carried, but the quantity $q + W$ is a state function. This is because $q + W = \Delta E$ and ΔE is a state function.

4.5.3 Limitations of the First Law of Thermodynamics

The first law of thermodynamics indicates that the exact equivalence between various forms of energy involved but it does not provide any information about the following:

(i) *The feasibility of the process, i.e., whether the process is possible or not:* We know that it is not possible to regenerate petrol from the products obtained when it is burnt in the engine of an automobile. The first law fails to tell us whether it is feasible or not.

(ii) *The conditions of the change:* It is unable to explain why the physical and chemical processes occur in a natural direction. For example, ice melts at room temperature but water does not freeze at room temperature.

(iii) *The direction of change:* It does not about the direction of flow of heat. For example, ice may be used to heat water by extracting heat from ice at lower temperature and supplying it to water at high temperature, without any external aid. The law tells us that the increase in the energy of the system should be equal to the decrease in energy of the surroundings and vice versa. The law does not indicate the direction of the process, since according to it, all processes are possible in which energy is conserved. For example, ice cannot be used for heating a body since energy always flows from a body at a higher energy level to a body at a lower energy level.

(iv) *The extent of the change:* The heat absorbed cannot be completely converted into work without producing some other change in some part of the system or the surroundings.

The arguments show the insufficiency of the first law of thermodynamics. The second law of thermodynamics provides the explanation of the above points.

MEMORY FOCUS

1. The first law is simply the law of conservation of energy which states that energy can be neither created nor destroyed.
2. Energy whenever disappears in one form, an equal amount of energy in some other form must appear.
3. It is impossible to construct a machine which can produce energy without spending energy.

(*Continued*)

MEMORY FOCUS (*Continued*)

4. Total mass and energy of an isolated system remains unchanged (modified in the light of Einstein's law, $E = mc^2$).
5. Mathematical form of first law of thermodynamics is: $\Lambda E = q + w$
   If w is work of expansion $w = -P\Lambda V$, then
   $$\Lambda E = q - P\Lambda V$$
   or $$q = \Lambda E + P\Lambda V$$
6. When certain heat energy is supplied to a system, a part of it is used up to increase its internal energy and another part to do work.
7. In any phenomenon, if the change in a property depends only on the initial and final states of the system, that property is called as a state function. State function does not depend on the path of the process.
8. Internal energy and enthalpy are state functions; however, work and heat energy are not state functions.

NOTEWORTHY POINTS

1. Both work and heat appear only during a change of state.
2. q is +ve when system absorbs heat and –ve when system evolves heat.
3. W is +ve when work is done on the system and –ve when system performs a work.
4. Work, energy and heat have same units. Whereas energy is a thermodynamic property of the system, work and heat are not.
5. It should be noted that the work done during the isothermal reversible process is greater than adiabatic process.
6. I cal > 1 J > 1 erg.
7. Both work and heat appear only at the boundary of the system.
8. Internal energy is an extensive property.
9. Internal energy is depends on temperature, pressure, volume and quantity of matter.
10. The internal energy (or intrinsic energy) cannot be calculated. By convention, the internal energy of an element in most stable form is zero.
11. There is no change in internal energy in a cyclic process.

Numerical Problems

Example 1 **Calculate the work performed in erg when 2 moles of hydrogen expand isothermally and reversibly at 25°C from 15 to 50 l.**

Solution:

The work done is given by the equation

$$w_{\max} = nRT \ln \frac{V_2}{V_1}$$

$$= 2.303 nRT \times \log_{10} \frac{V_2}{V_1}$$

Here $n = 2$ moles, $T = 25 + 273 = 298$ K, V_1=15 l and V_2=50 l

Also $R = 8.3 \times 10^7$ erg/°C/mole

Therefore, $w_{max} = 2.303 \times 2 \times 8.3 \times 10^7 \times 298 \log_{10}\dfrac{50}{15} = \mathbf{5.96 \times 19^{10}}$ **erg**

Example 2 **What is the maximum work obtainable from the isothermal expansion of 1mole of an ideal gas at 18°C from a pressure of 4 atm to a pressure of 1 atm? Also calculate the quantity of heat evolved in the process.**

Solution: The maximum work obtainable is given by the equation,

$$w_{max} = nRT\ln\frac{P_1}{P_2}$$

$$= 2.303nRT^2\ln\frac{P}{P_1}$$

Here $n = 1$ mole, $T = 18°C = 18 \times 373 = 291$ K, $P_1 = 4$ atm., $P_2 = 1$ atm.

Also $R = 8.314 \times 10^7$ erg/°C/mole

Therefore, $w_{max} = 2.303 \times 1 \times 8.314 \times 10^7 \times 291 \times \log_{10}\dfrac{4}{1}$

$= \mathbf{3.36 \times 10^{10}}$ **erg**

$$q = w_{max} = \frac{3.36\times10^{10}}{4.18\times10} = \mathbf{804\ cal}$$

Example 3 **Calculate the minimum work necessary to compress 2 moles of an ideal gas from 10 to 5 l at 27°C. How much heat is absorbed in the compression process?**

Solution: The minimum work necessary to compress the gas will be equal to the maximum work done by the gas if it expands, but opposite in sign. The maximum work obtainable under reversible and isothermal conditions is given by,

$$W_{max} = 2.303\ nRT\log\frac{V_2}{V}$$

Here, $n = 2$ moles, $R = 1.987$ cal/°C/mole, $T = 273 + 27 = 300$°K, $V_2 = 5$ l, $V_1 = 10$ l.

$$w_{max} = 2.303 \times 2 \times 1.987 \times 300 \times \log\frac{5}{10}$$

$$= -2.303 \times 2 \times 1.987 \times 300 \times 0.3010$$

$$= -826.4 \text{ cals.}$$

Therefore, the minimum work required to compress the gas is 826.4 cals.

Further in an isothermal process

$$W = q$$

Therefore $q = +826.4$ cals.

Therefore, the amount of heat absorbed = **826.4 cals.**

4.6 THE HEAT CONTENT OR ENTHALPY OF A SYSTEM

Processes at constant pressure are more common in chemistry because most of the operations are carried out in open vessel. Under conditions of constant pressure, the gas expands so that the work of expansion will be

$$w = P\Delta V$$

where ΔV is the increase in volume and P is the constant pressure. Putting this value in the first law of thermodynamics,

$$q = \Delta E + w \text{ or } q_p = \Delta E + P\Delta V \tag{4.7}$$

where subscript 'p' stands for the process at constant pressure.

Suppose when the system absorbs q_p calories of the heat, its internal energy increases from E_1 to E_2 and its volume increases from V_1 to V_2, we have

$$\Delta E = E_2 - E_1 \text{ and } \Delta V = V_2 - V_1$$

$$q_p = (E_2 - E_1) + P(V_2 - V_1)$$

$$q_p = (E_2 + PV_2) - (E_1 + PV_1)$$

Now as P and V are the properties of the state of the system only, it follows that the quantity $E + PV$, like the internal energy E, is dependent only on the thermodynamic state and not on its previous history. *The thermodynamics property $E + PV$ is called heat content or enthalpy of the system and is represented by the symbol H, i.e., by definition.*

$$H = E + PV \tag{4.8}$$

Thus if H_2 is enthalpy of the system in the final state and H_1 the enthalpy in the initial state, then equation (4.8) becomes

$$q_p = H_2 - H_1 = \Delta H$$

where $\Delta H = H_2 - H_1$ is the change in enthalpy. Thus the change in enthalpy of a system is equal to the heat absorbed at constant pressure (provided the work done is only the work of expansion, i.e., pressure–volume type).

Further combining equations (4.7) and (4.8), we get

$$\Delta H = \Delta E + P\Delta V \tag{4.9}$$

i.e., the change in enthalpy at constant pressure is equal to the increase in internal energy plus external work of expansion.

Enthalpy function is quite significant since most of the processes in the laboratory are carried out at constant pressure as for example reactions in an open beaker or flask take place under atmospheric pressure, i.e., at constant pressure. Absolute value of enthalpy like energy cannot be known. We are concerned only with the changes in enthalpy for most of the processes rather than its absolute values at various states. The changes in enthalpy may be measured by taking any suitable reference state of the elements. The processes in which heat is absorbed by the system are endothermic whereas the processes in which heat is evolved are exothermic reactions. In the first case the ΔH is positive and in the latter case the ΔH is negative.

Enthalpy a change accompanied by a certain process is called the enthalpy of that process. Thus enthalpy of evaporation means the change in the enthalpy involved when one mole of the substance gets converted from the liquid state into the vapour state. The enthalpies of fusion and sublimation have similar meanings.

4.7 HEAT CAPACITIES AT CONSTANT PRESSURE AND AT CONSTANT VOLUME

The heat capacity of a system, between any two temperatures, is the quantity of heat required to raise the temperature of the system from the lower to the higher temperature divided by the temperature difference, i.e., it is the amount of heat required to raise the temperature of the system through 1°C.

If the temperature of the system is raised from T_1 to T_2, its heat capacity may be expressed as

$$C_{(T_2,T_1)} = \frac{q}{T_2 - T_1} = \frac{q}{\Delta T}$$

where q is the heat absorbed and C is the capacity of system.

Since the heat capacity varies with temperature, it may be considered over a very narrow temperature range and may be represented by the differential equation

$$C = \frac{dq}{dT}$$

where dq is the small amount of heat absorbed by the system which raises its temperature by dT (say from T to $T+dT$).

We know q is not a state function and depends upon the path followed and hence C is also not a state function. It depends upon the path, i.e., whether the heat is being supplied at constant volume or at constant pressure. Thus there are two types of heat capacities of a system, particularly a gaseous system, the heat capacity of a gaseous system determined at constant volume is different from that of determined at constant pressure. *Thus the heat required to raise the temperature of 1 mole of a substance by 1°C at constant volume is denoted by C_v and the corresponding quantity at constant pressure is called molar heat at constant pressure, denoted by C_p.*

4.7.1 Heat Capacity at Constant Volume

In this case, the whole quantity of heat supplied is utilized in raising the temperature and no work is done.

At constant volume, $w = P\Delta V = 0$.

The first law of thermodynamics, $q = \Delta E + w$, becomes $q = \Delta E$.

Heat capacity of the system at constant volume is

$$C_V = \left(\frac{q}{T_2 - T_1}\right)$$

Substituting the value of $q C_V = \left(\frac{\Delta E}{T_2 - T_1}\right)_v$, we get

$$C_v = \left(\frac{dE}{dT}\right) \text{ or } C = \left(\frac{dE}{dT}\right)$$

Thus, *the heat capacity of a system at constant volume (C_v) is numerically equal to the increase in internal energy of the system per degree rise of temperature at constant volume.*

4.7.2 Heat Capacity at Constant Pressure

Work is done by the heat energy supplied to change the volume. Suppose the volume changes by ΔV, then $w = P\Delta V$. According to the first law of thermodynamics, $q = \Delta E + w = \Delta E + P\Delta V$. Therefore, heat capacity of the system at constant pressure is given as

$$C_p = \left(\frac{q}{T_2, T_1}\right)_p = \left(\frac{\Delta E = P\Delta V}{T - T}\right)_p$$

We know, $\Delta H = \Delta E + P\Delta V$

Hence
$$C_p = \left(\frac{\Delta H}{T_2 - T_1}\right)_p \text{ or}$$

$$C_p = \left(\frac{\partial H}{\partial T}\right)_p \text{ or } C_p = \frac{dH}{dT}$$

Thus, *the heat capacity of a system at constant pressure (C_p) is equal to the increase in heat content (or enthalpy) of the system per degree rise of temperature at constant pressure.*

4.7.3 Relationship Between C_p and C_v

Relationship between C_p and C_v is derived on the basis of differential calculus. We know that

$$C_v = \frac{dE}{dT} \text{ and } C_p = \frac{dH}{dT}$$

Again $\quad H = E + PV$ but $PV = RT$

Hence $\quad H = E + RT$

Differentiating it with respect to temperature T, we have

$$\frac{dH}{dT} = \frac{dE}{dT} + R \tag{4.10}$$

Putting the values of $\frac{dH}{dT}$ and $\frac{dE}{dT}$ in Eq. (4.10), we have

$$C_p = C_v + R, \text{ and } C_p - C_v = R \tag{4.11}$$

This is the desired relationship.

4.8 JOULE–THOMSON EFFECT

It was observed by Joule and Thomson that *'when a real gas at a certain pressure expands adiabatically through a porous plug into a region of low pressure, it is accompanied by cooling' (expect for hydrogen and helium which get warmed up).* This phenomenon is known as 'Joule–Thomson effect'.

A simple explanation of Joule–Thomson effect is based upon the fact that the molecules of the gas are moving and so possess a definite amount of kinetic energy. When the gas expands, it does some work in overcoming the van der Waals' forces of attraction existing between the molecules. Thus, a part of the energy is used up in doing this work and, therefore, the energy of the system and, as a result, the temperature also falls.

Experimental setup: The experimental setup of Joule and Thomson is illustrated schematically in Fig. 4.5. A tube thoroughly insulated (so as to be nearly under adiabatic conditions) was fitted with a porous plug, as indicated. The pressure on the left and right sides of a porous plug were kept at P_1 and P_2, respectively ($P_1 > P_2$). By applying pressure on the piston A on the left slowly enough (so as not to change the pressure P_1), a volume of gas V_1 was forced slowly through the porous plug and then allowed to expand to the pressure P_2 and volume V_2 by moving the piston B on the right outward. While the expansion was taking place, accurate temperature readings were taken on the gas in the two chambers. A fall in temperature was observed in all cases expect for hydrogen and helium.

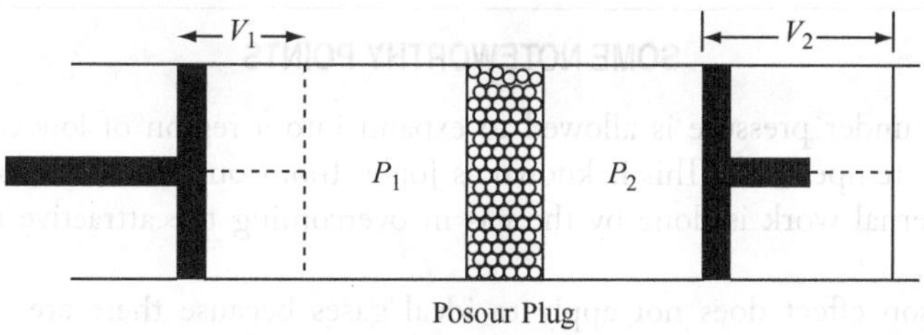

Figure 4.5 Joule Thomson effect

Explanation: The work done on the system at the left piston is P_1V_1 and the work done by the system at the right piston is P_2V_2. Hence, the net work done by the system is:

$$w = P_2V_2 - P_1V_1$$

Since the process was conducted adiabatically,

$$q = 0$$

Applying just law of thermodynamics

$$\therefore \quad \Delta E = -w$$

$$E_2 - E_1 = -(P_2V_2 - P_1V_1)$$

or $$E_2 + P_2V_2 = E_1 + P_1V_1$$

i.e., $$H_2 = H_1$$

$$\Delta H = 0$$

Thus, *during the free expansion of real gas, the heat content or enthalpy remains constant.* Under these conditions, the extent of the temperature change is found to depend both on the initial *t* temperature and the pressure of the gas. Joule and Thomson defined a term, μ, known as *Joule–Thomson co-efficient* as

$$\mu \text{ J.T.} = \left(\frac{\partial T}{\partial P}\right) \text{H}$$

Thus, μ may be defined as *the temperature change in degrees experienced by a gas in free expansion by a pressure drop of 1 atm under conditions of constant enthalpy.* For cooling, μ is positive, while for all observed heating, μ is negative. The temperature at a particular pressure at which the gas neither heats nor cools on expansion, i.e., where $\mu = 0$ is called an **inversion temperature.** It is clear from the above formula that:

(i) For cooling effect, μ will be positive because ∂T and ∂P both will be negative.
(ii) For heating effect, μ will be negative because ∂T is positive while dP is negative.
(iii) For no effect, μ will be zero because $\partial T = 0$ for any value of ∂P.

Since most of the gases at room temperature are below their inversion temperature, hence they undergo cooling on adiabatic expansion. Hydrogen and helium, on the other hand, are still above their upper inversion temperatures and hence they show heating. It may be expected that these gases would also

SOME NOTEWORTHY POINTS

1. When a gas under pressure is allowed to expand into a region of low constant pressure, it suffers fall in temperature. This is known as Joule–Thomson effect. The cooling is due to the fact that internal work is done by the gas in overcoming the attractive forces between the molecules.
2. Joule–Thomson effect does not apply to ideal gases because there are no attractive forces between the molecules.
3. At the inversion temperature, Joule–Thomson coefficient is zero.
4. The pressure at which no heating or cooling takes place on expansion is known as **inversion point** at which Joule–Thomson coefficient is zero.

exhibit cooling if their temperatures were lowered below their upper inversion point at a given pressure. This is actually the case.

4.9 REVERSIBLE-ISOTHERMAL EXPANSION OF AN IDEAL GAS

The isothermal expansion of a gas can be carried out reversibly by placing the cylinder of the gas in a thermostat and adjusting the external pressure so as to be less than the pressure of the gas by an infinitesimally small amount. As the gas expands, however, its own pressure decreases, since the temperature is maintained constant. Hence, if the process is to be thermodynamically reversible, it must be supposed that the external pressure is continuously adjusted so as to be always infinitesimally less than the pressure of the gas. The expansion will then take place extremely slowly so that the system is always in thermodynamic equilibrium. If at any instant during expansion, the external pressure was increased so that it became greater than the gas pressure by an extremely small amount, the process would be reversed; compression would take place and every stage would be exactly the reverse of that occurring during expansion. A general expression for the work of expansion in an isothermal reversible process may be easily derived as follows:

Suppose P is the pressure within a system undergoing an isothermal expansion by a reversible process. Then from the discussion above, it follows that the external pressure must be $P–dP$, where dP is a very small quantity. The small amount of the work, dw, done by the system when it increases its volume by an infinitesimal amount dV is equal to the product of the external pressure and the volume change, i.e.,

$$dw = (P-dP)dV = P\,dV - dP\,dV$$

Neglecting the term $dP\,dV$ which is extremely small, the equation becomes,

$$dw = PdV$$

Therefore, the total work done by the gas when its volume increases form V_1 to V_2 will be given by

$$\int dw = \int_{V_2}^{V_1} Pdv$$

or

$$w = \int_{V_2}^{V_1} Pdv \qquad \text{[this expression is valid for any gas]}$$

Now for n moles of an ideal gas, the equation of state is

$$PV = nRT, \text{ so that } P = \frac{nRT}{V}$$

For an isothermal process, T is constant. Also n and R are constant quantities. Therefore, putting the value of P in the above equation,

$$w - nRT\int_{V_1}^{V_2}\frac{dV}{V} \quad \text{or} \quad w = nRT\ln\frac{V_2}{V_1} \tag{4.12}$$

or

$$w = 2.303nRT\log_{10}\frac{V_2}{V_1}$$

Further as the process is isothermal, i.e., the temperature remains constant; therefore, by Boyle's law, we have

$$P_1V_1 = P_2V_2 \text{ or } \frac{V_2}{V_1} = \frac{P_1}{P_2}$$

where P_1, V_1 are the initial pressure and volume and P_2, V_2 are the final pressure and volume of the system. Putting this value in Eq. (4.12), we get

$$w = nRT\ln\frac{P_1}{P_2} \tag{4.13}$$

or

$$w = 2.303nRT\log_{10}\frac{P_1}{P_2}$$

This is another expression which gives the work of expansion in an isothermal reversible process. It may be noticed that although the expressions have been derived for the work due to expansion but the same expression is valid for compression (i.e., the work done by the system during compression will be negative as $\ln = \frac{V_2}{V_1}$ will be negative).

For the ideal gas, energy depends on temperature and hence ΔE for such a process will be zero and hence for an isothermal process

$$\Delta E = 0$$

and

$$q = \Delta E + w \qquad \text{[First law of thermodynamics]}$$

$$q = w \qquad \text{[Since temperature remains constant]}$$

Hence heat absorbed,

$$q = 2.303\, nRT\log_{10}\frac{V_2}{V_1}$$

For 1 mole of gas,

$$q = 2.303\, RT\,\frac{V_2}{V_1}$$

and

$$q = 2.303\, nRT\log_{10}\frac{V_2}{V_1}$$

For 1 mole of gas,

$$q = 2.303\, RT\,\frac{V_2}{V_1}$$

4.9.1 Maximum Work

It must be mentioned here that the work done in an isothermal reversible expansion of ideal gas is the *maximum* possible for the given increase of volume. This is because the work done by the expanding gas is the product of the external pressure and the increase in volume.

$$dw = (P - dp)dV$$

Hence, the only way to get more work for a given increase in volume is to increase the external pressure. It is evident that the smaller the value of dp, the greater is the work done or the work which can be obtained from the system is maximum when the external pressure differs infinitesimally from the internal pressure of the gas.

It follows, therefore, that the work of expansion in reversible isothermal process is maximum for a given change in the thermodynamic state. This is, therefore, sometimes written as w_{max}. Thus the work of isothermal expansion done by a system in an irreversible process is always less than for the same increase in volume when the change is carried out reversibly. In case of an isothermal compression, the reversible work of expansion is (numerically) equal to the maximum amount of work which must be done by the surrounding on the system in order to bring about the specified change in volume (or pressure).

i.e., $$w_{rev} > w_{irr}$$

4.10 SECOND LAW OF THERMODYNAMICS

The second law of the thermodynamics provides an explanation to the limitations of the first law. It tells us about the direction and the extent of change that can take place. For example, ice cannot be used for heating a body since heat energy always flows from a body at a higher energy level to a body at a lower energy level.

Some important statements of the second law are:

(i) It is impossible to convert heat completely into equivalent amount of work without producing some other change in some other part of the system.
(ii) All natural and spontaneous processes take place in one direction only and are thermodynamically irreversible.
(iii) It is impossible to take heat from a reservoir and convert it into work by a cyclic process, without transferring it to a cold reservoir.
(iv) Heat cannot pass by itself from cold body to hot body.
(v) It is impossible for a self-acting machine unaided by an external source to convey heat from a cold body to a hot body.

It may be noted that various statements of the second law appear to be quite different. However, all the statements are modifications of the same fundamental concept, viz., *all forms of energy can be converted into heat but no process is available for converting the heat so obtained completely to other forms of energy.*

Advantages of the Second Law

1. It helps in predicting the feasibility of a chemical reaction or a process. It can be decided whether a process is possible or not. This can be done by determining the free energy change (ΔG) for that process. For any process to take place, G must be negative. $\Delta G = \Delta H - T\Delta S$. The details have been discussed later on in the look.
2. Applying this law, we can calculate the maximum fraction of heat that can be converted into work. The efficiency of a heat engine can be determined from Carnot's theorem; according to which, the efficiency of a heat engine is given by the relation,

$$n = \frac{T_2 - T_1}{T_2} = 1 - \frac{T_1}{T_2}$$

where T_2 and T_1 are the temperature of the source and that of sink, respectively.

It is clear that $(T_2-T_1)/T_2$ is always less than one. Hence, the efficiency of the heat engine is always less than one. No heat engine has so for been constructed which can show an efficiency equals to unity.

3. The law determines the direction of heat flow.

4.10.1 Spontaneous Processes and Reactions (Basis of Second Law)

All naturally occurring processes without the assistance of external agencies are called spontaneous processes.

Some of the examples of the spontaneous processes taking place by themselves are:

(i) Flow of heat from a hot body to a cold body as soon as they are brought in contact.
(ii) Water runs down a hill.
(iii) A stone falls to the ground if it is dropped from some distance above.
(iv) A stretched piece of elastic contracts when released.
(v) Diffusion of gas from high pressure side to low pressure side.

All spontaneous processes are irreversible processes. The spontaneous processes do not proceed in the reverse direction by themselves. The water will not go uphill, the stone does not rise from the ground by itself and the elastic will never stretch by itself.

Spontaneous processes may, however, be reversed with the help of some external assistance, i.e., when energy from outside is supplied. Water can be lifted from a lower level to an upper level with the help of a pump. Gas molecules can be made to move from a region of low pressure to a region of high pressure by applying pressure on them. Heat can be made to flow from a colder body to a hotter body by doing external work on the system, e.g., refrigerator. It may, thus, be concluded that *spontaneous processes are irreversible and can be reversed only when work is done on the system.* All non-equilibrium situations tend to shift towards equilibrium situations on their own or in a natural way or spontaneously, but a change from an equilibrium state of a system to a non-equilibrium state cannot occur without external help to the system from the surroundings.

4.10.2 Spontaneous Reactions

The reaction which proceeds by itself once it has been properly initiated, if necessary, is known as spontaneous or feasible reaction. The term spontaneous does not represent that the reaction or the process occurs instantaneously (quickly). It merely represents that the reaction or the process has an urge to take place. The rate of reaction or the process may vary from slow to extremely fast.

Examples of some spontaneous reactions are given below, which take place on initiation.

(i) On passing electric spark through a mixture of hydrogen and oxygen, liquid water is formed. It is an exothermic reaction.

$$H_2(g) + \frac{1}{2}O_2(g) \rightarrow H_2O(l); \qquad \Delta H = -285.8 \text{ kJ}$$

The reaction is initiated by passing electric spark.

(ii) On ignition, graphite (carbon) undergoes almost complete combustion to give carbon dioxide. It is an exothermic reaction.

$$C(s) + O_2(s) \rightarrow CO_2(g); \qquad \Delta H = -393.5 \text{ kJ}$$

The reaction is initiated by ignition.

(iii) As already mentioned, the initiation is not always necessary. For example, nitric oxide and oxygen combine readily to form nitrogen dioxide without any initiation.

$$NO(g) + \frac{1}{2}O_2(g) \rightarrow NO_2(g); \qquad \Delta H = -56.52 \text{ kJ}$$

All the reactions cited above are exothermic reactions. However, it does not mean that only those reactions for which ΔH has a negative sign are spontaneous. In actual practice, it is found that even endothermic reactions are also spontaneous. For example,

(i) $H_2O\ (s) \longrightarrow H_2O\ (l);$ $\quad \Delta H = +6.0$ kJ
(ii) $H_2O\ (l) \longrightarrow H_2O\ (g);$ $\quad \Delta H = +40.0$ kJ
(iii) $N_2\ (g) + O_2\ (g) \longrightarrow 2NO\ (g);$ $\quad \Delta H = +180.5$ kJ

Thus, both the exothermic and endothermic reactions may be spontaneous. The basis of second law of thermodynamics is the spontaneity of processes.

MEMORY FOCUS

1. It is impossible to convert heat completely into equivalent amount of work without producing some other change in some other part of the system.
2. All natural and spontaneous processes take place in one direction only and are thermodynamically irreversible.
3. It is impossible to take heat from a reservoir and convert it into work by a cyclic process, without transferring it to a cold reservoir.
4. Heat cannot pass by itself from cold to hot body.
5. It is impossible for a self-acting machine unaided by an external source to convey heat from a cold body to a hot body.
6. All forms of energy can be converted into heat but no process is available for converting the heat so obtained completely to other forms of energy.
7. A process which has a natural tendency to take place either by itself or after initiation under a given set of conditions is known as spontaneous process.
8. A process that has no natural tendency to take place is called a non-spontaneous process.
9. The force which is responsible for the spontaneous nature or spontaneity of a process is called the driving force.
10. The tendency towards maximum randomness or disorder is another factor which determines the spontaneity of a process.
11. The overall tendency for a process to occur, i.e., driving force, can be expressed because of the two tendencies which are (a) tendency to acquire a state of minimum energy and (b) tendency to acquire a state of maximum randomness.

4.11 ENTROPY

It is defined as the property of a system which measures the randomness or the amount of disorder of a system, quantitatively. Like internal energy and enthalpy, entropy is also a state function.

It is a thermodynamic measure for randomness and represented by letter S. The entropy of a system is equal to the sum of the components of that system.

The greater the randomness or disorder in a system is, the higher is its entropy, e.g.,

$$\underset{\text{Less entropy}}{H_2O(l)} \longrightarrow \underset{\text{More entropy}}{H_2O(g)}$$

$$\underset{\text{Less entropy}}{CaCO_2(s)} \longrightarrow \underset{\text{More entropy}}{CaO(s) + CO_2(g)}$$

Unlike internal energy, enthalpy and free energy, the absolute entropy can be calculated.

The concept of entropy as a measure of disorder of randomness has led to the conclusion that all substances in their normal crystalline state at the absolute zero of temperature would be in the condition of maximum orderly arrangement because all motion has essentially ceased at 0 K. In other words, the entropy of a substance at 0 K is minimum. This led Plank to put forward another universal empirical law, the third law of thermodynamics, which states that, 'The entropy of a pure perfectly crystalline substance is zero at absolute zero (0 K) of temperature'.

Applying third law of thermodynamics, the absolute entropy of solids can be calculated at a temperature T can be calculated from the formula

$$S_T = \int_{T=0}^{T=T} C_p \frac{dT}{T} = \int_0^T C_p (d\ln T) \tag{4.14}$$

Absolute entropy of substance in the liquid as well as gaseous states at a particular temperature can also be calculated by taking into account the entropy phase changes.

4.11.1 Mathematical Explanation of Entropy

For a reversible Carnot Engine working between a higher temperature T_2 and a lower temperature T_1 with the heat absorbed q_2 and heat evolved q_1, we have

$$\frac{q_2 - q_1}{q_2} = \frac{T_2 - T_1}{T_2}$$

or

$$1 - \frac{q_1}{q_2} = 1 - \frac{T_1}{T_2}$$

or

$$\frac{q_1}{T_1} = \frac{q_2}{T_2}$$

or

$$\frac{q_2}{T_2} - \frac{q_1}{T_1} = 0 \tag{4.15}$$

Then, term q/T is called the reduced heat. Thus, for a reversible Carnot cycle, the algebraic sum of the reduced heats is equal to zero.

For an irreversible cycle, we have

$$\frac{q_2 - q_1}{q_2} < \frac{T_2 - T_1}{T_2}$$

or
$$\frac{q_2}{T_2} - \frac{q_1}{T_1} < 0 \quad (4.16)$$

Therefore, for an irreversible cycle, the sum of q/T terms is less than zero.

For an infinitely small reversible cycle, Eq. (4.9) can be written as

$$\frac{\delta q_2}{T_2} - \frac{\delta q_1}{T_1} = 0 \quad (4.17)$$

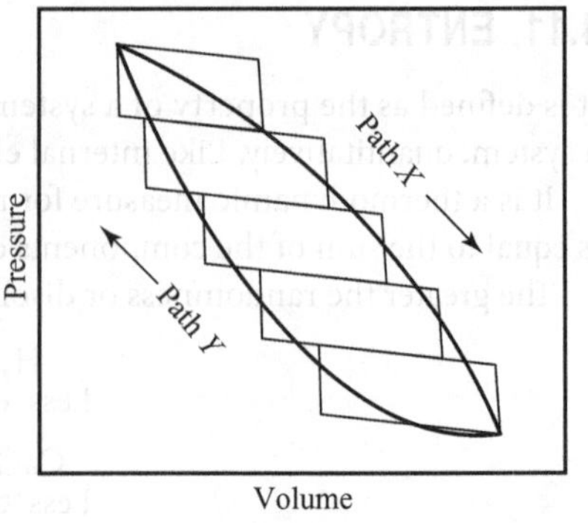

Figure 4.6 A reversible cycle is composed of a large number of reversible carnot cycles

Since a reversible cycle is made of a large number of infinitely small reversible Carnot cycles (Fig. 4.6), therefore, for a reversible cycle, Eq. (4.11) should be written as

$$\int \frac{\delta q_2}{T_2} - \int \frac{\delta q_1}{T_1} = 0$$

or
$$\int \frac{\delta q}{T} = 0 \quad (4.18)$$

where $\oint$ represents a closed or contour integral.

For an irreversible cycle, from Eq. (4.10), we have

$$\oint \frac{\delta q}{T} < 0 \quad (4.19)$$

The expression represented by Eq. (4.15) is called the *Clausius inequality.*

Suppose the reversible cycle is so performed that the system passes from state A to state B by path x and then from state B to state A by path y. For such a cycle, Eq. (4.14) can be written as

$$\int_A^B \frac{\delta q(x)}{T} + \int_B^A \frac{\delta q(y)}{T} = 0$$

or
$$\int_A^B \frac{\delta q(x)}{T} = -\int_B^A \frac{\delta q(y)}{T} = \int_A^B \frac{\delta q(y)}{T} \quad (4.20)$$

Therefore, the term $\int \frac{\delta q}{T}$ is the same for a change from state A to state B whether the system follows path x or path y. This implies that the term $\int \frac{\delta q}{T}$ represents a function of state. This function is termed as entropy and is represented by symbol S. Therefore,

$$\int_A^B \frac{\delta q_{\text{rev}}}{T} = S_B - S_A = \Delta S$$

Consequently,
$$dS = \frac{\delta q_{\text{rev}}}{T} \quad (4.21)$$

Hence, the change in entropy dS may be defined as equal to the heat absorbed reversibly by the system divided by the temperature (in kelvin) at which the heat is being absorbed.

4.11.2 Entropy Change in Chemical Reaction

The entropy change ΔS for a chemical change is equal to the sum of the entropies of the products of that change minus the sum of entropies of the reactants.

Let us consider a hypothetical reaction,

$$m\text{A} + n\text{B} \longrightarrow x\text{C} + y\text{D}$$

The entropy change,

$$\text{S} = \Sigma\text{S products} - \Sigma\text{S reactants}$$

$$= [x \times S_C + y \times S_D] - [m \times S_A + n\, S_B]$$

where S_A, S_B, S_C and S_D are the entropies of A, B, C and D, respectively.

A positive value for ΔS ($\Delta S > 0$) indicates an increase in randomness or disorder. A negative value for ΔS ($\Delta S < 0$) indicates a decrease in randomness or a decrease in order.

4.11.3 Units of Entropy

Entropy is usually expressed in terms of calories per degree, sometimes called as entropy units (e.u.). Entropy is an extensive property. It, therefore, mentions the quantity of the substance involved. The quantity is usually taken as 1 mole of the substance. Therefore, the entropy units are calories per degree per mole (cal K^{-1} mol^{-1}) in C.G.S. system and joule per degree per mole (JK^{-1} mol^{-1}) is S.I. units.

4.11.4 Physical Significance of Entropy

Entropy has been regarded as a measure of disorder or randomness of a system. Thus, when a system goes from a more orderly state to less orderly state, there is an increase in its randomness and hence entropy of the system increases. Conversely, if the change is one in which there is an increase in orderliness, there is decrease in entropy. For example, when a solid changes to a liquid, an increase in entropy takes place because with the breaking of the orderly arrangement of the molecule in the crystal to the less orderly liquid state, the randomness increases. Conversely, the process of solidification brings about more orderly state and consequently there is a decrease in entropy. The process of vaporization produces an increase in entropy.

Evidently, the expansion of a gas in vacuum and the diffusion of a solute from a concentrated solution to a dilute solution when the two are connected together, etc. are accompanied by the increase of disorder (more of randomness) and hence there is an increase in entropy.

4.11.5 Entropy Change Accompanying Change of Phase

A process of change of state, e.g., melting of solid or vaporization of liquid, may be carried out at constant temperature reversibly as the two phases are in equilibrium for all times during the change. Suppose the process of change of state of one mole of a substance is carried out reversibly, the amount of heat absorbed will be equal to the molar heat of fusion or vaporization and the temperature will be the melting point or the boiling point.

1. **Entropy of fusion:** *It is defined as the change in entropy when one mole of a solid substance transforms into the liquid from at its melting point.* The molar entropy change accompanying a fusion process is given by,

$$\Delta S_{\mathrm{f}} = \frac{\Delta H_{\mathrm{f}}}{T_{\mathrm{f}}} \tag{4.22}$$

where ΔS_{f} is the molar entropy change (for 1 mole of a solid) accompanying fusion process, ΔH_{f} is the molar heat of fusion and ΔT_{f} is the freezing or melting point on the absolute scale.

2. **Entropy of vaporization:** Similarly, the molar entropy change ΔS_{v} accompanying a vaporization process (for 1 mole of a liquid) is given by,

$$\Delta S = \frac{\Delta H_{\mathrm{v}}}{T_{\mathrm{b}}} \tag{4.23}$$

where ΔH_{v} is the molar heat of vaporization and T_{b} is the boiling point on the absolute scale.

Since ΔH_{f} and ΔH_{v} are both positive, an increase in entropy takes place during the processes of fusion and vaporization.

In the change of state from vapour to liquid or from liquid to solid, ΔH_{f} will be negative and during the processes of condensation of vapour or freezing of a liquid decrease in entropy takes place.

4. **Entropy of sublimation:** It is defined as the entropy involved in the transformation of a solid directly into vapour at a particular temperature called sublimation temperature.

It is represented by ΔS_{sub} and is given as

$$\Delta S_{\mathrm{vap}} = S_{\mathrm{vap}} - S_{\mathrm{liquid}} = \frac{\Delta H_{\mathrm{subs}}}{T} \tag{4.24}$$

where ΔH_{sub} is the enthalpy of sublimation per mole of a solid at temperature T (in kelvin).

5. **Entropy of transition:** It may be defined as the change in entropy when one mole of a solid changes from crystalline state into the other at the transition temperature.

$$\Delta S = \frac{\Delta H_{\mathrm{t}}}{T_{\mathrm{t}}} \tag{4.25}$$

where ΔH_{t} is the molar heat of transition and T is the transition point.

4.11.6 Entropy Changes in Reversible Processes

When a system gains heat, the surroundings lose heat and hence their entropies also change. The total change in entropy will, therefore, be equal to the algebraic sum of entropy changes of the system and the surroundings, i.e. $\Delta S = \Delta S_{\mathrm{sys}} + \Delta S_{\mathrm{sur}}$. Consider an isothermal reversible expansion of an ideal gas. The gas absorbs Q_{rev} heat at temperature T. The entropy change of the gas, i.e., system is given by,

$$\Delta S_{\mathrm{sys}} = \frac{Q_{\mathrm{rev}}}{T}$$

Since the expansion is reversible, the system is in equilibrium with its surroundings at all times during the expansion. Hence, the surroundings lose an amount of heat equal to Q_{rev}, or we can say that the surroundings absorb $-Q_{\mathrm{rev}}$, of heat. The entropy changes of the surrounding are given by,

$$\Delta S_{\mathrm{sur}} = -\frac{Q_{\mathrm{rev}}}{T}$$

Therefore, the total entropy changes of the universe:

$$\text{DS} = \text{DS}_{sys} + \text{DS}_{sur} = \frac{Q_{rev}}{T} - \frac{Q_{rev}}{T}$$

or $$\Delta S = 0$$

It is concluded that for a reversible process, the total entropy change of the universe is zero.

4.11.7 Entropy Changes in Irreversible Processes

When any part of the process is irreversible, the process as a whole becomes irreversible. Suppose that the heat lost by the surroundings is Q_{irrev}. This amount of heat will be absorbed by the system; however, the entropy change of the heat absorbed reversibly (Q_{rev}), and not upon the heat actually absorbed. If the heat is absorbed isothermally by the system at temperature T, the entropy change or the system will be

$$\Delta S_{sys} = \frac{Q_{rev}}{T}$$

This is due to the fact that entropy is a state function and hence entropy increase of the system will be the same; no matter whether the change has been brought in a reversible or an irreversible manner. However, the heat actually lost by the surroundings is Q_{irrev}. The loss of heat by the surroundings to the system can be adjusted to takes place reversibly and, therefore, the decrease in entropy is given by,

$$\Delta S_{sur} = -\frac{Q_{irrev}}{T}$$

∴ The net change in entropy is given by,

$$\Delta S = \Delta S_{sys} + \Delta S_{sur}$$

$$\Delta S = \frac{Q_{rev}}{T} - \frac{Q_{irrev}}{T}$$

We know that $$Q_{irrev} < Q_{rev}$$

or $$\Delta S > 0$$

Thus, in an irreversible process, there is a net increase in entropy of the universe (system + surroundings).

Entropy Change of the Universe

It is interesting to note that since all the natural spontaneous processes (such as, the transfer of heat from a hot body to a cold body) are irreversible and hence they must be associated with a net increase of entropy. The result about the spontaneous processes is so important that it is sometimes taken as the definition of the second law of thermodynamics.

Hence, the second law may be stated as,

All spontaneous processes are accompanied by a net increase of entropy.

OR

The net entropy of the universe tends to increase.

Clausius summed up the statements of the first and second laws of thermodynamics as below:

The energy of the universe remains constant; the entropy of the universe tends towards a maximum.

4.11.8 Entropy as Criterion of Spontaneity

From the above discussion, we conclude that the entropy of the system and its surroundings taken together increases in a thermodynamically irreversible process at a constant pressure; however, it remains constant in a thermodynamically reversible process.

This provides us a criterion of distinction between an irreversible and a reversible processes. Mathematically,

$$\Delta S_{sys} + \Delta S_{sur} = 0 \quad \text{(for a reversible process)} \tag{4.26}$$

$$\Delta S_{sys} + \Delta S_{sur} > 0 \quad \text{(for an irreversible process)}$$

$$\Delta S_{sys} + \Delta S_{sur} \geq 0 \tag{4.27}$$

Hence, 'equal to' sign refers to a reversible process and the 'greater than' sign refers to an irreversible process. These conclusions are quite significant, as these help us to predict whether a given process is thermodynamically feasible or not, i.e., whether it can take place or not.

4.11.9 Entropy Changes for an Ideal Gas

Consider a mole of an ideal gas contained in a cylinder fitted with a frictionless piston. Suppose an infinitesimal small amount of heat (q_{rev}) is supplied reversibly at constant temperature T, to the system which would undergo expansion involving pressure – volume work. From first law of thermodynamics,

$$q_{rev} = dE + PdV$$

We know that $$C_v = \frac{dE}{dT} \quad \text{or} \quad dE = C_v\, dT$$

and $$PV = RT \quad \text{or} \quad P = \frac{RT}{V}$$

Putting the values in the above equation, we get

$$q_{rev} = C_v dT + RT\frac{dV}{V}$$

Dividing throughout by T, we get

$$\frac{q_{rev}}{T} = C_v\frac{dT}{T} + R\frac{dV}{V}$$

However, $$\frac{q_{rev}}{T} = dS$$

Therefore, $$dS = C_v\frac{dT}{T} + R\frac{dV}{V}$$

This gives the total entropy change for one mole of an ideal gas when the changes in both temperature and volume take place reversibly. When the temperature changes from T_1 to T_2 and the volume from V_1 to V_2, then the entropy change accompanying the process would be

$$\Delta S = S_2 - S_1 = C_v\int_{T_1}^{T_2}\frac{dT}{T} + R\int_{V_1}^{V_2}\frac{dV}{V}$$

$$= C_v \ln \frac{T_2}{T_1} + R \ln \frac{V_2}{V_1} \tag{4.28}$$

It is assumed that C_v is constant over the temperature limit T_1 and T_2. Again for an ideal gas $P_1V_1 = RT_1$ and $P_2V_2 = RT_2$ for the initial and final states, respectively. Therefore,

$$\frac{V_2}{V_1} = \frac{T_2P_1}{T_1P_2}$$

Putting this value in the above equation, we get

$$\Delta S = C_v \ln \frac{T_2}{T_1} + R \ln \frac{T_2P_1}{T_1P_2}$$

Knowing $\quad C_p - C_v = R \quad$ or $\quad C_v = C_p - R$

$$\therefore \quad \Delta S = (C_p - R) \ln \frac{T_2}{T_1} + R \ln \frac{T_2P_1}{T_1P_2}$$

$$= C_p \ln \frac{T_2}{T_1} - R \ln \frac{T_2}{T_1} + R \ln \frac{T_2}{T_1} + R \ln \frac{P_1}{P_2}$$

$$= C_p \ln \frac{T_2}{T_1} + R \ln \frac{P_1}{P_2}$$

$$= C_p \ln \frac{T_2}{T_1} - R \ln \frac{P_2}{P_1} \tag{4.29}$$

Equation (4.24) gives the entropy change for one mole of an ideal gas when temperature and volume are varied, whereas Eq. (4.25) gives the entropy change when temperature and pressure are varied. These equations can be modified depending upon the conditions of constancy of temperature, pressure and volume as under:

Case I (for an isothermal process)

Since there is no change in temperature, i.e., $T_1 = T_2$

The above two equations take the form

$$(\Delta S)_T = R \ln \frac{V_2}{V_1} = -R \ln \frac{P_2}{P_1} = R \ln \frac{P_1}{P_2}$$

Case II (for an isobaric process)

For a process taking place at constant pressure, $P_1 = P_2$, the Eq. (4.25) becomes

$$(\Delta S)_p = C_p \ln \frac{T_2}{T_1} \tag{4.30}$$

Case III (for an isochoric process)

For a process taking place at constant volume, $V_1 = V_2$, the Eq. (4.24) becomes

$$(\Delta S)_v = C_v \ln \frac{T_2}{T_1} \tag{4.31}$$

MEMORY FOCUS

1. Entropy is the quantity which tells us whether a reaction or a physical change can occur in a system or not.
2. Mathematically *d*S = *d*q/T, where '*d*S' is difference in entropy and '*d*q' is heat supplied in a reversible process. T is absolute temperature.
3. Units of entropy are joule/degree or calories/degree.
4. Entropy cannot be felt like temperature and pressure.
5. Entropy is a measure of the unavailable energy. i.e. Unavailable energy = Entropy × Temperature.
6. Entropy is a measure of the disorder or randomness of the system.
7. According to Clausius, the entropy of the universe is increasing.
8. All processes in nature tend to occur only with an increase in entropy and the direction of change is always such as to lead to an increase in entropy.
9. Spontaneous processes are accompanied by the increase in entropy (randomness) of the system.
10. For all pure crystals, entropy is taken as zero.
11. The decreasing order of entropies is: gas > liquid > solid
12. Entropy is a state function and an extensive property.

Numerical Problems

Example 4 Calculate the entropy change accompanying the complete conversion of one mole of liquid water at 373 K to steam. Latent heat of vaporization of water is 2.257 kJ g^{-1}.

Solution: Heat required to vaporize one mole of water

$$q_{rev} = H_{vap} = 2.257 \text{ kJ/g}$$
$$= 2.25718 \text{ kJ mol}^{-1}$$
$$= 40.626 \text{ kJ mol}^{-1}$$
$$\Delta S_{(vap)} = \frac{\Delta H_{(vap)}}{T_b} = \frac{40.626}{373} = 0.1089 \text{ kJK}^{-1}\text{ mol}^{-1} = \mathbf{108.9\ JK^{-1}\ mol^{-1}}$$

Example 5 Calculate the total entropy change when 5 moles of ice is converted into water at 0 °C. The latent heat of fusion of ice is 80 cal. gm^{-1}.

Solution: We know that molar entropy change $\Delta S_f = \dfrac{\Delta H_f}{T_f}$

Here, the latent heat of ice = 80 cal. gm^{-1}
The molar heat of fusion H_f = 80 × 18 cal. mol^{-1}

$$T_f = 273 \text{ K.}$$
$$\Delta S_f = \frac{80 \times 18}{273} = 5.274 \text{ cal. deg}^{-1}$$

The entropy change per mole = 5.274 cal. deg^{-1}
The total entropy change for 5 moles = 5 × 5.274

$$= \mathbf{26.37\ cal.\ deg^{-1}}$$

Example 6 **Calculate the entropy for the conversion of 1 mole of ice at 274 K and 1 atm into steam at 350 K and 1 atm, the molar heat of fusion of ice $\Delta H_f = 7$ kJ mol^{-1} and molar heat of vaporization of water, $\Delta H_v = 45$ kJ mol^{-1} and $C_p = 76$ JK^{-1} (assume CP is constant within this temperature).**

Solution: (i) Entropy change for 1 mole of ice at 274 K

$$\Delta S_1 = \frac{\Delta H_t}{\Delta H_t} = \frac{7000}{274} = 25.54 \text{ JK}^{-1}\text{mol}^{-1}$$

(ii) Entropy change of *m* mole of water at 274 to water at 350 K at 1 atm.

$$\Delta S_2 = 2.303 \times C_p \log \frac{T_2}{T_1}$$

$$\Delta S_2 = 2.303 \times 76 \times \log \frac{350}{274} = 18.61 \text{ JK}^{-1}\text{mol}^{-1}$$

(iii) Entropy change of 1 mole of water at 350 K to steam at 1 atm.

$$\Delta S_3 = \frac{\Delta H_v}{\Delta T_v} = \frac{42000}{350} = 120 \text{ JK}^{-1}\text{mol}^{-1}$$

Total entropy change $= \Delta S_1 + \Delta S_2 + \Delta S_3$

$$= \mathbf{164.15\ JK^{-1}mol^{-1}}$$

Example 7 **One mole of an ideal gas at 300 K expands reversibly from 310^{-2} m^3 to 510^{-2} m^3. Calculate the entropy change for the gas.**

Solution: The entropy change $(\Delta S)_T$ for one mole of an ideal gas during isothermal reversible expansion is given as

$$(\Delta S)_T = 2.303 \text{ R} \log \frac{V_2}{V_1}$$

Here, $R = 1.987$ calories, $V_1 = 3 \times 10^{-2}$ m^3 and $V_2 = 5 \times 10^{-2}$ m^3

$$(\Delta S)_T = 2.303 \times 1.987 \log \frac{5 \times 10^{-2}}{3 \times 10^{-2}}$$

$$= 2.303 \times 1.987 \times [\log 5 - \log 3]$$

$$= \mathbf{1.0154\ calories/degree/mole}$$

Example 8 **Calculate the change in entropy when 2 moles of a gas is heated from 27 °C to 727 °C at constant pressure of 1 atm. The molar heat capacity of the gas is 23.7 JK^{-1} mol^{-1}.**

Solution: The entropy change $(\Delta S)_P$ for *n* moles of an ideal gas at constant pressure is given as

$$(\Delta S)_P = 2.303\, nC_p \log \frac{T_2}{T_1}$$

Substituting the given value, we have

$$(\Delta S)_P = 2.303 \times 223.7 \log \frac{1000}{300}$$

$$= 2.303 \times 223.7 \times [\log 10 - \log 3]$$

$$= \mathbf{27.081\ JK^{-1}mol^{-1}}$$

Example 9 **One mole of an ideal monoatomic gas expands reversible from a volume of 10 dm³ at temperature 300 K to a volume of 25 dm³ at temperature 230 K. Assuming that $C_p = \frac{3}{2}R$, calculate the entropy change for the process.**

Solution: For one mole of an ideal gas, the entropy change in the process involving the variables V and T is given by

$$\Delta S = 2.303\ C_v \log_{10}\frac{T_2}{T_1} + 2.303 R \log_{10}\frac{V_2}{V_1}$$

Given $T_1 = 300$ K, $T_2 = 230$ K, $V_1 = 10$ dm³, $V_2 = 25$ dm³, $R = 8.314$ JK⁻¹, $C_v = \frac{3}{2}R$

Substituting the values, we have

$$\Delta S = 2.303 \times \frac{3}{2} \times 8.314 \times \log_{10}\frac{230}{300} + 2.303 \times 8.314 \log_{10}\frac{25}{10}$$

$$= -3.314 + 7.619 = \mathbf{4.305\ JK^{-1}\ mol^{-1}}$$

Example 10 **Five moles of an ideal gas expands thermally and reversibly at 27 °C from an initial volume of 5–50 dm³ against a pressure that is gradually reduced. Calculate ΔG and ΔS for the process ($LR = 8.314$ JK⁻¹).**

Solution: We have that $(\Delta S)_T = 2.303\ nR \log\frac{V_2}{V_1}$

Putting the given values, we get

$$(\Delta S)_T = 2.303 \times 5 \times 8.314 \times \log\frac{50}{5} = 98.74\ JK^{-1}$$

We also know that $\Delta G = 2.303\ nRT \log\frac{V_1}{V_2}$

Putting the values, we get

$$\Delta G = 2.303 \times 5 \times 8.314 \times \log\frac{5}{50} \times 300 = -28720\ j\ mol^{-1} = \mathbf{28.72\ j\ mol^{-1}}$$

REVISION QUESTIONS

1. (a) What are the various thermodynamic process.
   (b) Distinguish between reversable process and irreversable process.
   (c) State and explain first law of thermodynamic.
   (d) Discuss absolute scale of temperature.
2. (a) What are the limitations of first law of thermodynamics?
   (b) State the second law of thermodynamics in as many ways as possible.
3. State second law of thermodynamics and explain the conditions under which heat can be converted into work.

4. (a) Explain the term 'entropy'. Show that it is a state function. Give its units.
   (b) Why is the internal energy of a system a state function?
5. (a) Show that for an irreversible process: $\Delta S_{sys} + \Delta S_{sur} > 0$; whereas, for a reversible process, there is no net entropy change.
   (b) Explain the terms: (i) Steady state (ii) Isotherm (iii) Absolute temperature
   (c) Define second law of thermodynamics.
6. (a) Show that for an isothermal expansion of an ideal gas: $\Delta S = R \text{ In } V_2/V_1$.
   (b) Write expression for entropy of melting, vaporization and sublimation.
   (c) Why does entropy of a solid increase on fusion?
7. (a) Briefly explain the physical significance of the term 'Entropy'.
   (b) Select the most random system of the following:
      (i) Sugar cubes in a box (ii) Sugar cubes scattered on a table
      (iii) Sugar dissolved in water (iv) Finely powdered sugar
8. (a) What is entropy? Derive an expression for entropy change during reversible isothermal expansion of an ideal gas.
   (b) Show that for an irreversible process: $\Delta S_{sys} + \Delta S_{surr} > 0$
9. (a) Prove that in a reversible process the entropy of the universe is constant while it increases in an irreversible process.
   (b) Derive an expression for the calculation of the entropy change of an ideal gas when the temperature changes from T_1 to T_2 and volume changes from V_1 to V_2.
   (c) Derive an expression for the calculation of the entropy change of an ideal gas when the temperature changes from T_1 to T_2 and the pressure changes from P_1 to P_2.
10. (a) State and explain (at least in two forms) the second law of thermodynamics.
    (b) One mole of an ideal gas at 300 K expands reversibly from 3×10^{-2} m³ to 5×10^{-2} m³. Calculate the entropy change for the gas ($R = 1.987$ cal).
    (c) Predict the magnitude of ΔS in the following:
       (i) $H_2(g) + Cl_2(g) + aq \longrightarrow 2\,HCl\,(aq)$ (ii) $H_2O\,(l) \xrightarrow{\Delta} H_2O\,(g)$
       (iii) $N_2(g) + 3H_2(g) \longrightarrow 2NH_3(g)$
       Justify your answer with suitable explanations.
11. (a) Rank the following in order of increasing entropy:
       (i) 1 mole of H_2O (e) at 25 °C and 1 atm. pressure
       (ii) 2 mole of H_2O (s) at 0 °C and 1 atm. pressure
       (iii) 1 mole of H_2O (v) at 100 °C and 1 atm. pressure
       (iv) 1 mole of H_2O (e) at 0 °C and 1 atm. pressure
    (b) An endothermic reaction D $\longrightarrow$ K proceeds virtually to the end at a certain temperature T. Determine:
       (i) Sign of S for this reaction
       (ii) Sign of G for the reaction: K $\longrightarrow$ D at T
       (iii) The possibility of reaction: K $\longrightarrow$ D proceeding at a low temperature.
12. Prove that

    (a) $\Delta S = nc_v \ln\frac{T_2}{T_1} + nR\ln\frac{V_2}{V_1}$ (b) $\Delta S = c_p \ln\frac{T_2}{T_1} - R\ln\frac{P_2}{P_1}$

13. (a) Predict with justification the magnitude of S for the following processes:
    (i) Protein (helical) $\longrightarrow$ Protein (random coil) (ii) Dry ice $\longrightarrow$ CO_2 (g)
    (iii) Normal egg $\longrightarrow$ Boil egg
    (b) Although the molar mass of He ($4\ g\ mol^{-1}$) is greater than the molar mass of H_2 ($2\ g\ mol^{-1}$), yet the entropy of H_2 is greater than the entropy of He. Explain.
14. Define entropy. Write its unit. Give its mathematical treatment. Why the efficiency of an engine is always less than unity? What is its physical significance?
15. (a) State and explain second law of thermodynamics. What are its advantages?
    (b) Write various applications of clapeyron-clausius equations.
    (c) Define the following:
    (i) Heat of neutralisation
    (ii) Heat of formation

MULTIPLE CHOICE QUESTIONS

1. The enthalpy of formation of KNO_3 is + 36.5 kcal. In this reaction
   (a) energy is liberated
   (b) energy is absorbed
   (c) KNO_3 is stable
   (d) KNO_3 is unstable
2. For a process to be spontaneous,
   (a) ΔG must be –ve
   (b) ΔG should be +ve
   (c) ΔH must be –ve
   (d) ΔS must be –ve
3. For a reaction to occur spontaneously,
   (a) $(\Delta H - T\Delta S)$ must be –ve
   (b) $(\Delta H + T\Delta S)$ must be –ve
   (c) ΔH must be –ve
   (d) ΔS must be –ve
4. The heat of formation of the compound in the following reaction is:
   $H_2(g) + Cl_2(g) \rightarrow 2HCl(g) + 44$ kcal
   (a) $-44\ cal\ mol^{-1}$
   (b) $-22\ kcal\ mol^{-1}$
   (c) $11\ kcal\ mol^{-1}$
   (d) $-88\ kcal\ mol^{-1}$
5. Which relation is correct?
   (a) $\Delta H = \Delta G + T\Delta S$
   (b) $\Delta G = \Delta H + T\Delta S$
   (c) $\Delta G = T\Delta S - \Delta H$
   (d) $\Delta G = \Delta H - S\Delta T$
6. The free energy change, $\Delta G = 0$ when
   (a) the reactants are completely consumed
   (b) a catalyst is added
   (c) the system is at equilibrium
   (d) the reactants are initially mixed
7. The total change in entropy of a system and surrounding for an irreversible process is always
   (a) zero
   (b) negative
   (c) positive
   (d) not known
8. Which of the following is the correct equation?
   (a) $\Delta U = Q - W$
   (b) $\Delta W = \Delta U + \Delta Q$
   (c) $\Delta U = \Delta W + \Delta Q$
   (d) none of these
9. The heat of neutralization of 0.5 mole of HNO_3 with 0.5 mole of NaOH is
   (a) 57.0 kJ
   (b) 34.9 kJ
   (c) 28.5 kJ
   (d) 11.4 kJ
10. The enthalpy change of a reaction does not depend on
    (a) the state of reactants and products
    (b) the nature of reactants and products

(c) different intermediate reaction
(d) the initial and final enthalpy change of a reaction

11. $C + \frac{1}{2} O_2 \rightarrow CO; \Delta H = -42$ kcal

$C + \frac{1}{2} O_2 \rightarrow CO; \Delta H = -42$ kcal. The heat of formation of CO_2
(a) −16 kcal
(b) +66 kcal
(c) +16 kcal
(d) −66 kcal

12. The heat of combustion is
(a) always –ve
(b) +ve
(c) Both (a) and (b)
(d) 0

13. $S + O_2 \rightarrow SO_2 + 2x$ kcal
$SO_2 + 1/2\, O_2 \rightarrow SO_3 + y$ kcal (find out the heat of formation of SO_3)
(a) $(2x - y)$
(b) $(2x + y)$
(c) $(x + y)$
(d) $2x/y$

14. First law of thermodynamics is represented by the equation
(a) $\Delta E = Q + W$
(b) $\Delta E = Q - W$
(c) $W = Q + \Delta E$
(d) All of these

15. If 0.75 mole of an ideal gas expands isothermally at 27°C from 15 litres, the maximum work obtained is
(a) 8.40 J
(b) 9.43 J
(c) 9.96 J
(d) 10.43 J

16. In which of the following cases, reaction is spontaneous at all temperature
(a) $\Delta H > 0, \Delta S > 0$
(b) $\Delta H < 0, \Delta S > 0$
(c) $\Delta H < 0, \Delta S < 0$
(d) $\Delta H < 0, \Delta S = 0$

17. If a gas expands at constant temperature and pressure, then its
(a) internal energy decreases
(b) internal energy remains the same
(c) internal energy increases
(d) entropy increases and then decreases

18. One mole of ideal gas at 300 K is expanded isothermally from an initial volume of 1 litre to 10 litres. The change in energy for this process is ($R = 2$ cal mol^{-1} K^{-1})
(a) 163.7 cal
(b) 850.2 cal
(c) 1381.1 cal
(d) zero

19. The enthalpy of vaporization of water is 386 kJ; what is entropy of vaporization of water?
(a) 0.5 kJ
(b) 1.03 kJ
(c) 22.05 kJ
(d) 1.5 kJ

20. The enthalpy change of reaction does not depend on
(a) the state of reactants and products
(b) the nature of reactants and products
(c) different intermediate reaction
(d) the initial and final enthalpies of a reaction

ANSWERS

1. (b)	6. (c)	11. (d)	16. (b)
2. (a)	7. (c)	12. (a)	17. (b)
3. (a)	8. (a)	13. (b)	18. (d)
4. (b)	9. (c)	14. (a)	19. (b)
5. (a)	10. (c)	15. (b)	20. (c)

Adsorption and Catalysis

5

CHAPTER OBJECTIVES

5.1 Adsorption

5.2 Types of Adsorption

5.3 Factors Affecting Adsorption of Gases on Solids

5.4 Adsorption Isobar (Effect of Temperature on Adsorption)

5.5 Adsorption Isotherm (Effect of Pressure)

5.6 Theory of Adsorption

5.7 Gibbs' Adsorption Equation

5.8 Applications of Gibbs' Adsorption Equation

5.9 Equation for Multi-layer Adsorption (B.E.T. Equation)

5.10 Catalysis

5.11 Homogeneous and Heterogeneous Catalyses

5.12 How does a Catalyst Work?

5.13 Mechanism of Homogeneous and Heterogeneous Catalyses

Adsorption finds extensive applications both in research laboratories and in industries. A few applications are: (i) production of high vacuum, (ii) in gas masks, (iii) for desiccation or dehumidification, (iv) removal of colouring matter from solution, (v) heterogeneous catalysis, (vi) separation of inert gases, (vii) softening of hard water, (viii) deionization of water, (ix) in curing diseases, (x) cleaning agents, (xi) froth floatation process, (xii) in adsorption indicators, (xiii) chromatographic analysis and (xiv) in dyeing.

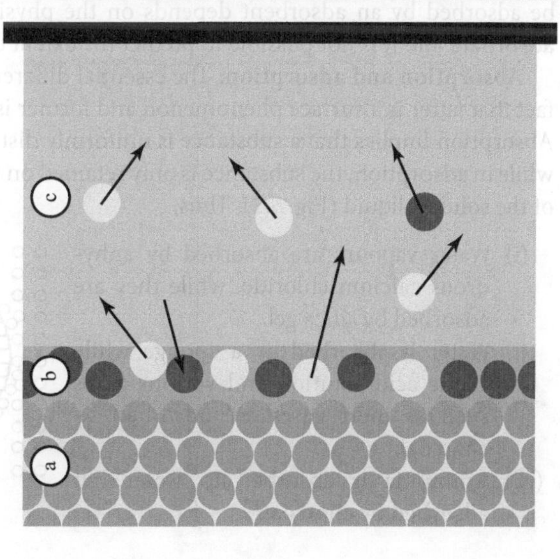

5.1 ADSORPTION

It is a well-known fact that molecules at the surface of a liquid and also of a solid are under strain and are experiencing a net inward pull due to the attractive forces of molecules in the bulk. As a result of these residual forces, the surface of a solid or a liquid tends to attract and retain the molecules of gases and other substances with which they come in contact. This results in higher concentration of molecules on the surface of a solid or a liquid than in the bulk of the medium. This phenomenon is known as adsorption. Thus, the phenomenon of the increase in concentration at a surface brought about by the action of molecular surface forces is called adsorption.

The substance upon whose surface the change of concentration occurs is called the adsorbent. The substance which is taken up on the surface is called adsorbate. The common surface between the two phases where the adsorbed molecules concentrate is called the interface. Figure 5.1 shows the phenomenon of adsorption.

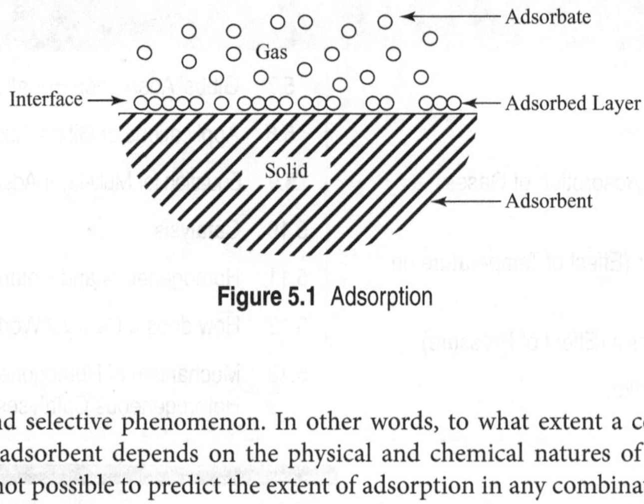

Figure 5.1 Adsorption

It is a specific and selective phenomenon. In other words, to what extent a certain substance will be adsorbed by an adsorbent depends on the physical and chemical natures of both adsorbent and **adsorbate** and it is not possible to predict the extent of adsorption in any combination.

Absorption and adsorption: The essential difference between absorption and adsorption lies in the fact that latter is a surface phenomenon and former is concerned with the whole mass of the absorbent. Absorption implies that a substance is uniformly distributed throughout the body of a solid or a liquid, while in adsorption, the substance is only retained on the surface but does not go into the bulk or interior of the solid or liquid (Fig. 5.2). Thus,

(i) Water vapours are absorbed by anhydrous calcium chloride, while they are adsorbed by silica gel.
(ii) Water is absorbed by a sponge while acetic acid solution and various gases such as inert gases are adsorbed by charcoal.
(iii) Ammonia is absorbed in water but adsorbed by charcoal.

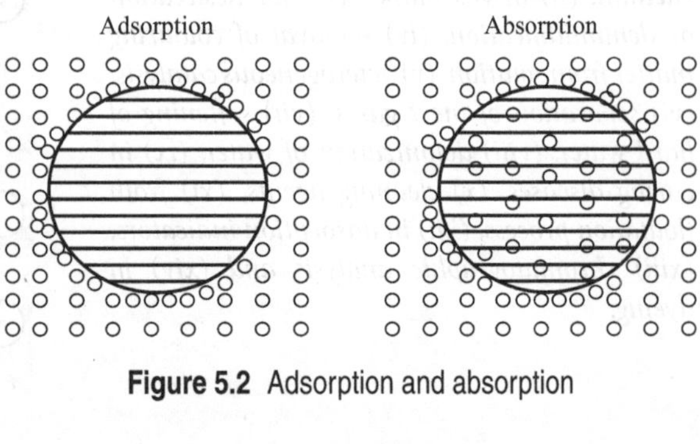

Figure 5.2 Adsorption and absorption

Both absorption and adsorption often take place side by side. It is thus difficult to distinguish the two processes experimentally. M.C. Bain introduced the general term sorption to include both the processes.

5.2 TYPES OF ADSORPTION

Adsorption normally means the increase in the concentration of adsorbate more at the surface as compared to its concentration in the bulk phase. This is called positive adsorption. Sometimes, in case of liquids, it is seen that the concentration of the solute is less at the surface layer than in the bulk part of the solution. This is also a case of adsorption due to difference in concentration. Such an adsorption is called negative adsorption.

Adsorption may be classified according to the nature of binding forces. Thus,

(i) **Physical adsorption:** Here, the adsorbate is held on the surface by van der Waals' forces. It is also called **physiosorption or van der Waals' adsorption.** This type of adsorption is temporary and can be reduced by increasing the temperature or reducing the pressure of the gas.

(ii) **Chemical adsorption:** Here, the adsorbate is held by forces which are strong as in usual chemical bonding. It is also called **chemisorption or Langmuir adsorption.** Compared to physical adsorption, it is more permanent and may or may not be decreased by increasing the temperature or reducing the pressure.

Adsorption is accompanied by the evolution of heat (exothermic). This is known as **heat of adsorption or enthalpy** (ΔH) and is equal to the heat evolved for the adsorption of one mole of adsorbate on the surface of adsorbent. Heat of adsorption for chemisorption is more than that of physiosorption (Table 5.1).

Table 5.1 Comparison of physical and chemical adsorptions.

Physical Adsorption	Chemical Adsorption
1. The forces responsible for the physical adsorption are very weak and they are known as van der Waals' forces.	1. The forces responsible for the chemical adsorption are quite strong and they are known as chemical forces.
2. It occurs at low temperature.	2. It occurs at high temperature.
3. Heat of adsorption is low and is in the range of 20–40 kJ mol^{-1}.	3. Heat of adsorption is high and is in the range of 40–400 kJ mol^{-1}.
4. It is not specific in nature and occurs in case of more easily liquefiable gases.	4. It is highly specific and occurs in the gases which can interact chemically with the adsorbent.
5. It is a reversible process.	5. It is an irreversible process.
6. In this, multi-layer adsorption occurs and thus adsorbed layer is several molecules thick.	6. In this, single-layer adsorption occurs.
7. No appreciable activation energy is involved in this process.	7. Activation energy may be involved in this adsorption process.
8. The rate of adsorption increases with the increase of pressure of the adsorbate.	8. The rate of adsorption decreases with the increase of pressure of the adsorption.
9. The amount of adsorption on a surface of an adsorbent is more a function of the adsorbate than the adsorbent.	9. The amount of adsorption is characteristic of both adsorbate and adsorbent.

MEMORY FOCUS

1. Molecules at the surface of a solid, a metal or a liquid experience a net inward force of attraction with free valencies.
2. The phenomenon of attracting and retaining the molecules of substance on the surface of a solid or a liquid resulting in the higher concentration of the molecules on the surface is called **adsorption**.
3. The substance adsorbed on the surface is called **adsorbate** and the substance on which it is adsorbed is called **adsorbent**.
4. The reverse process, i.e., the removal of adsorbed substance from the surface is called **desorption**.
5. The adsorption of gases on the surface of metals is called **occlusion**.
6. The term **sorption** is employed when adsorption and absorption take place simultaneously.
7. Adsorption is a **surface phenomenon,** whereas absorption is a **bulk phenomenon**. Adsorption occurs only at the surface of adsorbent, whereas absorption occurs throughout the body of the material.
8. When the concentration of the adsorbate is more on the surface of the adsorbent than in the bulk, it is called **positive adsorption**.
9. When the concentration of the adsorbate is less relative to its concentration in the bulk, it is called **negative adsorption**.
10. When a gas is adsorbed at the surface of a solid by weak forces (van der Waals' forces), it is called **physical adsorption**.
11. When a gas is held on the surface of a solid by forces similar to those of a chemical bond, it is called chemical adsorption or **chemisorption**.
12. Adsorption is accompanied by the evolution of heat. The amount of heat evolved when one mole of a gas is adsorbed on a solid, it is known as **molar heat of adsorption**. Its magnitude depends upon the nature of the gas.

5.3 FACTORS AFFECTING ADSORPTION OF GASES ON SOLIDS

Adsorption is accompanied by the decrease in entropy of the system, as the adsorbate changes from more random (gaseous or solution) state to less random state on the surface of the solid adsorbent. Thus, $T\Delta S$ is negative. Adsorption is thus accompanied by decrease in enthalpy (as ΔH is negative in

Learning Plus

Preferential Adsorption

Whenever a mixture of gases is *allowed to come in contact with a particular adsorbent under the* same *conditions, the more strongly adsorbable adsorbate is adsorbed to a greater extent irrespective of its amount.* For example, air, besides moisture, contains a number of gases such as O_2 and N_2, yet moisture is adsorbed more strongly on silica than the other gases of the air. This shows that different adsorbates are preferentially adsorbed.

It is also observed that *a preferentially adsorable adsorbate can displace a weakly adsorbed substance from the surface of the adsorbent.* For example, charcoal used in the mask contains O_2, N_2 etc. of the air already adsorbed in it. However, when it comes in contact with the poisonous gases which are preferentially adsorbable on charcoal, O_2, N_2 etc. are displaced by these gases.

exothermic process) and entropy of the system. The free energy change (ΔG) for a spontaneous process is given by the thermodynamic equation $\Delta G = \Delta H - T\Delta S$ and should be negative.

Since both ΔH and ΔS are negative, it is clear that ΔH should have a higher negative value, so that the net value of ΔG is negative. This is actually so. However, experiments show that the heat of adsorption goes on decreasing or becomes less negative as adsorption proceeds further. Ultimately, the values of ΔH and $T\Delta S$ become equal and ΔG become zero. Thus is the stage when adsorption stops and equilibrium is attained.

5.4 ADSORPTION ISOBAR (EFFECT OF TEMPERATURE ON ADSORPTION)

A graph drawn between the amount of substance absorbed per gram (x/m) and the temperature (T) at a constant pressure is called adsorption isobar. The adsorption isobars for physical adsorption and chemisorption are shown in Fig. 5.3. These are different from each other. While a physical adsorption isobar shows a decrease in x/m when the temperature raises, the isobar of chemisorption shows an increase in the beginning and then a decrease when the temperature increases. The initial increase is due to the fact that, like chemical reactions, chemisorption also requires activation energy.

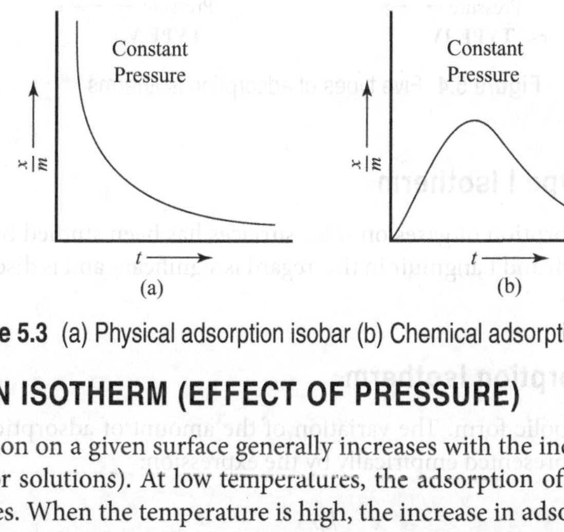

Figure 5.3 (a) Physical adsorption isobar (b) Chemical adsorption isobar

5.5 ADSORPTION ISOTHERM (EFFECT OF PRESSURE)

The extent of adsorption on a given surface generally increases with the increase in pressure (for gases) and concentration (for solutions). At low temperatures, the adsorption of a gas increases very rapidly when the pressure rises. When the temperature is high, the increase in adsorption is relatively less.

The relation between the amount adsorbed by an adsorbent and the equilibrium pressure (or concentration for solutions) of the adsorbate at a constant temperature is called the adsorption isotherm. Five different types of isotherms have been observed in the adsorption of gases on solids.

These isotherms depend upon the nature of the absorbent and the adsorbate. The five types of adsorption isotherms are shown in Fig. 5.4. In these plots, $p°$ represents the minimum pressure at which the gas condenses to liquid state and further study is not possible.

Type I. Adsorption of nitrogen on charcoal at 90 K.
Type II. Adsorption of nitrogen on iron gel at 78 K.
Type III. Adsorption of iodine on silica gel at 356 K.
Type IV. Adsorption of benzene vapours on ferric oxide gel at 323 K.
Type V. Adsorption of water vapours on charcoal at 373 K.

All the five types of isotherms are observed in the case of physical adsorption, while one Type I curve is observed in the case of chemisorption.

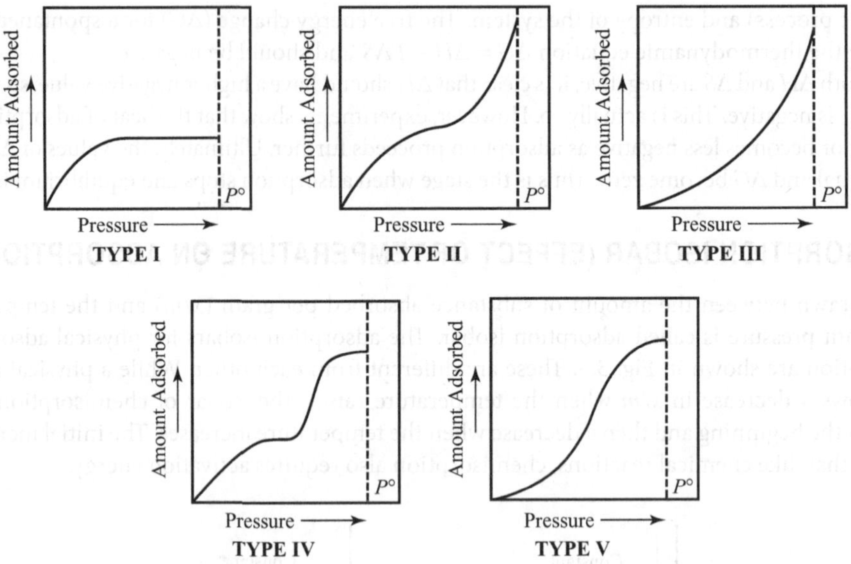

Figure 5.4 Five types of adsorption isotherms

5.5.1 Explanation of Type I Isotherm

The phenomenon of chemisorption of gases on solid surfaces has been studied by various workers, but the contribution of Freundlich and Langmuir in this regard is significant and is discussed in the following sections.

5.5.2 Freundlich Adsorption Isotherm

The Type 1 curve is of parabolic form. The variation of the amount of adsorption per unit mass with pressure for gases is often represented empirically by the expression:

$$\frac{x}{m} = \frac{kp^1}{n} \tag{5.1}$$

where x g of gas are absorbed by m g of adsorbent at pressure p, so that x/m denotes the amount of gas adsorbed by unit mass of the adsorbent, n and k are constants depending upon the nature of adsorbate and absorbent, and temperature, n is always less than unity, which means that the amount of absorption does not rise as rapidly as the pressure. Taking logarithms of both sides, Eq. (5.1) becomes

$$\log\frac{x}{m} = \log k + \frac{1}{n}\log p \tag{5.2}$$

This relation (5.1) was extensively used by Freundlich in the measurements of adsorption and is often called **Freundlich adsorption isotherm**.

A graphical representation of log x/m against log p would, therefore, give a straight line with slope equal to '1/n' and ordinate intercept equal to log k (Fig. 5.5). The experimental values, when plotted, however, show some deviation from linearity, especially at high pressures. This relation is hence considered as an approximate one and is suitable at lower pressures.

This shows that this equation does not have general applicability in reproducing adsorption of gases by solids.

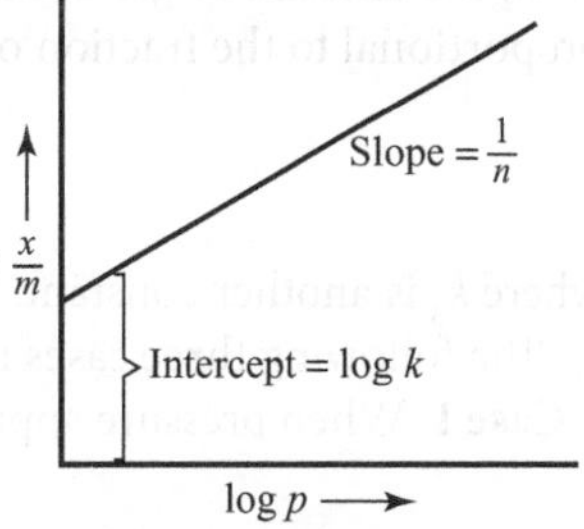

Figure 5.5 Freundlich adsorption isotherm. A plot between log *x*/*m* and log *p*

5.5.3 The Langmuir Adsorption Isotherm

A model for the chemisorption process was presented by Langmuir in 1916. The main points of his model are as follows:

(i) The chemisorption leads to the formation of a monomolecular layer over the surface of the adsorbent and an equilibrium is set up between the gas phase and partially formed monolayer.

(ii) The equilibrium state can be interpreted in terms of the dynamic equilibrium that results from an equal rate of evaporation of the adsorbed material and the rate of condensation of the gas-phase molecules.

(iii) The rate of evaporation can be taken to be proportional to the fraction of the surface covered and can be expressed, therefore, as $k_1\theta$, where k_1 is some proportionality constant and θ is the fraction of the surface that is covered.

(iv) The rate of condensation is taken to be proportional to the gas pressure P, which according to the kinetic molecular theory determines the number of molecular collisions per unit area, per unit time, and to the fraction of the bare surface, i.e., not already covered by adsorbed molecules, i.e., to $1-\theta$.

Rate of condensation $\propto (1-\theta)P$

$$= k_2(1-\theta)P$$

where k_2 is constant of proportionality. Equating the expressions deduced for the rate of evaporation and the rate of condensation

$$k_1\theta = k_2P(1-\theta) \tag{5.3}$$

Rearranging Eq. (5.3), we get

$$\theta = \frac{k_2P}{k_1 + k_2P} \tag{5.4}$$

Reversing

$$\frac{1}{\theta} = \frac{k_1 + k_2P}{k_2P}$$

$$= \frac{k_1}{k_2P} + 1$$

Introduction of $a = k_2 / k_1$ allows this result to be expressed as

$$\theta = \frac{aP}{1 + aP} \tag{5.5}$$

Now, the amount of gas adsorbed per unit area or per unit mass of adsorbent, x/m, must obviously be proportional to the fraction of surface covered and hence

$$\frac{x}{m} = k_3\theta = \frac{k_3 aP}{1+aP} \tag{5.6}$$

where k_3 is another constant.

The following three cases may arise:

Case I When pressure approaches zero (P→40), the expression (5.6) becomes:

$$\frac{x}{m} = \frac{k_3 aP}{1}$$

or
$$\frac{x}{m} \propto P \tag{5.7}$$

or the magnitude of adsorption is directly proportional to pressure.

Case II When pressure is very large, 1 can be neglected as compared to aP and Eq. (5.6) becomes:

$$\frac{x}{m} = k_3 = k_3\,(P^{\circ}) \tag{5.8}$$

or the magnitude of adsorption is independent of pressure or depends on its zeroth power.

Case III At intermediate range of pressure, the magnitude of adsorption depends on pressure to a power intermediate between zero and one or

$$\frac{x}{m} \propto P^{1/n} \tag{5.9}$$

or
$$\frac{x}{m} = k_3 P^{1/n}$$

The expression is identical with the Freundlich equation and this lends proof for the unimolecular theory of Langmuir.

5.5.4 Verification

Equation (5.6) may also be described as

$$\frac{x}{m} = \frac{bP}{1+aP} \tag{5.10}$$

where the constant b has been written for the product $k_3 a$.

Equation (5.10) is the Langmuir adsorption isotherm. The constants a and b depend upon the system under study and are calculated from experimental data. Their magnitude also depends upon the temperature.

At anyone temperature, the Langmuir isotherm can be verified by first dividing Eq. (5.10) by P on both sides and the then taking the reciprocals. The result is

$$\frac{P}{x/m} \quad \text{or} \quad \frac{P}{y} = \frac{1}{b} + \left(\frac{a}{b}\right)P \tag{5.11}$$

where
$$\frac{x}{m} = y$$

Adsorption Isostere

We know that adsorption increases with increase of pressure and decreases with increase of temperature. It means that by increasing the temperature and decreasing the pressure to certain values, we can keep the amount of adsorption constant, it means that for the same amount of adsorption, the plot between pressure and temperature will be linear. Such a plot of temperature versus pressure, keeping the amount of adsorption constant, is called adsorption isostere.

Learning Plus: Since a and b are constants, a plot of P/y vs. P should give a straight line with slope equal to a/b and an ordinate intercept, equal to $1/b$. Figure 5.6 gives such a plot. The excellent straight line confirms the Langmuir adsorption equation and shows its superiority to the Freundlich isotherm. This confirms the assumption that adsorbent is covered only by a unimolecular layer of gas molecules.

Limitation of Langmuir's theory: Langmuir's theory of unimolecular adsorption is valid only at comparatively low pressure and high temperatures. As the pressure is increased or temperature is lowered, additional layers are formed. This has led to the modern concept of multi-layer adsorption.

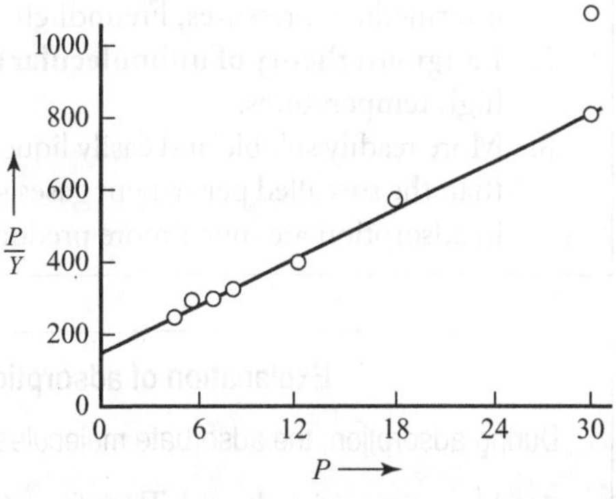

Figure 5.6 Application of Langmuir equation to adsorption of N² on mica at 90 K

5.5.5 Explanation of Type II and III Isotherms

Although Langmuir adsorption isotherm can satisfactorily explain Type I isotherm, it altogether fails to explain all the five types, which was proposed that in Types II and III, multi-layer formation takes place especially when the gas pressure is increased.

5.5.6 Explanation of Type IV and V Isotherms

To explain Type IV and V isotherms, it was suggested that in addition to the multi-layer adsorption, condensation of the gas molecules also takes place in the small pores and capillaries of the adsorbent even at pressures below $p°_L$. The distinction between these two types is again based on the relative magnitudes E_1 and E_L. If $E_1 > E_L$, Type IV isotherms are observed and if $E_1 < E_L$, Type V isotherms are observed.

MEMORY FOCUS

1. The magnitude of gaseous adsorption depends upon temperature, pressure, nature of the gas, and the nature of adsorbent.
2. Adsorption decreases with the increase in temperature, since it is accompanied by evaluation of heat.
3. The adsorption increases with the increase in pressure, since the adsorption of gas leads to the decrease in pressure.
4. The variation of adsorption with pressure at a constant temperature is called **isotherm**.

(Continued)

MEMORY FOCUS (*Continued*)

5. The relationship between the magnitude of adsorption and pressure is known as **Freundlich adsorption isotherm,** which can be expressed as $\frac{x}{m} = kp^1/n$.
6. At low pressure, the amount of the gas adsorbed per unit quantity of adsorbent is proportional to the pressure. At high pressure, the amount of adsorbed gas is independent of pressure. At intermediate pressures, Freundlich adsorption isotherm is expected to hold.
7. **Langmuir theory of unimolecular adsorption** is valid only at comparatively low pressures and high temperatures.
8. More readily soluble and easily liquefiable gases such HCl, Cl_2, SO_2 and NH_3 are adsorbed more than the so-called permanent gases such as H_2,O_2 and N_2 because van der Waals' forces involved in adsorption are much more predominant in the former gases than the latter category of gases.

Explanation of adsorption in terms of Gibbs–Helmholtz equation

During adsorption, the adsorbate molecules are attracted on the surface and as a result:

1. Heat energy is released. Therefore, ΔH is always negative and favours the process of adsorption.
2. The process of adsorption is accompanied by the decrease in the entropy of the system, i.e., ΔS is negative and opposes the process of adsorption.

   In spite of the two opposing factors, the adsorption does take place. According to Gibbs–Helmholtz equation, $\Delta G = \Delta H - T\Delta S$.

   Adsorption can take place only if ΔG has negative value which is possible if ΔH is greater than $T\Delta S$. Thus, adsorption occurs as long as ΔH is greater than $T\Delta S$ and the process of adsorption stops as soon as $\Delta H = T\Delta S$, so that $\Delta G = 0$. At this stage, adsorption equilibrium is said to be attained.

5.6 THEORY OF ADSORPTION

Adsorption is basically a surface phenomenon and is governed by the amount of surface made available by the adsorbent. It means that adsorption depends upon the interfacial area. Whenever the concentration of a particular substance in the interfacial layers between two different phases is greater than in the bulk of either phase, the substance is said to be adsorbed at the interface. The adsorption of gases and solutions on solid surfaces is well known. For example, moisture is adsorbed on the surface of glass, porcelain and organic substances are decolourized due to the adsorption of colouring matter on active charcoal. The adsorption of solutes from solutions on the surface of solids occurs in the same way as that of gases. The process of adsorption is governed by two factors, which are listed below.

(i) Decrease in the interfacial tension leads to the adsorption of the adsorbate. According to Gibbs' adsorption theorem, those substances, which lower the interfacial tension, tend to become concentrated at the interface. Therefore, solutes which lower the interfacial tension, between a solid and a solvent, will be concentrated at the surface of the solid and adsorbed on it.

(ii) The effect of solute and the surface tension of the solvent in which they are dissolved have been studied by Gibbs. It has been observed that those solutes which lower the surface tension of solvent have a greater concentration in the surface layer. Many organic solutes such as carboxylic acids, phenols and amines, which lower the surface tension of water, have a greater concentration in the interfacial layer than in the bulk. These solutes, which increase the surface tension, have a lower concentration in the surface layer. Some inorganic salts, which increase the surface tension of water, have smaller concentration in the interfacial layer than in the bulk. Gibbs derived the following equation for the adsorption of gases by solids.

$$-S = \frac{p}{RT}\frac{d\gamma}{dp} \tag{5.12}$$

where S is excess of solute in the surface layer when compared to the bulk, p is the partial pressure of the gas, γ is the surface tension, T is the absolute temperature and R is the gas constant, and $d\gamma/dp$ is the variation of surface tension with pressure.

The corresponding equation for the adsorption from solutions on solid surfaces would be

$$-S = \frac{c}{RT}\frac{d\gamma}{dp} \tag{5.13}$$

where c is concentration of solute in solution and other terms have same significance. The variation of surface tension with concentration can be obtained by plotting γ against c and finding the slope of the curve at the required point. Equations (5.12) and (5.13) are referred to as Gibbs' adsorption law.

Some important conclusions drawn from the studies on adsorption are summarized as:

(i) The amount of solute absorbed by 1 g of the adsorbent (x/m) depends on the concentration of the solute in the solution and it increases with this concentration.

(ii) This adsorption is reversible and can be decreased by decreasing the solute concentration in the solution or can be washed out by a prolonged washing, provided that the solute has not suffered a chemical change through adsorption.

(iii) Freundlich adsorption isotherm equation is applicable to the adsorption of solutes from solutions on solid surface (refer Section 5.4).

(iv) A given substance is more strongly adsorbed from solvents in which it is less soluble. Picric acid is strongly adsorbed on charcoal from aqueous solutions, but not from alcoholic solutions. Alcohol can elute out picric acid and from charcoal after its adsorption from the aqueous solution.

(v) If we have more than one solute in a solution, these will compete for the surface of the adsorbent. In fact, more strongly adsorbable substances can displace less strongly adsorbable substances from the surface of an adsorbent. Thus, quinine sulphate, a more strongly adsorbable substance, was used to elute out vitamin B from the surface of fuller's earth.

5.7 GIBBS' ADSORPTION EQUATION

We know that the adsorptions of solutes from solutions on solids are dependent upon the change of surface tension due to the dissolution of solute in a solvent. A mathematical relation between the amount of adsorption and the change in surface tension was derived by J.W. Gibbs which is referred to as Gibbs' adsorption equation. The derivation is based upon an important concept in thermodynamics, namely chemical potential, which is the partial molar free energy change of any system.

It is defined as the change in free energy produced on the addition of one mole of the substance at constant temperature to a large bulk of the mixture, so that the composition of the mixture does not change.

For a pure substance or a mixture of substances having constant composition, the variation in free energy is expressed by the following fundamental equation:

$$dG = VdP - S\,dT \tag{5.14}$$

If temperature is kept constant, i.e., $dT = 0$, we get

$$(dG)_T = (VdP)_T$$

or

$$\left(\frac{\partial G}{\partial P}\right)_T = V \tag{5.15}$$

If pressure is kept constant, i.e., $dP = 0$, we get

$$(dG)p = -\ (SdT)p$$

or

$$\left(\frac{\partial G}{\partial P}\right)_P = -S \tag{5.16}$$

If, however, the composition of the mixture is variable, then free energy (G) becomes a function of n_1, n_2, n_3 etc. the number of moles of the components besides being a function of T and P. Thus, for a many component system,

$$G = f\,(T, P, n_1, n_2, n_3 \ldots n_u)$$

and its total differential is given by

$$dG = \left(\frac{\partial G}{\partial P}\right)_{P,n_1,n_2} dT + \left(\frac{\partial G}{\partial P}\right)_{T,n_1,n_2} dP + \left(\frac{\partial G}{\partial P}\right)_{T,P,n_2} dn_1 + \left(\frac{\partial G}{\partial P}\right)_{T,P,n_1} dn_2 \tag{5.17}$$

But

$$\left(\frac{\partial G}{\partial P}\right)_{T,P,n_2} = \mu_1 \quad \text{and} \quad \left(\frac{\partial G}{\partial P}\right)_{T,P,n_1} = \mu_2$$

$$\therefore \quad dG = \left(\frac{\partial G}{\partial P}\right)_{P,n_1,n_2} dT + \left(\frac{\partial G}{\partial P}\right)_{T,n_1,n_2} dP + \mu_1 dn_1 + \mu_2 dn_2 \tag{5.18}$$

Substituting the values from Eqs (5.15) × (5.16) in Eq. (5.19), we get

$$dG = -SdT + VdP + \mu_1 dn_1 + \mu_2 dn_2 \tag{5.19}$$

The increase in free energy due to the increase in the surface area should also be taken into account. If dA is the increase in surface area and γ is the surface tension, then free energy change due to the increase in surface area becomes γdA.

Therefore, total free energy change would be given as

$$dG = -SdT + VdP + \mu_1 dn_1 + \mu_2 dn_2 + \gamma dA \tag{5.20}$$

For a two-component system consisting of solute and solvent only, suppose n_1 and n_2 represent the number of moles and μ_1 and μ_2 represent their respective chemical potentials, the free energy of the system would be given by,

$$G = \mu_1 n_1 + \mu_2 n_2$$

On mixing the two components, there is a change in the surface area of the solution which results in the change of surface energy of the system. Surface energy is equal to the work done to increase the surface area by 1 cm² and it depends upon the surface tension of the solvent. Suppose γ is the surface tension of the solvent, i.e., the surface energy per square centimetre and A is the surface area, then

$$\text{Surface free energy} = \gamma A$$

Total free energy of the system would include the surface energy also. Thus, we get

$$G = \mu_1 n_1 + \mu_2 n_2 + \gamma A$$

Differentiating this equation completely, we get

$$dG = \mu_1 dn_1 + n_1 d\mu_1 + \mu_2 dn_2 + n_2 d\mu_2 + \gamma dA + A d\gamma \tag{5.21}$$

Comparing Eqs (5.20) and (5.21), we get

$$A d\gamma + n_1 d\mu_1 + n_2 d\mu_2 = -SdT + VdP$$

or

$$SdT - VdP + Ad\gamma + n_1 d\mu_1 + n_2 d\mu_2 = 0$$

However, at constant temperature and pressure

$$dT = 0 \quad \text{and} \quad dP = 0$$

∴

$$A d\gamma + n_1 d\mu_1 + n_2 d\mu_2 = 0 \tag{5.22}$$

Let us imagine the system under consideration to be divided into two parts, one part being the surface phase and the other is the bulk phase. The surface phase is under the influence of surface forces, i.e., surface tension, whereas the phase does not involve such forces. Let n_1 and n_2 be the moles of two components (solvent and solute, respectively) in the surface corresponding to n^0_1 and n^0_2 in the bulk phase. Neglecting the surface forces, Eq. (5.22) for the bulk phase would take the form

$$n_1^0 d\mu_1 + n_2^0 d\mu_2 = 0$$

Multiplying this equation by $\frac{n_1}{n_1^0}$, we get

$$n_1 d\mu_1 + \frac{n_1 n_2^0}{n_1^0} d_2\mu = 0$$

Subtracting the above equation from Eq. (5.22), we get

$$A d\gamma + n_2 d\mu_2 - \frac{n_1 n_2^0}{n_1^0} d\mu_2 = 0$$

$$A d\gamma + \left(n_2 - \frac{n_1 n_2^0}{n_1^0} \right) d\mu_2 = 0$$

$$A\,d\gamma = -\left(n_2 - \frac{n_1 n_2^0}{n_1^0}\right) d\mu_2$$

or

$$-\frac{d\gamma}{d\mu_2} = \left(n_2 - \frac{n_1 n_2^0}{n_1^0}\right) A \qquad (5.23)$$

The expression on the R.H.S. of the equation represents the excess concentration of the solute in the surface phase per unit area of the surface. Here, n_2 is the number of moles associated with n_1 moles of the solvent in the surface phase and the term $n_1 n_2^0 / n_1^0$ is the corresponding number of moles associated within n_1^0 moles of the solvent in the bulk phase. If the factor on the R.H.S. of Eq. (5.10), i.e., excess concentration of solute per unit area of the surface, is represented by S_2, we get

$$-\frac{d\gamma}{d\mu_2} = S_2 \qquad (5.24)$$

We know that chemical potential (μ_2) of component 2 (i.e., solute) is given by,

$$\mu_2 = \mu_2^0 + RT \ln a_2$$

where a_2 is the activity of the solute and μ_2^0 is its chemical potential in the standard state, i.e., its activity is unity. Differentiating at constant temperature, we get

$$d\mu_2 = RT\, d\ln a_2$$

$$d\mu_2 = \frac{RT}{a_2} da_2$$

Substituting the value in Eq. (5.24), we get

$$S_2 = -\frac{d\gamma}{RT da_2} a_2$$

or

$$S_2 = -\frac{a_2}{RT}\frac{d\gamma}{da_2}$$

The above equation has been derived for the component 2 (i.e., solutes), but it is equally applicable to either of the two components of a binary system. Therefore, the equation in its general form can be expressed as

$$S = -\frac{a}{RT}\frac{d\gamma}{da} \qquad (5.25)$$

For dilute solutions, the activity may be replaced by concentration (c) and the above equation is expressed as

$$S = -\frac{C}{RT}\frac{d\gamma}{dc} \qquad (5.26)$$

where $d\gamma/dc$ represents the rate of variation of surface tension of the solution with concentration.

Equations (5.25) × (5.26) are referred to as Gibbs' adsorption equations.

5.8 APPLICATIONS OF GIBBS' ADSORPTION EQUATION

(i) **Predicting the nature of adsorption.** The rate of variation of surface tension of any solution with concentration can be conveniently determined graphically. From the plots of surface tension (γ) vs. concentration (C), the expression $d\gamma/dc$ under the desired conditions can be evaluated. The numerical value of $d\gamma/dc$ will predict the nature (or adsorption for a particular system), i.e., whether it is a case of positive or negative adsorption. When $d\gamma/dc$ comes out to be negative, i.e., when the surface tension of the solution decreases with concentration, S must be positive (from Eq. (5.26)), i.e., the concentration of the solute in the surface be greater than that of the bulk of the solution. This observation predicts that the adsorption would be positive adsorption. Surface-active substances such as soaps and non-electrolytes in aqueous solution exhibit this type of behaviour. Similarly, it can be concluded that when $d\gamma/dc$ is positive, it would be a case of negative adsorption. Many electrolytes in dilute aqueous solutions exhibit this type of behaviour.

(ii) **Calculation of adsorption of a gas (vapour) on the surface of a liquid in which it is not soluble.** Consider the adsorption of vapours of a pure substance, say, saturated hydrocarbon (component 2), on the surface of water (component 1).

The Gibbs' adsorption equation, in its simplest form, i.e., Eq. (5.24), derived above is

$$-\frac{d\gamma}{d\mu_2} = S_2$$

The chemical potential of the vapours (ideal gas) whose pressure is p can be expressed as:

$$\mu_2 = \mu_2^0 + RT\ln p$$

$$\therefore \qquad d\mu_2 = RTd\ln p = \frac{RT}{p}dp$$

Substituting the value of $d\mu$ in Eq. (5.24), we get

$$S_2 = -\frac{d\gamma}{RT\,dp}p$$

$$S_2 = -\frac{p}{RT}\frac{d\gamma}{dp} \tag{5.27}$$

Knowing the variation of surface tension of non-dissolving liquid 1 (water) with the vapour pressure of component 2 (hydrocarbon vapour) on the surface of a liquid. The variation of σ with p can be found out graphically as shown in the case of adsorption of n-pentane on water at 288 K.

Knowing the value of $\partial\gamma/\partial p$ from the plots, the values of Gibbs' adsorption S at different pressures can be calculated. The variation of adsorption with pressure is shown in Fig. 5.7(b). Because of the weak interactions of the non-polar saturated hydrocarbons with water (adsorbate-adsorbent) as compared to high mutual interactions between the molecules of hydrocarbon (adsorbate-adsorbate), the nature of the isotherm obtained is concave.

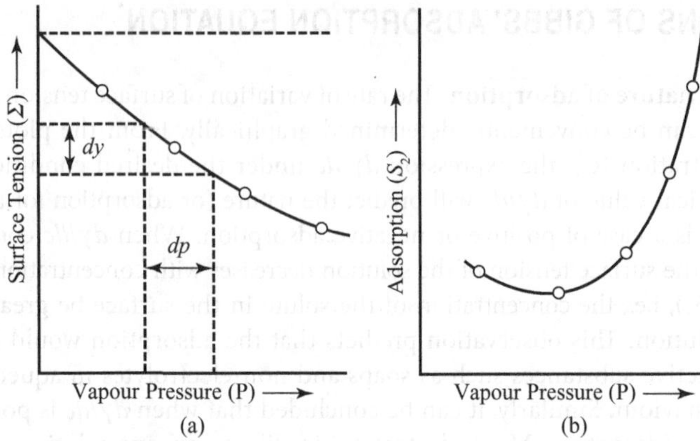

Figure 5.7 (a) Surface tension isotherm (b) Adsorption isotherm of n-Pentane gas on water surface

(iii) **Calculation of adsorption of a substance on the surface of a liquid in which it is soluble.** Let us consider the adsorption of alcohol on water surface, where the second component dissolves in the bulk phase of first component (water). In this case, the chemical potential of water changes. However, it is possible to use Gibbs' equation (5.26).

$$S_2 = -\frac{C_2}{RT}\frac{d\gamma}{dc} \tag{5.28}$$

From this equation, the value of Gibbs' adsorption S_2 can be calculated at different concentrations of c_2.

From the values of Gibbs' adsorption calculated, as explained above, and the variation of these values with the concentration of the solvent in the bulk phase, different substances have been divided into following two categories.

(a) **Substances forming aqueous solutions with positive deviations from Raoult's law:** The molecules of organic substances have non-polar part [a hydrocarbon radical] and a polar part, such as –OH (alcohol), –COOH (acid), and $-NH_2$ (amines), due to these polar parts, these organic molecules can be dissolved in water. The interaction between the molecules of water is greater than the interactions between the molecules of water and the molecules of these organic substances. As a result, the molecules of these substances are pushed to the surface and get adsorped. The concentration of these substances at the surface decreases the molecular interaction in the surface layer, i.e., the surface tension falls with increase in the concentration of e_2 organic substances.

With the increase in the length of carbon chain, the molecules are pushed up with greater force and reduce the surface tension to a greater extent. The limiting value of the reduction in the surface tension with concentration, i.e., the quantity $d\gamma/dc_2$, is called the surface activity. Substances which reduce the surface tension of a solvent are called surface-active substances or surfactants. For these substances, $d\gamma/dc_2 < 0$ and $S_2 > 0$.

Surfactants can concentrate on the surface of water to such an extent that they cover it almost completely. The surface tension is consequently decreased to a considerable extent. For example, the long-chain alcohols and acids arc first flatly oriented on the water surface at low concentration. As the concentration, *c*, increases, the molecules in the surface layer change their orientation and become packed closer with their polar group facing the water and their non-polar hydrocarbon radical being arranged vertically (Fig. 5.8).

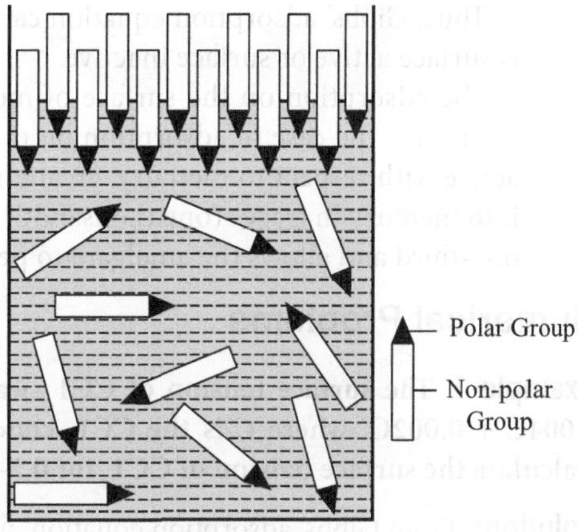

Figure 5.8 Adsorption of surfactant molecules from an aqueous solution

(b) **Substances forming aqueous solution with the negative deviation, from Raoult's law:** Inorganic salts, for example, strong electrolytes, dissolve in water giving ions. The interactions of the ions with the molecules of water are greater than that between the molecules of water themselves. Therefore, the ions are distributed in the bulk of the solution. Here, $d\gamma/dc_2 > 0$ and $S_2 < 0$. Such substances, which increase the surface tension of the solvent and are adsorbed negatively, are called the surface-inactive ones.

The surface-inactive substances change the concentration in the surface layer very slightly, because even the ions getting into the surface layer are highly hydrated. Therefore, these substances increase the surface tension only slightly and their adsorption (negative) is small. Figure 5.9(a) shows the variation of surface tension with concentration and Fig. 5.9(b) shows the adsorption isotherm for aqueous solution of a surfactant (butyric acid) and surface-inactive substance (KNO_3).

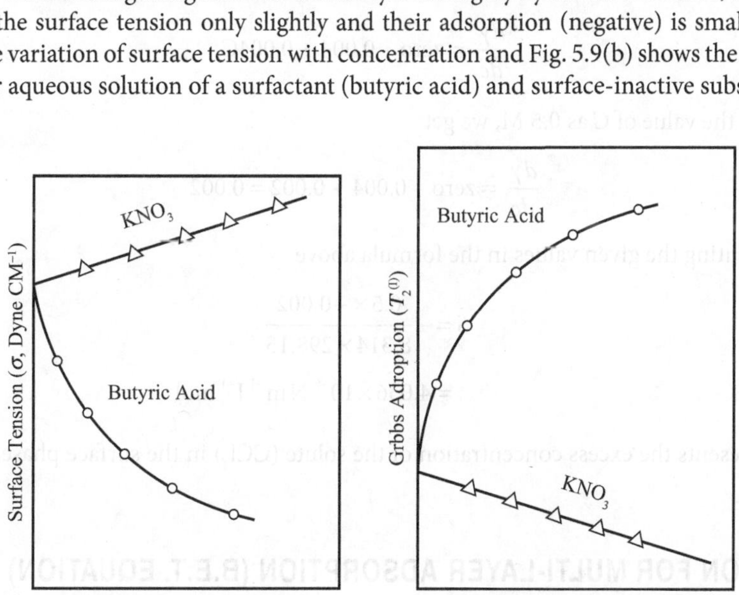

Figure 5.9 (a) Surface tension isotherms (b) Adsorption isotherms of butyric acid and KNO_3

Thus, Gibbs' adsorption equation can be employed to decide whether a particular substance is surface active or surface inactive.

The adsorption on the surface of molten metal melts is of great importance. For example, let us take the case of adsorption on the surface of amalgams. The alkali metals, being surface active with respect to mercury, greatly reduce its surface tension. The introduction of caesium into mercury in traces (one thousandth of an atomic percentage) reduces the surface tension by one-third and allows the amalgam to penetrate into very thin capillaries.

Numerical Problems

Example 1 **The surface tension of $CC1_4$–water mixture follows the equations: $\gamma(Nm^{-1}) = 1.072 - 0.004C + 0.002C^2$ where C is the $CC1_4$ concentration in mol/dm³. The temperature is 298.15 K. Calculate the surface tension of $CC1_4$ for 0.5-M solutions. Mention its merits.**

Solution: From Gibbs' adsorption equation, we have

$$S = -\frac{C}{RT}\frac{d\gamma}{dc}$$

Here, $d\gamma/dc$ represents the rate of variations of surface tension with concentration. This factor can be calculated from the following equation.

$$\gamma = 1.072 - 0.004C + 0.002C^2$$

Differentiating w.r.t. concentration, we get

$$\frac{d\gamma}{dc} = \text{zero} - 0.004 + 0.004C$$

Substituting the value of C as 0.5 M, we get

$$\frac{d\gamma}{dc} = \text{zero} - 0.004 + 0.002 = 0.002$$

Now, substituting the given values in the formula above

$$S = -\frac{0.5 \times -0.002}{8.314 \times 298.15}$$

$$= 4.036 \times 10^{-8}\,\text{N}\,\text{m}^{-1}\,\text{J}^{-1}\,\text{mol}$$

Here, S represents the excess concentration of the solute (CCl_4) in the surface phase per unit area of the surface.

5.9 EQUATION FOR MULTI-LAYER ADSORPTION (B.E.T. EQUATION)

Langmuir's theory of unimolecular adsorption is valid only at low pressure and high temperatures. As the pressure is increased or temperature is lowered, additional layers are formed. This has led to the modern concept of multi-layer adsorption, which would not be explained by Langmuir's theory. Brunauer, Emmett and Teller extended the Langmuir's approach and proposed a theory for multi-layer adsorption. They derived an equation based upon certain assumptions which is called B.E.T. equation. The main postulates on which the multi-layer adsorption theory is based are listed below:

(i) Condensation forces are taken as the main forces in uni- as well as multi-layer adsorptions.
(ii) The rate of condensation, is equal to the rate or evaporation.
(iii) The rate of evaporation from each succeeding layer is equal to the rate of condensation on the preceding layer.
(iv) The heat of adsorption in each layer other than the first layer is taken to be equal to the heat of liquefaction of the bulk adsorbate.
(v) The heat of adsorption is involved exponentially in each of the equilibrium rate expression.

The B.E.T. equation, based on the above assumptions, is given below.

$$\frac{P}{v(p_0+p)}=\frac{1}{v_m C}+\frac{(C-1)}{(v_m C)}\frac{p}{p_0} \tag{5.29}$$

where v and v_m are the volumes (reduced to standard conditions, i.e., STP) at pressure p and temperature T, respectively, when the surface is covered with a monomolecular layer; P_0 is the saturated vapour pressure of the liquefied gas at a given temperature T and C is a constant given by the expression

$$C=\text{Exponential}^{(E_1-E_L)/RT}$$

i.e. $$C=e^{(E_1-E_L)/RT} \tag{5.30}$$

where E_1 is the heat of adsorption of the first layer and E_L is the heat of condensation of the gas to liquid.

Equation (5.29) can be tested by plotting, $\dfrac{p}{v(p_0-p)}$ versus $\dfrac{p}{p^0}$.

The plot should be a straight line with slope $\dfrac{c-1}{v_m C}$ and intercept $\dfrac{1}{v_m C}$ as shown in Fig. 5.10. From the values of the slope and intercept, v_m and C can be calculated.

When p/p^0 in Eq. (5.29) in replaced by f, the equation takes the following form:

$$\frac{f}{v(1-f)}=\left(\frac{1}{V_m C}\right)+\frac{(c-1)f}{V_m C}=\frac{1+(c-1)f}{V_m C}$$

The validity of BET equation can also be tested by plotting $\dfrac{f}{V(1-f)}$ against f. A straight line should be obtained with a slope equal to $\dfrac{(C-1)}{V_m C}$ and the intercept on the Y-axis equal to $\dfrac{1}{V_m C}$.

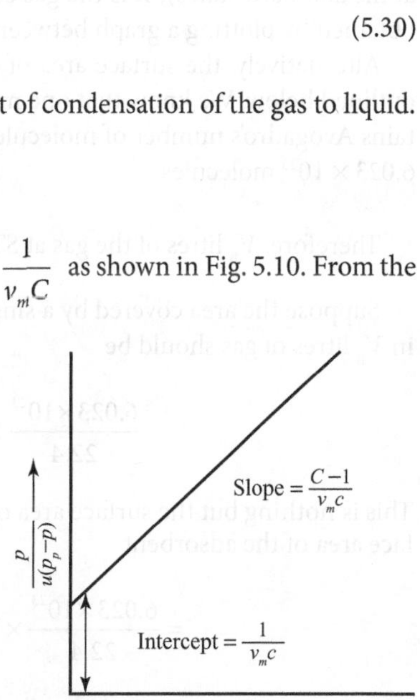

Figure 5.10 Isotherm plot according to BET equation

5.9.1 Determination of Surface Area of the Adsorbent

We know that

$$PV = nRT \qquad \text{[For } n \text{ molcs of an ideal gas]}$$

$$n = \frac{PV}{RT}$$

where P, V, T and R have their usual significance.

Avogadro's number (N) represents the number of molecules in one mole of any material.

Therefore, the number of molecules in n moles $= nN = \frac{PV}{RT}N$

Suppose S represents the area of one molecule surface, then, the surface of n molecules,

$$= \frac{PV}{RT}NS$$

If A represents the area of the solid per definite weight of adsorbent, it will be given as

The total area
$$A = \frac{P_0 V_B}{RT_0} NS \tag{5.31}$$

where A is the area in Angstroms² (Å²), $P^0 = 1$ arms, $T = 273$ K (the values of pressure and temperature at the standard states), R is the gas constant and V_B is the volume at point of inflection (B) in the curve obtained by plotting a graph between Y and P/P^0 as shown in Fig. 5.11.

Alternatively, the surface area of the adsorbent may be calculated directly from the value of VB as outlined below. We know that one mole of any gaseous substance at STP occupies 22.4 litres and it contains Avogadro's number of molecules, i.e., 6.023×10^{28} molecules. 22.4 litres of the gas at STP contains 6.023×10^{23} molecules.

Therefore, V_B litres of the gas at STP will contain $\frac{6.023 \times 10^{23}}{22.4}$ molecules.

Suppose the area covered by a single molecule is S, then the area covered by the molecules contained in V_B litres of gas should be

$$\frac{6.023 \times 10^{23}}{22.4} \times V_B \times S$$

This is nothing but the surface area of the adsorbent. Hence, the surface area of the adsorbent

$$= \frac{6.023 \times 10^{23}}{22.4} \times V_B \times S = \frac{NV_B S}{22.4}$$

where V_B is the volume in litres usually the volume is expressed in millilitre, i.e., c.c., so the expression would be $\frac{NV_B S}{22400}$.

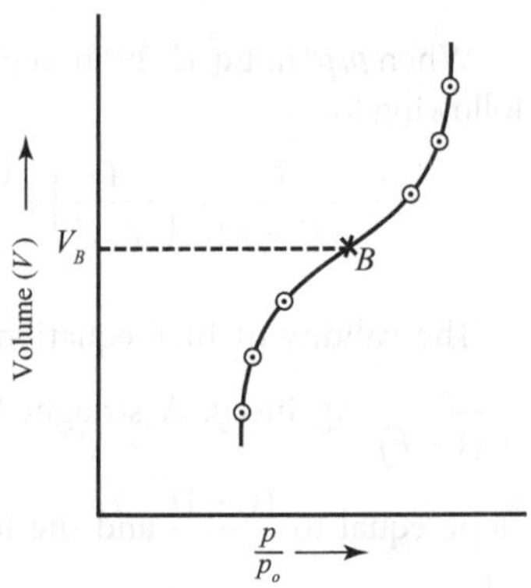

Figure 5.11 Calculation of V_B

5.10 CATALYSIS

Catalyst is a substance which changes (usually increases) the rate of a reaction but is recovered chemically unchanged after the completion of the reaction and the phenomenon itself is called **catalysis.** Although a catalyst is not consumed, it does enter into chemical reaction but is subsequently regenerated.

5.10.1 Positive and Negative Catalyses

When a catalyst increases the rate of a reaction, it is called a positive catalyst and the phenomenon is called positive catalysis.

Example or positive catalysis.

(i) Decomposition of potassium chlorate using MnO_2 as catalyst.

$$2KClO_3(s) \xrightarrow{MnO_2(s)} 2KCl(s) + 3O_2(g)$$

(ii) Manufacture of ammonia by Haber's process using finely divided iron as catalyst.

$$N_2(g) + 3H_2(g) \xrightarrow{Fe(s)} 2NH_3(g)$$

(iii) Hydrolysis of ethyl acetate using an acid or an alkali as catalyst.

$$CH_3COOC_2H_5 + H_2O \xrightarrow{H^+ \text{ or } OH^{-1}} CH_3COOH + C_2H_5OH$$

When a catalyst decreases the rate of a reaction, it is called **a negative catalyst** and the phenomenon is called **negative catalysis.**

Examples of negative catalysis.

(i) Decomposition of H_2O_2 is retarded in the presence of acetanilide.

$$2H_2O_2(l) \xrightarrow{\text{acetanilide(s)}} 2H_2O(l) + O_2(g)$$

(ii) Oxidation of chloroform can be retarded in the presence of ethyl alcohol.

$$\underset{\text{Chloroform}}{2CHCl_3(l)} + O_2(g) \xrightarrow{C_2H_5OH(l)} \underset{\text{Phosgene}}{2COcl_2(g)} + 2HCl(l)$$

5.11 HOMOGENEOUS AND HETEROGENEOUS CATALYSES

(A) A reaction in which the catalyst and the reacting substances are present in the same phase is called **homogeneous catalysis.**

(i) Oxidation of SO_2 to SO_3 in the presence of nitric oxide (NO) as catalyst.

$$2SO_2(g) + O_2(g) \xrightarrow{NO(g)} 2SO_3(g)$$

(ii) Hydrolysis of methyl acetate in the presence of an acid or base.

$$\underset{\text{Methyl acetate}}{CH_3COOCH_3(aq)} + H_2O(l) \xrightarrow[\text{or } OH^-(aq)]{H^+(aq)} \underset{\text{Acetic acid}}{CH_3COOH(aq)} + \underset{\text{Methyl alcohol}}{CH_3OH(aq)}$$

(B) A reaction in which the catalyst is present in a phase different from that of the reacting substances is called **heterogeneous catalysis.** All the reactions in this case occur at the surface of the catalyst. Hence, it is also known as or **contact catalysis.**

Example of heterogeneous catalysis.

(i) Manufacture of ammonia by Haber's process using finely divided iron as a catalyst.

$$N_2(g) + 3H_2(g) \xrightarrow[\text{divided Fe(s)}]{\text{finely}} 2NH_3(g)$$

(ii) Manufacture of SO_2 and O_2 in contact process using vanadium pentoxide or finely divided platinum as catalysis.

(iii) Manufacture of methyl alcohol from CO and H_2 using ($ZnO + Cr_2O_3$) as catalyst.

$$CO(g) + 2H_2(g) \xrightarrow{(ZnO+CuO+Cr_2O_3)} CH_3OH(g)$$

Differences between homogeneous and heterogeneous catalysis are listed in Table 5.2.

Table 5.2 Homogeneous catalysis versus heterogeneous catalysis.

Homogeneous Catalysis	Heterogeneous Catalysis
1. Reactants, catalyst and products are all in the same physical state, e.g., chamber process for the manufacture of H_2SO_4. $2SO_2(g) + O_2(g) \xrightarrow[\text{(Catalyst)}]{NO(g)} 2SO_3(g)$	Physical state of catalyst is different from that of reactants and products, e.g., contact process for the manufacture of H_2SO_4. $2SO_2(g) + O_2(g) \xrightarrow[\text{(Catalyst)}]{V_2O_5(s)} 2SO_3(g)$
2. It is not a surface phenomenon.	It is a surface phenomenon.
3. The reactant molecules on collision with catalyst from intermediate compound, which then yields products.	The reactant molecules are adsorbed on the surface of the solid catalyst, where reaction occurs due to close proximity of adsorbed molecules and then desorption of the product molecules from the surface of catalyst takes place, leaving the catalyst surface for fresh action.
4. Catalytic activity is independent of the size of catalyst molecule.	The catalytic activity increases when the particle size of catalyst is small or surface area is large.

5.12 HOW DOES A CATALYST WORK?

A catalyst decreases the activation energy of the reaction by lowering the energy barrier. Hence, it increases the rate of the reaction by allowing the reaction to follow a new path (Fig. 5.12).

The energy of activation in the new path (E'_a) is lesser when compared to that of the original path. That is, $E'_a < E_a$ lesser the energy of activations required for the reaction to take place, the faster would be the rate of the reaction. However, in case of reversible reactions, the catalyst decreases the activation of the forward and the backward reactions to the same amount. Hence, it increases the rate of the forward and the backward reactions to the same extent. Hence, the equilibrium is attained quickly but is not disturbed.

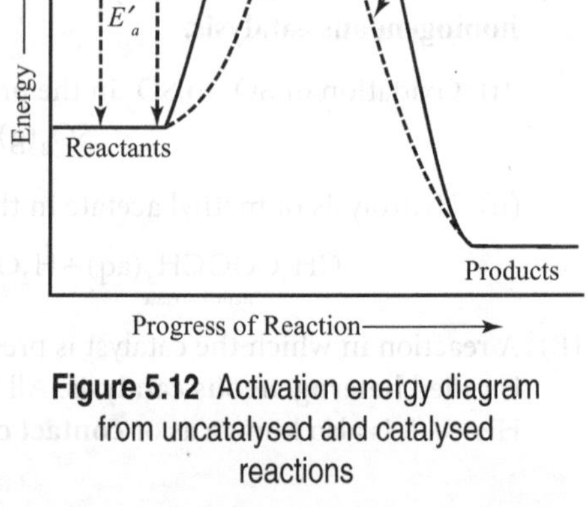

Figure 5.12 Activation energy diagram from uncatalysed and catalysed reactions

MEMORY FOCUS

1. A foreign substance which does not under go any chemical change but alters the reaction rate is called **catalyst** and this phenomenon is called catalysis.
2. The term catalyst was introduced first by Berzelius.
3. A substance which increases the reaction rate is called **positive catalyst.**
4. A substance which decreases the reaction rate is called a **negative catalyst** or inhibitor or anticatalyst.
5. Rate of decomposition of H_2O_2 is increased by the presence of finely divided metals and is decreased by adding acetanilide, acid etc.
6. The phenomenon in which one of the products catalyzes the same reaction is called **autocatalysis.** For example,
   (a) Mn^{2+} for the reaction of $KMnO_4$ with oxalic acid.
   (b) oxides of nitrogen for the reaction of Cu with HNO_3.
7. When the reactants and the catalyst are in the same phase, it is called homogeneous catalyst; otherwise, it is called **heterogeneous** catalysis.
8. A phenomenon in which one reaction catalyzes the other is called **induced catalysis.**
9. A substance which enhances the action of a catalyst is called **promoter.**
10. Substances which destroy the efficiency of the catalyst are called catalytic poisons.
11. Zeolites are the most important metal oxide catalysts. These catalysts are widely used in the petrochemical industries for (i) cracking of hydrocarbons and (ii) aromatization.
12. Shape selectivity is the moist remarkable feature of zeolite catalysis. The reactions depend on the size of the cavities (cages) and pores (apertures) 260 pm and 760 pm. Hence, depending upon the size of the reactant and product molecules, when compared to the size and the shape of the cavities and the pores of the zeolite, the reactions can move in a specified manner.
13. A special zeolite catalyst, called ZSM-5, is used to convert alcohol to gasoline or petrol.

5.12.1 Characteristics of Catalytic Reactions

(i) **A catalyst remains unchanged in mass and chemical composition at the end of the reaction:** It may undergo a physical change. Thus, granular manganese dioxide (MnO_2) used as a catalyst in the thermal decomposition of potassium chlorate is left as a fine powder at the end of the reaction.

(ii) **A small quantity of catalyst is generally needed to bring about the reaction:** For example, mass of finely divided platinum is needed to catalyze the decomposition of hydrogen peroxide. On the other hand, there are catalysts which are present in relatively large amount to be effective. Thus, in Friedel–Crafts reaction,

$$C_6H_6 + C_2H_5Cl \xrightarrow{AlCl_3} C_6H_5C_2H_5 + HCl$$

anhydrous aluminium chloride acts as a catalyst effectively only when present in sufficient amounts.

(iii) **A catalyst is more effective when present in the finely divided state:** Lump of platinum will have much less catalytic activity than colloidal or platinized asbestos. Finely divided nickel is a better catalyst than lumps of solid nickel.

(iv) **A catalyst is specific, i.e., selective in its action:** While a particular catalyst works for one reaction, it will not necessarily work for another reaction. For example, ethanol (C_2H_5OH) gives ethane (C_2H_4) when passed over hot aluminium oxide,

$$C_2H_5OH \xrightarrow{Al_2O_3} \underset{\text{ethene}}{C_2H_4} + H_2O \qquad \text{[Dehydration]}$$

but with hot copper, it gives ethanol (CH_3CHO).

$$C_2H_5OH \xrightarrow{Cu} \underset{\text{ethanal}}{CH_3CHO} + H_2 \qquad \text{[Dehydrogenation]}$$

Just as a key can open one lock only, similarly a catalyst can bring about a particular reaction.

(v) **A catalyst cannot, in general, initiate a reaction:** It only accelerates a reaction already in progress. However, there are certain reactions where the reactants do not combine for very long period. For example, a mixture of hydrogen and oxygen, which remains unchanged almost indefinitely at room temperature, can take place by the catalyst platinum black at once

$$H_2 \xrightarrow{\text{room temp.}} \text{No reaction}$$

$$2H_2 + O_2 \xrightarrow{\text{Pt black}} 2H_2O$$

Thus, it is now considered that the catalyst can initiate a reaction.

(vi) **A catalyst does not affect the final position of equilibrium, although it shortens the time required to establish the equilibrium:** In case of reversible reactions, it means that in a reversible reaction, the catalyst hastens the forward and the reverse reactions equally. Thus, the ratio of the rates of two opposing reactions, i.e., the equilibrium constant, remains unchanged.

(vii) **Temperature affects the catalyzed and uncatalyzed reactions in a similar way:** Some catalysts are, however, physically altered by a rise in temperature and hence their catalytic activity may be decreased. This is particularly true with colloidal solutions like that of platinum, since a rise in the temperature may cause their coagulation. In such a case, the rate of reaction is maximum at a particular temperature called the optimum temperature.

(viii) The activity of a catalyst may sometimes be increased by addition of a small quantity of another substance. This substance is not a catalyst itself for the reaction; however, it acts as a catalyst for the catalyst and is called **catalyatic promoter.**

(ix) The activity of the catalyst is reduced by the presence of small amounts of impurities in the reactants and such substances are called **catalytic poisons.**

5.12.2 Acid Base Catalysis

A number of homogeneous catalytic reactions are catalyzed by acids or bases; these are often referred to as **acid base catalysts.**

Examples of acid base catalysis.

(i) **Inversion of cane sugar:** $\underset{\text{cane sugar}}{C_2H_{22} + H_2} \xrightarrow{H^+} \underset{\text{glucose}}{C_6H_{12}O_6} + \underset{\text{fructose}}{C_6H_{12}O_6}$

(ii) **Hydrolysis of an ester:** $\underset{\text{ethyl acetate}}{CH_3COOC_2H_5} \xrightarrow[\text{or OH}]{H^+} \underset{\text{acetic acid}}{CH_3COOH} + \underset{\text{ethanol}}{C_2H_5OH}$

Recently, it has been found that not only H^+ and OH^- ions but all Bronsted bases acids (proton donors and acceptors) cause acid catalysis. Thus, general acid catalysts are:

H^+, undissociated acids (CH_3COOH), cations of weak bases (NH^+_4) and water (H_3O^+)

Thus, the general base catalysts are:

OH^-, undissociated bases, anions of weak acids (CH_3COO^-) and water (H_2O)

The catalysis brought about by general acids and bases is termed general acid–base catalysis. For elucidation, the decomposition of nitramide is also catalyzed by acetate ions (CH_3COO^-).

$$NH_2NO_2 \xrightarrow{CH_3COO^-} N_2O + H_2O$$

Mechanism of acid base catalysis

(a) In acid catalysis, the H^+ (or a proton donated by Bronsted acid) forms an intermediate complex with the reactant, which then reacts to give back the proton. For example, the mechanism of ketoenol tautomerism of acetone is

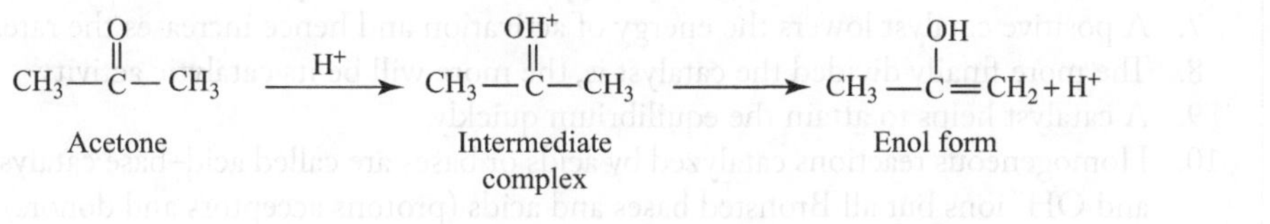

(b) In base catalysis, the OH^- ion (or any Bronsted base) accepts a proton from the reactant to form an intermediate complex which then reacts or decomposes to regenerate the OH^- (or Bronsted base). For example, the decomposition of nitramide by OH^- ions may be explained as follows:

$$NH_2NO_2 + OH \longrightarrow H_2O + \underset{\text{Intermediate complex}}{NHNO_2}$$

$$NHNO_2 \longrightarrow N_2O + OH^-$$

5.12.3 Enzyme Catalysis

Many organic reactions are taking place in the body of animals and plants to maintain the life process. These reactions being slow are remarkably catalyzed by the organic compounds known as enzymes.

Enzymes: Enzymes have been found to be complex protein molecules. The catalysis brought about by enzymes is known as enzyme catalysis. An enzyme is produced in a particular living cell to catalyze a reaction occurring in that cell. Many enzymes have been obtained in pure crystalline state from the cells to which they belong. Some examples of enzyme catalysis are given below:

(i) Inversion of cane sugar ($C_{12}H_{22}O_{11}$) by invertase present in yeast,

$$C_{12}H_{22}O_{11} \xrightarrow{\text{invertase}} C_6H_{12}O_6 + C_6H_{12}O_6$$

(ii) Conversion of glucose into ethanol by zymase present in yeast,

$$C_6H_{12}O_6 \xrightarrow{\text{zymase}} 2C_2H_5OH + 2CC$$

(iii) Hydrolysis of urea ($H_2N - CO - NH_2$) by urease present in soya bean,

$$H_2N-\overset{\overset{\displaystyle O}{\|}}{C}-NH_2 + H_2 \xrightarrow{\text{urease}} 2NH_3 + CO_2$$

MEMORY FOCUS

1. A catalyst remains unchanged in mass and chemical composition.
2. A catalyst cannot start a reaction. It can only change the rate of a reaction.
3. A catalyst has no effect on the position of an equilibrium in a reversible reaction.
4. Action of catalyst is highly specific, i.e., generally for a reaction there will be only one catalyst.
5. Generally, the catalyst can change neither the nature of products nor their yield.
6. Heat of a reaction is unchanged by the presence of a catalyst.
7. A positive catalyst lowers the energy of activation and hence increases the rate.
8. The more finally divided the catalyst is, the more will be its catalytic activity.
9. A catalyst helps to attain the equilibrium quickly.
10. Homogeneous reactions catalyzed by acids or bases are called acid–base catalysis. Not only H^+ and OH^- ions but all Bronsted bases and acids (protons acceptors and donors) cause this type of catalysis.
11. The mechanism of acid–base catalysis (homogeneous catalysis) is explained by intermediate compound formation theory.
12. Many organic reactions taking place in the body of animals and plants, which control the life process, are catalyzed by organic compounds called enzymes.

5.13 MECHANISM OF HOMOGENEOUS AND HETEROGENEOUS CATALYSES

(A) Mechanism of homogeneous catalysis

Usually, homogeneous catalysis involves a two-step mechanism, through intermediate compound formation.

First step: The catalyst reacts with the reactant to form an intermediate complex.

$$\underset{\text{Reactant}}{A} + \underset{\text{Catalyst}}{C} \rightarrow \underset{\text{Intermediate complex}}{AC}$$

Second step: The intermediate can then either decompose to form the product and regenerate the catalyst nor react with another reactant to from the product and regenerate the catalyst.

$$\underset{\text{Intermediate complex}}{AC} \rightarrow \underset{\text{Product}}{P} + \underset{\text{Catalyst}}{C}$$

or

$$\underset{\text{Intermediate complex}}{AC} + \underset{\text{Second reactant}}{B} \rightarrow \underset{\text{Product}}{P} + \underset{\text{Catalyst}}{C}$$

For example, the reaction $2SO_2(g) + O_2(g) \xrightarrow{NO(g)} 2SO_3(g)$ follows a two-step mechanism as shown below.

First step: $2NO + O_2 \xrightarrow{\text{Fast}} 2NO_2$

Second step: $2NO_2 + 2SO_2 \xrightarrow{\text{Fast}} 2SO_3 + 2NO$

Both the steps are fast. Therefore, the overall reaction is also fast while the uncatalyzed reaction is slow.

(B) Mechanism of heterogeneous catalysis

Heterogeneous catalysis involves the following steps

(i) The reactant molecules diffuse to the surface of the catalyst.

(ii) The reactant molecules are adsorbed at the active sites on the surface of the solid catalyst.

(iii) A chemical reaction occurs on the surface of the catalyst forming a product by weak van der Waals' forces (physical adsorption) or by partial weak chemical bonds (chemical adsorption); the product formed is called activated complex.

(iv) Decomposition of the activated complex to form the products.

(v) Desorption of the product molecules away from the catalyst surface.

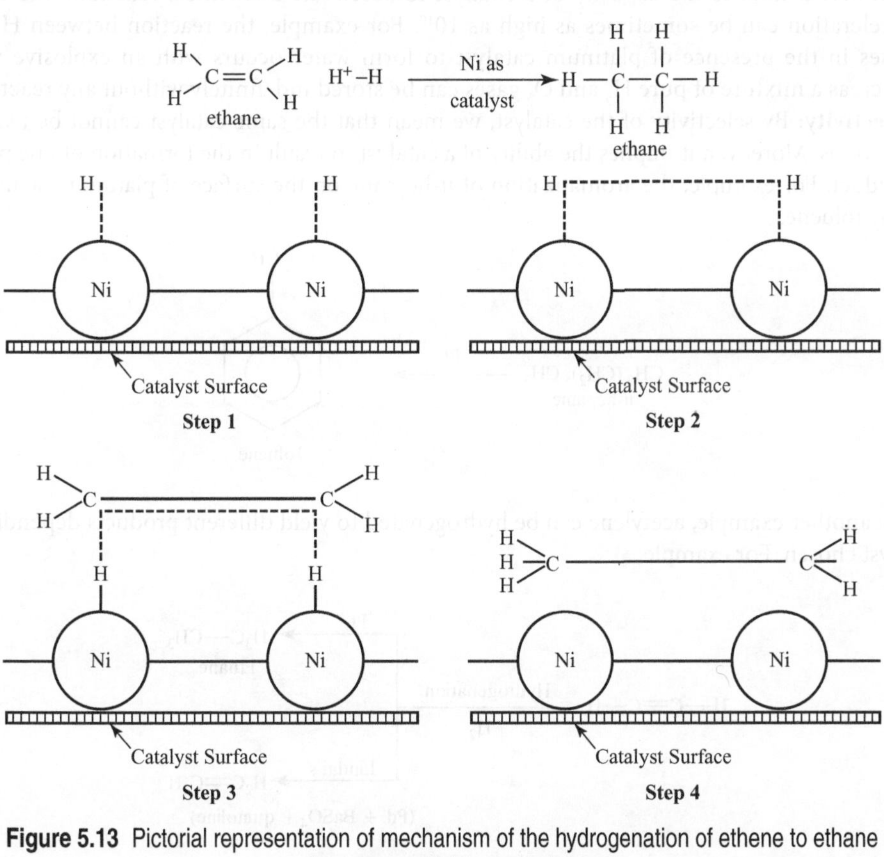

Figure 5.13 Pictorial representation of mechanism of the hydrogenation of ethene to ethane in presence of Ni as catalyst

For instance, the following stepwise mechanism is contemplated for the hydrogenation of ethane to ethene in presence of Ni catalyst (Fig. 5.13).

Mechanism

Step 1. Adsorption of hydrogen molecules on the catalyst surface due to residual valence bonds of Ni atoms.

Step 2. Since the bond length of H–H bond is lesser than the bond length of Ni–Ni bond, the H–H bond is stretched, weakened, and hence broken into atoms which are held to the catalyst surface by chemical bond.

Step 3. The chemisorbed H atoms are attached to ethene molecules by partial chemical bonds, thus forming an unstable activated complex.

Step 4. The unstable activated complex is decomposed to yield the product (ethane) and the catalyst surface is released for a fresh cycle of the above steps.

5.13.1 Significant Characteristics of Heterogeneous Catalysis

The two important aspects of heterogeneous catalysis are *activity* and *selectivity*.

(i) **Activity:** It implies the capacity of a catalyst to accelerate a chemical reaction. The degree of acceleration can be sometimes as high as 10^{10}. For example, the reaction between H_2 and O_2 gases in the presence of platinum catalyst to form water occurs with an explosive violence, whereas a mixture of pure H_2 and O_2 gases can be stored indefinitely without any reaction.

(ii) **Selectivity:** By selectivity of the catalyst, we mean that the same catalyst cannot be used for all reactions. Moreover, it implies the ability of a catalyst to result in the formation of one particular product. For example, the aromatization of *n*-heptane on the surface of platinum catalyst yields only toluene.

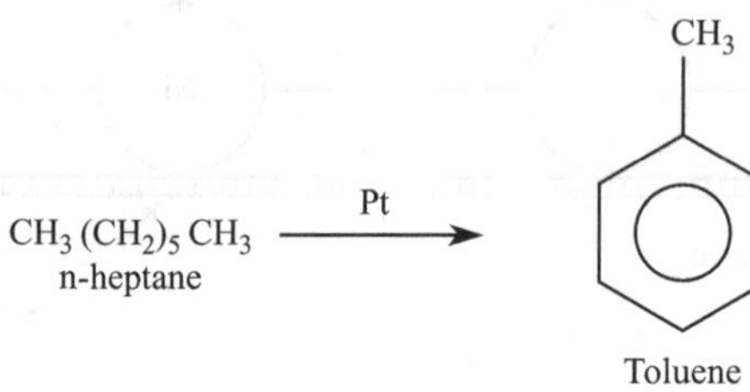

To cite another example, acetylene can be hydrogenated to yield different products depending upon the catalyst chosen. For example

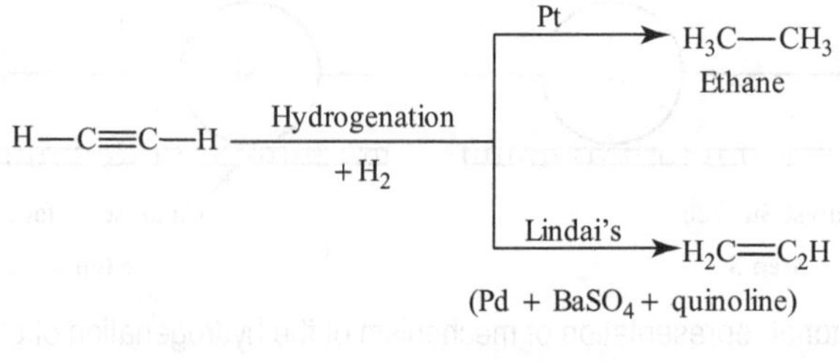

5.13.2 Facts Explained by Adsorption Theory

The adsorption theory reasonably explains the various facts of catalytic activity as discussed in the following sections.

5.13.2.1 Efficiency of Catalyst in Colloidal or Finely Divided State

The catalyst is more efficient when present in colloidal or in finely divided state than when present in lumps.

In heterogeneous catalysis, the rate of reaction is increased because of the increase in the concentration of the reactants on the surface of the solid catalyst due to adsorption at the active centres. When the catalyst is present in a finely divided state, the surface area is increased and thereby, the free valencies or active centres on the catalyst surface are increased. This accounts for the enhanced activity of the catalyst when present in finely divided state when compared to that in the lump form. The finer the catalyst subdivided is, the greater will be its activity (Fig. 5.14).

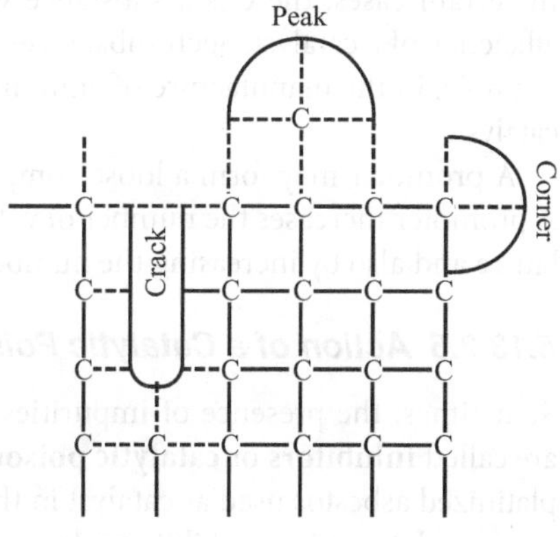

Figure 5.14 Illustration of the increase in the free valence bonds or active centers on the surface of catalyst when it is more finely divided

5.13.2.2 Selectivity of a Catalyst

Selectivity of a catalyst usually means the efficiency of the catalyst to bring about a desired reaction, whereas the activity of a catalyst denotes the overall conversion of the reactants into products. In heterogeneous or contact catalysis, the reaction process is initiated by the adsorption of the reactant molecules on the surface of the solid catalyst. The adsorption process results from the residual forces on the catalyst surface. These forces may lead to the formation of a chemical bond or just the polarization of the adsorbed molecule. In this process, the adsorbed molecule may undergo dissociation. The adsorbed species are in a more reactive form and hence they can undergo reactions more readily than in the absence of the catalyst. The catalytic selectivity is the consequence of the specific manner in which bond formation, polarization etc. take place.

5.13.2.3 Specificity of a Catalyst

The affinity of different catalysts for different reactant molecules is different. The adsorption occurs only when there is a strong affinity of the catalyst for a given reactant or a set of reactant.

5.13.2.4 Active Centres on a Rough Catalyst Surface

A catalyst surface contains unbalanced or free chemical bonds and it is on these that the gaseous reactant molecules are adsorbed on the catalyst.

These free valency bonds on the surface of the catalyst are not uniformly distributed. They are found to be greater in number of cracks, peaks and corners of the catalyst. The catalytic activity due to adsorption is found to be height at these spots, which are called active centres. The active centres accelerate the rate of reaction not only by increasing the concentration of the reactants at these centres but also by activating the adsorbed reactant molecule adsorbed at two such adjacent centres by stretching.

5.13.2.5 Action of a Promoter

In certain cases, there is a substance which is itself not necessarily a catalyst but it can increase the efficiency of a catalyst. Such substances which can act as catalyst for a catalyst are called **promoters**. For example, in the manufacture of ammonia by Haber's process, molybdenum acts as a promoter for iron catalyst.

A promoter may form a loose compound with the catalyst which increases the adsorption capacity. A promoter increases the number of valence bonds on the surface of the catalyst by changing the crystal lattice and also by increasing the number of peaks, cracks and corner on the catalyst surface.

5.13.2.6 Action of a Catalytic Poison

Sometimes, the presence of impurities in the reactants renders the catalyst inactive. Such substances are called **inhibitors** or **catalytic poisons**. For example, arsenious oxide inhibits the catalytic activity of platinized asbestos used as catalyst in the manufacture of H_2SO_4 by contact process.

A catalytic poison inhibits or destroys a catalyst by blocking the free valence bonds of its surface by preferential adsorption or by chemical combination.

MEMORY FOCUS

1. Function of a catalyst can be explained by:
   (a) intermediate compound theory, (b) acid–base theory and (c) adsorption theory
2. Modern theory of catalysis is adsorption theory.
3. Heterogeneous catalysis can be explained by the adsorption theory, while homogeneous catalysis can be explained by the intermediate compound theory.
4. The reactive sites of the catalyst are called active centres.
5. By dividing the catalyst into fine powder, there will be increase in
   (a) surface area, (b) active centres, (c) free valencies and (d) adsorbing capacity
6. A catalyst poison reduces the number of active centres by adsorbing on the free valencies, while a promoter increases the active centres.

REVISION QUESTIONS

1. Define the followings:
   (a) Absorption (b) Adsorption (c) Desorption
   (d) Positive adsorption (e) Negative adsorption (f) Adsorption isotherm
   (g) Adsorption isobar
2. Distinguish physical and chemical adsorptions.
3. State Freundlich adsorption isotherm and discuss the significance of the various terms involved.
4. Discuss briefly Langmuir's theory of adsorption. Deduce an expression for Langmuir unimolecular adsorption isotherm. Give importance and limitations of this theory.
5. Derive an expression for the Langmuir adsorption isotherm and discuss its behaviour at very low and high pressures.
6. Discuss Langmuir adsorption equation.
7. Discuss in brief the theory of adsorption.

8. (a) Discuss the applications of adsorption in industry and everyday life.
   (b) What is the effect of temperature on the adsorption of a gas on a solid surface?
9. Discuss the application of adsorption with special reference to catalysis.
10. (a) Distinguish adsorption and absorption.
    (b) Discuss the factors which affect the adsorption of a gas on a solid adsorbent.
    (c) Discuss in brief the type of adsorption isotherms commonly observes for the adsorption of gases on a variety of adsorbents as different temperatures.
11. (a) What is the theory of adsorption?
    (b) What is adsorption isotherm?
    (c) Distinguish physical and chemical adsorptions.
12. (a) Distinguish chemisorption and physical adsorption.
    (b) Derive Langmuir adsorption isotherm and prove its validity.
    (c) Adsorption explains the catalytic action of a metal in gaseous reactions. Explain.
13. Explain the following terms:
    (a) Catalyst
    (b) Homogeneous catalysis
    (c) Heterogeneous catalysis
    (d) Promoters
    (e) Catalytic poisons
14. What is meant by the term catalyst? Give general characteristics of catalytic reaction.
15. (a) Give examples of heterogeneous catalysis in which reactants taking part are
    (i) solid (ii) liquid (iii) gaseous state
    (b) Discuss the theory of heterogeneous catalysis and illustrate with examples.
16. Write a note on enzyme catalysis.
17. (a) Discuss intermediate compound formation theory of catalysis.
    (b) Define the following:
    (i) Catalytic promoter (ii) Catalytic inhibitor (iii) Catalytic poison
18. (a) What is heterogeneous catalysis? Explain by giving one example.
    (b) What mechanism would you suggest for heterogeneous catalytic reaction? Classify it with one example.
    (c) Why should a solid catalyst work best as a five powder?
19. (a) What is the function of catalytic promoters and inhibitors in chemical reaction?
    (b) Explain the important characteristics of enzyme catalysis.
    (c) Why is the reaction speeded up in the presence of a catalyst?
20. Explain the following with example:
    (a) Negative catalyst
    (b) Autocatalyst
    (c) Catalytic poison
    (d) Promoters
    (e) Specificity of a catalyst
21. (a) Give an example each for enzyme and acid–base catalyses.
    (b) Describe the theory of homogeneous and heterogeneous catalyses.
22. Explain the difference between homogeneous and heterogeneous catalyses. Discuss the theories of catalysis.

23. Explain, with proper examples, homogeneous and heterogeneous catalyses.
24. (a) What are the various factors influencing adsorptions? Discuss it.
    (b) What are the various types of adsorption isotherms? Explain the Gibbs adsorption isotherm for adsorption from solution

MULTIPLE CHOICE QUESTIONS

1. The phenomenon of concentrations of molecules of a gas or liquid at a solid surface is called
   (a) absorption (b) adsorption
   (c) catalysis (d) none of these
2. The adsorption of gases on metal surfaces is called
   (a) catalysis (b) occlusion
   (c) adsorption (d) absorption
3. The process of adsorption is
   (a) exothermic
   (b) endothermic
   (c) sometimes exothermic and sometime endothermic
   (d) none of the above
4. Physical adsorption is a ________ process.
   (a) reversible (b) irreversible
   (c) exothermic (d) none of these
5. Multi-molecular layers are formed in
   (a) absorption
   (b) physical adsorption
   (c) chemisorptions
   (d) reversible adsorption
6. The relationship between equilibrium pressure of gas and its amount adsorbed on the solid adsorbent at constant temperature is called
   (a) chemisorptions
   (b) adsorption isobars
   (c) adsorption isotherms
   (d) none of these
7. Freundlich isotherms is not applicable at
   (a) high pressure (b) low pressure
   (c) 273 K (d) room temperature
8. At low pressures, the amount of the gas adsorbed is ____________ proportional to the pressure
   (a) directly
   (b) inversely
   (c) sometimes directly and sometimes inversely
   (d) none of these
9. In gas masks, the poisonous gases are removed by the adsorbent by the process of
   (a) absorption (b) adsorption
   (c) catalysis (d) none of these
10. Heat of adsorption is defined as the energy liberated when of a gas is adsorbed on the solid surface
    (a) 1 molecule (b) 1 gm
    (c) 1 gm mole (d) 1 kg
11. In physical adsorption, the gas molecules are held to the solid surface by
    (a) hydrogen bond
    (b) sigma bond
    (c) pi bond
    (d) van der Waals' forces
12. The adsorption of hydrogen on charcoal is
    (a) physical adsorption
    (b) chemical adsorption
    (c) sorption
    (d) none of these
13. Adsorption takes place with
    (a) decrease in enthalpy of the system
    (b) increase in enthalpy of the system
    (c) no change in enthalpy of the system
    (d) none of these
14. The heat of adsorption in physical adsorption lies in the range
    (a) 1–10 kJ mol^{-1} (b) 10–400 kJ mol^{-1}
    (c) 40–100 kJ mol^{-1} (d) 10–400 kJ mol^{-1}
15. Which out of the following is incorrect?
    (a) Chemisorption is reversible in nature.
    (b) Physical adsorption is reversible in nature.

(c) ΔH is small in physical adsorption.
(d) ΔH is large in chemical adsorption.

16. Which of the following is incorrect?
(a) Chemisorption is caused by bond formation.
(b) Chemisorption is specific in nature.
(c) Chemisorption is reversible.
(d) Chemisorption increases with the increase in temperature.

17. Langmuir isotherms hold at low pressure but fail at
(a) low temperature
(b) high pressure
(c) intermediate pressure
(d) none of these

18. A catalyst
(a) does not react
(b) reacts and is produced in one of the later steps of reaction
(c) reacts in an early step and is produced in a later step
(d) reacts but is not produced

19. A catalyst can be described as a substance that
(a) undergoes change to accelerate the rate of the reaction
(b) increases the kinetic energy of the reactants
(c) provides a path of lower activation energy for the reaction
(d) lowers the potential energy of the products with the respect to the energy of the reactants

20. The minimum amount of energy needed to bring about a non-spontaneous change is the
(a) free energy (b) activation energy
(c) entropy (d) enthalpy

21. Which one of the following statements is incorrect?
(a) Catalysts speed up a reaction by lowering the activation energy.
(b) Heterogeneous catalysts are often found on a solid in contact with gaseous reactants.
(c) An automotive catalytic converter uses a catalyst to burn gasoline better for higher gas mileage.
(d) Enzymes are biological catalysts.

22. Enzymes are
(a) substances synthesized by chemists to decrease the reaction rate
(b) highly porous substances to activate acids and bases
(c) extremely poor in catalytic activity
(d) catalysts found in organism

23. Which one of the following statement is not true?
(a) Enzymes require optimum temperature.
(b) Enzymes require optimum pH.
(c) Enzymes increase activation energy.
(d) Enzymes are specific in nature.

24. A catalyst becomes ineffective by the catalytic poison because
(a) it combines chemically with one of the reactants
(b) it combines chemically with the catalyst
(c) it absorbs reactant molecules
(d) it is preferentially adsorbed on the catalyst

25. A promoter is a substance which
(a) lowers the kinetic energy of reactants
(b) lowers the activation energy of reaction
(c) enhances the activity of the catalyst
(d) enhances the concentration of the catalyst

26. Tetraethyl lead, $Pb(C_2H_5)_4$, when added to petrol acts as
(a) a catalyst
(b) a promoter
(c) an inhibitor
(d) an autocatalyst

27. An example of acid–base catalysis is
(a) inversion of cane sugar
(b) keto-enol tautomerism
(c) decomposition of nitramide
(d) all the above

28. In intermediate compound formation theory of catalysis, the intermediate complex formed has
    (a) lesser activation energy than uncatalyzed complex
    (b) more activation energy than uncatalyzed complex
    (c) lesser kinetic energy than the reactants
    (d) lesser kinetic energy than the products
29. Which of the following processes is an example of heterogeneous catalysis?
    (a) contact process
    (b) Haber's process
    (c) Hydrogenation of vegetable oils
    (d) all of those
30. The phenomenon of negative catalysis is also known as
    (a) autocatalysis
    (b) self-catalysis
    (c) inhibition
    (d) enzyme catalysis

ANSWERS

1. (b)	6. (c)	11. (d)	16. (c)	21. (c)	26. (c)
2. (b)	7. (a)	12. (a)	17. (b)	22. (d)	27. (d)
3. (a)	8. (a)	13. (a)	18. (c)	23. (c)	28. (a)
4. (a)	9. (b)	14. (a)	19. (c)	24. (d)	29. (d)
5. (c)	10. (c)	15. (a)	20. (b)	25. (c)	30. (c)

Photochemistry 6

CHAPTER OBJECTIVES

6.1 Introduction

6.2 Thermochemical and Photochemical Reactions

6.3 Laws Governing Light Absorption — Lambert's Law and Beer's Law

6.4 Laws of Photochemistry

6.5 Quantum Efficiency

6.6 Study of Some Photochemical Reactions

6.7 Fluorescence and Phosphorescence

Photochemistry provides a comprehensive overview of the concepts and reactions fundamental to the field, illustrating important applications in biological, medical and environmental sciences.

Grazyna Stochel

6.1 INTRODUCTION

The reactions which occur with the absorption of radiations (photons) are called photochemical reactions. Photon supplies the necessary energy to the reactants, enabling them to react to yield products. It may be emphasized that a photochemical reaction requires the absorption of a definite energy by the reactants to proceed to products.

Although the term 'radiation' includes a variety of rays, the radiations important from the standpoint of photochemistry lie, almost exclusively, in the visible and ultraviolet (UV) regions, i.e., the wavelength ranges from 2000 to 8000 Å (200–800 nm). ***Photochemistry** is, therefore, mainly concerned with rates and mechanism of reactions resulting from the exposure of a system to radiant energy which lies in the visible and UV regions, i.e., the wavelength ranges between 2000 and 8000 Å (200–800 nm).*

Photochemical reactions are usually cleaner and more efficient than other type of reactions. Important applications of photochemistry like photo polymerization, photohalogenation, nitrozation, pharmaceutical synthesis (e.g. synthesis of vitamin D) and photolethography (photo-resistant technology) for micro-electronics industries etc. have led to the development in the area of Photochemistry. A few examples of such photochemical reactions are the conversion of oxygen into ozone, the decomposition of hydrochloric acid and ammonia and the photosynthesis of carbohydrates in plants. In all these reactions, a part of the radiant energy supplied to the system is converted into free energy of the products. Moreover, these reactions are important because these reactions are several times faster than thermal reactions. Reactions as fast as 10^{-9}S and the associated processes as fast as 10^{-15}S are often observed.

6.2 THERMOCHEMICAL AND PHOTOCHEMICAL REACTIONS

Thermochemical reactions take place either by the absorption or by the evolution of heat energy and are referred to as endothermic and exothermic reactions. Some examples of these reactions are discussed in the following sections.

Endothermic reactions: The reactions take place with the absorption of heat energy, i.e., enthalpy change (H) in their case is positive. For example, the following chemical reactions proceed with the absorption of heat energy.

1. $H_2\,(g) + I_2\,(g) \rightarrow 2\,HI\,(g)$, $\quad \Delta H = 49.7$ kJ
2. $N_2\,(g) + O_2\,(g) \rightarrow 2\,NO\,(g)$, $\quad \Delta H = 180.5$ kJ
3. $C\,(s) + 2\,S\,(s) \rightarrow CS_2\,(l)$, $\quad \Delta H = 92.00$ kJ

Exothermic reactions: These reactions take place with the evolution of heat energy, i.e., enthalpy change (ΔH) in their case is negative. For example, combustion reactions proceed with the evolution of heat energy.

1. $C\,(s) + O_2\,(g) \rightarrow CO_2\,(g)$, $\quad \Delta H = -393.5$ kJ
2. $H_2\,(g) + ½\,O_2\,(g) \rightarrow H_2O\,(I)$, $\quad \Delta H = -285.9$ kJ
3. $CH_4\,(g) + 2\,O_2\,(g) \rightarrow CO_2\,(g), + 2\,H_2O\,(I)$, $\quad \Delta H = -890.3$ kJ

Ordinary chemical reactions are referred to as *thermal* or *dark* reaction in order to distinguish them from photochemical reactions.

Photochemical reactions: These take place only after the absorption of light radiations from the visible and UV regions of the spectrum. Some typical photochemical reactions include:

(a) Dissociation, e.g., $\quad 2\,HBr \rightarrow H_2 + Br_2$

(b) Rearrangement, e.g., $\quad$ Fumaric acid → Maleic acid

(c) Addition reactions, e.g., $\quad Br_2 + (C_6H_5)_2C = C\,(C_6H_5)_2 \rightarrow (C_6H_5)_2\,C\,Br - C\,Br\,(C_6H_5)_2$

(d) Polymerization, e.g., $\quad nC_2H_2 \rightarrow (C_2H_2)n$ (polyacetylene)

$\quad 2\,C_{14}H_{10} \rightarrow C_{28}\,H_{20}$ (dianthracene)

(e) Photocatalytic reactions, e.g., $\quad CO_2 + H_2O[\text{chlorophyll}] \rightarrow \frac{1}{n}(CH_2O_n) + O_2 + [\text{chlorophyll}]$

(f) Combination, e.g., $H_2 + Cl_2 \rightarrow 2\,HCl$
(g) Decomposition, e.g., $2\,O_3 \rightarrow 3\,O_2$
(h) Substitution, e.g., $C_6H_{12} + Br_2 \rightarrow C_6H_{11}Br + HBr$; etc.

Important observations about photochemical reactions: Some of the important observations of photochemical reactions are:

(i) Light has to be absorbed by the reactants.
(ii) If coloured light is used, the reactants may not be initiated by all colours. A photon of red light is less energetic than a photon of blue light. Usually reactions that are initiated by blue light cannot be initiated by red light. Reactions that are initiated by red light can often be initiated by green, yellow, blue and violet lights.
(iii) The rate of reaction depends on the intensity of light.
(iv) In some cases, the molecule that absorbs light may transfer its extra energy to another molecule which may undergo a reaction. This process is called photosensitization.

The main points of differences between these two types of reactions are given in Table 6.1.

Table 6.1 Differences between photochemical and thermochemical reactions.

Photochemical Reactions	Thermochemical Reactions
(i) These reactions involve only absorption of light.	(i) These reactions involve absorption or evolution of heat.
(ii) The presence of light is the primary requisite for the reaction to takes place.	(ii) They can take place in the dark as well as light.
(iii) Temperature has very little effect on the rate of a photochemical reaction.	(iii) Temperature has significant effect on the rate of a thermochemical reaction.
(iv) The free energy change (ΔG) of a photochemical reaction may be negative or positive. A few examples of photochemical reactions for which ΔG is positive and still they are spontaneous are: (a) The synthesis of carbohydrates in plants. (b) The decomposition of HCl to H_2 and Cl_2.	(iv) The free energy change (ΔG) of a thermochemical reaction is always negative.
(v) The rate of these reactions depends upon the intensity of light being absorbed.	(v) The intensity of light has no effect on the rate of these reactions.
(vi) Some of these are initiated by the presence of a photosensitizer. A photosensitizer acts in a different way than a catalyst.	(vi) They are accelerated by the presence of a catalyst.

6.3 LAWS GOVERNING LIGHT ABSORPTION — LAMBERT'S LAW AND BEER'S LAW

Light incident upon a system can be transmitted, refracted, scattered or absorbed. The fraction of incident light absorbed by a system depends on the thickness of the medium that is traversed. The intensity of absorption at a particular wavelength can be determined by passing a monochromatic beam of light through a sample of known thickness and measuring the intensity of transmitted light which is found to decrease. The decrease of intensity of incident light at any given wavelength on passing through a transparent absorbing substance is given by **Lambert's law**, according to which,

The rate at which the intensity decreases with thickness of the medium is proportional to the intensity of the incident light.

Mathematically, it may be expressed as

$$-\frac{dI}{dx} \propto I \text{ or } -\frac{dI}{I} = \alpha dx \qquad (6.1)$$

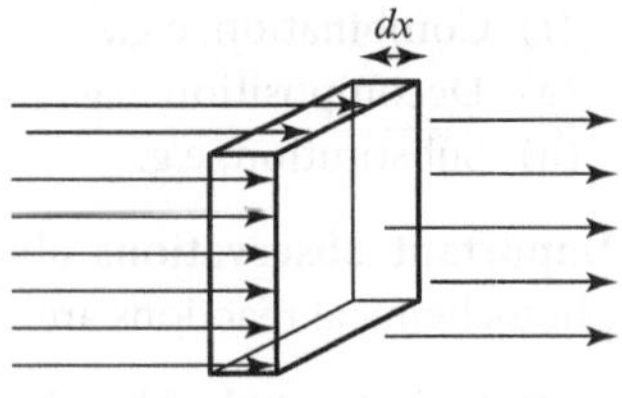

Figure 6.1 A beam of intensity I passes through medium of thickness *dx*, the intensity of the beam is reduced to I – *dI*

where I is the intensity of light of a particular wavelength (energy per unit area per unit time); dI is the change in light intensity produced by absorption radiation on passing through of a thin layer of thickness dx (Fig. 6.1); α is called absorption coefficient or extinction coefficient. The minus sign is introduced because there is reduction in intensity.

Equation (6.1) can be also written as

$$+\frac{dI}{I} = -\alpha dx \qquad (6.2)$$

Integrating Eq. (6.2) and applying boundary conditions, when $x = 0, I = I_0$ and when $x = x, I = I$.

$$\int_{I_0}^{I} \frac{dI}{I} = \int_{x=0}^{x=x} -\alpha dx$$

or
$$\ln\frac{I}{I_0} = -\alpha x \quad \text{or} \quad \frac{I}{I_0} - e^{-\alpha x} \qquad (6.3)$$

or
$$I = I_0 e^{-\alpha x} \qquad (6.4)$$

Equation (6.3) can also be rewritten as

$$2.303 \log\frac{I}{I_0} = -\alpha x \quad \log\frac{I}{I_0} = -\frac{\alpha}{2.303} \times \log\frac{I}{I_0} = -\alpha' x \qquad (6.5)$$

or
$$\frac{I}{I_0} = 10^{-\alpha' x} \qquad (6.6)$$

or
$$I = I_0 10^{-\alpha' x} \qquad (6.7)$$

The equantity $\alpha' = \frac{\alpha}{2.303}$ is called **absorption coefficient or absortivity** of the substance.

(α' and α both represent the same term i.e. absorption coeffiecient, but the term only differ in their numerical values)

Equation (6.5) can be rewritten as

$$\log\frac{I_0}{I} = \alpha' x \quad \text{or} \quad \alpha' = \frac{1}{x}\log\frac{I_0}{I} \qquad (6.8)$$

The factor $\log\frac{I_0}{I}$ is called optical density or **absorbance (A)** and $\frac{I}{I_0}$ is termed as **transmittance (T)**.

when
$$\frac{I_0}{I} = 10, \ \log\frac{I}{I_0} = 1 \Rightarrow \alpha' = \frac{1}{x}$$

This implies that absortivity is equal to the reciprocal of thickness of that layer in which the intensity of light falls to one tenth of its original value.

Note: If I_t is the intensity of transmitted light and I_0 is the intensity of incident beam of light, then, the intensity of absorbed light, I_a, is given by,

$$I_a = I_0 - I_t \tag{6.9}$$

Lambert–Beer's Law: Later, Beer found that the absorption coefficient (α) was proportional to the concentration (c) of solution and combined it with Lambert's law, this is known as Lambert–Beer's law and is stated as

On passing a beam of monochromatic radiation through a solution, the rate of decrease of intensity with the thickness of the solution is proportional to the intensity of incident radiation as well as concentration of the solution.

Mathematically,
$$-\frac{dI}{dx}\alpha cI \text{ or } -\frac{dI}{I} = \in' cdx \tag{6.10}$$

where $\in'$ is the absorption or extinction coefficient. Integration of Eq. (6.10) gives

$$\int -\frac{dI}{I} = \int \in' cdx \tag{6.11}$$

or
$$\ln I = \in' cx + \text{constant of integration} \tag{6.12}$$

At $x = 0$, $I = I_0$ and At $x = x$, $I = I$

Substituting these boundary conditions, we get

$$-\ln\frac{I}{I_0} = \in' cx \tag{6.13}$$

or
$$I = I_0 e^{-\in' cx} \tag{6.14}$$

Equation (6.14) is one of the well-known forms of Lambert–Beer's law, changing natural logarithm in Eq. (6.13) to the logarithm to the base 10, we get

$$-2.303\log\frac{I}{I_0} = \in' cx \tag{6.15}$$

or
$$-\log\frac{I}{I_0} = \frac{\in'}{2.303}cx \tag{6.16}$$

or
$$\log\frac{I_0}{I} = \frac{\in'}{2.303}cx = \in cx = \text{A}, \quad \text{where } \in = \frac{\in'}{2.303} \tag{6.17}$$

The quantity $\log\frac{I_0}{I}$ is known as the **absorbance (A)** and $\in$ is called the **molar absorption coefficient.**

The Eq. (6.17) tells that *the absorbance A is directly proportional to the concentration of the solution and the length of the path.* The length of the path can be kept constant and hence absorbance measurement can be used for measuring unknown concentrations of the coloured solutions. The molar absorption coefficient depends on temperature, *T*, at which the solution is kept and the wavelength, λ, of the incident light.

From Eq. (6.17),
$$\log\frac{I_0}{I} = \in cx \tag{6.18}$$

If $C = 1$ and $\log \dfrac{I_0}{I} = 1$ then $1 = \in x$

or
$$\in = \frac{1}{x} \quad \left(\log \frac{I_0}{I} = 1, \text{ show } \frac{I_0}{I} = 10 \text{ i.e. } I = \frac{1}{10} I_0 \right)$$

This gives us the definition of molar absorption coefficient. Thus, molar absorption coefficient is the reciprocal of that thickness of the solution layer of one molar solution which decreases the intensity of light to one-tenth of its original value.

Relation Between Transmittance and Absorbance of Solution

The absorbance of solution is denoted by A and is given as

$$A = \log_{10} \frac{I_0}{I} \quad \text{or} \quad A = -\log_{10} \frac{I_0}{I} \quad \text{or} \quad A = -\log_{10} T \tag{6.19}$$

6.3.1 Limitations of Lambert–Beer's Law

(i) The law is not obeyed if radiation used is not monochromatic.

(ii) The temperature of the solution should not be allowed to increase, as the absorption band in UV–visible spectra will shift towards longer wavelength.

(iii) The law is applicable for dilute solutions only; as in concentrated solution, the refractive index of the solution changes and high concentration of solute shows strong interionic interactions which can disturb the ability of a solute to absorb a given wavelength of the incident radiation.

MEMORY FOCUS

1. The reaction which takes place by the absorption of the visible and UV radiations (2000–8000 Å) is called a **photochemical reaction**, whereas the reactions which are caused by heat in the absence of light are called **thermal or dark reactions**.
2. Photochemical reactions involve the absorption of light radiations, whereas thermal reactions involve the absorption or evolution of heat.
3. For photochemical reactions, ΔG may be +ve or –ve, whereas for thermal reactions, ΔG is always negative.
4. The intensity of radiation can be defined as the number of photons that pass across a unit area in unit time.
5. **Lambert law** states that 'the rate at which the intensity decreases with thickness of the medium is proportional to the intensity of the incident light'.
   Mathematically, it may be expressed as
   $$-\frac{dI}{dx} \propto I \quad \text{or} \quad -\frac{dI}{I} = \alpha\, dx$$
   $$\log \frac{I_0}{I} = \alpha' x \quad \text{or} \quad \alpha' = \frac{1}{x} \log \frac{I_0}{I}$$
   where I_0 and I stand for the intensity of incident light and the intensity of light at a particular wavelength, α' is called absorption or extinction coefficient.

(*Continued*)

MEMORY FOCUS (*Continued*)

6. **Absorption or extinction coefficient** is equal to the reciprocal of thickness of that layer in which the intensity of light falls to one-tenth of its original value.
7. If I_t is the intensity of transmitted light and I_0 is the intensity of incident beam of light, then intensity of absorbed light, I_a, is given by

$$I_a = I_0 - I_t$$

8. **Lambert-Beer law** states that: 'On passing a beam of monochromatic radiation through a solution, the rate of decrease of intensity with thickness of the solution is proportional to the intensity of incident radiation as well as concentration of the solution.'
Mathematically, it may be expressed as

$$-\ln\frac{I}{I_0} = \in' cx \quad \text{or} \quad I = I_0 e^{-e'cx}$$

9. **Molar absorption coefficient ( )** is the reciprocal of that thickness of the solution layer of one molar solution which decreases the intensity of light to one-tenth of its original value.

Numerical Problems

Example 1 **When an incident beam of wavelength 3000 Å was allowed to pass through 2-mm thick Pyrex glass, the intensity of the radiation was reduced to one-tenth of its initial value. What part of the same radiation will be absorbed by 1 mm thick glass?**

Solution: We are given that $\frac{I_t}{I_0} = \frac{1}{10}$ and $x = 2$ mm = 0.2 cm

According to Lambert's law, we have

$$\log\frac{I_t}{I_0} = -\alpha x \quad \text{or} \quad \log\frac{1}{10} = -\alpha \times 0.2 \quad \text{or} \quad \alpha = \frac{1}{0.2} = 5$$

Again, we are given that $x = 1$ mm = 0.1 cm and $K = 5$

Applying Lambert's law, again $\log\frac{I_t}{I_0} = -\alpha x = -5 \times 0.1 = -0.5$

$$\therefore \quad \frac{I_t}{I_0} = \text{antilog}(-0.5) = \text{antilog}(\bar{1}.5) = 0.3162$$

We know $\quad I_\alpha = I_0 - I_t$

$$= I_0 - 0.3162, \qquad I_0 = \mathbf{0.6838}$$

I_0 **= 68.38% of the intensity of incident light.**

Example 2 **A 1-M solution in a cell of 0.200 mm thickness transmits 10% of a beam of light. Neglecting the absorption by a solvent and cell windows, what percentage of the initial intensity will be transmitted by a 2-M solution in a cell of 0.150 mm thick?**

Solution: According to Beer's law, we have $\log\frac{I_t}{I_0} = -\in' c \cdot x$

It is given that; transmittance $T = \frac{10}{100}$

$$\Rightarrow \quad \frac{I_t}{I_0} = \frac{1}{10}, \quad c = 1 \text{ mol dm}^{-3} \text{ and } x = 0.20 \text{ mm} = 0.02 \text{ cm}$$

$$\therefore \quad \log\frac{1}{10} = -\in' \times 1 \times 0.2 \quad \text{or} \quad \in' \frac{1}{0.2} = 50 \text{ dm}^3 \text{ mol}^{-1} \text{ cm}^{-1}$$

Applying Beer's law again, we have

$$\log\frac{I_t}{I_0} = -\in' c \cdot x$$

We know $c = 2$ M and $x = 0.15$ mm $= 0.015$ cm

$$\log\frac{I_t}{I_0} = -50 \times 2 \times 0.015 \qquad \therefore \log\frac{I_t}{I_0} = \text{antilog}(-15) = \text{antilog}\,\bar{2}.5 = 0.0316$$

or $\quad I_t = 0.0316$; $I_0 =$ **3.16% of the intensity of incident light.**

Example 3 What is the absorption coefficient of a solute which absorbs 90% of a certain wavelength of light beam passed through 1 cm cell containing 0.25-M solution?

Solution: Given that 90% of the incident light is absorbed

or $\quad I_0 = 100 \quad \text{and} \quad I_a = 90$

We know $\quad I_\alpha = I_0 - I_t \quad \text{or} \quad 90 = 100 - I_t \quad \text{or} \quad I_t = 100 - 90 = 10$

According to Lambert–Beer's law:

$$\ln\frac{I_t}{I_0} = -\in' cs \quad \text{or} \quad \log\frac{I_t}{I_0} = -\frac{\in'}{2.303} cx$$

where $\in'$ is molar absorption coefficient, c the molar concentration and x is the thickness of the solution.

$$\therefore \quad \log\frac{10}{1000} = \frac{\in'}{2.303} \times 0.25 \times 1$$

$$\therefore \quad \in' = \frac{2.303}{0.25} \times \log\frac{100}{10}$$

Hence, $\in' = 9.21$ and absorption coefficient = **9.21.**

Example 4 Light of definite wavelength was passed through a cell of 4 cm thickness containing 0.02-M solution of a given substance. If the molar extinction coefficient is 10.0 litre mole^{-1} cm^{-1}, calculate the optical density and percentage transmittance.

Solution: According to Lambert–Beer's law:

A, the optical density $\quad = \log\frac{I_0}{I_t} = \in' cx$

where $\quad \in' =$ molar extinction coefficient;

$c =$ concentration in mole litre^{-1} and

$x =$ thickness of the optical path.

Given that, A = 10.0 litre mole^{-1} × 0.02 mole litre^{-1} × 4 cm = 0.80

Now, $\quad A = \log\frac{I_0}{I_1} = -\log T = -\log\frac{I}{I_0}$

where I_0= the intensity of light in quanta per cm^2; I_t = the intensity of transmitted light in quanta per cm^2 and T = transmittance.

$$\therefore \qquad \log T = -0.80 = \bar{1}.20 \qquad \text{or} \qquad T = 0.1585$$

Hence, the percentage transmittance = 100 T = 100×0.17 = **17%.**

6.4 LAWS OF PHOTOCHEMISTRY

6.4.1 Grotthus–Drapper Principle of Photochemical Activation: (First Law of Photochemistry)

Grotthus–Drapper principle states that *only the light that is absorbed by a substance is effective in producing a photochemical change. This is known as the first principle of photochemistry.*

If I_0 is the incident light, I_t is the transmitted light and I_a is the absorbed light, then

$$\begin{aligned} I_a &= I_0 - I_t \\ &= I_0 - I_0 e^{-\epsilon c x} = I_0\left(1 - e^{-\epsilon c x}\right) \end{aligned} \tag{6.20}$$

Alternatively, according to Grotthus and Draper,

When light falls on a body, only that fraction of incident light which is absorbed can bring about a chemical change. This implies that reflected and transmitted light do not produce any chemical change.

It is important to note that it is not necessary that all the light which is absorbed will bring about chemical change. It is possible that atoms and molecules may absorb light and then they may re-emit in the form of line or band spectrum of same or different frequency. Sometimes, the light absorbed may bring about phenomena such as *fluorescence* and *phosphorescence*. The absorbed light can also be converted into thermal energy.

This law is a qualitative one and does not give any correlation between the amount of light absorbed by a system and the number of molecules which have reacted.

6.4.2 Stark–Einstein's Law of Photochemical Equivalence — The Second Law of Photochemistry

This law states that each molecule which takes part in the photochemical reaction absorbs one quantum of light energy which induces the reaction. Thus, one molecule absorbs the one quantum to excite one molecule, e.g. $A + h\nu \rightarrow A^*$.

The asterisk (*) here means that the molecule A* is in the excited or activited state after absorption of one quantum of energy by the molecule A in the normal state.

The energy acquired by a single molecule (or atom) in absorbing one quantum is dependent on the wavelength (or frequency) of irradiating light. A mole of photons is frequently referred to as an einstein.

The energy (E) absorbed by one mole of the reacting substance is thus given by

$$E = N_A h\nu \tag{6.21}$$

where N_A = Avogadro number and h = Plank's constant and ν is the frequency of the radiation absorbed.

or

$$E = N_A h \frac{C}{\lambda} \tag{6.22}$$

where C is velocity of light and λ is the wavelength.

Substituting, $N_A = 6.022 \times 10^{23}$, $h = 6.626 \times 10^{-34}$ Js, $C = 3.0 \times 10^8$ ms $^{-1}$

$$E = \frac{(6.022\times10^{23})(6.626\times10^{-34})(3.0\times10^{8})}{\lambda}$$

$$\mathbf{E = \frac{0.1197}{\lambda} J\ mol^{-1}} \tag{6.23}$$

Therefore, the energy possessed by one mole of photons or the energy absorbed by one mole of the reacting species is given by,

$$E = \frac{0.1197}{\lambda} \text{J mol}^{-1} \text{ in S.I. units.}$$

This is also known as an 'einstein' (one einstein (Nhv) is the energy absorbed by one mole of the reacting substance).

However, when the wavelength is expressed in nanometre,

$$\mathbf{E = \frac{0.1197}{\lambda_{(nm)}} J\ mol^{-1} \times 10^{9}} \tag{6.24}$$

However, in terms of C.G.S. units,

$$N_A = 6.022 \times 10^{23}, \quad h = 6.626 \times 10^{-27} \text{ ergs s}, \quad C = 3.0 \times 10^{10} \text{ cm/s}$$

$$E = \frac{(6.022\times10^{23})(6.626\times10^{-27})(3.0\times10^{10})}{\lambda}$$

$$E = \frac{0.1197\times10^{6}}{\lambda} \text{ergs mol}^{-1}; \quad \lambda \text{ is in centimetre} \tag{6.25}$$

Since $\quad 4.184 \times 10^7 \text{ ergs} = 1 \text{ cal}$

$$\therefore \quad E = \frac{119.7\times10^{6}}{4.184\times10^{7}\lambda} \text{cal mol}^{-1} \quad \text{or} \quad E = \frac{2.86}{\lambda} \text{cal mol}^{-1} \tag{6.26}$$

where λ is expressed in Å

$$\mathbf{E = \frac{2.86\times10^{8}}{\lambda(A)} cal\ mol^{-1}} \tag{6.27}$$

It is clear from Eq. (6.27) that the energy per einstein is inversely proportional to the wavelength. Therefore, the energies per einstein of X-rays radiations will be larger than that of UV or visible radiations, since the wavelengths associated with UV or visible radiation are larger than that of X-rays. It may be noted that the shorter the wavelength is, the greater is the energy per einstein. For example, the order of wavelength of some radiations is

X-rays < UV < Visible < Infrared

Therefore, the order of energies per einstein of these radiations would be:

X-rays > UV > Visible > Infrared

The energy, E, can also be expressed in electron volt (eV) using the relation,

$$1 \text{ eV} = 1.6022 \times 10^{-19} \text{ J}$$

The photochemical equivalence law applies only to the absorption or **primary photochemical process.** When as a result of primary absorption only one molecule decomposes and the products enter no further reaction, the number of molecules reacting will be equal to the number of energy quanta absorbed. More frequently, however, a molecule activated photochemically initiates a sequence of thermal reactions called secondary reactions, as a result of which several or many reactant molecules will undergo chemical change. However, in certain processes; less than one molecule may react per quantum. The deviations from the law are due to secondary reaction, which may lead to an increase or decrease in the quantum yield from unity.

Numerical Problems

Example 5 **Calculate the energy of one photon of light of wavelength 2450 Å. Will it be able to dissociate a bond in diatomic molecule which absorbs this photon and has a bond energy equal to 95 kcal per mole?**

Solution: Energy of one photon = hν

$$h\nu = \frac{hc}{\lambda}$$

Putting the values of Plank's constant (h) = 6.623×10^{-27} erg sec and the velocity of light (c) 3×10^{10} cm per second and the given wavelength of the photon in centimetres we get

$$\text{Energy of one photon} = \frac{(6.623\times10^{-27})\times3\times10^{10}}{2450\times10^{-8}}\text{ergs} = 8.11\times10^{-12}$$

$$= \frac{8.11\times10^{-12}}{4.184\times10^{-7}}\text{cals} = 1.938\times10^{-19}\text{ cals}$$

Energy required for the dissociation of one bond

$$= \frac{9500}{6.02\times10^{23}}\text{cals} = \mathbf{1.578\times10^{-19}\ cals}$$

This energy of photon is greater than the energy required to break the bond. Hence, the bond will be dissociated.

Example 6 **Calculate the value of einstein of energy corresponding to wavelength 6000 Å.**

Solution: We know, energy of einstein $= \frac{2.86\times10^{5}}{\lambda}$ kcal/ mole

$$= \frac{2.86\times10^{5}}{6000}\text{kcals/mole} = \mathbf{47.66\ kcals\,/\,mole}$$

Example 7 **Calculate the value of an einstein of energy in electron volts for radiation of frequency 3×10^{15} Hz.**

Solution: We are given that $\nu = 3 \times 10^{15}$ and we know that $c = 3 \times 10^{10}$ cm per second.

$$\lambda = \frac{c}{\nu} = \frac{3\times10^{10}}{3\times10^{15}} = 10^{-5}\text{ cm} = 10^{3}\text{ Å} \qquad [1\text{ Å} = 10^{-10}\text{m}]$$

We also know that the energy of an einstein is given by the relation:

$$E = \frac{2.86 \times 10^5}{\lambda} \text{kcals/mole}$$

$$= \frac{2.86 \times 10^5}{10^3} \text{kcals/mole} = \frac{286}{23.06} = \mathbf{12.4\,eV}$$

Example 8 **If the value of an einstein is 72 kcal, calculate the wavelength of the light.**

Solution: We are given that $E = 72$ kcal.

From the equation, $E = \frac{2.86 \times 10^5}{\lambda}$ kcals/mole, $\lambda = \frac{2.86 \times 10^5}{E}$ Å $= \frac{2.86 \times 10^5}{72} = \mathbf{3972.2\,Å}$

6.5 QUANTUM EFFICIENCY

Quantum efficiency or quantum yield may be defined as the number of moles reacting per einstein of the light absorbed. It is denoted by and is expressed as 'quantum efficiency or quantum yield may be defined as the number of moles reacting per einstein of the light absorbed'. It is denoted by ϕ and is expressed as:

$$\phi = \frac{\text{Number of molecules reacting in given time}}{\text{Number of quanta of light absorbed in the same time}} \tag{6.28}$$

If we measure the rate of formation of substance A in mole per second, dN_A/dt, the quantum yield ϕ_A is given by

$$\phi_A = \frac{dN_A/dt}{\text{Number of quanta absorbed / second}} = \frac{dN_A/dt}{I_a} \tag{6.29}$$

The law of photochemical equivalence is applicable only to *primary photochemical process*. It should be borne in mind that primary processes which are a direct result of the absorption of light, in many cases, are marked by *secondary reactions*. Some of the secondary reactions occur in long chains and deviations observed with respect to the applicability of the law are many.

The quantity of radiation absorbed by a system can be determined experimentally. It is usually done by measuring the intensity of radiation before and after passing through the reacting system. If the rate of chemical change also is determined by suitable method, it is possible to calculate the amount of chemical change produced by the absorption of a definite amount of radiation. In other words, the number of molecules reacted by the absorption of one quantum of light, i.e., quantum efficiency, can be calculated.

The quantum yields of some well-studied photochemical reactions are given in Table 6.2.

6.5.1 Explanation of the Unexpected Behaviour

In many cases, the quantum yield is higher and in some cases, the quantum yield is lower than what may be expected from the law. *The higher yield may be explained by the supposition that the quantum of energy absorbed in the primary process may be more than required to bring about the primary reaction and the excess may be passed on to many other reactants.*

The lower quantum yield, on the other hand, is supposed to be due to the following reasons:

(i) The recombination of some of the molecules of the reactants.
(ii) The deactivation of the excited molecules before they react.

(iii) The insufficiency of the absorbed quantum to activate some of the reacting molecules.
(iv) The loss of excitation energy of excited molecules in collision with non-excited molecules. The law of photochemical equivalence is strictly true only in case of a few reactions.

Table 6.2 Effective frequencies of radiations for a few common photochemical reactions and their quantum yields.

S.No.	Reactions	Effective Wavelength Å	ϕ (Quantum yield)
(i)	Maleic acid $\rightarrow$ fumaric acid	2000–2800	0.04
(ii)	$CH_3COCH_3 \rightarrow CO + C_2H_6$	3000	0.1
(iii)	$2\,NH_3 \rightarrow N_2 + 3\,H_2$	2100	0.2
(iv)	$2\,H_2O_2 \rightarrow 2\,H_2O + O_2$	3100	0.7
(v)	$SO_2 + Cl_2 \rightarrow SO_2Cl_2$	4200	1.0
(vi)	$H_2S \rightarrow H_2 + S$	2080	1.0
(vii)	$ClCH_2COOH + H_2O - HOCH_2COOH + HCl$	2537	1.0
(viii)	$2\,NO_2 \rightarrow 2\,NO + O_2$	3600	1.5
(ix)	$2\,HBr \rightarrow H_2 + Br_2$	2070–2530	2.0
(x)	$2\,HI \rightarrow H_2 + I_2$	2070–2820	2.0
(xi)	$(CH_3)_2N \rightarrow C_2H_6 + \frac{1}{2}\,N_2$	3360	2.0
(xii)	$3\,O_2 \rightarrow 2\,O_3$	1700–1900	3.0
(xiii)	$CO + Cl_2 \rightarrow COCl_2$	4000–4360	10^3
(xiv)	$H_2 + Cl_2 \rightarrow 2\,HCl$	4000–4360	10^6

6.5.2 Classification of Photochemical Reactions (Based on Their Quantum Efficiencies)

According to quantum yield, photochemical reactions may be classified into the following three categories:

(1) ***Those reactions in which ϕ (quantum yield) is a simple integer*** *such as 1, 2, 3 etc.* For example, the dissociation of HI and HBr, the ozonization of oxygen and the combination of SO_2 and Cl_2.
(2) ***Those reactions in which ϕ (quantum yield) is extremely low (less than 1).*** For example, the dissociation of acetone vapour, H_2O_2, NH_3 etc.
(3) ***Those reactions in which ϕ (quantum yield) is extremely high.*** For example, the combination of hydrogen and chlorine, carbon monoxide and chlorine, etc.

In order to explain these deviations from the law of photochemical equivalence, it is believed that photochemical reactions involve two distinct processes.

1. **Primary processes** are those in which light radiation is absorbed by atoms or molecules of reactants to give excited atoms or molecules or to cause dissociation of molecules yielding atoms (some in the excited state) or free radicals.

$$\underset{\text{(atom or molecules)}}{A} + \underset{\text{quantum of energy}}{h\nu} \rightarrow \underset{\text{excited atom or molecule}}{A^*}$$

It may be pointed out that the law of photochemical equivalence is applicable only to the primary processes in photochemical reaction.

2. **Secondary processes** involve the excited atoms, molecules or free radicals produced in the primary process. They may take place in dark also and irrespective of the light radiation. The light radiations are essential only for primary processes.

MEMORY FOCUS

1. Grotthus Draper law states that '*It is only the absorbed light radiations that are effective in producing a chemical reaction.*'
2. Stark–Einstein law states that '*In a photochemical reaction, each molecule of the reacting substance absorbs a single photon of radiation causing the reaction and is activated to form the products.*'
3. The overall photochemical reaction may consist of (A) primary reaction and (B) secondary reactions.
4. A primary reaction proceeds by the absorption of radiation whereas a secondary reaction is a thermal reaction which occurs subsequent to the primary reaction.
5. The energy possessed by an Avogadro number (N) of photons is referred to as one Einstein.

$$E = N_A hV$$

$$E = \frac{0.1197}{\lambda} \text{ J mol}^{-1}$$

$$E = \frac{0.1197 \times 10^6}{\lambda} \text{ ergs mol}^{-1}; \qquad \lambda \text{ is in cm}$$

$$\therefore \quad E = \frac{119.7 \times 10^6}{4.184 \times 10^7 \lambda} \text{ cal mol}^{-1} \quad \text{or} \quad E = \frac{2.86}{\lambda} \text{ cal mol}^{-1}$$

6. The higher the wavelength is, the smaller will be the energy per einstein.
7. The number of molecules reacted or formed per photon of light absorbed is termed quantum yield (ϕ).

$$\phi = \frac{\text{Number of molecules reacting in given time}}{\text{Number of quanta of light absorbed in the same time}}$$

8. For a reaction that obeys strictly the Einstein law, one molecule decomposes per photon, the quantum yield $\phi = 1$. When two or more molecules are decomposed per photon, $\phi > 1$ and reaction has a **high quantum yield**. If the number of molecules decomposed is less than one per photon, the reaction has **a low quantum yield.**
9. **Cause of high quantum yield:**
   The main reason for high quantum yield is:

   The higher yield may be explained by the supposition that the quantum of energy absorbed in the primary process may be more than required to bring about the primary reaction and the excess may be passed on to many other reactants.
10. **Causes of low quantum yield**:
    The main reasons for low quantum yield are:

    (i) recombination of some of the molecules of the reactants,
    (ii) deactivation of the excited molecules before they react;
    (iii) insufficiency of the absorbed quantum to activate some of the reacting molecules and
    (iv) loss of excitation energy of excited molecules in collision with non-excited molecules.

    The law of photochemical equivalence is strictly true only in case of a few reactions.

Numerical Problems

Example 9 **A certain system absorbs 3×10^{18} quanta of light per second. On irradiation for 20 minutes, 0.003 mole of the reactant was found to have reacted. Calculate the quantum yield for the process (Avogadro's number = 6.024×10^{23}).**

Solution: Energy absorbed per second = 3×10^{18} quanta

The number of moles reacting in 20 min = 0.003

The number of molecules reacting in 20 min = $0.003 \times 6.024 \times 10^{23}$

The number of molecules reacting per second = $\dfrac{0.003 \times 6.204 \times 10^{23}}{20 \times 60} = 1.506 \times 10^{18}$

Therefore, the quantum efficiency is:

$$(\phi) = \frac{\text{Number of molecules reacting per sec}}{\text{Number of quanta absorbed per sec}} = \frac{1.506 \times 10^{18}}{3 \times 10^{18}} = \mathbf{0.502}$$

Example 10 **In the photochemical reaction between Cl_2 and H_2, 2×10^{-10} moles of chlorine have been converted to hydrochloric acid by irradiation with light of wavelength 4000 Å. What is the quantum efficiency of reaction if the light energy absorbed in the system is 7.16×10^{-15} kcal?**

Solution: Energy of one photon = $h\nu = \dfrac{hc}{\lambda}$ where c is the velocity, λ is the wavelength, ν is the frequency and h stands for Planck's constant. Substituting these values:

$$\text{Energy of photon} = \frac{6.623 \times 10^{-27} \times 3 \times 10^{10}}{4000 \times 10^{-8}} = 4.97 \times 10^{-12} \text{ ergs}$$

Total energy absorbed = 7.16×10^{-15} kcal = $7.16 \times 10^{-15} \times 10^{-3} \times 4.18 \times 10^{7}$ ergs

= $7.16 \times 4.18 \times 10^{-5}$ ergs

$$\text{The number of photon} = \frac{\text{Total energy absorbed}}{\text{Energy of a photon}}$$

$$= \frac{1.16 \times 4.18 \times 10^{-5}}{4.97 \times 10^{-12}} = 6.02 \times 10^{-7}$$

The amount of chlorine reacted = 2.0×10^{-10} moles.

∴ The number of molecules of chlorine reacted = $2.0 \times 10^{-10} \times 6.02 \times 10^{23} = 12.04 \times 10^{13}$ molecules.

$$\text{Quantum efficiency} = \frac{\text{Number of molecules reacted}}{\text{Number of photons absorbed}}$$

$$= \frac{12.04 \times 10^{13}}{6.02 \times 10^{7}} = \mathbf{2 \times 10^{6}}$$

Example 11 **The quantum efficiency of the photochemical reaction $H_2(g) + Cl_2(g) \xrightarrow{h\nu} 2HCl(g)$ is 1×10^{6} with a wavelength of 480 nm. What amount of HCl (g) would be produced under these conditions when one joule of radiant energy is absorbed?**

Solution: We know that $E = Nh\nu = \dfrac{Nhc}{\lambda}$

$$E = \frac{6.023 \times 10^{23}\ \text{mol} \times 6.623 \times 10^{-34}\ \text{Js} \times 3 \times 10^{8}\ \text{m/s}}{480 \times 10^{-9}\ \text{m}} \qquad [1\ \text{nm} = 10^{-9}\ \text{m}]$$

$$= 2.494 \times 10^{-5}\ \text{Joules mol}^{-1}$$

It means that the number of quanta of energy absorbed $= \frac{1J}{2.494 \times 10^{-5}\ \text{J mol}^{-1}} = 0.40 \times 10^{-5}$

We know that: $\phi = \frac{\text{Number of } Cl_2 \text{ moles reacted}}{\text{Number of quanta absorbed}}$

$$1 \times 10^{6} = \frac{\text{Number of } Cl_2 \text{ moles reacted}}{0.40 \times 10^{-5}}$$

∴ The number of Cl_2 moles reacted = 1 × 106 × 0.40 × 10^{-5} = **4 moles**

It is clear that one mole of chlorine on reacting with one mole of hydrogen gives two moles of HCl.

Hence, the amount of HCl produced = 2 × 4 = **8 moles**

Example 12 **A certain system absorbs 2.0×10^{16} quanta of light per second. At the end of 20 minutes, it is observed that 0.002 moles of the irradiated substance has reacted. What is the quantum efficiency of the process? ($N = 6.023 \times 10^{23}$)**

Solution: The amount of energy absorbed in 20 min $= 2.0 \times 10^{16} \times 60 \times 20$ quanta.

The number of molecules reacted in 20 min $= 0.002 \times 6.023 \times 10^{23}$

$$\therefore \text{ Quantum efficiency} \quad \phi = \frac{\text{Number of molecules reacted}}{\text{Number of quanta absorbed}} = \frac{0.002 \times 6.023 \times 10^{23}}{20 \times 10^{16} \times 60 \times 20} = 50.0$$

6.6 STUDY OF SOME PHOTOCHEMICAL REACTIONS

The mechanism and rate law of some photochemical reactions are explained in the following sections.

(A) Kinetics of the photosynthesis of HBr: The quantum efficiency of photosynthesis of HBr from H_2 and Br_2 is very low, i.e., 0.01. The following chain mechanism has been suggested.

$$H_2 + Br_2 \xrightarrow{h\nu} 2HBr$$

Step	Processes		Rate
1.	$Br_2 + h\nu \xrightarrow{k_1} 2Br$	(Chain initiation)	$k_1 I_{abs}$
2.	$Br + H_2 \xrightarrow{k_2} HBr + H$	(Endothermic)	$k_2 [Br] [H_2]$
3.	$H + Br_2 \xrightarrow{k_3} HBr + Br$	(Chain propagation)	$k_3 [H] [Br_2]$
4.	$H + HBr \xrightarrow{k_4} Br + H_2$	(Chain inhibition)	$k_4 [H] [HBr]$
5.	$Br + Br \xrightarrow{k_5} Br_2$	(Chain termination)	$k_5 [Br]^2$

In this photochemical reaction, chain is initiated in the presence of light (Step 1) and this is also a primary process. Steps 2–5 are secondary processes. Step 2 is *endothermic*, i.e., this reaction takes place

to a small extent. The Step 3 depends on the formation of H atoms (Step 2); therefore, it also takes place to a very small extent. In Step 4, HBr is not formed, rather is consumed. Therefore, the quantum yield of this reaction is very low (i.e., 0.01). However, the quantum yield of this reaction increases with the increase in temperature, as Step 2, which is endothermic, takes place more efficiently at higher temperature and as a result Step 3 also occurs to a greater extent and hence result in more formation of HBr and thereby causing high quantum efficiency for this reaction.
The rate of formation of HBr is given by

$$\frac{d[\mathrm{HBr}]}{dt} = \mathrm{k}_2[\mathrm{H}_2][\mathrm{Br}] + \mathrm{k}_3[\mathrm{H}][\mathrm{Br}_2] - \mathrm{k}_4[\mathrm{H}][\mathrm{HBr}] \tag{6.30}$$

Applying steady-state condition to H and Br atom (i.e., when the concentration of the short-lived intermediate is very low suggested that the rate at which reactive intermediates are formed is equal to their rate of being used in the reaction).

$$\frac{d[\mathrm{H}]}{dt} = \mathrm{k}_2[\mathrm{H}_2][\mathrm{Br}] - \mathrm{k}_3[\mathrm{H}][\mathrm{Br}_2] - \mathrm{k}_4[\mathrm{H}][\mathrm{HBr}] = 0 \tag{6.31}$$

and

$$\frac{d[\mathrm{Br}]}{dt} = \mathrm{k}_1 I_{\mathrm{abs}} - \mathrm{k}_2[\mathrm{Br}][\mathrm{H}_2] - \mathrm{k}_3[\mathrm{H}][\mathrm{Br}_2] - \mathrm{k}_4[\mathrm{H}][\mathrm{HBr}] - \mathrm{k}_5[\mathrm{Br}]^2 = 0 \tag{6.32}$$

Adding Eqs (6.31) and (6.32), we get

$$\mathrm{k}_1 I_{\mathrm{abs}} - \mathrm{k}_5[\mathrm{Br}]^2 = 0$$

$$[Br] = \left(\frac{\mathrm{k}_1}{\mathrm{k}_5} I_{\mathrm{abs}}\right)^{1/2} \tag{6.33}$$

From Eq. (6.31), we get

$$[\mathrm{H}] = \frac{\mathrm{k}_2[\mathrm{Br}][\mathrm{H}_2]}{\mathrm{k}_3[\mathrm{Br}_2] + \mathrm{k}_4[\mathrm{HBr}]} \tag{6.34}$$

Substituting the value of [Br]

$$[\mathrm{H}] = \frac{\mathrm{k}_2\left(\frac{\mathrm{k}_1}{\mathrm{k}_5} I_{\mathrm{abs}}\right)^{1/2}[\mathrm{H}_2]}{\mathrm{k}_3[Br_2] + \mathrm{k}_4[\mathrm{HBr}]}$$

Putting the values of [H] and [Br] in Eq. (6.30), we get

$$\frac{d[\mathrm{HBr}]}{dt} = \mathrm{k}_2[\mathrm{H}_2]\left(\frac{\mathrm{k}_1}{\mathrm{k}_5} I_{\mathrm{abs}}\right)^{1/2} + \mathrm{k}_3\left[\frac{\mathrm{k}_2(\mathrm{k}_1/\mathrm{k}_5 I_{\mathrm{abs}})^{1/2}[\mathrm{H}_2]}{\mathrm{k}_3(\mathrm{Br}_2) + \mathrm{k}_4(\mathrm{HBr})}\right][\mathrm{Br}_2] - \mathrm{k}_4\left[\frac{\mathrm{k}_2(\mathrm{k}_1/\mathrm{k}_5 I_{\mathrm{abs}})^{1/2}[\mathrm{H}_2]}{\mathrm{k}_3(\mathrm{Br}_2) + \mathrm{k}_4(\mathrm{HBr})}\right] \times [\mathrm{HBr}]$$

when $T = 0 \Rightarrow [\mathrm{HBr}] = 0$
On solving the above equation, we get

$$\frac{d[\mathrm{HBr}]}{dt} = 2\mathrm{k}_2 \frac{\mathrm{k}_1^{1/2}}{\mathrm{k}_5^{1/2}}[\mathrm{H}_2]\, I_{\mathrm{abs}}^{1/2}$$

$$= k\,[H_2]\,I_{abs}^{1/2} \quad \left(\text{where } k = 2k_2 \frac{k_1^{1/2}}{k_5^{1/2}}\right)$$

This expression derived, based on the mechanism proposed, agrees with the experimental expression for the rate of formation of HBr found by Bodenstein and Lutkemeyer.

(B) Kinetics of the photosynthesis of HCl: In this reaction, as many as 10^6 molecules of HCl are formed when one quanta of light is absorbed. Thus, the quantum yield following chain mechanism has been proposed.

$$H_2 + Cl_2 \xrightarrow{h\nu} 2HCl$$

Steps	Processes		Rate
1.	$Cl_2 + h\nu \xrightarrow{k_1} 2\,Cl$	(Chain initiation)	$k_1 I_{abs}$
2.	$Cl + H_2 \xrightarrow{k_2} HCl + H$	(Exothermic)	$k_2\,[Cl]\,[H_2]$
3.	$H + Cl_2 \xrightarrow{k_3} HCl + Cl$	(Chain propagation)	$k_3\,[H]\,[Cl_2]$
4.	$Cl + Wall \xrightarrow{k_4} ½\,Cl_2$	(Chain termination at the wall of container)	$k_4\,[Cl]$

Unlike H_2–Br_2 reaction, Step 2 in which Cl combines with H_2 to form HCl and H is an exothermic step, i.e. even at room temperature, this reaction occurs to a large extent and thus Step 3 is also carried out efficiently because of the large availability of H atom. This explains extremely large quantum yield for this reaction as compared to H_2–Br_2 reaction. The rate of formation of HCl is given by

$$\frac{d[HCl]}{dt} = k_2[Cl]\,[H_2] + k_3[H]\,[Cl_2] \tag{6.35}$$

Applying steady-state condition to [H] and [Cl] atom, which are short-lived intermediates,

Rate of formation of [H] or [Cl] = Rate of disappearance of [H] or [Cl]

Or, net rate of formation of [H] or [Cl] = 0

$$\therefore \quad \frac{d[H]}{dt} = k_2\,[Cl]\,[H_2] + k_3\,[H]\,[Cl_2] = 0$$

i.e.

$$k_2\,[Cl]\,[H_2] = k_3\,[H]\,[Cl_2] \tag{6.36}$$

Similarly

$$\frac{d[Cl]}{dt} = k_1 I_{abs} - k_4\,[Cl] = 0 \tag{6.37}$$

$$\therefore \quad [Cl] = \frac{k_1}{k_4} I_{abs} \tag{6.38}$$

Substituting Eq. (6.38) in Eq. (6.35), we get

$$\frac{d[HCl]}{dt} = 2k_2[Cl][H_2] \quad \text{or} \quad \frac{d[HCl]}{dt} = 2k_2 \frac{k_1}{k_4} I_{abs}[H_2] \tag{6.39}$$

or

$$\frac{d[HCl]}{dt} = kI_{abs}[H_2]$$

Bodenstein and Unger also experimentally found that the rate of formation of HCl from H_2 and Cl_2 is also given as k I_{abs} $[H_2]$. There is a strong support in favour of chain mechanism:

(i) The combination of hydrogen and chlorine can be initiated in the absence of radiations by introducing either chlorine atom or hydrogen atom into the gas.

(ii) The quantum yield of this reaction can be decreased to a large extent by working in capillary tubes, since the chains are terminated at the walls. The capillary tubes provide large areas of walls and hence support chain mechanism for H_2–Cl_2 reaction.

(iii) In the presence of even small amounts of oxygen, the quantum efficiency is lowered. O_2 has inhibiting effect. The various steps, in the presence of O_2, can be depicted as in the following table.

Steps	Primary Process	Rate
1.	$Cl_2 + h\nu \xrightarrow{k_1} 2Clp$	$k_1 I_{abs}$
2.	$Cl_2 + H_2 \xrightarrow{k_2} HCl + H$	k_2 [Cl] $[H_2]$
3.	$H + Cl_2 \xrightarrow{k_3} HCl + Cl$	k_3 [H] $[Cl_2]$
4.	$H + O_2 \xrightarrow{k_4} HO_2$	k_4 [H] $[O_2]$
5.	$Cl + O_2 \xrightarrow{k_5} ClO_2$	k_5 [Cl] $[O_2]$
6.	$Cl + X \xrightarrow{k_0} ClX$	k_5 [Cl] [X]

Here, *X* is any substance which removes oxygen. Steps 4 and 5, which occur in the presence of O_2, remove H and Cl and hence decrease the rate of formation of HCl.

6.7 FLUORESCENCE AND PHOSPHORESCENCE

As discussed earlier, when a molecule is excited by light, it can lead to a chemical reaction or energy transfer by collision. However, if both fluorescence and phosphorescence do not occur, then the molecule will return to the ground state with the release of energy which is either a short-lived emission, known as fluorescence, or a long-lived emission, known as phosphorescence.

6.7.1 Fluorescence

The absorbed photoenergy sends electron from lower energy level to higher energy level and the excited molecule immediately or instantaneously re-radiates or emits a part of the absorbed energy at a greater wavelength, as the electron returns to or reverts back to its normal state within a short time ($=10^{-8}$ s). 'Florescence' is instantaneous and starts immediately after the absorption of light and stops the moment when the radiation is cut off.

Examples of fluorescence are solutions of fluorescein and eosin. When their solutions are placed in light, they show fluorescence from green to violet colour; Uranyl sulphate, chlorophyll (green pigment of photosynthetic organisms such as plants and algae), fluorspar (CaF_2) and petroleum also show this phenomenon. It may be pointed that different substances show the phenomenon of fluorescence in the light of different wavelengths. Thus, fluorspar shows fluorescence with blue light, chlorophyll with red light, uranium glass with green light and so on.

The phenomenon of fluorescence finds wider application such as:

(i) Fluorescence is used for lighting purposes in fluorescent tubes. Mercury arc producing large proportion of UV light is used in a tube coated with fluorescent salts which gives visible light; the intensity of which is not far different from day light.

(ii) By mixing fluorescent dyes with coloured paints, the fluorescence of the dye helps the light reflected by the paint to produce extraordinary brightness and lustre. These materials are used as road signs. For example, Brilliant Sulpho Flavine FF and Rhodamine 6G dispersed in a special kind of plastic are used for this purpose.

(iii) It is used in the industry for testing and identifying materials, e.g., rubber industry.

(iv) In analysis, the concentration of riboflavin (vitamin B_2) in chloroform has been examined.

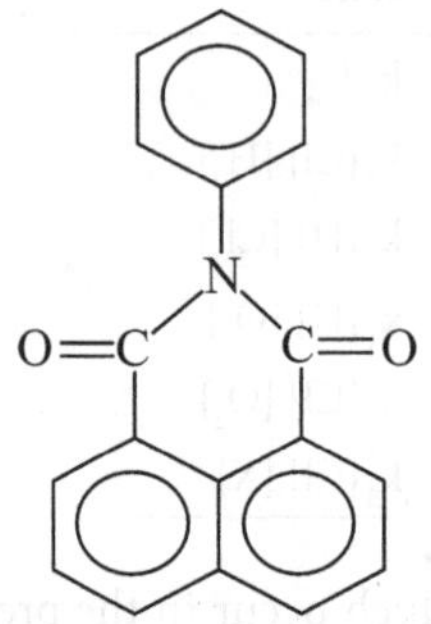

Brilliant Sulpho Flavine – FF

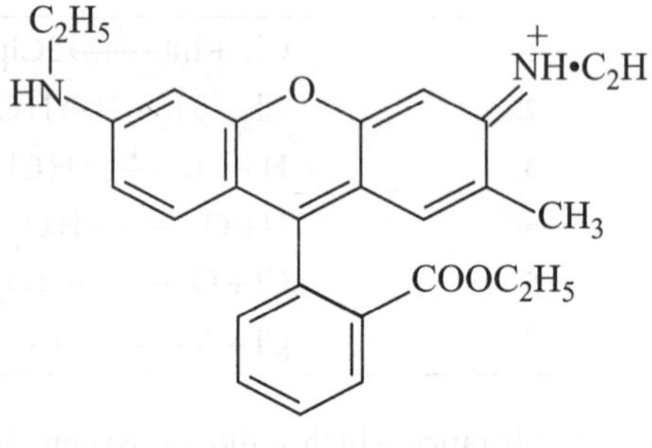

Rhodamine – 6G

(v) In television, the cathode stream of the photoelectric effect is made visible in the cathode ray tube by adding ZnS to which a little Ni is added to cut off phosphorescence which otherwise makes the picture blurred.

(vi) The use of fluorescent microscopes and fluoroscope used in X-ray diagnosis help in testing the condition of food stuff, detecting ring worms etc.

(vii) A fluorescent dye is used as whitener for washing cloth when mixed with washing powder.

6.7.2 Phosphorescence

In this process, the absorbed light energy raises the electron to a higher level and the excited electron reverts back to the original state not instantaneously but after some time lag and consequently absorbed light is re-radiated or emitted after some time. Thus, phosphorescence may be regarded as slow fluorescence. In this phenomenon, the emission of light of different wavelengths continues even after the source of light radiation has been cut off.

The familiar examples of phosphorescence are BaS or SrS containing about 2.5% alkali chlorides (NaCl or KCl) and a trace of heavy metal sulphide. Such a mixture is generally used for painting watch dials, electric switches etc.

Many dyestuffs which are fluorescent in aqueous solution become phosphorescent when dissolved in glycerol and cooled.

6.7.3 Photophysical Process—Consequence of Light Absorption (Jablonski Diagram)

In order to understand the mechanism of these photophysical processes, namely fluorescence and phosphorescence, we have to learn certain terms such as *spin multiplicity* and *singlet and triplet states.* In the ground state, most of the molecules having an even number of electrons have paired spin. The quantity (2S + 1) is known as spin multiplicity of a state, where 8 is the total electron spin. In the ground state, the two paired electrons (↑↓) have their spin in the opposite direction and the total spin (S) is taken as zero. This can be represented as

$$S_1 = \frac{1}{2}; \qquad S_2 = -\frac{1}{2}; \qquad \text{so that} \qquad S = S_1 + S_2 = \frac{1}{2} - \frac{1}{2} = 0$$

Hence, 2S + 1 = 1 and the molecule is said to have a *singlet ground state* Fig. 6.2(a). The spin multiplicity of the molecule in the ground state is 1.

When a photon absorbs suitable energy (h*ν*), one of the paired electrons goes to higher energy level and the molecule is said to be in the *excited state.* The spin orientations of the two single electrons may be either parallel or anti-parallel and they are represented in Fig. 6.2(b) and 6.2(c), respectively.

When the spins are parallel, as in Fig. 6.2(b), then

$$S = S_1 + S_2 = \frac{1}{2} + \frac{1}{2} = 1$$

Hence, $$2S + 1 = 3$$

The spin multiplicity of the molecule is 3 and the molecule is said to have the *triplet excited state.*

When the spins are anti-parallel, as in Fig. 6.2(c), then

$$S = S_1 + S_2 = \frac{1}{2} - \frac{1}{2} = 0$$

Hence, $$2S + 1 = 1$$

The spin multiplicity of the molecule is 1 and the molecule is said to have the *singlet excited state.* Depending upon the energy of the photon absorbed, the electron may jump to any of the higher electronic states 2, 3, 4…. Thus, we may get a series of singlet excited states represented as S_1, S_2, S_3 etc. and triplet excited states represented as T_1, T_2, T_3 etc. According to quantum mechanics, a singlet excited state has higher energy than the corresponding triplet excited state. The energy sequence is as stated in the following equation:

$$E_{S_1} > E_{T_1};\ E_{S_2} > E_{T_2};\ E_{S_3} > E_{T_3};\ \text{and so on}$$

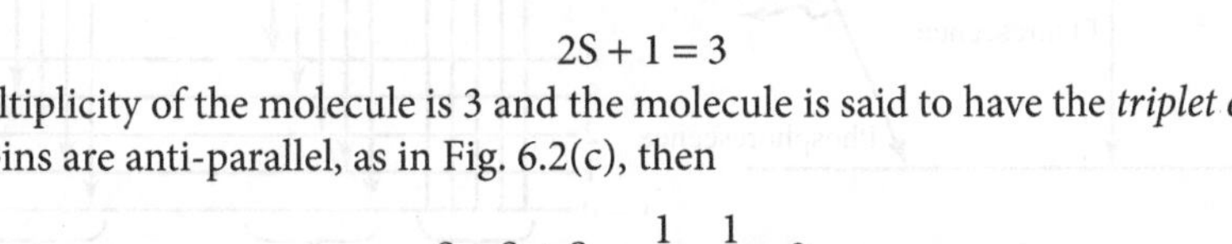

Figure 6.2 Parallel and anti-parallel spin orientations

When light is absorbed, the electron of the absorbing molecule jumps from S_0 to S_1, S_2 or S_3 singlet excited state depending upon the amount of photons absorbed as shown in Jablonski diagram (Fig. 6.3). For each singlet state (S_1, S_2, S_3 etc.), there is a corresponding triplet excited state (T_1, T_2, T_3 etc.). Whatever may be the excited state, singlet or triplet, the molecule of the substance in the ground state (A_0) is said to be in the excited state (A). Thus

$$A_0 + h\nu \rightarrow A^*$$

The *activated* molecule (A^*) returns to the ground state by dissipating its energy through the following types of transitions.

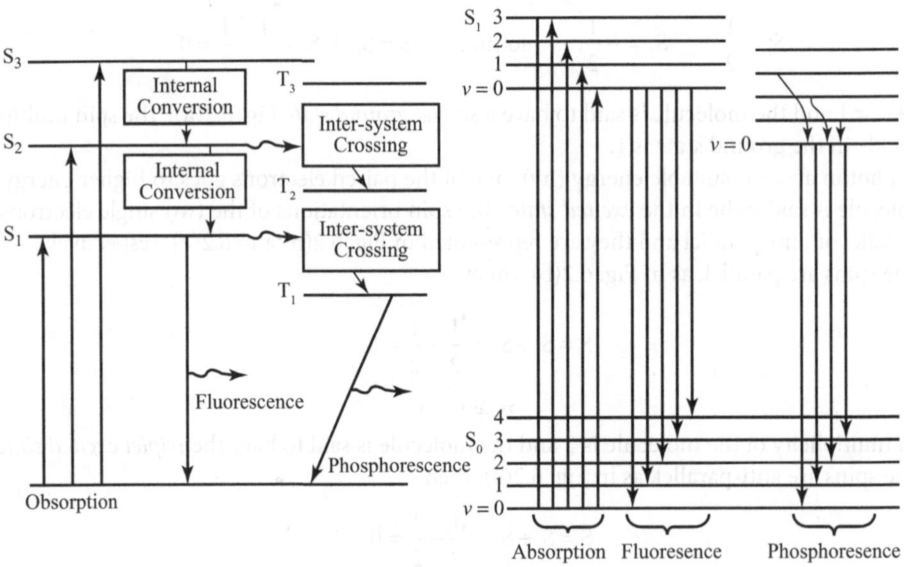

Figure 6.3 Jablonski diagram

6.7.3.1 Non-radiative Transition

These transitions include the return of the excited electron within the energy levels of same multiplicity, i.e., either from S_3, S_2 to the first excited S_1 or from T_3, T_2 to the first excited state T_1. These transitions do not *involve* the emission of any radiations and are thus referred to as *non-radiative or radiationless transitions.* The energy of the activated molecule is dissipated in the form of heat through molecules collisions. The process is called internal conversion (IC) and takes place in less than about 10^{-11} s. The *activated* molecule may also lose energy by another process called intersystem crossing (ISC). This process *involves* transitions between states (energy levels) of different multiplicity, for instance from S_2 to T_2 or S_1 to T_1, which are also *non-radiative or radiationless.* Such transitions are spectroscopically forbidden, but they do take place at a slower rate comparatively.

6.7.3.2 Radiative Transitions

These transitions include the return of activated molecule within the singlet state directly from S_3, S_2, S_1 to S_0 and also indirectly from singlet excited triplet state (T_1) to the ground state (S_0). Such transitions are accompanied by the emission of the radiation. These transitions from S_1 to S_0 are spectroscopically allowed and take place in about 10^{-8} s. This type of emissions of radiation is called *fluorescence.*

The transitions from T_1 to the ground state S_0 are forbidden spectroscopically and take place rather slowly. The lifetime is much longer being of the order of 10^{-3} s or greater. It is because the transitions involve the inversion of spin which requires time. The emission of radiation from triplet excited state to the ground state, i.e. from T_1 to S_0, is called phosphorescence.

Both the emitted radiations (fluorescence and phosphorescence) have shorter frequency (greater wavelength) when compared to the absorbed light used for excitation. This is clear because some part of absorbed light energy is dissipated as heat energy due to non-radiative transitions.

6.7.3.3 Photochemical Reaction

The molecule after activation may lose energy by undergoing chemical reaction with other molecules. The activated molecule in the singlet excited state does not get a chance to react chemically, since they return quickly to the ground state. However, molecules in the triplet excited state return to get a chance to undergo chemical reaction. Therefore, it is necessary for a molecule to be in the activated triplet excited state before it can undergo a chemical reaction.

6.7.4 Mechanism of Fluorescence and Phosphorescence

When a molecule absorbs a photon, it gets excited and after excitation, it can end up in any of vibration level/levels in the first (or second) excited states and these excited states are unstable; the molecule can directly return to ground electronic state with the emission of light which stays for 10^{-9}–10^{-6} s. This phenomenon is referred to as *fluorescence.* When the excited molecule radiates, the frequency is lower than that of the exciting radiation because the energy gap between the levels is smaller.

Since the vibrationally excited molecules in the ground state can undergo collision with other molecules, they lose energy and return to the zero point level. Fluorescence emission spectrum is the mirror image of the absorption spectrum (Fig. 6.3).

As phosphorescence involves a triplet state, the activated molecules in the singlet excited state must undergo intersystem conversion (ISC), involving loss of energy. Under these conditions, the transition between the states with different multiplicities will take place, although this is forbidden spectroscopically. The triplet state has a long lifetime and hence phosphorescence radiation is emitted slowly. The energy of a triplet state is lower than that of exiled singlet state because in the triplet state, the electrons having the same spin tend to avoid each other and as the electrons are at a large distance and hence there is smaller repulsion and thus the energy is lower in the triplet state. That is why the phosphorescence occurs at frequencies lower than fluorescence.

6.7.5 Difference Between Fluorescence and Phosphorescence

(1) Phosphorescence has much longer decay period (10^{-4} to 100 s) than fluorescence (10^{-6} to 10^{-9} s).
(2) Phosphorescence spectrum is not the mirror image of the absorption spectrum whereas fluorescence spectrum is the mirror image of the absorption spectrum.
(3) Phosphorescence is the radiation emitted in a transition between states of different multiplicity whereas fluorescence is the radiation emitted in a transition between states of same multiplicity.
(4) Phosphorescence is not observed in solutions at room temperature whereas fluorescence can be observed in solutions at room temperature.
(5) Phosphorescence is rarely observed in gases. Fluorescence is exhibited by some elements in the vapour state, e.g., sodium, iodine and mercury vapours.

MEMORY FOCUS

1. The quantum efficiency of the photosynthesis of HBr from H_2 and Br_2 is very low (0.01). This is due to the reason that most of the bromine atoms produced in the primary process recombined to give back Br_2 molecules. Thus, HBr molecules are less formed and efficiency becomes low.
2. The quantum efficiency of photosynthesis of HCl from H_2 and Cl_2 is very high (10^4 to 10^6). This is due to the reason that this reaction is self-propagating chain reaction.
   (i) $Cl_2 + h\upsilon \rightarrow 2\ Cl$
   (ii) $Cl + H_2 \rightarrow HCl + H$
   (iii) $H + Cl_2 \rightarrow HCl + Cl$
   Step (ii) and (iii) are repeated again and again till the available H_2 and Cl_2 molecules are consumed.
3. Certain molecules, when exposed to light radiation of short wavelength (higher frequency), emit light of longer wavelength. The process is called **fluorescence,** e.g., a solution of chlorophyll in ether shows blood-red fluorescence.
4. When a substance absorbs radiation of higher frequency and emits light even after the incident radiation is cut off, the process is called **phosphorescence,** e.g., barium sulphate, calcium sulphate and strontium sulphate exhibit phosphorescence.
5. The transitions in which the excited electron within the energy levels of same multiplicity, i.e. either from S_3, S_2 return to the first excited state (S_1) and do not involve the emission of any radiation are called **non-radiative transitions.**
6. The transitions in which the activated molecule within singlet state returns directly from S_3, S_2, S_1 to S_0 and also indirectly from singlet excited triplet state (T_1) to the ground state (S_0) and involves the emission of radiation are called **radiation transitions.**
7. Molecules in the triplet excited state return to get a chance to undergo chemical reaction. Therefore, it is necessary for a molecule to be in the activated triplet excited state before it can undergo a chemical reaction.

REVISION QUESTIONS

1. How do photochemical reactions differ from thermal reactions?
2. (a) Explain Lambert's and Beer's laws.
   (b) What do you understand by 'absorption coefficient' and 'molar extinction coefficient'?
3. State and explain Einstein's law of photochemical equivalence.
4. Derive the relationship between einstein (in calories) and the wavelength of the radiation (in angstrom units).
5. What is Einstein's law of photochemical equivalence? Is it applicable to all cases of photochemical reactions?
6. What do you understand by the term 'quantum yield' or 'quantum efficiency'? What does it signify?
7. How would you explain very high and very low quantum efficiencies of some photochemical reactions?

8. Give one example of photochemical reaction in which quantum yield is very high. Briefly explain the reasons for same.
9. What is meant by primary and secondary processes in photochemistry?
10. Explain photosynthesis of HBr. What is the quantum yield of this process?
11. Write short notes on 'fluorescence' and 'phosphorescence'. What is the difference between them?
12. How do you explain that in fluorescence the wavelength of the light emitted is usually greater than that of the light absorbed?
13. State and explain Einstein's law of photochemical equivalence. What are quantum yields? What are the causes of high and low quantum yields?
14. (a) Discuss the reaction $H_2(g)+Cl_2 \xrightarrow{h\nu} 2HCl(g)$
    (b) Define and calculate einstein of energy.
15. (a) Define the law of photochemical equivalence. Is there any exception to this law?
    (b) Discuss the Lambert's law and Beer's law. Is Beer's law applicable in the case of concentrated solution of light absorbing substances?
16. Explain the following.
    (a) Fluorescence spectrum is the mirror image of absorption.
    (b) Phosphorescence is the radiation emitted in a transition between a triplet state and a singlet state.
    (c) Phosphorescence has much longer decay periods than fluorescence.
17. Explain the mechanism of (i) fluorescence, (ii) phosphorescence and (iii) photosensitization.
18. (a) What is the importance of photochemistry in research? Explain the consequence of absorption.
    (b) Explain the Lawbert-Beer Law.

MULTIPLE CHOICE QUESTIONS

1. Photochemistry deals with the study of
   (a) photons
   (b) photos
   (c) reactions which proceed with the absorptions of UV light
   (d) reactions which proceed with the absorption of IR light
2. The wavelength of UV and visible regions of electromagnetic spectrum is
   (a) less than 2000 Å
   (b) more than 8000 Å
   (c) 2000–8000 Å
   (d) none of these
3. Which of the following statements about the photochemical reactions is true?
   (a) The presence of light is the primary requirement for reactions to take place.
   (b) Temperature has a very little effect on the rate of photochemical reactions.
   (c) ΔG for the photochemical spontaneous reactions is +ve or –ve.
   (d) All of the above.
4. 'It is only the absorbed light radiations that are effective in producing a chemical reaction.' This is the statement of

(a) Lambert's law
(b) Lambert–Beer's law
(c) Grotthus–Draper's law
(d) Stark Einstein's law

5. One einstein of energy is
(a) $E = \frac{2.859}{\lambda} \times 10^5$ ca mol^{-1}
(b) $E = \frac{2.859}{\lambda} \times 10^5$ kcal mol^{-1}
(c) $E = \frac{2.859}{\lambda} \times 10^5$ J mol^{-1}
(d) $E = \frac{2.859}{\lambda} \times 10^5$ kJ mol^{-1}

6. Which of the following statement is true?
(a) It is the second reaction in which the absorption of radiation takes place.
(b) It is the primary reaction in which the absorption of radiation takes place.
(c) The absorption of radiation takes place in both the primary and the secondary reactions.
(d) None of the above.

7. The number of molecules reacted or formed per photon of light absorbed is called
(a) yield of the reaction
(b) quantum efficiency
(c) quantum yield
(d) quantum productivity

8. In some photochemical reactions, low quantum yield is obtained. It is due to
(a) the deactivation of reacting molecules
(b) the occurrence of reverse primary reaction
(c) the recombination of dissociated fragments
(d) all of these

9. The energy associated with a photon is given by the equation
(a) $\varepsilon = h \times \lambda$
(b) $\varepsilon = h \times \nu$
(c) $\varepsilon = h \times c$
(d) $\varepsilon = h \times C^2$

10. One einstein is given by (N is Avogadro's number)
(a) $E = \frac{NhC}{\lambda}$
(b) $E = \frac{NhC}{\lambda}$
(c) $E = \frac{NhC}{\lambda}$
(d) $E = \frac{Nh}{c\lambda}$

11. A species which can both absorb and transfer radiant energy for activation of the reactant molecule is called
(a) radioactive substance
(b) an ionizer
(c) a photochemical substance
(d) a photosensitizer

12. The glow of fireflies is due to the aerial oxidation of luciferin. It is example of
(a) fluorescence
(b) phosphorescence
(c) chemiluminescence
(d) none of these

13. A glass of certain thickness is found to have a transmission of 70% of light. If the thickness of glass is reduced, the transmission of light would
(a) decrease
(b) increase
(c) remains the same
(d) reduce to zero

14. The energy of an einstein of radiation of wavelength 400 nm is ____________ that of radiation of 300 nm
(a) lesser than
(b) greater than
(c) equal to
(d) none of these

15. Photochemical decomposition of a substance is called
(a) thermal dissociation
(b) thermolysis
(c) photolysis
(d) none of the above

16. The substances which initiate a photochemical reaction, but itself does not undergo any chemical change is called
    (a) catalyst
    (b) fluorescent
    (c) sensitizer
    (d) none of these

17. Organic dyes such as eosin, chlorophyll and ultramarine show __________ in the visible or UV region
    (a) fluorescence
    (b) phosphorescence
    (c) chemiluminescence
    (d) none of these

ANSWERS

1. (c)	4. (c)	7. (c)	10. (c)	13. (a)	16. (c)
2. (c)	5. (b)	8. (d)	11. (d)	14. (a)	17. (a)
3. (d)	6. (b)	9. (b)	12. (c)	15. (c)	

Chemical Kinetics 7

CHAPTER OBJECTIVES

7.1 Introduction

7.2 Rate of a Reaction

7.3 Rate Constant and Rate Equation

7.4 Order of a Reaction

7.5 Molecularity of a Reaction

7.6 Zero-order Reactions

7.7 Intergrated Rate Law Equation for First-order Reactions

7.8 Second-order Reactions

7.9 Third-order Reaction

7.10 Reactions of Higher Order

7.11 Determination of Rate Law, Rate Constant and Order of a Reaction

7.12 Some Complications in Determination of Order of Reaction

7.13 Temperature Dependence of Reaction Rates

7.14 Mechanism of a Reaction (Concept of Molecularity and Order of a Reaction)

7.15 Theories of Reaction Rates

The rate of the observed reaction depends on the number of collisions per second which take place between active molecules of the participating substances.

W.C. McCLewis

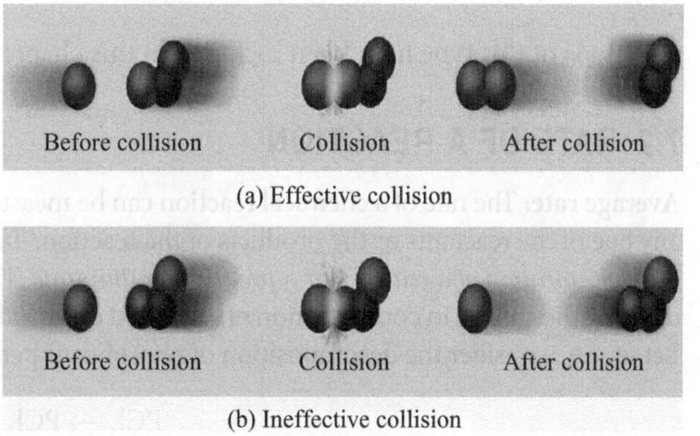

(a) Effective collision

(b) Ineffective collision

7.1 INTRODUCTION

Chemical kinetics deals with the study of rates of chemical reactions and the factors which influence them.

The main objectives of chemical kinetic include the following:

(i) To predict the time taken for the completion of a reaction
(ii) To find out the optimum conditions for a reaction
(iii) To study the mechanism of a reaction
(iv) To study the effect of other external factors on the rate of a reaction
(v) The most important objective of chemical kinetics is determination of the order of the reaction and thereby suggests a possible mechanism for that reaction. The suggested mechanism has to be confirmed by experiments. Thus, the study is useful in deciding the path through which the reaction proceeds.

It is well known that different reactions proceed at different speeds. Based on their speeds, the reactions can be broadly classified into three different types:

(i) The *instantaneous reactions*, i.e., which occur at once and are very fast. For example, ionic reactions such as acid–base neutralization and precipitation reactions.

$$KOH\ (aq) + HCl\ (aq) \dashrightarrow KCl\ (aq) + H_2O\ (l)$$
$$NaCl\ (aq) + AgNO_3 (aq) \dashrightarrow AgCl\ (s) + NaNO_3\ (aq)$$

Such type of reactions proceeds so quickly that their rates cannot be determined by common methods.

(ii) *Reactions, which take place at extremely slow* rate. Sometimes even years are required so that some appreciable change may take place. Rusting of iron is an example of such type of reactions. A large number of geological changes in nature also belong to this category.

(iii) *Reactions, which proceed with moderate rates.* In between these two extremes, there are reactions which proceed at rates which are neither too slow nor too fast and which can be easily determined by available methods. The decomposition of hydrogen peroxide, hydrolysis of an ester and hydrolysis of sugar etc. belong to this category.

$$H_2O_2 \longrightarrow H_2O + \frac{1}{2}O_2$$
$$CH_3COOC_2H_5 + H_2O \longrightarrow CH_3COOH + C_2H_5OH$$
$$C_{12}H_{11}O_{11} + H_2O \xrightarrow{H^+} C_6H_{12}O_6 + C_6H_{12}O_6$$

Reactions of this type have been included in this chapter.

7.2 RATE OF A REACTION

Average rate: The rate of a chemical reaction can be measured with respect to the change in concentration of any one of the reactants or the products of the reaction. *The rate of a reaction may be defined as the change in the concentration of a reactant or a product per unit time.* The average rate of a reaction can be determined by dividing the change in concentration of a reactant or a product by the time in which that change has occurred. Let us, e.g., consider the decomposition of phosphorus pentachloride to form trichloride and chlorine:

$$PCl_5 \rightarrow PCl_3 + Cl_2$$

The rate of this reaction can be expressed in various but all equivalent ways. Thus, we have

$$\text{Average rate in terms of reactants} = \frac{\text{Decrease in concentration of } PCl_5}{\text{Time interval}} = \frac{\Delta[R]}{\Delta t}$$

$$\text{Average rate} = \frac{[\text{Reactant}]-[\text{Reactant}]}{\text{Time interval}} = \frac{\Delta[R]}{\Delta t}$$

$$\text{Average rate in terms of products} = \frac{\text{Increase in conc. of either } PCl_3 \text{ or } Cl_2}{\text{Time interval}} = \frac{\Delta[P]}{\Delta t}$$

or

$$\text{Average rate} = \frac{[\text{Product}]-[\text{Product}]}{\text{Time interval}} = \frac{\Delta[P]}{\Delta t}$$

where $\Delta[R]$ or $\Delta[P]$ represent the change in concentration of a reactant or a product and Δt signifies change in time.

Units of rate of a reaction: The unit of concentration is moles per litre (mol L^{-1}) while that of time is seconds or minutes. Hence, the unit of rate of a reaction is moles per litre per second (mol L^{-1} s^{-1}) or moles per litre per minute (mol L^{-1} min^{-1}).

Instantaneous rate: According to the law of mass action, the rate of a reaction is governed by the molar concentrations of the reactants. Now, as the reaction proceeds, reactants are being used up and their concentration falls. Consequently, the rate of the reaction goes on decreasing. Thus, as the time passes, the reaction gets slower and slower continuously. *Therefore, it is not possible to obtain the rate of a reaction just by dividing the change in concentration of the substance by the time taken for the change to occur*. Therefore, the determination of average rates of chemical reactions does not serve any useful purpose.

Since the rate of a reaction varies from time to time, it is expressed at a particular instance of time. Therefore, a chemist is always interested in the rate of a reaction at a specified concentration of the reactant or a product, which is called the **instantaneous rate**. Hence, the **instantaneous rate** *of a reaction may be defined as the rate of change in concentration of any one of the reactants or products at a particular moment of time; i.e., at a specified concentration of a reactant or a product.* It is expressed as

$$+\frac{d[P]}{dt} \text{ or } -\frac{d[R]}{dt}$$

where $d[P]$ and $d[R]$ are change in concentration of products and reactants, respectively. The plus sign shows the increase in the concentration of the products while the negative sign denotes the decrease in the concentration of the reactants.

Instantaneous rate of reaction in general is written as dx/dt, where dx and dt are the small change in concentration in a small interval of time. The rate of a reaction is always positive, the –ve or +ve sign is written with the reaction rates to indicate decrease or increase of the concentration of reactants or products, respectively.

7.2.1 Measurement of Rate of a Reaction

While determining the rates of reactions experimentally, the progress of the reaction is monitored by observing from time to time the change in concentration of any one of the reactants or products.

A graph is plotted between concentration (x) against time (t) and a curve of the type as shown in Fig. 7.1 is obtained. To determine the rate of reaction at any time t, a tangent is drawn to the curve at the point corresponding to time t, as shown in Fig. 7.1. The slope of this tangent gives the rate of a reaction at that particular time. To determine the slope, the tangent is extended to cut the two axes. The distance along the ordinate gives the decrease in concentration, Δx, whereas the distance along the abscissa gives the corresponding time interval Δt, the ratio $\Delta x/\Delta t$ is equal to the slope of tangent and gives the rate of the reaction at time t.

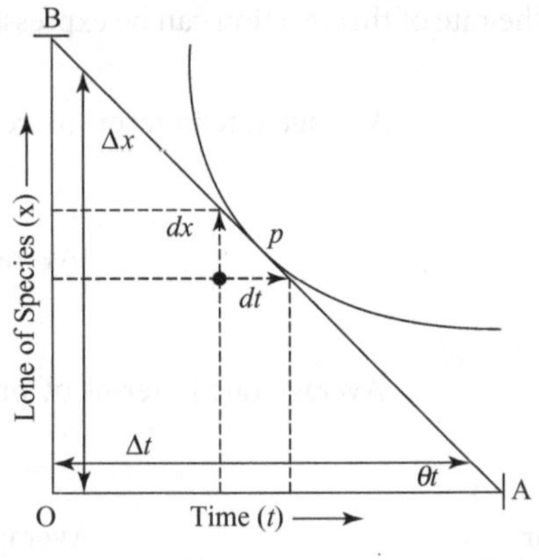

Figure 7.1 Measurement of rate of reaction

i.e., Rate of reaction = Slope of the tangent

$$= \tan\theta = \frac{\Delta x}{\Delta t} = \frac{\text{Change in concentration}}{\text{Time}}$$

7.2.2 Expressing the Rate of a Reaction

The rate of a reaction may be expressed in terms of decrease in the concentration of the reactants or terms of increase in the concentration of the products per unit time. It may be pointed out here that both these ways are otherwise exactly equivalent. Thus, if c is the concentration of a reactant at any time t, the rate can be expressed as $-dc/dt$ or if x is the concentration of the product at any time t, the rate of the reaction is given by dx/dt. Here, dt represents an infinitesimally small interval of time in which the concentration of the reactant decreases by an infinitesimally small amount dc and dx is the infinitesimally small amount of the product formed in time dt. Thus, for a reaction

$$N_2 + 3H_2O \rightarrow 2NH_3$$

We have

$$\text{Rate of reaction} = -\frac{d[N_2]}{dt} = -\frac{1}{3}\frac{d[H_2]}{dt} = \frac{1}{2}\frac{d[NH_3]}{dt}$$

$$\text{Rate of reaction} = -\frac{dc}{dt} = \frac{dx}{dt}$$

It may be mentioned here that the rate of reaction is always positive. *The negative sign in the above-mentioned expression merely indicates the decrease in concentration of the reactant with time while the positive sign indicates that the concentration of the product is increasing with time.*

Consider a general reaction, such as

$$aA + bB \rightarrow pP + qQ$$

where a moles of A react with b moles of B to produce p moles of P and q moles of Q.

The reaction rate for the above reaction may be expressed as the rate of change of any species per mole with time, i.e.

$$\frac{dx}{dt} = -\frac{d[A]}{dt}\times\frac{1}{a} = -\frac{d[B]}{dt}\times\frac{1}{b}$$

$$= +\frac{d[P]}{dt}\times\frac{1}{p} = +\frac{d[Q]}{dt}\times\frac{1}{q}$$

Taking into consideration the decomposition of phosphorus pentachloride

$$PCl_5 \rightarrow PCl_3 + Cl_2$$

We have, therefore

$$\text{Rate of reaction} = -\frac{d[PCl_5]}{dt} = +\frac{d[Cl_2]}{dt} = +\frac{d[PCl_3]}{dt}$$

7.2.3 Factors Influencing Rate of a Reaction

The rate of a reaction is influenced by a number of factors. These include the following:

(i) Nature of the reactants
(ii) Concentration of the reactants
(iii) Surface area of the reactants
(iv) Temperature
(v) Presence of a catalyst
(vi) Radiations

7.3 RATE CONSTANT AND RATE EQUATION

According to the law of mass action, the rate of a reaction is proportional to the product of active masses of the reactants.

Thus, for a general reaction $A + B \rightarrow \text{Products}$

The rate is given by, $\text{Rate} = \frac{dx}{dt}$ or $\frac{d[A]}{dt} \propto [A][B]$

$$\text{Rate} = \frac{dx}{dt} = -\frac{d[A]}{dt} = K[A][B]$$

If concentrations of A and B are unity, then

$$\text{Rate} = K \times 1 \times 1 = K$$

where K is a constant and is known as the **rate constant** or **velocity constant** or **specific reaction rate.** Numerically, *it is equal to the rate of the reaction when the concentration of all the reactions is taken as unity at a fixed temperature.* Each reaction is characterized by its own rate constant, the value of which depends on the nature of the reaction and temperature. Knowing the value of the rate constant, it is possible to determine the rate of reaction under particular concentration conditions. It may be mentioned here that the mathematical expression which represents observed or actual rate of a reaction in terms of concentration of the reacting species which influence the rate of a reaction is known as the **rate law or rate equation.**

For example

$$CH_3COOC_2H_5 + NaOH \rightarrow CH_3COONa + C_2H_5OH$$

The rate equation for this reaction is

$$\text{Rate} = K\,[CH_3COOC_2H_5][NaOH]$$

7.3.1 Differences Between Rate of a Reaction and Rate Constant

The differences are summarized in Table 7.1.

Table 7.1 The differences between the rate of a reaction and the rate constant are summarised.

Rate of a Reaction	Rate Constant
1. It may be defined as an instantaneous rate of change of concentration of any of the reactants or products at a particular instance.	1. It is proportionality constant and is equal to the rate of reaction when the concentration of each reactant is unity.
2. It depends upon concentration of the reactant as well as temperature.	2. It is independent of concentration of reactants but depends on temperature.
3. The units of the rate of a reaction are mol L^{-1} s^{-1}.	3. The units of rate constant depend upon the order of the reaction, e.g., for first order the unit is s^{-1}.

MEMORY FOCUS

1. The study of rates of chemical reactions and their mechanism is called chemical kinetics.
2. The rate of change in concentration of a particular reactant or product is called the rate of a reaction.

$$\text{Rate} = \frac{dc}{dt} = \frac{\text{change in concentration}}{\text{interval of time}}$$

3. The rate of any chemical reaction decreases with time as the concentration of reactants decreases with time.
4. The rate of a reaction depends on concentration of the reactants at that time.
5. The rate of a reaction has the units (mole/litre) time^{-1}.
6. The rate of a reaction is influenced by the following factors:
   (i) Concentration of the reactants
   (ii) Nature of the reactants
   (iii) Temperature
   (iv) Surface area of the reactants
   (v) Catalyst
   (vi) Exposure to radiation
7. The effect of concentration on the reaction rate is explained by the law of mass action.
8. According to the law of mass action, the rate of a chemical reaction is directly proportional to the product of active masses of the reactants.
9. For a reaction of the type $mA + nB \rightarrow$ products, the rate of $[A]^m[B]^n$ or rate $= K[A]^m[B]^n$ (K is a proportionality constant).
10. The rate of reaction when the concentration of all the reactants is taken as unity is called the specific rate or rate constant or velocity constant.
11. The rate constant is a constant which is independent of initial concentrations of the reactants and depends on temperature.
12. Coal burns rapidly in oxygen than in air as the concentration of oxygen (reactant) is increased.

(*Continued*)

MEMORY FOCUS (*Continued*)

13. $KMno_4$ reacts slowly with oxalic acid but rapidly with ferrous sulphate at room temperature due to the difference in the number of bonds to be broken.
14. Powdered coal burns faster than a lump of same weight because of increase in surface area.
15. In the presence of a catalyst, the rate of a chemical reaction is altered.
16. Decomposition of H_2O_2 proceeds rapidly in the presence of finely divided metals but is retarded when small quantity of acetanilide is added to it.
17. Hydrogen and chlorine combine rapidly in sunlight but very slowly in its absence.
18. In general, increases in temperature make many reactions to proceed rapidly.

7.4 ORDER OF A REACTION

Order of a reaction: *The order of a reaction may be defined as the sum of the powers to which the concentration terms are raised in the rate equation in order to express observed or actual rate of a reaction.*

For a general reaction $aA + bB \rightarrow \text{Products}$

If the rate or reaction is given by, $\text{Rate} = K[A]^m[B]^n$

Then, the order of the reaction is $(m + n)$, where the exponents 'm' and 'n' are determined experimentally. It may be mentioned here that there is no relationship between the exponent's m, n and the corresponding coefficients a, b in the balanced equation. However, in some cases m may be equal to a and n may be equal to b.

Thus, the order of a reaction is an **experimental quantity** *and cannot be deduced from the chemical equation.* Further, the order of a reaction may be a whole number, zero or a fraction.

(i) Zero-order reaction: A reaction whose rate is not influenced by the change in the concentration of reacting species is known as a *zero-order reaction.*

Examples:

(1) Photochemical reaction

$$H_2(g) + Cl_2(g) \xrightarrow{\text{Sunlight}} 2HCl(g)$$

(2) Decomposition of ammonia

$$2NH_3(g) \xrightarrow{\text{Mo or W}} N_2(g) + 3H_2(g)$$

The rate equation is $\text{Rate} = K$

(ii) Examples of first-order reactions:

(1) Decomposition of nitrogen pentoxide

$$N_2O_5 \longrightarrow N_2O_4 + \tfrac{1}{2}O_2$$

The rate equation is $\text{Rate} = K[N_2O_5]$

(2) Decomposition of ammonium nitrite

$$NH_4NO_2 \longrightarrow N_2 + 2H_2O$$

The rate equation is $\text{Rate} = K[NH_4NO_2]$

(iii) Examples of second-order reactions:

(1) Decomposition of hydrogen iodide.

$$2HI \longrightarrow H_2 + I_2$$

The rate equation is $\quad$ Rate = $K\,[HI]^2$

(2) Alkaline hydrolysis (sponification) of ethyl acetate.

$$CH_3COOC_2H_5 + NaOH \longrightarrow CH_3COONa + C_2H_5OH$$

The rate equation is $\quad$ Rate = $K[CH_3COOC_2H_5][NaOH]$

(3) Oxidation of HBr.

$$4HBr + O_2 \xrightarrow{673\text{ to }873\,K} 2Br_2 + 2H_2O$$

Experimentally it has been observed that the rate equation is

$$\text{Rate} = K\,[HBr][O_2]$$

(4) Combination of NO_2 and F_2.

$$\text{Rate} = K\,[NO_2][F_2]$$

(iv) Examples of third-order reactions:

(1) $2FeCl_3 + SnCl_2 \longrightarrow 2FeCl_2 + SnCl_4$

The rate equation is $\quad$ Rate = $K[FeCl_3]^2[SnCl_2]$

(2) Oxidation of nitric oxide $2NO + O_2 \longrightarrow 2NO_2$

The rate equation is $\quad$ Rate = $K[NO]^2[O_2]$

(v) Examples of fractional-order reactions:

Decomposition of acetaldehyde.

$$CH_3CHO \xrightarrow{720\,K} CH_4 + CO$$

Experimentally, it has been observed that the rate equation is Rate = $K\,[CH_3CHO]^{3/2}$. Thus, the order of the reaction is 3/2. It is important to differentiate the term 'order of a reaction' from the molecularity of a reaction.

7.4.1 Units for Rate Constant or Specific Reaction Rate

The units of rate constant depend upon the order of the reaction. The units of the rate constant of reactions of different orders are different.

The rate equation for a reaction of nth order is given by,

$$\frac{dx}{dt} = K[\text{Reactant Conc.}]^n$$

or

$$\text{Rate constant, } K = \frac{dx}{dt} \times \frac{1}{[\text{Conc.}]^n}$$

$$= \frac{\text{Conc.}}{\text{Time}} \times \frac{1}{(\text{Conc.})^n} = \frac{1}{\text{Time}} \times \frac{1}{(\text{Conc.})^{n-1}}$$

Thus

(i) Substituting $n = 0$, for a **zero-order reaction**, we get

$$K = \frac{1}{\text{Time}} \times \frac{1}{[\text{Conc.}]^{-1}} = \frac{\text{Conc.}}{\text{Time}} = \frac{\text{Moles per litre}}{\text{Time}}$$

$$K = \text{mol L}^{-1}\ \text{time}^{-1}$$

(ii) Substituting $n = 1$, for **a first-order reaction**, we get

$$= \frac{1}{\text{Time}} \times \frac{1}{[\text{Conc.}]^0} = \frac{1}{\text{Time}} = \text{Time}^{-1}$$

(iii) Substituting $n = 2$, for a **second-order reaction**, we get

$$K = \frac{1}{\text{Time}} \times \frac{1}{[\text{Conc.}]^{2-1}} = \frac{1}{\text{Time}} \times \frac{1}{\text{Conc.}}$$

$$= \frac{1}{\text{Time}} \times \frac{1}{\text{mol L}^{-1}} = \text{L mol}^{-1}\text{time}^{-1}$$

(iv) Substituting $n = 3$, for a **third-order reaction**, we get

$$K = \frac{1}{\text{Time}} \times \frac{1}{[\text{Conc.}]^{3-1}} = \frac{1}{\text{Time}} \times \frac{1}{(\text{Conc.})^2}$$

$$= \frac{1}{\text{Time}} \times \frac{1}{(\text{mol L}^{-1})^2}$$

$$= \text{L}^2\text{mol}^{-2}\text{time}^{-1}$$

In general, for a reaction of nth order, the units of rate constant are $(\text{mol L}^{-1})^{1-n}\ \text{s}^{-1}$, and for gaseous reaction of nth order, K has units of $(\text{atm})^{1-n}$.

7.5 MOLECULARITY OF A REACTION

In case of simple reactions, very often it is found that the molecularity of a reaction is the same as its order. For example, the formation of hydrogen iodide

$$H_2 + I_2 \longrightarrow 2HI$$

is a second-order reaction because the rate is given by

$$\text{Rate} = K[H_2][I_2]$$

It is also a bimolecular reaction since it involves two molecules, one each of hydrogen and iodine. However, this is not the case always, especially for reactions in which one of the reactants is present in large excess, e.g.

In the inversion of cane sugar

$$C_{12}H_{22}O_{11} + H_2O \longrightarrow C_6H_{12}O_6 + C_6H_{12}O_6$$

The molecularity of the reaction is two since two reactant molecules of cane sugar and water are involved. Since water is present in excess, its concentration remains practically unchanged during the course of the reaction. Thus, the rate of a reaction depends only on the concentration of sugar. The rate equation is, hence, given as

$$\text{Rate} = K\,[C_{12}H_{22}O_{11}]$$

Thus, the order of the reaction is only one and the reaction is referred to as '**pseudounimolecular reaction**.'

Hence, *molecularity of a reaction may be defined as the total number of species (i.e., atoms, ions or molecules) which participate in the step leading to the chemical change.*

7.5.1 Differences Between Order and Molecularity of a Reaction

The main differences between the order and molecularity of a reaction are summarized in Table 7.2.

Table 7.2 Difference between order and molecularity.

Order	Molecularity
1. It is the sum of the powers to which the concentration terms are raised in the rate equation for the reaction.	1. It is the total number of atoms, ions or molecules which participate in the step leading to the chemical change.
2. It may be a whole number, a fraction or even zero.	2. It is always a whole number.
3. It has to be determined experimentally and cannot be deduced from the balanced equation.	3. It is obtained from the simple balanced equation.
4. It belongs to the reaction as a whole no matter how many intermediate steps are involved in its completion.	4. It belongs to one-step reactions or a single step in case of complex reactions.
5. For a complex reaction, the overall order is the order of the slowest step involved in it.	5. The overall molecularity of a complex reaction has no significance. Each single step has its own molecularity.

MEMORY FOCUS

Order of a Reaction

1. The sum of powers of the concentration terms of the reactants which appear in the rate law is called the order of the reaction.
2. The number of atoms or molecules whose concentrations determine the rate of a reaction is called the **order of the reaction**.
3. The order of a reaction can be determined only experimentally.
4. If rate law for a reaction rate = $K[A]^m[B]^n$, its order is equal to $[m + n]$.
5. Order of a reaction may be a whole number, a fraction or even zero.
6. Chemical reactions are characterized in terms of order of reactions.
7. If the rate of a reaction does not change with time it is called a zero-order reaction.

(Continued)

MEMORY FOCUS (*Continued*)

8. If the rate of a reaction is influenced by only one concentration term, it is called a first-order reaction.
9. If the rate of a reaction depends upon two concentration terms, that reaction is said to be of second order.

Molecularity of a Reaction

1. The minimum number of molecules taking part in the process leading to the formation of products is called the **molecularity** of the reaction.
2. The number of atoms or molecules, which collide together at one and the same time for the reaction to take place is called **molecularity**.
3. The number of molecules, atoms or ions of the reactants used in the formation of the activated complex is called **molecularity.**
4. Molecularity of a reaction is always a whole number. It is never zero or fraction.
5. The description of various steps that lead to final products in a chemical reaction is called mechanism and the individual steps are called elementary processes.
6. The rate of a chemical reaction is the rate of the **slowest step** in it, which is called the **rate-determining** step.
7. The number of molecules involved in the rate-determining step of a reaction is called its molecularity.
8. Decomposition of N_2O_5 to N_2O_4 and O_2 is a unimolecular reaction.

7.6 ZERO-ORDER REACTIONS

Reactions in which the rate is independent of the concentrations of all the reactant species are called zero-order reactions. Accordingly, the rate remains constant. This is because throughout the progress of the reaction, it does not change with time. Many photochemical and some heterogeneous reactions are zero-order reactions. Well-known photochemical reactions, which take place when an equimolecular mixture of hydrogen and chlorine is exposed to sunlight to form hydrogen chloride, are a reaction of zero order

$$H_2(g) + Cl_2(g) \rightarrow 2HCl(g)$$

The reaction can be studied by enclosing H_2 and Cl_2 gases over water in a tube. In the presence of light, gases combine to form HCl gas. The rate of the reaction can be observed by the rise in water level due to dissolution of HCl gas formed. The water rises at the same rate at which H_2 and Cl_2 disappear. As a result, the concentration of gases in the gaseous phase does not change with time. Although the quantities of H_2 and Cl_2 decrease, their amount per unit volume i.e., their concentration remains the same.

The decomposition of hydrogen iodide and ammonia on the surfaces of gold and tungsten are some other examples of zero-order reactions. In these reactions, gold metal acts as a catalyst. At high pressure the catalyst surface is completely covered by the reactant molecules and on further increasing the pressure (or concentration) of the gas phase has no effect on the amount of reactant at the surface. The concentration of the reactant, therefore, remains constant although the product is being formed.

The **mathematical expression for the zero-order reaction** can be derived as explained below: Suppose a is the initial concentration of the reactant, the rate of the reaction can be expressed as

$$\frac{dx}{dt} \propto a^{\circ} = Ka^{\circ} \tag{7.1}$$

$$= K \qquad [\because a^{\circ} = 1]$$

where K is the rate constant of the reaction and the above expression may be written as

$$dx = Kdt$$

On integration,

$$\int dx = \int Kdt$$

$$x = Kt + I$$

where I is a constant of integration

when $t = 0, \quad x = 0 \qquad [\therefore I = 0]$

Hence $\qquad x = Kt$

or

$$K = \frac{x}{t} \quad \text{or} \quad x \propto t \tag{7.2}$$

That is, the amount of substance reacted is proportional to time.

7.6.1 Characteristics of a Zero-order Reaction

(i) Evaluation of rate constant (K): If the progress of the reaction is noted after different intervals of time and the value of K is evaluated from Eq. (7.2), a constant value of K is obtained always for the reactions of zero order.

The value of K can also be evaluated if the amount of reactant consumed at two different intervals of time is known. For example, let x_1 and x_2 be the amount of reactant which has reacted after time t_1 and t_2, respectively.

From Eq. (7.2) we have

$$x_1 = Kt_1 \quad \text{and} \quad x_2 = Kt_2$$

on subtracting,

$$x_2 - x_1 = K(t_2 - t_1)$$

or

$$K = \frac{x_2 - x_1}{t_2 - t_1} \tag{7.3}$$

The value of rate constant can also be evaluated graphically. The equation for zero-order reactions is $x = Kt$. The progress of the reaction is noted at different intervals of time experimentally. A graph is potted between the amount of reactant which have reacted (x) and time (t) (Fig. 7.2). A straight line passing through the origin is obtained for all reactions of the zero order. The value of K is obtained from the slope of the straight line.

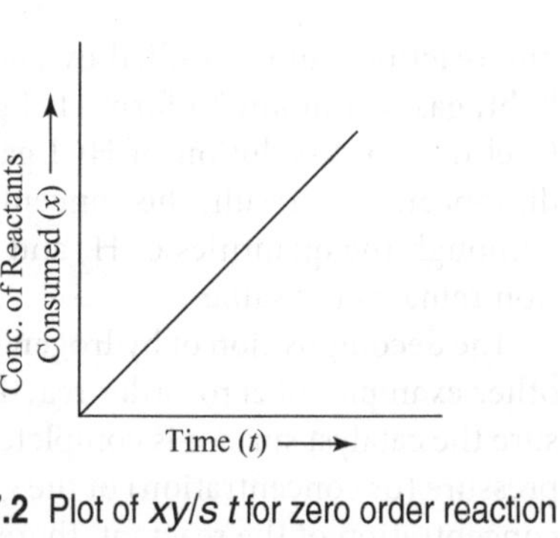

Figure 7.2 Plot of *xy/s t* for zero order reaction

(ii) Units of rate constant:

$$K = \frac{x}{t} = \frac{\text{moles/litre}}{\text{time}} = \text{mole l}^{-1}\ \text{time}^{-1}$$

If the concentration is expressed in mole per litre and time in seconds, K is expressed as mole l^{-1} s^{-1}.

(iii) Effect of change in concentration units on the value of *K*: Let the value of concentration in new units be n times those of the initial units, then

$$nx = Kt \qquad K = \frac{nx}{t}$$

The above expression shows that for zero-order reaction the change in units' concentration will change the numerical value of rate constant.

(iv) Time for fractional change: Let us calculate the time for the completion a definite fraction of the reactions of zero order. The time taken for the completion of half the reaction is called half-life period, $t_{1/2}$, and it can be calculated as shown below.

When

$$x = \frac{a}{2}, \quad t = t_{1/2}$$

$$\frac{a}{2} = Kt_{1/2}$$

or

$$t_{1/2} = \frac{a}{2K} \quad \text{or} \quad t_{1/2} \propto a$$

Hence half-life period is directly proportional to initial concentration of reactants.

Numerical Problems

Example 1 **The progress of a photochemical reaction between hydrogen and chlorine has been studied. The concentration of hydrogen reacted at different intervals of time is given below:**

Time (min)	2	4	7	11
Conc. of hydrogen consumed (mole/litre)	2.8	5.6	9.9	15.3

Show that this is a zero-order reaction.

Solution: The equation for zero-order reactions is $K = \frac{x}{t}$.

x (moles/litre)	t (min)	$K = \frac{x}{t}$ (moles $l^{-1}min^{-1}$)
2.8	2	$K = \frac{2.8}{2} = \mathbf{1.40}$
5.6	4	$K = \frac{5.6}{4} = \mathbf{1.40}$
9.2	7	$K = \frac{9.9}{7} = \mathbf{1.41}$
15.3	11	$K = \frac{15.3}{11} = \mathbf{1.39}$

A nearly constant value of K shows that the reaction is of zero order.

MEMORY FOCUS

Zero-order Reaction

1. Rate expression for these reactions is Rate = K or $\frac{dx}{dt} = K$ or $x = Kt$.
2. Units of rate constant (K) are moles l^{-1} (time units)$^{-1}$.
3. The concentration of products increases linearly with time.
4. Plot of concentration of product(s) with time is a straight line passing through the origin.
5. Time taken for completion of a definite fraction of the reaction is directly proportional to the initial concentration of the reactant.
6. Examples
   (i) Photochemical combination of H_2 and Cl_2 over water surface
   (ii) Thermal decomposition of HI on gold surface
   (iii) Thermal decomposition of NH_3 gas on tungsten or molybdenum
   (iv) Many enzymatic reactions

7.7 INTERGRATED RATE LAW EQUATION FOR FIRST-ORDER REACTIONS

A reaction is said to be of the first order if its rate depends on the first power of concentration of the reactant. Consider a general first-order reaction represented by

$$A \longrightarrow \text{Products}$$

Let the initial concentration of the reactant A be a moles per litre. Let x moles per litre of it decompose into products in time t. Thus after time t, the concentration of the reactant A will be $(a - x)$ moles per litre. If the rate is expressed in terms of reactant concentration, then

$$\frac{dx}{dt} = K(a - x) \tag{7.4}$$

Rearranging

$$\frac{dx}{(a - x)} = Kdt$$

On integrating this equation, we get $\quad -\text{In}(a - x) = Kt + 1$

where L is the constant of integration. When $t = 0$, $x = 0$, we have $1 = -\ln a$.
Substituting this in the above equation, we have $-\ln(a - x) = Kt - \ln a$

or

$$\ln a - \ln(a - x) = Kt$$

or

$$\ln \frac{a}{a - x} = Kt$$

or $$K = \frac{l}{t}\ln\frac{a}{a-x}$$ [taking natural logo]

or $$K = \frac{2.303}{t}\log\frac{a}{a-x} \tag{7.5}$$

Equation (7.5) is referred to as the *rate equation* of the first-order reactions. The constant K is called the first-order rate constant or the *specific rate* of the first-order reaction.

7.7.1 Characteristics or Significance of First-order Reaction

(i) Evaluation of rate constant (K): All the first-order reactions must obey the rate Eq. (7.2)

$$K = \frac{2.303}{t}\log\frac{a}{a-x}$$

On substituting the value of a and $a-x$ at different intervals of time, if a constant value of k is obtained, the reaction is of the first order

(ii) Calculation of rate constant (K): The value of K can be evaluated if the amount of reactant consumed at two different intervals of time is known.

If x_1 and x_2 are the g moles/litres of A converted to the product up to times t_1 and t_2, then

$$Kt_1 = 2.303\log\frac{a}{a-x_1}$$

$$Kt_2 = 2.303\log\frac{a}{a-x_2}$$

$$K(t_2 - t_1) = 2.303\left(\log\frac{a}{a-x_2} - \log\frac{a}{a-x_1}\right) = 2.303\log\frac{a-x_1}{a-x_2}$$

$$K = \frac{2.303}{(t_2 - t_1)}\log\frac{a-x_1}{a-x_2} \tag{7.6}$$

For graphical evaluation of K, Eq. (7.5) can be written as

$$\log\frac{a}{a-x} = \frac{Kt}{2.303}$$

$$\log a - \log(a-x) = \frac{Kt}{2.303}$$

or $$\log(a-x) = \frac{Kt}{2.303} + \log a \tag{7.7}$$

It is evident that a graph of $\log(a-x)$ versus t will result in a straight line with a negative slope being equal $-K/2.303$ (Fig. 7.3) since a, the initial concentration is constant for a given experiment. The intercept of the line on the ordinate (at $t = 0$) is simply the initial concentration of reactant. From the slope of the straight line obtained in the graph, the value of K can be determined.

(iii) Units of rate constant (*K*):

From Eq. (7.5), we have $K = \frac{2.303}{t} \log \frac{a}{a-x}$

Since the expression contains ratio of two concentration terms, *K* is independent of units in which concentration of the reactants is given as

$$K = \text{time}^{-1} \log \frac{\text{moles/litre}}{\text{moles/litre}} = \text{time}^{-1}$$

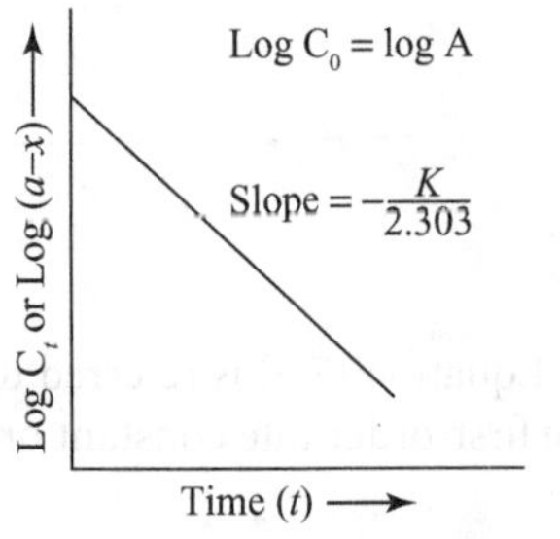

Figure 7.3 Plot of log C_t and time (t) of first order kinetics

The first-order rate constant *K* has units of reciprocal time i.e., time^{-1}. It is expressed in seconds, *K* is expressed in s^{-1} and *t* is expressed in min^{-1}.

(iv) Effect of change in concentration on the value of K: Let the value of concentration in new units be *n* times those of the initial units, then

$$K = \frac{2.303}{t} \log \frac{na}{n(a-x)} = \frac{2.303}{t} \log \frac{a}{a-x}$$

The expression is the same as the original unimolecular reaction equations. Therefore, a change in concentration units will not change the rate constant.

(v) Time for fractional change: Time for the completion of any fraction of the reaction (say, half-life period) can be calculated from the equation for the first-order reactions.

When $$t = t_{1/2}, x = \frac{a}{2}$$

$$\therefore \quad K = \frac{2.303}{t_{\frac{1}{2}}} \log \frac{a}{a - \frac{a}{2}}$$

$$t_{1/2} = \frac{2.303}{K} \log 2 = \frac{0.693}{K} \quad (7.8)$$

Equation (7.8) does not involve any concentration term. It, therefore, follows that for a first-order reaction, *the half-life period or time for the completion of any fraction of a reaction is independent of the initial concentration of the reactants.*

7.7.2 Examples of the Reactions of First Order

(i) Decomposition of hydrogen peroxide:

$$\underset{\text{Solution}}{H_2O_2} \xrightarrow{Pt} H_2O + \tfrac{1}{2} O_2$$

The progress of this reaction can be followed in two ways as explained below:

1. By measuring the volume of oxygen gas evolved during the decomposition of H_2O_2 at definite intervals, when the volume of oxygen evolved is proportional to the amount of H_2O_2 decomposed.
2. By titrating, a definite amount of reaction mixture (solution) with standard $KMnO_4$ at definite intervals of time. The volume of $KMnO_4$ at definite intervals of time is proportional to unchanged

H_2O_2, i.e., $(a - x)$. As in the beginning no H_2O_2 is decomposed, the volume of $KMnO_4$ used at the commencement of the reaction corresponds to the initial concentration of H_2O_2.

If V_0 and V_t represent the volume of $KMnO_4$ used at the start of the reactions and at any time t, $a \propto V_0$ and $a - x \propto V_t$.

Therefore,
$$K = \frac{2.303}{t} \log \frac{V_0}{V_t}$$

Numerical Problems

Example 2 **The decomposition of hydrogen peroxide was studied by titrating it at different intervals with $KMnO_4$. Calculate the velocity constant for it from the following data, assuming it to be of first order.**

(Sec)	0	600	1200
$KMnO_4$(cm³)	22.8	13.8	8.2

Solution: Initial concentration of H_2O_2 corresponds to 22.8 cc of $KMnO_4$.

Time	$(a-x)$ corresponds to litre value	$K = \frac{2.303}{t} \log \frac{a}{a-x}$
600	13.8	$K = \frac{2.303}{600} \log \frac{22.8}{13.8} = \mathbf{0.000837}$
1200	8.2	$K = \frac{2.303}{1200} \log \frac{22.8}{8.2} = \mathbf{0.000852}$

$$\text{Average value of } K = \frac{0.000837 + 0.000852}{2} = 0.000844 \times 10^{-4}\ s^{-1}$$

Hence, the velocity constant of the reaction is = **8.44 10^{-4} s^{-1}**.

(ii) Decomposition of ammonium nitrite: An aqueous solution of ammonium nitrite decomposes on warming as follows:

$$NH_4NO_2 \longrightarrow N_2 + 2H_2O$$

The progress of the reaction is studied by measuring the volume of the nitrogen gas liberated at regular intervals of time. The amount of ammonium nitrite decomposed at any time i.e., x is proportional to the volume of nitrogen liberated. The value of the initial concentration a, is given by the volume of the nitrogen evolved at the end of the reaction. The volume is often known as the volume collected at infinite time. Thus if V_t be the volume of N_2 evolved at any time t and V_∞ is the volume of N_2 evolved when the decomposition is complete (i.e., at infinite time), K is given by

$$K = \frac{2.303}{t} \log \frac{V_\infty}{V_\infty - V_t}$$

Example 3 **The decomposition of an aqueous solution of ammonium nitrite was studied at a particular temperature. The volume of nitrogen gas collected at different intervals of time was as follows:**

Time	10	15	20	°
Vol. of N_2 (cm)³	6.25	9.00	11.40	35.05

From the above data, prove that the reaction is of the first order.

Solution: If the reaction is of the first order, it must obey the equation

$$K = \frac{2.303}{t}\log\frac{a}{a-x} = \frac{2.303}{t}\log\frac{V_\infty}{V_\infty - V_t}$$

In the present case, $V_\infty = 35.05\text{cc}$

The value of K at each instance of time can be calculated as follows:

(*t*) (min)	(*t*)	$V_\infty - V_t$	$K = \frac{2.303}{t}\log\frac{V_\infty}{V_\infty - V_t}$
10	6.25	35.05 – 6.25 = 28.80	$K = \frac{2.303}{10}\log\frac{35.05}{28.80} = \mathbf{0.0196}$
15	9.00	35.05 – 9.00 = 26.05	$K = \frac{2.303}{15}\log\frac{35.05}{26.05} = \mathbf{0.0198}$
20	11.40	35.05 – 11.40 = 23.65	$K = \frac{2.303}{20}\log\frac{35.05}{23.65} = \mathbf{0.0196}$

Since the value of K comes out to be nearly constant, the given reaction is of the first order.

(iii) Decomposition of nitrogen pentoxide: Nitrogen pentoxide is a volatile solid, which decomposes in the gaseous state or when dissolved in some organic solvent like carbon tetrachloride. The total reaction is represented by the equation

$$N_2O_5 \longrightarrow N_2O_4 + \frac{1}{2}O_2$$
$$N_2O_4 \rightleftharpoons 2NO_2$$

When the reaction is carried out in carbon tetrachloride, N_2O_4 remains in solution and thus the progress of the reaction can be noted by collecting the oxygen evolved and measuring its volume from time to time. If V_t be the volume at any time and V_∞ is the measure of the original concentration (a) of N_2O_5 and $(V_\infty - V_t)$ is the measure of the undecomposed nitrogen pentoxide $(a - x)$ left at time t. Thus

$$K = \frac{2.303}{t}\log\frac{a}{a-x} = \frac{2.303}{t}\log\frac{V_\infty}{V_\infty - V_t}$$

Example 4 From the following data for the decomposition of N_2O_5 in CCl_4 at 321 K, show that the reaction is of first order.

Time (minutes)	10	15	20	25	°
Vol. of O_2 (in mol)	6.30	8.93	11.40	13.5	34.75

Solution: If it is a unimolecular reaction, $K = \frac{2.303}{t}\log\frac{V_\infty}{V_\infty - V_t}$

In this case, $V_\infty = 34.75$

Time	$V_\infty - V_t$	$K = \frac{1}{t}\log\frac{V_\infty}{V_\infty - V_t}$
10	$34.75 - 6.30 = 28.45$	$K = \frac{1}{10}\log\frac{34.75}{28.45} = \mathbf{0.00868}$
15	$34.75 - 8.95 = 25.80$	$K' = \frac{1}{15}\log\frac{34.75}{25.80} = \mathbf{0.00862}$
20	$34.75 - 11.40 = 23.35$	$K' = \frac{1}{20}\log\frac{34.75}{23.35} = \mathbf{0.00863}$
25	$34.75 - 13.50 = 21.25$	$K' = \frac{1}{25}\log\frac{34.75}{21.25} = \mathbf{0.00834}$

Since the value of K' is almost constant, it is a reaction of the first order.

7.7.3 Pseudo First-order Reaction

(iv) Inversion of cane sugar of sucrose (hydrolysis of cane sugar in dilute solution):

$$\underset{\text{Sucrose}}{C_{12}H_{22}O_{11}} + H_2O \longrightarrow \underset{\text{Glucose}}{C_6H_{12}O_6} + \underset{\text{Fructose}}{C_6H_{12}O_6}$$

The reaction is monomolecular because the molar concentration [active mass] of water does not undergo an appreciable change in dilute solution and its reaction velocity is proportional to the active mass of cane sugar, as all the constituents of the reaction mixture are optically active. The progress of the reaction can be studied with the help of a polarimeter. Cane sugar and glucose are dextrorotatory and fructose is laevorotatory. The laevorotatory power of fructose is greater than the dextrorotatory power of glucose at the same concentration. Therefore, as the reaction proceeds, the rotation (dextrorotation) of the solution decreases becomes zero and then reaches negative value. Because of this, the reaction is known as ***inversion of cane sugar***.

The polarimeter reading at the start of the experiment when $t = 0$ is θ_0 and after infinite time, ($t = \infty$) when inversion is almost complete, it is θ_∞. The total inversion that has taken place is $\theta_0 - \theta_\infty$ and this is evidently proportional to *a*, i.e., initial concentration of sugar.

Now, $$K = \frac{2.303}{t}\log_{10}\frac{a}{a-x}$$

Substituting the value in the above equation, $$K = \frac{2.303}{t}\log_{10}\frac{\theta_0 - \theta_\infty}{\theta_t - \theta_\infty}$$

Example 5 **A 20% solution of cane sugar having a dextrorotation of 34.50Υ is inverted by 0.5 N lactic acid at 250 ΥC. The rotations determined polarimetrically are as follows:**

T (min)	0	14.35	111.360	°
Rotation	34.50Υ	31.10Υ	13.980Υ	−10.77Υ

The value of *K* in both the cases is almost the same and hence inversion of cane sugar is a first-order reaction. Show that the inversion of sugar is a unimolecular reaction.

Solution: Rotation θ_0 when t is $0 = 34.5°$

Rotation θ_∞ after infinite time $= -10.77°$

Now, initial concentration of sugar (a), when $t = 0$

$$\theta_\infty - \theta_0 = +34.5 - (-10.77) = 34.5 + 10.77 = 45.27$$

Substituting the value in the equation for a unimolecular reaction, we obtain

Time	$\theta_t - \theta_\infty$ (corresponding $a - x$)	$K = \frac{2.303}{t} \log_{10} \frac{\theta_0 - \theta_\infty}{\theta_1 - \theta_\infty}$
14.35	$31.10 - (10.77) = 41.87°$	$K = \frac{2.303}{14.35} \log_{10} \frac{45.27}{41.87} = \mathbf{2.303 \times 0.00236}$
111.360	$13.98 - (-10.77) = 24.75°$	$K = \frac{2.303}{111.360} \log_{10} \frac{45.27}{24.75} = \mathbf{2.303 \times 0.00235}$

(v) Hydrolysis of ethyl acetate, $CH_3COOC_2H_5$:

$$CH_3COOC_2H_5 + H_2O \xrightarrow[\text{acid}]{H^+} H_3CCOOH + C_2H_5OH$$

The reaction is unimolecular, since water is present in large excess, its molar concentration i.e., active mass remains almost constant throughout hydrolysis. To accelerate the rate of hydrolysis of ester with water, mineral acid is added to the reaction mixture.

Free acetic acid is formed during hydrolysis of the ester; the progress of the reaction can be studied by titrating definite volume of the reaction mixture against standard alkali solution at definite intervals of time.

At the commencement of the experiment, the amount of hydrochloric acid added is either known or can be determined by titrating the reaction mixture (definite volume) with standard alkali solution, at definite interval of time t, the amount of acid (x) formed can thus be determined and hence the amount of acetic acid formed at the end of the reaction corresponds to the initial concentration (a) of the ester. If V_0, V_t and V_∞ are the volumes of standard alkali solution needed to neutralize a definite amount of reaction mixture in the beginning of the reaction, after t minutes and at the end of the reaction, the amount (x) of acetic acid formed or amount of the ester hydrolyzed in t minutes will correspond to $V_t - V_0$ and the initial concentration of ester corresponds to $V_\infty - V_0$. Therefore

$$a - x = (V_\infty - V_0) - (V_t - V_0) = V_\infty - V_t$$

Now, $$K=\frac{2.303}{t}\log\frac{a}{a-x}$$

Substituting the value in the above equation, $K=\frac{2.303}{t}\log\frac{V_\infty - V_0}{V_\infty - V_t}$

Example 6 **1 cc of methyl acetate was added to a flask containing 20 cc of N HCl maintained at a temperature of 25 ΥC. 2 ccs of the reaction mixture was withdrawn at different intervals and titrated with a standard alkali.**

Time/min	**0**	**75**	**119**	**183**	**∘**
cc of alkali used	**19.24**	**24.20**	**26.60**	**29.32**	**42.30**

From the above data, show that the hydrolysis of methyl acetate is a unimolecular reaction.

Solution: $V_\infty - V_0 = a, V_\infty - V_t = (a-x)$

Initial concentration of ester a, $(V_\infty - V_0) = 42.03 - 19.24 = 22.24 = 22.79$

Time	**$V_\alpha - V_t$ corresponding to $a-x$**	$K=\frac{2.303}{t}\log_{10}\frac{V_\infty - V_0}{V_\infty - V_1}$
75	$42.03 - 24.20 = 17.83$	$K=\frac{2.303}{t}\log_{10}\frac{22.79}{17.83} = \mathbf{0.0003273}$
119	$42.03 - 26 = 15.43$	$K=\frac{2.303}{119}\log_{10}\frac{22.79}{15.43} = \mathbf{0.0003277}$
183	$42.03 - 29.32 = 12.71$	$K=\frac{2.303}{183}\log_{10}\frac{22.79}{12.71} = \mathbf{0.003191}$

Since the value of K is constant, hence the reaction is a first-order reaction.

MEMORY FOCUS

1. Rate expression for these reactions Rate = K[A] or $K=\frac{2.303}{\tau}\log\frac{a}{(a-x)}$
   (a = initial concentration; $(a-x)$ concentration at time t)
2. Units of the rate constant (K) are time and is independent of method of expressing the concentration.
3. A plot of $\log\left(\frac{a}{a-x}\right)$ with t is a straight line passing through origin with slope = K/2.303.
4. A plot of $\log(a-x)$ with time is a straight line with slope = –K/2.303.
5. Time taken for completion of a definite fraction of the reaction is independent of initial concentration of the reactant.
6. Half change time period for a first order reaction is independent of the initial concentration
   $$t_{1/2}=\frac{0.693}{K}$$

(Continued)

MEMORY FOCUS (*Continued*)

7. Example
   (i) decomposition of gases H_2O_2 to H_2O and O_2^-
   (ii) Decomposition of gaseous N_2O_5 and $SO_2Cl_2^-$
   (iii) Decomposition of gaseous H_2O_2 in solution with Pt as catalyst
   (iv) Decomposition of NH_4NO_2 to water and N_2^-
   (v) Ester hydrolysis in acid medium.
   (vi) Radioactive disintegrations.
   (vii) Inversion of cane sugar.

7.8 SECOND-ORDER REACTIONS

Reactions in which the rate of a reaction is determined by the change of two concentration terms are known as second-order reactions. Second-order reactions may be of two types as explained below.

In some cases, the rate of a reaction is determined by the change of two concentration terms of the same reactant. For example, $2HI \longrightarrow H_2 + I_2$

In general, we may represent such a case as: $2A \longrightarrow$ Products

There are other cases where the rate of a reaction is determined by the change of two concentration terms of two different reactants, e.g.

$$CH_3COOC_2H_5 + NaOH \longrightarrow CH_3COONa + C_2H_5OH$$

In general, such a case may be represented as follows:
Let us calculate the rate constants of these two types of reactions separately.
Equation of the second order for reactions of first type.

Consider a general reaction. $\qquad 2A \longrightarrow \text{Products}$

or $\qquad A + A \longrightarrow \text{Products}$

Let a moles/litre be the concentration of each of the reactant to start with and $(a - x)$ moles per litre be their concentration after time t. The rate of the reaction is given by

$$\frac{dx}{dt} \propto (a-x)^2$$

or

$$\frac{dx}{dt} = K(a-x)^2 \qquad (7.9)$$

where K is called the rate constant for a second-order reaction.

Rearranging the above equation, $\qquad \dfrac{dx}{(a-x)^2} = Kdt$

On integration of this equation, $\qquad \dfrac{1}{a-x} = Kt + I$

where I is the constant of integration.

When $t = 0$, $x = 0$

and hence $$I = \frac{1}{a}$$

Substituting this in equation in the above given integration equations, we have

$$\frac{1}{a-x} = Kt + \frac{1}{a}$$

$$Kt = \frac{1}{a-x} - \frac{1}{a} \quad \text{or} \quad K = \frac{x}{a(a-x)}$$

or $$K = \frac{1}{t}\frac{x}{a(a-x)} \tag{7.10}$$

Equation (7.10) is called the rate equation for the second order and K is called the second-order rate constant.

7.8.1 Characteristics of Second-order Reaction

7.8.1.1 Evaluation of Rate Constant (K)

All the reactions of second order must obey the following equations depending upon the type of reaction.

$$K = \frac{1}{t}\frac{x}{a(a-x)}$$

and $$K = \frac{2.303}{t(a-x)}\log\frac{b(a-x)}{a(b-x)}$$

If we put the values of x, a and b in the above equations in different intervals of time, a constant value of K must be obtained for reactions of the second order.

The value of K can also be evaluated if the amount of the reactants consumed at two different intervals of time is known. If x_1 and x_2 are the gram mole/litre of A and B converted to products up to time t_1 and t_2, then

$$Kt_1 = \frac{x_1}{a(a-x_1)} \quad \text{and} \quad Kt_2 = \frac{x_2}{a(a-x_2)}$$

$$Kt_2 - Kt_1 = \frac{x_2}{a(a-x_2)} - \frac{x_1}{a(a-x_1)}$$

$$K(t_2 - t_1) = \frac{1}{a}\left(\frac{x_2}{a-x_2} - \frac{x_1}{a-x_1}\right)$$

or $$K = \frac{1}{a(t_2 - t_1)}\left(\frac{x_2}{a-x_2} - \frac{x_1}{a-x_1}\right) \tag{7.11}$$

For graphical evaluation of K, Eq. (7.11) can be written as

$$Kt = \left(\frac{1}{a-x} - \frac{1}{a}\right)$$

or
$$\frac{1}{a-x} = Kt + \frac{1}{a}$$

It is evident that a graph of $\left[\frac{1}{a-x}\right]$ versus t will result in straight line since initial concentration a is constant for an experiment. The slope of this line is equal to K and intercept on the concentration axis $(t = 0)$ is equal to $\left[\frac{1}{a}\right]$ (Fig. 7.4).

Thus, the value of K can be determined by plotting $\frac{1}{a-x}$ versus (t).

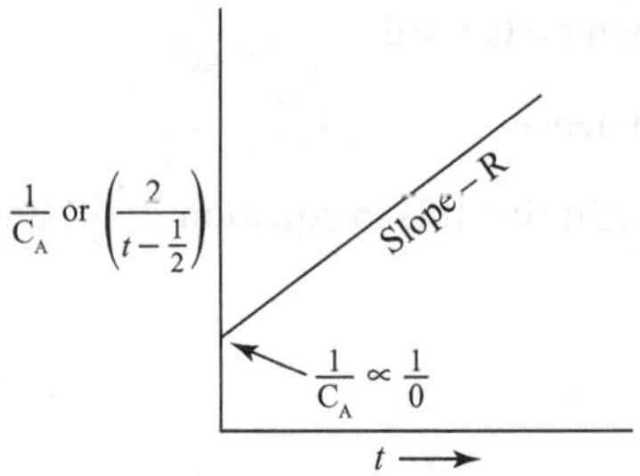

Figure 7.4 Linear plot for second order kinetics for $2A \rightarrow$ products or $A + B \rightarrow$ product where, $C_A = C_B$ or $(a = b)$

From the slope of the straight line obtained in the graph, the value of K can be determined.

7.8.1.2 Units of Rate Constant (K)

From Eq. (7.12) of second-order rate constant, we have

$$K = \frac{1}{t}\left[\frac{1}{a-x} - \frac{1}{a}\right]$$

or
$$K = \frac{1}{t}\left[\frac{x}{a(a-x)}\right]$$

Looking at the above equations we find that there are two concentration terms in the denominator, whereas there is only one concentration term in the numerator. Therefore, the second-order rate constant depends on the units in which the units of the reactants are expressed.

Dimensionally,
$$K = \frac{1}{\text{sec}} \times \frac{\text{moles/litre}}{\text{moles/litre} \times \text{moles/litre}}$$
$$= \text{litre moles}^{-1}\text{sec}^{-1}$$

7.8.1.3 Effect of Change in Concentration Units on the Value of K

Let the value of concentration in the new units be n times in the initial units:

$$a' = na \quad \text{or} \quad x' = nx \quad \text{and} \quad a' - x' = na(a-x)$$

where a', x' and $(a' - x')$ are molar concentrations in new units.

$$K = \frac{1}{a't}\left(\frac{x'}{a'-x'}\right) = \frac{1}{nat}\left[\frac{nx}{na-nx}\right]$$

$$= \frac{1}{nat}\left[\frac{nx}{n(a-x)}\right] = \frac{1}{nat}\left(\frac{x}{(a-x)}\right)$$

The new value of K' is $\frac{1}{n}$ of the original value.

7.8.1.4 Time for Fractional Change

Time for the completion of any fraction of the reaction (say, the half-life period) can be calculated from the equation for the second-order reaction.

When $$t = t_{1/2},\ x = \frac{a}{2}$$

$\therefore$ $$x = \frac{1}{t_{1/2}} \frac{\frac{a}{2}}{a\left(a - \frac{a}{2}\right)} = \frac{1}{t_{1/2}a}$$

$\therefore$ $$t_{1/2} = \frac{1}{Ka} \qquad (7.12)$$

It is clear from Eq. (7.12) that time for the completion of same fraction of change is inversely proportional to the initial concentration.

7.8.2 Example of the Second-order Reaction

(i) Hydrolysis of an ester by an alkali.

Initial molar concentration = a

$$CH_2COOC_2H_5 + NaOH \rightarrow CH_3COONa + C_2H_5OH$$

Final molar concentration = $(a - x)$

To follow the progress of this reaction, a dilute solution of ethyl acetate in distilled water is mixed with an equivalent quantity of caustic soda solution. The reaction mixture is kept at a constant temperature throughout the experiment.

Measured volumes of the liquids are then withdrawn at regular intervals and at once titrated against an acid solution of known strength. The volume of acid used in each case corresponds to the amount of unreacted NaOH or $CH_3COOC_2H_5$ at that time, i.e., $(a - x)$ gram moles. The volume of acid required initially corresponds naturally to the original concentration a.

Numerical Problems

Example 7 **From the following data show that the saponification of ethyl acetate using equal concentration of ester and alkali is a bimolecular reaction.**

Time (min)	**0**	**4.89**	**10.37**	**28.18**
Acid (used)	**447.65**	**38.92**	**32.62**	**20.58**

Solution: The bimolecular reaction equation is given as

$$K = \frac{1}{at} \times \frac{x}{a - x}$$

Initial concentration of ester a corresponding to the ccs of acid used when t is 0 = 47.65.

Time (min)	$a-x$	$x=a-[a-x]$	$K=\frac{1}{at}\times\frac{x}{a-x}$
4.89	38.92	$47.65-38.92=8.73$	$K=\frac{1}{47.65\times 4.89}\times\frac{8.73}{38.92}=\mathbf{0.00096}$
10.37	32.62	$47.65-32.62=15.03$	$K=\frac{1}{47.65\times 10.37}\times\frac{15.03}{32.62}=\mathbf{0.00093}$
28.18	20.58	$47.65-20.58=27.07$	$K=\frac{1}{47.65\times 28.18}\times\frac{27.07}{20.58}=\mathbf{0.00098}$

Since the value of K in these cases is almost constant, the reaction is bimolecular.

SECOND-ORDER REACTION

1. Rate expression for these reactions can be

   (a) rate = $K|A|^2$ or $K=\frac{1}{\tau}\left[\frac{1}{(a-x)}-\frac{1}{a}\right]$;

   or $$K=\frac{1}{\tau}\left[\frac{x}{a(a-x)}\right]$$

   (b) rate = $K|A||B|$

   or $$K=\frac{2.303}{\tau(b-a)}\log\frac{a(b-x)}{b(a-x)}$$

2. Units of rate constant are mole lit^{-1} time^{-1}.
3. (a) A plot of $\log\frac{a(b-x)}{b(a-x)}$ with time will be a straight line passing through origin whose slope is equal to $K(b-a)/2.303$.

   (b) A plot of $\frac{1}{(a-x)}$ with time is a straight line with slope equal to K.
4. Time taken for the completion of a definite fraction of the reaction is inversely proportional to the initial concentration.
5. Half change time period ($t_{1/2}$) for a second-order reaction is inversely proportional to the initial concentration (a).$t_{1/2}$ $\left[t_{1/2}=\frac{1}{Ka}\right]$
6. Examples

   (i) Conversion of ammonium cyanate to urea
   (ii) Benzoin condensation
   (iii) Saponification reaction (hydrolysis of esters in basic medium)
   (iv) Formation of HI from the elements
   (v) Decomposition of ozone
   (vi) Decomposition of Cl_2O and N_2O to the elements

(Continued)

SECOND-ORDER REACTION (*Continued*)

HALF-LIFE

1. Time required for completion of half of the reaction is called half-life of the reaction.
2. The half-life period varies inversely as the $(n - 1)$th power of the initial concentration of the reactants.

$t_{1/2}\ \alpha \left(\frac{1}{a}\right)^{n-1}\quad t_{1/2}\alpha(a)^{1-n}$ $\qquad$ (n = order of the reaction, a = initial concentration)

3. For zero-order reactions, the half life is directly proportional to the initial concentration

$$(t_{1/2} \infty\, a \quad \text{or} \quad t_{1/2} = ak)$$

4. For first-order reactions the half life is independent of initial concentration of the reactant $\left(t_{1/2} = \frac{0.693}{k} = \text{constant}\right)$

5. For second-order reactions the half life is inversely proportional to the initial concentration of the reactant

$$\left(t_{1/2}\ \alpha \frac{1}{a} \text{ or } t_{1/2} = \frac{1}{Ka}\right)$$

7.9 THIRD-ORDER REACTION

Reactions in which the rate of a reaction is determined by the change of three concentration terms are known as **third-order reactions.**

There can be many ways for the reactions to be of third order. For instance,

(i) A reaction may be of third order with respect to a single species,

$$3A \longrightarrow \text{Products}$$

or

$$A + A + A \longrightarrow \text{Products}$$

(ii) A reaction may be of first order with respect to one species and second order with respect to another species

$$A + 2B \longrightarrow \text{Products}$$

or

$$2A + B \longrightarrow \text{Products}$$

(iii) It may be of first order with respect to three different species

$$A + B + C \longrightarrow \text{Products}$$

Let us consider the rate expression of only a simple case where there is only one reactant say A

$$3A \longrightarrow \text{Products}$$

Let the initial concentration be a. Let x moles per litre of each reactant decompose in time t. Then at any given time, concentration of each reactant will be $(a - x)$. The rate of reaction is then given by

$$\frac{dt}{dx} \propto (a - x)^3$$

or

$$\frac{dt}{dx} = K(a - x)^3$$

where K is called the rate constant for a third-order reaction. Rearranging, we get

$$\frac{dt}{(a-x)^3} = Kdt$$

On integration, we get
$$\frac{1}{2(a-x)^2} = Kt + I$$

where I is a constant of integration. When $t = 0$, $x = 0$ and hence $I = \frac{1}{2a}$. Substituting this in the above equation, we get

$$\frac{1}{2(a-x)^2} = Kt + \frac{1}{2a^2}$$

$$\frac{1}{2}\left[\frac{1}{(a-x)^2} - \frac{1}{a^2}\right] = Kt$$

$$K = \frac{1}{2t}\left[\frac{1}{(a-x)^2} - \frac{1}{a^2}\right] \tag{7.13}$$

$$K = \frac{1}{2t}\left[\frac{a^2 - (a-x)^2}{a^2(a-x)^2}\right]$$

$$K = \frac{1}{2t}\left[\frac{x(2a-x)}{a^2(a-x)^2}\right] \tag{7.14}$$

Equations (7.13) and (7.14) are the rate equations for a third-order reaction when the initial concentrations of the three reactants are the same.

7.9.1 Characteristics of a Third-order Reaction

7.9.1.1 Evaluation of Rate Constant (K)

All the reactions of third order must obey Eq. (7.14):

$$K = \frac{1}{2t}\left[\frac{x(2a-x)}{a^2(a-x)^2}\right]$$

If we put the value of x and a in the above equation at different intervals of time, a constant value of K must be obtained for the reaction of the third order.

For graphical evaluation of K, Eq. (7.13) can be written as

$$K \times 2t = \frac{1}{(a-x)^2} - \frac{1}{a^2}$$

or
$$\frac{1}{(a-x)^2} = K \times 2t + \frac{1}{a^2}$$

It is evident that a graph of $\left[\frac{1}{(a-x)^2}\right]$ vs. t will result in a straight line since the initial concentration, (a), is constant for a given experiment. The slope of this line is equal to $2K$ and an intercept is equal to $1/a^2$. This plot is illustrated in Fig. 7.5.

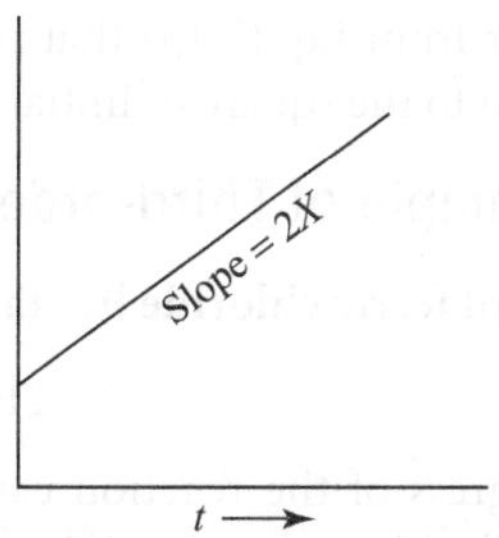

Figure 7.5 Typical linear plot for third order reaction

7.9.1.2 Units of Rate Constant (K)

From Eq. (7.14)

$$K = \frac{1}{2t} \times \frac{x(2a-x)}{a^2(a-x)^2}$$

Looking at the third-order rate equation, we find that there are three concentration terms in the denominator whereas there is only one concentration term in the numerator. Therefore, the third-order rate constant, unlike the first-order rate constant depends on the units in which the units of the reactants are expressed.

$$K = \frac{1}{\text{sec}} \times \frac{\text{mole/litre} \times \text{mole/litre}}{(\text{mole/litre})^2 \times (\text{mole/litre})^2}$$

$$= \frac{1}{\text{sec}} \times \frac{1}{(\text{mole/litre})^2} = \text{litre}^2\text{mole}^{-2}\,\text{sec}^{-1}$$

7.9.1.3 Effect of Change in Concentration Units on the Value of K

Let the value of concentration in the new units be n times in the initial units:

$$a' = na \quad \text{and} \quad a' - x' = n(a-x)$$

where a', x' and $(a' - x')$ are molar concentrations in new units

$$K = \frac{1}{2t}\left[\frac{nx(2na-nx)}{a^2n^2(an-nx)^2}\right] = \frac{1}{2tn^2}\left[\frac{x(2a-x)}{a^2(a-x)^2}\right] = \frac{1}{2n^2}K$$

The new value of K' is $\frac{1}{n^2}$ of the original value.

7.9.1.4 Time for Fractional Change

Time for the completion of any fraction of the reaction (say, half-life period) can be calculated from the equation of the third-order reaction.

When $$t = t_{1/2},\ x = \frac{a}{2}$$

$$K = \frac{1}{2t_{1/2}}\left[\frac{a/2(2a-a/2)}{a^2(a-a/2)^2}\right] = \frac{1}{2t_{1/2}} \times \frac{3}{a^2}$$

$$t_{1/2} = \frac{3}{2K}\frac{1}{a^2} \tag{7.15}$$

It is clear from Eq. (7.15) that time for the completion of the same fraction of change is inversely proportional to the square of initial concentration.

7.9.2 Example of Third-order Reactions

Reduction of ferric chloride by stannous chloride.

$$2FeCl_3 + SnCl_2 \longrightarrow 2FeCl_2 + SnCl_4$$

The progress of the reaction can be studied by mixing equivalent solution of ferric chloride and stannous chloride and maintaining the reaction mixture at constant temperature. Some amount of the mixture is withdrawn at definite intervals of time and the mercury chloride solution is added to it to remove excess of $SnCl_2$ and ferrous ions are estimated by titration against standard $K_2Cr_2O_7$.

The volume of $K_2Cr_2O_7$ used at each interval corresponds to the amount of ferrous chloride formed or the amount of ferric chloride reduced, i.e., x and the volume of dichromate used at infinite time (when the experiment is almost complete) is proportional to the total amount of $FeCl_2$ formed which corresponds to the total amount of $FeCl_2$ initially present, i.e., ***a***.

$$K = \frac{1}{2t} \times \frac{x(2a-x)}{a^2(a-x)^2}$$

Numerical Problems

Example 8 The reduction of ferric chloride by stannous chloride was studied by Nayas. The reactions were taken in equivalent amounts and the progress of the reaction was followed by titrating definite volume of the reaction mixture against standard potassium dichromate solution after adding mercuric chloride to remove the excess of stannous chloride. The results obtained are given below.

T (min)	1	3	7	∞
ccs of $K_2Cr_2O_7$ used	7.17	13.32	18.06	31.25

Show that the reaction is of the third order.

Solution: $K = \frac{1}{2t} \times \frac{x(2a-x)}{a'(a-x)^2}$

Now, initial concentration of $FeCl_3$ corresponds to the volume of dichromate used at infinite time $a = 31.25$, $2a = 62.50$ and volume used at a particular interval is equivalent to the amount of $FeCl_3$ which has undergone reduction i.e., x.

t	*x*	$a-x$	$2a-x$	$K = \frac{1}{2t} \times \left[\frac{x(2a-x)}{a^2(a-x)^2}\right]$
1	7.17	$31.50 - 7.17 = 24.08$	$62.50 - 7.17 = 55.33$	$K = \frac{1}{2\times1} \times \frac{7.17 \times 55.33}{(31.25)^2 \times (24.08)^2} = \mathbf{3.50 \quad 10^{-4}}$
3	13.32	$31.25 - 13.32 = 47.93$	$62.50 - 13.52 = 49.18$	$K = \frac{1}{2\times3} \times \frac{13.32 \times 49.18}{(31.25)^2 \times (47.93)^2} = \mathbf{3.47 \quad 10^{-4}}$
7	18.06	$31.25 - 18.06 = 13.19$	$62.52 - 18.06 = 44.44$	$K = \frac{1}{2\times7} \times \frac{18.06 \times 44.44}{(31.25)^2 \times (13.19)^2} = \mathbf{3.38 \quad 10^{-4}}$

The value of rate constant is fairly constant and hence the reaction is of third order.

7.10 REACTIONS OF HIGHER ORDER

Reactions involving four or more molecules (i.e., reactions with high molecularity) are rare. According to the kinetic theory, the rate of a chemical reaction is proportional to the number of collisions taking place between the reacting molecules. Obviously, the possibility of simultaneous collision of more than three molecules is very rare. It is, therefore, believed that such reactions can take place through a series of steps involving lesser number of molecules. Such reactions, which do not take place in one step, but take place in a series of steps, are known as complex reactions. In such cases, the slowest out of the various possible steps of the reaction steps becomes the rate-determining step. The order of the slowest step thus becomes the overall order of the complex reaction.

MEMORY FOCUS

1. Rate expression for the third order reactions is

$$\text{Rate} = k(a-x)^3 \quad \text{or} \quad K = \frac{1}{2x}\left[\frac{x(2a-x)}{a^2(a-x)^2}\right]$$

2. Units of rate constant (K) are litre^2 mole^{-2}s^{-1}.
3. A plot of $\left[\frac{1}{(a-x)^2}\right]$ vs time (t) will be a straight live with slope of the line being equal to 2k. The value of rate constant (K) can be calculated graphically.
4. The time for the completion of same fraction of change (say half time) is inversely proportional to the square of the initial concentration.
5. The half life period $[t_{1/2}]$ can be calculated as under:

$$t_{1/2} = \frac{3}{2K}\frac{1}{a^2}$$

6. A common example of 3rd order reaction is reduction of ferric chloride by stannous chloride

$$2\,FeCl_3 + SnCl_2 \longrightarrow 2FeCl_2 + SnCl_4$$

7.11 DETERMINATION OF RATE LAW, RATE CONSTANT AND ORDER OF REACTION

Different methods employed for these determinations are given below:

(i) Graphical method: This method is employed when there is only one reactant involved in the reaction under consideration. Let us consider the decomposing of H_2O_2 solution in water and the decomposition of N_2O_5 dissolved in carbon tetrachloride as an example for illustrating this method

$$2H_2O_2(l) \xrightarrow{Pt} 2H_2O(l) + O_2(g)$$

$$2H_2O_2(g) \longrightarrow 4NO_2(g) + O_2(g)$$

The progress of the reaction is followed by noting the amount of reactants consumed or products formed (x) at regular intervals of time (details already discussed). These values are then plotted against time (t). The plots of the curve are shown in Fig. (7.6) from the curve, the value of the rate of the reaction at different instants [dx/dt] can be determined from the slope of the curve at a particular time. The slope of the curve is equal to $\tan\theta$. Another graph is now plotted between the value of [dx/dt] at a particular time and $(a-x)$, $(a-x)^2$ or $(a-x)^3$, respectively at that time. The order of the reaction is determined from the graph which gives a straight line. If a straight line is obtained by plotting [dx/dt] against $(a-x)$, the reaction would be of the first order. Similarly, if the plot of [dx/dt] against $(a-x)^2$ comes out to be a straight line, the reaction would be of the second order.

Numerical Problems

Example 9 The decomposition of N_2O_5 in CCl_4 solution follows the first-order rate law. The concentration of N_2O_5 measured at different intervals of time is given below:

Time in seconds (t)	0	80	160	410	600	1130	1720
N_2O_5	5.5	5.0	4.8	4.0	3.4	2.4	1.6

Calculate its rate constant at $t = 410$ s and $t = 1130$ s. What do these results show?

Solution: The equation for the first-order reaction is $K = \frac{2.303}{t} \log \frac{a}{a-x}$

(i) At $t = 410$ s, $K = \frac{2.303}{410} \log \frac{5.5}{4}$

$$K = 0.0056\ (1.7404 - 1.6021) = 7.7\quad 10^{-4}\ s^{-1}$$

(ii) At $t = 1130$ s, $K = \frac{2.303}{123s^{-1}} \log \frac{5.5}{2.4}$

$$K = 0.0024\ s^{-1}\ (1.74.4 - 1.3802) = 7.7\quad 10^{-4}\ s^{-1}$$

This proves that the reaction is of the first order.

(ii) Initial rate method: This method is employed when more than one reactants are involved in the reaction. For instance, let us consider the reaction: $aA + bB \xrightarrow{} \text{Products}$.

In this method, the change in the concentration of one reactant (say A) is a measure as time passes, while that of the B is kept constant. A graph between concentration of A and time is plotted and a tangent at a point corresponding to $t = 0$ on the curve is drawn. The slope of the tangent at a point corresponding to $t = 0$ on the curve is hence drawn. The slope of the tangent gives initial rate of the first set of reaction.

In another set of the same reaction, the initial concentration of one reactant (B) is changed while that of the other (A) is kept constant. The procedure similar to the first set is followed to get the initial rate. In this way, different initial rates for different initial concentration are measured. Such data are used to solve the different rate equation. Rate = $K[A]^x\ [B]^y$.

There are three quantities, namely K, x and y to be evaluated; therefore, three sets of initial rates would be sufficient to evaluate all the three.

By starting with certain concentrations of A and B, the rate of reaction, $\Delta x_1/\Delta t$ for short interval of time Δt *has been* calculated. The [A] is double, keeping [B] the same, and the value of the rate of the reaction, $\Delta x_2/\Delta t$ has been calculated:

$$\frac{\Delta x_2/\Delta t}{\Delta t/\Delta t} = \frac{k[2A]^x\,[B]^y}{k[A]^x\,[B]^y} = 2^x$$

The value of x gives the order of the reaction w.r.t. A.

Similarly, the procedure is repeated by keeping [A] the same and doubling [B] and the value of y is determined which is the order of the reaction w.r.t. B. The overall order of the reaction is their sum, i.e., $x + y$.

Example 10 **From the following data, determine the order of the reaction and the rate constant for the reaction**

S. No.	[A] (mol L^{-1})	[B] (mol L^{-1})	Rate (mol L^{-1} s^{-1})
1.	0.35	0.64	0.008
2.	0.35	1.28	0.032
3.	1.05	0.64	0.024

Solution: Let the order w.r.t. A be x and w.r.t. B be y. Then, the rate law for the reaction is given by

$$\text{Rate} = k[A]^x\,[B]^y$$

Substituting the values at S. No. 1, we get

$$\text{Rate} = k(0.35)^x\,(0.64)^y = 0.008 \qquad \text{(i)}$$

Likewise, for the values at S. No. 2, we get

$$\text{Rate} = k(0.35)^x - (1.28)^y = 0.032 \qquad \text{(ii)}$$

Further, for values at S. No. 3,

$$\text{Rate} = k(1.05)^x - (0.64)^y = 0.024 \qquad \text{(iii)}$$

Dividing Eq. (ii) by Eq. (i) we get $\dfrac{k(0.35)\times(1.28)y}{k(0.35)\times(0.64)y} = \dfrac{0.032}{0.008} = 4$

$$2^y = 4 = 2^2,\ \therefore\ y = 2$$

i.e., the order with respect to B is 2. Similarly, dividing Eq. (iii) by Eq. (ii), we get

$$\frac{k(1.05)\times(0.64)y}{k(0.35)\times(1.28)y} = \frac{0.024}{0.008} = 3$$

$$3^x = 3 = 3^1, \qquad [\therefore\ x = 1]$$

i.e., the order with respect to A is 1. Hence, the overall order of the reaction

$$= x + y = 1 + 2 = 3$$

Rate law for the reaction

$$\text{Rate} = k[A][B]^2$$

Now rate constant $$(K) = \frac{\text{Rate}}{[A][B]^2}$$

From values at S. No. 1

Rate constant $$(k) = \frac{0.008}{0.35 \times (0.64)^2} = \mathbf{0.0558\ mol^{-2}\,L^2\,s^{-1}}$$

From values at S. No. 2

Rate constant $$(k) = \frac{0.032}{0.35 \times (1.28)^2} = \mathbf{0.058\ mol^{-2}\,L^2\,s^{-1}}$$

Comment. It may be concluded from the above results that the rate of a reaction depends upon the concentration but the rate constant is independent of the concentrations.

(iii) Integrated rate law method: The differential rate law equation for a reaction $\alpha A \rightarrow$ Products, can be written as

$$\frac{-d[A]}{dt} = k[A]^x,$$

where x is the order of the reaction.

If we start with 'a' moles per litre of A and after time t, x moles per litre have reacted, so that the concentration after time t is $(a - x)$ moles per litre. Depending upon the order of the reaction the differential rate law equations would become

$$\frac{dx}{dt} = K(a - x)^0 \qquad \text{[For a zero-order reaction]}$$

$$\frac{dx}{dt} = K(a - x) \qquad \text{[For a first-order reaction]}$$

$$\frac{dx}{dt} = K(a - x)^2 \qquad \text{[For a second-order reaction]}$$

$$\frac{dx}{dt} = K(a - x)^3 \qquad \text{[For a third-order reaction]}$$

These equations can be integrated and the integrated rate law equations for reaction depending upon there are given as

$$K = \frac{x}{t} \qquad \text{[For zero order]}$$

$$K = \frac{2.303}{t} \log \frac{a}{a - x} \qquad \text{[For first order]}$$

(for derivation, refer to article 7.5)

$$K = \frac{1}{t}\frac{x}{a(a - x)} \qquad \text{[For second order]}$$

$$K = \frac{1}{2t}\left[\frac{x}{(a - x)2} - \frac{1}{a}\right] = \frac{1}{2t}\left[\frac{x(2a - x)}{a^2(a - x)^2}\right] \qquad \text{[For third order]}$$

In this method, the known amounts of the reactants are taken and the progress of the reactions is followed after regular intervals of time. The experimental results are substituted in the rate equations for zero-, first-, second- and third-order reactions. The rate equation, which gives a constant value of K, gives the order of the reaction.

(iv) Half-life period: Half-life period of a reaction is defined as the time in which half of the reactants originally present get converted into products. In other words, it is the time in which the concentration of the reactants falls to half of the initial value.

When $$t = t_{1/2},\ x = \frac{x}{2}$$

Substituting this in the rate equation for the first order, i.e.,

$$K = \frac{2.303}{t}\log\frac{a}{(a-x)}$$

We have $$K = \frac{2.303}{t_{1/2}}\log\frac{a}{(a-x)}$$

or $$t_{1/2} = \frac{2.303}{K}\log 2 = \frac{2.303\times 0.3010}{K} \quad \text{or} \quad t_{1/2} = \frac{0.693}{K}$$

The above equation does not involve any concentration term. It, therefore, follows that for a first-order reaction, the half period is independent of the initial concentration of the reactants.

In general, the half-life period varies inversely as the $(n-1)$th power of the initial concentration of the reactants $t_{½}\ \alpha\left(\frac{1}{2}\right)^{n-1}$.

Starting with two different initial concentrations of the reactants, we get

$$t_1\,(1/2)\alpha\left(\frac{1}{a^1}\right)^{n-1} \quad \text{and} \quad t_2\,(1/2)\alpha\left(\frac{1}{a^2}\right)^{n-1}$$

Example 11 **1.0×10^{-6} g of radioactive iodine is injected into the blood of a patient. How long will it take for radioactivity to fall to 10% of the initial value? The half-life period of radioactive iodine is 8.05 days.**

Solution: We know that $K = \frac{0.693}{t_{1/2}} = \frac{0.693}{8.05\text{ days}} = 0.086\text{ days}^{-1}$

$$t = \frac{2.303}{K}\log\frac{a}{a-x}$$

$$t = \frac{2.303}{0.086}\log\frac{1.0\times 10^{-6}\,g}{0.5\times 10^{6}\,g} = \frac{2.303}{0.086\text{ days}}\log^2$$

$$t = \frac{2.303\times 0.3010}{0.086} = \mathbf{8.0604\ days}$$

Example 12 **75% of a reaction of the first order was completed in 32 min. When was it half completed?**

Solution: The reaction is 75% completed in 32 min.

When $$t = 32:\ x = \frac{3}{4}\alpha$$

Substituting these values in the rate equation of the first order, we get

$$K = \frac{2.303}{t}\log\frac{a}{a-x}$$

$$K = \frac{2.303}{32}\log\frac{a}{a-\frac{3a}{4}}$$

or $$K = \frac{2.303}{32}\log 4 = \frac{2.303}{32}\times 0.6021\ \text{min}^{-1}$$

Let the reaction be half completed in 't' minutes.

i.e., when $t = t'$; $x = \frac{x}{2}$

$$t' = \frac{2.303}{K}\log\frac{a}{a-\frac{a}{2}}$$

or $$t' = \frac{2.303\times 32}{K2.303\times 0.6021}\log 2 = \frac{2.303\times 32}{K2.303\times 0.6021}\times 0.3010 = \mathbf{16\ min}$$

Hence, the reaction was completed in 16 min.

7.12 SOME COMPLICATIONS IN DETERMINATION OF ORDER OF A REACTION

7.12.1 Consecutive Reactions

It has been observed that in some cases the order of a reaction as determined experimentally does not conform to the one suggested by chemical equation which represents the reactions. This is because the reaction is not a single-step reaction as the chemical equation suggests, but the final products are formed as a result of two or more reactions. Chemical reactions such as $A \xrightarrow{K_1} B \xrightarrow{K_2} C$ which proceed from reactants to products through one or more intermediate stages are called **consecutive reactions.**

In these reactions, each stage has its own rate and its own velocity constant. The overall rate depends on the relative magnitudes of the various stages. It is a well-established fact that in any sequence of reaction of varying speed, the one that is slowest will determine the rate of the overall reaction. Consequently, if, in a reaction such as that represented above $K_1 >> K_2$, the conversion of B to C will determine the rate of formation of the product. On the other hand if, $K_2 >> K_1$, the formation of B from A will control the rate, and C will result from B as soon as the latter appears. However, when K_1 and K_2 are comparable in magnitude, the rate of the overall reaction depends on both constants, and it may effect some complication in the determination of order of the reaction.

For example, the thermal decomposition of acetaldehyde is a fractional order (1.5) reaction.

$$H_3C-CHO \xrightarrow{450\,°C} CH_4 + CO$$

An instance of a consecutive reaction is the decomposition of sodium hypochlorite in alkaline solutions. The overall reaction is

$$3NaClO \longrightarrow NaClO_3 + 2NaCl$$

Although the reaction appears to be of third order, yet it has been determined experimentally to be of the second order. The reaction proceeds in two steps

$$2NaClO \longrightarrow NaCl + NaClO_2 \quad (7.16)$$

and

$$NaClO_2 + NaClO \longrightarrow NaCl + NaClO_3 \quad (7.17)$$

It is found that step (7.16) is two times slower than step (7.17), so the order of the reaction is two. Thus, the slowest reaction among the consecutive reactions is the rate-determining step.

Some more examples of the consecutive reactions are discussed under mechanism of reaction.

(i) Reversible reactions or opposing reactions:

Opposing or reversible reactions are those reactions in which both forward and backward reactions take place simultaneously.

Let us consider the simplest case where both the forward and the backward reactions follow first-order kinetics and are in equilibrium:

$$A \underset{K_b}{\overset{K_f}{\rightleftharpoons}} B \quad (7.18)$$

where K_f and K_b are the rate constants for the forward and backward reactions, respectively. The overall rate of reaction is given by

$$\text{Rate of reaction} = r_f - r_b$$

i.e.,

$$\frac{dx}{dt} = -\frac{d[A]}{dt} - \frac{d[B]}{dt} - K_f[A] - K_b[B] \quad (7.19)$$

where r_f and r_b refer to the rates of forward and backward reactions, respectively.

(ii) Dependence of reaction rate on species other than reactants:

The rate of the reaction

$$CH_3COOH_3 + I_2 \xrightarrow{H^+} H_3CCOCH_2I + HI$$

depends up on H^+ ions, the catalyst used.

The rate is given as Rate $= K\,[CH_3COCH_3]\,[H^+]$

(iii) Negative order of the reaction:

Certain reactions have negative order w.r.t. reactants/products. For example, during decomposition of ozone to oxygen, $2O_3 \longrightarrow 3O_2$, the order w.r.t. O_2 is −1.

This indicates that the rate of reaction decreases as the concentration of O_2 increases.

1. The difference in the potential energy of the reactants and that of the activated complex is the activation energy (E_a). The reactants must have this minimum energy to undergo the reaction through the transition state.
2. For exothermic reactions, the potential energy of the products is less than that of the reactants, whereas in case of endothermic reactions it is reversed.
3. The activation energy of the reaction is the additional energy which the reactant(s) molecules required to form the activated complex i.e., the difference between the energy of the activated complex and the reactant molecules.

MEMORY FOCUS

1. The chemical reactions, which proceed from reactants to products through one or more intermediate stages are called **consecutive reactions.**
2. The **slowest reaction** among the consecutive reactions **is the rate-determining step.** The order of the slowest reaction becomes the overall order of the reaction.
3. A **reversible reaction** is the one in which both the forward and the backward reactions take place simultaneously. The overall rate of such reactions is equal to the rate of forward reaction minus the rate of backward reaction.
4. In some cases, the reaction rate depends on species other than reactants. For instance,

$$CH_3COOH_3 + I_2 \xrightarrow{H^+} H_3CCOCH_2I + HI$$

The rate is given as: Rate = $K[CH_3COCH_3]$ $[H^+]$
Here, $[H^+]$ is the molar concentration of the catalyst.

5. Certain reactions have negative order w.r.t. reactants/products. For example, during decomposition of ozone to oxygen

$$2O_3 \rightarrow 3O_2 \text{ (order w.r.t. } O_2 \text{ is } -1)$$

This indicates that the rate of reaction decreases as the concentration of O_2 increases.

7.13 TEMPERATURE DEPENDENCE OF REACTION RATES

Temperature has a marked influence on reaction rates. An increase in temperature, in general, results in an increase in the rate of a reaction. In many homogeneous reactions, the rate is almost doubled by rising the temperature only by 10 ; however, in some cases the increase may be even higher. The effect of temperature is generally expressed in terms of temperature coefficient which may be defined as 'the ratio of the velocity constants of a reaction at two temperatures differing by 10 °C, usually 25 °C and 35 °C.'

For most of the homogeneous reactions, the rate constant is doubled or tripled for each 10° rise in the temperature.

Thus, $$\text{temperature coefficient} = \frac{K_{t+10}}{K_t} = 2 \text{ to } 3$$

where K_t is the velocity constant at t °C and K_{t+10} is the velocity constant at 10° higher.

Many reactions of industrial importance, like the Haber's process for manufacture of NH_3 are carried out at a higher temperature to speed up the reaction.

7.13.1 Explanation of Effect of Temperature

The increase in the rate of a reaction with increase in temperature may be explained on the basis of collision theory. A chemical reaction is the result of collisions between the reacting molecules, but all the collisions are not effective collisions.

For a collision to be effective, the reactants molecules must be associated with a certain minimum amount of energy known as the threshold energy. Thus, the threshold energy may be defined as the minimum amount of energy which must be possessed by the reactant molecules so that their collisions results in actual chemical change.

However, most of the molecules have energy lower than the threshold energy. Thus, for a reaction to occur, the reacting molecules must be activated before they can react to get converted in to products. In other words, some extra energy should be supplied to the reactants so that their collisions become effective and the reaction occurs.

This additional energy required by the reactants molecules so that their collisions may result in an actual chemical change is called activation energy. This is commonly denoted by E_a, Thus, we have

Activation energy = Threshold energy – Energy actually possessed by the reactant molecules.

i.e., $$E_a = E_{threshold} - E_{reactants}$$

Due to the tremendous number of collisions, there occurs a wide range of distribution of velocities and kinetic energy of molecules. The distribution of kinetic energies is shown in Fig. 7.5. The shaded area under the curves at temperature T_2 and T_1 represents the fraction of molecules undergoing effective collisions. The area under the curve at temperature T_2 is more than double than that a lower temperature, T_1. It is seen that if lower temperature is raised, there occurs a large increase in the number of molecules having energy higher than threshold energy. Thus, at a higher temperature, the number of effective collisions is much higher and hence the rate of reaction increases.

7.13.2 Arrhenius Equation

The effect of temperature on the rate constant of *a* can be described by means of an equation proposed by Arrhenius and is commonly known as Arrhenius equation.

$$K = A_e^{-Ea/RT}$$

where K is the rate constant, A is a constant called the frequency factor and is characteristic of the reaction, E_a is the activation energy of the reaction, T is the Kelvin temperature, R is the gas constant in J K^{-1} mol^{-1} and $e^{-Ea/RT}$ is the fraction of activated molecules. Taking log, the above equation becomes

$$\ln K = \ln A - \frac{E_a}{RT}$$

This may be written as $$\log K = \log A - \frac{E_a}{2.303RT}$$

Thus, if $\log K$ is plotted against $1/T$, a straight line should be obtained whose slope is equal to $-E_a/2.303R$.

Knowing the slope, E_a can be calculated. Alternatively, the value of E_a can be calculated as

$$K = A_e^{-Ea/RT}$$

Comparing the Ks at two different temperatures T_1 and T_2, we have

$$\frac{K_2}{K_1} = e\left[(-E_a/R)\left(\frac{1}{T_2} - \frac{1}{T_1}\right)\right]$$

Taking logarithm, we get $$\ln \frac{K_2}{K_1} = -\left(\frac{Ea}{R}\right)\left(\frac{1}{T_2} - \frac{1}{T_1}\right)$$

Converting to common logarithm

$$\log_{10}\frac{K_2}{K_1} = -\frac{E_a}{R\times 2.303}\left(\frac{1}{T_2}-\frac{1}{T_1}\right)$$

$$= \frac{E_a}{R\times 2.303}\left(\frac{1}{T_2}-\frac{1}{T_1}\right)$$

$$= \frac{E_a}{R\times 2.303}\left(\frac{T_2-T_1}{T_1\,T_2}\right)$$

$$E_a = 2.303\mathrm{R}\left[\frac{T_1T_2}{T_2-T_1}\right]\left[\log\frac{K_2}{K_1}\right]$$

MEMORY FOCUS

1. In most of the reactions, for every 10 °C rise in temperature, the rate of the reaction doubles.
2. Ratio of the rate constants of a reaction at two temperatures which differ by 10° (usually 35 °C and 25 °C) is called temperature coefficient.
3. Temperature coefficient $= \dfrac{K_T+10}{K_T} = 2$ to 3 (for many).
4. Variation of rate constant with temperature is given by the Arrhenius equation $K = \mathrm{A}e^{-Ea/RT}$ [A = frequency factor constant, E_a = energy of activation, R = gas constant]
5. The term $e^{-Ea/RT}$ is called Boltzmann factor.
6. Boltzmann factor represents the fraction of molecules having energy in excess of activation energy.
7. Arrhenius equation may also be written as

$$\log K = \log \mathrm{A} - E_a/2.303\ \mathrm{RT}$$

8. Effective collisions are those in which the colliding molecules must have
   (a) energy equal to or greater than the threshold energy
   (b) proper orientation

Numerical Problems

Example 13 **The rate constant, *K*, for the first-order gas phase decomposition of ethyl iodide, $C_2H_5I \times C_2H_4 + HI$ is $1.60 \times 10^{-5}\ s^{-1}$ at 600 K and $6.36 \times 10^{-3}\ s^{-1}$ at 700 K. Calculate the energy of activation of this reaction ($R = 8.314\ J\ mol^{-1}\ K^{-1}$).**

Solution: According to Arrhenius equation

$$\log\frac{K_2}{K_1} = \frac{E_a}{2.303R}\left[\frac{1}{T_1}-\frac{1}{T_2}\right]$$

$$\log \frac{6.36\times10^{-3}\, s^{-1}}{1.60\times10^{-5}\, s^{-1}}$$

$$= \frac{E_a}{2.303\times8.314\,\text{J mol}^{-1}\text{K}^{-1}}\left[\frac{1}{600\,K}-\frac{1}{700\,K}\right]$$

$$\log 397.5 = \frac{E_a}{2.303\times8.314\,\text{J mol}^{-1}}\times\frac{1}{4200}$$

$$2.5993 = \frac{E_a}{2.303\times8.314\times4200\,\text{J mol}^{-1}}$$

$$E_a = 2.5993x\ 2.303x\ 8.314x\ 4200\ \text{J mol}^{-1}$$

$$E_a = 2.09\times10^5\ \text{J mol}^{-1} = \mathbf{209\ kJ\ mol^{-1}}$$

Example 14 **Calculate the activation energy of a reaction whose reaction rate at 27 ΥC gets doubled for 10 ΥC rise in temperature (R = 8.314 J K^{-1} mol^{-1}).**

Solution: It is given that

When $T_1 = 27 + 273\ \text{K} = 300\ \text{K}$

When $T_2 = 27 + 10\ °\text{C} + 273 = 310\ \text{K}$

$K_2 = 2K_1$

Substituting these values in the equation

$$\log\frac{K_2}{K_1} = \frac{E_a}{2.303R}\left(\frac{T_2-T_1}{T_1\,T_2}\right),$$

we get

$$\log\frac{2K_1}{K_1} = \frac{E_a}{2.303\times8.314R}\times\left(\frac{310-300}{300\times310}\right)$$

or

$$\log 2 = \frac{E_a}{19.15}\times\frac{10}{300\times310}$$

we get

$$E_a = 53611.9\ \text{J mol}^{-1} = \mathbf{53.612\ kJ\ mol^{-1}}$$

7.14 MECHANISM OF A REACTION (CONCEPT OF MOLECULARITY AND ORDER OF A REACTION)

The most important application of chemical kinetics is that it helps to study the manner in which the reactant molecules are changed to products. In a reaction, the formation of products results directly from a suitable collision between the reactant molecules. However, in most of the reactions, the overall change from the reactants into the products is brought about in a number of step reactions, each step being called an elementary reaction. A series of step reactions proposed for the completion of an overall reaction is called the mechanism of the reaction.

While proposing a reaction mechanism on the basis of kinetic studies, the following two facts must be considered.

(i) The rate of formation of the ultimate products of a reaction cannot be greater than the rate of the slowest step in the sequence. The slowest step is then called the rate-determining step and if it is much slower than the others, it exclusively determines the rate and order of the overall reaction.
(ii) The rates of formation and utilization of the intermediate substance are equal and fast.

Let us now analyze some reactions with a view to propose their reaction mechanism.

(i) **Reaction between NO_2 and F_2**

The reaction $2NO_2 + F_2 \rightarrow 2NO_2F$ follows a second-order rate law. It obeys the following function(s):

$$\text{Rate} = \frac{dx}{dt} = k[NO_2][F_2]$$

The rate law indicates that both the NO_2 and F_2 are involved in the rate-determining step but stoichiometry shows that any reaction between NO_2 and F_2 must produce something besides NO_2F. These two facts suggest that the most likely mechanism for the reaction is

1. $NO_2 + F_2 \rightarrow NO_2F + F$
2. $F + NO_2 \rightarrow NO_2F$ (fast)

The first bimolecular process is the rate-determining step. Its rate law confirms that the overall reactions are of second order and the molecularity is also two.

(ii) **Reaction between H_2 and ICI**

The stoichiometric equation for this reaction is

$$H_2 + 2ICI \rightarrow 2HCl + I_2$$

Rate law, $$\frac{dx}{dt} = K[ICI][H_2]$$

Order = 2 and molecularity is also 2. The rate law suggest that if no other complication exists, the rate-determining step should involve one molecule each of H_2 and ICI.

$$ICI + H_2 \rightarrow \text{Activated complex} \rightarrow HCl + HI \text{ (slow)} \quad (7.20)$$

Since HI is not a product of the reaction, its disappearance should occur by a fast reaction such as

$$ICI + HI \rightarrow I_2 + HCl \text{ (fast)} \quad (7.21)$$

Combination of reactions (7.20) and (7.21) gives the stoichiometry and rate law for the reaction.

(iii) **Gas phase reaction between NO and H_2**

The reaction

$$2NO\,(g) + 2H_2\,(g) \rightarrow N_2(g) + 2H_2O\,(g)$$

Rate law, $$\frac{dx}{dt} = K[NO]^2\,[H_2]$$

Order: 3 Suggested mechanism

$$2NO\,(g) + H_2\,(g) \rightarrow N_2(g) + H_2O\,(g) \text{ (slow)}$$

$$H_2O_2\,(g) + H_2\,(g) \rightarrow 2H_2O \text{ (fast)}$$

(iv) Reaction between NO_2 and CO

The reaction

$$NO_2\,(g) + CO\,(g) \rightarrow CO_2\,(g) + NO\,(g)$$

Rate law, $$\frac{dx}{dt} = K[NO_2]^2$$

Order: 2 Suggested mechanism

1. $NO_2 + NO_2 \rightarrow NO + NO_3$ (slow)
2. $NO_3 + CO \rightarrow CO_2 + NO_2$ (fast)

The first step being slow is the rate-determining step. The reaction (i) is bimolecular or the reaction is of second order.

(v) Thermal decomposition of N_2O_5

$$2N_2O_5 \rightarrow 4\,NO_2 + O_2$$

According to rate law, rate of reaction = $K[N_2O_5]$

The suggested mechanism is given below:

1. $N_2O_5 \rightarrow NO_2 + NO_3$ (slow)
2. $N_2O_5\,NO_3 \rightarrow 3NO_2 + O_2$

The first reaction is the slowest one involving one molecule of N_2O_5. Thus, the molecularity and order of the reaction is one.

MEMORY FOCUS

1. A series step reactions proposed for the completion of an overall reaction is called the mechanism for the reactions.
2. The slowest step is the rate-determining step and it determines the overall order of the reactions.
3. For determining the mechanism for a reaction, its order is determined experimentally and the rate law equation is accordingly written. The sequence of step for the overall order of the reaction must be the order of the slowest step of the proposed sequence of steps.
4. For example, the reaction between NO_2 and F_2, i.e.

   $2NO_2 + F_2 \rightarrow NO_2F + F$ follows a second order rate law equation

   $$\frac{dx}{dt} = K[NO_2][F_2]$$

   The proposed mechanism is

   (i) $NO_2 + F_2 \rightarrow NO_2F + F$

   (ii) $NO_2 + F \rightarrow NO_2F$

   The first step is the rate-determining step.

7.15 THEORIES OF REACTION RATES

There are two important theories of reaction rates, namely

(i) The collision thcory
(ii) The transition state theory or theory of absolute reaction rates
(iii) Lindeman theory

Let us discuss these theories in sequence.

(i) The collision theory: According to this theory, molecules must collide to bring about the reaction. The rate of the reaction is proportional to the number of colliding molecules per unit time. It is understood that molecules cannot react if they are kept at a distance from one another. However, collision between molecules cannot be the only factor involved in determining the rate of a reaction.

Chemical reactions involve redistribution of atoms; and redistribution of atoms requires rupturing of the bonds holding the atoms together in the original molecules and the formation of new bond in the product molecules.

For example, for the reaction

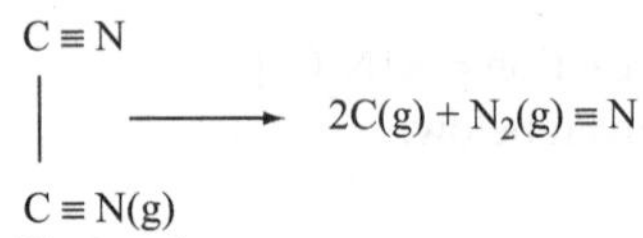

(Cynogen)

to occur, the bond holding the C atom to the N atom must be ruptured and bond must form between N atoms. We thus postulate that molecules react only in a collision, if they possess an energy equal to or greater than a certain critical value known as the **energy of activation** E_a, which is characteristic for each reaction.

Thus the salient hypothesis of classical collision theory is that *if the energy of the colliding molecules is less than E_a, no reaction occurs; if energy of the colliding molecules is equal to or greater than E_a, reaction occurs.*

Then, reaction rate = number of colliding molecules per unit volume per unit time × fraction of colliding molecules with energy E_a or greater. Only a very small fraction of the total number of collision actually results in a reaction.

For simplicity, it is further assumed that molecules utilize only their kinetic (translational) energy to acquire E_a. This kinetic energy converted to vibrational energy upon impact causes the atoms in the molecules to vibrate violently; the large vibrational amplitudes weaken the bonds.

The rate constant of bimolecular reaction from collision theory is given as follows

$$K = Pze^{\frac{-E_a}{RT}}$$

where P is the **probability factor** or **steric factor**, z is the number of colliding molecules per cc per sec in a reacting system which contains unit concentration of the reactant (i.e., /mole/litre). If x be the fraction of activated molecules possessing threshold energy i.e., the minimum energy required for effective collision that can bring about a chemical reaction, then the rate constant of a bimolecular reaction is given as $(K) = zx$.

$$\text{The fraction of activated molecules } (x) = \frac{\text{Number of activated molecules}}{\text{Total number of molecules}} = \frac{\Delta N}{N}$$

Comparison with Arrhenius equations:
Comparison of the collision theory equation, modified with the steric factor

$$K = Pze^{-E_a/RT}$$

and the **Arrhenius** equation

$$K = Ae^{-E_a/RT}$$

shows that we can identify the frequency factor A with the term *Pz*, i.e., A = *Pz*. *P is called the steric factor.*

The value of *P* normally would be less than unity or exactly one if there were no steric requirements. Even this refinement, however, cannot adequately handle some reactions that appear to have steric factors greater than unity. In addition to this inadequacy, there is a more fundamental lack of predictive character in collision theory. There is no way to deduce the activation energy except by direct experimental measurement.

Apparently, not all collision molecules possessing the required energy of activation lead to a reaction. The manner in which they collide is also important. By analogy, the damage in flight (reaction) resulting from the collision of two zero planes depends not only on their speeds (energy) but also on their relative positions of orientation. Thus, some molecules must be oriented in a very specific manner for reaction to occur. Other molecules may react on colliding whatever be their orientation (Fig. 7.6).

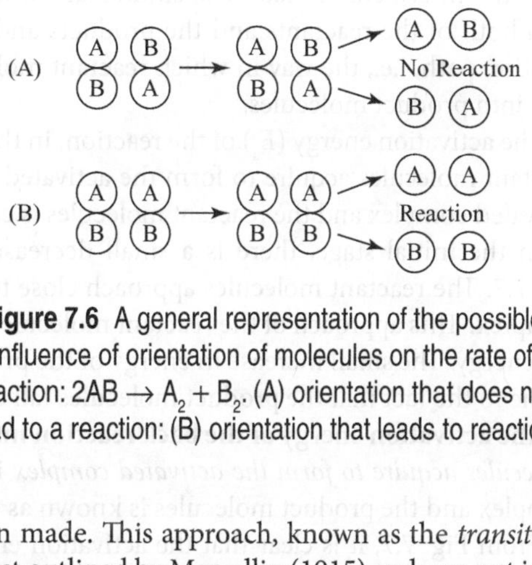

Figure 7.6 A general representation of the possible influence of orientation of molecules on the rate of reaction: $2AB \rightarrow A_2 + B_2$. (A) orientation that does not lead to a reaction: (B) orientation that leads to reaction

(ii) The transition-state theory: Another approach, as an alternative to the collision theory, to explain the mechanism of reaction has been made. This approach, known as the *transition-state theory or* ***the absolute reaction rate theory*** was first outlined by Marcellin (1915) and was put in its developed present form in 1935 by **Henry Eyring, Polany and Plezer**. This approach is also known as **activated complex theory**.

According to this theory 'when reactants are converted into the products, there exists a state which is intermediate between the two and is known as **activated complex**.' Therefore, the reactant molecules are first transformed into an intermediate energy-rich activated complex or transition state. This activated complex may be formed by some loose association or bonding of the reactant molecules with necessary rearrangement of valency bonds and energy. The activated complex, though unstable and has a transient existence, is treated formally as a definite molecule with an independent entity. The activated complex subsequently breaks up to produce the resultants at a definite rate.

Schematically, this is represented below:

$$\text{Reactants} \rightarrow \text{Activated complex} \rightarrow \text{Products}$$

The dissociation of HI into H_2 and I_2 is depicted below:

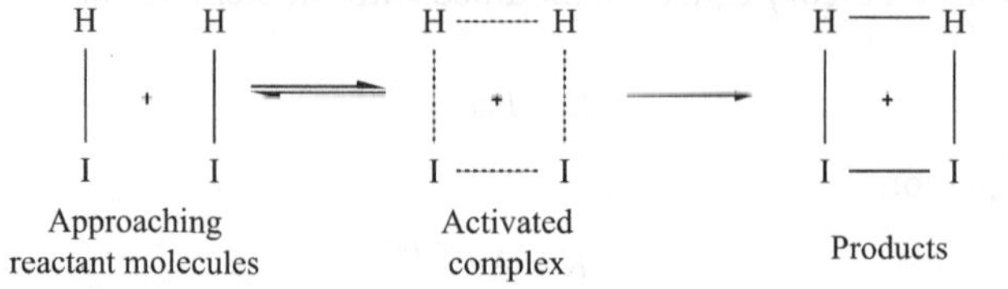

The transition state does not represent an observable substance. In principle, it cannot be isolated. Nevertheless, it is assumed to possess properties common to real molecules, such as a molecular weight, interatomic distances, a definite enthalpy, ability to rotate and vibrate. Further, it is endowed with theoretical properties not possessed by real molecules. Once the transition state is formed, it can do only one of two things: (i) either it returns to the initial reactants or (ii) it proceeds to form products. To simplify the application of the theory, it is commonly assumed that once the transition state is formed, it will decompose into the products.

In the intermediate state (transition state or activated complex state), the potential energy is higher than both of the reactants and the products and corresponds to a maximum along the lowest energy reaction path, i.e., the way in which reactant molecules approach and rearrange themselves for conversion into product molecules.

The activation energy (E_a) of the reaction, in the light of this theory, is the additional energy which the reactant molecules acquire to form the activated complex, i.e., the difference between the energy of the activated complex and the reactant molecules is called the energy of activation of the reaction (Fig. 7.7).

In the initial stage, there is a small decrease in the potential energy of the reactants as seen in Fig. 7.7. The reactant molecules approach close to one another before they react to form the activated complex. This approach of the reactant molecules, close to one another, results in the decrease of potential energy. The small increase in energy of the product molecules in the final stage (as shown in Fig. 7.7) is due to the fact that the product molecules fall apart after the reaction.

The activation energy of the back reaction may be defined *as the additional energy which the product molecules acquire to form the activated complex* i.e., the difference between the energy of the activated complex and the product molecules is known as the energy of the back reaction.

From Fig. 7.7, it is clear that the activation energy of the back reaction is less than the activation energy of the forward reaction. This is due to the fact that the products possess higher energy than the reactants and the reaction is *exothermic*. In case the product is associated with the lower energy than the associated with the reactants, the forward reaction will be *endothermic*.

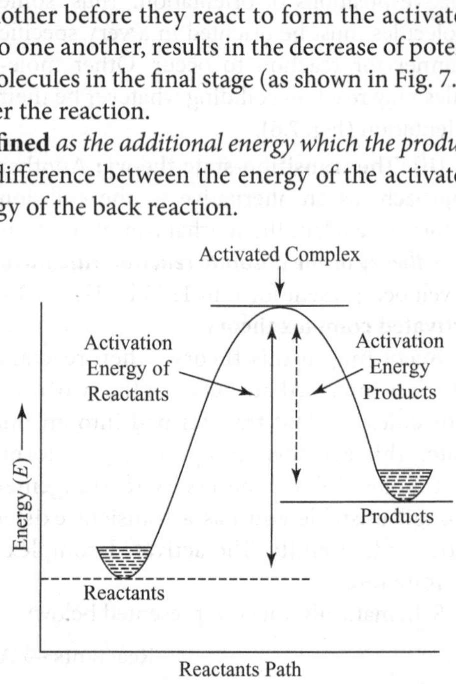

Figure 7.7 Energy variation along the reaction path

Derivation of rate constant for biomolecular or second order reaction. The influence of the transition state in a bimolecular or second-order reaction can be represented as follows:

$$A + B \rightleftharpoons X^* \longrightarrow \text{Products}$$

Reacting Molecules — Activated Complex (Transition State)

The overall process consists of equilibrium between the reactant molecules and the activated complex [or transition state] X and subsequent decomposition of the activated complex to give products at a definite rate. The conversion of the activated complex into the products is considered to be so slow that it does not affect the equilibrium between the reactant molecules and the activated complex. *In this theory, the intermediate is considered formally to be involved in a thermodynamic equilibrium.* The advantage of assuming the existence of equilibrium is that its profitable use can be made of in the various thermodynamic relations involving the equilibrium constant. All our thermodynamic relations involve the equilibrium constant. All functions associated with the transition state are designated by the superscript.

(iii) Lindeman's theory of unimolecular reactions: There are certain reactions whose order of reaction cannot be justified based on collision theory as well as transition state theory. In the year 1922, Lindeman explained this anomaly by putting forth his theory for the reaction rates. Let us take the dissociation of nitrogen pentoxide as an example:

$$2N_2O_5(g) \longrightarrow 4NO_2(g) + O_2(g)$$

The experimentally determined order for this reaction has been found to be a first-order reaction. It is not possible to account for it on the basis of the above said theories. According to his theory, **'there exists a time lag between activation and the actual reaction between the molecules.'** In between this time lag, either of the two processes may take place.

(i) The reactions may proceed directly as explained on the basis of the above mentioned two theories.

(ii) The molecules may be deactivated. For instance

(i) $X + X \quad X + X$ ....Activation
(ii) $X + X^* \quad X + X$ ....Deactivation
(iii) X^* Products........................Reaction

In case the time lag is sufficient, step (iii) i.e., the formation of products would be slow and the reaction should follow first-order kinetics. In case X reacts as soon as it is formed step (ii) would be of the slowest and the reaction would be of the second order. We know that the slowest step is always the rate-determining step i.e., the order of the reaction always depend upon the slowest step of reaction.

This theory can be proved by studying the effect of change of pressure on the dissociation of N_2O_5. At high atmospheric pressure, the chance of the activated molecule(s) to change into product(s) directly would be higher and the reaction would be of the first order.

Nitrogen pentoxide is a volatile solid, which decomposes as such or when dissolved in some organic solvent like carbon tetrachloride as shown below:

$$N_2O_5 \quad N_2O_4 + 1/2O_2(g)$$
$$2NO_2(g)$$

The progress of the reaction can be monitored by measuring the volume of oxygen evolved from the time to time.

If V_t be the volume of O_2 at any time t and V_∞ the final volume of oxygen when the reaction is complete, the V_∞ is a measure of the initial concentration of N_2O_5 and $(V_\infty - V_t)$ is a measure of undecomposed $N_2O_5(a - x)$ remaining at time t. Thus

$$K = \frac{2.303}{t} \log_{10} \frac{V_\infty}{V_\infty - V_t}$$

On substituting the values of V_∞, $(V_\infty - V_t)$ at different time intervals, t, the value of K is found to be constant. Thus, the reaction is of the first order.

At sufficiently low pressure, all the activated molecules will react before they can be deactivated. In such a case, the reaction kinetics would change from first order to second order with decreasing pressure. In case the above reaction is carried out at a pressure lesser than atmospheric pressure, it has been experimentally observed that the same reaction takes place through a different mechanism and the order of the reaction comes out to be 2. In other words, the slowest step becomes that in which two species, the activated molecule react. Not only the above cited example, but also several other gases show this type of behaviour.

MEMORY FOCUS

1. The molecules react in a collision only if they possess energy equal to greater that a certain value known as the **energy of activation (E_a)**.
2. According to collision theory: Reaction rate = Number of colliding molecules per unit volume per unit time fraction of colliding molecules with energy E_a or greater.
3. Rate constant of a bimolecular reaction $K = Pze^{-E_a/RT}$ where P is the probability factor or steric factor, z is the number of colliding molecules per cc per sec, if x is the fraction of activated molecules possessing threshold energy i.e., minimum energy required for effective collision.
4. The rate of bimolecular reaction = zx

   The fraction of activated molecules $(x) = \frac{\Delta N}{N} = \frac{\text{Number of activated molecules}}{\text{Total number of molecules}}$
5. According to the transition theory, during the collision, the reactant molecules form a **transition state or activated complex** which decomposes to give the products.
6. In the process of formation of activated complex, the kinetic energy of the two molecules is converted into potential energy.
7. When the molecules come into contact, the interpenetration of their electron clouds occurs which allows the rearrangement of valence electrons.
8. The difference in the potential energy of the reactants and that of the activated complex is the activation energy (E_a). The reactants must have this minimum energy to undergo the reaction through the transition state.
9. For exothermic reactions, the potential energy of the products is less than that of reactants, whereas in case of endothermic reactions it is reversed.
10. The activation energy of the reaction is the additional energy which the reactant molecules require to form the activated complex i.e., the difference between the energy of the activated complex and the reactant molecules.
11. The activation energy of the backward reaction is the additional energy which the product molecules acquire to form the activated complex i.e., the difference between the energy of the activated complex and the product molecules.
12. In this theory, the intermediate is considered formally to be involved in a thermodynamic equilibrium with reactants and the products.

REVISION QUESTIONS

1. (a) What do you understand by rate of a reaction? What are its units?
   (b) How is the rate of a reaction expressed and measured?
   (c) Why the rate of a reaction cannot be measured by dividing the total amount of a substance by the total time taken?
2. Define the following terms:
   (a) molecularity (b) order of reaction
   (c) reaction velocity (d) pseudomolecular reactions
3. (a) Write a note on 'zero-order reactions.'
   (b) Derive an expression for rate constant (K) of a zero-order reaction. What will be the expression for half-life period? How long will it take for the reaction to complete?
4. (a) Derive an expression for the rate constant of a first-order reaction. Show that the half-life period is independent of the initial concentration of the reactant.
   (b) Show that in a first-order reaction the rate constant is independent of the change of the concentration units.
5. (a) Derive an expression for the time required for completion of one-third of a first-order reaction.
   (b) Derive an expression for the velocity constant of the inversion of cane sugar, when water is used in excess.
6. Deduce the expression for the specific rate constant for a reaction of the type $2A \rightarrow$ Product.
7. Show that the half-life period of a second-order reaction is inversely proportional to the initial concentration of the reaction.
8. (a) Derive an expression for a reaction ($3A \rightarrow$ Products) of third order. Give two examples of a third-order reaction.
   (b) Prove that half-life period of a third-order reaction is inversely proportional to the square of the initial concentration of the reactants.
9. (a) Why are reactions of the higher order rare?
   (b) What is meant by order and molecularity of a reaction? How is the order of a reaction related to its half-life period?
10. Explain why the hydrolysis of ethyl acetate in the presence of dilute acid follows first-order kinetics while in the presence of dilute alkali follows a second-order kinetic.
11. Write short notes on the following:
    (a) consecutive reactions (b) side reactions or parallel reactions
12. Give a brief account of collision theory of reaction rates. How would you account for a bimolecular reaction using this theory?
13. Give a brief description of absolute theory of reaction rates. How can you explain the bimolecular reactions using this theory?
14. What are the weaknesses of collision theory and how have these been improved upon by transition state theory?
15. (a) State and explain the difficulties which come in the way of determination of order of a reaction. How can they be overcome?
    (b) Derive an expression for the rate constant of the reaction: $2A \rightarrow$ Products.

16. Describe any one method of determining order of a reaction.
17. Write short notes on the following:
    (a) order of a reaction (b) molecularity of a reaction
    (c) collision theory of reaction rate
18. (a) Discuss order of reaction.
    (b) What do you understand by first order kinetics in a reaction.
    (c) Write about second order reactions.
    (d) Derive the equation for second order reaction.
19. (a) Discuss the kinetics of complex reactions.
    (b) Write a short note on second order reactions.
    (c) What are acid base and enzyme catalysis

MULTIPLE CHOICE QUESTIONS

1. Which three factors affect the rate of a chemical reaction?
   (a) temperature, pressure and humidity
   (b) temperature, reactant concentration and catalyst
   (c) temperature, reactant concentration and pressure
   (d) temperature, product concentration and container volume
2. For first-order reactions the rate constant, k, has the unit(s)
   (a) L mol^{-1} (b) $time^{-1}$
   (c) $(mol/L)^{-1}$ $time^{-1}$ (d) time mol L^{-1}
3. Which of the following is the correct set of units for the reaction rate, if concentration are measured in molarity and the time in seconds
   (a) mol Ls (b) mol L-1 s^{-1}
   (c) mol^{-1} L^{-1} s^{-1} (d) none of these
4. In a series of reaction, which is the rate-determining step?
   (a) the simplest reaction
   (b) the main reaction involving the major reactant
   (c) the slowest reaction
   (d) the fastest reaction
5. Rate law for chemical reactions are determined.
   (a) by examining the coefficient in the balance chemical equation
   (b) from the equilibrium constant
   (c) from the rates of the forward and reverse reaction of the system at equilibrium
   (d) By Experiment
6. For a reaction, A→product, a graph of [A] versus time is found to be a straight line. What is the order of this reaction?
   (a) a zero order (b) first order
   (c) second order (d) third order
7. Which concentration plot is a linear for a first-order equation? (A is one of the reactants).
   (a) [A] versus time
   (b) square root of [A] versus time
   (c) In [A] versus time
   (d) $[A]^2$ versus time
8. As the temperature increases, the reaction rate__________
   (a) decrease than increase
   (b) decreases
   (c) increases
   (d) remains the same
9. The half-life of a first-order process
   (a) depends on the reactant concentration raised to the first power

(b) is inversely proportional to the square of the reactant concentration
(c) is inversely proportional to the reactant concentration
(d) is totally independent of the reactant concentration

10. The reaction, $2NO(g) \rightarrow N_2(g) + O_2(g)$, proceeds in a single elementary step. This reaction is thus
(a) the molecularity can't be determined from the given information
(b) termolecular
(c) bimolecular
(d) unimolecular

11. For a certain reaction, the rate = k $[NO]^2[O_2]$, when the initial concentration of NO is tripled, the initial rate
(a) decreases by a factor of nine
(b) increases by a factor of three
(c) increases by a factor of six
(d) increases by a factor of nine

12. What happens when molecules collide with less than the activation energy needed for the reaction?
(a) they stick together but do not react
(b) they react, but more slowly
(c) they react if the bonds are arranged in the correct orientation
(d) they do not react : they simply bounce off each other

13. According to chemical kinetic theory, a reaction can occur
(a) if the reactant collide with the proper orientation
(b) if the reactant possess sufficient energy of collision
(c) if the reactant are able to form a correct transition state
(d) all of the above

14. The half-life of radioactive sodium is 15.0 hours. How many hours would it take for a 64 g sample to decay to one-eigth of its original concentration?
(a) 3 (b) 15
(c) 30 (d) 45

15. If the reaction below is carried out in a closed, rigid container, how could you best follow the rate of the reaction?
$2SO_2(g) + O_2(g) \rightarrow 2\ SO_3(g)$
(a) monitor density of the mixture
(b) monitor the temperature
(c) monitor the pressure
(d) monitor the change in concentration

16. Which of the following does not affect the rate of a chemical reaction
(a) Enthalpy of the reaction
(b) Enthalpy of reactants
(c) temperature
(d) surface area

17. The reaction A → P is a second-order process. When the initial concentration of A is 0.50 M, the half-life is 8.0 minutes. What is the half-life if the initial concentration of A is 0.10 M?
(a) 1.6 minutes (b) 8.0 minutes
(c) 40.0 minutes (d) 14.0 minutes

18. The hydrolysis of ester in alkaline medium is a
(a) first-order reaction with molecularity 1
(b) second-order reaction with molecularity 2
(c) first-order reaction with molecularity 2
(d) second-order reaction with molecularity 1

19. An endothermic reaction A→B has an activation energy of x KJ mol^{-1}. If the energy change of the reaction is y KJ, the activation energy of the reverse reaction is
(a) $-x$ KJ (b) $(x-y)$ KJ
(c) $(x+y)$ KJ (d) $(y-x)$ KJ

20. Half-life period of a zero-order reaction is
(a) proportional to initial concentration of reactants
(b) independent of initial concentration of reactants
(c) inversely proportional to initial concentrations of reactants
(d) inversely proportional to the square of initial concentrations of reactants

21. The rate of a gaseous reaction is equal to $K[A][B]$. The volume of the reaction vessel

containing these gases is suddenly reduced to one-fourth the initial volume. The rate of the reaction would be
(a) 1/16 (b) 16/1
(c) 1/8 (d) 8/1

22. In the presence of an acid, the initial concentration of methyl acetate was reduced from 0.1 M to 0.5 M in 5 h and 0.25 M in 10 h. The reaction must be of
(a) first order
(b) zero order
(c) second order
(d) third order

ANSWERS

1. (b)	6. (a)	11. (d)	16. (d)	21. (b)
2. (b)	7. (c)	12. (d)	17. (c)	22. (a)
3. (b)	8. (c)	13. (d)	18. (b)	
4. (c)	9. (d)	14. (d)	19. (b)	
5. (d)	10. (c)	15. (c)	20. (a)	

Quantum Mechanics 8

CHAPTER OBJECTIVES

8.1 Introduction
8.2 Classical Mechanics and its Limitations
8.3 Origin of Quantum Mechanics
8.4 Black Body Radiations
8.5 Kirchoff's Law
8.6 Stefan–Boltzmann Fourth Power Law
8.7 Wein's Displacement Law
8.8 Planck's Radiation Law
8.9 Postulates of Quantum Mechanics
8.10 Operators in Quantum Mechanics
8.11 Schrödinger Wave Equation
8.12 Eigenvalues and Eigenfunctions (or Wave Functions)
8.13 Normalized and Orthogonal Eigenfunctions
8.14 Concept of Atomic Orbital
8.15 Probability Distribution Curves
8.16 Radial Probability Distribution Curves

Combining the two basic ideas—de Broglie's dual nature of matter and Heisenberg's uncertainty principle—Erwin Schrödinger introduced the differential equation that governs matter waves and developed a three-dimensional mechanical model of the atom. Thus, many previously unsolved problems were successfully resolved.

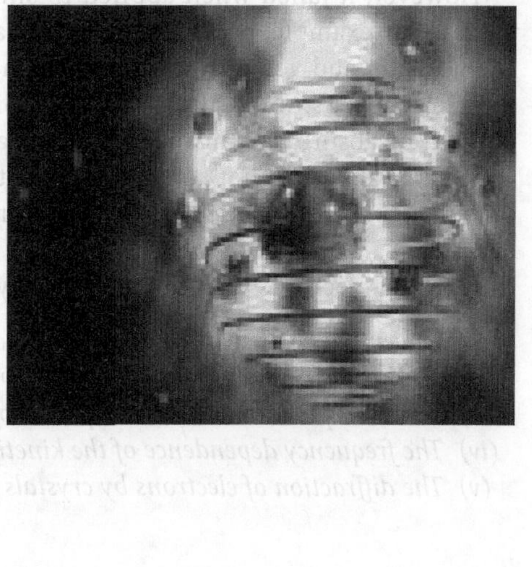

8.1 INTRODUCTION

The phenomena associated with large size objects can be easily explained by **classical mechanics**, which is also called *Newtonian mechanics*, as it was put forward by Newton. Like the continuous motion associated with a falling stone or motion of the planets, classical mechanics assume that energy is emitted or absorbed continuously. However, classical mechanics fails when it is applied to small particles such as atoms and molecules.

The mechanics that satisfactorily explains the properties associated with the movement of small particles such as atoms and molecules is called **quantum mechanics or wave mechanics**. This is so called because it assumes that the energy is emitted or absorbed not continuously but discontinuously in the form of discrete packets of energy called *quanta*. In addition, it takes into account the particle nature as well as the wave nature of the moving particles.

The basis for handling all problems in quantum mechanics is the Schrödinger wave equation. In this chapter, we will discuss the basics of quantum mechanics.

8.2 CLASSICAL MECHANICS AND ITS LIMITATIONS

Earlier, most of the phenomena connected with the motion of bodies and transfer of energy could be explained by Newton's laws of motion and Maxwell's electromagnetic wave theory, which postulated that the energy is emitted or absorbed continuously. Newton's laws of motion could explain the motion of planets and the falling of stones while Maxwell's electromagnetic wave theory could explain the transfer of energy,

The branch of science based on Newton's laws of motion and Maxwell's electromagnetic wave theory employed for explaining the phenomena connected with the motion and energy is known as classical mechanics.

8.2.1 Limitations

Classical mechanics could explain the phenomena associated with large-size objects moving at speeds much below the speed of light, e.g., motion of celestial bodies and other objects on the earth.

However, it failed when applied to small particles such as electrons, protons, atoms and molecules. For example, according to classical mechanics, it is possible to determine simultaneously the position and velocity (or momentum) of a moving particle. However, this is contradicted by the Heisenberg's uncertainty principle. Similarly, classical mechanics assumes that the energy is emitted or absorbed continuously whereas Planck's quantum theory postulates that the energy is emitted or absorbed not continuously but discontinuously in the form of packets of energy called *quanta*. Further, the concept of quantum numbers was introduced arbitrarily to explain the atomic spectra.

Classical mechanics fails to explain many other phenomena, a few of which are as follows:

(i) The energy density distribution or black body radiation.
(ii) The heat capacities of mono-atomic solids such as copper metal.
(iii) The absorption and emission spectra of atoms and molecules, especially the line spectra of it.
(iv) The frequency dependence of the kinetic energy of emitted electrons in the photoelectric effect.
(v) The diffraction of electrons by crystals in a manner similar to that observed for X-rays.

8.3 ORIGIN OF QUANTUM MECHANICS

The laws of Newtonian mechanics explained successfully the motion of celestial bodies and other stationary objects. These laws explained successfully the physical behaviour and the bulk properties of matter. When the same laws were applied to particles such as electrons, protons, neutrons and nuclei, which are too small to be seen through the highest powered microscopes, serious difficulties were encountered. It means that the classical concepts are not valid in the region of atomic dimensions. In the microscopic world of atoms and molecules, the energy and moments do not have the same meaning as in the case of classical dynamics, On the other hand, these dynamic variables have a discrete value in a particular state of an atom and do not undergo change in a continuous manner from one state to another as one would be expecting under classical laws. Thus, to include new concepts, the Newtonian laws of macroscopic world were modified.

In view of the failure of classical mechanics to explain the phenomena associated with the small particles, a new mechanics was put forward by Schrödinger in 1926 to explain these phenomena. This is called wave mechanics. It is based on de Broglie concept of dual character of matter and thus, takes into account the particle as well as wave nature of the material particles. It is also called particle mechanics or quantum mechanics.

The phenomena of chemistry cannot be understood thoroughly without a firm understanding of the principal concepts of quantum mechanics. The same is true of virtually all the spectroscopic techniques that are now so important for investigating the composition and structure. Present-day techniques for studying chemical reactions have progressed to the point where the quantum mechanics has to be used in its interpretation. In fact, the structures of atoms and molecules cannot be discussed without making use of quantum mechanical concepts.

8.3.1 Classical Mechanics Versus Quantum (or wave) Mechanics

The differences between classical mechanics and quantum mechanics are as follows:

S. No.	Classical Mechanics	Quantum (or wave) Mechanics
1.	It is applicable to macroscopic particles.	It is applicable to microscopic particles.
2.	Newton's laws are the fundamental principles on which it is based.	Heisenberg's uncertainty principle and de Broglie's concept of dual nature of matter (particle nature and wave nature) are its foundations.
3.	Maxwell's electromagnetic wave theory is its basis. According to this theory, any amount of energy may be absorbed or emitted continuously.	Plank's quantum theory is its basis. According to this theory, energy is absorbed or emitted discontinuous only in packets, i.e., quantum.
4.	The state of a system is defined with certainty. To define the state, it is required to specify all the forces acting on the particles and their positions with their velocities.	It provides the possibilities of finding the particles at different locations of space.

However, under certain conditions, the quantum mechanics as well as classical mechanics provide same results. The following are some of the examples:

(i) Quantum (wave) mechanics yields the same results as Newton's classical mechanics when the macroscopic body is moving at a velocity much less than the velocity of light (i.e., $v << c$).

(ii) When the Planck's constant approaches the limit 0 (i.e., $h \rightarrow 0$), the time-independent Schrödinger equation (to be discussed later) is reduced to Newton's second law.

(iii) Both wave mechanics and classical mechanics give the same results when the system is in highly excited state, i.e., when the value of principal quantum number n is very high.

SOME NOTEWORTHY POINTS

1. The maximum number of electrons in a major (or main) energy level (or shell) is given by $2n^2$.
2. The number of subshells in a major energy level is equal to the value of n.
3. The number of orbitals in a subshell is equal to the number of m (magnetic quantum number) as given by $(2l+1)$.
4. The total number of orbitals in a shell is given by n^2.
5. An orbital can accommodate not more than two electrons and that too with opposite spin.
6. All substances have magnetic properties due to the magnetic moment associated with electron spin and with orbital angular momentum of the electron.
7. The two electrons in different atomic orbitals are far apart and suffer less electron–electron repulsion than when these occupy the same orbital with opposite spins.
8. The energy of atomic orbitals for H atom is (as the energy depends on the value of n):
$$1s < 2s = 2p < 3s = 3p = 3d < 4s = 4p = 4d = 4f$$
9. The energy of orbitals other than the H atom depends upon the values of $(n+1)$ and varies as
$$1s < 2s < 2p < 3s < 3p < 4s < 3d < 4p < 5s < 4d < 5p < 6s < 4f < 5d < 6p < 7s < 5f < 6d$$

8.4 BLACK BODY RADIATIONS

When a radiation falls on an object, a part of it is reflected, a part absorbed and the remaining transmitted. This is due to the fact that no object is a perfect absorber. Contrastingly, we may visualize a black body which completely absorbs all the radiations that are incident on it and retains all the radiant energy that strikes it. For experimental purposes, a black body is generally a blackened metallic surface where the energy is almost completely absorbed.

A perfectly black body is defined as one which absorbs all the heat radiations (corresponding to all wavelengths) incident on it.

It practice, a perfectly black body is not available. A body showing close approximation to a perfectly black body can be constructed. A good approximation to a black body—a body capable of emitting and absorbing all frequencies (or wavelengths) of radiations—is a pinhole in an empty container maintained at a constant temperature because the radiations leaking out of the hole has been absorbed and re-emitted inside the container so many times that it has come to thermal equilibrium with the walls. It is in thermal equilibrium in the sense that the temperature of the radiation-filled space inside the container is the same as that of the walls. It means the radiation incident on the walls is completely absorbed.

The black body designed by Ferry shown in Fig. 8.1 is the most commonly used one. It consists of a double walled metallic sphere with a conical projection P facing the hole H.

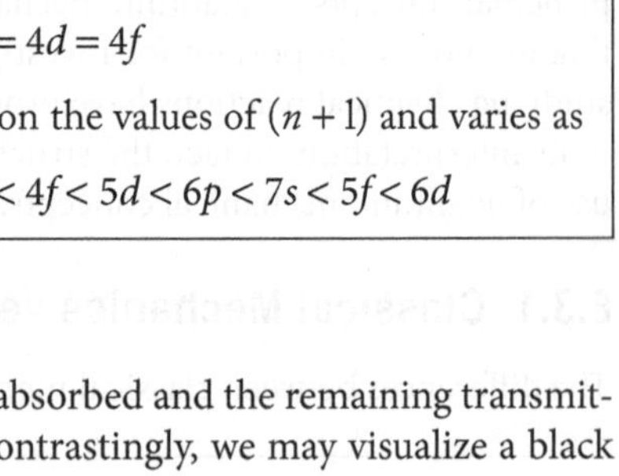

Figure 8.1 Ferry's black body

The inside of the sphere is coated with lamp black. The projection helps in avoiding any direct reflection since the body will not behave as a black body otherwise.

When such a body is heated, it emits radiations of all types of wavelengths. The radiations are independent of the nature of the substance. Such heat radiations in a uniform temperature enclosure are known as black body radiations. In addition, the black body completely absorbs heat radiations of all wavelengths. Hence, the black body completely emits the radiations of all wavelengths at that temperature. *A black body is not only a perfect absorber of radiant energy but also a prefect radiator and emitter.*

8.5 KIRCHOFF'S LAW

Black body is a perfect absorber as well as a perfect radiator (or emitter) of energy. However, all other bodies absorb the radiant energy incident on them and also emit it. The relationship between the energy absorbed and the energy emitted by a body can be qualitatively studied with the help of the following experiment:

Two balls—one black (B) and the other white (W)—are suspended with a thread in a hollow enclosure with its walls maintained at uniform temperature and thermally insulated to keep the temperature constant (Fig. 8.2). To start with, the whole space inside the enclosure is filled with the radiations emitted by the walls. The walls of the enclosure absorb as well as emit radiations continuously.

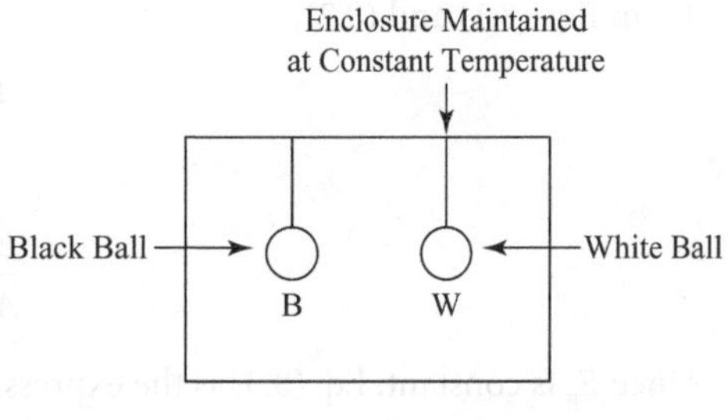

Figure 8.2 Qualitative verification of Kirchoff's law

Since both the balls (black, B as well as white, W) absorb as well as emit radiations, it is observed that they also acquire the temperature of the enclosure. Their temperature remains constant as long as the enclosure is maintained at constant temperature.

It is further found that both the balls are at the same temperature even when black body is a better absorber. Obviously, this can be so only because the black ball is also a better radiator. It may, therefore, be concluded that a good absorber is a good radiator while a poor absorber is a poor radiator. Thus, substances such as glass or polished metals which are very poor absorbers are also seen to emit very little radiation.

Kirchoff's law gives a quantitative relationship between the energy absorbed and the energy emitted by a body, which is stated as follows:

The ratio of the emissive power (or emittance) of a body to the absorptive power (or absorbance or absorptivity) at a given temperature is constant independent of the nature of the body and is equal to the emissive power (emittance) of a perfectly black body.

The *emissive power* or *emittance* is the energy emitted by the surface per unit time per unit area and *absorptive power* or *absorbance* or *absorptivity* is the fraction of the incident light absorbed by the surface per unit area per unit time.

Then, according to Kirchoff's law,

$$\frac{E_S}{A_S} = E_B$$

where E_S is the emissive power of a substance, A_S is the absorptive power of a substance and E_B is the emissive power of a perfectly black body.

The Kirchoff's law can be derived as follows:

Let Q be the amount of radiation incident on a body per unit area per second.

If A_S is the absorptive power of the surface, the amount of radiation absorbed per unit area per second = $A_S \times Q$.

If E_S is the emissive power of the surface, the amount of radiation emitted by the body per unit area per second = E_S.

When the body is in thermal equilibrium with the enclosure, amount of radiation emitted by the surface per unit area per second = Amount of radiation absorbed by the body per unit area per second.

$$\therefore \quad E_S = A_S \times Q \tag{8.1}$$

For a perfectly black body, $A_S = A_B = 1$ and E_S may be replaced by E_B when A_B and E_B are the absorptive power and the emissive power, respectively, of a perfectly black body.

$$\therefore \quad E_B = Q \tag{8.2}$$

From Eqs (8.1) and (8.2),

$$E_S = A_S \times E_B$$

or

$$\frac{E_S}{A_S} = E_B \tag{8.3}$$

Since E_B is constant, Eq. (8.3) is the expression for the Kirchoff's law.

8.5.1 Spectral Distribution of Black Body Radiation

Electromagnetic radiation is a wave of electric and magnetic fields travelling through space. Such waves are generated by the acceleration of electric charge, as in oscillating motion of electrons in the antenna of a radio transmitter. A hot object emits electromagnetic radiations because its atoms and electrons are continuously being accelerated. The atoms vibrate around their mean positions and their electrons are moved from one location to another. At high temperature, an appreciable proportion of the radiation is in the visible region of the spectrum. A higher proportion of short wavelength blue light is generated as the temperature is raised. We observe this behaviour when an iron bar glowing red hot becomes white hot when heated further. This is because, on further heating the red hot iron bar, more blue light mixes with the red light and changes the perceived colour towards white. At low temperatures, the emission is mainly in the infrared region, but as the temperature is raised, the spectral range of the radiation widens, opening up fresh regions of the spectrum. In addition, if the temperature gets high enough, such as white heat, a range of waves covering the visible light spectrum are emitted besides the radiations of lower wavelengths that have started at a lower temperature.

Thus, the total radiation emitted by a hot body is composed of electromagnetic waves of varying range of wavelengths constituting a spectrum. Thus, the total intensity of radiation and its distribution among the various wavelengths of the spectrum are different at different temperatures, and may also be quite different for different objects even at the same temperature. If, however, the radiation emerging through a small hole in the wall of an isothermal hollow body is measured, it is found that the total intensity of the radiation and its distribution among wavelengths are independent of the material, size and shape of the enclosure. This radiation is called black body radiation.

The energy of the black body radiation radiated per unit area per unit time increases with increase in temperature accompanied by the changes in colour from red hot to white hot through the intervening colours. It may also be noted that at the same temperature, the rate of emission of energy is far less for a bright platinum metal surface than for black surface, e.g., an incandescent carbon filament at a temperature of the order of much above 3500 K.

8.6 STEFAN-BOLTZMANN FOURTH POWER LAW

The law, which was discovered experimentally by Stefan in 1879 and derived using the principles of thermodynamics by Boltzmann in 1884, states that *the total amount of energy* E *(i.e., the intensity of total radiation) radiated by a perfectly black body radiation is proportional to the fourth power of its absolute temperature*, that is,

$$E \propto T^4 \quad \text{or} \quad E = \sigma T^4 \tag{8.4}$$

where σ is the Stefan's constant

$$= 5.669 \times 10^{-5} \text{ erg s}^{-1} \text{ cm}^{-2} \text{ K}^{-4}$$

$$= 5.669 \times 10^{-8} \text{ W (W} = 1 \text{ J s}^{-1}) \text{ m}^{-2} \text{ K}^{-4}$$

This law is called *Stefan–Boltzmann fourth power law.* Boltzmann's name is included to this law because he explained it theoretically by applying the laws of thermodynamics.

An alternative form of the law is in terms of the ***excitance****, M, the power emitted by a region of surface divided by the area of the surface.* Because the excitance is proportional to the energy density in the container, M is also proportional to T^4, and we can write

$$M = \sigma T^4; \qquad \sigma = 5.669 \times 10^{-8} \text{ W m}^{-2} \text{ K}^{-4}$$

where power is the rate of supply of energy. Its SI units are Watts, W = 1 J s^{-1}.

The Stefan–Boltzmann law implies that 1 cm^2 of the surface of a black body at 1000 K radiates about 6 W when all wavelengths of emitted light radiation are taken into account.

8.7 WIEN'S DISPLACEMENT LAW

The energy emitted by a black body at a given temperature (as given by Stefan–Boltzmann law) does not consist of a single frequency (or wavelength) and is also not uniformly distributed along the spectrum. Lummer and Pringsheim studied the distribution of energy among different wavelengths of black body spectrum which is described in the following.

The black body radiations were obtained by electrically heating a carbon tube. The radiations obtained were passed through slit S_1 followed by making the radiations parallel with the help of a concave reflector mirror M_1. The rays were then made to fall on the prism made of fluorspar to obtain the spectrum. The rays after reflection through the prism were focussed by another concave mirror M_2 and slit S_2 on to a linear bolometer (an extremely sensitive instrument for measuring heat radiations) as shown in Fig. 8.3.

The radiations corresponding to different wavelengths were focussed on to the bolometer one by one and then intensities measured by rotating the prism gradually. By maintaining the black body (carbon tube) at different temperatures, a number of observations were made in the temperature ranges between 621 and 1645 K. The curves were obtained by plotting the monochromatic emittance E_λ along the ordinates and

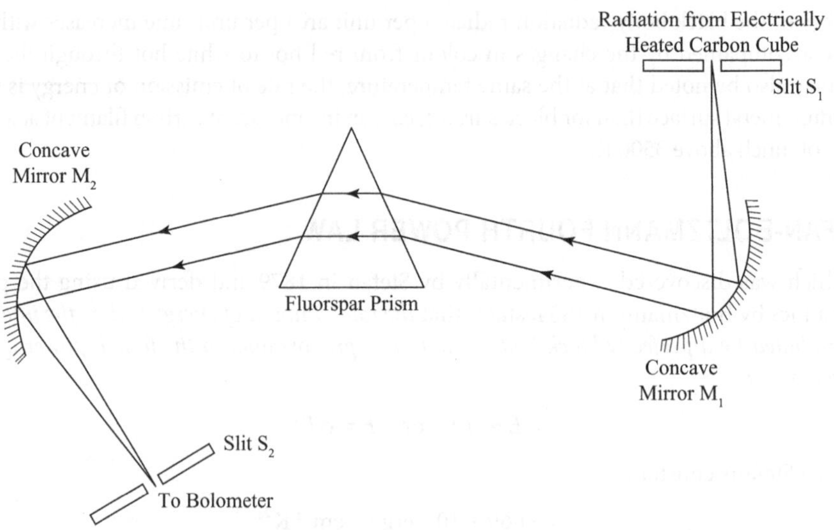

Figure 8.3 Experimental study of spectral distribution of black body radiation by Lummer and Pringsheim

the same lengths along the abscissae for each temperature. Monochromatic emittance (E, corresponding to wavelength λ is the amount of energy emitted per unit area per unit time within a unit wavelength interval around λ, i.e., between $(\lambda - 1/2)$ and $(\lambda + 1/2)$. Here, the unit wavelength interval is 1 Å.

The typical curves from a black body at different temperatures are as shown in Fig. 8.4.

Some important conclusions drawn from the curves shown in Fig. 8.4 are as follows:

(i) At a given temperature, the energy is not uniformly distributed among the different wavelengths in the radiation spectrum obtained from the black body.

(ii) At a given temperature, the intensity of radiation increases with increase in wavelength and at a particular wavelength λ_m, its value is maximum. With further increase in wavelength, the intensity of heat radiations decreases. Hence, for each temperature, there is a wavelength λ_{max} for which the energy radiated is maximum.

(iii) With increase in temperature, λ_{max} decreases. λ_{max} *is the wavelength at which the emission of energy takes place.* The points on the dotted line are representing λ_{max} at various temperatures. This means that with increase in temperature, E_m increases but corresponding λ_{max} decreases.

(iv) Higher the temperature, more pronounced is the maximum in the graph.

(v) The area under the curve at a particular temperature represents the total energy emitted by the black body per unit area per second for the complete spectrum corresponding to all the wavelengths at that temperature (i.e., from λ_0 to $\lambda = \infty$)

(vi) Since the area under the curve represents the total energy emitted by the black body per unit area per second, therefore, $E_p \propto T^4$ (Stefan–Boltzmann law).

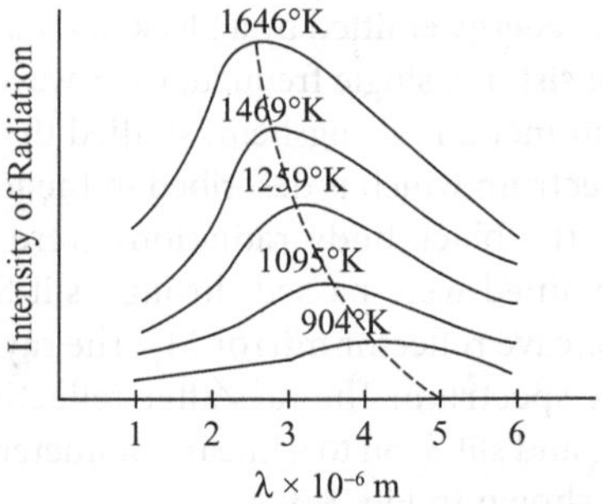

Figure 8.4 Distribution of energy in the spectrum of a black body at different temperatures

(vii) It is observed from the curves that λ_{max} is inversely proportional to the temperature T, i.e.,

$$\lambda_{max} T = \text{constant} = 0.2892 \text{ cm K}$$

$$\therefore \qquad \lambda_{max_1} T_1 = \lambda_{max_2} T_2 = \lambda_{max_3} T_3 \ldots \tag{8.5}$$

Obviously, the wavelength λ_{max} for which the emittance of a black body is maximum is inversely proportional to its absolute temperature. This is known as *Wien's displacement law.*

This law shows that with the increase in temperature, λ_{max} decreases. The law explains the chromaticity in the colours of visible light radiation from red (greater wavelength) to yellow (lower wavelength) with the increase in the temperature of the hot body.

8.8 PLANCK'S RADIATION LAW

Planck studied black body radiation from the viewpoint of thermodynamics. In 1900, he could account for the experimental observations by proposing that energy of each electromagnetic oscillator is limited to discrete values and cannot be varied arbitrarily. This proposal is quite contrary to the viewpoint of classical physics; in which, all possible energies are allowed. This limitation of energies to discrete values is called the *quantization of energy*. In particular, Planck found that he could account for the observed distribution of energy if he supposed that permitted energies of electromagnetic oscillator of frequency v (nu) are integer multiples of hv.

$$E = nhv \tag{8.6}$$

where $n = 0, 1, 2, \ldots$ and h is the fundamental constant.

It is known as *Planck's constant*. Planck's assumptions may be summed up as follows:

Energy is emitted or absorbed not continuously, but discontinuously in the form of packets of energy called quanta. The energy of each quantum is given by the relation $E = hv$, where v is the frequency of the radiation and h is called the Planck's constant.

The total energy emitted or absorbed is either unit quantum, i.e., hv, or a whole number multiple of hv, i.e., nhv. Based on this assumption, Planck was able to derive an expression for the energy E_λ radiated by a black body at wavelength λ. This is known as Planck's radiation law or Planck's distribution law.

$$E_\lambda = \frac{8\pi hc}{\lambda} \frac{1}{e^{hc/kT\lambda} - 1}$$

This expression fits the experimental curve very well at all wavelengths (Fig. 8.4), and the value of h, which is an undetermined parameter in the theory, may be obtained by varying its value until a best fit is obtained. The currently accepted value for $k = 6.62608 \times 10^{-34}$ J s.

The Planck's radiation law also known as *Planck's distribution* accounts for *Stefan–Boltzmann law*. Stefan–Boltzmann law can be obtained from Planck's distribution by integrating the emittance, i.e., emission power of the black body over all the wavelengths from $\lambda = 0$ to $\lambda = \infty$, and then converting the definite integral into a standard formula. The result is

$$E = \frac{2\pi^2 k}{15c^2 h^3} T^4 \tag{8.7}$$

Equation (8.7) is exactly of the same form as the expression for Stefan–Boltzmann law $E = \sigma T^4$.

$$\therefore \qquad \sigma = \frac{2\pi^2 k}{15c^2 h^3} \tag{8.8}$$

The numerical evaluation of the constant σ from Equation (8.5) accords with the exponential value of σ. *The Planck's radiations law (Planck's distribution)* accounts for *Wien's law.*

If λT is large, $e^{hc/kT\lambda} >> 1$ such that 1 in the denominator may be neglected. This gives rise to

$$E_\lambda = \frac{8\pi hc}{\lambda^5} e^{-hc/kT\lambda}, \tag{8.9}$$

which is *Wien's law.*

Planck's radiation law resembles *Rayleigh–Jeans law*. If λT is large, $e^{hc/kT\lambda}$ may be expanded by the exponential theorem of $e^x = 1 + x + x^2 + \ldots$ such that Planck's radiation law equation reduces to the form

$$E_\lambda = \frac{8\pi hc}{\lambda^5} \times \frac{kT\lambda}{hc} = \frac{8\pi}{\lambda^4} kT, \tag{8.10}$$

which is *Rayleigh–Jeans law.*

Hence, Planck's radiation law reduces to Wien's law for short wavelength and to Rayleigh–Jeans law for long wavelengths. Plots corresponding to these three radiation laws are shown in Fig. 8.5.

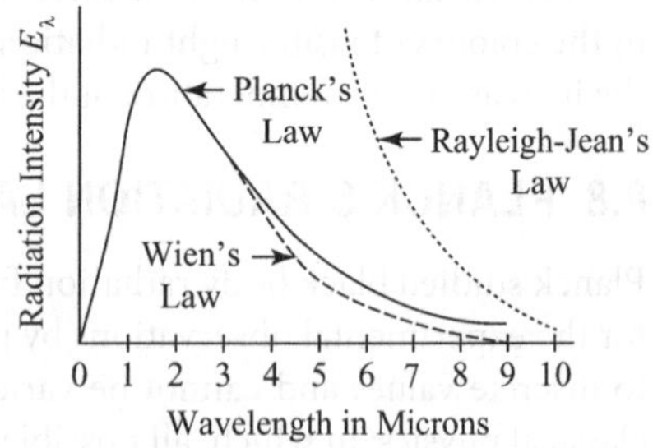

Figure 8.5 Curves corresponding to the three radiation laws

MEMORY FOCUS

1. The mechanics that satisfactorily explains the properties associated with the movement of small particles like atoms, molecules, etc. is called *quantum mechanics* or *wave mechanics.*
2. This branch of science based on Newton's laws of motion and Maxwell's electromagnetic wave theory employed for explaining the phenomena connected with motion and energy is known as classical mechanics.
3. The classical concepts are not valid in the region of atomic dimensions.
4. In the microscopic world of atoms and molecules, the energy and moments do not have the same meaning as in the case of classical dynamics.
5. Schrödinger combined the idea of dual nature of microscopic objects in motion given by de Broglie and Heisenberg's principle of uncertainty and put forward a three-dimensional wave mechanical model of an atom.
6. A perfectly black body is defined as one which absorbs all the heat radiations (corresponding to all wavelengths) incident on it.
7. A black body is not only a perfect absorber of radiant energy but also a prefect radiator and emitter.
8. Kirchhoff's law states "*The ratio of the emissive power (or emittance) of a body to the absorptive power (or absorbance or absorptivity) at a given temperature is constant independent of the nature of the body and is equal to the emissive power (emittance) of a perfectly black body.*"
9. Stefan's law states that the total amount of energy E (i.e., the intensity of total radiation) radiated by a perfectly black body is proportional to the fourth power of its absolute temperature, that is

$$E \propto T^4$$

(*Continued*)

MEMORY FOCUS (*Continued*)

10. The wavelength λ_{max} for which the emittance of a black body is maximum is inversely proportional to its absolute temperature. This is known as *Wien's displacement law*.
11. Energy is emitted or absorbed not continuously but discontinuously in the form of packets of energy called quanta. The energy of each quantum is given by the relation $E = hv$, where v is the frequency of the radiation and h is called the Planck's constant.

8.9 POSTULATES OF QUANTUM MECHANICS

Schrödinger wave mechanics is based upon certain assumptions called 'postulates'. The validity of these postulates lies in the results obtained after their application. For example, in case of hydrogen atom, energy for the system as calculated using wave mechanics was found to be same as obtained experimentally and also obtained by N. Bohr using classical mechanics and quantization concepts. In general, it is observed that there is exact correspondence between the results due to classical mechanics and wave mechanics under certain conditions. This is called ***correspondence principle***.

The following are the postulates of quantum (wave) mechanics.

Postulate 1: *Every time-independent state of system is described completely by a function ψ called wave function which is single valued, continuous and finite throughout the configuration space.*

Postulate 2: *Every physically measurable or observable quantity in classical mechanics, such as position, velocity, momentum and energy, of a system has a corresponding operator in quantum mechanics. The nature of the operator depends upon the classical expression for the observable quantity.*

For example,

(i) The operator corresponding to a position coordinate is multiplication by the value of that coordinate, i.e., the operator for a position coordinate x is the multiple x.

(ii) The operator of evaluating the derivative with respect to x is represented by d/dx when this operator is applied to function x^n, we obtain a new function as

$$\frac{d}{dx}(x^n) = nx^{n-1} \tag{8.11}$$

(iii) The operation representing the momentum (p) in the direction of any coordinate q is the differential operator.

$$\frac{h}{2\pi i}\frac{\partial}{\partial q} \quad \text{or} \quad -\frac{ih}{2\pi}\frac{\partial}{\partial q} \quad \text{or} \quad -\frac{\hbar}{2i}\frac{\partial}{\partial q} \quad (\hbar \text{ is read as } h \text{ cross}) = \frac{h}{2\pi}$$

where h is Planck's constant.

To understand the significance of setting up and using the operator corresponding to a given observable can be learnt by taking as an example of the operator for linear momentum parallel to x-axis which is

$$\hat{p} = \frac{h}{2\pi i}\frac{\partial}{\partial x} \tag{8.12}$$

To find out the linear momentum of a particle parallel to x-axis, the following eigenvalue equation is used

$$\hat{p}\psi = p\psi \tag{8.13}$$

Table 8.1 Few classical mechanical observables and their corresponding quantum operators (for one-dimensional system).

Observables		Operators	
Name	**Symbol**	**Symbol**	**Operation**
Position	x	$\hat{x}$	Multiplication by x
Position squared	x^2	$\hat{x}^2$	Multiplication by x^2
Momentum	p_x	$\hat{p}_x$	$\frac{h}{2\pi}\frac{\partial}{\partial x}$
Momentum squared	p^2x	$\hat{p}^2_x$	$\frac{-h^2}{4\pi}\frac{\partial^2}{\partial^2 x}$
Kinetic energy	$T=\frac{p^2_x}{2m}$	$\hat{T}_x$	$\frac{-h^2}{8\pi^2 m}\frac{\partial^2}{\partial x^2}$
Potential energy	$V(x)$	$\bar{V}(x)$	Multiplication by $V(x)$
Total energy	$E=T_x * V(x)$	$\hat{H}$	$\frac{-h^2}{8\pi^2 m}\frac{\partial^2}{\partial x^2}+V(x)$

It means that value of momentum p can be known by differentiating the wave function ψ with respect to x and multiplying the result with $h/2\pi i$ followed by the use of the eigenvalue equation. A few classical mechanical observables and their quantum mechanical operators are listed in Table 8.1.

For three-dimensional systems, the operators are obtained by the summation of the corresponding operators. For example,

(i) For momentum,
$$\hat{p}=\frac{h}{2\pi i}\left(\frac{\partial}{\partial x}+\frac{\partial}{\partial y}+\frac{\partial}{\partial z}\right) \tag{8.14}$$

(ii) For kinetic energy,
$$\hat{T}=\frac{-h^2}{2\pi^2 m}\left(\frac{\partial^2}{\partial x^2}+\frac{\partial^2}{\partial y^2}+\frac{\partial^2}{\partial z^2}\right)=\frac{-h^2}{2m}\nabla^2 \tag{8.15}$$

(iii) For total energy,
$$\hat{H}=\frac{-h^2}{2\pi m}\nabla^2+V(x,y,z) \qquad [\because E=T+V] \tag{8.16}$$

Postulate 3: *The only possible values that can be observed of a physical property, such as angular momentum or energy, are the eigenvalues a, in the operator equation known as eigenvalue equation:* $\hat{A}\psi = a\psi$ *where* $\hat{A}$ *is the operator for the physical quantity and* ψ *is the well-behaved eigenfunction.*

However, we shall see later that the form of operators for various physical quantities can be derived from the direct physical considerations and then by solving the above eigenvalue equation, the eigenvalues and eigenfunctions can be found.

Since the observable physical quantities have obviously real magnitude, the eigenvalues, a, must be real.

Postulate 4: *The only possible measured values of an observable are the eigenvalues obtained from the eigenvalue equation (from Postulate 3) above. Hence, if the wave function ψ of the system is an eigenfunction of the operator for the observable, then measurement of that observable gives eigenvalue. When the wave function ψ of the system is not an eigenvalue, the value of the observable obtained on measurement has a probability that can be calculated from the wave function.*

Postulate 5: *When a great many measurements of any observable represented by an operator $\hat{A}$ are made on a system characterized by a function ψ, the average value of the quantity (represented by $\bar{a}$) is given by*

$$\bar{a} = \frac{\oint \psi * \hat{A} \psi d\tau}{\oint \psi * \psi d\tau} \tag{8.17}$$

where $\oint$ represents integration over the whole of the configuration space.

Postulate 6: *The time development of a given wave function is given by*

$$\frac{ih}{2\pi} \frac{\partial \psi}{\partial t} = \hat{H} \psi \tag{8.18}$$

which is Schrödinger's equation, where H is the Hamiltonian operator.

If the wave function ψ of a physical system is given at a certain instant, not only all the properties of the system at that instant are described but also its behaviour at all subsequent instants is determined. The mathematical expression of this fact is that $(d\psi/dt)$ (derivative of the wave function with respect to time) at any given instant must be determined by the value of the function itself at that instant.

8.10 OPERATORS IN QUANTUM MECHANICS

An operator is a mathematical operation or procedure which when carried out on a function changes it into another function. In other words, *(Operator)×(Function) = Another function*

The function on which the operation is carried out is known as *operand*. The left-hand side of the above equation should not mean that a function has been multiplied with the operator. The operator written alone is meaningless. Let us consider a few examples:

(i) $\sqrt{5} = 2.236$. Here, the symbol $\sqrt{\ }$ stands for square root and hence is an operator. It conveys that one is to take the square root of 5 which is the operand and 2.236 is the result.

(ii) $\frac{d}{dx}(x^3) = 3x^2$. Here, $\frac{d}{dx}$, which stands for differentiation with respect to *x*, is the operator, x^3 is the operand and $3x^2$ is the resultant function.

(iii) $\int x^3 dx = \frac{x^2}{4} + C$. Here, $\int x dx$ which stands for integration with respect to *x* is the operator, x^3 is the operand while $\frac{x^2}{4} + C$ is the resultant function.

When the symbol for the operator is not self-explanatory, a suitable letter or symbol for the operator is used with the symbol (^) over it, i.e., a carat is put over the symbol.

8.10.1 Types of Operators

Following are the important types of operators:

(i) **Linear operators:** An operator is said to be linear if it is operated on the sum of two or more functions and gives the result as sum of the results of operation on each function.

For example, if $\hat{A}$ is applied on the sum of two functions $\psi(x)$ and $\phi(x)$, linear operator gives

$$\hat{A}[\psi(x)+\phi(x)] = \hat{A}\psi(x)+\hat{A}\phi(x) \tag{8.19}$$

Operator square root is not linear because

$$\sqrt{\psi}(x)+\phi(x) \neq \sqrt{\psi(x)}+\sqrt{\phi(x)}$$

Similarly, an operator which squares the function or raises the power by three or any number is not linear. However, the differential operator is linear, i.e.,

$$\frac{\partial}{\partial x}[\psi(x)+\phi(x)] = \frac{\partial}{\partial x}\psi(x)+\frac{\partial}{\partial x}\phi(x)$$

(ii) **Eigen operators:** An operator, say A, is said to be an eigenoperator if it operates on a function, say $\psi(x)$, giving a result which is the original function multiplied by some constant.

The following quality should exist.

$$\hat{A}\psi(x) = a\psi(x) \tag{8.20}$$

The constant a is called the eigenvalue and the function $\psi(x)$ is called eigenfunction or *characteristic function or proper function of the operator $\hat{A}$.*

For example, when an operator d^2/dx^2 is carried out on the function cos $2x$.

$$-\frac{d^2}{dx^2}(\cos 2x) = 4(\cos 2x)$$

In this case, cos $2x$ is the eigenfunction and 4 is the eigenvalue of the operator $-d^2/dx^2$.

(iii) **Laplacian operator:** It is represented by ∇^2 (del squared) and is defined as

$$\nabla^2 = \frac{\partial^2}{\partial x^2}+\frac{\partial^2}{\partial y^2}+\frac{\partial^2}{\partial z^2} \tag{8.21}$$

This is common operator in quantum mechanics.

The Schrödinger equation

$$\frac{\partial^2\psi}{\partial x^2}+\frac{\partial^2\psi}{\partial y^2}+\frac{\partial^2\psi}{\partial z^2}+\frac{8\pi^2 m(E-V)}{h^2}\psi = 0$$

in terms of Laplacian operator may be written as

$$\nabla^2\psi+\frac{8\pi^2 m(E-V)}{h^2}\psi = 0 \tag{8.22}$$

Besides this, a few more commonly employed operators are Hermitian operator, unitary operator and Hamiltonian operator.

MEMORY FOCUS

The following are the postulates of quantum (wave) mechanics:

Postulate 1. Every time-independent state of system is described completely by a function *y* called wave function which is single valued, continuous and finite throughout the configuration space.

Postulate 2. Every physically measurable or observable quantity in classical mechanics, like position, velocity, momentum, energy, etc., of a system has a corresponding operator in quantum mechanics. The nature of the operator depends upon the classical expression for the observable quantity.

Postulate 3. The only possible values that can be observed of a physical property like angular momentum or energy are the eigenvalues a, in the operator equation known as eigenvalue equation: $\hat{A}y = ay$ where $\hat{A}$ is the operator for the physical quantity and *y* is the well-behaved eigenfunction.

Postulate 4. The only possible measured values of an observable are the eigenvalues obtained from the eigenvalue equation (from Postulate 3) above. Hence, if the wave function y of the system is an eigenfunction of the operator for the observable, then measurement of that observable gives eigenvalue. When the wave function y of the system is not an eigenvalue, the value of the observable obtained on measurement has a probability that can be calculated from the wave function.

Postulate 5. When a great many measurements of any observable represented by an operator $\hat{A}$ are made on a system characterized by a function *y*, the average value of the quantity (represented by $\bar{a}$) is given by

$$\bar{a} = \frac{\oint \psi * \hat{A} \psi d\tau}{\oint \psi * \psi d\tau}$$

Postulate 6. The time development of a given wave function is given by

$$\frac{ih}{2\pi}\frac{\partial \psi}{\partial t} = \hat{H}\psi$$

which is Schrödinger's equation, where $\hat{H}$ is the Hamiltonian operator.

Following are the important types of operators:

1. **Linear operators:** An operator is said to be linear if it is operated on the sum of two or more functions and gives the result as sum of the results of operation on each function.
2. **Eigen operators:** An operator, say *A*, is said to be an eigenoperator if it operates on a function say, $\psi(x)$, giving a result which is the original function multiplied by some constant.
3. **Laplacian operator.** It is represented by ∇^2 (del squared) and is defined as

$$\nabla^2 = \frac{\partial^2}{\partial x^2} + \frac{\partial^2}{\partial y^2} + \frac{\partial^2}{\partial z^2}$$

This is the common operator in quantum mechanics.

8.11 SCHRÖDINGER WAVE EQUATION

Keeping in view the de Broglie concept of particle nature and wave nature of small particles (put forward in 1924), Schrödinger in 1926 proposed an equation for the wave motion of the electron in the three dimensions. This equation is called Schrödinger wave equation and is given as follows:

$$\frac{\partial^2\psi}{\partial x^2}+\frac{\partial^2\psi}{\partial y^2}+\frac{\partial^2\psi}{\partial z^2}+\frac{8\pi^2 m}{h^2}(E-U)\psi=0 \tag{8.23}$$

The various symbols in the above equation have the following meanings:

(i) ψ represents the amplitude of the wave function associated with the electron.
(ii) m is the mass of the electron.
(iii) h is the Plank's constant.
(iv) E represents the total energy of the electron, i.e., kinetic energy + potential energy.
(v) U is the potential energy of the electron.
(vi) x, y and z are the usual Cartesian coordinates of the electron at a point where amplitude wave function is ψ.

The above equation is usually written in the form

$$\nabla^2\psi+\frac{8\pi^2 m(E-U)}{h^2}\psi=0$$

where ∇^2 (del. squared) $=\frac{\partial^2\psi}{\partial x^2}+\frac{\partial^2\psi}{\partial y^2}+\frac{\partial^2\psi}{\partial z^2}$ is called the Laplacian operator. (The term operator is a mathematical procedure which transforms one function into another).

8.11.1 Derivation of Schrödinger Wave Equation

To develop Schrödinger wave equation, let us first consider the simplest type of wave motion like that of stretched string travelling along the x-axis between two fixed points.

Suppose that a wave is allowed to travel along the string in one direction, so that when the wave reaches the fixed point, it is reflected in the opposite direction. If the two waves travelling in the opposite direction have the same wavelength, speed and amplitude, the string will be divided into a number of vibrating segments, separated by stationary points of zero amplitude. Such a vibrating motion is called *stationary* or *standing wave*. The points of zero amplitude (points *a*) are called *nodes*. The distance from one node to the next is equal to one half of the wavelength (λ) (see Fig. 8.6). Evidently, for such a wave-length of the string (l) = $n(\lambda/2)$.

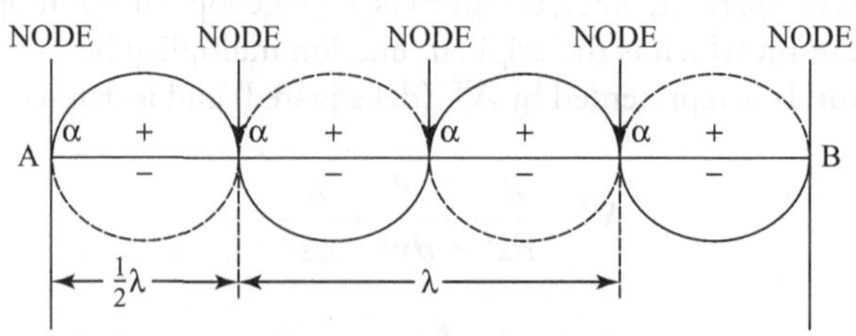

Figure 8.6 Stationary wave in a string

It can be shown that for this type of wave motion, the equation can be written as

$$\frac{\partial^2 f(x)}{\partial x^2} = -\frac{4\pi^2}{\lambda^2} f(x) \tag{8.24}$$

where $f(x)$, called the amplitude function, is a function of the displacement x from the origin and λ is the wavelength of the wave.

Equation (8.24) represents the wave motion in only one direction. For extending this equation to the wave motion in three directions, $f(x)$ for one coordinate may be replaced by $\psi(x, y, z)$; i.e., amplitude function for three coordinates which may be simply written as ψ for the sake of brevity. Thus, the equation for wave motion in three directions will be

$$\frac{\partial^2 \psi}{\partial x^2} + \frac{\partial^2 \psi}{\partial y^2} + \frac{\partial^2 \psi}{\partial z^2} = -\frac{4\pi^2}{\lambda^2}\psi \tag{8.25}$$

Equation (8.25) is applicable to all particles such as electrons, nuclei, atoms, molecules and photons. Substituting the de Broglie equation, i.e.,

$$\lambda = \frac{h}{mv} \tag{8.26}$$

where m is the mass of the particle, v is its velocity and h is Planck's constant, Equation (8.25) becomes

$$\frac{\partial^2 \psi}{\partial x^2} + \frac{\partial^2 \psi}{\partial y^2} + \frac{\partial^2 \psi}{\partial z^2} = -\frac{4\pi^2 m^2 v^2}{\lambda^2}\psi$$

Further, we know that,

Total energy (E) of the particle = Kinetic energy + Potential energy.

i.e.,

$$E = \frac{1}{2}mv^2 + U$$

or

$$\frac{1}{2}mv^2 = E - U \tag{8.27}$$

where U represents the potential energy of the particle.

Substituting the value of mv^2 from Equation (8.26) in Equation (8.27), we get

$$\frac{\partial^2 \psi}{\partial x^2} + \frac{\partial^2 \psi}{\partial y^2} + \frac{\partial^2 \psi}{\partial z^2} + \frac{8\pi^2 m(E-U)}{\lambda^2}\psi = 0$$

or

$$\frac{\partial^2 \psi}{\partial x^2} + \frac{\partial^2 \psi}{\partial y^2} + \frac{\partial^2 \psi}{\partial z^2} = -\frac{8\pi^2 m(E-U)}{\lambda^2}\psi$$

or

$$\frac{\partial^2 \psi}{\partial x^2} + \frac{\partial^2 \psi}{\partial y^2} + \frac{\partial^2 \psi}{\partial z^2} + \frac{8\pi^2 m}{\lambda^2}(E-U)\psi = 0 \tag{8.28/8.23}$$

This equation which represents the wave function of any particle in three directions is called the **Schrödinger wave equation.**

8.12 EIGENVALUES AND EIGENFUNCTIONS (OR WAVE FUNCTIONS)

A look at Equation (8.8) above shows that for a stationary (or standing) wave in a stretched string, the amplitude function $f(x)$ can have significance only for certain definite values of A. Obviously, these functions will be those which satisfy the following conditions:

(i) $f(x)$ must be equal to zero at each end of the string, as the string is fixed at these points and the amplitude of vibration will be zero.
(ii) $f(x)$ must be single-valued and finite, i.e., at every point on the vibrating string, the amplitude has a definite value at any instant.

Similarly, Schrödinger wave equation, being a second-order differential equation, can have many solutions for the amplitude function ψ but only those values of ψ are acceptable (or have significance) which correspond to some definite values of the total energy E. Such values of the total energy E are called *eigenvalues* (which correspond to discrete energy levels as postulated by Bohr). The corresponding values of the function ψ are called *eigenfunctions* or *wave functions.*

Just as in the case of stretched string, the eigenfunctions of the Schrödinger equation will be those which satisfy the following conditions:

(i) ψ must be single-valued and finite, i.e., for each value of the variables x, y and z, there is only one definite value of the function y.
(ii) ψ must be continuous, i.e., there must not exist any sudden changes in y as the variables are changed.
(iii) ψ must become zero at infinity.

8.12.1 Physical Significance of the Wave Function

If the particle under consideration is an electron, the wave function ψ represents the amplitude of the electron wave at any instant of time when the coordinates of the electron are x, y and z.

However, just as in case of light or sound, the square of the amplitude of the wave at any point gives the intensity of the light or sound at that point. Similarly, the square of the amplitude of the electron wave, i.e., ψ^2, at any point gives the intensity of the electron wave at that point. However, keeping in the view the Heisenberg's uncertainty principle, the intensity of the electron at any point may be interpreted as the probability of finding the electron at that point. Thus, ψ^2 at any point gives the probability of finding the electron at that point. However, the probability is directly related to the density (concentration) of the electron cloud. Thus, ψ^2 gives the electron density at any point. Since the region around the nucleus which represents the electron density at different points is called an orbital, the wave function ψ for an electron in an atom is called *orbital wave function* or simply *atomic orbital.*

8.13 NORMALIZED AND ORTHOGONAL EIGENFUNCTIONS

The integral of $\psi\psi^* \, d\tau$ over whole of the configuration space is proportional to the total probability of finding the particle somewhere in space. It must be finite and equal to unity for whole of the configuration space. That is,

$$\int_{-\infty}^{\infty} \psi\psi^* \, d\tau = 1$$

where $\int_{-\infty}^{\infty}$ represents the integration over the whole space. This is also represented as ϕ. The eigenfunction satisfying the above relation is said to be normalized to unity or simply normalized.

If the result of integration is not equal to unity but is equal to some constant c, then,

$$\int_{-\infty}^{\infty} \psi\psi^* d\tau = c \qquad \text{or} \qquad \frac{1}{c}\int_{-\infty}^{\infty} \psi\psi^* d\tau = 1$$

8.14 CONCEPT OF ATOMIC ORBITAL

In the light of Heisenberg's uncertainty principle, Bohr's concept that the electrons move along definite circular paths lost its significance. Now we can talk of only the probability of finding the electron at different points around the nucleus. The probability of finding the electron of a given energy E at any point can be calculated using Schrödinger wave equation, i.e., by solving the equation for ψ corresponding to the coordinates of that point and then finding them.

By calculating the probabilities at different points, it is found that there is a three-dimensional space around the nucleus within which the probability of finding the electron of a given energy is maximum. This space is called an atomic orbital.

Hence an atomic orbital may be defined as a three-dimensional space around the nucleus within which the probability of finding the electron of a given energy is maximum.

If dots are used to represent the electron probabilities, i.e., a dot represents a point where there is a probability for the electron to be present, then the orbital corresponding to 1*s* electron may be represented as shown in Fig. 8.7.

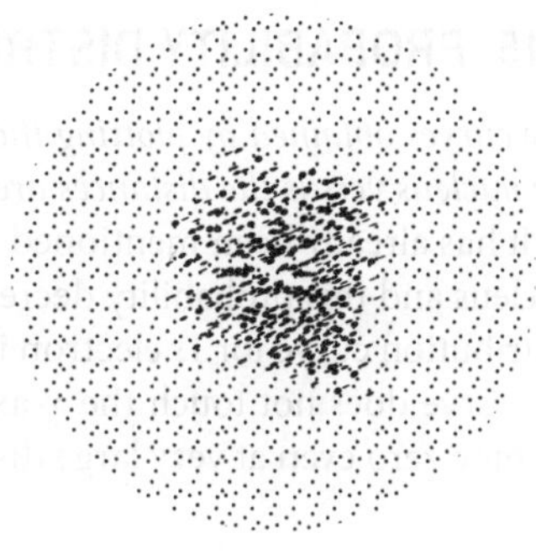

Figure 8.7 Representation of 1*s* orbital

Such a representation of an orbital is called the electron cloud picture. It may be noted that the intensity of dots which represents the electron probability is greater near the nucleus and fall off as the distance from the nucleus increases. Nowhere is the probability of finding the electron equal to zero. Even at large distances from the nucleus, there is a finite, though small, probability of finding an electron of a given energy. This means that electron clouds do not have sharp boundaries. However, for the sake of pictorial clarity and for convenience of representation, a boundary surface may be drawn which encloses a certain volume of the space around the nucleus within which the probability of finding the electron of a given energy is maximum, i.e., 90–95%.

It may be noted that it is not possible to determine the exact position of the electron at any instant of time (in accordance with Heisenberg's uncertainty principle). Hence, the probabilities are calculated in small segments of volume (called volume elements) in space around the nucleus.

SOME NOTEWORTHY POINTS

1. de Broglie's equation and uncertainly principle are applicable to all moving bodies and they are significant only in case of microscopic particles like atomic and subatomic particles such as electrons, protons, etc.
2. de Broglie's equation and uncertainly principle have no significance for macroscopic particles and can be neglected in everyday life.
3. de Broglie's equation explains only systems in which the energy of particle is described only in terms of speed. In the case of electrons where energy is partly due to speed and partly due to position, the de Broglie's theorem is used as an approximation.

(Continued)

SOME NOTEWORTHY POINTS (*Continued*)

4. The wave function of an electron (ψ) in the field of nucleus of an atom is called atomic orbital. It is a three-dimensional amplitude of electron wave.
5. $\psi^2 dv$ is the probability of finding the electron in a small volume element dv surrounding the nucleus of an atom.
6. The particular values of ψ are called eigenfunctions and the values of energies which correspond to these are called eigenvalues.
7. The eigenfunction for n electron is called an atomic orbital.
8. Wave equation is applicable to both atoms and also molecules, but the treatment is complicated.
9. On solving the equations, we find regions in space where ψ is positive as well as negative, therefore it has no significance. But ψ^2 (probability) is always positive and hence is a significant term.

8.15 PROBABILITY DISTRIBUTION CURVES

The curves obtained by plotting the probabilities (ψ^2) of finding the electron at different distances (r) from the nucleus versus the distances are called probability distribution curves.

It has already been mentioned above that the probability of finding 1*s* electron is maximum near the nucleus and this probability decreases as the distance from the nucleus increases. Hence the probability distribution curve for 1*s* electron is as shown in Fig. 8.8(a). From this curve, it may be further noted that the curve does not touch the *x*-axis even at very large distances showing that the probability does not become zero even at very large distances from the nucleus.

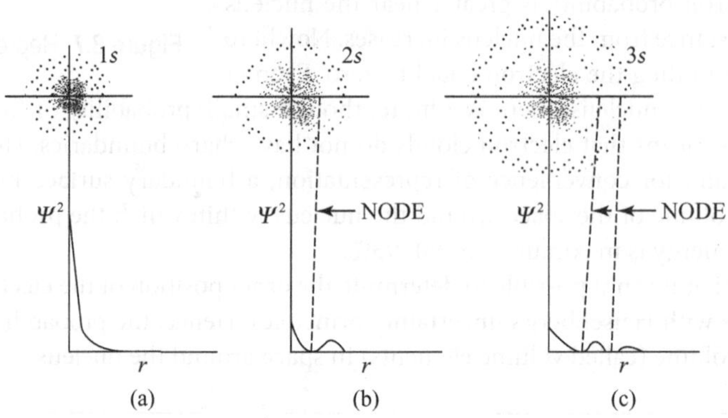

Figure 8.8 Probability distribution curves for (a) 1*s*, (b) 2*s* and (c) 3*s* orbitals

Similarly, the probability distribution curves for 2*s* and 3*s* electrons are shown in Fig. 8.8(b) and 8.8(c) respectively. It may be noted that for 2*s* electron, ψ^2 goes to zero and then increases again in value before finally approaching zero at large values of *r*. The intermediate regions where ψ^2 goes to zero are called nodal surfaces or simply nodes. The number of nodes increases with the increasing value for the principal quantum number (*n*). The 3*s* orbital possesses two nodes as shown in Fig. 8.8(c).

8.16 RADIAL PROBABILITY DISTRIBUTION CURVES

Imagine a large number of spherical shells of very small thickness present concentrically around the nucleus (see Fig. 8.9). The probability of finding the electron in a spherical shell of thickness *dr* at a distance *r* from the nucleus is called *radial probability*. It is equal to the product of the volume of the shell of thickness dr and the probability (ψ^2) per unit volume, i.e.,

$$\text{Radial probability} = 4\pi r^2\, dr \times \psi^2$$

The volume of a spherical shell of thickness dr at a distance r from the centre can be calculated as follows:

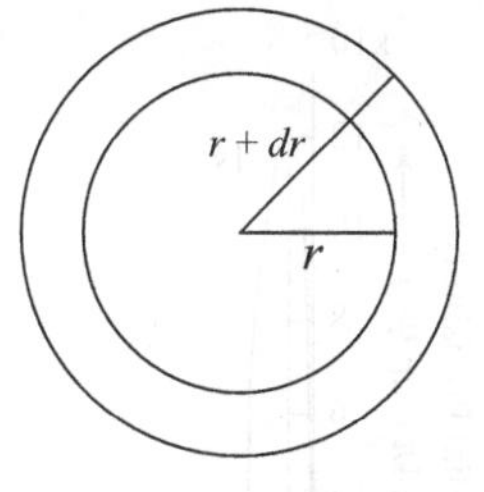

Figure 8.9 Concentric spherical shells around the nucleus

Required volume = [Volume of sphere with radius $r + dr$] – [Volume of sphere with radius r]

$$= \frac{4\pi}{3}(r+dr)^3 - \frac{4\pi}{3}r^3$$

$$= \frac{4\pi}{3}(r^3 + 3r^2dr + 3rdr^2 + dr^3) - \frac{4\pi}{3}r^3$$

$$= \frac{4\pi}{3}(3r^2dr) = 4\pi r^2 dr$$ (dr and its higher powers are neglected because dr is very small)

It is more useful to study the radial probability at different distances from the nucleus. The plots of radial probability versus distance from the nucleus are called radial probability distribution curves. The radial probability distribution curves for 1s, 2s, 2p, 3s, 3p and 3d orbitals for the hydrogen atom are shown in Fig. 8.10.

The main features of these curves are briefly described in the following sections:

8.16.1 Radial Probability Distribution Curve for 1s Orbital

The probability (ψ^2) of finding 1s electron is maximum near the nucleus (as shown in Fig. 8.8(a)). However, the volume $4\pi r^2 dr$ is very small because r is very small near the nucleus. Hence, the radial probability ($4\pi r^2\, dr\, \psi^2$) is very small near the nucleus. As the distance r from the nucleus increases, ψ^2 goes on decreasing whereas the volume $4\pi r^2\, dr$ goes on increasing. In the beginning, as the decrease in the value of ψ^2 is not so large as compared to the increase in the value of r (and hence $4\pi r^2\, dr$), the radial probability goes on increasing, reaches a maximum value at a distance of 0.529 Å and then begins to decrease abruptly and finally becomes negligible at large distances.

The distance of maximum probability (0.529 Å) is same as Bohr's radius for the first orbit. However, there is a great difference between the two viewpoints. According to Bohr, the electron is revolving at a constant distance of 0.529 Å from the nucleus whereas wave mechanical model predicts that the probability for the electron to be present is maximum at 0.529 Å but the electron may be present at distances less as well as greater than this distance.

Further, it appears that the energy of 1s electron should be different for different positions of the electron with respect to nucleus. However, this is not so. The energy of 1s electron is fixed. This is because the wave mechanical model does not deal with the fixed positions of the electron; rather, it is opposed to the concept of fixed positions.

8.16.2 Radial Probability Distribution Curves for other s Orbitals

The radial probability distribution curve for 2s orbital consists of two maxima, separated by a region where the radial probability is zero. This region is called a node. Thus, the 2s orbital has one node. Similarly, 3s orbital has three maxima and two nodes. In general, for any ns orbital, there are maxima and (n – 1) nodes. Further, it may be seen from the curves that as n increases, the distance of maximum radial probability increases but the average electron density decreases as shown by the heights of the humps.

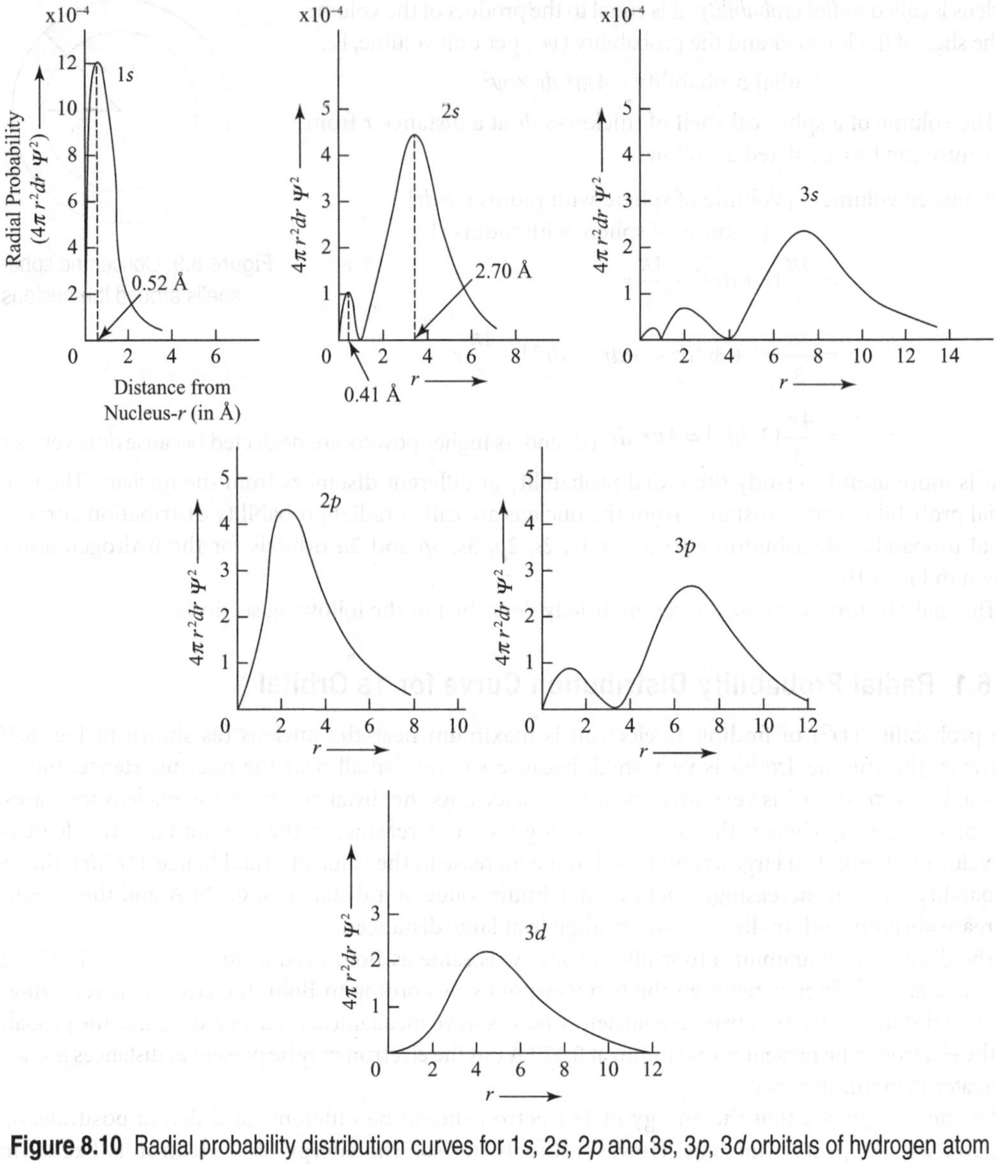

Figure 8.10 Radial probability distribution curves for 1*s*, 2*s*, 2*p* and 3*s*, 3*p*, 3*d* orbitals of hydrogen atom

8.16.3 Comparison of Radial Probability, Distribution Curves for 1*s* with other *s* Atomic Orbitals

(i) 1*s* orbital has only one maximum radial probability while the other *s* orbitals have more than one maximum. In general, any *ns* orbital has *n* maxima.

(ii) 1*s* orbital does not have any node whereas other *s* orbitals have nodes. In general, any *ns* orbital has $(n - 1)$ nodes.

(iii) As n increases, the distance of maximum radial probability increases but the average electron density decreases as shown by the heights of the humps.

8.16.4 Radial Probability Distribution Curves for *p* Orbitals

A look at the radial probability distribution curve for $2p$ orbital shows that the probability of $2p$ electron increases as the distance from the nucleus increases, reaches a maximum value at equal to 2.1 Å and then decreases to a minimum.

The radial probability distribution curve for $3p$ has two maxima separated by a region of zero probability, i.e., one node. Similarly, $4p$ has three maxima and two nodes. In general, any np orbital has $(n - 1)$ maxima and $(n - 2)$ nodes.

In view of the discussion given above, it may be noted that the number of nodes in any ns or np orbital is $(n - l - 1)$ where n represents the main shell and l represents the sub-shell ($l = 0, 1, 2, 3$ for s, p, d and f sub-shells, respectively).

8.16.5 Comparison of Radial Probability Distribution Curves for 2*s* and 2*p* Orbitals

(i) $2s$ orbital has two maxima at distances of 0.4 Å and 2.70 Å from the nucleus separated by one node whereas $1p$ orbital has only one maximum at a distance of 2.1 Å and there is no node.

(ii) The first hump in the case of $2s$ indicates that $2s$ electron spends some of its time close to the nucleus more than $2p$. Hence, $2s$ has lower energy than $2p$.

(iii) The larger distance of the bigger peak in case of $2s$ than $2p$ indicates that the electron of $2s$ (being spherically symmetrical) extends a little beyond that of $2p$ (Fig. 8.11).

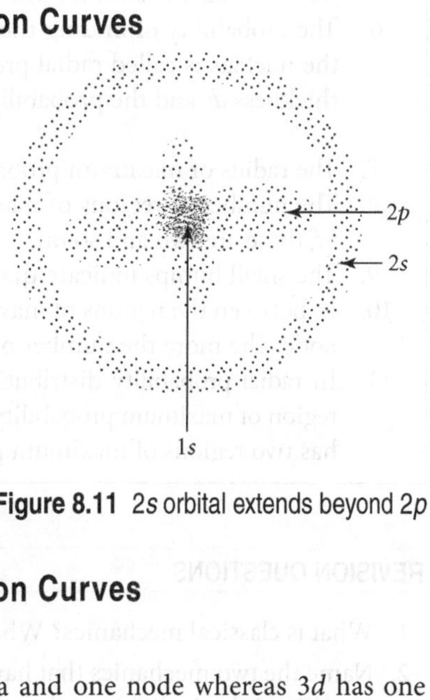

Figure 8.11 *2s* orbital extends beyond *2p*

8.16.6 Comparison of Radial Probability Distribution Curves for 3*s*, 3*p* and 3*d* Orbitals

(i) $3s$ has three maxima and two nodes, $3p$ has two maxima and one node whereas $3d$ has one maximum and no node.

(ii) The distance of maximum radial probability for $3s$ is greater than that of $3p$ which in turn is greater than that of $3d$. Thus, the sizes of the electron clouds of $3s$, $3p$ and $3d$ are of the order $3s > 3p > 3d$.

(iii) Two humps in $3s$, one hump in $3p$ and no hump in $3d$ indicate that $3s$ is attracted to the nucleus more than $3p$ which in turn is attracted to the nucleus more than $3d$. Thus, their energies are in the order $3s < 3p < 3d$.

MEMORY FOCUS

1. The Schrödinger wave equation is given as follows:

$$\frac{\partial^2\psi}{\partial x^2}+\frac{\partial^2\psi}{\partial y^2}+\frac{\partial^2\psi}{\partial z^2}+\frac{8\pi^2 m}{h^2}(E-U)\psi=0$$

The various symbols used in the equation have the useful meanings.
2. Only those values of ψ which correspond to some definite values of the total energy E are acceptable (or have significance). Such values of the total energy E are called *Eigen values.* The corresponding values of the function ψ are called *Eigen functions* or *wave functions.*
3. The wave function ψ for an electron in an atom is called *orbital wave function* or *simply atomic orbital.*
4. An atomic orbital may be defined as a three-dimensional space around the nucleus within which the probability of finding the electron of given energy is maximum.
5. *The curves obtained by plotting the probabilities (ψ^2) of finding the electron at different/distances (r) from the nucleus versus the distances are called probability distribution curves.*
6. The probability of finding the electron in a spherical shell of thickness dr at a distance r from the nucleus is called radial probability. It is equal to the product of the volume of the shell of thickness dr and the probability (ψ^2) per unit volume, i.e.,

   Radial probability $= 4\pi r^2\, dr \cdot \psi^2$
7. The radius of maximum probability of 1s electron is 0.529 Å.
8. The number of regions of maximum probability for $1s$, $2p$, $3d$ and $4f$ is one. For $2s$, $3p$, $4d$ and $5f$, these are two and so on.
9. The small humps indicate that the electron has a tendency to penetrate closer to the nucleus.
10. In between the regions of maximum probability, there is a region of zero electron density called node. The more the number of nodes are, more is the energy of an orbital.
11. In radial probability distribution curves, the first orbital of each type ($1s$, $2p$, $3d$, $4f$) has one region of maximum probability and no node, whereas the first orbital of each type ($2s$, $3p$, $4d$, $5f$) has two regions of maximum probability and one node and so on.

REVISION QUESTIONS

1. What is classical mechanics? What are its limitations?
2. Name the two mechanics that have been proposed to overcome the limitations of classical mechanics. What is basic difference between them?
3. Explain the failure of classical mechanics based on the phenomenon of black body radiation.
4. Write notes on (i) Stefan–Boltzmann law and (ii) Wien's displacement law.
5. State and explain Kirchoff's law about black body radiation.
6. Why does classical mechanics fail to explain the spectral distribution of black body radiation? Give Planck's radiation law.
7. Comment on the statement that Wien's law and Rayleigh–Jeans law follow Planck's radiation law.
8. Give main points of difference between classical mechanics and quantum mechanics. How are the two related?

9. What is an operator? Write quantum operators for the following.
   (a) Linear momentum P (b) Angular momentum (c) Kinetic energy
10. What do you understand by
    (a) Linear operator (b) Laplacian operator
11. Derive Schrödinger wave equation. What is the significance of ψ in it?
12. What do you understand by eigenvalues and eigenfunctions? What are the conditions which the eigenfunctions must satisfy?
13. Based on the postulates of quantum mechanics, derive Schrödinger wave equation.
14. Derive Schrödinger wave equation for hydrogen-like atoms in terms of spherical polar co-ordinates.
15. Write a short note on concept of operators in quantum mechanics.
16. (a) Derive Schrödinger equation for hydrogen-like atoms of spherical polar co-ordination.
    (b) What do Laplacian and Hamiltonian operator stand for?
17. What is zero point energy? Write its significance.
18. What are the different postulates of quantum mechanics? Discuss them in detail.
19. (a) Discuss the significance of a wave function.
    (b) Define an operator.
    (c) Write the postulates of quantum mechanics.
    (d) Derive Schrödinger wave equation based on the above postulates.
20. (a) What is a linear operator?
    (b) Explain what do you understand by the term 'zero point energy'?
    (c) Write Schrödinger wave equation for hydrogen-like atom in spherical polar coordinates.
21. (a) How does quantum mechanics differ from classical mechanics (or Newtonian mechanics)?
    (b) What is the basis of quantum mechanics?
22. Write Schrödinger wave equation. What do various symbols signify? In addition, write the equation in terms of Laplacian operator.
23. Write the Schrödinger wave equation for the wave motion of a particle in one direction. From it, deduce the equation for three dimensions.
24. Briefly describe the physical significance of the wave function ψ in the Schrödinger wave equation.
25. What do you understand by probability distribution curves? Discuss the probability distribution curves for 1s, 2s and 3s orbitals.
26. What do you understand by a node? How many nodes are present in (i) 2s orbital and (ii) 3s orbital?
27. What is meant by radial probability? Write its mathematical expression. Draw the radial probability distribution curves for 1s, 2s, 3p, 3s, 3p and 3d orbitals for the hydrogen atom.
28. Draw radial probability distribution curve for 1s electron. How do you interpret it?
29. The distance of maximum probability for 1s electron is same as Bohr's radius. Then, how does the wave mechanical model differ from Bohr's model?
30. Draw the radial probability distribution curves for 1s, 2s and 3s orbitals. How do they differ from each other?
31. Draw the radial probability distribution curve for 2p orbital. Discuss its important features.
32. (a) What are the various operators used in quantum mechanics?
    (b) Derive the Schrodinger wave equations and explain its significance

MULTIPLE CHOICE QUESTIONS

1. In Schrödinger wave equation, the symbol ψ represents the
   (a) wavelength of the spherical wave
   (b) amplitude of the spherical wave
   (c) frequency of the spherical wave
   (d) none of these
2. The energy of electron in an atomic orbital is always
   (a) different (c) infinite
   (b) zero (d) same
3. The principal quantum number n represents
   (a) average size of the electron cloud
   (b) average energy of the electron
   (c) average distance of the electron from the nucleus
   (d) all of the above
4. Out of the following, which is the correct set of quantum numbers for the outermost electron of potassium ($Z = 19$)?

	n	l	m	s
(a)	4	3	2	−1/2
(b)	4	2	0	−1/2
(c)	4	1	0	+1/2
(d)	4	0	0	−1/2

5. Wave nature of electron was demonstrated by
   (a) Schrödinger (c) Davisson
   (b) de Broglie (d) Heisenberg
6. Which of the following statement is false?
   (a) ψ^2 gives the electron density at any point
   (b) ψ^2 at any point give the probability of finding the electron at that point
   (c) ψ^2 stands for an atomic orbital with definite energy
   (d) ψ^2 is square of the amplitude of electronic wave
7. Which of the following condition is not satisfied by the eigenfunction of the Schrödinger wave equation?
   (a) ψ must be single valued and finite
   (b) ψ must be continuous
   (c) ψ must be discontinuous
   (d) ψ must become zero at infinity
8. Which of the following statement is false about electromagnetic radiations?
   (a) They do not require any medium for transmission
   (b) They have electric and magnetic fields at right angles to the direction of propagation
   (c) They are characterized by definite frequency and wavelength
   (d) They travel with velocity different from that of light
9. The expression Δ^2 represents which operator (del squared) $= \dfrac{\partial^2}{\partial x^2} + \dfrac{\partial^2}{\partial y^2} + \dfrac{\partial^2}{\partial z^2}$
   (a) Hermitian operator
   (b) Laplacian operator
   (c) Unitary operator
   (d) Hamiltonian operator
10. Quantum mechanics is applicable to
    (a) microscopic particles
    (b) macroscopic particles
    (c) Both microscopic and macroscopic
    (d) neither of the two
11. Which of the following statement is correct?
    (a) All electromagnetic waves do not possess the same velocity
    (b) Matter waves are associated with electric and magnetic fields
    (c) Energy of the electron in an atom is negative
    (d) Matter waves and electromagnetic waves are alike
12. Which of the following scientists developed the equation $E = hv$?
    (a) Einstein (c) Max Planck
    (b) Schrödinger (d) de Broglie
13. Assuming the velocity to be same, which of the following is associated with smallest de Broglie wavelength?
    (a) An electron
    (b) A proton
    (c) An α particle
    (d) All will have same value of wavelength

14. The total number of orbitals in a shell with principal quantum number n is
    (a) $2n$ (c) n^2
    (b) $2n^2$ (d) $n+1$
15. The quantum number which determines the shape of the orbital is
    (a) principal (c) magnetic
    (b) azimuthal (d) spin
16. Which of the following law gives quantitative relationship between the energy absorbed and the energy emitted by a body?
    (a) Stefan's law
    (b) Wien's displacement law
    (c) Planck's radiation law
    (d) Kirchoff's law
17. The statement *total amount of energy radiated by a perfectly black body is proportional to the fourth power of its absolute temperature* is
    (a) Planck's radiation law
    (b) Stefan's law
    (c) Wien's displacement law
    (d) Kirchoff's law
18. The symbol P_x stands for which of the following operators?
    (a) Position (c) Momentum
    (b) Position squared (d) Energy
19. Schrödinger wave equation is applicable to the wave motion of an electron in
    (a) one dimension
    (b) two dimensions
    (c) three dimensions
    (d) all the above dimensions
20. Which of following statement in respect of Schrödinger wave equation is wrong?
    (a) The symbol psi (ψ) represents the amplitude of the wave
    (b) h is Planck's constant
    (c) U stands for kinetic energy
    (d) E stands for total energy

ANSWERS

1. (b)	5. (b)	9. (b)	13. (c)	17. (b)
2. (d)	6. (c)	10. (a)	14. (c)	18. (c)
3. (d)	7. (c)	11. (c)	15. (b)	19. (c)
4. (d)	8. (d)	12. (c)	16. (d)	20. (c)

Ionic Equilibria 9

CHAPTER OBJECTIVES

- 9.1 Introduction
- 9.2 Arrhenius Theory of Ionization
- 9.3 Ionisation of Weak Electrolytes – Ostwald's Dilution Law
- 9.4 Arrhenius Concept of Acids and Bases
- 9.5 Ionisation Constant of Weak Acids and Bases (Arrhenius Concept)
- 9.6 Bronsted–Lowry Concept of Acids and Bases
- 9.7 Lewis Concept of Acids and Bases
- 9.8 Ionic Product of Water
- 9.9 pH Scale
- 9.10 Buffer Solution

Most of the fundamental ideas of science are essentially simple, and may, as a rule, be expressed in a language comprehensible to every one.

Albert Einstein

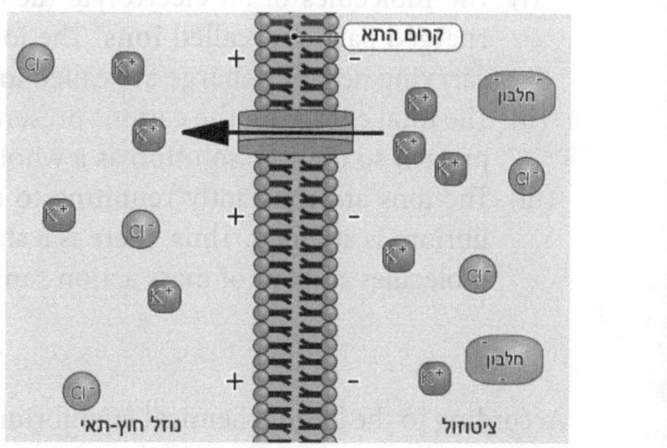

9.1 INTRODUCTION

There are some substances which can conduct electricity in molten or in their aqueous state like sodium chloride, whereas some substances can never conduct electricity in any state like sugar. The conduction of electricity is always due to the formation of ions. It means that there are two types of substances: One, which can ionize and second which can never ionize in any state. Based on this, there are two types of substances: (i) *Electrolytes* (ii) *Nonelectrolytes*

The electrolytes are further divided in two parts: (a) *Strong electrolytes* (b) *Weak electrolytes*

A substance that dissociates almost completely into ions in an aqueous solution is called a strong electrolyte. For example, NaCl, NaOH, KOH, HCl, H_2SO_4, etc., are some of the examples of such a class of compounds.

On the other hand, there are substances like acetic acid (CH_3COOH) where the degree of dissociation is low. Such substances, which dissociate to a small extent in an aqueous solution and hence conduct electricity to a small extent, are called weak electrolytes.

When dissolved in suitable solvents, electrolytes undergo dissociation to form charged chemical species known as ions. However, this dissociation into ions seldom takes place to completion due to reversible nature of the process. In fact, the extent of dissociation is different in different cases.

In the solutions of weak electrolytes, an equilibrium is established between the ions in solution and the undissociated molecules. This is known as **ionic equilibrium**. Therefore, ionic equilibrium *is the equilibrium, which is, established between the undissociated molecules and the ions in the solution of weak electrolytes*. Reactions involving ionic equilibrium may be classified as follows:

(i) Equilibria involving soluble ionic compounds such as weak acids and weak bases.
(ii) Equilibria involving sparingly soluble ionic compounds: Solubility product.

In this chapter, we discuss at length the first type of ionic equilibria as per the details of the syllabus.

9.2 ARRHENIUS THEORY OF IONIZATION

A satisfactory explanation of the behaviour of electrolytes in a solution was put forward by the Swedish physicist, Suante Arrhenius in 1887, in his theory, known as the theory of electrolytic dissociation. The main postulates of this theory can be summed up as follows:

(i) The molecules of an electrolyte (acid, base or salt), in a solution break up into two types of charged particles, called **ions**. The ions carrying positive charge are called **cations**, while those carrying negative charge are called **anions**.
(ii) The total charge on the cations present in the solution is equal but opposite to that on the anions present so that the solution as a whole is electrically neutral.
(iii) The ions are constantly reuniting to form undissociated or unionized molecules until an equilibrium is attained. Thus, there is a state of dynamic equilibrium between ionized and unionized molecules and law of mass action can be satisfactorily applied to this system.

$$A^+B^- \rightarrow A^+ + B^-$$

According to the law of chemical equilibrium, ionization constant will be $K = \dfrac{[A^+]\times[B^-]}{[AB]}$

(iv) The ions are free to move and under the influence of electricity are directed towards oppositely charged electrodes. Thus, only those substances, which ionize in solution, will be conductors of electricity. *Electric current does not produce ions, but it only has a directive effect on ions already present in the solution.*

One of the important points in the theory of electrolytic dissociation is that the decomposition of the electrolyte into ions is not affected by electric current, as previously thought, but occurs as soon as the electrolyte is dissolved in water. The solution of every electrolyte contains free ions, independent of whether it is undergoing electrolysis or not. The effect of the current is only directive. As soon as the electrodes are introduced in the solution, the positively charged cations move towards the cathode, whereas the anions move towards the anode.

(v) The electrolyte may not be completely ionized and *the proportion of the dissociated molecules to the total number of molecules is termed 'degree of dissociation' (or ionization)*. Electrolytes with high degree of dissociation (or ionization) are called **strong electrolytes**, while those with low degree of ionization are called **weak electrolytes.**

The degree of dissociation is quite considerable even at moderate dilutions and it increases with dilution until at infinite dilution complete dissociation takes place.

(vi) The properties of electrolytes are the properties of their ions, e.g., the properties of acids and bases are the properties of H^+ and OH^- ions, respectively.

(vii) These ions behave like molecules in depressing the freezing point, elevating the boiling point, lowering the vapour pressure and establishing osmotic pressure. So far, as the strong electrolytes are concerned, Arrhenius's theory is only of historical interest, because all strong electrolytes are known to ionize completely at all dilutions.

9.2.1 Degree of Dissociation or Ionization

Not all the molecules of an electrolyte undergo ionization in solution. The fraction of the molecules of an electrolyte, present as ions, is called degree of ionization. It is denoted by α. Thus

$$\alpha = \frac{\text{No. of dissociated molecules}}{\text{Total No. of molecules}}$$

The degree of dissociation depends on the following factors:

(i) **Nature of solute:** It is the nature of the solute that contributes most to the degree of ionization. Weak electrolytes possess lower degree of ionization and strong electrolytes possess higher degree of ionization. Mineral acids, alkalis and most of the salts are examples of strong electrolytes while ammonium hydroxide and most of the organic acids are examples of weak electrolytes.

(ii) **Nature of the solvent:** Solvents with high dielectric constant are highly ionizing. Water, we know, is a good ionizing solvent. Ethyl alcohol, acetone, etc., are also capable of ionizing but not to the extent of the solvent, water. There are instances of polar molecules. They have comparatively **high dielectric constant**. However, non-polar or feebly polar molecules having small value of the dielectric constant are very weak ionizers or are incapable of ionizing. Table 9.1 enables this comparison to be understood clearly.

Table 9.1 Dielectric constant and ionizing influence of solvents.

Solvent	Dielectric Constant	Ionization
Water	81.0	Strong ionizer
Ethyl alcohol	72.0	Moderate ionizer
Acetone	21.0	Very moderate ionizer
Chloroform	5.1	Very feebly ionizer
Carbon disulphide	2.6	Negligible ionizer
Benzenes	2.3	Negligible ionizer

(iii) **Presence of other solutes:** Presence of other ions also affects the degree of ionization. Thus, presence of same ions from another solute depresses the degree of ionization of the given solute.

(iv) **Concentration:** The lesser the concentration is, the greater will be the number of solvent molecules and hence greater will be the ionization. Therefore, degree of ionization is inversely proportional to concentration.

(v) **Temperature:** The increase in temperature results in an increase in the velocities and kinetic energies of molecules, which overcome the forces of attraction, and hence there is a greater ionization.

9.3 IONISATION OF WEAK ELECTROLYTES—OSTWALD'S DILUTION LAW

In an aqueous solution of a weak electrolyte like BA, a fraction of it dissociates into ions, and there exists an equilibrium between the free ions B^+ and A^-, and the undissociated electrolyte. Thus

$$BA \xrightleftharpoons{\text{water}} B^+ + A^-$$

Since ionization is a reversible process, therefore, the rate of formation of ions and their removal by recombination will be governed by the law of mass action.

Suppose that originally C mole of the binary electrolyte BA is dissolved in 1 L of solution and that its degree of dissociation is α, then at equilibrium

$$\underset{C(1-\alpha)}{\overset{C}{AB}} \xrightleftharpoons{\text{water}} \underset{C\alpha}{\overset{0}{B^+}} + \underset{C\alpha}{\overset{0}{A^-}} \quad \begin{matrix}\text{...Original}\\ \text{...At equilibrium}\end{matrix}$$

and the respective molar concentrations at equilibrium will be

$$[BA] = C(1-\alpha)$$

$$[B^+] = C\alpha$$

$$[A^-] = C\alpha$$

According to law of chemical equilibrium,

$$K = \frac{[B^+][A^-]}{[BA]}$$

where K is a constant and is known as dissociation or ionization constant.

or
$$K = \frac{C\alpha^2}{1-\alpha} \tag{9.1}$$

This mathematical expression is known as **Ostwald's dilution law.** If degree of dissociation is very small, as is usually the case for weak electrolyte at ordinary concentration, then $(1 - \alpha)$ will be approximately equal to unity, so the dilution law becomes

$$C\alpha^2 = K \qquad \text{or} \qquad \alpha = \sqrt{\frac{K}{C}} \tag{9.2}$$

that is, **the degree of dissociation of weak electrolyte is inversely proportional to the square root of the molar concentration of solution.** Thus, it follows that the degree of dissociation would increase with dilution and in case of very dilute solutions will be nearly equal to unity. This implies that even weak electrolytes are almost completely ionized in very dilute solutions.

Further, since C is reciprocal of V (volume)

$$\therefore \qquad \alpha = \sqrt{K \times V} \tag{9.3}$$

or α is directly proportional to the volume.

9.3.1 Verification of Ostwald's Dilution Law

The degree of dissociation (α) at different concentrations was determined by using the formula

$$\alpha = \lambda_V / \lambda_\infty$$

where λ_V is equivalent conductance at dilution, V, and λ_∞ is equivalent conductance at infinite dilution. Substituting the value of at different concentrations in Eq. (9.1), the value of ionization constant (K) was calculated. The value of concentration constant (K) calculated is observed to be constant for weak electrolytes. The law was found to hold good in the case of weak electrolytes, such as acetic acid and ammonium hydroxide. However, the law failed completely when it was applied to strong electrolytes such as hydrochloric acid, sodium hydroxide or potassium chloride.

MEMORY FOCUS

1. According to Arrhenius theory of ionisation, the mere process of dissolution breaks up the electrolytes into charged particles called **ions.**
2. The ions carrying positive charge are called **cations,** while those carrying negative charge are called **anions.**
3. There is a dynamic equilibrium between the ionised particles and unionised electrolytes and law of mass action can be applied.
4. Weak electrolytes are only partially ionised in water. The fraction of the electrolyte ionised is called **degree of ionisation.**
5. The degree of ionisation depends on the following factors.
   (i) Nature of solute (ii) Nature of solvent (iii) Presence of other solutes
   (iv) Concentration (v) Temperature.

(Continued)

MEMORY FOCUS (*Continued*)

6. Arrhenius theory explains the process of electrolysis. The decomposition of electrolytes into ions is not brought about by electric current. Current only directs the ions already present in the solution towards oppositely charged electrodes.
7. This theory also explains that properties of the electrolytes in solution are the properties of the ions that they give in solution.
8. Ostwalds applied the law of chemical equilibrium to the dissociation of electrolytes in solution and put forth a mathematical formula

   $K = K = \dfrac{C\alpha^2}{1-\alpha}$ refer to as Ostwald's dilution formula.
9. This formula is applicable to weak electrolytes only. It does not hold good for strong electrolytes.

9.4 ARRHENIUS CONCEPT OF ACIDS AND BASES

An acid is a substance, which furnishes hydrogen ions in its aqueous solution, e.g., HCl, HNO_3, CH_3COOH, etc., are regarded as acids as they furnish hydrogen ions in their aqueous solutions.

$$\underset{\text{Hydrochloric acid}}{HCl} \xrightleftharpoons{\text{water}} H^+ + Cl^-$$

$$\underset{\text{Nitric acid}}{HNO_3} \xrightleftharpoons{\text{water}} H^+ + NO_3^-$$

$$\underset{\text{Acetic acid}}{CH_3COOH} \xrightleftharpoons{\text{water}} H^+ + CH_3COO^-$$

Base: A base is a substance, which furnishes hydroxyl ions in its aqueous solution, e.g., NaOH, KOH, NH_4OH, etc., are regarded as bases as they furnish hydroxyl ions in their aqueous solution.

$$\underset{\text{Sodium hydroxide}}{NaOH} \xrightleftharpoons{\text{water}} Na^+ + OH^-$$

$$\underset{\text{Potassium hydroxide}}{KOH} \xrightleftharpoons{\text{water}} K^+ + OH^-$$

$$\underset{\text{Ammonium hydroxide}}{KOH} \xrightleftharpoons{\text{water}} NH_4^+ + OH^-$$

9.4.1 Limitation of Arrhenius Theory

Soon after Arrhenius had proposed his concept of acids and bases, it was realized that his concept has a limited scope only. This is because it attributes all acidic and basic properties to the formation of hydrogen and hydroxyl ions in *aqueous solution only*. However, there are substances such as NH_3, CaO, etc., which do not contain the hydroxyl group but can still act as bases since they are known to neutralize acids. Similarly, there are substances, such as carbon dioxide, which cannot, by themselves, dissociate to form hydrogen ions but even then act as acids in aqueous solution.

9.5 IONISATION CONSTANT OF WEAK ACIDS AND BASES (ARRHENIUS CONCEPT)

Like other electrolytes, different acids and bases dissociate to different extents in aqueous solutions. Depending upon the degree of dissociation, they furnish different amounts of H^+ (aq) or OH^- (aq) ions in solution. Accordingly, they are referred to as strong and weak acids and bases.

(i) The strength of an acid depends upon the number of free H^+ ions present in a solution. Since, on dilution, the ionization or the number of H^+ ions increases, the strength of an acid also increases. Moreover, at infinite dilution, the ionization in case of all acids is nearly complete, so that acids tend to be *equally strong at infinite dilution.*

(ii) The mineral acids like HCl and H_2SO_4, which are completely ionized at all dilutions, are said to be strong acids, whereas acids like acetic acid, phosphoric acid that ionize to a less extent are called *weak acids.*

A. Strengths of acids and their dissociation constants: Acids such as HCl, HNO_3 and H_2SO_4 *which are almost completely dissociated in aqueous solution are known as* strong acids. On the other hand, acids such as CH_3COOH and HCN, which are only slightly dissociated in aqueous solution, *are termed as* weak acids.

The relative strengths of weak acids can be easily compared in terms of their dissociation constants. In aqueous solution, the dissociation of a weak acid, say HA, may be represented as

$$HA + H_2O\ (l) \rightarrow H_3O^+(aq) + A^-(aq)$$

According to the law of chemical equilibrium, the equilibrium or dissociation constant for this reaction is given by the relationship

$$K_c = \frac{[H_3O^+][A^-]}{[HA][H_2O]}$$

Since water used as a solvent is present in large quantity, its concentration remains practically constant (say K)

$$\therefore \qquad K_a = K_c \times K = \frac{[H_3O^+][A^-]}{[HA]}$$

where K_a is known as dissociation constant of an acid. It is a constant, at a given temperature for a given acid. The higher the value of K_a the stronger is the acid.

For instance, the dissociation constant of acetic acid may be written as follows

$$CH_3COOH + H_2O\ (l) \rightarrow CH_3COO^-(aq) + H_3O^+(aq)$$

$$K_a = \frac{[CH_3COO^-][H_3O^-]}{[CH_3COOH]}$$

B. Strengths of bases and their dissociation constants: Just as in the case of acids, a base according to Arrhenius concept would be termed as strong if it is almost completely dissociated in solution. On the other hand, if a base is only slightly dissociated in solution, it would be called a weak base. Both NaOH and KOH are prominent examples of strong bases while NH_4OH, AgOH, etc., are weak bases.

The dissociation constants of bases provide a convenient practical measure of comparing their relative strengths. The dissociation of a weak base, say BOH, may be represented as

$$BOH(aq) + H_2O(l) \rightleftharpoons B^+(aq) + OH^-(aq)$$

$$\therefore \qquad K_c = \frac{[B^+][OH^-]}{[BOH][H_2O]}$$

Since water is used as a solvent, its concentration remains practically constant (say, K)

$$\therefore \qquad K_b = K_c \times K = \frac{[B^+][OH^-]}{[BOH]}$$

where K_b, is known as dissociation constant of a base. It is a constant, at a given temperature, for a given base. The higher the value of K_b is, the stronger will be the base.

For instance, the dissociation constant of NH_4OH may be written as follows:

$$NH_4OH \rightleftharpoons NH_4^+ + OH^-$$

$$K_b = \frac{[NH_4^+][OH^-]}{[NH_4OH]}$$

9.6 BRONSTED–LOWRY CONCEPT OF ACIDS AND BASES

According to this concept, *a substance (molecular or ionic) that donates a proton, (H^+) to another substance is known as an acid, whereas a substance (molecular or ionic), which* accepts a proton *from any other substance is called* a *base.* In other words, an acid is a proton donor and a base is a proton acceptor.

For example, consider the reaction between HCl and H_2O

$$HCl + H_2O \rightleftharpoons H_3O^+ + Cl^-$$

As HCl donates a proton to water, hence HCl is an *acid.* On the other hand, H_2O accepts a proton from HCl, hence it is a base. The main features of Bronsted–Lowry concept are given as follows.

(i) An acid–base reaction involves the transference of a proton from a proton donor (acid) to a proton acceptor (base). Thus, this concept is also called proton transfer theory of acids and bases.

(ii) Not only molecules but also ions are capable of acting as acids and bases. For example

$$\underset{\text{Acid}}{NH_4^+} + \underset{\text{Base}}{S^{2-}} \rightleftharpoons \underset{\text{Acid}}{NS^-} + \underset{\text{Base}}{NH_3}$$

$$\underset{\text{Acid}}{HSO_4^-} + \underset{\text{Base}}{CO_3^{2-}} \rightleftharpoons \underset{\text{Acid}}{HCO_3^-} + \underset{\text{Base}}{SO_4^{2-}}$$

(iii) For an acid to act as a proton donor, a base (proton acceptor) must be present to receive proton. For example, acetic acid or HCl acts as an acid in water but not in benzene because benzene does not accept proton.

$$\underset{\text{Acid}}{CH_3COOH} + \underset{\text{Base}}{H_2O} \rightleftharpoons \underset{\text{Acid}}{H_3O^+} + \underset{\text{Base}}{CH_3COO^-}$$

$$\underset{\text{Acid}}{HCl} + \underset{\text{Base}}{H_2O} \rightleftharpoons \underset{\text{Acid}}{H_3O^+} + \underset{\text{Base}}{Cl^-}$$

(iv) Whenever an acid and base reacts another pair of acid and base is formed, e.g.

$$\underset{\text{Acid}_1}{HCl} + \underset{\text{Base}_2}{NH_3} \rightleftharpoons \underset{\text{Acid}_2}{NH_4^+} + \underset{\text{Base}_1}{Cl^-}$$

$$\underset{\text{Acid}_1}{HNO_3} + \underset{\text{Base}_2}{F^-} \rightleftharpoons \underset{\text{Acid}_2}{NF} + \underset{\text{Base}_1}{NO_3^-}$$

(v) **Amphoteric compounds:** Certain molecules and ions may exhibit either acidic or basic behaviour under appropriate conditions. Such species are known as amphiprotic or amphoteric. For example, water may lose a proton to a base such as NH_3 or gain a proton from an acid such as HCl and so it is classified as amphiprotic.

$$\overset{\text{Acid}_1}{H_2O} + \overset{\text{Base}_2}{NH_3} \rightleftharpoons \overset{\text{Acid}_2}{NH_4^+} + \overset{\text{Base}_1}{OH^-}$$

$$\underset{\text{Acid}_1}{HCl} + \underset{\text{Base}_2}{N_2O} \rightleftharpoons \underset{\text{Acid}_2}{N_3O^+} + \underset{\text{Base}_1}{Cl^-}$$

Species like HSO_3^-, HS^-, HCO_3^-, NH_3 are also amphoteric in nature.

(vi) All Bronsted acids are Arrhenius acids but not all Bronsted bases are Arrhenius bases.

(vii) An acid–base reaction can take place even in the absence of a solvent.

9.6.1 Conjugate Acid Base Pairs

When an acid loses a proton, it forms what is called a conjugate base of the acid. For example

$$\text{Acid} + \text{Proton} \rightarrow \text{Conjugate base}$$

$$HCl \rightarrow H^+ + Cl^-$$

$$H_2SO_4 \rightarrow H^+ + HSO_4^-$$

$$CH_3COOH \rightarrow H^+ + CH_3COO^-$$

When a base accepts a proton, it forms what is called the conjugate acid of the base. For example

$$\text{Base} + \text{Protons} \rightarrow \text{Conjugate acid}$$

$$H_2O + H^+ \rightarrow H_3O^+$$

$$NH_3 + H^+ \rightarrow NH_4^+$$

$$S^{2-} + H^+ \rightarrow HS^-$$
$$CO_3 + H^+ \rightarrow HCO_3^-$$

Now, an acid does not form its conjugate base unless a second base (base$_2$) is present to accept a proton. When the second base (base$_2$) accepts the proton, it forms its conjugate acid, i.e., acid$_2$. For example

$$HCl + H_2O \rightleftharpoons H_3\overset{+}{O} + Cl^-$$

In the forward reaction, HCl donates a proton, thus behaves as an acid, whereas water accepts a proton, and thus behaves as a base. In the reverse reaction, $H_3\overset{+}{O}$ donates a proton to chloride ion and thus $H_3\overset{+}{O}$ is behaves as an acid while Cl^- behaves as a base. Such pairs of substances, which are formed from one another by gain or lose of a proton, are known as conjugated acid–base pairs. Thus, hydrochloric acid is a conjugate acid of chloride ion while water is a conjugate base of hydronium ion.

In general, each acid base reaction involves two pairs of conjugate acids and bases. These are labelled as 1 and 2 as shown below:

$$Acid_1 + Base_2 \rightleftharpoons Acid_2 + Base_1$$

Some examples in this regard are given below:

	Acid$_1$ + Base$_2$	⇌	Acid$_2$ + Base$_1$
(a)	$H_2O + NH_3$	⇌	$NH_4^+ + OH^-$
(b)	$HCl + NH_3$	⇌	$NH_4^+ + Cl^-$
(c)	$HNO_3 + H_2O$	⇌	$H_3O^+ + NO_3^-$
(d)	$HSO_4^- + H_2O$	⇌	$H_3O^+ + SO^{2}_{4}{}^-$
(e)	$NH_4^+ + CH_3COO^-$	⇌	$CH_3COOH + NH_3$
(f)	$H_2O + COO^{2}_{3}{}^-$	⇌	$HCO_3^- + OH^-$

9.6.2 Relative Strength of Acids and Bases

According to Bronsted–Lowry concept, the strength of an acid depends upon its tendency to lose protons while the strength of a base depends upon its tendency to gain protons.

(i) Let us consider the reaction of hydrochloric acid with water:

$$\underset{\text{Strong acid}}{HCl} + H_2O \rightleftharpoons H_3O^+ + \underset{\text{Weak conjugate base}}{Cl^-}$$

HCl, being a strong acid, is highly ionized and thus the equilibrium shifts towards the right-hand side. On the other hand, Cl^- (conjugated base of HCl) has a less tendency to accept a proton. Thus, a strong acid has a weak conjugate base:

Strong acid H^+ + Weak conjugate base

(ii) Now, let us consider the reaction of acetic acid with water:

$$\underset{\text{Weak acid}}{CH_3COOH} + H_2O \rightleftharpoons H_3O^+ + \underset{\text{Strong conjugate base}}{CH_3COO}$$

CH_3COOH, being a weak acid, is only slightly ionized and thus the equilibrium shifts towards the left-hand side. On the other hand, CH_3COO^- (conjugate acid of CH_3COOH) has a greater tendency to accept a proton. Thus, a weak acid has a strong conjugate base.

9.6.3 Limitation of Bronsted—Lowry Theory

The theory fails to explain acid–base reactions in which no proton transfer takes place, e.g., in the reaction given below:

$$CaO + SO_3 \rightarrow CaSO_4$$

involves the neutralization of basic oxide (CaO) with acidic oxide (SO_3) to form a salt ($CaSO_4$) without any transfer of proton between them. Similarly, reactions taking place in non-aqueous solvents cannot be explained by this theory, e.g.

$$SO_2 + SO_2(l) \rightleftharpoons SO^{2+} + SO_3^{2-}$$

9.7 LEWIS CONCEPT OF ACIDS AND BASES

An acid is a substance, which can accept a pair of electrons to form a coordinate bond, and a base is a substance, which can donate a pair of electrons to form a coordinate bond. Thus, the base is an electron pair donor and acid is an electron pair acceptor.

Examples of acid base reactions.

(i) *The molecule of a Lewis acid with a central atom having an incomplete octet, and a Lewis base with one or more unshared pairs of electrons,* e.g.

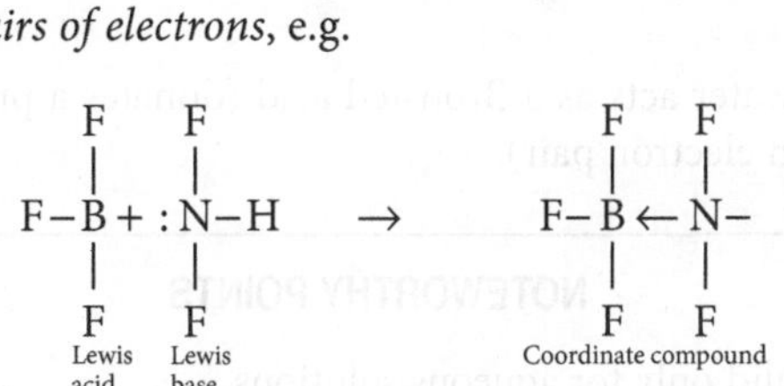

(ii) *The molecule of a Lewis acid containing multiple bonds, e.g.,* SO_3 has multiple bonds. It is not an Arrhenius or Bronsted acid. It is a Lewis acid and neutralizes bases.

$$\underset{\text{Lewis base}}{Ca^{2+}O^{2-}} + \underset{\text{Lewis acid}}{SO_3} \rightarrow [SO_4^{2-}]\,Ca^{2+}$$

(iii) *Simple cations like Al^{3+}, Ag^+, act as Lewis acids because they have empty orbitals in their outermost energy levels.*

$$\underset{\text{Lewis acid}}{Ag^+} + \underset{\text{Lewis acid}}{2NH_3} \rightarrow [H_3N:Ag:NH_3]^+$$

$$\underset{\text{Lewis acid}}{Al^{3+}} + \underset{\text{Lewis acid}}{6H_2O} \rightarrow \left[Al(H_2O)_6\right]^{3+}$$

(iv) ***The molecule of a Lewis acid having a central atom with empty d-orbitals:*** Sn in $SnCl_4$ has empty *d*-orbitals, which are used up to make $SnCl_4$ a Lewis acid, the base being a negative ion, Cl^-, F^-

$$\underset{\text{Lewis}}{SNCl_4} + \underset{\text{Lewis}}{2Cl^-} \rightarrow SnCl_6^{2-}$$

9.7.1 Limitations of Lewis Concept

(i) It is very general and includes all the coordination compounds and reactions. Protonic acids do not form coordinate bond with bases.
(ii) Acid–base reactions are quite fast. However, the reactions taking place forming coordinate compounds are very slow. Thus, they do not fit in a normal acid–base concept.
(iii) The catalytic action of acids in different reactions is due to H^+ ions. Lewis acids may not contain hydrogen, so, they cannot act as catalysts.
(iv) The essential requirement or the main feature of Lewis concept of acid is the formation of coordinate bond with bases. However, protonic acids (H_2SO_4, HCl) do not form coordinate bond with bases. This is against the tenets of Lewis theory.

Lewis bases are essentially Bronsted bases because all such substances, e.g., NH_3, CN^-, etc., are also capable of accepting a proton. However, Bronsted acids are not Lewis acids.

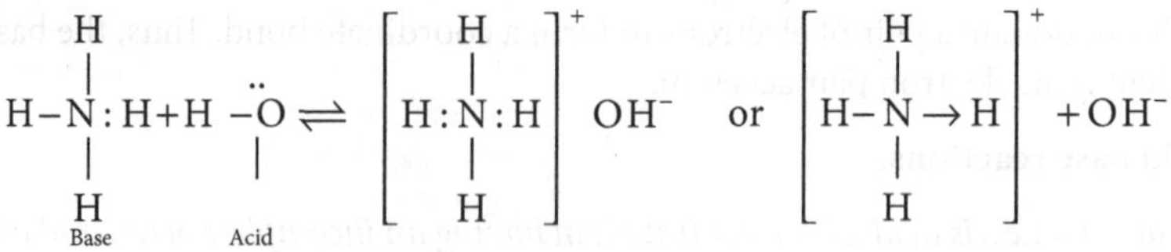

Water in the above reaction water acts as a Bronsted acid (donates a proton). However, water is not a Lewis acid (does not accept an electron pair).

NOTEWORTHY POINTS

1. Arrhenius theory is valid only for aqueous solutions.
2. In complex compounds, the metal atoms (or ions) act as Lewis acids while the ligands act as Lewis bases.
3. Acids like HCl, HBr, HI, HNO_3 and H_2SO_4 are strong acids, whereas H_2SO_3, H_2S, HCN, CH_3COOH, etc., are weak acids.
4. Many liquids are likely to undergo autoprotolysis as in the case of H_2O, e.g.,

$$NH_3 + NH_3 \rightleftharpoons NH_4^+ + NH_2^-$$

$$H_2SO_4 + H_2SO_4 \rightleftharpoons H_3SO_4^+ + HSO_4^-$$

$$CH_3OH + CH_3OH \rightleftharpoons CH_3OH_2^+ \rightleftharpoons CH_3O^-$$

(Continued)

NOTEWORTHY POINTS (*Continued*)

5. The strength of an acid depends upon the nature of a solvent.
   For example, HCl, HNO_3, H_2SO_4 and $HClO_4$ have some strength in water. It is due to levelling effect of water. With acetic acid as a solvent, the strength of above acids follows the order shown below:

   $$HClO_4 > H_2SO_4 > HCl > HNO_3$$

   Thus, acetic acid is called as a differentiating solvent.
6. Levelling effect of water is due to its high dielectric constant and strong proton-accepting tendency.
7. $HClO_4$ is the strongest while HCN is the weakest hydracids acids ever known. CsOH is the strongest base known.
8. Acetic acid behaves as a strong acid in ammonia and a base in anhydrous HF.
9. All Arrhenius acids are also Bronsted acids but not all Arrhenius bases are Bronsted bases.
10. A conjugate pair of an acid and a base differs by a proton only, i.e.

    $$\text{Conjugate acid} \rightleftharpoons \text{Conjugate base} + H^+$$

11. All Bronsted bases are also Lewis bases but all Bronsted acids are not Lewis acids.

9.8 IONIC PRODUCT OF WATER

Water is a very weak conductor of electricity. Therefore, it must ionize feebly, i.e., self-ionization must take place to a very small extent so that the following equilibrium is set up between water and its ions.

$$H_2O(l) + H_2O(l) \rightleftharpoons H_3O(aq) + OH^-$$

Applying the law of mass action, the equilibrium constant, K_c, for the above reaction can be written as

$$K_c = \frac{[H_3O^+][OH^-]}{[H_2O]^2}$$

Since the degree of ionization of water is very low, its ionization does not appreciably change the concentration of water molecules. Therefore, $[H_2O]^2$ may be regarded as a constant. Thus

$$\therefore \qquad K_c[H_2O]^2 = [H_3O^+][OH^-]$$

or $$K_c \times K = [H_3O^+][OH^-]$$

where K is another constant representing $[H_2O]^2$

or $$K_w = [H_3O^+][OH^-]$$

Where K_w is again a constant as it is the product of two constants K_c, and K.

The new constant K_w is called the dissociation constant or the ionic product of water. Its value is expressed in mol² L⁻². Because dissociation of water increases with rise in temperature, the value of K_w also increases with temperature.

Ionic product of water is generally taken as 1×10^{-14} which is the value at room temperature, 298 K. It is evident that in pure water the concentration of H_3O^+ and OH^- must be equal to one another. Thus, we have

$$[H_3O^+] = [OH^-]$$

But
$$[H_3O^+][OH^-] = 1\times10^{-14}$$

∴
$$[H_3O^+] = [OH^-] = \sqrt{1\times10^{-14}} = 1\times10^{-7}\ \text{mol L}^{-1}$$

9.8.1 Concentrations of H_3O^+ and OH^- ions in Aqueous Solutions of Acids and Bases

It has been mentioned in the above discussion that the ionization of water is an equilibrium process and the ionic product of water remains constant so long as temperature is kept constant. Thus, at a particular temperature, K_w has the same value regardless of the fact whether water is pure or some substance, such as an acid or a base, has been dissolved in it. For instance, if a strong acid, say HCl, is added to water, concentrations of H_3O^+ (aq) and OH^- (aq) will not remain equal as the concentration of H_3O^+ (aq) ions will become very high. Therefore, in accordance with Le Chatelier's principle, H_3O^+ (aq) ions will combine with OH^- (aq) ions so that K_w remains constant. Thus, the addition of an acid decreases the concentration of OH^- (aq) ions as given by the expression

$$[H_3O^+][OH^-] = K_w \quad \text{or} \quad [OH^-] = \frac{K_w}{[H_3O^+]}$$

Similarly, the addition of a base increases the concentration of OH^- (aq) ions and decreases the concentration of H_3O^+ (aq) ions as given by the expression

$$[H_3O^+] = \frac{K_w}{[OH^-]}$$

In both the cases, the self-ionization of water is suppressed due to the extra supply of H_3O^+ (aq) ions. The actual concentrations of H_3O^+ (aq) and OH^- (aq) ions in different, acidic or basic, solutions may be illustrated by considering a few examples.

Example 1 **What will be the concentrations of hydronium and hydroxyl ions in a 0.1 M solution of HCl at 298 K assuming HCl to be completely ionized?**

Solution:
$$HCl(aq) + H_2O(l) \rightarrow H_3O^+(aq) + Cl^-(aq)$$

Since HCl is completely ionized

$$[H_3O^+] = [HCl] = 0.1 = 1\times10^{-1}\ \text{mol L}^{-1}$$

But
$$[H_3O^+][OH^-] = K_w = 1\times10^{-14}$$

∴
$$[OH^-] = \frac{K_w}{[H_3O^+]} = \frac{1\times10^{-14}}{1\times10^{-1}} = 1\times10^{-13}\ \text{mol L}^{-1}$$

Example 2 **Calculate the concentrations of H_3O^+ and OH^- ions in a 0.0001 M solution of NaOH at 295 K, if NaOH is taken to be completely ionized.**

Solution:
$$NaOH(aq) + H_2O(l) \rightarrow Na^+(aq) + OH^-(aq)$$

As NaOH is completely ionized

$$[OH^-] = [NaOH] = 0.0001\,M = 1\times10^{-4}\,mol\,L^{-1}$$

Again $$[H_3O^+][OH^-] = K_w = 1\times10^{-14}$$

$\therefore$ $$[H_3O^+] = \frac{K_w}{[OH^-]} = \frac{1\times10^{-14}}{1\times10^{-4}} = 1\times10^{-10}\,mol\,L^{-1}$$

9.9 pH SCALE

As already pointed out, pure water has equal concentration of H^+ and OH^- ions, each having the value 1×10^{-7} moles/litre at room temperature. For this reason, water is regarded as neutral. However, if one of the ions is present in greater concentration than the other, the solution becomes acidic or basic. Thus, a solution having a concentration of hydrogen ions more than 1×10^{-7} moles/litre is said to be acidic. Similarly, a solution having a concentration of hydrogen ions less than 1×10^{-7} moles/litre is known as basic. Thus, the concept of ionic product of water enables us to classify a solution as acidic, neutral or basic by referring to the H^+ ion concentration or the OH^- ions concentration in them. However, it is customary to express acidity or basicity in terms of H^+ ion concentration only.

Sorensen (1909) suggested a logarithmic scale for expressing the hydrogen ion concentration. The scale has been found to be a very convenient form for expressing the hydrogen ion concentration and is known as the pH scale.

pH of a solution is defined as the negative logarithm of hydronium ion concentration

$$pH = -\log[H_3O^+] \quad \text{or} \quad pH = -\log\left[\frac{1}{H_3O^+}\right]$$

(i) **pH in acidic solutions:**
In an acidic solution, $[H_3O^+] > 10^{-7}$ e.g., 10^{-6}, 10^{-5}, etc.

$\therefore$ $$pH < 7$$

If pH lies between 0 and 7, the solution is acidic and the acidic character decreases as pH values changes from 0 to 7.

(ii) **pH in water or neutral solutions:**
In water or in a neutral solution $[H_3O^+] = 10^{-7}$ moles/litre

$$pH = -\log[H^+_{aq}]\ \log(10^{-7})$$
$$= +7\log 10 = 7\times1 = 7$$

(iii) **pH in basic solutions:**
In a basic solution

$$[H_3O^+] = 10^{-7}\ \text{moles/litre}$$

i.e., 10^{-8}, 10^{-9}, 10^{-10}, etc.

$\therefore$ $$pH > 7$$

If pH lies between 7 and 14, the solution is basic and the basic character increases as pH values change from 7 to 14. However, a variation pattern of $[H]^+$, pH and $[OH]^-$, pOH is shown in Table 9.2.

Table 9.2 Variation of $[H]^+$, pH, $[OH]^-$ and pOH.

$[H^+]$ M	pH		$[OH^-]$ M	pOH	
$1 \times 10^0 \cong 1$	0		1×10^{-14}	14	
1×10^{-1}	1	Acidic	1×10^{-13}	13	Acidic
1×10^{-2}	2		1×10^{-12}	12	
1×10^{-7}	7	Neutral	1×10^{-7}	7	Neutral
1×10^{-8}	8		1×10^{-6}	6	
1×10^{-10}	10	Basic	1×10^{-4}	4	Basic
1×10^{-14}	14		1×10^{-6}	0	

Thus, the acidity or basicity of a solution may be represented on a scale called the pH scale (Fig. 9.1).

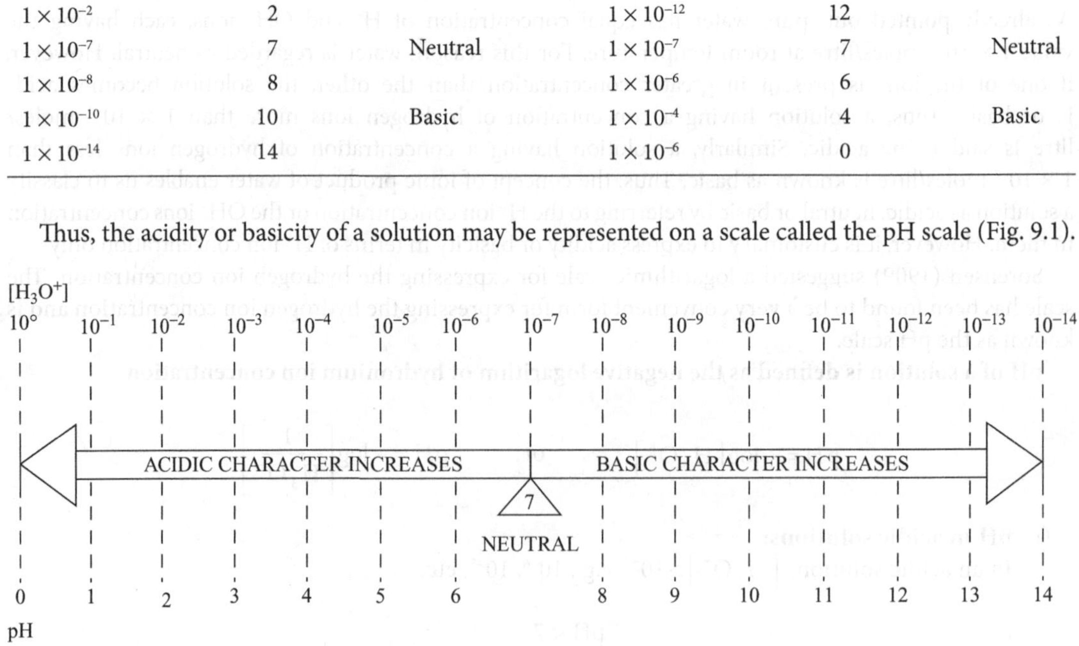

Figure 9.1 pH Scale

9.9.1 The pOH Scale

The pOH of a solution may be defined as the negative logarithm of the hydroxyl ion concentration.

Mathematically, $$pOH = -\log [OH^-]$$

Similarly, we may define pKw as the negative logarithm of the ionic product of water (K_w).
Now, we can write the ionic product equation for water in terms of pH, pOH and pK_w:

$$\therefore \quad [H^+]\,[OH^-] = K_w = 10^{-14}$$

Taking log on both sides

$$\log [H^+] + \log [OH^-] = \log K_w = -14 \log 10$$

or $$-\log[H^+] - \log[OH^-] = -\log K_w = 14$$

or $$pH + pOH = pK_w = 14$$

The variation of pH and pOH value along with H^+ and OH^- concentration has been summarized in Table 9.2. The actual calculation of pH value is illustrated with the help of the following examples.

NOTE WORTHY POINTS

1. A solution having pH = 0 is acidic.
2. pH of a solution decreases on heating as the dissociation of water is an endothermic process.
3. pH values greater than 14 are possible for concentrated strong base and negative pH values are possible for concentrated strong acids, but it is for dilute solutions that pH scale is most useful.
4. pH of solutions of acids, bases and salts of same concentration varies from one another. It depends on the extent to which a particular substance ionises in a solution.
5. pH of solutions can be calculated by pH meters (accurately) or by an electromotive force method or by a universal indicator (generally).
6. pH of boiling water is 6.5625, although it is neutral.
7. pH value is zero for 1 N solution of a strong acid. In case of 2 N, 3 N, 4 N, etc., solutions' pH values are negative.
8. The relationship between pH and acidity is inverse in nature. Thus, as pH goes up acidity comes down.
9. The relationship between pH and acidity is also a logarithmic one. For example, decrease of pH by one unit, corresponds to 10-fold increase in acidity. Similarly, when the pH goes down by two units, increase in the corresponding acidity is by a factor of 100.
10. K_w values do not change on the addition of a salt, acid or base at a particular temperature.
11. Degree of ionisation of pure water is 1.8×10^{-9} or $1.8 \times 10^{-7}\%$ at 25 °C which indicates that it is a very weak electrolyte.

Numerical Problems (On pH and pOH)

Example 3 **Calculate the pH and pOH values of the following solutions.**
(i) 0.01 M HCl (ii) 0.001 M NaOH
(iii) A solution containing 0.315 g of nitric acid in 500 cm³ of the solutions.

Solution:

(i) Since HCl is completely ionised

$$[H^+] = [HCl]$$

Hence, $$[H^+] = 0.01 = 1 \times 10^{-2}\ M$$

∴ $$pH = -\log[H^+] = -\log[1 \times 10^{-2}] = 2$$

$$pH + pOH = 14$$

$$2 + pOH = 14$$

$$\therefore \quad pOH = \mathbf{12}$$

(ii) Because NaOH is completely ionised

$$\therefore \quad [OH^-] = [NaOH]$$

or $$[OH^-] = 0.001\ M = 1 \times 10^{-3}\ M$$

Hence, $$[H^+] = \frac{K_w}{[OH^-]} = \frac{1\times10^{-14}}{1\times10^{-3}}$$

$$\left[\because K_w = [H][OH^-] = 1\times10^{-11}\ M\right]$$

$$pH = -\log [H^+] = -\log [1 \times 10^{-11}] = 11$$

$$pOH = -\log [OH^-] = -\log (1 \times 10^{-3}) = \mathbf{3}$$

(iii) Strength of HNO_3 solution

$$= 0.315\ g/500\ ml = 0.63\ g/litre$$

$$= \frac{0.63}{63} = 0.01\ M\ (1 \text{ mole of } HNO_3 = 63\ g)$$

$$\therefore \quad [H^+] = 0.01\ M = 1 \times 10^{-2}\ M$$

Hence, $$pH = -\log [H^+] = -\log (1 \times 10^{-2}) = 2$$

and $$pOH = 14 - pH = 14 - 2 = \mathbf{12}$$

Example 4 **Assuming calcium hydroxide is completely ionized, calculate the pH and pOH values of a 0.005 M solution of it.**

Solution: $$Ca(OH)_2 \rightarrow Ca^{2+} + 2OH^-$$

Since $Ca(OH)_2$ is completely ionized, one mole of $Ca(OH)_2$ furnishes 2 moles of OH^- ions.

$$[OH^- = 2[Ca(OH)_2]$$

$$= 2 \times 0.005\ M = 0.01\ M = 1 \times 10^{-2}\ M$$

$$pOH = -\log (1 \times 10^{-2}) = 2$$

$$[H^+] = \frac{K_w}{[OH^-]} = \frac{1\times10^{-14}}{1\times10^{-2}} = 10^{-12}$$

$$pH = -\log (1 \times 10^{-12}) = \mathbf{12}$$

Example 5 **What is the $[H^+]$ of a solution with a pH of 9.6?**

Solution: $$pH = -\log [H^+] = 9.6$$

$$\log [H^+] = -pH = -9.6 = 10.4$$

$$[H^+] = \mathbf{2.51 \quad 10^{-10}\ M}$$

Example 6 **How many grams of NaOH must be dissolved in 1 L of the solution to give it a pH value of 12?**

Solution: By definition

$$[H^+] = 10^{-pH}$$

$$[H^+] = 10^{-12}\ M$$

Moreover, $[H^+][OH^-]\ K_w$

However, $K_w = 1 \times 10^{-14}$

Substituting the values in Eq. (9.1)

$$10^{-12} \times [OH^-] = 1 \times 10^{-4}$$

$$[OH^-] = 1 \times 10^{-2}\ M$$

Since NaOH is a strong electrolyte, it undergoes complete ionization:

$$NaOH \rightarrow Na^+ + OH^-$$

$\therefore$ $[OH^-] = [NaOH] = 1 \times 10^{-2}\ M$

$\therefore$ Normality of NaOH $= 1 \times 10^{-2}$ [NaOH is a monoacid base]

Eq. wt. of NaOH $= 40.$

Strength of NaOH $=$ Eq. wt. $\times$ normality

$= 40 \times 10^{-2} = \mathbf{0.40\ g/L^{-1}}$.

Example 7 **The value of K_w is 9.55 10^-14^ at a certain temperature. Calculate the pH of water.**

Solution: Here, we are given that

$$K_w = 9.55 \times 10^{-14}$$

Now, as for water $[H^+] = [OH^-]$

$\therefore$ $K_w = [H^+] \times [OH^-] = [H^+][H^+] = [H^+]^2$

i.e., $[H^+]^2 = 9.55 \times 10^{-14}$

$\therefore$ $[H^+] = \sqrt{9.55 \times 10^{-14}} = 3.09 \times 10^{7}\ M$

$$pH = -\log\,[H^+] = -\log\,[3.09 \times 10^{-7}]$$

$$= -\log 3.09 - \log 10^{-7}$$

$$= -[-7 + 0.4900] = \mathbf{6.51}$$

Example 8 **Calculate the $[H_2O]^+$ and $[OH^-]$ and pH of 0·02 M solution of hydrocyanic acid, HCN. The ionization constant of HCN at 298 K is 7.2 10^-10^.**

Solution:

$$HCH_{(aq)} + H_2O_{(l)} \rightarrow H_3O^+_{(aq)} + CN^-_{(aq)}$$

$$K_a = 7.2 \times 10^{-10}$$

If x mole of HCN is ionized, it will give x mole of H_3O^+ and x mole of CN^- $0.2 = x$ mole of HCN will remain unionized,

$$K_a = \frac{x \times x}{(0.2 - x)} = \frac{x^2}{(0.2 - x)} = 7.2 \times 10^{-2}$$

Assuming x to be very small

$$\frac{x^2}{0.2} = 7.2 \times 10^{-2} \quad \text{or} \quad x = 1.2 \times 10^{-2}$$

$\therefore$ $$[H_3O]^+ = (CN^-) = 1.2 \times 10^{-5} \text{ mol L}^{-1}$$

$$[OH^-] = \frac{1.0 \times 10^{-14}}{1.2 \times 10^{-5}} = 8.3 \times 10^{-10} \text{ mol L}$$

$$pH = -\log(1.2 \times 10^{-5}) = \mathbf{4.92}$$

Example 9 **At what molar concentration, the pH of nitrous acid would be 2.0? (K_a of nitrous acid = 4.5 10^{-4})**

Solution: Let the molar concentration of NO_2 be M/x

$$HNO_2 \rightarrow H^+ + NO_2^-$$

$$K_a = \frac{[H^+][NO_2^-]}{[HNO_2]} = \frac{[H^+]^2}{[HNO_2]}$$

$$pH = 2.0$$

$\therefore$ $$[H^+] = 10^{-2} \text{ M}$$

$$4.5 \times 10^{-4} = \frac{(10^{-2})^2}{[HNO_2]}$$

$$[HNO_2] = \frac{10^{-4}}{4.5 \times 10^{-4}} = \frac{1}{4.5} \text{ M} = \mathbf{0.222 \text{ M}}$$

Thus, molar concentration of HNO_2 solution having pH 2.0 is **0.222 M.**

Example 10 **Calculate the pH of a solution made mixing 50 mL of 0.01 M $Ba(OH)_2$ solution with 50 mL of water.**

Solution: $M_1V_1 = M_2V_2$

$\therefore$ $$50 \text{ mL} \times 0.01 \text{M} = 100 \text{ mL} \times \text{M}$$

$$M_2 = 0.005 \text{ M}$$

or concentration of $Ba(OH)_2$ after mixing with water = 0.005 M

$$[OH^-] = 2[Ba(OH)_2] = 2 \times 0.005 = 0.01 \text{ M}$$

$$K_w = [H^+][OH^-] = 10^{-14}$$

$$[H^+] = \frac{10^{-14}}{[OH^-]} = \frac{10^{-14}}{10^{-1}} = \mathbf{10^{-13} \text{ M}}$$

A salt of some weak acid (CH_3COOH) with a weak base (NH_4OH) such as ammonium acetate ($H_3C\ COO\ NH_4$) may be sometime used as a buffer.

$$H_3C-COO\ NH_4 \rightleftharpoons H_3C\ COO^- + NH_4^+$$

The H^+ ions from any source combine with $H_3C\ COO^-$ ions to form weakly ionised acetic acid. Similary the OH^- ions are retained by NH_4^+ ions to form a weak base NH_4OH. But a common practice is to use mixture of two substances.

Example 11 Calculate the pH value of the solution obtained by mixing 50 mL of N–HCl and 30 mL N–NaOH solution assuming both to be completely dissociated in normal solution.

Solution: Total volume of solution after mixing = 50 + 30 = 80 mL

$$50 \text{ mL N–HCl} = 50 \text{ meq. HCl}$$

$$30 \text{ mLN–NaOH} = 30 \text{ meq. NaOH}$$

∴ HCl neutralized by NaOH = 30 meq
HCl left behind 50 −30 = 20 meq in 80 mL solution.

$$[HCl] = \frac{20}{80} \times 1000 \text{ meq L}^{-1}$$

$$= \frac{20}{80} \times \frac{1000}{1000} \text{g eq. L}^{-1} = 0.25 \text{ g eq. L}^{-1}$$

$$\therefore \quad [H^+] = 0.25\text{g eq. L}^{-1}$$

$$pH = -\log[H^+] = -\log[0.25] = \mathbf{0.602}$$

9.10 BUFFER SOLUTION

It is sometimes necessary that a solution of definite pH value is required to be prepared and stored. The preveration of such a solution is even more difficult than its preparation. If the solution comes in contact with air, it will absorb CO_2 and becomes more acidic because $CO_2 + H_2O \rightleftharpoons H_2CO_3$ (a weak acid). If the solution is stored in a glass bottle, the alkaline impurities present in glass may dissolve into the solution and may alter its pH. A buffer solution is defined as a solution which resists any change in its pH value i.e it helps in maintaining the pH value practically consistant even when small amounts of the acid or the base are added to it. Then those solutions which have reserve acidity and alkalinity are called buffer solutions.

Characterestics of a Buffer: It has the following charateristics:

(i) It possesses a definite pH value
(ii) Its pH value is not changed on keeping or on dilutions
(iii) Its pH value is not appreciably changed on the additions of small quantity of either an acid or a base to it.

Types of buffer solutions: Buffers are of two types:

(i) **Acidic buffer:** This is obtained by mixing equimolecular quantities of a weak acid and its salt with a strong base, e.g.,

$$CH_3COOH + CH_3COONa$$

(ii) **Basic buffer:** It is obtained by mixing equimolecular amounts of a weak base and its salt with a strong acid, e.g., $NH_4OH + NH_4CI$

9.10.1 Buffer Action of Buffer Solutions

The mechanism by which a buffer solution resists the change in pH value on adding a strong acid or a strong base is known as buffer action.

A. Buffer action of an acid buffer: Let us consider a solution of CH_3COOH and CH_3COONa in water. The species present in it are:

CH_3COOH and H_2O (l) are nearly unionized, while CH_3COONa is highly ionized.

$$CH_3COONa \rightarrow CH_3COO^- + Na^+$$

(highly ionized)

(i) **Addition of acid:** When a strong acid, say HCl, is added, CH_3COOH formed is nearly unionized. Therefore, there is no increase in H^+ ion concentration:

$$\underset{\text{(of buffer)}}{CH_3COO^-} + \underset{\text{(from added acid)}}{H^+} \rightarrow \underset{\text{(unionized)}}{CH_3COOH}$$

(ii) **Addition of base:** When a strong base, say NaOH, is added, then H_2O formed is nearly unionized. Therefore, there is no increase in the OH^+ ion concentration:

$$\underset{\text{(of buffer)}}{CH_3COOH} + \underset{\text{(from added base)}}{OH^-} \rightarrow CH_3COO^- + \underset{\text{(unionized)}}{H_2O(l)}$$

Thus, in both cases, pH remains almost the same. It is clear from the above mechanism that reserve acidity is due to the presence of CH_3COOH and reserve alkalinity is due to the presence of CH_3-COO^- ions.

B. Buffer action of a basic buffer: Let us consider a solution of NH_4OH and NH_4Cl. The species present in it are:

NH_4OH (aq) and water are nearly unionized while NH_4Cl is highly ionized.

$$NH_4Cl \rightarrow NH_4^+ + Cl^- \text{ (highly ionized).}$$

(i) **Addition of acid:** When a small amount of a strong acid, say HCl, is added, then H+ ions are consumed by the excess of NH_4OH to form H_2O, which is nearly unionized.

$$\underset{\text{(from added acid)}}{H^+_{(aq)}} + \underset{\text{(of buffer)}}{NH_4OH_{(aq)}} = NH_4^+ + \underset{\text{(unionized)}}{H_2O(l)}$$

Therefore, there is no increase in the H^+ ion concentration.

(ii) **Addition of Base:** When a small amount of a strong base is added, then OH^- ions are consumed by NH^+ ions to form nearly unionized NH_4OH.

$$NH_4Cl \rightarrow NH_4^+ + Cl^-$$

$$\underset{\text{(from butter)}}{NH_4} + \underset{\text{(from added base)}}{OH^-} \longrightarrow \underset{\text{unionized}}{NH_4OH}$$

Therefore, there is no increase in the OH^- concentration. Thus, pH in both the cases remains the same.

9.10.2 Applications of Buffer Solutions

(i) Buffer solutions are important in biological and chemical processes. For instance, the pH of blood is 7.4, which is maintained in the body by some buffer mechanism.
(ii) Buffers are also used in electroplating, dyes and photographic material.
(iii) They are also used in analytical chemistry. In volumetric analysis, e.g., Erio-chrome Black T (EBT) works as an indicator in the presence of a buffer.

MEMORY FOCUS

1. Those solutions, which have reserve acidity and alkalinity are called buffer solutions.
2. A good buffer has the following characteristics:
   (i) They possess a definite pH value.
   (ii) The pH value remains steady on keeping or on dilution.
   (iii) The pH value is not appreciably changed on the addition of suitable quantity of either an acid or a base to the buffer solution.
3. An acidic buffer is a mixture of weak acid and a salt of it with a strong base, e.g., mixture of acetic acid and sodium acetate. The pH of this type of buffer solution is always less than 7.
4. A basic buffer is a mixture of weak base and a salt of it with a strong acid, e.g., mixture of ammonium hydroxide and ammonium chloride. The pH of this type of buffer solution is always greater than 7.
5. **Buffer action of acid buffer**
   Let us take the following example:

$$H_3CCOOH \rightleftharpoons H_3CCOO- + H^+$$

$$H_3CCOONa \rightarrow H_3CCOO^- + Na^+$$

Due to common ion effect of H_3CCOO^- ion, the dissociation of the weak acetic acid is further suppressed. The small amount of H^+ ions from any source will combine with H_3CCOO^- ion to form H_3CCOOH whose dissociation has already been suppressed.

$$H_3CCOO^- + H^+ \rightarrow H_3CCOOH$$

The small amount of OH^- from any source will react with H_3CCOOH as under

$$H_3CCOOH + OH^- \rightarrow H_3C - COO^- + H_2O$$

(Continued)

MEMORY FOCUS (*Continued*)

Thus, the pH value remains constant.

$$NH_4OH \rightleftharpoons NH_4^+ + OH^-$$

$$NH_4Cl \rightleftharpoons NH_4^+ + Cl^-$$

Due to common ion effect of NH_4^+ ions, the dissociation of weak base NH_4OH is further suppressed.

The OH^- ions from any source will combine with NH_4^+ ions to form NH_4OH whose dissociation is already suppressed.

The H^+ ions from any source will react with NH_4OH as shown below:

$$NH_4OH + H^+ \rightarrow NH_4^+ + H_2O$$

The pH value of the solution remains constant.

7. Buffer solution contains a very small actual concentration of H^+ or OH^- ions, but they possess a large reserve of opposite ions. Hence, they are known as solutions with **reserve acidity and reserve alkalinity.**
8. The capacity of a buffer solution to resist a change in its pH value is known as buffer capacity.

$$\text{Buffer capacity} = \frac{\text{Number of moles of acid or base added to 1 litre}}{\text{Change in pH}}$$

REVISION QUESTIONS

1. Give salient features of Arrhenius theory of ionizations. How does this theory explain
   (a) Electrolysis (b) Ionic reactions (c) Colligative properties (d) Heats of neutralization.
2. (a) State Ostwald's dilution law. How it is experimentally verified?
   (b) Calculate the pH value of a solution whose hydrogen ion concentration is 0.006 gm ion/litre.
3. State and explain Ostwald's Dilution law. Can this be applied to the dissociation of HCl in aqueous solution?
4. What is Ostward's dilution law? How it is verified? What are its limitations?
5. Define and explain ionic product of water. Discuss the factors controlling the degree of ionization of an electrolyte.
6. The dissociation constant of a weak monobasic acid is 4×10^{-10}. Calculate the hydrogen ion concentration of 0.1 N solution of this acid.
7. Explain the following terms:
   (a) Arrhenius acids (b) Arrhenius bases
   (c) Bronsted acids (d) Bronsted bases
   (e) Lewis acids (f) Lewis bases
8. Compare the Lewis theory of acids and bases with the Bronsted–Lowry concept.
9. State and explain Lowry–Bronsted theory and Lewis theory of acids and bases. In what way, Lewis acid differs from Bronsted acid?
10. What are acids and bases? Explain the different theories of acids and bases.

11. (a) Discuss the Bronsted–Lowry concept and Lewis concept of acids and bases.
    (b) Write three methods for determining the strength of acids.
12. (a) Define Lewis acids and Lewis bases. Give one example of each.
    (b) What do you understand by conjugate acids and bases in the following reactions:

$$\text{(i) } NH_3 + H_2O \rightleftharpoons NH_4^+ + OH^-$$

$$\text{(ii) } HNO_3 + H_2O \rightleftharpoons H_3O^+ + NO_3^-$$

    (c) Calculate the pH of a solution having hydrogen ion concentration 2.5×10^{-3} M.
13. Calculate the pH of the following solution:
    (a) 0.0002 M HCl
    (b) 0.1 M NH_4OH $(K_b = 1.8 \times 10^{-5})$
14. Calculate the pH of a solutions made by mixing 50 ml of 0.01 M $Ba(OH)_2$ solution with 50 ml of water.

MULTIPLE CHOICE QUESTIONS

1. An acid is a compound that gives H^+ ions in water and a base is a compound that gives OH^- ions in water. This concept was given by
   (a) Arrhenius (c) Bronsted
   (b) Lewis (d) Lowry
2. According to Bronsted-Lowry concept, an acid is a substance that
   (a) accepts a proton
   (b) releases a proton
   (c) accepts an electron pair
   (d) releases an electron pair
3. Which one of the following is the limitation of Arrhenius concept?
   (a) free H^+ and OH^- ions do not exist in water
   (b) this concept is limited to water only
   (c) some bases do not contain OH^- ions
   (d) all of these
4. When calcium oxide is dissolved in water, the following reaction takes place

$$O^{2-} + H_2O \rightarrow 2OH^-$$

   The Bronsted acid is
   (a) O^{2-}
   (b) H_2O
   (c) OH^-
   (d) none of these
5. A substance accepts a proton. According to Bronsted–Lowry concept it is
   (a) an acid (c) a neutral substance
   (b) a base (d) amphoteric
6. When calcium oxide is dissolved in water, the following reaction takes place

$$O^{2-} + HCl \rightarrow H_3O^+ + Cl^-$$

   the species that acts as a Bronsted base is
   (a) H_2O (c) H_3O^+
   (b) HCl (d) Cl^-
7. A weak base has ______ conjugate acid and a weak acid has a ______ conjugate base
   (a) strong, strong (c) strong, weak
   (b) weak, strong (d) weak, weak
8. In the following hypothetical reaction

$$HA + B^- \rightleftharpoons HB + A^-$$

   The conjugate base of acid HA is
   (a) B^- (c) A^-
   (b) HA (d) none of these
9. In the reaction between NH_3 and HCl

$$HCl + NH_3 \rightleftharpoons NH_4^+ + Cl^-$$

   the conjugate acid of NH_3 is
   (a) HCl
   (b) NH_4^+

(c) Cl^-
(d) none of these

10. The strength of a Bronsted acid depends upon its tendency to __________ a proton
(a) gain (c) react with
(b) donate (d) none of these

11. Molecules or ions that can behave both as Bronsted acid and base called
(a) monoprotic acids
(b) polyprotic acids
(c) amphiprotic substances
(d) polyprotic base

12. HCO_3^- is an example of
(a) conjugate acid (c) amphiprotic ion
(b) conjugate base (d) amphoteric

13. According to Lewis concept, an acid is __________ acceptor
(a) proton (c) electron
(b) base (d) electron pair

14. All cations and molecules that are short of an electron pair act as ________
(a) Lewis acids (c) Bronsted acid
(b) Lewis bases (d) Bronsted bases

15. According to Lewis concept, a reaction between an acid and a base is the transfer of
(a) a proton from acid to a base
(b) OH^- ion from bases to acid
(c) electron pair from acid to base
(d) electron pair from base to acid

16. The strength of an acid depends upon
(a) the concentration of H^+ ions in aqueous solution
(b) the concentration of electrons pairs in aqueous solution
(c) the tendency to release the electron pair
(d) the tendency to gain the electron pair

17. The autoionization of water can be represented by
(a) $H_2O \rightleftharpoons H^+ + OH^-$
(b) $H_2O + H^+ \rightleftharpoons H_3O^+ + OH^-$
(c) $H_2O + H_2O \rightleftharpoons H_3O^+ + OH^-$
(d) $H_2O + HA \rightleftharpoons H_3O^+ + OH^-$

18. For a weak acid with α as its degree of dissociation, the value of dissociation constant is given by (C is concentration of acid in moles per litre)
(a) $K_a = C\alpha$ (c) $K_a = C^2\alpha$
(b) $K_a = C\alpha^2$ (d) $K_a\ C^2\alpha^2$

19. The dissociation constant K_b for a base BOH is given by the expression
(a) $K_b = \dfrac{[B^+][H^+]}{[BOH]}$
(b) $K_b = \dfrac{[B^+][OH^-]}{[BOH]}$
(c) $K_b = \dfrac{[B^+][OH^-]^2}{[BOH]}$
(d) $K_b = \dfrac{[B^+]^2[OH^-]}{[BOH]}$

20. Which is correct about pH?
(a) $pH = -\log[H^+]$ (c) $pH = 10^{-pH}$
(b) $pH = \log\dfrac{1}{H^+}$ (d) all of these

21. The concept of pH was introduced by
(a) Arrhenius
(b) Bronsted
(c) Lewis
(d) Sorensen

22. The pH of 0.001 M HCl is
(a) 0.001 (c) 3
(b) 10^{-3} (d) –3

ANSWERS

1. (a)	5. (b)	9. (b)	13. (d)	17. (c)	21. (d)
2. (b)	6. (b)	10. (b)	14. (a)	18. (b)	22. (c)
3. (d)	7. (a)	11. (c)	15. (d)	19. (b)	
4. (b)	8. (c)	12. (c)	16. (b)	20. (d)	

Distribution Law 10

CHAPTER OBJECTIVES

10.1 Introduction
10.2 Conditions for the Validity of the Distribution Law
10.3 Effect of Temperature on Distribution Coefficient
10.4 Thermodynamic Derivation of Distribution Law
10.5 Distribution Law and Molecular State of Solute
10.6 Applications of Distribution Law

The process of extraction is more efficient if the solvent is used in a number of small portions than in one whole lot. Thus, multiple extractions are preferred over simple extraction.

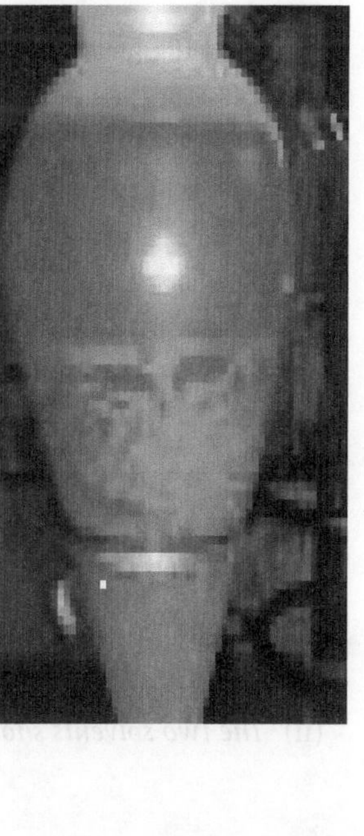

10.1 INTRODUCTION

It is a matter of common knowledge that when a solute (solid or liquid) is added to a mixture of two immiscible liquids, in both of which it is soluble; it distributes itself between the two liquids in such a way that the ratio of their concentration in the two liquids is always constant at a constant temperature. An example of this type of behaviour is furnished by shaking of iodine with a mixture of two immiscible liquids; for example, carbon tetrachloride and water. Berthelot and Jungfleish, in 1872, found that when solutions of iodine in tetrachloride having different concentrations were shaken with distilled water, the iodine distributed itself between the two solvents in such a way that the ratio of its concentrations in the two layers is constant at constant temperature. Mathematically, it may be expressed as

$$\frac{\text{Concentration of iodine in carbon disulphide } (C_1)}{\text{Concentration of iodine in water } (C_2)} = K$$

where K is a constant, which is called distribution coefficient or partition coefficient. However, soon Nernst (1891), based on a large number of experiments, pointed out that the ratio C_1/C_2 is constant only when the solute has the same molecular conditions, i.e., the same molecular mass in both the phases. On this basis, Nernst generalized his observations in the form of a law that is known as **Nernst's distribution law** or **partition law** (Fig. 10.1).

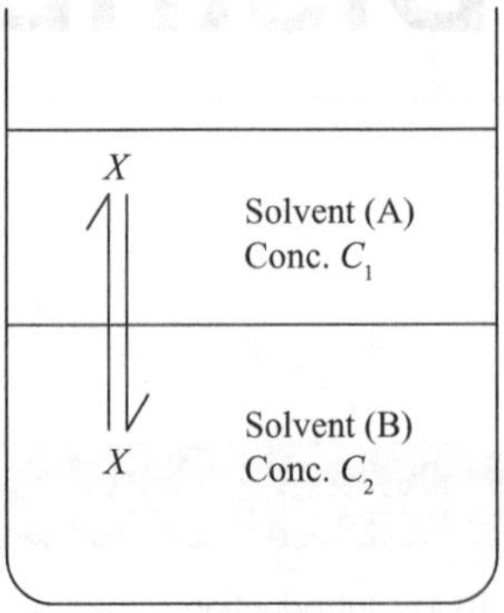

Figure 10.1 Distribution of a solute between two immiscible solvents

According to this law:

> *When a solute (solid or liquid) distributes itself between two immiscible solvents in contact with each other, there exists, for similar molecular species, at a given temperature, a constant ratio of distribution between the two solvents irrespective of any other molecular species present.*

Thus, if C_1 and C_2 are the concentrations of the solute in the similar molecular state in the two immiscible solvents, then

$$\frac{\text{Concentration of the solute in solvent A}}{\text{Concentration of the solute in solvent B}} = \frac{C_1}{C_2} = K \qquad (10.1)$$

where K is a constant that is known as **distribution coefficient or the partition coefficient.**

Some results of the distribution of oxalic acid distributed between ether and water and the distribution of iodine distributed between carbon disulphide and water (at 25 °C) are given in Table 10.1.

Table 10.1 Illustration of distribution law with two examples.

Oxalic Acid Between Ether and Water at 18 °C			Iodine Between Carbon Disulphide and Water at 25 °C		
Grams of Acid per 10 ml of H_2O (C_1)	Grams of Acid per 10 ml of Ether (C_2)	$K = C_1/C_2$	Grams of Iodine per 10 ml of CS_2 (C_1)	Grams of I_2 per 10 ml of H_2O (C_2)	$K = C_1/C_2$
0.473	0.053	9.2	1.74	0.0041	420
0.436	0.046	9.5	1.29	0.0032	400
0.304	0.031	9.8	0.66	0.0016	410
0.203	0.0205	9.9	0.41	0.0010	410

10.2 CONDITIONS FOR THE VALIDITY OF THE DISTRIBUTION LAW

Various conditions for the validity of distribution law are:

(i) *The temperature should remain constant.*
(ii) *The two solvents should be mutually immiscible or very sparingly soluble in each other.*

(iii) *The solutions are dilute, i.e., the amount of solute added should be small. Large concentration of added solute sometimes changes the mutual solubility of the two solvents.*
(iv) *The solute must have the same molecular state in both the solvents.*
(v) *The solute should not react with the solvent.*

10.3 EFFECT OF TEMPERATURE ON DISTRIBUTION COEFFICIENT

As already mentioned, the value of distribution or partition coefficient (K) is constant only at a constant temperature. Because, a change in temperature generally changes the solubility of the solute in two different solvents to different extents; therefore, the distribution coefficient is expected to vary with a variation in temperature. *The variation in the value at distribution coefficient per degree centigrade rise in temperature* is known as **temperature coefficient**.

Generally, the temperature coefficient is very small. In case of succinic acid distributed between water and ether, the temperature coefficient is 0.055 per degree, while for acetic acid distributed between water and ether, it is 0.0064 per degree.

10.4 THERMODYNAMIC DERIVATION OF DISTRIBUTION LAW

10.4.1 Principle

> *It is based upon a thermodynamic principle that if a substance is present in two phases in contact with each other, then, at equilibrium, its* ***chemical potential*** *must be the same in both the phases.*

Let us suppose that a solute X is present in two immiscible solvents A and B, which are in contact with each other. Let μ_A be the chemical potential of the solute X in solvent A and μ_B be its chemical potential in the solvent B.

We know from thermodynamics that the chemical potential (μ) of a substance in a solution is given by

$$\mu = \mu^\circ + \text{RT} \ln a \tag{10.2}$$

where μ° is the standard chemical potential and a is the activity of the solute in the solution.

Therefore, the chemical potential of the solute (X) in solvent A is given by

$$\mu_A = \mu^\circ_A + \text{RT} \ln a_A \tag{10.3}$$

and the chemical potential of the solute (X) in solvent B is given by

$$\mu_B = \mu^\circ_B + \text{RT} \ln a_B \tag{10.4}$$

When the two phases are in equilibrium, these chemical potentials will be equal to each other,

i.e., $$\mu_A = \mu_B$$

∴ $$\mu^\circ_A + \text{RT} \ln a_A = \mu^\circ_B + \text{RT} \ln a_B$$

or $$\text{RT} \ln a_A - \text{RT} \ln a_B = \mu^\circ_B - \mu^\circ_A$$

or $$\text{RT} \ln \frac{a_A}{a_B} = \mu^\circ_B - \mu^\circ_A$$

or $$\ln \frac{a_A}{a_B} = \frac{\mu^\circ_B - \mu^\circ_A}{\text{RT}}$$

At constant temperature, μ°_A and μ°_B are also constant for a given solute in the particular solvents. Since R is also constant (being the gas constant), it follows that at constant temperature:

$$\ln \frac{a_A}{a_B} = \text{constant}$$

or
$$\frac{a_A}{a_B} = \text{constant} \quad (10.5)$$

Since the solutions are dilute, they behave ideally; hence, Henry's law, according to which activity is proportional to mole fraction, can be applied to each phase.

$$\therefore \quad \frac{a_A}{a_B} = \frac{k_A x_A}{k_B x_B} \quad (10.6)$$

where x_A and x_B are the mole fractions of the solute in the two phases and k_A and k_B are the Henry's law constants, for the solute in the two phases.

Therefore, from Eqs (10.5) and (10.6), we have

$$\frac{k_A x_A}{k_B x_B} = \text{constant}$$

or
$$\frac{x_A}{x_B} = \text{constant} \quad \text{[At constant temperature]}$$

If the solutions are dilute, the mole fraction may be replaced by concentration. By doing so, we get

$$\frac{C_A}{C_B} = \text{constant} \quad \text{[At constant temperature]}$$

This is distribution law.

Numerical Problems

Example 1 **The concentration of iodine between water and carbon tetrachloride is as follows:**

Water layer	**0.000321**	**0.000503**	**0.00764**	**0.00115**
CCl_4 layer	**0.002745**	**0.0429**	**0.06541**	**0.0101**

Show that these results illustrate distribution law.

Solution: I. $$K = \frac{C_{water}}{C_{CO_4}} = \frac{0.000321}{0.002745} = 0.0117$$

II. $$K = \frac{C_{water}}{C_{CO_4}} = \frac{0.000503}{0.0429} = 0.0117$$

III. $$K = \frac{C_{water}}{C_{CO_4}} = \frac{0.00764}{0.06541} = 0.00117$$

IV. $$K = \frac{C_{water}}{C_{CO_4}} = \frac{0.00115}{0.0101} = 0.0114.$$

As the value of K is constant in all the cases, hence, the given data illustrates the distribution law.

Example 2 **In the distribution of succinic acid between ether and water at 15 °C, 20 ml of the ethereal layer contains 0.092 g of the acid. Find out the weight of the acid present in 50 ml of the aqueous solution in equilibrium with it if the distribution coefficient for succinic acid between water and ether is 5.2.**

Solution: Let the weight of succinic acid in aqueous layer be x g.

Concentration in aqueous layer, $C_{\text{water}} = \dfrac{x}{50}$ g/ml.

Concentration in ethereal layer, $C_{\text{ether}} = \dfrac{0.092}{20}$ g/ml.

We are given:

$$\frac{C_{\text{water}}}{C_{\text{ether}}} = 5.1$$

or

$$\frac{\dfrac{x}{50}}{\dfrac{0.092}{20}} = 5.1$$

or

$$x = 1.196 \text{ g}$$

Therefore, the weight of acid present in 50 ml of aqueous solution = **1.196 g.**

10.5 DISTRIBUTION LAW AND MOLECULAR STATE OF SOLUTE

Nernst pointed out that the distribution law is applicable only if the solution remains in the same molecular state in both the phases (solvents). If, however, the solute undergoes some association or dissociation or enters into chemical combination with one of the solvents, then the normal distribution law is not valid and it needs some modification. We shall now discuss modification of the distribution law in different cases.

10.5.1 Case I: When the Solute Undergoes Association in One of the Solvents

Suppose that X represents the molecular formula of the solute; let it remains in the normal molecular state in Phase I (Fig. 10.2), in which its concentration is C_1. Suppose that n molecules of it (solute) associate in Phase II to form a complex molecule $(X)_n$. The associated molecules will exist in equilibrium with single molecules in Phase II as follows:

$$(X)_n \rightleftharpoons nX \qquad (10.7)$$

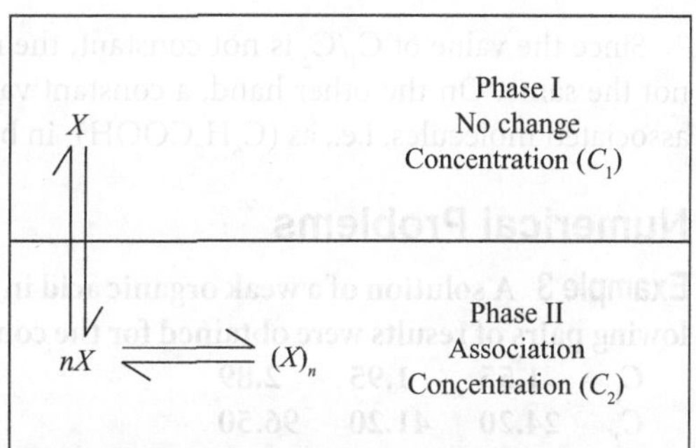

Figure 10.2 Association in one of the phase

Let C_2 be the total concentration of the solute X in Phase II.

Applying law of chemical equilibrium to the above equilibrium, we have

$$K = \frac{[X]^n}{[(X)_n]} \qquad (10.8)$$

or $$[X] = \sqrt[n]{K \times [(X)_n]} \tag{10.9}$$

$$= \text{constant} \times \sqrt[n]{[(X)_n]} \tag{10.10}$$

If the solute is assumed to exist mostly in associated form (which is generally true in dilute solution), the concentration of the associated molecules $[(X)_n]$ may be taken to be equal to C_2, i.e.,

From Eqs (10.10) and (10.11), we get

$$[X] = \text{constant} \times \sqrt[n]{C_2} \tag{10.11}$$

Applying the distribution law for the similar molecular species in both the phases, we have

$$\frac{C_1}{[X]} = \text{constant} \tag{10.12}$$

Substituting the value of [X] in Eq. (10.12), we have

$$\frac{C_1}{\text{constant} \times \sqrt[n]{C_2}} = \text{constant}$$

or $$\frac{C_1}{\sqrt[n]{C_2}} = \text{constant} \tag{10.13}$$

This relationship has been confirmed by studying the distribution of benzoic acid between water and benzene. The following results were obtained (See Table 10.2).

Table 10.2 Distribution of benzoic acid between water and benzene.

Grams of Benzoic Acid per 10 ml of Water (C_1)	Grams of Benzoic Acid per 100 ml of Water (C_2)	$K = \frac{C_1}{C_2}$	$K = \frac{C_1}{\sqrt{C_2}}$
0.0150	0.242	0.062	0.0305
0.0195	0.412	0.048	0.0304
0.0289	0.970	0.030	0.0293

Since the value of C_1/C_2 is not constant, the molecular state of benzoic acid in water and benzene is not the same. On the other hand, a constant value of $C_1/\sqrt{C_2}$ proves that benzoic acid exists mainly as associated molecules, i.e., as $(C_6H_5COOH)_2$ in benzene, but in normal state in water.

Numerical Problems

Example 3 **A solution of a weak organic acid in benzene was shaken with water. At equilibrium, the following pairs of results were obtained for the concentration (C_1) in water layer and (C_2) in benzene layer.**

C_1	**1.55**	**1.95**	**2.89**
C_2	**24.20**	**41.20**	**96.50**

What conclusion can be drawn from these results concerning the molecular condition of the organic acid in benzene?

Solution: The weak organic acid, in benzene, may be present either as normal or in the form of associated molecules. In the former case, C_1/C_2 should be constant while in the latter case, the ratio $\frac{C_1}{n\sqrt{C_2}}$ should be constant, giving the number of molecules of the acid which associate to give an associated molecule.

The value of C_1/C_2 for various sets are:

Set I $$\frac{C_1}{C_2} = \frac{1.50}{24.20} = 0.0619$$

Set II $$\frac{C_1}{C_2} = \frac{1.95}{41.20} = 0.0473$$

Set III $$\frac{C_1}{C_2} = \frac{2.89}{96.50} = 0.0299$$

Since the value of C_1/C_2 is not constant, the acid does not exist as single molecule in benzene layer.

Let us find out $C_1/\sqrt{C_2}$ in various sets.

Set I $$\frac{C_1}{\sqrt{C_2}} = \frac{1.50}{\sqrt{24.20}} = 0.305$$

Set II $$\frac{C_1}{\sqrt{C_2}} = \frac{1.95}{\sqrt{41.20}} = 0.304$$

Set III $$\frac{C_1}{\sqrt{C_2}} = \frac{2.89}{\sqrt{96.50}} = 0.303$$

A fairly constant value of $C_1/\sqrt{C_2}$ shows that the weak organic acid exists as a double molecule in benzene layer.

Example 4 **Succinic acid has its normal molecular weight in water. In the distribution of the acid between water and benzene, the concentration in the aqueous and the benzene layers was found to be C_1 and C_2, respectively.**

C_1 (g/100 ml)	**0.150**	**0.195**	**0.289**
C_2 (g/100 ml)	**2.420**	**4.120**	**9.700**

Calculate the molecular weight of succinic acid in benzene.

Solution: In the previous numerical, we used hit-and-trial method to determine the value of n. In this method, let us try logarithmic method for calculating the value of n.

According to logarithmic method,

$$n = \frac{\log C_2 - \log C_2'}{\log C_1 - \log C_1'}$$

Taking Sets I and II, we get

$$n = \frac{\log 4.12 - \log 2.42}{\log 0.195 - \log 0.150}$$

$$= \frac{0.6138 - 0.3820}{0.2900 - 0.1761} = \frac{0.2318}{0.1139} = 2.03 \approx 2$$

Taking Sets II and III, we get

$$n = \frac{\log 9.70 - \log 4.12}{\log 0.0289 - \log 0.195}$$

$$= \frac{0.9868 - 0.6138}{0.4609 - 0.2900} = \frac{0.3730}{0.1709} = 2.17 \approx 2$$

Thus, the value of $n = 2$. This leads to the conclusion that succinic acid exists as dimer in benzene. Hence, the molecular mass of succinic acid in benzene = 2 × 118 = **236**.

10.5.2 Case II: When the Solute Undergoes Dissociation in One of the Solvents

As before, suppose the solute X does not dissociate in Phase I, but dissociates into Y and Z in Phase II (Fig. 10.3). Let C_1 be the concentration of the solute in Phase I and C_2 be the total concentration of the solute in Phase II. Now, the following equilibria exist in the system.

(i) $$\underset{\text{Phase I}}{X} \rightleftharpoons \underset{\text{Phase II}}{X}$$

(ii) $$\underset{\text{Phase II}}{X} \rightleftharpoons \underset{\text{Phase II}}{Y} + \underset{\text{Phase II}}{Z}$$

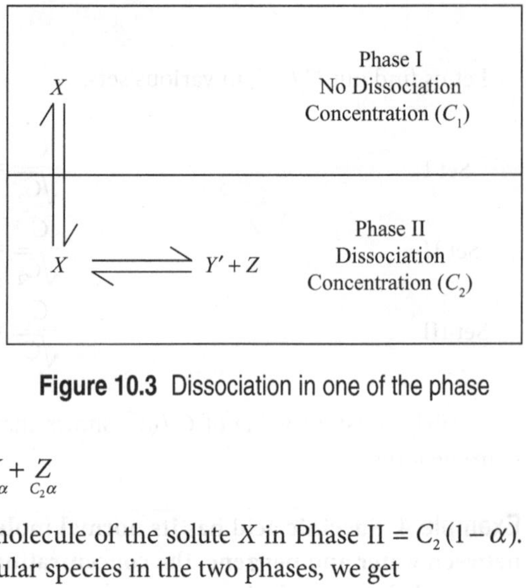

Figure 10.3 Dissociation in one of the phase

Suppose, α is the degree of dissociation of the solute X in Phase II.

$$\underset{C_2(1-\alpha)}{X} \rightleftharpoons \underset{C_2\alpha}{Y} + \underset{C_2\alpha}{Z}$$

As is evident, the concentration of the undissociated molecule of the solute X in Phase II = $C_2(1-\alpha)$. Then, by applying the distribution law to similar molecular species in the two phases, we get

$$\frac{C_1}{C_2(1-\alpha)} = \text{K} \tag{10.14}$$

In this case, the solute undergoes the same degree of dissociation in Phase I and not in Phase II. The relation will be reduced to

$$\frac{C_1(1-\alpha)}{C_2} = \text{K}$$

If solute undergoes dissolution, its both tile phase with the same degree of dissociation then the above relationship takes the form

$$\frac{C_1(1-\alpha)}{C_2(1-\alpha)} = K \tag{10.15}$$

10.5.3 Case III: When the Solute Enters into Chemical Combination with One of the Solvents

Suppose X represents the molecular formula of the solute. Let it remain as such in Phase I, but enters into chemical combination with solvent in Phase II (Fig. 10.4) according to the following equation:

$$X + nS \rightleftharpoons Xns \tag{10.16}$$

Let C_1 be the concentration of solute X in Phase I and C_2 be its total concentration in Phase II.

If α is the fraction of solute that enters into chemical combination with the solvent, then the concentration of solute that remains uncombined $X = C_2(1-\alpha)$ and the concentration of complex molecules Xns formed $= C_2\alpha$.

By applying law of chemical equilibrium to the Eq. (10.16), we get

$$K = \frac{C_2\alpha}{C_2(1-\alpha)[\text{solvent}]^n}$$

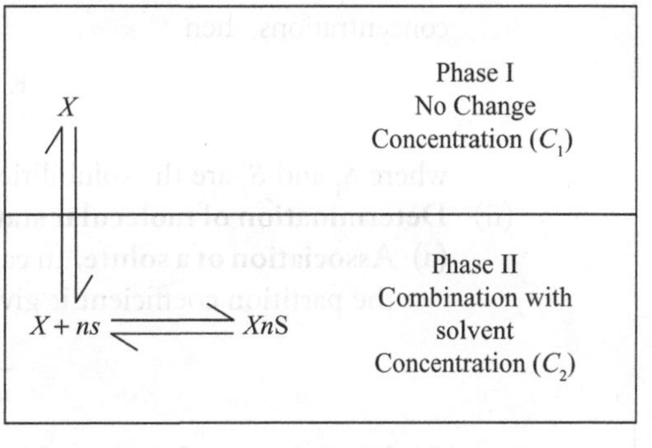

Figure 10.4 Combination with the solvent

However, the solvent is taken in large quantity; therefore, its concentration may be taken as constant.

$$\therefore \qquad \frac{C_2(\alpha)}{C_2(1-\alpha)} = \text{constant} \tag{10.17}$$

Applying the distribution law to similar molecular species in both the solvents, we have

$$\frac{C_1}{C_2(1-\alpha)} = \text{constant} \tag{10.18}$$

(since $C_2(1-\alpha)$ is the concentration of solute that remains uncombined in Phase II).

Dividing Eq. (10.15) by (10.14), we have

$$\frac{C_1}{C_2\alpha} = \text{constant} \tag{10.19}$$

As α, the fraction of the solute that combines with the same solvent, is also constant at a given temperature, therefore, Eq. (10.19) may be rewritten as

$$\frac{C_1}{C_2} = \text{constant} \tag{10.20}$$

Thus, we conclude that when solute enters into chemical combination with one of the solvents, there is no change in the fundamental equation of the distribution law except that the numerical value of partition constant is changed.

MEMORY FOCUS

1. The main applications of distribution law are:
   (i) **Determination of solubility of a solute in a solvent.**
   When a large quantity of the solute is shaken with two immiscible solvents in contact with each other, it is assumed that the solute forms saturated solution in both the solvents. In such a case, the concentration of the solute present in a particular solvent may be taken to be equal to their solubilities. If distribution law is supposed to be applicable at such concentrations, then
   $$K = \frac{C_1}{C_2} = \frac{S_1}{S_2}$$
   where S_1 and S_2 are the solubilities of the solute in solvents 1 and 2, respectively.
   (ii) **Determination of molecular state of solute in different solvents**
   (a) Association of a solute. In case a solute undergoes association in one of the solvents, the partition coefficient is given by
   $$\frac{C_1}{n\sqrt{C_2}} = K$$
   (b) Dissociation of a solute. If the solute undergoes dissociation in one of the solvents, then
   $$\frac{C_1}{C_2(1-\alpha)} = K$$
   (iii) Study of complex ions.
   (iv) Another important application of the distribution law is in the extraction of substance from mixtures.
   (v) The efficiency of extraction increases by increasing the number of extractions using only a small amount of the extracting solvent every time.
   (vi) It is not possible to extract the whole of the dissolved substance, however large the number of extractions may be.
   (vii) Multi-step extraction is more efficient than a single-step extraction.
   (viii) A modern technique of separating a mixture of small amounts of organic material called chromatography is based upon the distribution law.

10.6 APPLICATIONS OF DISTRIBUTION LAW

The distribution law has been quite helpful in solving many physicochemical problems of theoretical and practical natures. Some of them are discussed in the following sections.

10.6.1 Determination of Solubility of a Solute in a Solvent

When a large quantity of the solute is shaken with two immiscible solvents in contact with each other, it is assumed that the solute forms saturated solution in both the solvents. In such a case, the concentration of the solute present in a particular solvent may be taken to be equal to their solubilities. If the distribution law is supposed to be applicable at such concentrations, then

$$K = \frac{C_1}{C_2} = \frac{S_1}{S_2}$$

where S_1 and S_2 are the solubilities of the solute in solvents 1 and 2, respectively. Thus, if the value of partition coefficient and solubility of the solute in one of the solvents is known, then the other can easily be found.

10.6.2 Determination of Molecular State of Solute in Different Solvents

(a) Association of a solute: In case a solute undergoes association in one of the solvents, the partition coefficient is given by,

$$\frac{C_1}{n\sqrt{C_2}} = K$$

where C_1 and C_2 are the concentrations of the solute in a solvent in which it does not undergo association and C_2 is its concentration in the solvent in which it undergoes association. From here, the value of n can be calculated. By this method, we have been able to show that when acetic acid is distributed between benzene and water, the value of n is equal to 2, i.e., acetic acid exists as dimer in benzene.

(b) Dissociation of a solute: As already shown, if the solute undergoes dissociation in one of the solvents, then

$$\frac{C_1}{C_2(1-\alpha)} = K$$

where C_1 and C_2 are the concentrations of the solute in solvent 1 (in which it remains as such) and solvent 2 (in which it dissociates) and α is the degree of dissociation of the solute in second solvent. Knowing C_1, C_2 and K, α the degree of dissociation can easily be calculated.

10.6.3 Determination of Distribution Indicators

When iodine is added in water and CS_2 layers, it distributes considerably in CS_2 layer or in other words the partition coefficient of iodine is very high in favour of CS_2. As a result, considerable amount of I_2 is passed into CS_2 layer and it acquires a deep violet colour. This solution can be titrated against hypo solution until the violet colour in CS_2 layer disappears.

10.6.4 Study of Complex Ions

An important application of the distribution law is found in the study of complex actions of various types (See Fig. 10.5). For example,

$$I^- + I_2 \rightleftharpoons I_3^-$$

$$Br^- + Br_2 \rightleftharpoons Br_3^-$$

$$Cu^{2+} + 4NH_3 \rightleftharpoons \left[Cu(NH_3)_4\right]^{2+}$$

This is due to the fact that by distribution experiment, we can find the actual concentrations of the various reaction species at an equilibrium without disturbing the equilibrium. From the values, thus, the equilibrium constant along with the formula of the complex can be determined. Let us illustrate it by taking up the first equilibrium reaction.

$$I^- + I_2 \rightleftharpoons I_3^-$$

A study of this equilibrium involves the following steps.

Step (i) A solution of iodine in carbon disulphide is shaken with water to determine the value of partition coefficient. Let this be K.

Step (ii) A known amount of iodine is dissolved in a known volume of carbon disulphide and the solution so obtained is shaken with a potassium iodide solution of known concentration.

Step (iii) In the end, the concentration of iodine left in carbon disulphide solution is determined volumetrically.

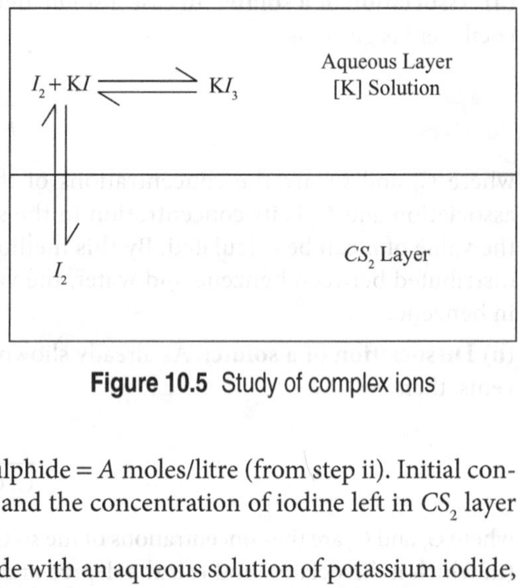

Figure 10.5 Study of complex ions

Let the initial concentration of iodine in carbon disulphide = *A* moles/litre (from step ii). Initial concentration of KI solution = *B* moles/litre (from step ii) and the concentration of iodine left in CS_2 layer after distribution = C moles/litre (from step iii).

After shaking a solution of iodine in carbon disulphide with an aqueous solution of potassium iodide, the concentration of iodine in aqueous layer must be very high.

Let us calculate the total concentration of iodine in aqueous layer. Total concentration of iodine (free + combined) in aqueous layer

= Initial concentration of iodine in CS_2 – Concentration of iodine left after distribution in CS_2 layer

= $(A - C)$ moles/litre

According to the distribution law,

$$= \frac{\text{Concentration of free iodine in aqueous layer}}{\text{Concentration of iodine in } CS_3 \text{ layer}} = K$$

∴ Concentration of free iodine in aqueous layer

$= \mathrm{K} \times$ Concentration of iodine in CS_2 layer

$= \mathrm{K} \times \mathrm{C}$ moles/litre

∴ Amount of iodine that has reacted with KI

= Total amount of iodine in aqueous layer – amount of free iodine in aqueous layer

$= (A - C) - \mathrm{KC}$ moles/litre

$= D$ moles/litre (say)

Let us now suppose that complex ion formed is I_3^-, then the following equilibrium should exist in the aqueous solutions:

$$I^- + I_2 \rightleftharpoons I_3^-$$

According to this equation

Number of moles I_2 combined = Number of moles of KI combined

= Number of moles of KI_3 obtained

Since, the number of moles of I_2 that has combined with KI = D

∴ Concentration of I_3^- (KI_3) in aqueous layer = D moles/litre (10.21)

Concentration of I^- (KI) in aqueous layer $(B - D)$ moles litre (10.22)

Concentration of I_2 in aqueous layer = KC moles/litre (10.23)

By applying law of chemical equilibrium to Eq. (10.18), we have

$$\mathrm{K} = \frac{[I_3^-]}{[I_2][I^-]}$$

By substituting the various values of $[I_3^-]$ and $[I_2]$ from Eqs (10.21), (10.22) and (10.23), calculate the value of K.

Calculate the value of K by starting with different initial concentration of iodine and potassium iodine solutions. It will be observed that the value of K is reasonably constant. This proves that the formula of complex ion is I_3^-.

10.6.5 In the Process of Extraction

Another important application of the distribution law is in the extraction of substance from mixture in the laboratory as well as in industry. In the laboratory, the process is often employed for the removal of dissolved substances from aqueous solution, with solvent such as benzene, chloroform and carbon tetrachloride while in industry, the process is frequently used to remove the impurities, from the end product. The method is also used in extracting organic material from plants and other natural resources.

It is based up on the principle that when an aqueous solution of the substance is shaken with an organic solvent, which is immiscible with water, the substance (present in aqueous layer) passes into the organic solvent because its partition coefficient is highly in favour of organic solvent. The organic solvent is, then, distilled to separate out the substance.

For extraction to be more efficient, the following two conditions must be observed.

(i) The value of partition coefficient (k) of the substance between water (or any other solvent in which the substance is present) and the extracting solvent should be highly in favour of the extracting solvent.

(ii) The given extracting solvent should not be used in one instalment but in maximum number of instalments. This fact can be proved as follows.

Suppose one litre of an aqueous solution contains x grams of an organic substance and that one litre of ether is available for extraction. Let the distribution coefficient of the substance between ether and water be 2.

Now, if whole quantity of the ether is used in one lot and x grams is the amount of the substance that passes (or get extracted) into ethereal layer, then

$$\frac{C_{\text{ether}}}{C_{\text{water}}} = 2$$

or

$$\frac{\frac{x}{1000}}{\frac{X - x}{1000}} = 2$$

or

$$x = \frac{2}{3}X = 0.67\,X$$

Thus, 66.7% of the total solute present in aqueous layer will be extracted. If the same amount of ether is used in two instalments of 500 ml each.

First instalment: After shaking with first instalment of 500 ml, let x_1 grams of the substance passes (or extracted) into the ether layer. Then, according to the distribution law,

$$\frac{\frac{x_1}{500}}{\frac{(X - x_1)}{1000}} = 2$$

or

$$x_1 = \frac{X}{2} = 0.5\,X$$

Thus, 50% of substance is taken out in the first operation.

Second instalment: Now, the amount of the substance which remains to be extracted in the aqueous layer $= X - X/2 = X/2$ grams. Now, one litre of the aqueous solution containing $X/2$ grams of the substance is shaken with the remaining 500 ml of ether. Let x_2 grams of the substance passes (or get extracted) into the ether layer, then, according to the distribution law,

$$\frac{\frac{x_2}{500}}{\frac{\left(\frac{X}{2} - x_2\right)}{1000}} = 2$$

or

$$x_2 = \frac{1}{4}X = 0.25\,X$$

Therefore, 25% of the solute comes out in the second operation. Thus, the total amount of the solute that gets extracted if the extracting solvent is used in two instalments of 500 ml each = 50 + 25 = 75%.

In other words, the total amount of the solute that is extracted increases from 66.7% to 75% if the same one litre of ether is used in two instalments instead of one. Similarly, it can easily be shown that if extraction is carried out in four instalments, the amount of the substance extracted will be 93.75%. Therefore, in general, it can be said that the process of extraction will be more efficient if it is carried out in a number of instalments rather than in one instalment.

A general formula can be derived which enables the calculation of the amount of the substance that is left unextracted after a given number of operations.

Let V ml of the solution containing W grams of the substance is repeatedly extracted with v ml of the extracting solvent. Let W_1 g be the weight of the substance that remains unextracted after first operation. Then, according to the distribution law,

$$\frac{\frac{W_1}{V}}{W - \frac{W_1}{v}} = \text{K}$$

or

$$\frac{W_1}{V} = \frac{(W - W_1)}{v}\text{K}$$

$$W_1 v = (W - W_1)\text{K}V$$

or

$$W_1(v + \text{K}V) = \text{K}VW$$

or

$$W_1 = \frac{\text{K}V}{(v + \text{K}V)}W \tag{10.24}$$

After second operation, let W_2 grams is the amount of the substance which remains unextracted. Consequently, the amount of the substance extracted will be given by $(W_1 - W_2)$ grams. Therefore, according to the distribution law:

$$\frac{\frac{W_2}{V}}{\frac{(W_1 - W_2)}{v}} = \text{K}$$

or

$$W_2 = \frac{\text{K}V}{(v + \text{K}V)}W_1 \tag{10.25}$$

Substituting the value of W_1 from Eq. (10.24) to Eq. (10.25), we get

$$W_2 = \left[\frac{\text{K}V}{v + \text{K}V}\right]\left[\frac{\text{K}V}{v + \text{K}V}\right]W$$

or

$$W_2 = \left[\frac{\text{K}V}{v + \text{K}V}\right]^2 W$$

In general, the amount of substance that remains unextracted at the end of n operations, W_n, is given by

$$W_n = \left[\frac{\mathrm{K}V}{v+\mathrm{K}V}\right]^n W \tag{10.26}$$

For the extraction to be more efficient, W_n should be as small as possible. This is possible only if the factor $\left[\frac{\mathrm{K}V}{v+\mathrm{K}V}\right]^n$ is as small as possible.

In $\left[\frac{\mathrm{K}V}{v}+\mathrm{K}V\right]^n$, denominator is greater than the numerator. Therefore, this factor will be smallest possible only if n is as large as possible. However, $n \times v$ is equal to the total volume of the extracting liquid available, which is constant. Thus, n can be made large by decreasing the value of v. In other words, the efficiency of extraction increases by increasing the number of extractions (n) using only a small amount of the extracting solvent (v) every time.

10.6.6 Application of Principle of Extraction to Desilverization of Lead

The principle of extraction studied above is highly useful in desilverization of lead by Parke's process. The given argentoferous lead is melted and heated to 800 °C; zinc, which is immiscible with lead, is then added and the system thoroughly mixed. Silver, which was originally present in lead, will pass into zinc layer because its partition coefficient between zinc and lead is 300 at 8,000, i.e., it is 300 times more soluble in zinc than in lead. Zinc layer, which contains more of silver now, is separated. By repeating this process three or four times, almost entire amount of silver can be extracted from lead.

Numerical Problems

Example 5 12 g of organic substance A is present in 100 ml of its aqueous solution. How much of it would be left behind after extracting the solution with two successive applications of 50 ml ether. The distribution coefficient of organic substance A between water and ether is 2 in favour of ether.

Solution: We are given:

Total volume of the aqueous solution (V) = 100 ml
Volume of the extracting solvent in each application (v) = 50 ml
Total weight of the organic substance A present (W) = 12 g
Number of extractions (n) = 2
Distribution coefficient (K) = 1/2
(The value of K is taken as 1/2 because in the derivation of the above relationship, K was taken to be in favour of aqueous solution.)

We know that
Amount of substance left unextracted, $W_n = \left[\frac{\mathrm{K}V}{v+\mathrm{K}V}\right]^n W$

Substituting the various values in the above relation, we get

$$W_n = \left[\frac{\frac{1}{2}\times 100}{50+\frac{1}{2}\times 100}\right]^2 \times 12 = \left(\frac{1}{2}\right)^2 \times 12 = 3$$

Therefore, the amount of organic substance A left behind after extraction = **3 g**

Example 6 **An aqueous solution contains 10 g of solute per litre. When one litre of the solution is treated with 100 ml of ether, 6 g of the solute is extracted. How much more of the solute would be extracted from the aqueous solution by further 100 ml ether? (Assume that the molecular state of the solute is the same in ether and water.)**

Solution: Total amount of solute in aqueous solution = 10 g

Amount of solute extracted in ether layer after first extraction = 6 g

Volume of ether taken for extraction = 100 ml

∴ Concentration of solute in ether layer = 6/100

Amount of solute left in aqueous layer = 10 – 6 = 4 g

Volume of aqueous solution = 1,000 ml

∴ Concentration of solute in aqueous layer = 4/1,000

According is the distribution law, distribution coefficient (K) is given by

$$K = \frac{C_{ether}}{C_{water}} = \frac{\frac{6}{100}}{\frac{4}{100}} = \frac{6}{100} \times \frac{100}{4} = 15.0$$

Suppose x grams of the solute can be extracted with further 100 ml of ether. In such a case,

concentration of solute in ether layer (C_{ether}) = $\dfrac{x}{100}$

concentration of solute in aqueous layer (C_{water}) = $\dfrac{4-x}{1000}$

(4 g of the solute was left in aqueous solution after first extraction).

According to the distribution law,

$$K = \frac{C_{ether}}{C_{water}} = \frac{\frac{x}{100}}{\frac{4-x}{1000}} \quad \text{or} \quad 15 = \frac{x}{100} \times \frac{1000}{4-x}$$

$$15(4-x) = 10x \quad \text{or} \quad 60 - 15x = 10x$$

$$25x = 60 \quad \text{or} \quad x = \frac{60}{25} = \mathbf{2.4\ g}$$

10.6.7 Determination of Degree of Hydrolysis

Distribution law can easily be applied for the determination of degree of hydrolysis of salts in aqueous solution, which will be discussed fully in further chapters.

MEMORY FOCUS

(1) Nernst's distribution law states that "if a solute x distributes itself between two immiscible solvents A and B at constant temperature and x is in the same molecular condition in both solvents."

$$\frac{\text{Concentration of } X \text{ in } A}{\text{Concentration of } X \text{ in } B} = K_D$$ where K_D is called distribution coefficient or partition coefficient.

(2) Validity conditions of distribution law are:
(i) The temperature is kept constant throughout.
(ii) The molecular state of the solute is the same in the two solvents.
(iii) The concentrations of the solute are noted after the equilibrium has been established.
(iv) The law does not hold when the concentrations are high.
(v) The two solvents are non-miscible or only slightly soluble in each other.

(3) Modification in the distribution law:

(i) When association occurs in one solvent, the distribution equation is modified as $\frac{C_1}{\sqrt[n]{C_2}} = K_D$.

(ii) The modified distribution law equation when there is dissociation in one of the solvents.
$$\frac{C_1}{C_2(1-x)} = K_D.$$

(iii) When solute enters into chemical combination with one of the solvents, there is no change in the fundamental equation of distribution law except that the numerical value of partition constant is changed.

REVISION QUESTIONS

1. Fill in the blanks with appropriate words

a. The essential condition for the distribution law to be obeyed is that the solute must have .... in both the phases.
b. The variation in the value of distribution coefficient per degree centigrade rises in temperature is known as ....
c. Benzoic acid exists as .... in benzene.
d. For efficient extraction, the given extracting solvent should .... be used in one instalment.
e. Principle of extraction can be used for desilverization of ....

2. Write the serial number of correct answer

a. In case of the distribution law, the two solvents should be
(i) mutually immiscible
(ii) mutually miscible
(iii) mutually miscible at high temperature

b. The relationship $\frac{C_1}{n\sqrt{C_2}}$ is applicable in case the solute
   (i) undergoes dissociation (b) undergoes association
   (ii) enters into chemical combination with one of the solvents (d) none of these
c. The partition coefficient of iodine between carbon tetrachloride and water is 85 in favour of carbon tetra chloride. This means that the solubility of iodine in ether is
   (i) higher than in water
   (ii) lower than in water
   (iii) same as in water
d. In case a solute enters into chemical combination with one solvent:
   (i) The fundamental equation of the distribution law and numerical value of partition coefficient remains the same.
   (ii) The fundamental equation of the distribution law remains the same but the numerical value of partition coefficient changes.
   (iii) Both fundamental equation as well as the value of partition coefficient changes.
e. For efficient extraction, the extracting solvent should be used
   (i) in one instalment only
   (ii) in two instalments
   (iii) as many instalments as possible

3. What is distribution law? Explain.
4. State and explain distribution law. Mention the conditions under which it is applicable.
5. What is temperature coefficient? Discuss the effect of temperature on partition coefficient of a solute between two immiscible solvents.
6. Derive the distribution law thermodynamically. On what principle this derivation is based? Define.
7. Derive a relation for the distribution law in case the solute undergoes association in one of the solvents. With its help, how will you find out the extent of complexity of the solute in the solvent?
8. Discuss the distribution law in case the solute undergoes dissociation in one of the solvents.
9. Prove that there is no change in the fundamental equation of the distribution law except a change in the numerical value of partition coefficient in case the solute reacts will one of the solvents.
10. What are the various applications of the distribution law? Discuss.
11. How distribution law helps in determining:
    (a) The solubility of a solute in a solvent?
    (b) Molecular state of a solute in a solvent?
12. Discuss the application of distribution law in the study of complex ions.
13. Comment on the statement: "For efficient extractions, the given solvent should be used in as many instalments as possible."
14. Discuss the importance of distribution law in the process of extraction.
15. Principle of extraction can be applied to the desilverization of lead. Explain.

MULTIPLE CHOICE QUESTIONS

1. The Nernst's distribution law is also known as
   (a) Partition law (c) Raoult's law
   (b) Henry's law (d) Equilibrium law
2. The Nernst's distribution law does not hold good if
   (a) The temperature throughout the experiment is not constant
   (b) Concentration of the solute in two solvents are high
   (c) There is association or dissociation of the solute in one of the solvents
   (d) All of the above
3. For the study of distribution law, the two solvents should be
   (a) miscible
   (b) non-miscible
   (c) volatile
   (d) reacting with each other
4. The modified distribution law for the solute undergoing dissociation in one of the solvent is

   (a) $K_D = \frac{C_1}{\sqrt{C_2}}$ (c) $K_D = \frac{C_1}{C_2(\alpha - 1)}$

   (b) $K_D = \frac{C_1}{C_2(1-\alpha)}$ (d) $K_D = \frac{C_1}{C_2^2}$
5. At a constant temperature, the solubility of a gas in a liquid is proportional to the pressure of the gas above it. This law is
   (a) Nernst's distribution law
   (b) Henry's law
   (c) Le Chatelier's principle
   (d) Raoult's law
6. The Nernst's distribution law equation $C_1/C_2 = K_D$ applies when
   (a) The molecular state of the solute is the same in both the solvents
   (b) The molecular state of the solute is different in both the solutes
   (c) The molecular state of the solute may be the same or different in both the solutes
   (d) None of the above
7. The Henry's law equation is
   (a) $C = k \times P$ (c) $P = C \times k$
   (b) $k = C \times P$ (d) $C = k \times \sqrt{P}$
8. Benzoic acid when shaken with mixtures of benzene and water undergoes dismerization in benzene. The distribution law applicable here is

   (a) $\frac{C_1}{C_2} = K_D$ (c) $\frac{C_1}{\sqrt[2]{C_2}} = K_D$

   (b) $\frac{C_1}{\sqrt[3]{C_2}} = K_D$ (d) none of these
9. The greater the distribution ratio in favour of the organic solvent is, the ________ will be the amount extracted in any one operation
   (a) lesser (c) equal
   (b) greater (d) none of these
10. With the help of distribution law, we can
    (a) determine the association or dissociation of the solute in one of the solvents.
    (b) determine the solubility of solute in various solvents.
    (c) deduce the formula of complex ions.
    (d) all of the above.
11. Multiple extractions are ________ efficient when compared to single extraction
    (a) more (c) the same
    (b) less (d) none of these
12. Is it possible to extract the whole of the dissolved substance by using large number of extractions?
    (a) Yes (c) Sometimes
    (b) No (d) None of these
13. In liquid–liquid chromatography technique, the component with ________ distribution coefficient is extracted first
    (a) lower (c) intermediate
    (b) higher (d) none of these
14. The Nernst's distribution law $K_D = C_1/C_2$ is not applicable if the solute undergoes
    (a) association in one of the solvents
    (b) dissociation is one of the solvents

(c) association or dissociation in one of the solvents
(d) none of the above

15. In the first extraction, the amount of the substance left unextracted is given by the formula (K is distribution coefficient, V millilitre of the aqueous solution contain A gram of an organic substance)
(a) $x_1 = A\dfrac{KV}{KV+v}$
(b) $x_1 = A\dfrac{KV}{KV+v2}$
(c) $x_1 = A^2KV/(KV+v^2)$
(d) $x_1 = A\dfrac{k2V}{KV+v^2}$

16. When a bottle of soda water is opened, the partial pressure of CO_2 ___________
(a) decreases
(b) increases
(c) remains the same
(d) none of these

17. When succinic acid or oxalic acid is shaken with ether and water, it ___________ in water.
(a) associates to form dimer
(b) dissociates into ions
(c) remains the same
(d) none of these

18. When the solute undergoes association in one of the solvent, the Nernst's distribution law is modified is
(a) $C_1 = \sqrt{C_2} \times K_D$
(b) $\sqrt{C_2}\, x\, C_1 = K_D$
(c) $C_1 = K_D\, x\, C^2_{\ 2}$
(d) $C_1 = K_D \times C^3_{\ 2}$

19. The distribution law is applied in the
(a) Haber's process for the manufacture of NH_3
(b) Parke's process for the extraction of Ag
(c) Contact process for the manufacture of H_2SO_4
(d) None of the above

ANSWERS

1. (a)	5. (b)	9. (b)	13. (b)	17. (b)
2. (d)	6. (a)	10. (d)	14. (c)	18. (a)
3. (b)	7. (a)	11. (a)	15. (a)	19. (b)
4. (b)	8. (c)	12. (b)	16. (a)	

Electrochemistry 11

CHAPTER OBJECTIVES

- 11.1 Introduction
- 11.2 Electrolysis
- 11.3 Electrolytic Conduction
- 11.4 Kohlrausch Law
- 11.5 Migration of Ions
- 11.6 Transport Number
- 11.7 Limitations of Arrhenius Theory
- 11.8 Modern Theory of Strong Electrolytes

This branch finds applications in almost every field of science and engineering. Today, under the heading of 'Electrochemistry' one can describe batteries, fuel cells, techniques of chemical analysis, corrosion of metals, manufacture of materials and many other topics.

Electrochemistry is the study of the processes involved in the interconversion of electrical energy and chemical energy.

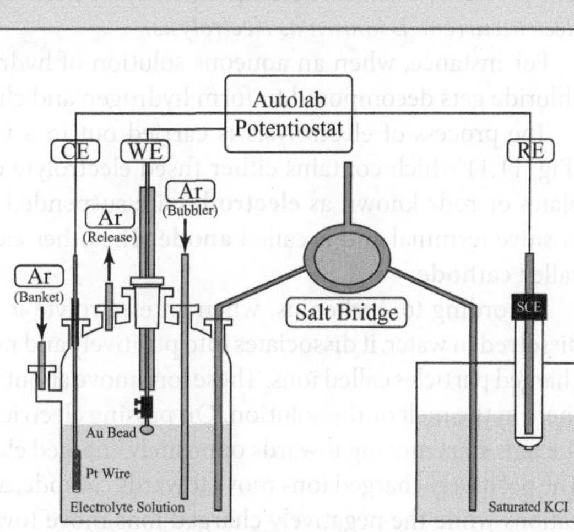

11.1 INTRODUCTION

Electrochemistry is that branch of chemistry which deals with chemical changes produced by electricity and production of electricity by chemical changes. Electrochemical changes are of two types.

(i) A chemical change is effected by passing electric current through electrolytes in the molten state or through their aqueous solutions. In these changes, electric current can be employed to bring about a chemical reaction. The phenomenon of decomposition of an electrolyte with the help of electric current is called ***electrolysis*** and a device in which electrolysis is carried out is called an ***electrolytic cell***.

(ii) A chemical change which produces electric current because of the electron transfer from one substance to another is a redox reaction. When a redox reaction is carried out directly, chemical energy in the form of heat is produced. However, when a redox reaction is carried out indirectly, chemical energy is produced in the form of electrical energy. A device employed to convert the chemical energy of a redox reaction into electrical energy is called an ***electrochemical cell***. The study of the principles involved in utilizing chemical changes for producing electrical energy is referred to as ***electromotive chemistry***. The details have been discussed in chapter 10.

The flow of current in the external circuits, i.e., through metallic conductors is nothing but the flow of electrons. However, the flow of electricity through solutions is due to the flow of ions. The passage of electricity through electrolytes in the molten state or through their solutions due to the movements of ions is called ***electrolytic conduction***. The knowledge of electrolytic conduction being of primary importance in the study of electrochemistry, this topic would be dealt with at the outset of this chapter.

Besides electrolytic conduction, we discuss in the chapter a very important law (Kohlrausch law) dealing with the amount of current carried in solution and transport number which relates to the current carried by an ion and its mobility. We also explain the failure of Ostwald dilution formula for strong electrolytes and explain the modern theory of strong electrolytes.

11.2 ELECTROLYSIS

'The process of chemical decomposition of an electrolyte (in the dissolved or molten state) by the passage of electric current' is known as electrolysis.

For instance, when an aqueous solution of hydrogen chloride is subjected to electrolysis, hydrogen chloride gets decomposed to form hydrogen and chlorine gases.

The process of electrolysis is carried out in a vessel known as electrolytic tank or electrolytic cell (Fig. 11.1) which contains either fused electrolyte or aqueous solution of the electrolyte. Two metallic plates or rods known as **electrodes** are suspended into the **electrolyte.** One electrode is connected to positive terminal and is called **anode.** The other electrode is connected to the negative terminal and is called **cathode.**

According to Arrhenius, when an electrolyte is fused or dissolved in water, it dissociates into positively and negatively charged particles called ions. These ions move about here and there in the melt or the solution. On passing electric current, the ions start moving towards oppositely charged electrodes. The positively charged ions move towards cathode, are called cations while the negatively charged ions move towards the anode, and are called **anions**. The movement of cations and anions is responsible for the electrical conductivity of the solutions of electrolytes. In addition to electrical conductivity, the movement of the ions also causes transport of matter from one part of the system to another. As the ions reach the respective electrodes, a chemical reaction takes place. Such a reaction that takes place at the electrode between the

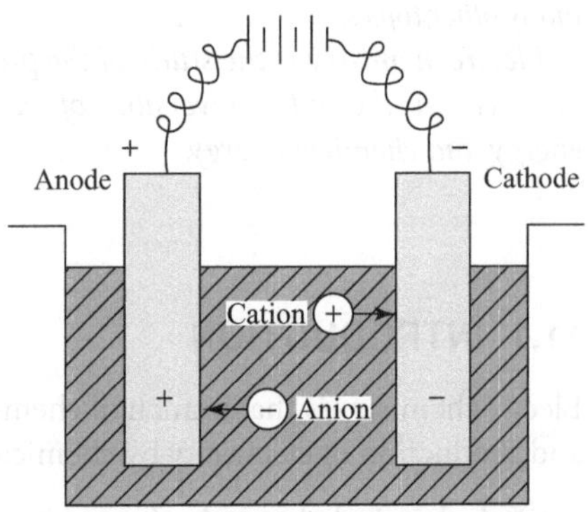

Figure 11.1 Electrolytic cell

electrolyte and electrode is called an electrochemical reaction. A few electrochemical reactions which may take place at cathode and anode are mentioned below.

(i) Reactions at Cathode: Reaction taking place at cathode is always **reduction**, which simply involves the gain of electrons by the cation, e.g.,

(i) $2H^+ + 2e^- \rightleftharpoons H_2$
(ii) $Cu^{2+} + 2e^- \rightleftharpoons Cu$
(iii) $Ag^+ + e^- \rightleftharpoons Ag$

(ii) Reactions at Anode: Reaction taking place at anode is always **oxidation** which simply involves the loss of electrons by a substance e.g.,

(i) $2Cl^- - 2e^- \rightleftharpoons Cl_2$
(ii) $4OH^- - 4e^- \rightleftharpoons 2H_2O + O_2$

11.2.1 Faraday's First Law of Electrolysis

It states that *the amount of any substance deposited at an electrode is directly proportional to the quantity of electricity passed through the electrolyte solution.*

Mathematically: $W \propto Q$

$$W \propto I \times t \quad \text{or} \quad W = z \times I \times t \tag{11.1}$$

where z is a constant of proportionality and is called ***electrochemical equivalent*** of the substance. If $I = 1$ ampere and $t = 1$ second, then, W = z. *Thus, electrochemical equivalent is defined as the weight of a substance deposited by the passage of 1-ampere current for one second (or one coulomb of electricity).*

Importance: This law is used to calculate:

(i) The values of electrochemical equivalents of different ions.
(ii) The weights of different ions deposited by passing different quantities of electricity through their electrolytes.

It has been found out experimentally that by passing **1 faraday, i.e., 96,500 coulombs** of electricity through an electrolyte, it results in the decomposition of one gram equivalent (i.e., equivalent weight expressed in grams) of the substance being deposited on the electrode. Therefore, **1 faraday or 96,500 coulombs deposit one gram equivalent of the substance.**

On the other hand, one coulomb of electricity deposits electrochemical equivalent (z) of the substance. Hence, we may conclude that

$$\text{Electrochemical equivalent (z)} \times 96{,}500 = \text{Gram equivalent}$$

11.2.2 Faraday's Second Law of Electrolysis

It states that *when the same quantity of electricity is passed through different electrolytes, the amounts of different substances produced at the electrodes are directly proportional to their equivalent weights.*

Let us explain the above law. Consider three cells, one contains water and a little HCl, the second cell contains copper sulphate solution and the third cell contains silver nitrate solution. Let these cells be connected in series with an electric battery (Fig. 11.2).

Since all the three cells have been connected in series, they will receive the same quantity of electricity during the same period. On passing electric current through these cells, hydrogen, copper and silver

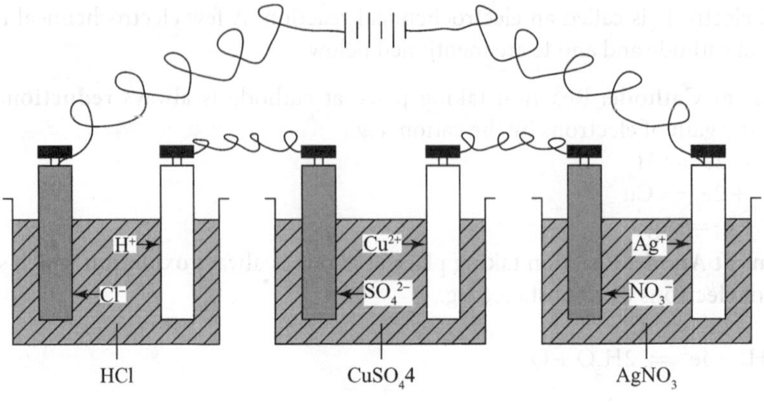

Figure 11.2 Passage of electricity through different cells connected in series

will be deposited at their respective cathodes in the ratio of their chemical equivalent, i.e., 1:31.75:108. In other words,

$$\frac{\text{Weight of hydrogen liberated}}{\text{Weight of copper liberated}} = \frac{\text{Eq. wt. of hydrogen}}{\text{Eq. wt. of copper}} = \frac{1}{31.75}$$

or

$$\frac{\text{Weight of hydrogen liberated}}{\text{Weight of silver liberated}} = \frac{\text{Eq. wt. of hydrogen}}{\text{Eq. wt. of silver}} = \frac{1}{108}$$

11.2.3 Application of Electrolysis

(i) Electroplating: *We may define electroplating as the process of depositing a superior metal like gold, silver or nickel over a baser metal like copper or iron with the help of electricity.*

The main objectives of carrying out electroplating are as follows:

(1) For preservation
(2) For decoration
(3) For repairs

Process: The cleaned article is then made the cathode by joining with negative terminal while anode is made of the metal to be deposited on the article. The cathode and the anode are placed in an electrolyte bath containing a soluble salt of the metal to be deposited. For example, if we wish to electroplate a copper fork with silver metal, then the fork is made the cathode and the silver plate is made the anode while the electrolyte used is a complex salt of silver, i.e., potassium argentocyanide K [Ag $(CN)_2$]. The process of electrolysis is carried out in an electrolytic tank which is made of glass or enamelled iron (Fig. 11.3).

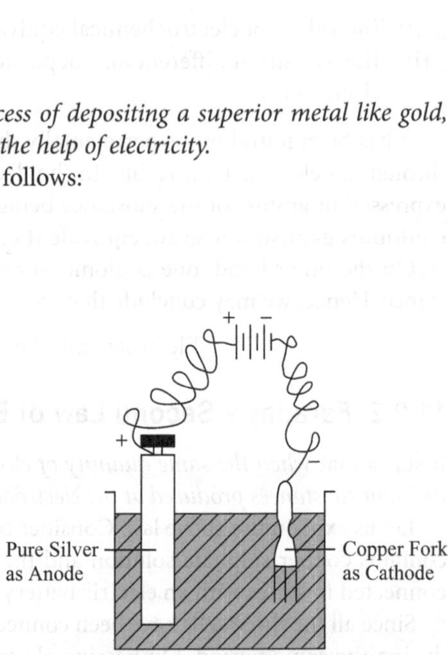

Figure 11.3 Electroplating

On passing electric current, silver from the electrolyte gets deposited on the fork (cathode) while equivalent amount of silver dissolves from the anode into electrolyte solution.

Keeping in view the above requirements, the electrolytes employed for various types of electroplating are as follows:

Copper plating : $CuSO_4$ solution containing dil. H_2SO_4
Gold plating: $K[Au(CN)_2]$ solution
Nickel plating: $NiSO_4(NH_4)_2 SO_4.6H_2O$ solution
Silver plating: $K[Ag(CN)_2]$ solution

Chemistry of silver plating: Deposition of silver in the above example is explained as follows:

$$K[Ag(CN)_2 \rightleftharpoons K^+ + [Ag(CN)_2]^- \quad \text{(ionization)}$$

$$[Ag(CN)_2]^- \rightleftharpoons Ag^+ + 2CN^- \quad \text{(ionization)}$$

At cathode: $Ag^+ + e^- \rightarrow Ag$

At anode, cyanide ions (CN^-) cause an equivalent amount of silver from the anode to dissolve forming AgCN and finally $K[Ag(CN)_2]$.

(ii) Electrorefining of metals: *We may define electrorefining as the process of removal of impurities from an impure metal by electrolysis.*

For this purpose, the impure metal is made the anode in an electrolyte bath containing the solution of a suitable soluble salt of the metal. The cathode is made of a thin plate of pure metal (Fig. 11.4). As electricity is passed, pure metal from the electrolyte deposits on the cathode and an equivalent amount dissolves from the anode. The impurities present generally fall at the bottom as anode mud.

(iii) Electrotyping: We may define electrotyping as the process of obtaining the impressions of letters with the help of electrolysis.

(iv) Electrometallurgy: We may define electrometallurgy as the process of extraction of metals from their ores by electrolysis.

For example, sodium metal is obtained by the electrolysis of either fused caustic soda or fused sodium chloride. Similarly, aluminium is extracted by the electrolysis of bauxite (i.e., Al_2O_3) in fused cryolite.

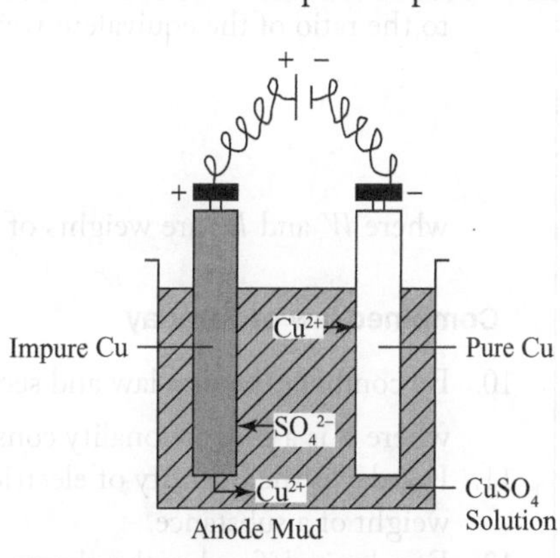

Figure 11.4 Electro-refining of metals

MEMORY FOCUS

First Law

1. The quantity of a substance decomposed or produced is directly proportional to the quantity of electric charge passed into the solution.
(where Z is a proportionality constant called electrochemical equivalent)
2. Electrochemical equivalent is characteristic of the substance reacting at the electrode.

(Continued)

MEMORY FOCUS (*Continued*)

3. Units of electrochemical equivalent are grams/coulomb or kg/coulomb.
4. The quantity of a substance decomposed or produced at the electrode by 1 coulomb of electricity is called its electrochemical equivalent.

$$\text{Electrochemical equivalent of a substance} = \frac{\text{gram equivalent of a substance}}{96{,}500}$$

5. When 96,500 coulombs of electricity is flowing through an electrode, it is equal to its gram equivalent weight.
6. Unit of electrical charge is coulomb, unit of current strength is ampere.
7. One ampere = one coulomb per sec.; quantity of electricity $(Q) = ct$
   where c = current strength in ampere and t = time in sec.

Second Law

8. When same quantity of electricity is passed through different electrolytes in a series, the weights of each substance reacted or deposited or liberated at the electrode is directly proportional to its equivalent weight, i.e., $w \propto E$.
   where E is the equivalent weight of the substance.
9. For two substances, the ratio of the weights of the substances reacted or deposited is identical to the ratio of the equivalent weights

$$\frac{W_1}{W_2} = \frac{E_1}{E_2}$$

   where W_1 and W_2 are weights of the substances and E_1 and E_2 are their equivalents.

Combined law of Faraday

10. By combing the first law and second law w a Q·E or w a ct E or $w = \frac{ct\,E}{\text{F}}$
    where F is a proportionality constant called Faraday.
11. Faraday is the quantity of electricity required to decompose or to liberate one gram equivalent weight of a substance.
12. Faraday is defined as the charge carried by one mole of electrons.
13. I Faraday = $6.023 \times 10^{23} \times 1.602 \times 10^{-19}$ coulombs = **96,500 coulombs.**

11.3 ELECTROLYTIC CONDUCTION

A substance which allows the electric current to flow through it is called a conductor while another which does not allow any electric current to flow through it is called a non-conductor or an insulator.

Conductors are of two kinds:

(i) **Metallic conductors** *which allow the electric current to flow through them without undergoing any chemical decomposition.* They remain unchanged even after the flow of electricity through them. Examples are copper, silver, etc. The conduction is due to the flow of electrons.

(ii) Electrolytic conductors (Electrolytes) *are substances whose molten or aqueous solutions conduct the electric current and which are decomposed by the passage of electric current.* Examples are acids, bases and salts. The conduction is due to the movement of ions.

A compound whose aqueous solution does not conduct electric current is called a **non-electrolyte.** Solutions of cane sugar, glycerine and alcohol do not conduct electricity and hence are examples of non-electrolytes.

11.3.1 Differences Between Metallic Conductor and Electrolytic Conductor

The main differences between metallic conductor and electrolytic conductors are as follows:

S. No.	Metallic Conductor	Electrolytic Conductor
1.	The flow of electricity through metallic conductor is due to the flow of electrons.	The flow of electricity through electrolytic conductor is due to the flow of ions.
2.	Flow of electricity does not cause the decomposition of metallic conductor.	Flow of electricity causes the decomposition of electrolytic conductor.
3.	Metallic conductance decreases with the rise in temperature.	Electrical conductance increases with the rise in temperature.
4.	Resistance of metals increases with rise in temperature.	Resistance of electrolytes decreases with rise in temperature.
5.	Flow of electricity does not cause the flow of matter.	Flow of electricity causes the flow of matter.
6.	Metals usually have high conductivity.	Electrolytic solutions have lower conductivity than metals.

11.3.2 Factors Affecting Electrolytic Conduction

The factors which influence the free movement of ions in solution and thus the conductance of the solution are given below:

(i) Nature of the electrolyte: *Strong electrolytes*, A substance that dissociates completely into ions in aqueous solution is called a strong electrolyte. For example, mineral acids such as, HCl, H_2SO_4, etc.; alkalies such as NaOH and KOH; salts such as NaCl, NH_4Cl, etc.

On the other hand, a weak electrolyte is one, which dissociates only partially into ions in an aqueous solution. For example, organic acids like HCOOH, CH_3COOH, and bases like $C_6H_5NH_2$, etc.

Strong electrolytes have higher value of conductance than those of weak electrolytes. This is because the number of ions in solution in the case of strong electrolytes (which are completely ionized) is larger than in the case of weak electrolytes (which are only partially ionized).

The above definition of strong and weak electrolytes is ambiguous, since in very dilute solutions all electrolytes are almost equally ionized. To remove the ambiguity, a convention is adopted. The criterion of dilution is fixed as one molar solution. If the substance is highly dissociated in 1 M solution, it is considered as strong electrolyte. If in 1 M solution it is slightly dissociated, it is considered as a weak electrolyte.

(ii) The interionic interactions: The forces existing between the oppositely charged ions of the electrolyte (solute) in solution are also called solute–solute interactions. In dilute solutions, the ions are far apart so that the forces of attraction between the oppositely charged ions are almost negligible. The ions are free to move and they make maximum contribution towards conductance.

However, in concentrated solutions, ions are close to each other and ionic forces of attraction are no longer negligible. Consequently, the mobility of the ions falls so that the conductance of the solution decreases.

(iii) **The solvation of ions:** The ions of the electrolyte in solution get associated with the molecules of the solvent. The hydration of Na^+ ions and Cl^- ions from NaCl in water is shown in Fig.11.5.

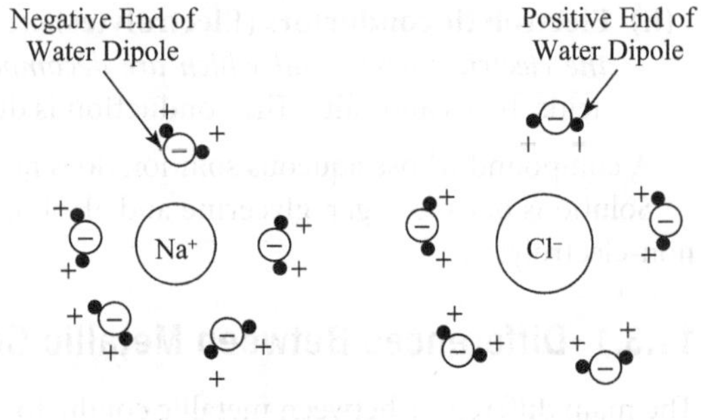

Figure 11.5 Hydration of ions

Thus, there exists an interaction between the ions of the solute and solvent molecules due to solvation. It affects mobility of the ions and hence conductance of the solution.

(iv) **Viscosity of the solvent:** The viscosity of the solvent depends upon the interaction amongst the molecules of the solvent, i.e., solvent–solvent interactions. Since the ions have to pass through the solvent, their mobility is affected by the viscosity of the solvent. The greater the viscosity of the solvent, the lower would be the mobility of the ions passing through it and lower would be the conductance of the solution.

(v) **Temperature:** With increase in temperature, the conductance of an electrolytic solution also increases, whereas the conductance of metallic conductors decreases with rise in temperature. The increase in conductance of solutions with temperature is due to increase in the average kinetic energy of ions carrying the current. Moreover, the viscosity of the solvent decreases with rise in temperature and thus the speed of ions increases.

It can be concluded that the conductance of an electrolyte in solution depends upon the following:

(i) The number of ions per unit volume of the solution.

(ii) The charge on the ions in solution. The greater the charge the higher would be the conduction. The conductance of $MgCl_2$ solution is greater than that of NaCl solution of the same concentration.

(iii) The mobility (speed) of the ions. The greater the mobility of the ions in solution, the higher would be its conductance.

11.3.3 Electrical Conductance

The term conductance of any conductor (metallic or electrolytic solution) is related to the resistance of the conductor. The resistance of the conductor can be understood through Ohm's law. The law states that,

'The strength of current (I) *passing through a conductor is directly proportional to the potential difference* (E) *applied across the conductor and is inversely proportional to the resistance* (R) *of the conductor'.*

Mathematically,

$$R \propto \frac{1}{C} \quad \text{or} \quad R = k\frac{1}{C} \tag{11.2}$$

The current strength is measured in amperes, the potential difference in volts and the resistance in ohms. The three units of electrical measurements are, therefore, related to each other by Ohm's law, as

$$R = \frac{E}{I} \quad \text{or} \quad \text{Ohms} = \frac{\text{Volts}}{\text{Amperes}}$$

Thus, if 1 ampere of current flows through a conductor when a voltage of 1 V is applied to it, the resistance of the conductor is taken as 1 ohm. Every substance offers resistance to electricity to a small or large extent. Ohm's law gives us the exact value of resistance. It holds good for metallic as well as electrolytic conductors.

In electrochemistry, the term conductance (C) is more commonly used. The term "C" implies the ease with which the current flows through the conductor.

It is a common observation that a substance with little resistance must possess a high conductance and vice versa. Therefore, if C is the conductance of a solution and R the resistance it offers to the passage of current, then

$$R \propto \frac{1}{C} \quad \text{or} \quad R = k\frac{1}{C}$$

where k is constant of proportionality.

when $k = 1$,
$$R = \frac{1}{C} \quad \text{or} \quad C = \frac{1}{R}$$

Hence, conductance is defined as the reciprocal of resistance.

Units: The units of electrical conductance are also the reciprocal ohm, i.e., **Ohm^{-1}** or **mho**. Its SI unit is siemens represented by symbol S. For example, if the resistance of an electrolyte is 10 ohms, its conductivity will be 1/10 mhos or ohm^{-1} or siemens or Ω^{-1}.

11.3.4 Specific Conductance

It is observed that resistance R of a conductor is

(i) directly proportional to its length (l) and
(ii) inversely proportional to its area of cross-section (a)

i.e.
$$R \propto \frac{1}{a} \quad \text{or} \quad R = \rho\frac{1}{a} \tag{11.3}$$

where ρ is a constant of proportionality called **specific resistance or resistivity**. Its value depends upon the material of the conductor.

If $l = 1$ cm and $a = 1$ sq. cm. then $\rho = R$, then specific resistance of a material is defined as the *resistance of material of a specimen of a unit length with unit area of cross-section. In other words, specific resistance is the resistance of one centimetre cube of material.*

'The reciprocal of specific resistance is known as conductivity or specific conductance.' It is denoted by k (kappa). Thus, if k is the specific conductance and ρ is the specific resistance, then the two terms are related as

$$k = \frac{1}{\rho}$$

We know
$$R = \rho \times \frac{l}{a} \quad \therefore \quad \rho = \frac{a}{l} \times R$$

or
$$k = \frac{1}{\rho} = \frac{l}{a} \times \frac{1}{R} \quad \therefore \quad k = \frac{1}{a} \times \text{Conductance}$$

Now, if $l = 1$ cm and $a = 1$ sq. cm, then $C = k$.

Hence, conductivity or specific conductance of a solution may be defined as the conductance of a solution of unit length with unit area of cross-section. Alternatively, it may be defined as conductance of one centimetre cube of the solution of electrolyte or it may be defined as the conductance of one centimetre cube of the material.

11.3.5 Equivalent Conductance and Molecular Conductance

Specific conductance (k) of a metallic conductor (of a fixed length and fixed area of cross-section) has a definite value at a given temperature. However, this is not so in the case of electrolytic solutions. In the electrolytic solution, the current is carried by ions. Thus, conductance of a solution depends upon the number of ions present in the given volume. The number of ions present per cubic centimetre of the solution varies with the concentration of the electrolyte. Therefore, the specific conductance of the electrolytic solution, amongst other factors, depends on its concentration also.

For comparison of the conductances of different electrolytic solutions, it is essential that equal volume of these solutions contain such a definite amount of the electrolytes which give ions carrying the same total charge of electricity. In other words, it is necessary to take solutions of equivalent concentrations.

For example, one mole of potassium chloride in its aqueous solution gives one mole of K^+ and one mole of Cl^- ions. One mole of potassium sulphate will give two moles of K^+ ion and one mole of ions. One mole of potassium chloride gives ions carrying a total charge of 2 faradays; one mole of potassium sulphate gives ions carrying a total charge of 4 faradays. Thus, in order to get comparable results, one-half mole of K_2SO_4 should be taken for every one mole of KCl. Therefore, consider the conductance of a certain volume of the solution in which definite amount of the solid (g. eq. wt. or g. mol. wt.) is dissolved. The conductance is called *equivalent conductance or molecular conductance* according as the weight dissolved in gram equivalent mass or gram molecular mass (Fig. 11.6)

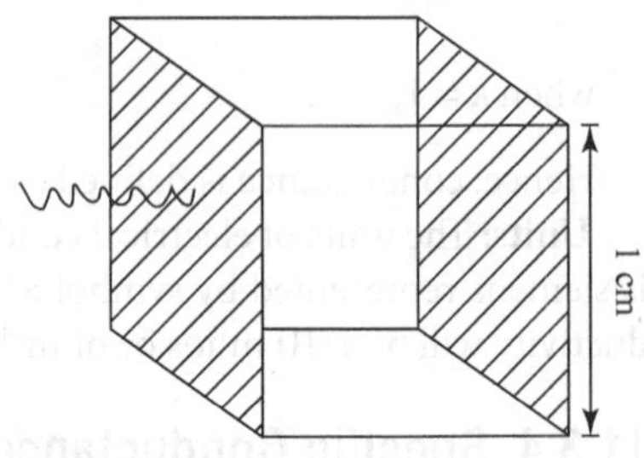

Figure 11.6 Diagramatic illustration of specific conductance

Hence, **equivalent conductance** is defined as the conductance of all the ions produced by 1 g equivalent of electrolyte dissolved in the solution. It is denoted by Λ_{eq}.

Its value is given by the product of the specific conductance k and the volume V in c.c. containing 1 g equivalent of the electrolyte.

11.3.6 Relation Between Specific Conductance and Equivalent Conductance

Consider 1 c.c. of a solution containing 1 g equivalent of an electrolyte placed between two large electrodes 1 cm apart. The area of cross-section of the solution will obviously be 1 sq. cm. Therefore, in such a case

$$\text{Conductance (C)} = \text{Specific conductance } (k) = \text{Equivalent conductance } (\Lambda_{eq})$$

If the solution is now diluted, say 9 c.c., the conductance of the solution will be nine times the specific conductance as there will be now nine cubes (Fig. 11.7).

However, even now, as the solution contains gram equivalent of the electrolyte between the electrodes, the conductance measured will be the equivalent conductance. Thus, in this case:

$$\text{Equivalent conductance} = 9 \times \text{Specific conductance}$$

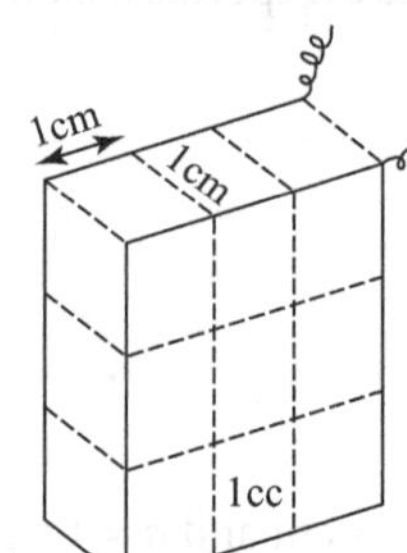

Figure 11.7 Illustration of Equivalent conduct

If the solution is now diluted to 1,000 c.c., the conductance of the solution will be 1,000 times the specific conductance as there will be now 1,000 cubes of the material. However, even now, as the solution contains only 1 g equivalent of the electrolyte between the electrodes, the conductance measured will be the equivalent conductance.

Hence, in general, $$\Lambda_{eq} = k \times V \tag{11.4}$$

where V is the volume of the solution in c.c. containing 1 g equivalent of the electrolyte.

If C gram equivalents of an electrolyte are present in 1 L of the solution or if C gram equivalents per litre are the concentration of the solution, the volume containing 1 g equivalent of the electrolyte will be

$$\frac{1}{C} \text{ litre} \quad \text{or} \quad \frac{1{,}000}{C} \text{ c.c}$$

Therefore, the equivalent conductance, $$\Lambda_{eq} = k \times \frac{1{,}000}{C_{eq}} \tag{11.5}$$

Molar conductance *may be defined as the conducting power of all the ions produced by dissolving one mole of an electrolyte in a solution. It is denoted by* Λ_m. Like equivalent conductance, the molar conductance of a solution is related to its specific conductance as

$$\Lambda_m = k_v \times V \tag{11.6}$$

where V is the volume in c.c., containing one mole (g. mol. mass) of the electrolyte.

or $$\Lambda_m = k_c \times \frac{1{,}000}{\text{molarity}} = k_c \times \frac{1{,}000}{C_m} \tag{11.7}$$

where C_m is the concentration of solution in moles per litre.

11.3.7 Experimental Measurement of Conductance

The electrical conductance of a solution can be determined by measuring its resistance because conductance is just the reciprocal of resistance. The **Wheatstone Bridge method** is employed for the purpose. The apparatus used is shown in Fig. 11.8. To determine the resistance of an electrolytic solution, specially designed cells known as conductivity cells are used.

AB is a uniform wire and a sliding contact point H moves over it. S represents the source of AC current and C is the conductivity cell containing the solution whose resistance is to be measured. R is the resistance box and P is the phone head to detect the current.

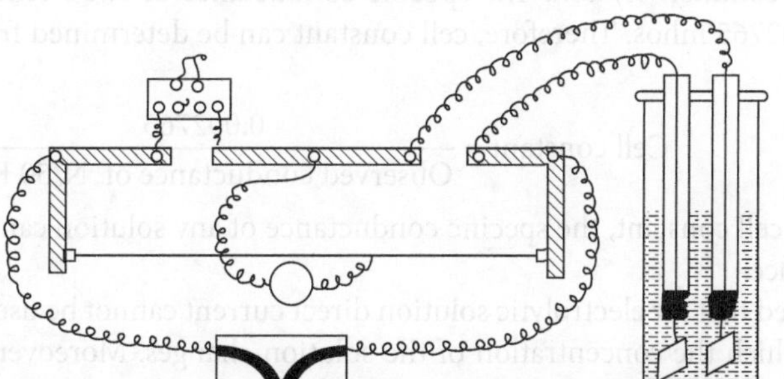

Figure 11.8 Circuit for finding the conductance of the solution

From the resistance box a known resistance R, which is of the same order (approximately) as that of solution under examination, is introduced. The sliding contact H is moved along the wire AB to detect the point of minimum sound. Under these circumstances

$$\frac{\text{Resistance of solution }(X)}{\text{Resistance R}} = \frac{\text{Length HB}}{\text{Length HA}}$$

The length HA and HB can be read from the scale, R is known, therefore, resistance of the solution X can be calculated.

Reciprocal of this resistance gives the conductance of the solution

$$\left[\mathrm{C} = \frac{1}{X}\right]$$

11.3.7.1 Calculations of Specific Conductance

We know that the specific conductance (k) is the reciprocal of specific resistance (ρ):

$$\text{Specific conductance} = \frac{l}{a} \times \text{Conductance}$$

or

$$k = \mathrm{C} \times \frac{l}{a}$$

where C is the conductance of the solution, l is the length of solution column, i.e., distance between the electrodes and a is the area of cross-section of electrons. The factor l/a is constant for a given conductivity cell. This factor is called '**cell constant**' and is expressed in cm^{-1}. Thus

$$\text{Specific conductance } (k) = \text{Conductance (C)} \times \text{Cell constant}$$

The value of the cell constant is not calculated from the value of l and a, as these are difficult to measure for a given cell from its dimensions. It is determined from the relation:

$$\text{Cell constant} = \frac{\text{Specific conductance }(k)}{\text{Observed conductance (C)}}$$

For this purpose, the conductance of a standard solution (say N/50 KCl) is determined at 25 °C by using the given conductivity cell. The specific conductance of N/50 KCl, solution at 25 °C has been found to be 0.002765 mhos. Therefore, cell constant can be determined from the following expression once for all:

$$\text{Cell constant} = \frac{0.002765}{\text{Observed conductance of N/50 KCl}} \tag{11.8}$$

By knowing the cell constant, the specific conductance of any solution can be calculated from the measured conductance.

It may be noted that for electrolytic solution direct current cannot be used because it causes electrolysis because of which the concentration of the solution changes. Moreover, the products of electrolysis may get accumulated at the electrodes and back electromotive force (EMF) may be set up. The resistance of the solution thus changes.

Numerical Problems

Example 1 **The resistance of a decinormal solution of an electrolyte in a conductivity cell was found to be 245 ohm. Calculate the equivalent conductance of the solution if the electrodes in the cell were 2 cm apart and each area of 3.5 sq. cm.**

Solution: We know that conductivity is the reciprocal of resistance.

Therefore, conductivity, $C = \frac{1}{R} = \frac{1}{245}\text{ohm}^{-1}$

Now specific conductance (k) = Observed conductance × Cell constant

$$= \text{Observed conductance} \times \frac{1}{a}$$

$$= \frac{1}{245} \times \frac{2}{3.5}\text{ohm}^{-1}\text{cm}^{-1}$$

Now, equivalent conductance $\Lambda_v = k \times V$

where V is the volume containing 1 g equivalent of the electrolyte.

As the solution is decinormal (i.e., 0.1 N), it means that 0.1 g equivalent of the electrolyte is present in 1,000 c.c. of the solution. Therefore, volume containing 1 g eq. of the electrolyte = 1,000/0.1 = 10,000 c.c.

$$\text{Eq. conductance } (\Lambda_v) = \frac{1}{245} \times \frac{2}{3.5} \times 10{,}000$$

$$= \mathbf{23.32\ ohms^{-1}cm^2eq^{-1}}$$

Example 2 **Specific conductance of N/50 KCI at 25 °C is 0.002768 ohms and it has a resistance of 520 ohms. An N/25 solution of a salt kept in the same cell was found to have a resistance of 300 ohms at 25 °C. Calculate (i) the cell constant and (ii) the equivalent conductance of the salt solution.**

Solution: (i) We know that

Specific conductance (k) = Observed conductance × Cell constant

or

$$\text{Cell constant} = \frac{\text{Specific conductance}}{\text{Observed conductance}}$$

$$= \text{Specific conductance} \times \text{Observed resistance}$$

Now, for the KCl solution, we are given that

Specific conductance = 0.002768 mhos

and observed resistance = 520 ohms

Therefore, cell constant = 0.002768 × 520 = **1.43936**

(ii) As before, we have for the salt solution

Conductivity (k) = Observed conductance × Cell constant

$$= \frac{\text{Cell constant}}{\text{Observed resistance}}$$

But, cell constant = 1.43936 [As calculated in (i)]

and observed resistance = 300 ohms (given)

$$\therefore \quad \text{Conductivity}(k) = \frac{1.43936}{300} \text{ ohms}$$

As the solution is N/25, therefore, volume containing 1 g equivalent of the salt will be

$$= 25 \times 1{,}000 = 25{,}000 \text{ c.c.}$$

Hence, equivalent conductance, $\Lambda_v = k - V$

$$= \frac{1.43936}{300} \times 25{,}000 \text{ mhos}$$

$$= \mathbf{119.945\ mhos\ cm^2\ g\ eq^{-1}}$$

Example 3 Calculate equivalent conductivity of 1 MH_2SO_4 solution if its specific conductivity is 2.6×10^{-1} ohm^{-1} cm^{-1}.

Solution: Equivalent mass of $H_2SO_4 = \frac{\text{Molar mass}}{\text{Basicity}} = \frac{98}{2} = 49$

Therefore, $1\ MH_2SO_4 = 2NH_2SO_4$

or concentration of the solution, $C_{eq} = 2 \text{ eq L}^{-1}$

Now, $$\Lambda_{eq} = \kappa \times \frac{1{,}000}{C_{eq}} = 2.6 \times 10^{-1} \times \frac{1{,}000}{2}$$

$$= \mathbf{130\ ohm^{-1}\ cm^2\ eq^{-1}}$$

Example 4 0.05 M NaOH solution offered a resistance of 31.6 ohm in a conductivity cell at 298 K. If the cell constant of the cell is 0.367 cm^{-1}, calculate the molar conductivity of the sodium hydroxide.

Solution: It is given that resistance of the solution = 31.6 ohm

Therefore, conductance $= \frac{1}{31.6} \text{ ohm}^{-1}$

Therefore, specific conduction (k) = Conductance cell constant

$$= \frac{1}{31.6} \text{ ohm}^{-1} \times 0.367 \text{ cm}^{-1}$$

$$= 0.0116 \text{ ohm}^{-1} \text{ cm}^{-1}$$

Molar conductance $$(\Lambda_m) = \frac{k \times 1{,}000}{C}$$

$$= \frac{0.0016 \times 1000}{0.05}$$

$$= \mathbf{232\ ohm^{-1}\ cm^2\ mol^{-1}}$$

11.3.8 Effect of Dilution on Conductance

Equivalent conductance of an electrolyte increases with dilution. This is due to the fact that on dilution the same amount of an electrolyte furnishes a large number of ions in solution. However, the increase in the number of ions on dilution is much less as compared to the increase in the volume of the solution. As a result, the total number of ions present per c.c., decreases on dilution; thus, the specific conductance decreases while equivalent conductance increases.

On progressive dilution, the values of equivalent conductance go on increasing and ultimately it becomes constant due to the fact that dissociation of electrolyte is complete. The maximum (limiting) value of equivalent conductance at infinite dilution (or zero concentration) is represented by Λ_∞ or Λ_0.

The ratio of the equivalent conductance at dilution V (Λ_v) to the equivalent conductance at infinite dilution (Λ_∞) gives the degree of dissociation a, i.e.,

$$\alpha = \frac{\Lambda_v}{\Lambda_\infty} \text{ or } \frac{\Lambda_c}{\Lambda_0} \tag{11.9}$$

Thus, when $\alpha = 1$, i.e., when dissociation is complete, then $\Lambda_c = \Lambda_{v.}$

The manner in which the equivalent conductance varies with dilution is shown in Fig. 11.14 in which the values of equivalent conductances of some electrolytes are plotted against the square root of concentration, $\sqrt{c}$. It follows from the curves obtained that there is a wide difference in the conductance behaviour of strong and weak electrolytes.

11.3.8.1 Conductance Behaviour of Strong Electrolytes

It is seen from Fig. 11.9 that for strong electrolytes like KCl, $BaCl_2$, etc., there is only a small increase in conductance with dilution. This is due to the fact that a strong electrolyte is completely dissociated in solution and hence there is no change in the number of ions. However, if the concentration is high, the movement of the ions is retarded by greater interionic attractive forces so that the conductance falls. With increase in dilution, the ions are far apart so that the interionic attractive forces are almost negligible and hence the conductance increases. It is noted that in case of strong electrolytes, the equivalent conductance has a tendency to attain a certain limiting value as the concentration approaches zero near infinite dilution. This is called equivalent conductance at infinite dilution and is commonly denoted by Λ_0. This limiting value represents complete dissociation of the electrolyte. In case of strong electrolytes, it is obtained by extrapolating the equivalent conductance graph to zero concentration as shown in Fig. 11.9.

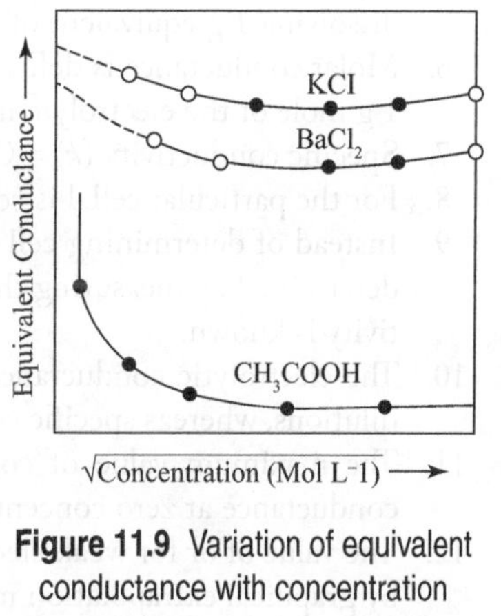

Figure 11.9 Variation of equivalent conductance with concentration

It has been observed that the variation of molar conductance with concentration in case of strong electrolytes at low concentrations is given by the expression:

$$\Lambda_m = \Lambda_m^\infty - b\sqrt{C}$$

where C is the molar concentration of the electrolyte and b is a constant depending upon the nature of the solvent and temperature. Thus, at infinite dilution or zero concentration, $\Lambda_m = \Lambda_m^\infty$.

11.3.8.2 Conductance Behaviour of Weak Electrolytes

It follows from Fig. 11.9 that in case of weak electrolytes (e.g., CH_3COOH) there is a sharp increase in equivalent conductance near infinite dilution.

This is due to the fact that at infinite dilution, the ionization of the electrolyte is complete and hence the increase in conductance is due to an increase in the number of ions in solution. Further, in case of weak electrolytes, the nature of the plot is a curve and not a straight line. Thus, it is not possible to obtain the limiting value of equivalent conductance (Λ_0) by the curve to zero concentration.

So, Λ_0 cannot be obtained by extrapolation in case of weak electrolytes. The equivalent conductance at infinite dilution in such cases is obtained by an indirect method based on Kohlrausch law.

MEMORY FOCUS

1. The reciprocal of resistance is called conductance, i.e., $C = \frac{1}{R}$.
2. Specific conductivity of a solution is the conductance of a solution of 1 cm length and having 1 sq. cm area of cross section.
3. Specific conductivity is also defined as the conductance of 1 cm^3 of a solution of the electrolyte.
4. Equivalent conductivity of a solution at dilution V is defined as the conductance of all the ions produced from 1 g equivalent of the electrolyte dissolved in Vcm^3 of the solution when the distance between the electrodes is 1 cm and area of the electrodes is so large that whole of solution is contained between them.
5. Equivalent conductance is also defined as the conducting power of all the ions produced by dissolving 1 g equivalent of the electrolyte in solution.
6. Molar conductance is defined as the conducting power of all the icons produced by dissolving 1 g mole of the electrolyte in solution.
7. Specific conductivity (k) = Conductance (C) × cell constant.
8. For the particular cell, l is constant and this constant is called cell constant.
9. Instead of determining cell constant from the dimensions of the cell, it is more conveniently determined by measuring the conductance of a standard KCl solution whose specific conductivity is known.
10. The electrolytic conductance as well as the equivalent and molar conductance increase with dilutions, whereas specific conductance of an electrolytic solution decreases with dilution.
11. The maximum value of conductance corresponding to zero concentration is called molar conductance at zero concentration () or at infinite dilution ().
12. The value of or for weak electrolytes such as NH_4OH and CH_3COOH cannot be determined by graphical extrapolation method.

11.4 KOHLRAUSCH LAW

We know that equivalent conductance of the electrolytes increases with dilution until at a limiting value it becomes constant. This limiting value (or maximum value) is called equivalent conductance at infinite dilution.

Kohlrausch observed a regularity in the equivalent conductance at infinite dilution of pairs of salts having a common ion. Table 11.1 shows equivalent conductances at infinite dilution of electrolytes at 298 K. Table 11.1 also shows that the difference in conductance at infinite dilution of potassium and sodium salts having a common anion gives the same value (23.41). It is independent of the anion. In the same manner, the difference in equivalent conductances at infinite dilution of pairs of salts having the anions bromide and chlorides (2.06) is independent of the cations. Kohlrausch observed that difference between the Λ_0 value for two salts, (which were strong electrolytes) having the same cation, is always constant.

Table 11.1 Equivalent conductances at infinite dilution at 298 K (in ohms^{-1} cm^2 eq^{-1}).

Pairs of Electrolytes Having Same Anions			**Pairs of Electrolytes Having Same Cations**		
Electrolytes	Λ_0	**Difference**	**Electrolyte**	Λ_0	**Difference**
KBr	151.92	23.41	NaBr	128.51	2.06
NaBr	128.51	—	NaCl	126.45	—
KCl	149.86	23.41	KBr	151.92	2.06
NaCl	126.45	—	KCl	149.86	—
KNO_3	144.96	23.41	HBr	428.22	2.06
$NaNO_3$	121.55	—	HCl	426.16	—

From these observations, Kohlrausch concluded that Λ_0 value of an electrolyte is the sum of two terms, one for the anion and other for the cation. These terms are known as ionic conductances of the ions concerned. This was first noticed by Kohlrausch and the following law was put forward by him. *At infinite dilution when electrolyte is fully dissociated each ion makes a definite contribution towards equivalent conductance of the electrolyte, irrespective of the nature of the other ion with which it is associated. In addition, that the value of equivalent conductance at infinite dilution of an electrolyte is equal to the sum of two values, one depending upon cation and other upon anion at a constant temperature.* Thus, mathematically

$$\Lambda_0 = \lambda_\alpha^0 + \lambda_c^0 \tag{11.10}$$

where λ_α^0 and λ_c^0 signify the ionic conductance of the anion and cation respectively at infinite dilution and are expressed in ohm^{-1} cm^2 or mhos cm^2. The law is also known as law of independent migration of ions or the law of independent ionic mobilities.

11.4.1 Applications of Kohlrausch's Law

(i) Calculation of equivalent conductance of weak electrolytes at infinite dilution (Λ_0): The equivalent conductance of a weak electrolyte cannot be obtained graphically, because a weak electrolyte is not completely ionized even when it is highly diluted. However, it may be calculated indirectly by employing Kohlrausch's law. For example, equivalent conductance of acetic acid at infinite dilution can be calculated from the equivalent conductances at infinite dilution of hydrochloric acid, sodium acetate and sodium chloride as follows.

The equivalent conductances at infinite dilution of the three strong electrolytes HCl, CH_3COONa and NaCl are experimentally determined. Then, on adding values of the first and second and subtracting that of the third gives the values of Λ_0 of acetic acid.

Thus

$$\Lambda_0 \, HCl = \lambda°H^+ + \lambda°Cl^{-1} \tag{11.11}$$

$$\Lambda_0\ CH_3COONa = \lambda°Na^+ + \lambda°CH_3COO^- \quad (11.12)$$

$$\Lambda_0\ NaCl = \lambda°Na^+ + \lambda°\ Cl^- \quad (11.13)$$

Adding Eqs (11.11) and (11.12) and subtracting Eq. (11.13), we have

$$\Lambda_0\ HCl + \Lambda_0\ CH_3COONa - \Lambda_0 NaCl$$

$$= \lambda°H^+ + \lambda°CH_3COO^-$$

$$= \Lambda_0 \text{ for } CH_3COOH$$

(ii) Calculation of degree of dissociation of weak electrolytes and their dissociation constants: The degree of dissociation, (α) may be defined as the fraction of the total electrolyte which dissociates to give ions. The degree of dissociation of an electrolyte is given by

$$\alpha = \frac{\text{Equivalent conductance at a given concentration}}{\text{Equivalent conductance at infinite dilution or zero concentration}}$$

i.e.,

$$\alpha = \frac{\Lambda_C}{\Lambda_O}$$

where, Λ_c is equivalent conductance at a given concentration while Λ_0 is the equivalent conductance at infinite dilution or that at zero concentration. The dissociation constant of a weak electrolyte is given by the relation:

$$K = \frac{Ca^2}{1-\alpha}$$

Putting the values of α and concentration (C), K can be calculated.

MEMORY FOCUS

1. Kohlrausch concluded that each ion makes a definite contribution towards the total molar or equivalent conductance of an electrolyte at infinite solution.
2. The molar conductivity of an electrolyte at infinite dilution is the sum of ionic conductance of the cations and the anions each multiplied with the number of ions present in one formula unit of the electrolyte. Mathematically, $\Lambda_m{}^{\circ}$ for $A_xB_y = x\,\lambda_A{}^{o+} + y\,\lambda_B{}^{o-}$
3. The equivalent conductivity of an electrolyte at infinite dilution is the sum of two values: one depending upon the cation and the other upon the anion.

$$\Lambda_{eq}{}^{\circ} = \lambda c^{o+} = \lambda a^{\circ}$$

4. Kohlrausch's law can be applied to calculate the molar or equivalent conductance at infinite dilution (Λ_o) for weak electrolytes.
5. The degree of ionisation of weak electrolytes can be calculated by the relation:
6. The dissociation constant of a weak electrolyte can be calculated from the formula, $k = \dfrac{C\alpha^2}{1-\alpha}$

11.5 MIGRATION OF IONS

When electricity is passed through the solution of an electrolyte, the ions present in solution are liberated at the electrodes, i.e., electrolysis occurs. It is seen that the ions in solution move and carry electricity with them. This movement of ions towards the oppositely charged electrodes is called migration of ions. The ions actually exist independently and move with the passage of current in solution towards the oppositely charged electrodes. It can be shown experimentally as follows:

Take a U-tube (Fig. 11.10). Fill its lower position by $Cu_2Cr_2O_7$ jelly obtained by mixing a 5% solution of agar and a mixture of $CuSO_4$ and $K_2Cr_2O_7$ and allow it to set. The jelly is green in colour. To mark the position of jelly, some charcoal is scattered over the green jelly. Over charcoal in both the limbs is added KNO_3 solution in agar agar, i.e., KNO_3 jelly, then aqueous solution of KNO_3. When current is passed, blue-coloured Cu^{2+} ions are seen moving towards negative electrode and yellow coloured $Cr_2O_7^{2-}$ ions are seen moving towards positive electrode. After some time, the two types of ions are seen moving with well-defined boundaries. The rise of blue colour due to the movement of Cu^{++} ions has been found to be more than that of yellow colour due to $Cr_2O_7^{2-}$ ion. This experiment clearly shows that ions move and they move with different speeds.

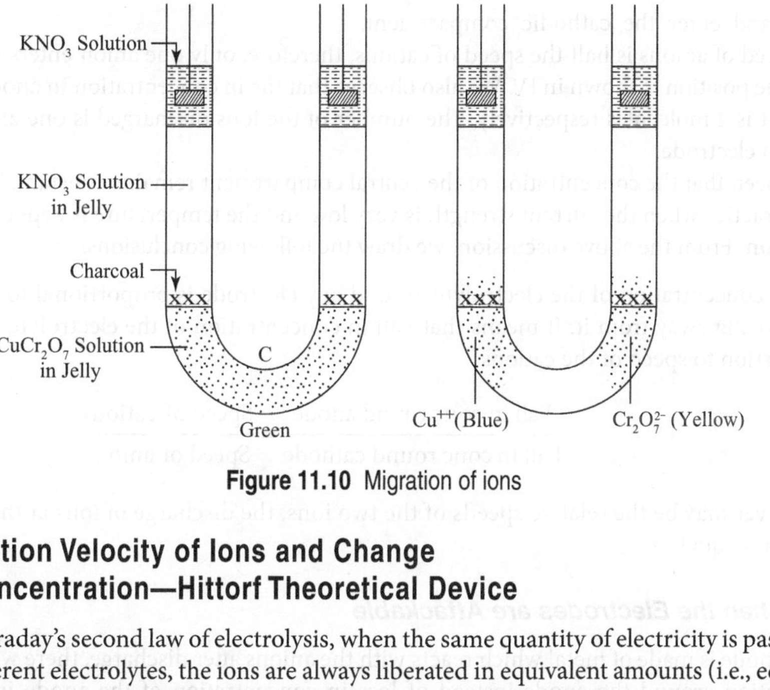

Figure 11.10 Migration of ions

11.5.1 Migration Velocity of Ions and Change in Concentration—Hittorf Theoretical Device

According to Faraday's second law of electrolysis, when the same quantity of electricity is passed through solutions of different electrolytes, the ions are always liberated in equivalent amounts (i.e., equal number of ions are discharged at respective electrodes). This, however, does not necessarily means that the ions move with the same speed. This was explained by theoretical device by Hittorf discussed below:

11.5.1.1 When the Electrodes are Non-Attackable

Consider a cell containing the solution and provided with the anode A and the cathode C (Fig. 11.11). Let the solution lying between the electrodes A and C be divided into three compartments (the anodic, the middle and the cathodic) with the help of the imaginary porous partitions a and c permeable to ions only (shown by dotted lines). Now, the following different cases may be considered. Let the electrodes be non-attackable. There is no change in concentration and no discharge of any of the compartment.

Case II. When only anions (–ive charge) move.

Although the cations have not still moved, still we see that thc ions liberated are the same (viz., 2) at both the electrodes. However, see that the concentration in anodic chamber has not altered, when in cathodic chamber it has fallen by two molecules.

Case III. When cations (+ive charge) and anions (–ve charge) at the same speed.

Let us assume that two cations move from anodic compartments to the cathodic compartment in a given time. Then, get position III. In this case, there is a fall concentration of two cubes in each compartment. However, we find that the number of liberated cubes at both the electrodes is the same, i.e., four.

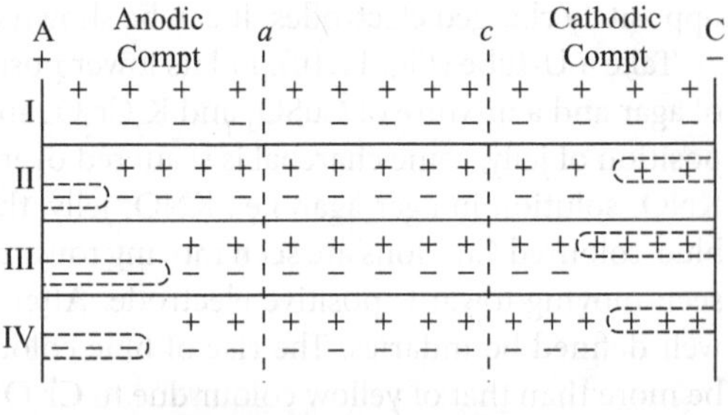

Figure 11.11 Migration velocity of ions-non-attackable electrodes

Case IV. When cations (+ive charge) move at double the speed of anions (–ve charge).

Suppose that two cations from anodic compartment leave and enter the cathodic compartment. Since the speed of anions is half the speed of cations, therefore, only one anion enters the anodic compartment. The position is shown in IV. We also observe that the in concentration in anodic and cathodic compartment is 2 molecules respectively. The number of the ions discharged is one and the same viz., three on each electrode.

We have seen that the concentration of the central compartment remains constant. This condition is realized in practice when the current strength is very low and the temperature is kept constant.

Conclusion: From the above discussion, we draw the following conclusions:

(i) Fall in concentration of the electrolyte around any electrode is proportional to the speed of the ion moving away from it. It means that Fall in concentration of the electrolyte around anode is proportion to speed of the cations.

i.e.,
$$\frac{\text{Fall in conc.round anode}}{\text{Fall in conc.round cathode}} = \frac{\text{Speed of cations}}{\text{Speed of anions}} \quad (11.14)$$

(ii) Whatever may be the relative speeds of the two ions, the discharge of ions at the two electrodes is always equal.

11.5.1.2 When the Electrodes are Attackable

In case the anode is made of metal which reacts with the anions after discharge, there will be an increase in concentration around the anode instead of loss in concentration of the anode instead of loss in concentration of the electrode around it.

The above relationship has been derived on the assumption that the ions do not react with the electrodes or the electrodes are non-attackable. However, in a silver voltameter (containing $AgNO_3$ solution and having silver electrodes) each NO_3^- ion that reaches the anodic compartment reacts with Ag^+ of the electrode forming $AgNO_3$ and instead of increase, there is increase of concentration in anodic compartment a silver voltameter, the above three cases are shown again in the e.g., Fig. 11.12.

It is seen that in every case, there is always an increase in concentration in the anodic compartment. In this case, the speed ratio is calculated as follows:

A silver voltameter is placed in series and amount of silver deposited (W) on the cathode is noted. In addition, an increase of concentration around (w) is noted.

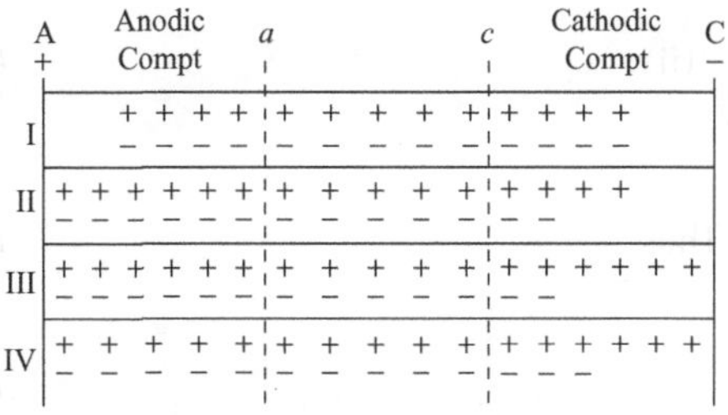

Figure 11.12 Migration of ions-attackable electrodes

Now,

fall of conc. around cathode = rise of conc. around anode

In addition, if no silver had migrated, increase around anode would have been equal to W. However, actual increase is w. The difference (W – w), therefore, gives the fall round anode. Therefore,

$$\frac{\text{Speed of } Ag^+}{\text{Speed of } NO_3^-} = \frac{W - w}{w} \tag{11.15}$$

11.6 TRANSPORT NUMBER

As seen earlier, electric current is carried by ions. However, each ion carries a definite fraction of the total current depending upon its speed. The fraction of the total current carried by an ion' is called its **transport number or Hittorf's number or Transference number.**

The greater the speed of the ion, the greater the amount of current it will carry with it and hence greater will be the transport number. Let u_a and u_c be the speed of the anion and the cation. Then

$$\text{Current carried by the anion} \propto u_a = k\,u_a$$
$$\text{Current carried by the cation} \propto u_c = ku_c$$

Since current is carried only by ions.

Total current carried = Current carried by anion + current carried by cation = $Ku_a + Ku_c = K(u_a + u_c)$.

Transport number of the anion, $$t_a = \frac{Kua}{K(u_a a + u_c)}$$

or $$t_a \text{ or } t^- = \frac{u_a}{u_a + u_c} \tag{11.16}$$

Similarly, transport number of cation will be

$$t_c \text{ or } t^+ = \frac{u_c}{u_a + u_c} \tag{11.17}$$

11.6.1 Important Relations Concerning Transport Number

(i) $$t^- + t^+ = \frac{u_a}{u_a + u_c} + \frac{u_c}{u_a + u_c} = \frac{u_a + u_c}{u_a + u_c} = 1$$

That is, sum of transport number of anion and cation of an electrolyte is always unity. Thus, no current is lost in any other function and the whole current is carried by the ions.

(ii)
$$t^- = \frac{u_a}{u_a + u_c} = \frac{1}{1 + \frac{u_c}{u_a}} = \frac{1}{1 + r}$$

Thus
$$t^- = \frac{u_a}{u_a + u_c} = \frac{1}{1 + r} \quad (11.18)$$

$$t^+ = \frac{u_c}{u_a + u_c} = \frac{r}{1 + r} \quad (11.19)$$

11.6.2 Factors Controlling Transport Number

The transport number of an ion is directly proportional to the speed of the ion, which in turn depends upon the following factors:

(i) Nature of the ion: The smaller the size of the ion the greater will be its speed and higher would be its transport number.

(ii) Nature of the counter ion present: The transport number of an ion depends not only upon the mobility of the alone but also on the mobility of counter ion associated with it in the electrolyte as is evident from the following example.

For example, transport number of chloride ion in sodium chloride is 0.604 and in hydrochloric acid, it is 0.166. This can be understood from the fact that the mobility of hydrogen ion is greater than that of sodium ion.

$$t_{cl^-} \text{ (in NaCl)} = \frac{u_{cl^-}}{u_{cl^-} - + u_{Na^+}}$$

$$t_{cl^-} \text{ (in HCl)} = \frac{u_{cl^-}}{u_{cl^-} - + u_{H^+}}$$

Obviously, the t_{cl^-} in NaCl is greater than t_{cl^-} in HCl as in the latter case, the denominator $(u_{Cl^-} + u_{H^+})$ is much greater than that in the former case $(u_{Cl^-} + u_{Na^+})$ while numerator u_{Cl^-}, i.e., the mobility of the chloride ion, is same in each case.

(iii) Hydration of the ions: Hydration of the ion affects the speed and thus the transport number of that ion. For instance, the size of Li^+ ions is smaller than Na^+ ion, but bit is more prone to hydration. The transport number of Li^+ is expected to be higher based on its size, but in practice, the value has been found to be less. This is due to its greater tendency towards hydration.

(iv) Concentration: Although the effect of concentration on transport number is very small, yet sometimes it becomes significant due to the formation of complex ion with change in concentration.

It has been observed that for some salts the values of the transport numbers are abnormal in the sense that they are even greater than unity for cation and so are negative for anion. A simple case is that of the solution of cadmium iodide in which the transport number of cations is greater than unity and for the anions a negative value is obtained. This is because of the formation of the complex ions, e.g.,

$$CdI_2 + \text{aqueous} \rightarrow Cd^{+2}(aq) + 2\,I^-(aq)$$

$$Cd\,^2(aq) + 4I\,(aq) \rightarrow Cd\,I^{-2}(aq)$$

At high concentration, cadmium exists as Cd^{-2} as well as CdI_4^{-2} so cadmium migrates towards both cathode and anode giving negative values for its transport numbers.

(v) **Temperature:** With increase of temperature, the ions tend to approach a 0.5 value of transport number. It has been observed that ions having transport number less than 0.5 at ordinary temperature show an increase in the value, whereas the other ion which has transport number more than 0.5 shows a decrease in the value, with rise of temperature. It implies that the cation and the anion tend to equalize the value of transport number with increase in temperature. The effect of temperature is very small when the value of transport number of the cation and the anion is already 0.5 as in the case of KCI.

(vi) **Interionic effects:** The speed of the ions and consequently their transport numbers are affected due to interionic effects as discussed.

11.6.3 Determination of Transport Numbers

There are a number of methods for the determination of transport number, e.g., (i) Hittorf's method (ii) Lodge's moving boundary method and (iii) EMF method. Out of these, the two methods are described below:

11.6.3.1 Hittorff's Method

Principle: The method is based upon the principle that the fall in concentration around an electrode is proportional to the speed of the ion moving away from it.

$$\frac{\text{Fall in concentration around cathode}}{\text{Fall in concentration around anode}}$$

$$= \frac{\text{Speed of anion}}{\text{Speed of cation}} = \frac{u_a}{u_c} \qquad (11.20)$$

$$\text{Since transport number of cation } t^+ = \frac{\text{Speed of the cation}}{\text{Speed of the cation} + \text{Speed of the anion}}$$

$$\text{Transport number of cation, } t^+ = \frac{\text{Fall around anode}}{\text{Fall around the anode}} + \text{Fall around the cathode}$$

$$= \frac{\text{Fall around anode}}{\text{Total fall around both the electrodes}}$$

If the concentrations are measured in terms of gram equivalent, then

Transport number of cation, t^+

Number of gram equivalents lost from the anodic compartment = Total number of gram equivalent lost from both the compartment.

Further, the total number of gram equivalent lost from both the compartments is equal to the gram equivalent deposited on each electrode. This is clear from Hittorf's theoretical device since in each case considering the total number of ions from both anodic and cathodic compartments is equal to the number of ions discharged or deposited on each electrode.

It is more convenient to use a copper coulometer in series with the experimental solution in the determination of transport number since the number of gram equivalents deposited shall be equal to the number of gram equivalents of copper deposited in copper coulometer.

Therefore, the transport number of cation, t^{+1}

$$= \frac{\text{Number of gram equivalents lost from the Anode compartment}}{\text{Number of gram equivalents of copper deposited in the copper coulometer}}$$

Now,
$$t^+ + t^- = 1 \qquad (11.21)$$

$$\text{Transport number of anion} = 1 - t^+$$

Thus, transport number of the cation (t^+) can be determined by measuring fall in concentrations around anode and total amount deposited on an electrode.

It may be pointed out here that in order to apply the above relationship, the electrolysis must be carried out so that intermixing of the solution around the two electrodes is avoided completely.

The above equation is valid only if the electrodes are not attacked by the ions in solutions. If the anode is attacked by the anions, there will be an increase in the concentration around the anode instead of decrease. However, this increase in concentration at anode is equal to decrease in concentration that would have taken place if the anode were inert. Examples are as in the case of electrolysis of $AgNO_3$ using silver anode.

11.6.3.2 Experimental Procedure

The most widely applicable form of Hittorf's apparatus is shown in Fig. 11.13. It consists of a two-limbed vessel connected by a narrow U-tube in the middle. Both limbs as well as the U-tube are provided with stop cooks at the bottom for the withdrawal of the solution. Two limbs are provided either with Pt electrodes (non-attackable) or with suitable attackable electrodes. Suppose we want to determine the transport number of Ag^+ in $AgNO_3$. Then, the electrodes are either of Pt or of pure Ag.

The apparatus in filled with a standard solution of $AgNO_3$ (usually 0-IN) and the cell is connected in series with a source of direct and a milliammeter as shown. A low current (10–15 mA) is passed for 2–3 h so as to produce an appreciable (but not too large change in concentrations). Low current is used to minimize thermal and resulting diffusion effects.

After the electrolysis has taken place, the solutions from anodic as well as central compartments are withdrawn separately by opening stopcock at the bottom and each analyzed by weighing and titration with standard ammonium thiocyanate (NH_4CNS) solution. The change in the amount of electrolyte in the anode compartment is thus obtained. There should be no change in the concentration of the solution taken out from the central compartment (i.e., U-tube). It is important that the amount of solution

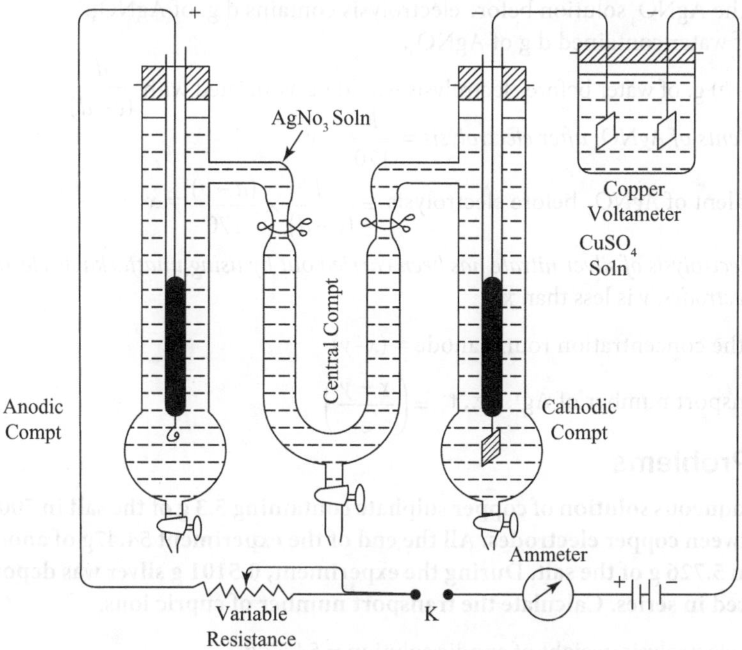

Figure 11.13 Hittorf's apparatus of measuring transport number

withdrawn from anode for analysis should include all the solution whose composition has changed. This can be checked by withdrawing a further portion of solution from the anode limb and finding whether it has the same composition as the original $AgNO_3$ solution or not. The weight of Cu deposited in the voltameter is accurately determined.

11.6.3.3 Observations and Calculations

With the change in concentration of a solution, the density of the solution also undergoes a change and consequently the volume also changes. The change in concentration is thus determined with reference to a definite weight of the solvent. Therefore, while determining the transport number of ions we take into account the known weight of the solution and not the known volume of the solution before and after the electrolysis.

(i) Let the weight of copper deposited on cathode of copper

$$\text{voltameter} = W \text{ g.}$$

the equivalent of the copper deposited in copper voltameter $= \dfrac{W}{31.75} = \text{z}$

(ii) Let a g. of anode solution after electrolysis contains b g. of $AgNO_2$.

$$\text{Wt. of water} = (\text{a} - \text{b}) \text{ g.}$$

(a – b) g. of water at the end contains b g. of $AgNO_3$.

(iii) Let c g of the $AgNO_3$ solution before electrolysis contains d g. of $AgNO_3$.
(c – d) g of water contained d g of $AgNO_3$.

Hence $(a-b)$ g. of water before electrolysis would be associated with $\frac{d}{(c-d)} \times (a-b) g.$ of $AgNO_3$

$\therefore$ *Equivalents of $AgNO_3$ after electrolysis* $= \frac{b}{170} = y$

and equivalent of $AgNO_3$ before electrolysis $= \frac{d}{(c-d)} \times \frac{(a-b)}{170} = x$

(a) *If the electrolysis of silver nitrate has been carried out by using unattackable electrodes, i.e., Platinum electrodes, y* is less than x.

$\therefore$ Fall in the concentration round anode = $(x-y)$

Hence transport number of Ag^+, *i.e.*, $t^+_{Ag} = \left(\frac{x-y}{z}\right)$

Numerical Problems

Example 5 An aqueous solution of copper sulphate containing 5.3 g of the salt in 500 g solution was electrolyzed between copper electrodes. All the end of the experiment 54.47g of anode solution was found to contain 5.726 g of the salt. During the experiment, 0.5101 g silver was deposited in a silver coulometer placed in series. Calculate the transport number of cupric ions.

Solution: After electrolysis, weight of anodic solution = 54.57 g

Weight of $CuSO_4$ in it = 5.726 g

Weight of water = 54.57 – 5.726 = 48.844 g

Weight of $CuSO_4$ per gram of water $= \frac{5.726}{48.844} = 0.1172$ g

Before electrolysis, weight of solution = 500 g

Weight of $CuSO_4$ = 5.3 g

Weight of water = 500 – 5.3 = 494.7 g

Weight of $CuSO_4$ per gram of water $= \frac{5.3}{494.7} = 0.0107$ g

Increase in the conc. of $CuSO_4$ around anode

$$= 0.1172 - 0.0107 = 0.1065 \text{ g } \frac{0.1065}{79.75} = 0.00136 \text{ g eq.}$$

Equation wt. of $CuSO_4 = \frac{63.5+32+64}{2} = 79.75$

Weight of Ag deposited in coulometer = 0.5101 g

$$= \frac{0.5101}{108} = 0.00472 \text{ g eq.}$$

Fall in conc. around anode due to migration of Cu^{+2} ions

$$= 0.00472 - 0.00136$$
$$= 0.00336 \text{ g eq.}$$

$$\text{Transport No of } Cu^{+2} \text{ ions} = \frac{\text{g eq. of } Cu^{+2} \text{ ions lost from anodic soln}}{\text{g eq. of } Ag^{+}\text{ions deposited}}$$

$$= \frac{0.00336}{0.00472} = \mathbf{0.712}$$

$$\text{Transport No of } SO_4^{2-} \text{ ions} = 1 - 0.712 = \mathbf{0.288}$$

Example 6 A solution of $AgNO_3$ containing 0.00739 g of the solute per gram of water was electrolyzed between Ag electrodes. During the experiment, 0.078 of Ag got deposited at the cathode. At the end of the experiment, the anode portion contained 23.14 of water and 0.236 g of $AgNO_3$. What are the transport numbers of Ag^+ and NO_3^- ions?

Solution: After electrolysis, 23.14 g of water contains 0.236 g of $AgNO_3$

$$1 \text{ g of water contains} = \frac{0.236}{23.14} = 0.0102 \text{ g of } AgNO_3$$

Before electrolysis, 1 g of water contains = 0.0074 g of $AgNO_3$
Increase of $AgNO_3$ around anode = 0.0102 – 0.0074

$$= 0.0028 \text{ g}$$

$$= \frac{0.0028}{170} = 0.0000165 \text{ g eq.}$$

Weight of Ag deposited at the cathode = 0.078 g

$$= \frac{0.078 \text{ g}}{170} = 0.0000458 \text{ g eq.}$$

Fall in conc. around anode due to migration of Ag^+ ions

$$= 0.000458 - 0.0000165 = 0.0000293 \text{ g eq.}$$

$$\text{Transport No. of } Ag^+ \text{ ions} = \frac{\text{g eq. of } Ag^{+}\text{ions from anodic solution}}{\text{g eq. of Ag deposited at the cathode}}$$

$$= \frac{0.0000293}{0.0000458} = \mathbf{0.6397}$$

$$\text{Transport number of } NO_3^- \text{ ions} = 1 - 0.6397 = \mathbf{0.3603}$$

11.7 LIMITATIONS OF ARRHENIUS THEORY

In the case of weak electrolytes, it is understandable to assume that the increase in conductance with dilution may result because of increasing, degree of ionization with dilution. The reason for such behaviour lies in the fact that when a weak electrolyte is dissolved in water, a small portion of it dissociates. Moreover, there is a dynamic equilibrium between the undissociated and the dissociated species i.e., the corresponding ions. Such equilibrium would lead to increased ionic concentration on dilution (Arrhenius theory). This idea of Arrhenius led to the suggestion that at infinite dilution, weak electrolytes (or all electrolytes in terms of Arrhenius concept) are completely dissociated. As such, the degree of dissociation, α of electrolyte (weak) can be obtained from the relation

$$\alpha = \frac{\Lambda_v}{\Lambda_0}$$

where Λ_v is the equivalent conductance at a given concentration. The result obtained in this way agreed with those calculated from other measurements, i.e., colligative properties only for weak electrolytes.

However, certain difficulties or anomalies appear evident when considering the case of strong electrolytes.

(i) The values of α calculated, in the above manner did not agree with those based on colligative properties.
(ii) The values of equilibrium constant based on Arrhenius concept varied with concentration of the electrolyte.
(iii) The increase of conductance (Λ) with dilution stand no justification on the argument of increasing charged species with dilution. This is because that X-ray studies of strong electrolytes, e.g., NaCl gives clear evidence that non-ionized molecule of NaCl does not exist even in the solid state. Thus, it is reasonable to assume, beyond any doubt, that a strong electrolyte like NaCl should be completely ionized at all concentrations in water.

11.8 MODERN THEORY OF STRONG ELECTROLYTES

A satisfactory explanation for the variation of equivalent conductance with concentration was dealt with quantitatively by *Debye–Huckel–Onsagar* about interionic attraction theory. Here, we shall deal with only the qualitative aspects of this theory of conductance. This theory is based on the realization of the following facts:

(i) Although the strong electrolytes are completely ionized, their dissociation will be incomplete. A complete dissociation visualizes that the ions of electrolyte are far apart and there is no interionic attractions. Thus, even strong electrolytes should be assumed to be completely dissociated only in very dilute solutions.
(ii) Because of attraction between positive and negative ions, each ion in solution is surrounded by ions of opposite charge, whose net charge is equal to that of the central ion. Thus, there is present an ionic atmosphere of opposite charge around any ion. In the absence of any external field applied, the ionic atmosphere has a central symmetry, i.e., atmosphere of opposite charges is spherically and symmetrically distributed about any ion (Fig. 11.14).

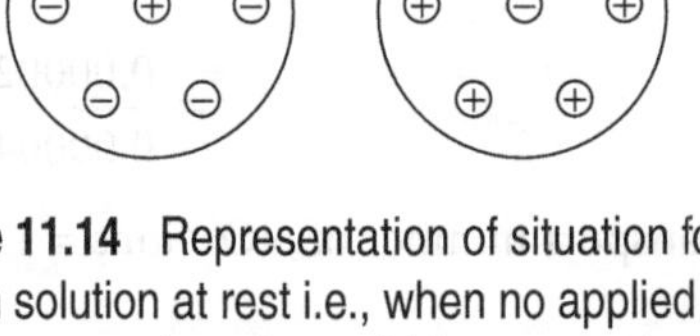

Figure 11.14 Representation of situation for the ions in solution at rest i.e., when no applied field. (a) ion atmosphere for +; (b) ion atmosphere for

(iii) When the potential is applied across two electrodes immersed in the solution of the electrolyte for conductance measurement, the above situation of the ions in solutions gets disturbed, thus giving rise to two-fold effects, namely (a) the relaxation or asymmetry effects, and (b) the electrophoretic effect. Both these effects finally result in the decrease of the speed of the ions and so their conducting power.

(a) Relaxation effect: This arises due to occurrence of the following situation. When the current is passed through the solution, positive ions will tend to move towards the cathode and, the negative ions towards the negative electrode, i.e., anode. Now, both the ions are surrounded by an atmosphere of opposite charge, so that under an applied potential, the central ion will tend to move in the direction opposite to the movement of the atmosphere. That is to say, when the positive central ion moves towards the cathode, its ionic atmosphere will move towards the anode. During its motion, the central ion will build up its ionic atmosphere again whereas the old one will die out. If this formation of the new atmosphere and the destruction of the previous one occur simultaneously, there will be no retarding force affecting the movement of the central ion. However, there is a time gap or the so-called time of relaxation between the formation of new ionic atmosphere and the breaking up of the old.

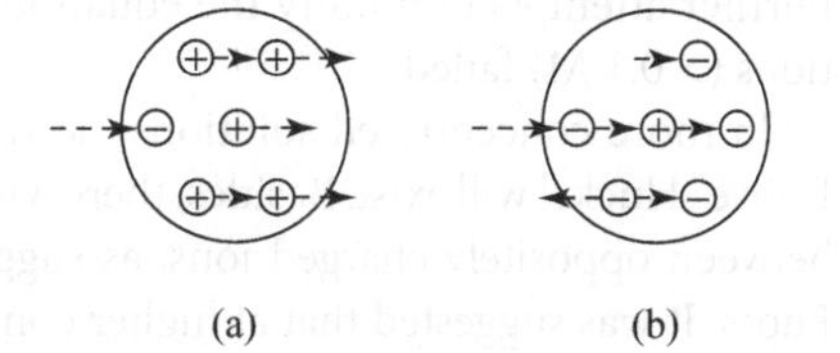

Figure 11.15 Representation of Asymmetric ionic atmosphere (a) –ve ion moving against +ve ionic atmosphere (b) +ve ion moving against –ve ionic

Thus, the symmetry of the ionic atmosphere will be disturbed and a new situation arises as shown in Fig. 11.15. The ionic atmosphere is no longer uniform around the central ion and consequently the ion experiences a retarding force against the direction of its motion and the ion slows down by these interionic attractions.

(b) Electrophoretic effect: This is another factor, which also opposes the motion of the moving ions. As the ions are solvated in solution and when the ions move, they carry these solvent molecules with them. In this way, any ion under the influence of applied field does not move through a stationary medium. Thus, a positive ion moving to the cathode has to make its way through the stream of the solvent molecules with the negative ions migrating towards the anode in the opposite direction. Similarly, the negative ions have to make their way through the solvent molecules carried by the positive ions in the opposite direction.

Debye–Huckel considered the above factors and obtained an equation relating the observed conductance to that at infinite dilution, thus showing

$$\Lambda = \Lambda_0 - a\sqrt{c} \tag{11.22}$$

quantitatively that the equivalent conductance depends on the square root of the concentration of the electrolyte. A is a constant and depends on the dielectric constant and viscosity of the medium. However, when the slopes of the straight lines in the plot of Λ vs. $\sqrt{c}$ were calculated based on the above equation, it agreed with the observed behaviour only up to concentrations of about 0.1 M. Onsagar modified the Debye–Huckel equation, which is known as **Debye–Huckel–Onsagar equation or simply as Onsagar equation.** For a univalent electrolyte such as KCl, it is

$$\Lambda_c = \Lambda_0 - \left[\frac{82.4}{(DT)^{1/2}\,n} + \frac{8.20\times10^5}{(DT)^{3/2}}\right]\sqrt{c} \tag{11.23}$$

where D is the dielectric constant, T is the temperature in Kelvin scale, Λ_c is the equivalent conductance at concentration c in moles/litre and Λ_0 is the equivalent conductance at infinite dilution.

D and n are constants for a particular solvent, so at constant temperature the above equation is reduced to

$$\Lambda_c = \Lambda_0 - [A + B\Lambda_0]\sqrt{c} \tag{11.24}$$

where A and B are constants depending on the temperature and the nature of solvent.

The validity of the equation is tested by plotting the experimental value of Λ against $\sqrt{c}$. Λ_0 obtained by extrapolation is then compared with the limiting value predicted by the equation. Some comparison between theory and experiments are shown in Fig. 11.16.

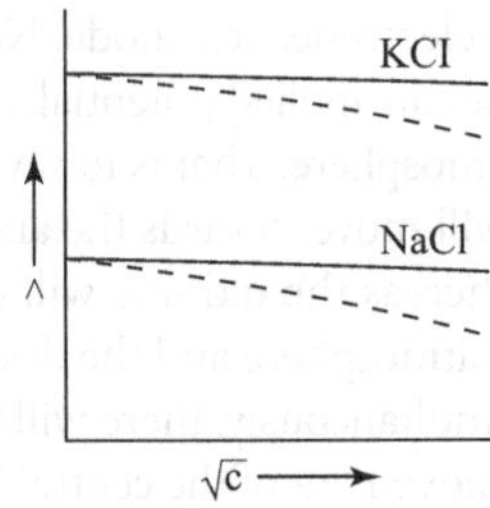

Figure 11.16 Experimental (solid lines) and theoretical (dashed lines) equivalent conductances

The agreement is usually good at very low concentration; however, in more concentrated solution, the conductivity is higher than predicted by the equation.

The modification of the Debye–Huckel by equation. Onsagar rendered it applicable to electrolytic solutions as high as 0.1 M. Further attempts to modify the equation for higher concentrations (> 0.1 M) failed.

In more concentrated solutions, no doubt, the effects of the Debye–Huckel will exist. Besides, there will be specific attractions between oppositely charged ions, as suggested by Bjerrum and Fuoss. It was suggested that at higher concentrations, oppositely charged ions would be tied up and present as ion pairs. For example, in a concentrated solution of NaCl, ion pairs of $Na^+ Cl^-$ may be present instead of individual Na^+ or Cl^- ions. The formation of these ion pairs will decrease conductance and other properties (colligative) of electrolyte. The concept of ion pair is of prime importance in solvent of low dielectric constant. Further, the interactions between the ions and solvent molecules should also be considered for a comprehensive structural model of the electrolyte. Thus, a satisfactory theory of electrolytes should deal with care, the extent of solvation, the nature of ion pair, or the formation of undissociated molecules, in conjunction with the Debye–Huckel effects. However, such a treatment is not possible in the absence of detailed data for the molecular world. Nevertheless, the theory of Debye–Huckel, Onsagar, and Bjerrum and Fuoss has led to the understanding of many experimental results.

Significance of Debye–Huckel Theory: This equation explains the fact that equivalent conductance of a strong electrolyte at any concentration (Λ_c) is less than the limiting value of Λ_0 at infinite dilution in spite of the complete ionization of strong electrolyte.

NOTEWORTHY POINTS

1. Both oxidation and reduction take place in the cell because electrical energy is being used.
2. In electrolytic cell, anode is positively charged whereas cathode is negatively charged.
3. During electrolysis of molten electrolyte, cations are liberated at cathode, while anions at the anode.
4. When two or more ions compete at the electrodes then the ion with higher reduction potential gets liberated at the cathode while the one with lower reduction potential at the anode.
5. Cell constant is equal to l/a having units of cm^{-1} and is determined with the help of conductivity bridge, where a standard solution of KCl is used.
6. Conductivity water: It is the highly purified water whose own conductance is very small. It is prepared by the demineralization of ordinary water by passing through cation and anion exchange resin.
7. Coulomb is the smallest unit of electricity.
8. Electrochemical equivalent of a metal = $\dfrac{\text{Equivalent weight of metal}}{96{,}500}$

   Chemical equivalent or equivalent weight = electrochemical equivalent (z) × 96,500.
9. Equivalent and molar conductance of a solution increases with the dilution whereas the specific conductance of the solution decreases.

(Continued)

NOTEWORTHY POINTS (*Continued*)

10. Equivalent conductivity increases with pressure because viscosity of a dilute solution decreases with increasing pressure.
11. Transport number: Transport number or transference number is defined as the fraction of the current carried by an ion.

$$\text{Transport number} = \frac{\text{Current carried by ion}}{\text{Total current carried}}$$

Transport number of cation + transport number of anion = 1

$$n_c + n_a = 1$$

REVISION QUESTIONS

1. Define the following terms
   (a) Electrolysis
   (b) Specific conductance
   (c) Equivalent conductance
   (d) Molar conductance
   (e) Degree of dissociation
   (f) Cell constant
2. How is the specific conductance of an electrolyte solution determined? Describe the experimental method.
3. (a) Explain the factors that affect the conductance of an electrolyte.
   (b) Calculate equivalent conductivity from the following data:
   (i) 0.1 N solution has a resistance 2.5×10^2 ohms, (ii) Cell constant = $1.15\ \text{cm}^{-1}$
4. (a) Write a short note on variation of equivalent conductance with dilution
   (b) Write down the methods for determination of equivalent conductance at infinite dilution for strong and weak electrolytes.
   (c) How would you measure the conductivity of an aqueous salt solution?
5. Define specific conductance, conductivity, equivalent conductivity and molar conductivity. What is the relationship between them? What effect do they produce upon dilution?
6. Define equivalent conductance and molar conductance. Establish the relation between specific conductance and equivalent conductance.
7. A solution of copper sulphate weighing 20 g was electrolyzed using 0.02 faraday of electricity. Calculate the weight of the resulting solution.
8. A current 4.0 amperes is passed for 8 h between nickel electrodes in 500 ml of 2 M solution of nickel nitrate. What will be the molarity of the solution at the end of electrolysis?
9. (a) Explain ionic conductance, transport number of an ion and absolute ionic velocity
   (b) The absolute velocity of Ag^+ is 0.00057 cm sec^{-1} and of the NO_3^- is 0.00069 cm sec^{-1}. Assuming complete dissociation, calculate the specific conductivity of 0.01 M solution of silver nitrate.
10. What do you understand by the terms: transport number and ionic mobility of an ion? How are these related to each other? Write their units.
11. Describe Hittorf's method for determining the transport number of Ag^+ and NO_3^- ions in a solution of $AgNO_3$ when the electrodes are made of silver.

12. Write a short notes on:
    (a) Transport number (b) Kohlrausch law of independent migration of ions
13. How does Kohlrausch's law helps in determining the dissociation constants of a weak acid by conductance measurements?
14. (a) Derive the expression for the determination of transport number by Hittorf's method when electrodes are not attacked?
    (b) Explain the factors on which the transport number depends.
15. Explain how is Kohlrausch's law helpful in determining the equivalent conductance of weak electrolyte at infinite dilution.

MULTIPLE CHOICE QUESTIONS

1. On passing electrical current through an electrolyte solution
   (a) cations move towards anode
   (b) anions move towards cathode
   (c) cations move towards cathode and anions towards anode
   (d) both cations and anions move in same directions
2. The conductivity of an electrolyte is due to the
   (a) presence of ions in the electrolyte
   (b) free movement of ions in the solutions
   (c) reunion of ions in the solution
   (d) release to heat energy due to ionization
3. The electrical conductivity of an electrolyte depends upon
   (a) the number of molecules in the electrolyte
   (b) the number of ions present in the electrolyte
   (c) the number of ions present in the solution
   (d) the number of molecules of the solvent
4. If λ_∞ is the equivalent conductance at infinite dilution and λv the equivalent conductance of the electrolyte at the dilution v, the degree of dissociation is given by
   (a) $\alpha = \dfrac{\lambda_\infty}{\lambda_v}$ (c) $\alpha = \lambda_\infty - \lambda_v$
   (b) $\alpha = \dfrac{\lambda_v}{\lambda_\infty}$ (d) $\alpha = \lambda_v - \lambda_\infty$
5. One faraday is equal to
   (a) 96.500 coulombs
   (b) 9,650 coulombs
   (c) 965.00 coulombs
   (d) 96,500 coulombs
6. The statement of Kohlrausch's law is
   (a) The equivalent conductance of an electrolyte at infinite dilution is equal to the products of equivalent conductances of the component ions.
   (b) The equivalent conductance of an electrolyte at infinite dilution is equal to the difference of equivalent conductances of the component ions.
   (c) The equivalent conductance of an electrolyte at infinite dilution is equal to the sum of equivalent conductances of the component ions.
   (d) None of the above.
7. Kohlrausch's law can be expressed as
   (a) $\lambda_\infty = \lambda_\infty - \lambda_c$ (c) $\lambda_\infty = \lambda_\infty + \lambda_c$
   (b) $\lambda_\infty = \lambda_c^- - \lambda_\infty$ (d) $\lambda_\infty = \lambda_\infty \times \lambda_c$
8. If the equivalent conductance of a certain solution of acetic acid is 39.07 ohm^{-1} cm^2 $eqvt^{-1}$. If λ_α of CH_3COOH is 390.7, the degree of dissociation of acetic acid is
   (a) 0.1 (c) 0.5
   (b) 0.2 (d) 0.75
9. The equivalent conductance at infinite dilution of NaCl, HCl and CH_3COONa at 25 °C is 126.0, 426.0 and 91.0 ohm^{-1} cm^2, respectively.

The equivalent conductance of acetic acid at infinite dilution at 25 °C will be

(a) 643.0 (c) 217.0
(b) 517.0 (d) 391.0

10. If the degree of dissociation of weak electrolyte at a certain temperature is 0.1, the percentage of the molecules undissociated will be
(a) 10% (c) 50%
(b) 20% (d) 90%

11. For strong electrolytes, the degree of dissociation is
(a) nearly equal to one
(b) nearly equal to zero
(c) nearly equal to infinity
(d) nearly equal to 0.5

12. The equivalent conductance at infinite dilution of NH_4Cl, NaOH and NaCl at 18°C is, respectively, 129.8, 227.4 and 108.9 $ohm^{-1}\ cm^2\ eqvt^{-1}$. The equivalent conductance of NH_4OH at infinite dilution at 18°C will be
(a) 128.8 (c) 227.4
(b) 108.9 (d) 238.3

13. The transport number of an ion depends on the mobility of that ion. It also depends upon the
(a) mobility of the solvent
(b) mobility of the electrolyte
(c) mobility of the other ion with which it is associated
(d) none of the above

14. On passing electrical current through an electrolyte solution, the cations
(a) move towards cathode with the speed equal to that of anions towards anode
(b) move with faster speed than that of anions
(c) move with different speed as compared to that of anions
(d) move with slower speed than that of anions

15. The Hittorf's rule state that
(a) The loss of concentration around any electrode is proportional to the speed of the ions moving towards is.
(b) The loss of concentration around any electrode is proportional to the speed of the ions moving away from it.
(c) The loss of concentrations around both the electrodes is proportional to the sum of speed of cations and anions.
(d) None of the above.

16. The Hittorf's rule can be represented by the expression

(a) $\dfrac{\text{Fall around anode}}{\text{Rise around cathode}} = \dfrac{\text{speed of cation}}{\text{speed of anion}}$

(b) $\dfrac{\text{Fall around anode}}{\text{Fall around cathode}} = \dfrac{\text{speed of anion}}{\text{speed of cation}}$

(c) $\dfrac{\text{Fall around anode}}{\text{Fall around cathode}} = \dfrac{\text{speed of cation}}{\text{speed of anion}}$

(d) $\dfrac{\text{Rise around cathode}}{\text{Fall around anode}} = \dfrac{\text{speed of anion}}{\text{speed of cation}}$

17. One faraday of electricity is passed through an HCl electrolyte solution. Select the correct electrode result
(a) 1 g of chloride ions is deposited at the anode.
(b) 1 g of hydrogen ions is deposited at the cathode.
(c) 5 g of hydrogen ions the deposited at the anode.
(d) 35 g of chloride ions are deposited at the anode.

18. One faraday will oxidize mole(s) of Cu to Cu^{2+} ions.
(a) 0 (c) ¼
(b) ½ (d) 1

19. Select the incorrect statement about the chemical activity at electrodes during electrolysis
(a) Anions give up electrons.
(b) Cations take up electrons.
(c) Oxidation occurs at the anode.
(d) Proton transfer occurs in the reactions.

20. How many grams of copper would be produced by the reduction of Cu^{2+} if 3.0 amperes

of current are passed through a copper (II) nitrate solution for one hour?
(a) 18.20 (c) 31.80
(b) 3.56 (d) 63.50

21. A current of 5.0 amperes is passed through molten magnesium chloride, $MgCl_2$ for 3.0 h. How many grams of magnesium, Mg (s), can be produced by this reduction? Faraday constant: 1 F = 96,485 C/mol; molar atomic mass (Mg) = 24.305 g/mol.
(a) 0.30 g (c) 17.6 g
(b) 6.80 g (d) 24.5 g

22. Specific conductance is the conductance of
(a) one centimetre cube of solution of an electrolyte
(b) one centimetre cube of a solid electrolyte
(c) one gram of the solution of an electrolyte
(d) one gram of the solid electrolyte

23. The units of specific conductance are
(a) ohm cm (c) ohm^{-1} cm
(b) ohm cm^{-1} (d) ohm^{-1} cm^{-1}

24. The equivalent conductance of a solution of an electrolyte
(a) increase with solution
(b) decreases with dilution
(c) does not vary with dilution
(d) none of these

25. The units of equivalent conductance are
(a) ohm cm eqvt (b) ohm^{-1} cm^{-1} $eqvt^{-1}$
(c) ohm^{-1} cm^2 $eqvt^{-1}$ (d) ohm^{-1} cm^{-2} $eqvt^{-1}$

26. The cell constant can be obtained by
(a) dividing specific conductance by observed conductance
(b) dividing observed conductance by specific conductance
(c) multiplying specific conductance by observed conductance
(d) multiplying specific conductance by equivalent conductance

27. The cell constant is the ratio of
(a) distance between electrodes to area of electrode
(b) area of electrode to distance between electrodes
(c) specific conductance to area of electrode
(d) specific conductance to distance between the electrodes

28. Which out of the following will decompose on passing electric current?
(a) glucose (c) silver nitrate
(b) urea (d) ethyl alcohol

29. On passing 0.1 faraday of electricity through $AlCl_3$, the amount of aluminium metal deposited on the cathode is (at mass of AI = 27)
(a) 0.27 g (c) 0.9 g
(b) 0.81 g (d) 135.g

30. On passing 96,500 coulombs of electricity through a dilute solution of an acid, the volume of hydrogen collected at STP is
(a) 1,120 ml (c) 2,240 ml
(b) 11,200 ml (d) 22,400 ml

ANSWERS

1. (c)	6. (c)	11. (a)	16. (c)	21. (b)	26. (a)
2. (b)	7. (c)	12. (d)	17. (b)	22. (a)	27. (d)
3. (c)	8. (a)	13. (c)	18. (b)	23. (d)	28. (c)
4. (b)	9. (d)	14. (c)	19. (d)	24. (a)	29. (c)
5. (d)	10. (d)	15. (b)	20. (b)	25. (c)	30. (b)

Electromotive Force and Oxidation–Reduction System 12

CHAPTER OBJECTIVES

- 12.1 Introduction
- 12.2 Single Electrode Potential
- 12.3 Standard Electrode Potential
- 12.4 Measurement of Single Electrode Potential
- 12.5 Reference Electrodes
- 12.6 Electrochemical Series
- 12.7 Cell Potential or EMF of a Cell
- 12.8 Derivation of Nernst Equation (Concentration Dependence of Electrode Potential)

Electromotive force is the energy per unit charge that is converted reversibly from chemical, mechanical, or other forms of energy into electrical energy in a battery or dynamo.

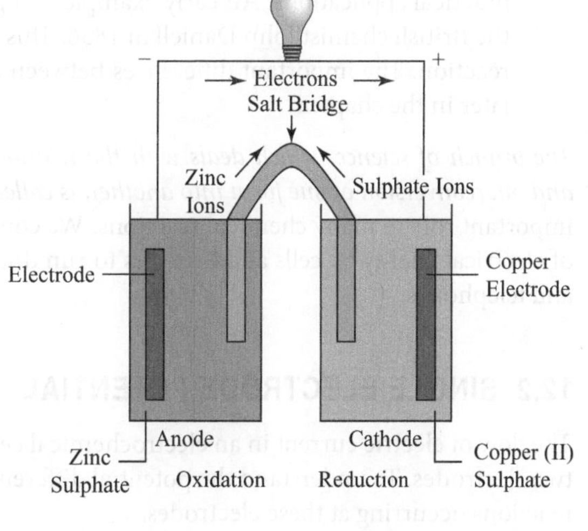

12.1 INTRODUCTION

We know that electric current can be used to bring about a chemical reaction (electrolysis). However, it is possible to carry out the reverse process, i.e., a chemical reaction can be used to generate electric current. The chemical reactions employed for generating electric current are oxidation–reduction reactions and are called **redox reactions**. In these reactions, electrons are lost (oxidation) by one substance and are accepted (reduction) by a second substance. In general, a chemical change, which is accompanied by flow of electrons, is known as an **electrochemical change**. The electrochemical changes are of two types and corresponding to these two types of changes there are two types of cells namely (i) electrolytic cells and (ii) electrochemical cells. We provide a brief description of these two types below:

(i) Electrolytic cells: There are chemical changes, which are brought about by passing electric current through electrolytes in the molten state, or through their aqueous chemical energy, i.e., current can be used to bring about a chemical reaction. *This phenomenon of decomposition of an electrolyte with the help of an electric current is called electrolysis. A device in which electrolysis is carried out is called an electrolytic cell.*

(ii) Electrochemical cells: There are chemical changes, which produce electric current as a result of the electron transfer from one substance to another. A device employed to convert the chemical energy of a redox reaction into electrical energy is called an electrochemical cell.

Electrochemical reactions can be accomplished by two methods: (a) direct redox reaction and (b) indirect redox reaction.

(a) Direct redox reaction: When a redox reaction is carried out directly, chemical energy in the form of heat is produced. For example, if we place a zinc rod in a solution of copper sulphate, the following reaction occurs and a rise in temperature of the solution is noticed.

$$Zn\ (s) + Cu^{2+}\ (aq) \longrightarrow Zn^{2+}\ (aq) + Cu\ (s)$$

(b) Indirect redox reaction: When a redox reaction is carried out indirectly, chemical energy is changed is converted into electrical energy. In this case, oxidation half-reaction and reduction half-reaction are carried out in separate beakers. These cells are very important, because of their many practical applications. An early example of a galvanic cell is a Daniell cell, which was invented by the British chemist John Daniell in 1836. This cell was constructed based on the above given redox reaction. The important differences between an electrolytic cell and a galvanic cell are discussed later in the chapter.

The branch of science, which deals with the relationship between electrical energy and chemical energy and interconversion of one form into another, is called ***electrochemistry***. Electrochemical energy plays an important role in many chemical reactions. We come across many processes in daily life such as supply of electrical energy by cells and batteries to run different appliances such as radios, transistors, watches and telephones.

12.2 SINGLE ELECTRODE POTENTIAL

The flow of electric current in an electrochemical cell indicates that a potential difference exists between two electrodes. To understand the potential difference between two electrodes, let us consider the redox reactions occurring at these electrodes.

When an electrode, say a metal, is immersed in a solution of its ions, then any of the following three possibilities can take place.

(i) The metal ions (M^{n+}) may collide with the electrode and do not undergo any change.

(ii) M^{n+} ions may collide with the electrode, gain electrons and get converted into metal atoms (i.e., the ions are reduced):

$$M^{n+} + ne^- \longrightarrow M \qquad \text{(reduction)}$$

(iii) M atoms on the electrode may lose electrons to the electrode and become M^{n+} ions and dissolve in the solution (i.e., oxidation occurs):

$$M \longrightarrow M^{n+} + ne^- \qquad \text{(oxidation)}$$

These possibilities are shown in Fig. 12.1.

Now, if the metal has relatively a high tendency to get oxidized, its atoms will lose electrons readily and M^{n+} ions would dissolve in the solution. The electrons lost would be accumulated on the metal electrode and hence the electrode acquires a slight negative charge with respect to the solution. Some of the M^{n+} ions in the solution will take up electrons and become M atoms. After some time, equilibrium will be established as shown below:

$$\underset{\text{electrode}}{M_{(s)}} \longrightarrow \underset{\text{solution}}{M^{n+}} + \underset{\text{on electrode}}{2ne^-}$$

When such equilibrium is attained, it results in the separation of charges (negative on the electrode with respect to the solution).

Similarly, if the metal ions have relatively greater tendency to be reduced, they will absorb electrons from the electrode. As a result, a net positive charge will be developed on the electrode with respect to the solution. This will also result in the separation of charges (positive on the electrode with respect to the solution).

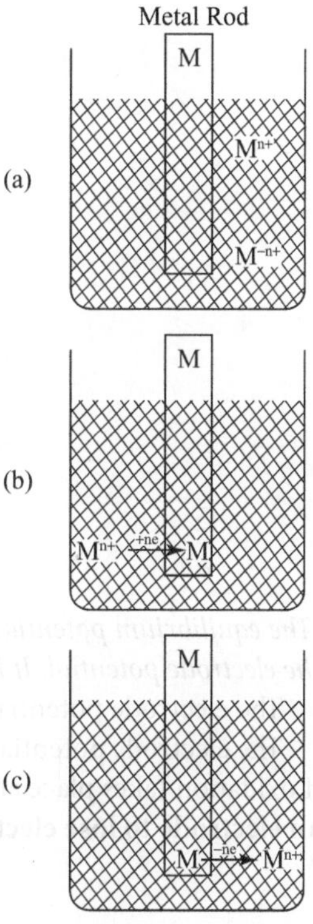

Figure 12.1 Electrode equilibrium

Metal ions collide but do not undergo any change. Metal ions gain electrons and get converted to metal, i.e., reduction occurs (electrons are lost by metal rod and it acquires +ve charge).

Metal atoms on electrode lose electrons to form matel ions, i.e., oxidation occurs. Electrons lost by metal accumulates on metal rod (metal rod acquires –ve charge). Due to separation of charges between the electrode and the solution, an electrical potential is set up between the metal electrode and its solution. *The electrical potential difference set up between the metal and its solution is known as an electrode potential. Thus, the electrode potential is a measure of tendency of an electrode in a half-cell to gain or lose electrons.*

12.2.1 Definition

When a metal is immersed in an aqueous solution containing the metal ions, the metal atoms have some tendency to pass into the solution as cations or metal ions start depositing on the metal surface (Fig. 12.2). In the beginning, both these changes occur with different speeds but soon equilibrium is established.

$$M \underset{\text{Reduction}}{\overset{\text{Oxidation}}{\longleftrightarrow}} M^{n+} + ne^-$$

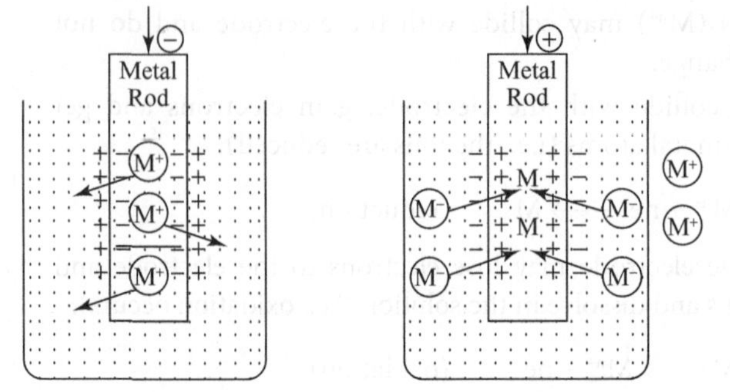

(a) Electrode Potential: Oxidation (De-electronation)

(b) Electrode Potential: Reduction (Electronation)

Figure 12.2 Oxidation and reduction potential

'The equilibrium potential difference between the metal electrode and the solution surrounding it' is called the electrode potential. It is also defined as the tendency of an electrode to lose or gain electrons.

The electrode potential is commonly expressed in two ways:

The electrode potential is ***oxidation potential and Reduction potential*** termed as oxidation potential, if oxidation takes place at the electrode. Oxidation involves loss of electrons. Therefore, **the tendency of an electrode to lose electrons is called oxidation potential.**

$$M \longrightarrow M^{n+} + ne^-$$

The metal ions dissolve in the solution and electrons remain on the metal rod making it negatively charged. The tendency of the metal to change into ions is also known as **electrolytic solution pressure.** The electrode potential is called reduction potential if reduction takes place at the electrode.

Reduction involves gain of electrons; so, **the tendency of an electrode to gain electrons is called its reduction potential.**

$$M^{n+} + ne^- \longrightarrow M$$

As metal ions start depositing on the metal surface, this developes a positive charge on the metal rod. Since oxidation is just a reverse of reduction, therefore, reduction potential is obtained from the oxidation potential by simply changing the sign.

In general, for an electrode

Oxidation potential = −Reduction potential

For example, in a zinc electrode, the standard oxidation potential is represented as

$$E^0_{Zn/Zn^{2+}} = 0.76\text{ V}$$

and standard reduction potential as

$$E^0_{Zn^{2+}/Zn} = -0.76\text{ V}$$

It is a common practice to express all the electrode potentials as reduction potentials.

The differences between oxidation potential and reduction potential may be summarized as follows:

Oxidation Potential	Reduction Potential
1. The tendency of an electrode to lose electrons when in contact with the solution of its own ions in a half-cell is known as oxidation potential.	1. The tendency of an electrode to gain electrons when in contact with solution of its own ions in a half-cell is known as reduction potential.
2. It is inversely proportional to the concentration of ions of the electrode.	2. It is directly proportional to the concentration of ions of the electrode.

MEMORY FOCUS

1. When a metal is dipped in a solution containing its own ion, the stress exerted on the metal surface is described in terms of electrode potential.
2. The tendency of an electrode to gain electrons is called its reduction potential.
3. The tendency of an electrode to lose electrons is called its oxidation potential.
4. Reduction potential (RP) denotes the ability of an electrode to get reduced.
5. Oxidation potential (OP) denotes the ability of an electrode to get oxidized.
6. Electrode potential depends on
   (a) concentration of electrolyte and
   (b) nature and purity of electrode
7. Electrode potential of pure metal in contact with its own ion of unit activity (concentration of 1 molar) is called standard electrode potential. It is denoted as E°.
8. Electrode potential of standard hydrogen electrode (SHE) is fixed as 0 V.
9. SHE is the one in which hydrogen gas is bubbled at 1 atm. pressure through a solution of 1 molar H^+ over a platinised surface.
   SHE is represented as Pt, H_2 (1 atm)/H^+ (1 M); E° = 0.00 V
10. For a given electrode, both RP and OP have the same value, but with opposite signs.
11. To avoid confusion, IUPAC has directed to use only standard reduction potential (SRP) values which are referred as electrode potentials.
12. When SHE is acting as a cathode in a cell, at the other electrode oxidation takes place spontaneously, such electrodes act as anodes.

12.3 STANDARD ELECTRODE POTENTIAL

The electrode potential of a half-cell, in which the metal is suspended in a solution of one molar (1 M) *concentration and the temperature is kept at* 25 °C (298 K) *is called standard electrode potential. It is usually denoted by E°.*

It is not possible to determine the absolute value of electrode potential of a single electrode. It is because a single electrode constitutes only a half-cell. A half-cell would not lose or gain electrons by itself. The loss or gain of electrons can take place only in a complete circuit containing two half-cells connected to each other. Therefore, the electrode potential can only be measured by using some electrode as a reference electrode: the reference electrode commonly employed is the standard or NHE (Fig. 12.3). The potential of NHE is taken as zero.

Hydrogen electrode: It is a familiar example of gas electrodes. It consists of a platinized platinum strip dipped in an acid solution through which hydrogen gas can be bubbled as shown in Fig. 12.3. In the NHEs, H_2 gas at 1 atmospheric pressure is passed into 1 M HCl at 298 K.

The reversible changes are represented as

$$2H^+_{(aq)} + 2e^- \longrightarrow H_2\,(g)$$

and

$$E^{\circ}_{H^+/H_2} = 0;\ E^{\circ}_{H_2/H^+} = 0$$

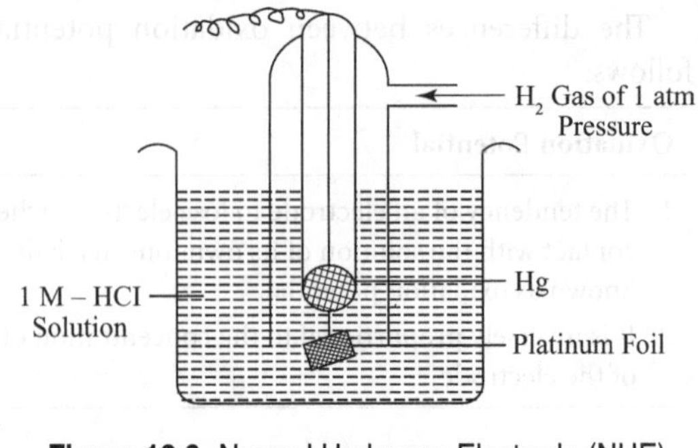

Figure 12.3 Normal Hydrogen Electrode (NHE)

The NHE is represented as Pt; H_2 (1 atm/HCl (1 M) when placed as anode and HCl (l M)/H_2 (1 atm); Pt when placed as cathode.

12.4 MEASUREMENT OF SINGLE ELECTRODE POTENTIAL

To measure the relative value of electrode potential of any given electrode, a cell is set up which consists of the given electrode as one of the electrode and the second electrode is NHE. The electromotive force (EMF) of the cell is measured. The measured EMF of the cell would give the electrode potential of the electrode under investigation because the electrode potential of NHE is taken as zero. The direction of flow of current further indicates whether oxidation takes place or the reduction take place at the electrode under investigation with respect to hydrogen electrode. Accordingly, the electrode potential is termed as oxidation or reduction potential.

12.4.1 Sign Conventions

According to IUPAC convention, potential is given a positive sign if reduction occurs and a negative sign if oxidation occurs at the electrode under investigation with respect to the hydrogen electrode. Accordingly, the electrode potential is termed as oxidation or reduction potential.

For measurement of standard electrode potential, a cell is set up in which one-half of the cell consists of metal and its ion at 1 M concentration and the other half of the cell consists of the hydrogen electrode. The temperature in both the half-cells is kept at 25 °C.

(i) **Measurement of electrode potential of Zn^{2+}/Zn electrode:** When coupled with NHE, some electrodes undergo spontaneous oxidation and permit loss of electrons. Zinc electrode when combined with NHE is oxidized to Zn^{2+} ions. In other words, the zinc electrode pushes electrons into the external circuit and forms the anode. By convention, the anode is shown on the left (Fig. 12.4) and it is said to possess a negative value of *reduction electrode potential* (–0.76 V).

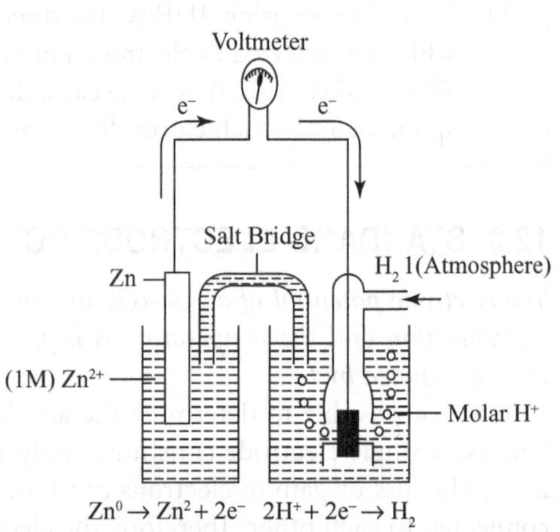

Figure 12.4 Zinc electrode acts as anode when it is connected to NHE

(ii) **Measurement of electrode potential of Cu^{2+}/Cu electrode:** In case the electrode coupled with NHE undergoes spontaneous reduction, it allows oxidation of NHE. When a copper electrode dipped in Cu^{2+} solution is coupled with NHE, Cu^{2+} ions are reduced to Cu(s) and at NHE, H_2 is oxidized to H^+ ion by pushing electrons into the external circuit (Fig. 12.5). Copper electrode acts as a cathode and it is said to have positive value of *reduction electrode potential* (0.34 V).

12.5 REFERENCE ELECTRODES

An electrode whose electrode potential in accurately known or whose electrode potential has been arbitrarily fixed is called a reference electrode. In other words, it is an electrode of standard potential with which we can compare the potentials of other electrodes.

It is combined with any electrode whose electrode potential has to be determined to form an electrochemical cell through a salt bridge. The EMF of the cell thus formed is noted and the electrode potential of the electrode under test can be known. In the case of NHE, the electrode potential is taken as zero. Therefore, the EMF of the cell itself becomes the electrode potential of the electrode concerned. A reference electrode should satisfy the following conditions:

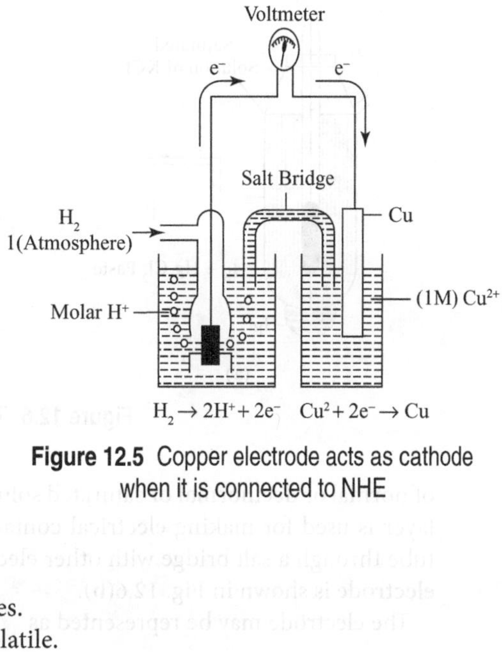

Figure 12.5 Copper electrode acts as cathode when it is connected to NHE

(i) It should be easily set up.
(ii) It should not be bulky or space occupying.
(iii) It should not be easily poisoned by the impurities.
(iv) It should be usable even when the solutes are volatile.
(v) Its electrode potential should not change even by change in pH of the solution.

Reference electrodes are divided into two types, which we discuss in the following sections.

12.5.1 Primary Reference Electrodes

Standard hydrogen electrode: For details, please refer to section 12.3. It is the primary reference electrode employed for cell potential measurements. Its potential is taken as zero. It is combined with the electrode whose electrode potential has to be measured to form a cell. The EMF of the cell so constituted is measured, which gives the potential of the electrode since potential of SHE is taken as zero. However, it is quite difficult to set up a hydrogen electrode. Moreover, it cannot be used in solutions containing redox solution. It is also affected by oxidizing agents such as MnO_4^-, $Cr_2O_7^{2-}$ and Fe^{3+} and some compounds of Hg, As and S. Consequently, this electrode cannot be used for solutions containing these ions. In such cases, some other electrodes are employed which are standardized using SHE. These electrodes are called as secondary reference electrodes.

12.5.2 Secondary Reference Electrodes

They include calomel electrode, Ag–AgCl electrode, glass electrode, quinhydrone electrode, etc.

(i) **Calomel electrode:** It consists of a tube, at the bottom of which there is a layer of mercury and mercurous chloride, Hg_2Cl_2 (Fig. 12.6). The remaining portion of the cell is filled with a solution

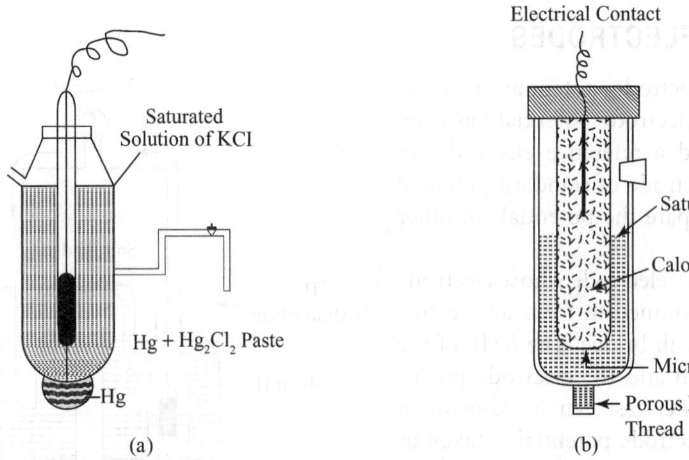

Figure 12.6 The calomel electrode

of normal or decinormal or saturated solution of KCl. A platinum wire dipping into the mercury layer is used for making electrical contact. The electrode is connected with the help of a side tube through a salt bridge with other electrode to make a complete cell. A compact form of the electrode is shown in Fig. 12.6(b).

The electrode may be represented as

$$\text{Hg; Hg}_2\text{Cl}_2\text{ (s) KCl (solution)}$$

When oxidation occurs at this electrode, the following reaction takes place

$$2Hg(l) \rightarrow Hg_2^{2+} + 2e^-$$
$$Hg_2^{2+} + 2Cl^- \rightarrow Hg_2Cl_2(s)$$
$$\overline{2Hg(l) + Hg_2Cl_2(s) + 2e^-}$$

When reduction occurs at this electrode, the following reactions take place

$$Hg_2Cl_2(s) \rightarrow Hg_2^{2+} + 2Cl^-$$
$$Hg_2^{2+} + 2e^- \rightarrow 2Hg(l)$$
$$\overline{Hg_2Cl_2(s) + 2e^- \rightarrow 2Hg(l) + 2Cl^-}$$

The electrode is thus, reversible with respect to chloride ion (anion). The equilibrium reaction may be represented as

$$2Hg + 2Cl^- \longrightarrow Hg_2Cl_2\,(s) + 2e^-$$

The potential of the calomel electrode depends upon the concentration of the potassium chloride solution used in the calomel electrode. If potassium chloride used is as saturated solutions, the electrode is known as **saturated calomel electrode (SCE)** and if potassium chloride solution is IN, the electrode is known as **normal calomel electrode (NCE)** while for 0.1 N potassium chloride solution the electrode is referred to as **decinormal calomel electrode (DNCE)**. The reduction potentials of calomel electrodes on hydrogen scale at 298 K are as follows:

Saturated KCl:	**0.2422 V**
1.0 KCl:	**0.2800 V**
0.1 N KCl:	**0.3338 V**

Calomel electrode is preferred to NHE as a reference electrode. Because it can be set up conveniently and once set, it can be used for long. In the case of NHE, it is difficult to maintain the supply of H_2 gas at 1 atm pressure for long.

(ii) Silver–Silver chloride electrode: The electrode is represented as

Ag: AgCl (s) KCl solution; or Ag + AgCl/Cl⁻ (ACl⁻)

Its electrode reaction is

$$Ag(s) + Cl^- \longrightarrow AgCl(s) + e^-$$

It is prepared by depositing a thin layer of AgCl electrolytically on a silver or platinum wire and is then immersed in a solution containing Cl^- ions (Fig. 12.7).

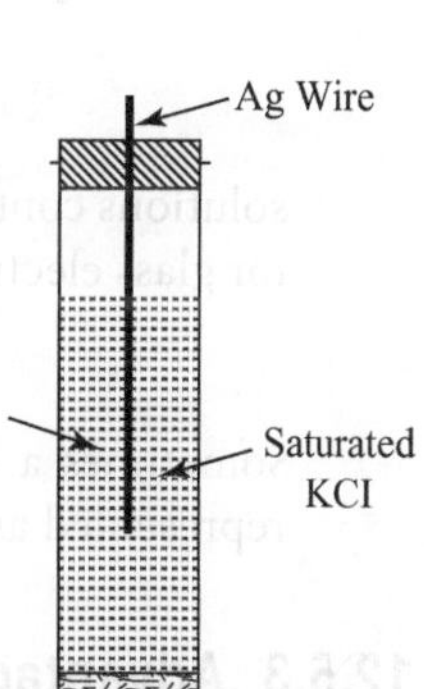

Figure 12.7 Ag-AgCl Elects Electro

The chloride ion concentration, i.e., of hydrochloric acid solution must be indicated, or at least any of the other sparingly soluble chloride solution. Other insoluble salts employed in making electrode of this type are AgBr, Agl, Hg_2Cl_2, sparingly soluble sulphates and oxalates for making other electrodes of Ag–AgCl type.

The Ag–AgCl electrode is reversible with respect to chloride ion. Its potential at 298 K with saturated KCl is 0.290 V and that with 0.1 N KCl is 0.199 V w.r.t. SHE.

(iii) Glass electrode: When two solutions of different pH are separated by a glass membrane, a potential is established across the membrane whose magnitude depends on the difference in pH of the two solutions, a potential is established between the glass and the solution. The value of the potential is a function of the particular glass used and is given by the equation:

$$E_g = E_g^\circ + \frac{RT}{F}\ln(H^+) = E_g^\circ - \left(\frac{2.303\ RT}{F}\right)$$

The electrode reaction is assumed to involve the reduction of H^+ ions. The glass electrode thus functions in the same manner as a reversible hydrogen electrode.

Construction of glass electrode: It consists of a thin walled glass bulb with a long neck and is made up of a special type of glass of low melting point and comparatively high conductivity [Figs. 12.8(a) and 12.8(b)]. A solution of known pH (usually 1 N HCD is taken in the bulb. A platinum electrode is introduced into the solution and a pinch of quinhydrone is added to it. A glass electrode is used for determining pH of the solution, especially for coloured solutions and

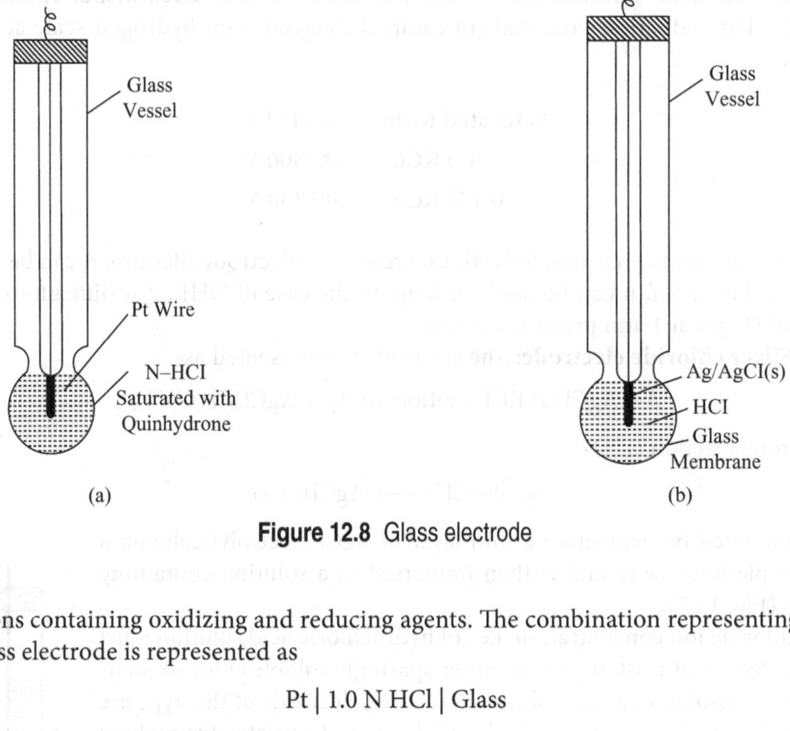

Figure 12.8 Glass electrode

solutions containing oxidizing and reducing agents. The combination representing the half-cell for glass electrode is represented as

$$\text{Pt} \mid 1.0\ \text{N HCl} \mid \text{Glass}$$

Sometimes a silver–silver chloride electrode instead of platinum is also employed. It is then represented as: Ag/AgCl(s) | 0.1 M HCl | Glass.

12.5.3 Advantages of Glass Electrode

The advantages of using a glass electrode are listed below:

(i) It is the most simple and convenient to use and gives accurate results.
(ii) It can be used in turbid, coloured and even colloidal solutions.
(iii) It can used in alkaline as well as in solutions containing oxidizing agents.
(iv) It is not easily poisoned by any kind of cations and anions present in the solutions.
(v) It can measure pH value of solutions available in small quantities (only a few millilitres of the solutions are required).

(vi) Equilibrium is attained rapidly.
(vii) It does not require the use of any salt bridge.

12.5.4 Limitations of Glass Electrode

(i) For EMF measurement using glass electrodes, ordinary potentiometers cannot be employed because of high-resistance glass. Either quadrant electrometers or vacuum tube voltameters (VTVMs) that require practically no current for their operations must be employed.
(ii) It can be used up to a pH of 13 but being sensitive to Na^+ ions above a pH of 9, the linear relationship between glass electrode and pH is altered. Therefore, an alkaline error is introduced beyond a pH of 9, which should be taken into consideration.
(iii) In strongly acidic solutions of pH < 1, the salt effect due to anions is observed.
(iv) It does not function properly in some organic solvents such as pure alcohol.

12.6 ELECTROCHEMICAL SERIES

The standard reduction potentials of a number of electrodes along with their corresponding half-cell reactions are given in Table 12.1. The table in which the reduction potentials of various electrodes have been arranged in the increasing order is called electrochemical series or activity series or electromotive series.

Table 12.1 Standard reduction electrode potentials at 298 K.

Electrode	Electrode Reaction	E° (Volts)	Electrode	Electrode Reaction	E° (Volts)
Li^+/Li	$Li^+ + e^- \rightleftharpoons Li$	−3.045	Sn^{2+}/Sn	$Sn^{2+} +.2e^- \rightleftharpoons Sn$	−0.136
K^+/K	$K^+ + e^- \rightleftharpoons K$	−2.925	Pb^{2+}/Pb	$Pb^{2+} + 2e^- \rightleftharpoons Pb$	−0.126
Cs^+/Cs	$Cs^+ + e^- \rightleftharpoons Cs$	−2.923	Pe^{3+}/Pe	$Pe^{3+} + 3e^- \rightleftharpoons Fe$	−0.036
Ba^{2+}/Ba	$Ba^{2+} + 2e^- \rightleftharpoons Ba$	−2.906	$H^+/H_2/Pt$	$H^+ + 2e^- \rightleftharpoons H_2$	0.0
Ca^{2+}/Ca	$Ca^{2+} + 2e^- \rightleftharpoons Ca$	−2.866	$Cu^{2+}, Cu^+/Pt$	$Cu^{2+} + 2e^- \rightleftharpoons Cu^+$	+0.153
Na^+/Na	$Na^+ + e^- \rightleftharpoons Na$	−2.71	Cu^{2+}/Cu	$Cu^{2+} + 2e \rightleftharpoons Cu$	+0.337
Mg^{2+}/Mg	$Mg^{2+} + 2e^- \rightleftharpoons Mg$	−2.36	$I_2/I/Pt$	$I_2 + 2e^- \rightleftharpoons 2I^-$	+0.534
$A1^{3+}/AI$	$A1^{3+} + 3e^- \rightleftharpoons AI$	−1.66	Ag^+/Ag	$Ag^+ + e^- \rightleftharpoons Ag$	+0.799
Zn^{2+}/Zn	$Zn^{2+} + 2e^- \rightleftharpoons Zn$	−0.76	Hg^{2+}/Hg	$Hg^{2+} + 2e^- \rightleftharpoons Hg$	+0.854
Cr^{3+}/Cr	$Cr^{3+} + 3e^- \rightleftharpoons Cr$	−0.74	Br_2/Br^- Pt	$Br_2 + 2e^- \rightleftharpoons 2Br^-$	+1.066
Fe^{2+}/Fe	$Fe^{2+} + 2e^- \rightleftharpoons Fe$	−0.44	Cl_2/Cl^- Pt	$Cl_2 + 2e^- \rightleftharpoons 2Cl^-$	+1.359
Cd^{2+}/Cd	$Cd^{2+} + 2e^- \rightleftharpoons Cd$	−0.403	Au^{3+}/Au	$Au^{3+} + 3e^- \rightleftharpoons Au$	+1.498
Co^{2+}/Co	$Co^{2+} + 2e^- \rightleftharpoons Co$	−0.28	F_2/F^- Pt	$F_2 + 2e^- \rightleftharpoons 2F$	+2.870
Ni^{2+}/Ni	$Ni^{2+} + 2e^- \rightleftharpoons Ni$	−0.25			

12.6.1 Applications of Electrochemical Series

We have studied the detail of various applications of electrochemical series +2 classes. A mention of these may be made again as given below:

(i) ***To predict the relative ease of oxidation and reduction of various elements:*** In the electrochemical series, F_2 has the higher reduction potential with +ve sign (or reduced most easily) and Li^+ has the lowest reduction potential with –ve sign (or reduced with greatest difficulty). In other words, F_2 is the strongest oxidizing agent or Li^+ ion is the weakest oxidizing agent. Thus, the substances, which have lower reduction potential, have the tendency to be oxidized easily while those, which have higher reduction potential, have the tendency to be reduced easily.

(ii) ***To calculate the EMF of an electrochemical cell:*** This is discussed in section 12.7.1.

(iii) ***To predict the feasibility of a redox reaction:*** This feasibility of a redox reaction is found from the E° values of the two electrodes.

A redox reaction is feasible in case the value of E° cell calculated from the E° values of the electrodes comes out to the positive. The species undergoing reduction have higher reduction potential than the species undergoing oxidation.

(iv) ***To predict whether a metal reacts with acid to give H2 gas or not:*** Let us take the reaction of H_2SO_4 with Ag metal.

$$Ag + H_2SO_4 \longrightarrow Ag_2SO_4 + H_2$$

or

$$Ag + 2H^+ \longrightarrow 2Ag^+ + H_2$$

The cell involving the above reactions is represented as

$$Ag \mid Ag^+ \parallel H^+ \mid H_2$$

The E°_{cell} can be calculated as

$$E^\circ_{cell} = E^\circ_{(right)} - E^\circ_{(left)}$$

$$= E^\circ_{H^+,H} - E^\circ_{Ag^+,Ag} = 0 - 0.80 = -0.80V$$

Since E°_{cell} of the supposed reaction comes out to be negative, so this reactions does not take place. Hence Ag metal does not liberate H_2 gas from H_2SO_4.

Numerical Problems Based on Electrochemical Series

Example 1 **Determine the standard EMF of the following cell.**

$$\mathbf{Cu, Cu^{2+} \parallel Ag^+, Ag}$$

$$\left(\mathbf{E^\circ_{Cu^{2+},Cu} = 0.34\ V\ and\ E_{Ag^+,Ag} = 0.80\ V}\right)$$

Solution: Given that

$$E^\circ_{Cu^{2+},Cu} = 0.34\ V$$

and

$$E_{Ag^+,Ag} = 0.80\ V$$

For the given cell

Standard EMF,

$$E^\circ_{cell} = E^\circ_{cathode} - E^\circ_{anode}$$

$$= 0.80 - 0.34$$

$$= 0.46\ V$$

Example 2 Can 1 M $CuSO_4$ solution be stored in a silver vessel?

$$\left(E^{o}_{Cu^{2+},Cu} = 0.34\ V \text{ and } E_{Ag^{+},Ag} = 0.80\ V\right)$$

Solution: Let us find out whether Ag(s) reacts with Cu^{2+} or not, i.e., whether the following reaction is feasible or not.

$$Cu^{+2}\ (aq) + 2\ Ag\ (s) \longrightarrow Cu\ (s) + 2\ Ag^{+}\ (aq)$$

The cell for the above redox reaction

$$Cu + Cu^{2}+ \,||\, Ag^{+} \,|\, Ag$$

$$E^{\circ}_{cell} = E^{\circ}_{cathode} - E^{\circ}_{anode}$$

$$= 0.80 - 0.34 = 0.46\ V$$

The E_{cell} being positive, the reaction is feasible. Therefore, $CuSO_4$ solution cannot be stored in an Ag vessel.

MEMORY FOCUS

1. The list of standard reduction potential values arranged in the increasing order is commonly called as the electrochemical series.
2. Since this series also determines the activity of respective elements, it is also called as activity series.
3. The element with the least SRP is lithium (–3.05 V). Lithium electrode tops the table of activity series.
4. The element with highest SRP is fluorine (+2.87 V). Lithium electrode tops the table of activity series.
5. The smaller the SRP value of an element, the greater is its ability to act as a agent in a solution.
6. The greater the SRP value of an element, the greater is its ability to act as an oxidising agent in a solution.
7. Based on the electrochemical data, lithium is the best reducing element and fluorine is the best oxidising element.
8. An element can reduce all those present under it and can oxidise all those present above it, in the electrochemical series.
9. Metals whose SRP values are negative can displace hydrogen from acids.
10. Metals can be displaced from their salt solutions using another metal, as a reagent, with more negative SRP. value.
11. Alkali and alkaline earth metals are placed near the top of the series and are good reducing agents.
12. Electropositivity of a metal gradually decreases, going down in the series (with increase in SRP value).
13. For storing a salt solution, metal container with higher SRP cannot be used.
14. An active metal cannot be obtained by electrolysis of an aqueous solution. This is because H^{+} ions are easily reduced to give hydrogen.

12.7 CELL POTENTIAL OR EMF OF A CELL

An electrochemical cell consists of two electrodes, which have different electrode potentials. *The difference in electrode potential, which causes the current to flow from the electrode at a higher potential to the electrode at lower potential, is known as the cell potential.*

The differences of potential between the electrodes of a cell can be measured by using a voltmeter. The cell is sending the current to the circuit due to the potential difference. This potential difference is called the EMF of the cell, if no or very little current is allowed to flow at the time of measurement. It is measured by a potentiometer.

The differences between EMF of a cell and potential difference are summarized below:

Cell Potential or EMF	Potential Difference
Definition	**Definition**
1. It is the difference between electrode potentials of two electrodes when no current is flowing through the circuit, i.e., open circuit.	1. It is the difference between electrode potentials of two electrodes when current is flowing through the circuit, i.e., closed circuit.
2. It is determined by means of the potentiometer.	2. It is determined by means of a simple voltmeter.
3. It is the maximum voltage obtainable from a cell.	3. It is less than maximum voltage obtainable from a cell.
4. The work calculated from EMF is the maximum work obtainable from a cell.	4. The work calculated from potential difference is less than maximum work obtainable from a cell.
5. It is responsible for the flow of steady current through the circuit.	5. It is not responsible for the flow of steady current through the circuit.

12.7.1 Calculation of EMF of a Cell

An electrochemical cell is always based upon a redox reaction, which can be split into two half-reactions namely the oxidation half-reaction and the reduction half-reaction. The standard EMF of a cell (E°_{cell}) is given by the relation

$$E^\circ_{cell} = \begin{bmatrix}\text{Standard oxidation potential}\\ \text{of oxidation half-reaction}\end{bmatrix} - \begin{bmatrix}\text{Standard oxidation potential}\\ \text{of reduction half-reaction}\end{bmatrix}$$

Since oxidation potential = –reduction potential

The above expression may be written as

$$E^\circ_{cell} = \begin{bmatrix}\text{Standard reduction potential}\\ \text{of reduction half-reaction}\end{bmatrix} - \begin{bmatrix}\text{Standard reduction potential}\\ \text{of oxidation half-reaction}\end{bmatrix}$$

12.7.1.1 Notation and Convention

In the representation of a cell, the electrode at which oxidation takes place is shown on the left-hand side and the electrode at which reduction takes place is shown on the right-hand side. Hence, the above expression is written as

$$E^\circ_{cell} = E^\circ_{(right\ electrode)} - E^\circ_{(left\ electrode)} - E^\circ_{(cathode)} - E^\circ_{(anode)}$$

[both the electrode potentials being reduction potentials]

Thus, the EMF of a galvanic cell ($E°_{cell}$) is calculated through the following steps:

Step I: The overall electrochemical cell reaction is split into two half-reaction, i.e., oxidation half-reaction and reduction half-reaction.

Step II: Equate the number of electrons in the two half-reactions by multiplying with a suitable number, if necessary. However, E° values are not multiplied.

Step III: Formulate the galvanic cell in such a way that the electrode on which oxidation takes place and acts as anode in shown on the left-hand side while the electrode on which reduction takes place and acts as cathode is shown on the right-hand side.

Step IV: The standard EMF of the cell is given by the relationship:

$$E°_{cell} = E°_{cathode} - E°_{anode}$$

$$= E°_{right} - E°_{left}$$

The electrode potentials of the cathode as well as anode should be in terms of reduction potentials.

Note: On calculation, if $E°_{cell}$ has a positive value, the cell reaction is feasible. On the other hand, if $E°_{cell}$ has a negative value, the cell reaction is not feasible.

MEMORY FOCUS

1. The extent of a electrochemical reaction can be obtained by the EMF of the cell.
2. The EMF of a galvanic cell is its ability to drive the electrons in to the circuit.
3. The EMF of a cell is accurately measured by potentiometer method or by null method. Null method is also known as Poggendorff's compensation method.
4. The EMF of a cell is also obtained from electrode potential.

$$EMF_{cell} = E_{right} - E_{left}$$

or $$EMF_{cell} = E_{cathode} - E_{anode}$$

5. The EMF of a cell is the difference between the ratio potentials of reduction half-cell and oxidation half-cell.
6. If the EMF of the cell is +ve, electricity is generated spontaneously.
7. If the EMF of the cell is –ve, the reverse reaction takes place.
8. The best galvanic cell is the one with the highest +ve EMF value.
9. If the EMF of the cell is zero, the cell stops functioning.

Numerical Problems Based on the Calculation of EMF

Example 3 **Determine the standard EMF of the cell:**

$$\mathbf{Cu + 2Ag^+ \longrightarrow Cu^{2+} + 2Ag}$$

Given $\mathbf{E^0_{Ag/Ag^+} \longrightarrow 0.80}$ **(oxidation potential)**

$\mathbf{E^0_{Cu^{2+}/Cu} \longrightarrow 0.34}$ **(reduction potential)**

Solution: The given cell reaction

$$Cu + 2Ag^+ \longrightarrow Cu^{2+} + 2Ag$$

The cell for the above cell reaction is represented as

$$Cu \mid Cu^{2+} \parallel Ag^+ + 2Ag$$

$$E^\circ_{cell} = E^\circ_{(right)} - E^\circ_{(left)} = E^\circ_{(cathode)} - E^\circ_{(anode)}$$

$$= 0.80 - 0.34 = 0.46$$

Therefore, the EMF of the given cell reaction = 0.46 V.

Example 4 Predict the feasibility of the given reaction.

$$\mathbf{2Ag + Cu^{2+} \longrightarrow 2Ag^+ + Cu}$$

Solution: Cell involving the given reaction is

$$Ag \mid Ag^+ \parallel Cu^{2+} \mid Cu$$

We know that

$$E^\circ_{cell} = E^\circ_{(right)} - E^\circ_{(left)}$$

$$= E^0_{Cu^{2+}/Cu} - E^0_{Ag^+/Ag}$$

$$= 0.34 - (0.80) = -0.46\ V$$

Since the value E°_{cell} comes out to be negative, the given reaction is not feasible.

Example 5 Can a solution of 1 M $CuSO_4$ be stored, in a vessel made of Ni metal? Given that $E^0_{Ni/Ni^{2+}}$ = +0.25 V and $E^0_{Cu^{2+}/Cu}$ = −0.34 V.

Solution: The reduction potentials of the two electrodes, i.e., $E^0_{Ni/Ni^{2+}} = +0.25$ V and $E^0_{Cu^{2+}/Cu} = -0.34$ V suggest that nickel will act as an anode and copper will act as a cathode. Now, we wish to know whether the following reactions take place or not.

$$Ni + CuS0_4 \longrightarrow NiSO_4 + Cu$$

or

$$Ni + Cu^{2+} \longrightarrow Ni^{2+} + Cu$$

The cell involving this reaction can be represented as

$$Ni \mid Ni^{2+} \parallel Cu^{2+} \mid Cu$$

We know that $E^\circ_{cell} = E^\circ_{(right)} - E^\circ_{(left)} = 0.34 - (-0.25) = 0.59$ V

Since E°_{cell} is positive, it means that $CuSO_4$ reacts with Ni. Hence, $CuSO_4$ cannot be stored in a Ni vessel.

Example 6 Predict whether silver reacts with 1 M H_2SO_4 to give out hydrogen gas or not. Given that standard reduction potential of Ag is 0.80 V.

Solution: If silver reacts with sulphuric acid, the following reaction should take place.

$$Ag + H_2SO_4 \longrightarrow Ag_2SO_4 + H_2$$

or $$2Ag + 2H^+ \longrightarrow 2Ag^+ + H_2$$

The cell involving the above reaction is represented as

$$Ag \mid Ag^+ \parallel H^+ \mid H_2$$

$$E^\circ_{cell} = E^\circ_{(right)} - E^\circ_{(left)}$$

$$= E^0_{H^+, H} - E^0_{Ag^+, Ag}$$

$$= 0 - 0.80 = -0.80\ V$$

Since E°_{cell} of the supposed reaction comes out to be negative, this reaction does not take place. Hence, silver cannot liberate hydrogen from the acid.

Example 7 **State whether the following representation of a galvanic cell is correct or not.**

$$\mathbf{Pb/Pb^{2+}\ (1\ M) \parallel Mg^{2+}\ (1\ M)/Mg}$$

Given that $$\mathbf{E^0_{Pb^{2+}, Pb} = -1.126 \text{ and } E^0_{Mg^{2+}/Mg} = -2.36\ V}$$

Solution: The representation is correct if E°_{cell} comes out to be positive.

$$E^\circ_{cell} = -2.363 - (-0.126) = -2.363 + 0.126 = -2.237\ V$$

Since E° is negative, the position of the cathode and anode must be interchanged, i.e., the representation is not correct.

Example 8 **Calculate ΔGϒ for the cell reaction at 25ϒC: $Zn + FeSO_4 \rightarrow Fe + ZnSO_4$.**
The standard reduction potentials are

$$\mathbf{E^0_{(Zn^{2+}, Zn)} = -1.76 \text{ and } E^0_{(Fe^{2+}, Fe)} = -0.44\ V}$$

Predict whether the reduction is spontaneous or not.

Solution: For the given cell reaction, the cell can be represented as

$$Zn \mid Zn^{2+} \parallel Fe^{2+} \mid Fe$$

$$\text{Standard EMF of the cell } (E^\circ_{cell}) = E^\circ_{cathode} - E^\circ_{anode} = -0.44\ V - (-0.76\ V) = 0.32\ V$$

Now, $\Delta G^\circ = -nFE^\circ = -2 \times 96500\ C \times 0.32\ V = -61760\ CV = -61760\ J$

Since the free energy change (ΔG°) is negative, the reaction is spontaneous.

12.8 DERIVATION OF NERNST EQUATION (CONCENTRATION DEPENDENCE OF ELECTRODE POTENTIAL)

The EMF of a cell depends both on (i) *Standard EMF of cell, i.e., its nature* (ii) *the concentrations of the species involved in the cell reaction and* (iii) *temperature.*

For a general cell reaction of the type,

$$aA + bB \longrightarrow lL + mM$$

In thermodynamics, the free energy (ΔG) accompanying the above general equation is given as

$$\Delta G = \Delta G^\circ + \text{RT} \ln \frac{[\text{L}]^l [\text{M}]^m}{[\text{A}]^a [\text{B}]^b}$$

where ΔG° is the free energy change of the reaction under standard conditions.

Since $\Delta G = -n\text{FE}$ and $\Delta G^\circ = -n\text{FE}^\circ$

$$\therefore \quad -n\text{FE} = -n\text{FE}^\circ + \text{RT} \ln \frac{[\text{L}]^l [\text{M}]^m}{[\text{A}]^a [\text{B}]^b}$$

or
$$\text{E} = \text{E}^\circ - \frac{\text{RT}}{n\text{F}} \ln \frac{[\text{L}]^l [\text{M}]^m}{[\text{A}]^a [\text{B}]^b}$$

or
$$\text{E} = \text{E}^\circ - \frac{2.303\ \text{RT}}{n\text{F}} \log \frac{[\text{L}]^l [\text{M}]^m}{[\text{A}]^a [\text{B}]^b}$$

where E is the EMF of the cell, E° is the standard EMF of the cell.

[A], [B], [L] and [M] are the molar concentration of the species A, B, L and M, respectively. *a, b, l and m are the stoichiometric coefficients in the chemical equation, n is the number of electrons involved, R is the gas constant, T is the absolute temperature, F is one Faraday* (96500 coulombs). The equation is known as Nernst equation.

As 25 °C, $R = 8.314$ Joules/Kelvin/mole, $T = 25\ °\text{C} + 273 = 298$ K (at standard conditions) and $F = 96500$ coulombs.

$$\therefore \quad \frac{2.303\ \text{RT}}{\text{F}} = \frac{2.303 \times 8.314 \times 298}{96500} = 0.0591$$

Substituting the value of $\frac{2.303\ \text{RT}}{\text{F}}$ in the equation above, we get

$$\text{E} = \text{E}^\circ - \frac{0.0591}{\text{n}} \log \frac{[\text{L}]^l [\text{M}]^m}{[\text{A}]^a [\text{B}]^b} \qquad (12.1)$$

or
$$\text{E} = \text{E}^\circ + \frac{0.0591}{n} \log \frac{[\text{A}]^a [\text{B}]^b}{[\text{L}]^l [\text{M}]^m}$$

It is seen from Nernst equation, that if the product of the molar concentration of products is more than the product of the reactants, EMF, (E) is less than the standard EMF, (E°). On the other hand, if the product of the molar concentration of the reactants is greater than that of the molar concentration of products, E is greater than E°.

12.8.1 Application of Nernst Equation

(i) Calculation of the EMF of a cell: Consider, e.g., the cell

$$\text{Zn} \mid \text{Zn}^{2+} \parallel \text{Cu}^{2+} \mid \text{Cu}$$

The cell reaction may be written as

$$\text{Zn} + \text{Cu}^{2+} = \text{Cu} + \text{Zn}^{2+}$$

Applying Nernst Eq. (12.1)

$$E = E - \frac{0.0591}{2}\log\frac{[Cu][Zn^{2+}]}{[Zn][Cu^{2+}]}$$

Here, number of electrons involved in the half-cell reaction is 2.

In the standard state, the activity (molar concentration) of a solid is taken as unity.

$$E = E^\circ - \frac{0.0591}{2}\log\left[\frac{Zn^{2+}}{Cu^{2+}}\right]$$

From this equation, it is obvious that E may be equal to E° or may be greater or less than E°. E depends on the ratio of the concentration of Zn^{2+} ions and Cu^{2+} ions.

(ii) Calculation of the single electrode potential: Although the standard electrode potential is fixed, the concentration of the ion in the half-cell affects its single reduction electrode potential. This is also given by Nernst's equation.

$$Zn^{2+} + 2e^- \longrightarrow Zn$$

$$E = E^\circ - \frac{RT}{nF}\ln\frac{[Zn]}{[Zn^{2+}]} = E^\circ - \frac{RT}{nF}\ln\frac{1}{[Zn^{2+}]} = E^\circ + \frac{RT}{nF}\ln[Zn^{2+}]$$

[∴ Zn being solid, its concentration is taken as unity]

$$E = E^\circ + \frac{RT}{nF}\ln[Zn^{2+}] = E^\circ + \frac{2.303\,RT}{nF}\log[Zn^{2+}] = E^\circ + \frac{0.0591}{n}\log[Zn^{2+}]$$

where E and E° are the single electrode potential and standard potential, *n*, R and F have the usual significance.

The general expression for electrode potential of a metal in contact with M^{n+} ions reaction involving the electrode may be written as

$$M^{n+} + ne^- = M$$

$$E = E^\circ + \frac{0.0591}{n}\log[M^{n+}]$$

(iii) Calculation of equilibrium constant: At the equilibrium concentration of the two metal ions, there is no potential difference between the two electrodes and the EMF of the cell becomes zero.

Consider a reaction at equilibrium

$$aA + bB \rightarrow cC + dD$$

$$K_C = \frac{[C]^c[D]^d}{[A]^a[B]^b}$$

The Nernst equation can be represented as

$$E_{cell} = E^\circ_{cell} - \frac{0.0591}{n}\frac{[C]^c[D]^d}{[A]^a[B]^b}$$

At the stage of equilibrium, $E^\circ_{cell} = 0$

Therefore, the above equation can be written as

$$\text{Zero} = E^{\circ}_{cell} - \frac{0.0951}{n}\log K_c$$

or

$$E^{\circ}_{cell} = \frac{0.0951}{n}\log K_c$$

where K_c, is the equilibrium constant for the reaction.

Knowledge of standard electrode potential can thus be employed for calculating the equilibrium constant (K_c) for the reaction.

MEMORY FOCUS

1. The EMF of a cell depends upon
   (i) its nature, i.e., standard EMF of the cell
   (ii) The concentrations of the species involved in the cell reaction and
   (iii) temperature
2. Nernst derived an equation combining the above three factors for a cell reaction of the type:

$$aA + bB = lL + mM$$

3. The Nernst equation for the above reaction is $E = E^{\circ} - \frac{RT}{nF}\ln\frac{[L]^l[M]^m}{[A]^a[B]^b}$

   where E is the EMF of the cell, E ° is the standard EMF of the cell. Other terms have their usual meaning.
4. Putting the values of n, the number of electron involved in the reaction, R is the gas constant (8.314 Joules/degree absolute/mole) $T = 25\,°C + 273 = 298$ K, $F_2 = 96500$ coulombs we get,

$$E = E^{\circ} - \frac{0.0591}{n}\log\frac{[L]^l[M]^m}{[A]^a[B]^b}$$

5. The important applications of Nernst equation are:
   (i) Calculation of EMF of a cell.
   (ii) Calculation of single electrode potential which is given as

$$E = E^{\circ} + \frac{0.0591}{n}\log[M^{n+}]$$

Numerical Problems Based on Nernst Equation

Example 9 **Calculate the half-cell potential at 298 K for the reaction:**

$$Zn^{2+}(aq) + 2e^- \longrightarrow Zn(s)$$

$$\textbf{If } [Zn^{2+}aq] = 0.1 \text{ M and } E^{\circ} = -0.76 \text{ V}$$

Solution: Given: $n = 2$, $[Zn^{2+}\ aq] = 0.1$ M and $E^{\circ} = -0.76$ V

According to Nernst equation:

$$E = E^\circ + \frac{0.0591}{n}\log[M^{n+}]$$

$$= -0.76 + \frac{0.059}{2}\log 0.1 = -0.7895\ V$$

Thus, half-cell potential is **0.7895 V.**

Example 10 **Calculate the EMF of the cell in which the reaction is:**

$$\mathbf{Mg(s) + 2Ag^+(aq) \longrightarrow Mg^{2+}(aq) + 2Ag(s)}$$

When $[Mg^{2+}] = 0.130$ M and $[Ag^+] = 1.0 \times 10^{-4}$ M,

Given that $E^0_{Ag/Ag^+} = 0.80$ V and $E^0_{Mg^{2+}/Mg} = -2.37$ V

Solution: The cell is represented as

$$Mg(s)\,|\,Mg^{2+}(0.130)\,||\,Ag^+(1.00 \times 10^{-4}\ M)\,|\,Ag$$

Electrode reactions are

$$Mg(s) \longrightarrow Mg^{2+}(aq) + 2e^- \quad \text{(oxidation)}$$

$$2Ag^+(aq) + 2e^- \longrightarrow 2Ag(s) \quad \text{(reduction)}$$

The overall reaction is

$$Mg(s) + 2Ag^+(aq) \longrightarrow Mg^{2+}(aq) + 2Ag(s)$$

Since the reaction involves two moles of electrons, therefore $n = 2$ mol. The cell's EMF expression is written as

$$E = E^\circ + \frac{0.0591}{2}\log\frac{[Ag^+]^2}{[Mg^{2+}]}$$

$$= E^\circ_{Ag/Ag^+} - E^\circ_{Mg^{2+}/Mg} + \frac{0.0591}{2}\log\frac{(1.0\times10^{-4})^2}{(0.130)}$$

$$= [0.80 - (-2.37\ V)] - \frac{0.0591}{2}\log\frac{1.30\times10^{-1}}{1.00\times10^{-8}}$$

Example 11 **For the concentration of Ag^+ (aq) will the EMF of the given cell be zero at 25ϒC, if concentration of Cu^{2+}(aq) is O.IM?**

Cu(s) I Cu^{2+} (O.IM) II Ag^+ (aq) I Ag(s)

(Given E: $Ag^+/Ag = +0.80$ V; $E_{Cu^{2+}/Cu} = 0.34$ V

Solution: We know that

$$E^\circ_{cell} = E^\circ_{cathode} - E^\circ_{anode} = 0.80\ V - 0.34\ V = 0.46\ V$$

The cell reaction can be written as

$$Cu(s) + 2Ag^+ \longrightarrow Cu^{2+} + 2Ag$$

According to Nernst

$$E_{cell} = E^\circ_{cell} + \frac{2.303RT}{nF}\log\frac{[Ag^+]^2}{[Cu^{2+}]}$$

$$0 = 0.46 + \frac{0.0591}{2}\log\frac{(x)^2}{0.1}$$

or $$-0.46 = 0.0295 \times 2 \log x - 0.0295 \log 0.1$$

or $$-0.4895 = 0.059 \log x$$

$$\log x = \frac{-0.4895}{0.059} = -8.2969$$

$$x = \text{Antilog } (-8.2969)$$

Antilog of 8.2969 = 1.981 × 108 M = **5.04 10⁹ M**

Example 12 **Calculate the cell potential of the given cell at 25 ϒC**

[$R = 8.31$ JK^{-1}mol^{-1}, $F = 96500$ C mol^{-1}].

Ni(s) | Ni^{2+} (0.01 M) || Cu^{2+} (0.1 M) | Cu(s)

(Given $E^\circ_{Cu^{2+}/Cu} = +0.34V$; $E^\circ_{Ni^{2+}/N} = -0.25V$)

Solution: For the given cell,

$$E^\circ_{cell} = E^\circ_{cathode} - E^\circ_{anode} = 0.34 \text{ V} - (-0.25 \text{ V}) = 0.59 \text{ V}$$

$$E_{cell} = E^\circ_{cell} + \frac{2.303 \text{ RT}}{nF}\log\frac{[Cu^{2+}]}{[Ni^{2+}]}$$

$$= 0.59 \text{ V} + \frac{0.0591}{2}\log\frac{0.1M}{0.01M}$$

$$= 0.59 \text{ V} + 0.0295 \log 10 = 0.59 \text{ V} + 0.0295 \text{ V} = \mathbf{0.6195 \text{ V}}$$

Example 13 **Calculate the voltage generated in the following cell:**

Fe | Fe^{2+} || Mn^{2+} | Mn

when an iron rod is immersed in 1.0 M $FeSO_4$ solution and Mn rod is immersed in 0.1 M $MnSO_4$ solution. Given Eϒ for Fe^{2+}/Fe is –0.40 V and Mn^{2+}/Mn is – 1.18 V.

Solution:

$$E^\circ_{cell} = E^\circ_{cathode} - E^\circ_{anode} = E^\circ_{right} - E^\circ_{left}$$

$$= (-1.18 + 0.40) \text{ V} = -0.78 \text{ V}$$

According to Nernst equation:

$$E_{cell} = E^\circ_{cell} + \frac{0.591}{2}\log\frac{[Mn^{2+}]}{[Fe^{2+}]}$$

$$= -0.78 \text{ V} + 0.0296 \log (0.1/1.0)$$
$$= -0.78\text{V} + 0.0296 (-1) = -.8096 \text{ V}$$

Minus sign shows that the actual cell should be represented as

$$\text{Mn} + \text{Mn}^{2+} \,||\, \text{Fe}^{2+} \,|\, \text{Fe, whose EMF is 0.8096 V}$$

Example 14 **Calculate the equilibrium constant for the reaction:**

$$\mathbf{Fe^{2+}(aq) + Ag^{+}(aq) \rightleftharpoons Fe^{3+}(aq) + Ag(s) \text{ at } 25ϒC}$$

If $$\mathbf{E^{\circ}_{Fe^{3+}/Fe^{2+}} = 0.77V \text{ and } E^{\circ}_{Ag^{+}/Ag} = 0.80V}$$

Solution: The cell for the given cell reaction is represented as

$$\text{Fe}^{2+} \,|\, \text{Fe}^{3+} \,||\, \text{Ag}^{+} \,|\, \text{Ag}$$

Now, $$E^{\circ}_{cell} = E^{\circ}_{cathode} - E^{\circ}_{anode} = 0.80 \text{ V} - (0.77 \text{ V}) = 0.03 \text{ V}$$

We know that, $$E^{\circ}_{cell} = \frac{0.059}{n} \log K_C \quad \text{or} \quad \log K_C = \frac{n \times E^{\circ}_{cell}}{0.059}$$

or $$\log K_C = \frac{1 \times 0.03 \text{V}}{0.059} = 0.50847$$

or $$K_C = \text{Antilog } 0.50847 = 3.2245$$

Example 15 **What is the concentration of Ni^{2+} in the cell at 25ϒC, if the EMF is 0.601 V?**

$$\mathbf{Ni(s) \,|\, Ni^{2+} (a = ?) \,||\, Cu^{2+} (0.75M) \,|\, Cu(s)}$$

$$\mathbf{\text{Given } E_{Ni/Ni^{2+}} = 0.25 \text{ V, and } E_{Cu^{2+}/Cu} = 0.34 \text{ V}}$$

Solution: We know that $E^{\circ}_{cell} = E^{\circ}_{right} - E^{\circ}_{left}$ where E°_{right} and E°_{left} are reduction potentials of cathode and anode, respectively.

Given $E^{\circ}_{Ni/Ni^{2+}} = 0.25$ V is the oxidation potential.

Therefore, the reduction potential

$$E^{\circ}_{Ni^{2+}/Ni} = -0.25 \text{ V}$$

$$E^{\circ}_{cell} = 0.34 \text{ V} - (-0.25 \text{ V}) = 0.59 \text{ V}$$

According to Nernst equation:

$$0.601 \text{ V} = 0.59 \text{ V} + 0.0296 \log \frac{0.75}{[\text{Ni}^{2+}]}$$

$$0.011 = 0.0296 \log \frac{0.75}{[\text{Ni}^{2+}]}$$

or $$\log \frac{0.75}{[\text{Ni}^{2+}]} = \frac{0.011}{0.0296} = 0.37162$$

$$\therefore \qquad \frac{0.75}{[Ni^{2+}]} = \text{Antilog } 0.37162 = 2.3529$$

or $$[Ni^{2+}] = \frac{0.75}{2.3529} = 0.3188\,M$$

REVISION QUESTIONS

1. Give two important differences between EMF and potential difference.
2. Describe the construction and working of the following electrodes:
   (a) hydrogen electrode (b) calomel electrode
3. (a) Explain the meaning of an electrode potential being positive or negative with respect to hydrogen electrode.
   (b) What is a reference electrode? Describe briefly.
4. (a) How is electrode potential experimentally determined? Distinguish between single electrode potential and standard electrode potential.
   (b) How is single electrode potential determined?
   (c) Represent a concentration cell according to the IUPAC convention.
5. Derive Nernst equation for single electrode potential and explain the terms involved in it. Write its application.
6. (a) Differentiate between the characteristics of an electrolytic cell and those of a galvanic cell. What do your mean by standard electrode potential? How can it be measured?
7. (b) What are reference electrodes? Describe the construction of calomel electrode.
8. (a) How is glass electrode constructed? Write the cell reactions.
   (b) What is the effect of electrolyte concentration on electrode potential?
   (c) Explain the construction of Ni–Cd cell.
9. (a) The electrode potential of Zn is assigned a negative value, whereas that of Cu is assigned a positive value. Give reasons.
   (b) For the Daniell cell involving the cell reaction:

   $$Zn\,(s) + Cu^{2+} \rightarrow Zn^{2+}_{eq} + Cu_{(s)},$$

   the standard free energies of formation for $Cu^{2+}_{(eq)}$ and $Zn^{2+}_{(eq)}$ are 15.66 kcals mol^{-1} and –35.14 kcal mol^{-1}, respectively. Calculate E° of the cell.
   (c) Calculate the EMF of the Zn–Ag cell at 25 °C when $[Zn^{2+}] = 0.01$ M and [Ag+] = 10 M, E°_{cell} at 25 °C = 1.56 V.
   (d) Write a short note on calomel electrode
10. (a) What do you understand by electrochemical series?
    (b) Write down the cell for which the overall reaction is

    $$Cd + Cu^{2+}_{(a=1)} \rightarrow Cd^{2+}_{(a=2)} + Cu$$

11. (a) Derive Nernst equation for the calculation of cell's EMF.
    (b) The standard electrode potential for Hg_2^{2+}/Hg and Hg^{2+}/Hg is 0.799 V and 0.855 V, respectively. Calculate at 298 K the equilibrium constant for the reaction.

(c) What is the cell reaction and cell EMF at 298 K of the cell?

$$Zn/Zn^{2+}_{(a=1)} \,||\, Pb^{2+}_{(a=1)} \,|\, Pb$$

12. What is standard electrode potential? A galvanic cell consists of a. copper plate immersed in a 10.0 M solution of copper sulphate and an iron plate immersed in a 1.0 M solution of ferrous sulphate and 298 K. Calculate the voltage for the cell. Given E °Cu^{++} + Cu = 0.34 V, E °F^{++} Fe = –0.44 V
13. What is single electrode potential? Derive Nernst equation for single electrode potential?
14. Define standard electrode potential and derived Nernst equation for single electrode potential.
15. (a) Explain the construction and working of a calomel electrode. Indicate the role of calomel electrode.
    (b) What are ion-selective electrodes?
16. (a) Define single electrode potential and standard electrode potential? Derive the Nernst equation for electrode potential.
    (b) Explain the construction and working of a calomel electrode. Mention its advantages.
    (c) Write a short note on ion-selective electrode.
17. (a) Discuss the origin of electrode potential and derive Nernst equation for electrode potential.
    (b) What are ion-selective electrodes? Give the construction of glass electrode and explain the experimental method of determining pH glass electrode.
    (c) An electrochemical cell consists of iron electrode dipped in 0.1 M $FeSO_4$ and silver electrode dipped in 0.05 M $AgNO_3$. Write the cell representation, cell reaction and calculate EMF of the cell at 298 K. Given that the standard reduction potentials of iron and silver electrodes are –0.44 and +0.80 V, respectively.
18. (a) Derive Nernst equation for single electrode. Explain the determination of single electrode potential using standard hydrogen electrode.
19. (b) Write brief notes on:
    (i) Calomel electrode (ii) Glass electrode
    (c) Write the electrode reactions and calculate the EMF of the following cell at 298 K given $E°_{cell}$ n = 1.3 V

$$Cu(s)\ Cu^{2+} + (1 \times 10^{-2}\ M)\ 11\ Ag^{+}\ (1 \times 10^{-1}\ M)\ Ag(s)$$

MULTIPLE CHOICE QUESTIONS

1. The chemical reaction taking place at anode is
   (a) oxidation (c) ionization
   (b) reduction (d) hydrolysis
2. In the cell, $Zn|Zn^{2+}||Cu^{2+}|Cu$, the negative terminal is
   (a) Cu (c) Zn
   (b) Cu^{2+} (d) Zn^{2+}
3. 96500 C of electricity liberates from $CuSO_4$ solution
   (a) 63.5 g of Cu (c) 96500 g of Cu
   (b) 31.75 g of Cu (d) 100 g of Cu
4. An unknown metal M displaces nickel from nickel (II) sulphate solution but does not displace manganese from manganese sulphate solution. Which order represents the correct order or reducing power?
   (a) Mn>Ni>M (c) Mn>M>Ni
   (b) Ni>Mn>M (d) M>Ni>Mn

5. Electrode potentials (E°_{red}) of four elements A, B, C and D are −1.36, −0.32, 0 and −1.26 V, respectively and the order of these elements is:
(a) A, D, B and C (c) B, D, C and A
(b) C, B, D and A (d) C, A, D and B

6. When a copper wire is dipped into silver nitrate solution, the solution gradually becomes blue. The reason for this is
(a) oxidation of silver
(b) reduction of copper
(c) oxidation of copper
(d) oxidation of nitrate ion

7. Reduction potentials for the following half-cell reactions are

$$Zn \rightarrow Zn^{2+} + 2e^- \; E^\circ = +0.76 \text{ V}$$

$$Fe \rightarrow Fe^{2+} + 2e^- \; E^\circ = +0.44 \text{ V}$$

The EMF for the cell reaction

$$Fe^{2+} + Zn \rightarrow Zn^{2+} + Fe \text{ will be}$$

(a) −1.20 V (c) −0.32 V
(b) +0.32 V (d) +1.20 V

8. The standard reduction potentials of Cu^{2+}/Cu and Cu^{2+}/Cu^+ are 0.337 V and 0.153 V, respectively. The standard electrode potential of Cu^+/Cu half-cell is
(a) 0.184 V (c) 0.521 V
(b) 0.827 V (d) 0.490 V

9. During the electrolysis of NaCl solution, the reaction at cathode is
(a) $Na^+ + e^- \rightarrow Na$ (c) $2H^+ + 2e^- \rightarrow H_2$
(b) $Cl^- - e^- \rightarrow Cl$ (d) $2H^- - 2e^- \rightarrow H_2$

10. The standard reduction potentials at 298 K for different metal ions having = 2 charge is
(a) $E^\circ = -0.76$ V (c) $E^\circ = 0.0$ V
(b) $E^\circ = -0.74$ V (d) $E^\circ = +0.77$ V

11. Certain current liberates 0.504 g of H_2 in 2 h. How many grams of copper can be liberated by the same amount of current flowing for the same time in $CuSO_4$ solution?
(a) 12.7 g (c) 31.8 g
(b) 15.9 g (d) 63.5 g

12. What is the amount of chlorine evolved, when 2 A of current is passed for 30 min in an aqueous solution of NaCl?
(a) 1.32 g (c) 9.81 g
(b) 4.56 g (d) 12.6 g

13. For the cell reaction:

$$Cu^{2+}(C_1\, aq) + Zn(s) = Zn^{2+}(C_2\, aq) + Cu(s)$$

Of an electrochemical cell, the change in free energy, at a given temperature is a function of
(a) $in(C_1)$ (c) $in(C_1 + C_2)$
(b) $in(C_2)$ (d) $in(C_2/C_1)$

14. At 25 °C the electrode potential of Mg^{2+}/Mg electrode in which concentration of Mg^{2+} ions is

$$0.12 \text{ M } (E^\circ M^{2+}g/Mg = -2.36 \text{ V},$$
$$R = 8.314 \text{ J}^{-1}\text{l}, F = 96500 \text{ coulomb mol}^{-1})$$

(a) −2.39 (c) −3.95
(b) −2.95 (d) −4.75

15. Which of the following will not displace hydrogen?
(a) Ba (c) Hg
(b) Pb (d) Sn

16. Which of the following reaction occurs at the cathode during the charging of lead storage battery?
(a) $Pb^{2+} + 2e^- \rightarrow Pb$
(b) $Pb \rightarrow Pb^{2+} + 2e^-$
(c) $Pb^{2+} + SO^{2-}_4 \rightarrow PbSO_4$
(d) $Pb\,SO_4 + 2H_2O \rightarrow 2PbO_2 + 4H^+ + SO^{2-}_4 + 2e^-$

17. Electrolytic cells are electrochemical cells in which ____________ reactions are forced to occur by the input of electrical energy
(a) spontaneous
(b) non-spontaneous
(c) exothermic
(d) endothermic

18. In any electrochemical cell, the cathode is always__________
(a) a nonmetal
(b) attached to a battery

(c) the electrode at which some species gain electrons
(d) the electrode at which some species lose electrons

19. $Zn \rightarrow Zn^{2+} + 2e^-$, e° = +0.76V, $Cr^{3+} + 3e^- \rightarrow Cr$, e° = –0.74 V
(a) Zn (c) Zn^{2+}
(b) Cr (d) Cr^{3+}

20. Which of the following statements associated with electrochemical cells is incorrect?
(a) the function of a salt bridge in an electrochemical cell is to complete the circuit
(b) cell potential is the potential difference in a voltaic cell
(c) a Bronsted–Lowery acid–base reaction can be the basis of the net reaction in a chemical cell
(d) a half-reaction corresponds to one electrode in a voltaic cell

21. Which of the following statements associated with batteries is incorrect?
(a) in a dry cell, the reaction $Zn \rightarrow Zn^{2+}$ continues to occur even when the battery is not being used
(b) secondary batteries are rechargeable
(c) the cell reaction in a primary battery is not reversible
(d) electrodes with greater surface area give a greater potential

22. What is indicated when a chemical cell's voltage (E°) has dropped to zero?
(a) the concentration of the reactants has increased
(b) the concentration of the products has decreased
(c) the cell reaction has reached equilibrium
(d) the cell reaction has completely stopped

23. Standard cell potential is
(a) measured at a temperature of 25°C
(b) measured when ion concentrations of aqueous reactants are 1:00 M
(c) measured under the conditions of 1:00 atm for gaseous reactants
(d) all of the above

24. The standard reduction potentials in volts for Pb^{2+} and Ag^+ are –0.13 and +0.80, respectively. Calculate E° in volts for a cell in which the overall reaction is $Pb + 2Ag^+ \rightarrow Pb^{2+} + 2Ag$
(a) 1.73 (c) 0.93
(b) 0.67 (d) 1.47

25. Predict the products in the electrolysis of aqueous potassium bromide
(a) hydrogen and bromine
(b) potassium metal and oxygen
(c) oxygen and bromine
(d) potassium metal and bromine

26. In the lead storage battery
(a) a reversible reaction can occur to recharge the battery
(b) lead is oxidized to create a flow of electrons
(c) lead forms the cathode when it is being reduced
(d) all of the above

27. The advantages of fuel cells include
(a) they can be recharged by the addition of more material to be oxidized and/or reduced
(b) they can be made to produce little or no harmful pollutants
(c) they can be made to run very quietly
(d) all of the above

28. Calculate the voltage (E) of a cell with E° = 1.1 V, if the copper half-cell is at standard conditions but the zinc ion concentration is only 0.001 molar and the temperature is 25 °C. The overall reaction is

$$Zn + Cu^{+2} \rightarrow Cu + Zn^{+2}$$

(a) 0.39 V (c) 6.19 V
(b) 1.43 V (d) 1.19 V

29. Calculate the potential (in volts) for the following voltaic cell at 25 °C

$$Cr/Cr^{3+}(0.10\ M)IICu^2 + (0.0010\ M)/Cu$$

(a) 1.25 V (c) 1.41 V
(b) 1.33 V (d) 1.57 V

30. The salt bridge in the electrochemical cell serves to
    (a) increase the rate at which equilibrium is attained
    (b) increase the voltage of the cell
    (c) maintain electrical neutrality
    (d) increase the oxidation/reduction rate

ANSWERS

1. (a)	6. (c)	11. (c)	16. (d)	21. (d)	26. (d)
2. (c)	7. (b)	12. (a)	17. (b)	22. (c)	27. (d)
3. (b)	8. (a)	13. (d)	18. (c)	23. (d)	28. (d)
4. (c)	9. (c)	14. (a)	19. (a)	24. (c)	29. (b)
5. (a)	10. (a)	15. (c)	20. (c)	25. (a)	30. (c)

Solid State (Crystalline State) 13

CHAPTER OBJECTIVES

13.1 Introduction
13.2 Crystalline and Amorphous Solids
13.3 Some Terms Used in Crystal Structure
13.4 Crystal Lattice and Unit Cell
13.5 Elements of Symmetry
13.6 Crystallographic Designations
13.7 Laws of Crystallography
13.8 Crystal Systems
13.9 Types of Unit Cells in Crystal System (Bravais Lattice)
13.10 Bragg's Method of Crystal Analysis
13.11 Types of Crystalline Solids

Perfection is unattainable. However, the orderly arrangement of atoms/ions/molecules, a typical of crystalline solids, is very close to perfection.

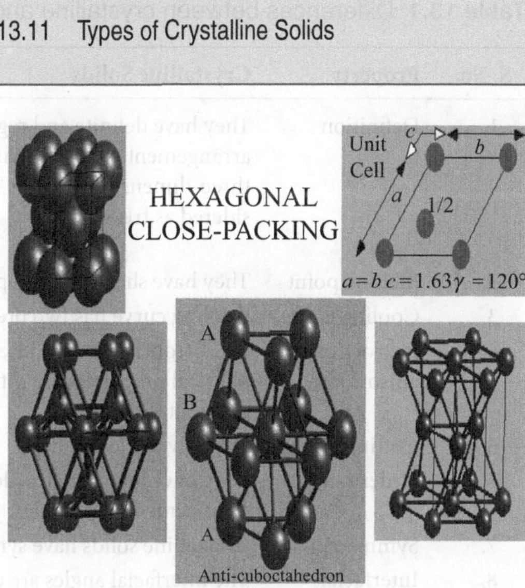

13.1 INTRODUCTION

The solid state represents the state of matter in which the constituent particles (atoms, ions or molecules) are closely packed and are held together by strong intermolecular forces of attraction. These strong forces of attraction in the structural units of solids do not allow any type of motion except that particles only vibrate about their mean positions. The main characteristics of the solids are listed below:

(i) **Rigidity and definite shape:** Solids are rigid and have definite shape and volume.
(ii) **Low compressibility:** Solids have extremely low compressibility due to non-availability of vacant spaces.
(iii) **High density:** They generally have high density due to close packing of the structural units.
(iv) **Melting point:** Generally, solids have definite melting points which depend on the magnitude of attractive forces between the particles. On supplying sufficient energy to solids, the particles move away from their fixed positions, resulting into formation of liquid state.
(v) **Diffusion:** In solids, the particles either diffuse slowly or do not diffuse at all.
(vi) **Vapour pressure:** Vapour pressure of solids is much less than that of liquids at a given temperature. Some particles near the solid surface may acquire sufficient energy to get converted into vapour state.
(vii) **Crystal formation:** Most of the solids can form crystals.

In this chapter, we will discuss the difference between crystalline and amorphous states of solids (Table 13.1), geometry and symmetry of crystals, types of crystals, elements of symmetry, Weiss and Miller's indices, laws of crystallography, and crystal diffraction.

13.2 CRYSTALLINE AND AMORPHOUS SOLIDS

Based on the arrangement of particles (atoms, molecules, ions), solids are classified as crystalline or amorphous solids. Liquids change to solid state on cooling. When the process of cooling slows down to

Table 13.1 Differences between crystalline and amorphous solids.

S. No.	Property	Crystalline Solids	Amorphous Solids
1.	Definition	They have definite and regular arrangement of particles in three-dimensional space. They are considered as true solids.	They do not have any definite arrangement of particles and do not have any definite geometrical shape. They are considered as pseudo-solids or super-cooled liquids.
2.	Melting point	They have sharp melting points.	They do not have sharp melting points.
3.	Cooling curve	Cooling curve has two breaks.	It is smooth curved without any break.
4.	Isotropy and anisotropy	Anisotropic is nature, i.e., their physical properties are different in different directions.	Isotropic in nature. Their physical properties are similar in all directions.
5.	Cutting	They give a clean cut.	They give irregular cuts.
6.	Order range	They have long-range order in the arrangement of particles.	They have either short-range order or do not have any order.
7.	Symmetry	Crystalline solids have symmetry.	They do not have symmetry.
8.	Interfacial angles	The interfacial angles are definite crystal line substances.	Amorphous substances do not have definite interfacial angles.
9.	Rigidity	They are rigid.	They are not rigid.
10.	Heat of fusion	They have definite value of heat of fusion.	They do not have definite value.

allow the constituent particles to arrange themselves in a very regular and orderly fashion, crystalline solids having definite geometrical shape with flat faces and sharp edges are formed.

Crystalline solids exist in well-defined crystalline forms, with their atoms or molecules or ions arranged in a certain definite geometrical pattern, throughout the three-dimensional network of the crystal [Fig. 13.1(a)]. Examples of crystalline solids are NaCl, CuCl, ZnS, diamond etc.

If cooling is very slow, there is a chance that the particles get frozen in a chaotic (disorder) state. The resulting solid with constituent particles arranged in a random fashion is an amorphous solid meaning without form.

Amorphous solids have completely random particles (the arrangement of atoms or molecules) and lack a long-range order in their structures as shown in Fig. 13.1 (b). In fact, amorphous solids can be regarded as ***super-cooled liquids***, having small structural units with a short-range order. Examples of amorphous solids are glass, rubber, plastics, carbon blocks and allotropic forms of elements such as selenium and antimony.

The two forms of solids differ from each other in the following respects.

(i) Cooling curve: Cooling curve for a crystalline substance has two breaks '*a*' and '*b*' which correspond to the beginning and end of process of crystallization.

As shown in Fig. 13.2 (a), temperature remains constant during crystallization. The process of crystallization is accompanied by some liberation of energy, which compensates for the loss of heat and causes the temperature to remain almost constant. Cooling curve for an amorphous substance is smooth as shown in the Fig. 13.2 (b).

(ii) Melting point: A crystalline substance has a sharp melting point, i.e., it changes abruptly into liquid state at a particular temperature, e.g., crystal NaCl melts at 1093 K.

An amorphous substance does not have definite melting point; however, it softens over a wide range of temperatures, e.g., if glass is heated gradually, it softens and starts without a definite and abrupt change.

(iii) Cutting: When we try to break/cut a crystal with a sharp edged tool, it split or cleaves along preferred directions and characteristics shapes are formed resulting into pieces as shown in Fig. 13.3 (a).

The amorphous substances, when broken, exhibit a typical conchoidal fracture which is not planar but a combination of curved surfaces as shown in Fig. 13.3 (b).

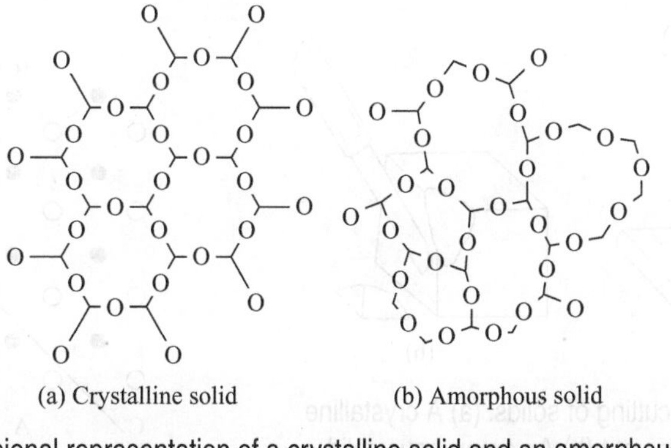

(a) Crystalline solid (b) Amorphous solid

Figure 13.1 Two-dimensional representation of a crystalline solid and an amorphous solid. Each substance has the formula A_2O_3, where A is shown by • and O is shown by O.

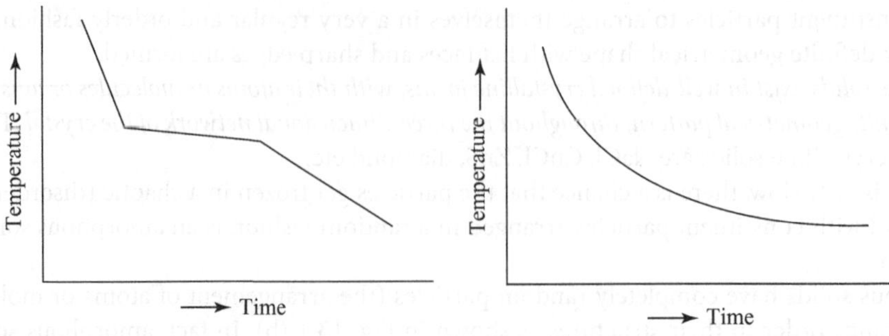

Figure 13.2 (a) Cooling curve of crystalline substances (b) Cooling curve of amorphous substances

(iv) **Isotropy and anisotropy:** Crystalline substances are anisotropic, i.e., their physical properties such as refractive index, electrical conductivity, thermal conductivity, and coefficient of thermal expansion are different in different directions, within the crystal; e.g., coefficient of thermal expansion of silver iodide crystal is positive in one direction and negative in another direction. It is because of the difference in arrangement of the constituent units of crystals in different directions as shown in Fig. 13.4.

Two-dimensional arrangement of only two different types of atoms has been depicted; if the properties are measured along AB and CD, the properties measured along CD will be different, from those measured long AB. Along AB, there is only one type of particles; however, along CD, there are two different types of particles. This type of situation is exhibited by crystals; therefore, they are anisotropic.

In amorphous substances (also in gases and liquids), the particles are disordered and random. Therefore, all directions are equivalent and hence such properties remain same in all directions and they are isotropic.

(v) **Order range:** Crystalline substances exhibit long-range order throughout the body of the crystal. Amorphous substances either do not exhibit short-range order as in liquids. Even when a regular arrangement exists, it extends only up to a few angstroms.

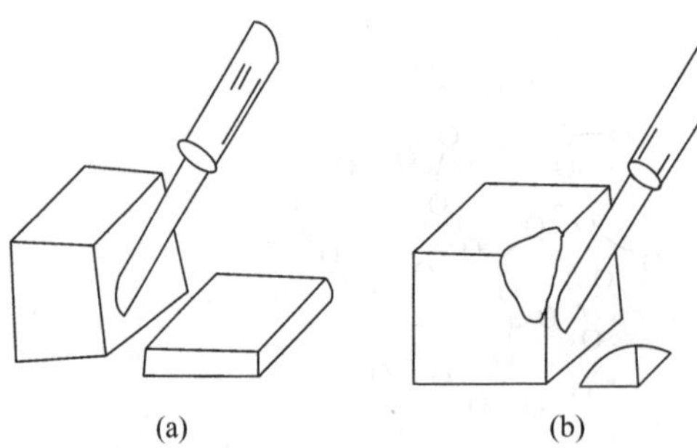

Figure 13.3 The cutting of solids: (a) A crystalline solid gives a clean cut (b) An amorphous solid gives an irregular cut

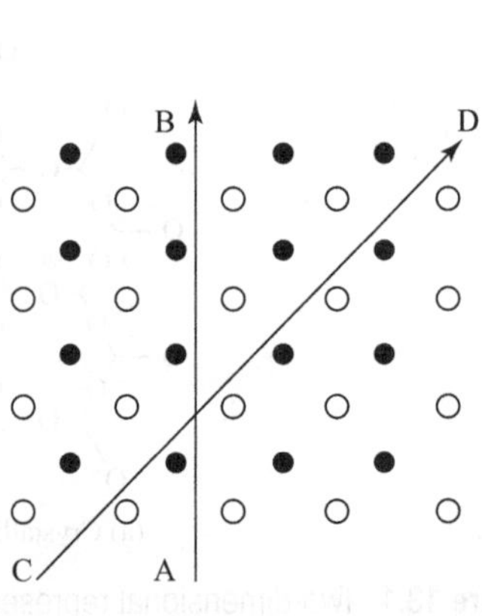

Figure 13.4 Anisotropy in crystals

(vi) **Symmetry:** (a) Crystalline forms have definite symmetry such as cubic, tetragonal or hexagonal. (b) Amorphous forms do not have symmetry; hence, they do not have any definite geometrical shape.

(vii) **Interfacial angles:** Crystals have plane faces. The angle between two faces is called interfacial angle (Fig. 13.5). In crystals, this angle is definite and always remains constant no matter how the faces develop.

When a crystalline solid is hammered, it breaks up into smaller crystals which are of the same geometric shape.

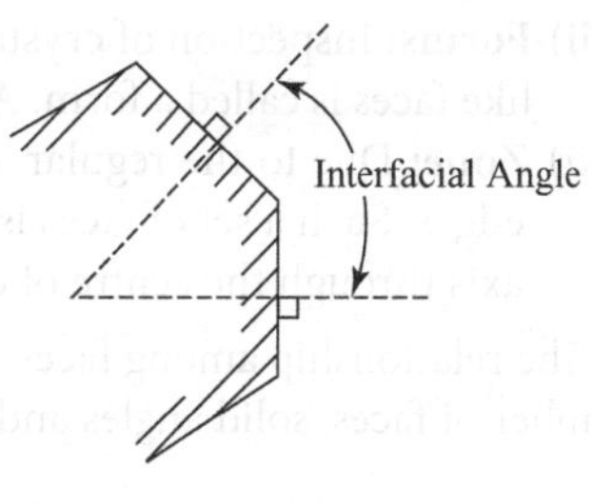

Figure 13.5 Interfacial angle between two faces of a crystal

Amorphous substances do not have well-defined planes. The pieces, formed by breaking of an amorphous substance, are generally not flat and intersect at random angles.

(viii) **Rigidity:** Crystalline substances are rigid and so their shape is not distorted by mild deforming forces. Amorphous substances are not very rigid. They can be distorted by compressing forces.

(ix) **Heat of fusion:** The crystalline solids have definite value of heat of fusion, whereas amorphous substances do not have definite value of the same.

13.3 SOME TERMS USED IN CRYSTAL STRUCTURE

(i) **Crystal:** A crystal is a solid which is composed of particles (atoms, ions or molecules) arranged in an orderly repetitive way. In other words, a crystal is defined *as "a homogeneous anisotropic substance having a definite geometric shape with surfaces that are usually planer and have sharp edges."*

As already explained, the physical properties, such as refractive index and electrical conductivity, of anisotropic substance are different in different directions.

(ii) **Crystallography:** Crystallography is the branch of science which deals with the geometry, properties and structure of crystals and crystalline substances.

(iii) **Crystal face:** A crystal is a solid composed of atoms, ions or molecules arranged in orderly repeated manner. The crystal of a substance is bound by plane surfaces called faces.

(iv) **Edges:** An edge is fanned by the intersection of two adjacent faces.

(v) **Crystal angle or interfacial angle:** The angle between the faces of a crystal is called crystal angle or interfacial angle. It may be of interest to note that although the size and the shape of a crystal of a given substance may vary with conditions of its formation, the crystal angles are always constant (Fig. 13.6). The measurement of angles is important in the study of crystals.

(vi) **Solid angle:** A solid angle is formed by the crystal faces and crystal angles intersection of three or more faces (Fig. 13.6).

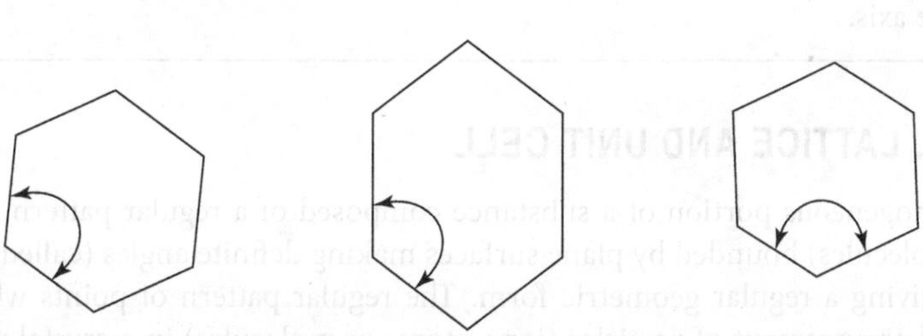

Figure 13.6 Crystal faces and crystal angles

(vii) **Forms:** Inspection of crystals shows that there are a number of like faces present in it. Each set of like faces is called a form. A crystal made entirely of like faces is called a simple form.

(viii) **Zone:** Due to the regular shape of a crystal, faces are found to occur in sets meeting in parallel edges. Such a set of faces is called a zone. Each zone forms a complete belt around the crystal. An axis through the centre of crystal and parallel to the edges of a zone is termed as zone axis.

The relationship among faces, edges and solid angles may be given as $f + c = e$ where f, c and e are the number of faces, solid angles and edges, respectively.

MEMORY FOCUS

Crystallography. It is the branch of science which deals with the geometry, properties and structure of crystals.

Crystal. A homogenous solid body having definite geometric form bounded by plane surfaces formed by the crystallization of certain liquid or gaseous substances is called a crystal.

Salient features of crystals.

1. **Faces.** Crystals are bounded by a number of surfaces which are perfectly flat, which are called faces. Faces are of two types:
   (i) **Like faces.** In fluorspar and alum, all the faces of the crystal are alike in all respects.
   (ii) **Unlike faces.** Galena crystals exhibit combination of the cube and octahedron. Such type of crystals are said to possess unlike faces.
2. **Form.** All the faces corresponding to a crystal are said to constitute a form of the crystal. A simple form of crystal is made up of like faces and a crystal with two or more simple forms is known as combination.
3. **Edge.** The intersection of two adjacent faces gives rise to the formation of an edge.
4. **Interfacial angle.** The angle between the normals to the two intersecting faces is called interfacial angle.

   The crystals are bounded by plane faces, straight edges and interfacial angles. The relationship between these elements can be expressed as

   $$f + c = e + 2,$$

   where f is number of faces, c is number of angles and e is number of edges.
5. **Zones and zone axis.** The faces of a crystal occur in sets called zones which meet in parallel edges or would do so if the planes of the faces are extended.

   A line drawn through the centre of a crystal in a direction parallel to the edges of a zone is called zone axis.

13.4 CRYSTAL LATTICE AND UNIT CELL

A crystal is a homogeneous portion of a substance composed of a regular pattern of structural units (ions, atoms or molecules) bounded by plane surfaces making definite angles (called interfacial angles) with each other giving a regular geometric form. The regular pattern of points which describes the three-dimensional arrangement of particles (ions, atoms or molecules) in a crystal structure is known as the **crystal lattice or space lattice**. The space lattice has to be thought of as extending in all directions throughout the entire crystal.

Unit cell: It is the smallest portion of the space lattice which shows the pattern of the whole lattice. Such unit cells are repeated over and over again in three dimensions and result into whole of the space lattice of the crystal. *It may be defined as the small portion of the space lattice, which, when moved a distance equal to its own dimensions; in various directions, generates the whole lattice.* The space lattice and the unit cell (marked by bold lines) are shown in Fig. 13.7. Actually, it is not possible for a single unit cell to exist alone since the boundaries of a unit cell are also the boundaries of the neighbouring unit cells. If a crystal has the same unit cell containing the same lattice points throughout the whole crystal, it is called an ideal-crystal.

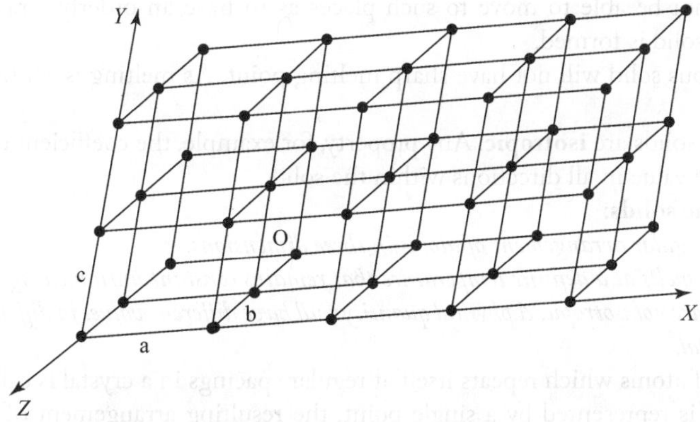

Figure 13.7 Space lattice and unit cell (Marked by bold lines)

It may be noted that it is the points and not the lines which represent the space lattice. The lines drawn represent three axes which describe the relative positions of the points.

In order to describe a unit cell, we ought to know the distances a, b, c, i.e., the lengths of the edges of unit cell and the angles between the three imaginary axes. Broadly speaking, the following three different types of unit cells are known to exist in a cubic system (Fig. 13.8).

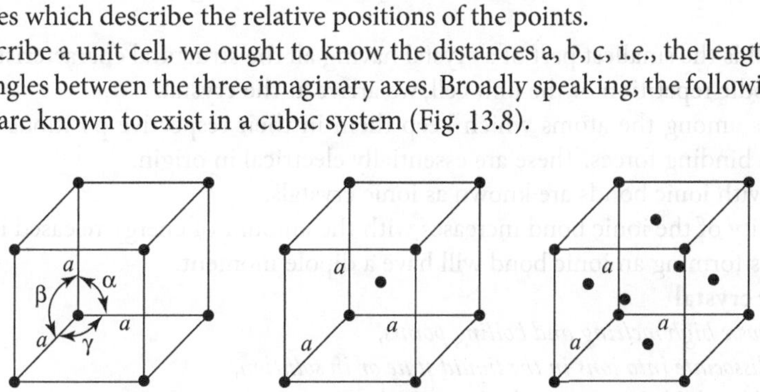

Figure 13.8 Types of unit cells in a cubic system

(i) Simple: It is that arrangement in which the atoms, ions or molecules are present only at the corners of the unit cell.

(ii) Face-centred: When in the unit cell, besides the points at the corners, there is one point present in the centre of each face, it is called face-centred arrangement.

(iii) Body-centred: In this arrangement, in addition to the points at the corners, there is one point at the centre within the body of the unit cell.

(iv) Body-face centred: In this arrangement, in addition to the points at the corners, the points are located at the centre of any two parallel faces of the unit cell. Such unit cells are encountered in monoclinic hexagonal and orthorhombic crystals.

MEMORY FOCUS

1. Solids are characterized by definite shape, size and ability to resist shear.
2. Solids are of two kinds—amorphous and crystalline.
3. The atoms in an **amorphous solid** are randomly situated without any orderly arrangement. Example: glass.
4. As a liquid solidifies from the molten state, because of high viscosity and sudden cooling, the atoms will not be able to move to such places as to have an orderly arrangement. Thus, an amorphous solid is formed.
5. An amorphous solid will not have sharp melting point. As melting is continued, temperature changes.
6. Amorphous solids are **isotropic**. Any property, for example, the coefficient of linear expansion has the same value in all directions within the solid.
7. **In crystalline solids**:
   (a) *There is regular arrangement of atoms in three dimensions.*
   (b) *A crystal melts at a definite temperature that remains constant until melting is completed.*
   (c) *Crystals are not isotropic. A physical quantity will have different values in different directions within a crystal.*
8. The group of atoms which repeats itself at regular spacings in a crystal is called the **basis**.
9. If the **basis** is represented by a single point, the resulting arrangement of points in space is known as **lattice.**
10. The environment surrounding every lattice point is identical.
11. If the group of atoms forming the basis repeats itself at regular intervals in space, we get the crystal.
12. **A unit cell** is the smallest part of a crystal having all the structural characteristics of the crystal. By a periodic repetition of the unit cell, we arrive at the crystal.
13. The forces among the atoms which keep them in their respective positions in a unit cell are termed as binding forces. These are essentially electrical in origin.
14. Crystals with ionic bonds are known as ionic crystals.
15. The stability of the ionic bond increases with the amount of energy released in its formation.
16. The atoms forming an ionic bond will have a dipole moment.
17. **The ionic crystal**
   (a) *will have high melting and boiling points,*
   (b) *will dissociate into ions in the liquid state or in solution,*
   (c) *is highly soluble in solvents that cause ionization,*
   (d) *is not conducting due to lack of free electrons and*
   (e) *is brittle and breaks along cleavage planes.*
18. A crystal structure is fully defined by the lattice constants a, b and c and the interfacial angles α, β and γ.
19. The simple cubic crystals may be of three possible types called simple cubic lattice, body-centred lattice and the face-centred lattice.
20. The number of equidistant neighbours that an atom has in the given structure is called the **co-ordination number**.

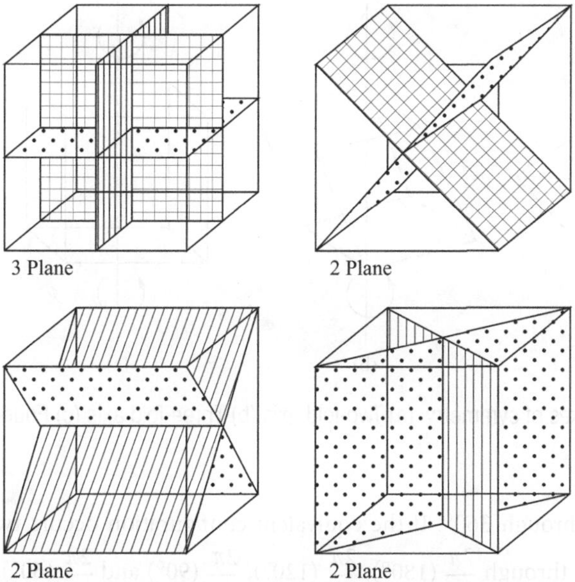

Figure 13.9 Plane of symmetry in a cube

13.5 ELEMENTS OF SYMMETRY

A symmetry element is a geometrical entity, such as a line (or axis), of a plane or a point with respect to which one or more symmetry of operations may be carried out.

13.5.1 Plane of Symmetry and Reflections

If a crystal can be divided by an imaginary plane passing through its centre into two equal portions, each of which is a mirror image of the other, the crystal is said to possess a plane of symmetry. In other words, a plane of symmetry divides the crystal into two identical and similarly placed halves.

Any line drawn perpendicular to this plane intersects the surface of crystal at equal distance on either centre. The plane of symmetry can be illustrated by taking the example of a cube. Figure 13.9 shows dissected planes in a cube. In all, there are nine planes of symmetry in a cube. The inspection of a brick shows that there are only three planes of symmetry in it.

The following examples illustrate the plane of symmetry (Fig. 13.10). The standard rotation for a plane of symmetry is σ that is also symbol for the operation of reflecting through the plane.

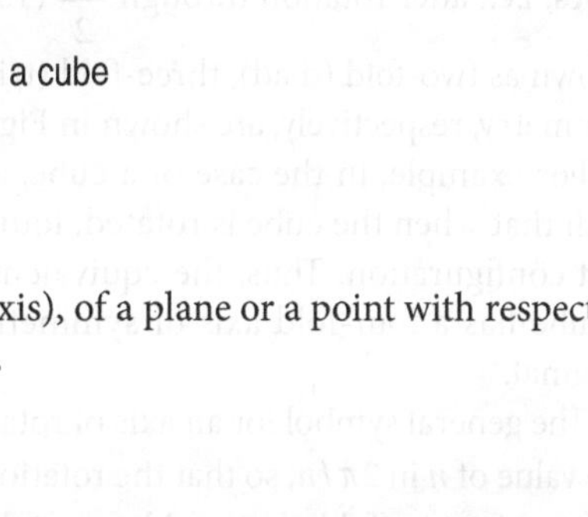

Figure 13.10 Example of plane of symmetry

13.5.2 Axis of Symmetry or Axis of Rotation

It is an imaginary line through the centre of the crystal about which the crystal may be rotated, so that it represents the equivalent configuration more than once in the course of

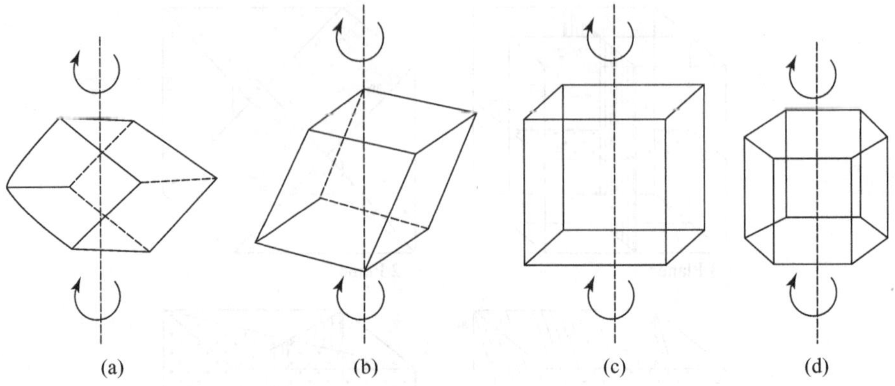

Figure 13.11 Various axes of symmetry: (a) Two-fold axis (b) Three-fold axis (c) Four-fold axis (d) Six-fold axis

a complete revolution through 360°. If the equivalent configuration occurs twice, thrice, four and six times, i.e., after rotation through $\frac{2\pi}{2}$ (180°), $\frac{2\pi}{3}$ (120°), $\frac{2\pi}{4}$ (90°) and $\frac{2\pi}{6}$ (60°), the axes of rotation are known as two-fold (diad), three-fold (triad), four-fold (tetrad) and six-fold (hexad) rotations; the axes of symmetry, respectively, are shown in Fig. 13.11.

For example, in the case of a cube, an axis passing perpendicularly through the centre of a face is such that when the cube is rotated, four different faces come into the same position giving the equivalent configuration. Thus, the equivalent configuration is repeated four times in the revolution. Thus, a cube has a four-fold axis of symmetry. The equivalent configuration is indistinguishable from the original.

The general symbol for an axis of rotation is C_n when n denotes the order of the axis. By order is meant the value of n in $2\pi/n$, so that the rotation through $2\pi/4$ gives an equivalent configuration. Thus, c_2, c_3, c_4, c_6 stand for two-fold, three-fold, four-fold and six-fold axes, respectively.

A two-fold axis of symmetry (or diad) is that axis about which if a crystal is rotated, then-it occupies the same position two times in a complete rotation. An axis of two-fold symmetry for a cube is shown in Fig. 13.11(a). It intersects at the middle point of the two opposite edges of the cube. As there are 12 edges in a cube, there will be in all six axes of two-fold symmetry.

A three-fold axis of symmetry (or triad) is that axis about which if crystal is rotated, then it occupies the same position three times in a complete rotation. Figure 13.11(b) shows an axis of three-fold symmetry for a cube. This axis intersects at the two opposite solid angles of the cube. As there are eight solid angles in a cube, there will be in all four axes of three-fold symmetry.

A four-fold axis of symmetry (or tetroid) is that axis about which if crystal is rotated, then it occupies the same position four times in a complete rotation. Figure 13.11(c) shows an axis of four-fold symmetry for a cube. It intersects at the middle points of the two opposite faces of a cube. As there are six faces in a cube, there will be in all three axes of four-field symmetry.

A six-fold axis of symmetry is that axis about which if a crystal rotated, then it occupies the same position six times in a complete rotation. It intersects at the middle points of the two faces of a hexagon.

13.5.3 Centre of Symmetry or Inversion Centre

A crystal is said to possess a centre of symmetry if every face has an identical face at an equal distance on the opposite side of this centre. If we invert the object though this centre and see the result, we find the object is again indistinguishable from the original. For example, the dimer of NO_2 (N_2O_4) in the solid phase possesses a centre of symmetry. A line drawn between symmetrically equivalent nitrogen atoms is bisected by the inversion centre.

The centre of symmetry may be defined as *an imaginary point inside the crystal such that any straight line drawn through it will intersect the faces, edges and solid angles of the crystal at equal distances on opposite sides.*

In addition, it may be defined as *a centre inside the crystal around which like faces, edges and solid angles are arranged in pairs on opposite side are equal distances.*

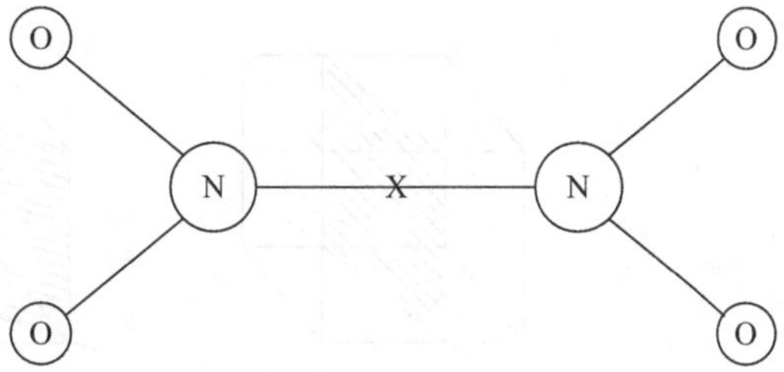

Figure 13.12 More elements showing the inversion centre by a cross X

A crystal may have any number of planes or axes of rotation but it has only one centre of symmetry. The above three elements of symmetry in a cube are shown in Fig. 13.12.

13.5.4 Improper Axis or Rotation Reflector Axis and Improper Rotation

This symmetry element has associated with it a rotation operation in that about the axis of rotation plus a reflection operation through plane perpendicular to the axis.

13.5.5 Axis of Rotation Inversion

The symmetry element has associated with it a rotation like that about the axis of rotation plus an inversion operation through the centre of symmetry.

The symbol of axis of rotation inversion is $\bar{C}_n$, i.e., a bar over C_n or $\bar{2}$, $\bar{3}$, $\bar{4}$, $\bar{6}$ etc. Generally, in a molecule, the number of axes of rotation and axes of rotation inversion are the same.

13.5.6 Total Elements of Symmetry

The important elements of symmetry which very important from the viewpoint of the crystal structure are (i) plane of symmetry, (ii) axis of symmetry and (iii) centre of symmetry.

Now, a crystal may have different number of each of symmetry. The number of planes, axes and centre of symmetries possessed by a crystal is termed as the elements of symmetry of the crystal. Let us explain this term by taking a specific case of a cubic crystal, e.g., NaCl. This crystal possesses 23 elements of symmetry as in clear from Fig. 13.13. These elements of symmetry are:

(i) **Rectangular plane of symmetry:** This is shown in Fig. 13.13 (a). There can be two more such planes, each of which will be perpendicular to the plane already shown. Thus, there are three such rectangular planes of symmetry.

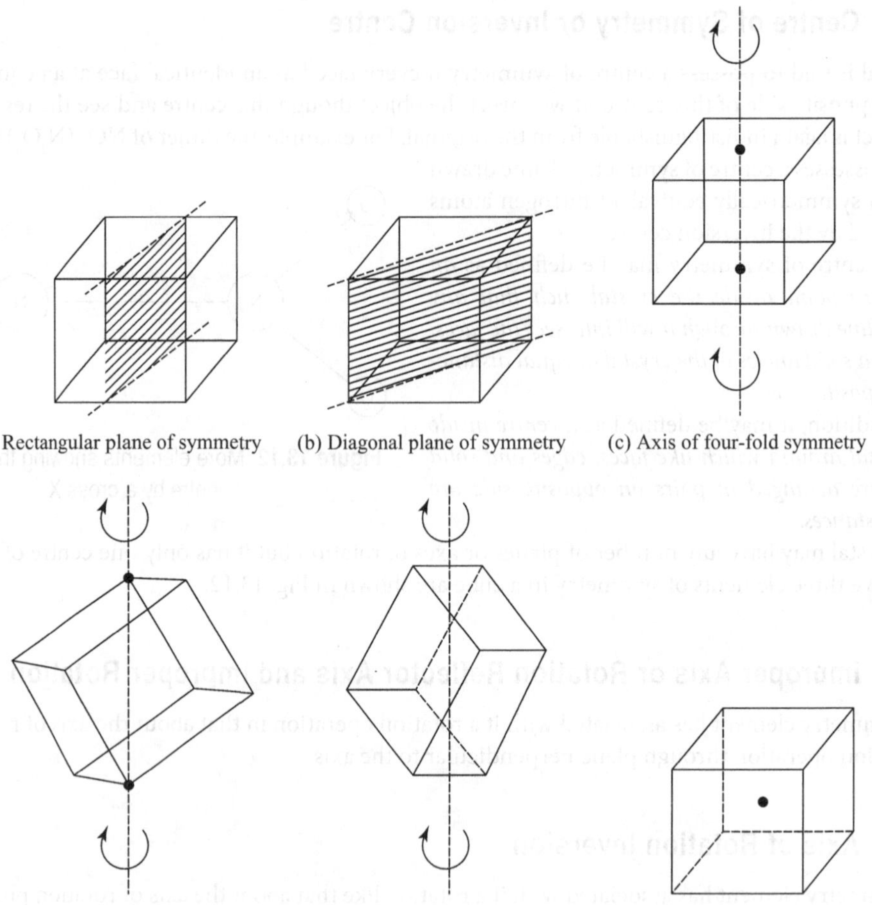

(a) Rectangular plane of symmetry (b) Diagonal plane of symmetry (c) Axis of four-fold symmetry

(d) Axis of three-fold symmetry (e) Axis of two-fold symmetry (f) Crystal of symmetry

Figure 13.13 Various elements of symmetry in a cubic crystal

(ii) Diagonal plane of symmetry: This is shown in Fig. 13.13 (b). There can be a total of six such planes passing diagonally through the cube, as a little imagination will prove.

(iii) Axes of four-fold symmetry: One such plane is shown in Fig. 13.13 (c). There can be a total of three such axes mutually perpendicular.

(iv) Axes of three-fold symmetry: One such plane is shown in Fig. 13.13 (d). There can a total of four such axes.

(v) Axes of two-fold symmetry: One such axis is shown in Fig. 13.13 (e). There can be evidently six such axes.

(vi) Centre of symmetry: There is only one centre of the cube Fig. 13.13 (f).

Thus, the elements of symmetry of various types are:

(i) Planes of symmetry = 3 + 6 = 9 elements.
(ii) Axes of symmetry = 3 + 4 + 6 =13 elements.
(iii) Centre of symmetry = 1 element.

Therefore, total elements of symmetry = 23.

13.6 CRYSTALLOGRAPHIC DESIGNATIONS

A concise method of showing the parameters relations of the crystal face to the crystallographic axes is termed as crystallographic designations. There are a number of crystallographic designations used by Miller, Weiss, Leny, Naumam and Goldschmidt. Out of them, two chief systems are very often used, which are discussed in the following sections.

13.6.1 Weiss Indices (Parameter System of Weiss)

The ratios of the distances from the origin at which a face intersects the crystallographic axes are called the parameters of the crystal face or Weiss indices.

For example, ABC is the face of a crystal which intersects the crystallographic axes *a*, *b* and *c* at the distance OA, OB and OC respectively from the origin O. The parameters of the face ABC are then the ratios of OA, OB and OC. In the parameter system of Weiss, it is essential to close a face whose intercepts on the crystallographic axes are taken as unit lengths for expressing the intercepts of any other face. Such a face is called a unit face.

The selection of the unit face is arbitrary and depends upon the nature of the crystal. The ratio of the intercepts that the unit face makes with the crystallographic axes is called axial ratios. It is the convention to write the intercept on *b*-axis as unity in expressing the axial ratio, e.g., the axial ratio for gypsum is a, b, c = 0.6899 : 1 : 0.4124. In obtaining the parameters of any other face in a gypsum crystal, we shall take 0.6899 as our unit of measurement along *a*-axis, 1 along *b*-axis and 0.4124 along *c*-axis.

For example, let us take face ABC as unit face and we are to find out the Weiss parameters of the face ADE. From Fig. 13.14, it is seen that the face ADE cuts the crystallographic axis *a* at the same distance as that of unit face ABC, but it cuts *b*-axis such that OD = 2 OB and cuts *c*-axis such that OE = ½OC. Thus, the intercepts of the face ADE on the crystallographic axes, *a*, *b* and *c* are 1, 2 and ½ in terms of the intercepts of unit face ABC on the corresponding axes. In the parameter system of Weiss, these intercepts are written before the corresponding axis. Thus, the Weiss symbol for the face ADE is $(a : 2b : \frac{1}{2} c)$. If a face is parallel to a particular axis, then it is considered to cut the axis at infinity. Hence, the sign of infinity, ∞, is put as its parameter before the corresponding axial letter, e.g., *a* face cutting *b* and *a* axes at 1 and 3 units, respectively, compared to the unit face and is parallel to *c*-axis have Weiss symbol as $(a : 3b : \infty c)$.

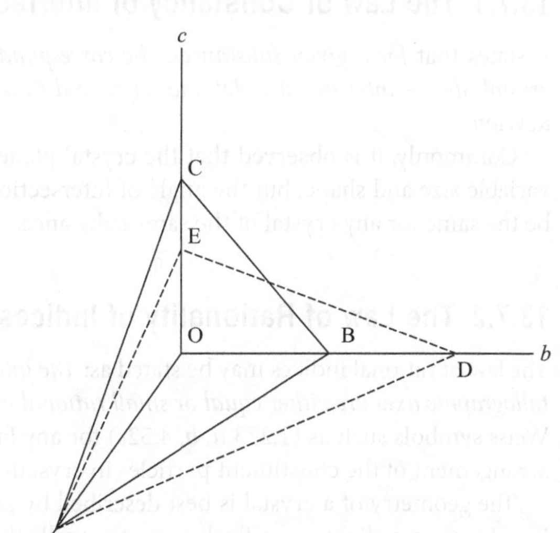

Figure 13.14 Parameters

13.6.2 Index System of Miller (Miller Indices)

The reciprocals of parameters are called the indices (plural index). In the index system of Miller, these indices are written in the order of axes *a*, *b* and *c* and are given in the simplified form by clearing the fractions, if any. As for example, consider the crystal face ADE which has the Weiss symbol ($a : 2b : 1/2\ c$). The reciprocals of the parameters, i.e., indices are 1, 1/2 and 2 in the axial order a, b and c. Now, the indices are multiplied by 2 to clear the fraction. Thus, we get 2, 1 and 4.

Thus, the Miller symbol for the face ADE is, therefore, (214) which is read as two one four. In the same way, if the Weiss symbol for a face is (a: $3b$: ∞ c), then its Miller symbol is (310) which is read as three one naught.

As considered earlier, the unit face of gypsum crystal cut the three axes at unequal distance, even its Miller symbol is (111). It is due to the fact that the unit face has been defined as cutting each axis at unit length for the purpose of measuring the intercepts of other faces. Thus, the Miller symbol for the unit face' ABC is also (111) (Fig. 13.8).

It may be noted that a face parallel to an axis will contain 0 (naught), the reciprocal of infinity at the appropriate position in the, Miller symbol. Again, as the Miller symbols are based on reciprocals of parameter, the larger the figure in the Miller symbol is, the nearer to the origin will the face intersect the corresponding axis. In the same way, the smaller the figure in the Miller symbol is, greater will be the distance from the origin at which the face will cut the axis. *Thus, Miller indices of given face are inversely proportional to the intercepts of the face on the various axes.*

13.7 LAWS OF CRYSTALLOGRAPHY

Crystallography is based on three fundamental laws, namely:

(i) The law of constancy of interfacial angles
(ii) The law of rationality of indices
(iii) The law of symmetry

13.7.1 The Law of Constancy of Interfacial Angles

It states that *for a given substances, the corresponding faces or planes forming the external surface of a crystal always intersect at a definite angle and thus this angle remains constant no matter how the faces develop.*

Commonly, it is observed that the crystal planes are unequally developed so as to produce faces of variable size and shape, but the angle of intersection of any two corresponding faces is always found to be the same for any crystal of the same substance.

13.7.2 The Law of Rationality of Indices

The law of rational indices may be stated as: *The intercepts that any face of the crystal makes with the crystallographic axes are either equal or small rational multiplies of the intercepts made by unit face.* Thus, the Weiss symbols such as (1.273 a, b, 4.52c) for any face are impossible. This law is a result of the ordered arrangement of the constituent particles in crystals.

The geometry of a crystal is best described by proposing some crystallographic axes which are some imaginary co-ordinate axes. Such axes are usually three and sometimes four; all meeting at a point and are so chosen as to bear a definite relation to the characteristic feature of the crystal.

For any crystal, a set of three coordinate axes can be chosen, so that all the faces of the crystal will either intercept these axes at definite distances from the origin or be parallel to some of the axes, in which case intercepts are at infinity.

Consider a three-dimensional co-ordinate system with axes X, Y and Z having origin at the point O. Let ABC be the unit cell and LMN the face of the crystal under study (Fig. 13.15). The intercepts are OA (= a), OB (= b) and OC (= c). For the face under study, the intercepts will be OL (= ma), OM (= nb) and ON (pc) where m, n and p are either simple whole numbers including infinity or fractions of whole numbers.

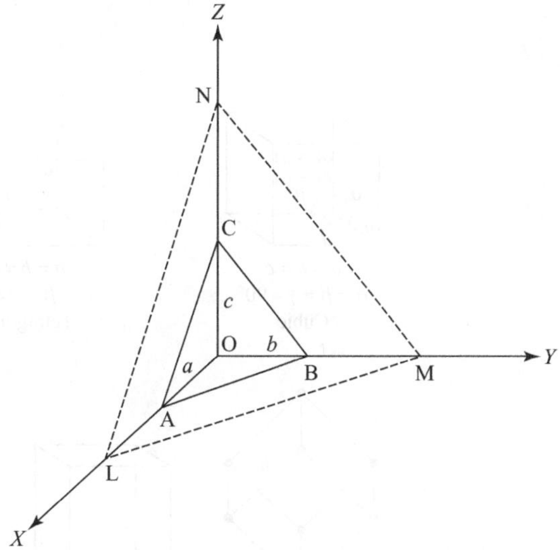

Figure 13.15 Illustration of the law of rationality of indices

The law may be illustrated with data on crystals of the mineral Topaz, $Al_2(FOH)_2\ (SiO_4)$ for which four different planes have the parameters,

1. $m = 1$ $n = 1$ $p = 1$
2. $m = 1$ $n = 1$ $p = \infty$
3. $m = 1$ $n = 1$ $p = 2/3$
4. $m = 2$ $n = 1$ $p = \infty$

and hence the intercept ratios are:

1. $a : b : c$ 2. $a : b : \infty c$

3. $a : b : 2/3c$ 4. $2a : b : \infty c$

For any particular plane, these ratios characterize the plane and may consequently be used to represent it. The co-efficient a, b and c (as discussed earlier) are known as the Weiss indices of a plane. However, Weiss indices are rather awkward in use and have consequently been replaced by Miller indices.

The Miller indices of a plane are obtained by taking the reciprocals of the Weiss coefficients and multiplying through by the smallest number that will express all the four planes mentioned above for topaz are respectively (111), (120), (223) and (120) in the Miller system of crystal face notation.

13.7.3 The Law of Symmetry

This law states that *all crystals of the same substances possess the same element of symmetry. There are three possible types of symmetry: (i) plane of symmetry, (ii) axis of symmetry and (iii) centre of symmetry.*

These have been discussed earlier. The total number of plane, axis and centre symmetries possessed by a crystal is termed as elements of symmetry of the crystal. Different types of crystal systems are shown in Fig. 13.16.

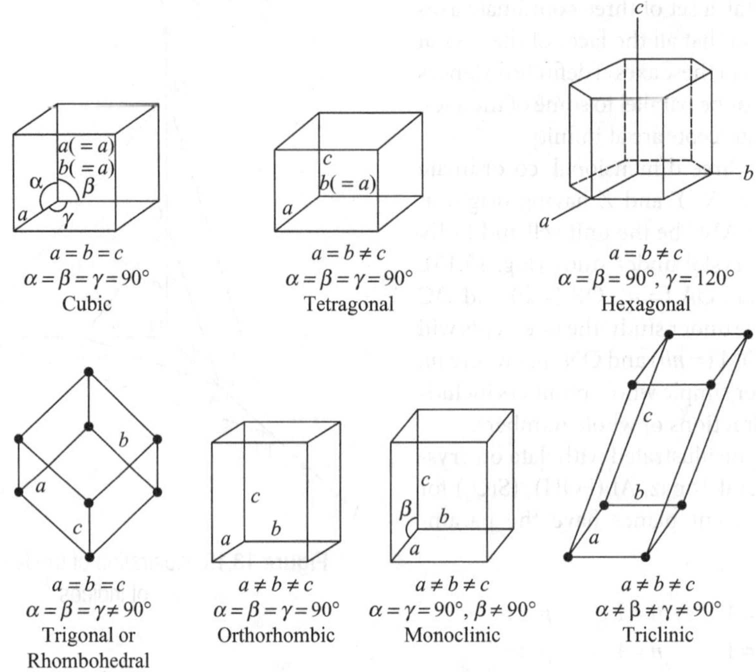

Figure 13.16 Crystal systems

MEMORY FOCUS

I. Laws of crystallography. Crystallography is based upon three fundamental laws:

1. **The law of constancy of interfacial angles:** The angle between the corresponding faces or planes forming the external surface of a crystal remains constant for a given substance, no matter how the face is developed.

   A substance may crystallize under different conditions to give crystals with faces of variable size and shapes. The angle of intersection of any two corresponding faces will be same.

   For example, Interfacial angles in all sodium chloride crystals are found to be 90°, irrespective of their shape and size.
2. **Hauy's law of rationality of intercepts or indices:** The intercepts of any face or plane of a crystal on suitable crystallographic axes can be expressed by small multiples of three unit distances or their simple multiple of three distance or their functional multiples.
3. **Law of constancy of symmetry:** All the crystal of the same substance possess the same elements of symmetry. There are three possible type of symmetry:
   (i) **Plane of symmetry.** A crystal can be divided by an imaginary plane passing through its centre into two equal portions, each of which is a mirror image of the other; the crystal is said to possess a **plane of symmetry**.

   The standard notation for a plane of symmetry is represented by a. Cubic crystals (e.g., NaCI) possess **three rectangular** planes of symmetry and six diagonal planes of symmetry.

(Continued)

MEMORY FOCUS (*Continued*)

(ii) Axis of symmetry. It is an imaginary straight line about which the crystal may be rotated, so that it represents the same appearance more than once during the complete revolution.

The axis of symmetry are called diad, triad, tetrad or hexad axes of symmetry depending upon same appearance of crystal two times, three times, four times and six times, respectively, i.e., after the rotation of 180°, 120°, 90° or 60°.

(iii) Centre of symmetry. A crystal is said to have centre of symmetry if every face has an identical face at an equal distance on the opposite side. Thus, the centre of symmetry is a point that any line drawn through it will meet the surface of crystal at equal distances on either side.

Note. It is important to note that a crystal may possess any number of planes or axes of symmetry, but it can have only one centre of symmetry.

II. Element of symmetry in a cube. The total number of plane, axis and centre of symmetry possessed by a crystal is called **element of symmetry**.

Thus, in a cubic crystal, there are three elements of symmetry as

1. Plane of symmetry

3	+			6	=	9
Rectangular planes				diagonal plane		

2. Axis of symmetry

3	+	4	+	6	=	13
(four-fold)		(three-fold)		(two-fold)		

3. Centre of symmetry. 1 (one) = 1

Total = 23

13.8 CRYSTAL SYSTEMS

All the crystals have been classified into seven crystal habits or spectrums illustrated in Table 13.2.

Table 13.2 Classification of crystal habits and spectrums.

S. No.	Systems	Symmetry Elements	Axes	Angles	Examples
1.	Cubic	Four three-fold axes	$a = b = c$	$\alpha = \beta = \gamma = 90°$	NaCl, Cs, Cl, CaO
2.	Orthorhombic	Three two-fold axes	$a \neq b \neq c$	$\alpha = \beta = \gamma = 90°$	K_2SO4, $HgCl_2$ rhombic sulphur
3.	Tetragonal	One four-fold axes	$a = b \neq c$	$\alpha = \beta = \gamma = 90°$	SnO_2, TiO_2
4.	Monoclinic	One two-fold axes	$a \neq b \neq c$	$\alpha = \gamma = 90° \neq \beta$	Gypsum, monoclinic sulphur, borax, $KClO_3$
5.	Hexagonal	One six-fold axes	$a = b \neq c$	$\alpha = \beta = 90°, \gamma = 120°$	Quartz, ZnO, AgI
6.	Rhombohedral or trigonal	One three-fold axes	$a = b = c$	$\alpha = \beta = \gamma \neq 90°$	Calcite, ice, graphite
7.	Triclinic	None	$a \neq b \neq c$	$\alpha \neq \beta \neq \gamma \neq 90°$	$CuSO_4$, $5H_2O$, $K_2S_2O_3$

Numerical Problems

Example 1 **Identify the crystal system to which each of the following belongs from the given data.**

a	b	c	α	β	γ
9 Å	9 Å	9 Å	90°	90°	90°
5.3 Å	8.0 Å	7.9 Å	90°	90°	90°
11.9 Å	11.9 Å	11.9 Å	90°	90°	90°
11.9 Å	11.9 Å	15.6 Å	90°	90°	90°
9.8 Å	19.8 Å	6.9 Å	90°	120°	90°
5.3 Å	8.0 Å	7.9 Å	95°	85°	125°

Solution:

(i) In this case, $a = b = c = 9$ A, $\alpha = \beta = \gamma = 90°$, the crystal belongs to **cubic system**.
(ii) Since, $a \neq b \neq c$, but $\alpha = \beta = \gamma = 90°$, the crystal belongs to **ortho-rhombic system**.
(iii) Since, $a = b = c = 11.9$ A and $\alpha = \beta = \gamma = 90°$, the crystal belongs to **cubic system**.
(iv) Since, $a = b \neq c$, but $\alpha = \beta = \gamma = 90°$, the crystal belongs to **tetragonal system**.
(v) Since, $a \neq b \neq c$, and $\alpha = \gamma = 90°$, $\beta = 120°$, the crystal belongs to **monoclinic system**.
(vi) Since, $a \neq b \neq c$, and $\alpha \neq \beta \neq \gamma \neq 90°$, the crystal belongs to **Triclinic system.**

Example 2 **How many atoms are assigned to the unit cell of (i) a body-central cube, (ii) a face-centred cube and (iii) a simple cube.**

Solution: **(i) In a body-centred cube**, there are eight corner atoms and one in the centre. Since the contribution of each of the corner atom of the unit cell is 1/8. Eight such corner atoms contribute $(8 \times 1/8) = 1$, atom to the unit cell. There is one atom in the centre of the unit cell in a body-centred cube. Hence, the total number of atoms assigned for body-centred cube is eight corner atoms. $(8 \times 1/8) + 1 = 2$. Therefore, the total is two atoms.

(ii) In the face-centred cube, there are eight corner atoms and one in each face. Hence, for eight corner atoms, the contribution is $8 \times 1/8 = 1$ atom each. The contribution of face-centred atoms is 1/2. Hence, for six face atoms the total contribution is $6 \times 1/2 = 3$. The total is $3 + 1 = 4$ atoms.

(iii) In the simple cube, there are only eight corner atoms. Hence, the, contributions for the unit cell is $8 \times 1/8 = 1$ atom.

13.9 TYPES OF UNIT CELLS IN CRYSTAL SYSTEM (BRAVAIS LATTICE)

It has been shown on geometrical basis that there are only 14 kinds of unit cells known as Bravais lattice, as shown in Fig. 13.17 corresponding to seven crystal systems. It may be mentioned that as per the structures shown in the diagram, there are only points and no line. The lines have been used simply to represent the complete shape. These lines do not actually exist in the unit cells. The various types of unit cells are called simple or primitive face-centred, body-centred and end-face-centred. These are summarized in Table 13.3.

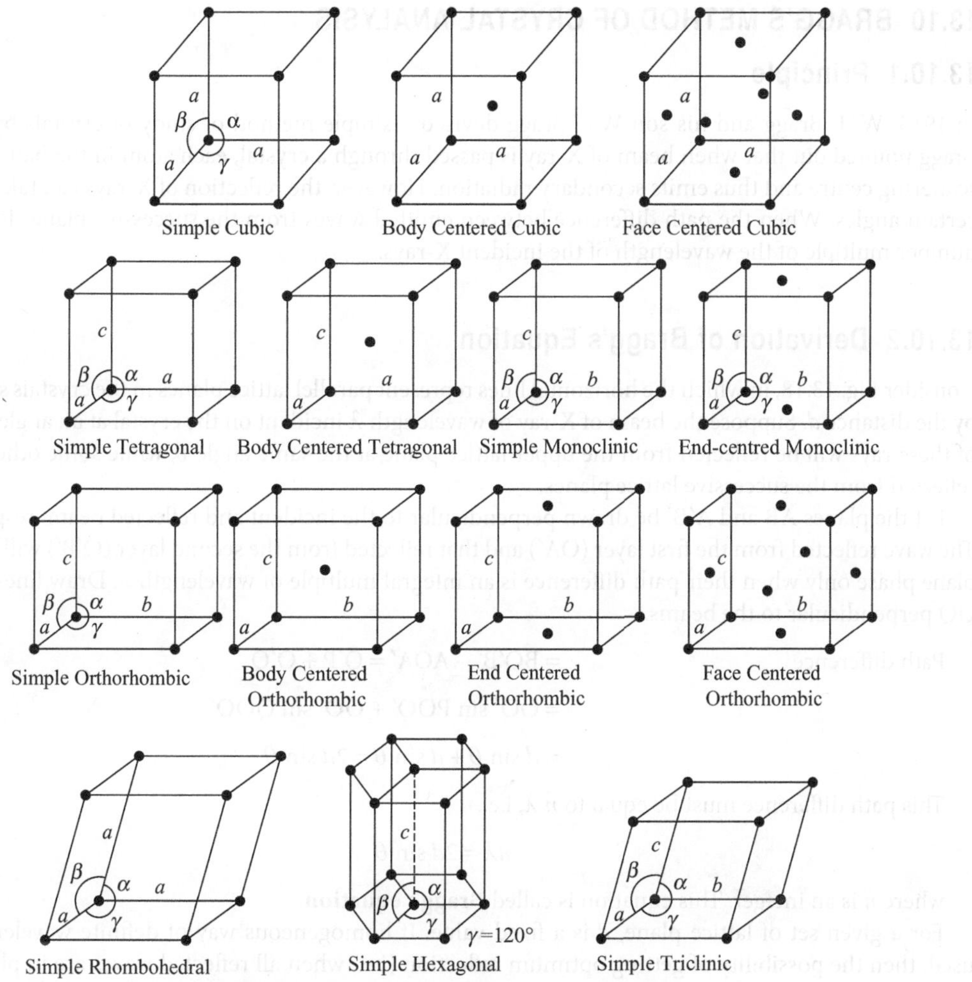

Figure 13.17 Types of unit cells (Bravais Lattice)

Table 13.3 Distribution of lattice types in crystal systems.

S. No.	Systems	Distribution of Lattice Types	Total Number of Lattice Types
1.	Cubic	Simple, body-centred and face-centred	3
2.	Orthorhombic	Simple, end-face-centred, body-centred and face-centred	4
3.	Tetragonal	Simple and body-centred	2
4.	Monoclinic	Simple and end face-centred	2
5.	Hexagonal	End-face-centred only	1
6.	Rhombohedral	Simple only	1
7.	Triclinic	Simple only	1

13.10 BRAGG'S METHOD OF CRYSTAL ANALYSIS

13.10.1 Principle

In 1915, W.H. Bragg and his son W.L. Bragg devised a simple method of study of crystals by X-rays. Bragg pointed out that when beam of X-ray is passed through a crystal, each atom in the path acts as a scattering centre and thus emits secondary radiation. However, the reflection of X-rays can take place at certain angles. When the path difference between emitted waves from the successive planes is a whole number multiple of the wavelength of the incident X-rays.

13.10.2 Derivation of Bragg's Equation

Consider Fig. 13.18, in which the horizontal lines represent parallel lattice planes in the crystals separated by the distance d. Suppose the beam of X-ray of wavelength λ incident on the crystal at an angle θ. Some of these rays will be reflected from the upper lattice plane at the same angle θ, while some other will be reflected from the successive lattice planes.

Let the planes AB and A′B′ be drawn perpendicular to the incident and reflected beam, respectively. The wave reflected from the first layer (OA′) and that reflected from the second layer (O′B′) will be in the plane phase only when their path difference is an integral multiple of wavelength λ. Draw lines OP and OQ perpendicular to the beams.

$$\text{Path difference} = BO'B' - AOA' = O'P + O'Q$$

$$= OO' \sin POO' + OO' \sin QOO'$$

$$= d \sin \theta + d \sin \theta = 2d \sin \theta$$

This path difference must be equal to $n\,\lambda$, i.e.,

$$n\lambda = 2d \sin \theta$$

where n is an integer. This equation is called **Bragg's equation.**

For a given set of lattice plane, d is a fixed value. If homogeneous way of definite wavelength λ is used, then the possibility of getting optimum reflection (i.e., when all reflected waves are in phase with

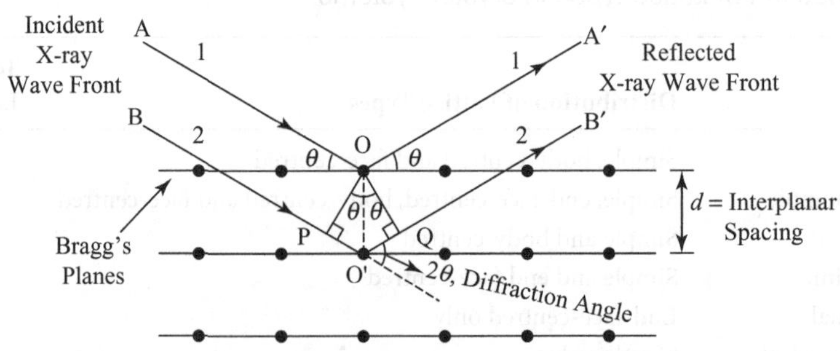

Figure 13.18 Bragg's Equation Derivation

one corner) depends on θ. If θ, the angle which the rays make with the plane of the crystal, is regularly increased, a number of positions corresponding to $n = 1, 2, 3$ etc. will be found at which the reflections will be maximum.

The reflection corresponding to $n = 1$ is called *first-order reflection*. The reflection corresponding to $n = 2$ is the *second-order reflection* and so on.

In experiments of X-ray reflection, it is customary to set $n = 1$. If λ is known, the value of d, the distance between the successive lattice planes in the crystal, can be determined, by measuring θ. On the other hand, if d is known, λ can be evaluated.

Example 3 **The second-order reflection from a crystal at KCl (which utilizes in the face-centred cubic (lattice) occurs at 29°–22′. Calculate the distance between successive 100 planes.**

Solution: From Bragg's equation, $n\lambda = 2d \sin\theta$

where n = order of reflection = 2; λ = wavelength of X-rays = 1.54 Å = 1.54×10^{-8} cm and θ = angle of reflection = 29°22′ ($\sin\theta = 0.4904$).

Substituting the values in the equation

$$2 \times 1.54 \times 10^{-8}\ \text{cm} = 2d \times 0.4904$$

$$\therefore \qquad d = \frac{1.54 \times 10^{-8}\ \text{cm}}{0.4904} = 3.14 \times 10^{-8}\ \text{cm}$$

The distance between the successive 100 planes in KCl crystal

$$= 3.14 \times 10^{-8}\ \text{cm} = 3.14\ \text{Å}$$

13.10.3 Bragg's Method for Determining Crystal Structure

The apparatus used by Bragg is shown in Fig. 13.19. X-ray generated in X-ray tube is passed through a narrow slit to get narrow beam of homogeneous wavelength. The beam is then projected on the surface of the crystal mounted at the centre of rotating table, fitted with scale. The reflected rays are passed into an ionizing chamber which can also be rotated. The extent of ionization produced by the reflected beam is measured by means of an electrometer.

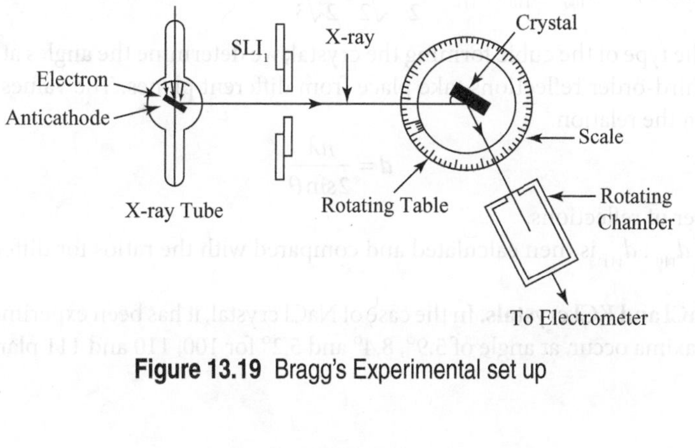

Figure 13.19 Bragg's Experimental set up

The crystal is rotated gradually to increase the glancing angle θ and intensities of reflected rays (as measured by the electrometer), which is plotted against the angles of incidence. The angles at which sudden increase in ionization occurs are noted. The lowest angle at which maximum reflection occurs corresponds to $n = 1$. The process is related for each plane of the crystal.

13.10.4 Applications of Bragg's Equation

13.10.4.1 Calculation of the Distance Between Lattice Planes

From Bragg's equation, we know that

$$n\lambda = 2d \sin\theta$$

$$d = \frac{n\lambda}{2\sin\theta}$$

where the terms have then usual significance. The angle θ, measured at the maxima, is determined experimentally for $n = 1, 2, 3$. Knowing the value of the wavelength (A) of the X-rays used, we can calculate the value of d.

13.10.4.2 In Deciding Bravais Lattice

Different crystal systems have different Bravais lattices. For a cubic system, there are three types of the lattices, e.g., simple, face-centred and body-centred. They crystal faces are the planes which can pass through the atoms or (100), (110) and (111). Let the distance between the 100 planes of the simple cubic lattice beam. Therefore, the distances between the planes of the three types can be calculated, as given below.

$d_{100} : d_{110} : d_{111}$ for simple cubic, face-centred cubic and body-centred cubic lattices will be:

Simple cubic: $$d_{100} : d_{110} : d_{111} = a : \frac{a}{\sqrt{2}} : \frac{a}{\sqrt{3}} = 1 : 0.707 : 0.577$$

Face-centred cubic: $$d_{100} : d_{110} : d_{111} = \frac{a}{2} : \frac{a}{2\sqrt{2}} : \frac{a}{\sqrt{3}} = 1 : 0.707 : 1.154$$

Body-centred cubic: $$d_{100} : d_{110} : d_{111} = \frac{a}{2} : \frac{a}{\sqrt{2}} : \frac{a}{2\sqrt{3}} = 1 : 1.414 : 0.577$$

To determine the type of the cubic forming the crystal, we determine the angles at which the first-, the second- and the third-order reflections take place from different planes. The values of d_{100}, d_{110} and d_{111} are calculated from the relation:

$$d = \frac{n\lambda}{2\sin\theta}$$

where n is the order of reflections.

The radio $d_{100} : d_{110} : d_{111}$ is then calculated and compared with the ratios for different types of lattices as given above.

Structure of NaCl and KCl crystals. In the case of NaCl crystal, it has been experimentally seen that first order reflection maxima occur, at angle of 5.9°, 8.4° and 5.2° for 100, 110 and 111 planes, respectively.

We know that for the first-order reflection,

$$d = \frac{n\lambda}{2\sin\theta} \propto \frac{1}{\sin\theta}$$

$$\therefore \quad d_{100} : d_{110} : d_{111} = \frac{1}{\sin\theta_1} : \frac{1}{\sin\theta_2} : \frac{1}{\sin\theta_3}$$

$$= \frac{1}{\sin 5.9^\circ} : \frac{1}{\sin 8.4^\circ} : \frac{1}{\sin 5.2^\circ}$$

$$= 9.33 : 6.84 : 11.04$$

$$= 1 : 0.704 : 1.14$$

$$= 1 : \frac{1}{\sqrt{2}} : \frac{2}{\sqrt{3}}$$

From the result, it is dear that NaCl crystal has centred cubic structure.
From KCl crystal, the experimental values of

$$d_{100} : d_{110} : d_{111} = 1 : \frac{1}{\sqrt{2}} : \frac{2}{\sqrt{3}}$$

Therefore, potassium chloride crystal has a simple cubic structure.

13.11 TYPES OF CRYSTALLINE SOLIDS

We have already classified solid substances as crystalline and amorphous solids and described the main points of difference in their behaviour. Now, we will classify the crystalline solids based on the nature of constituent particles and the major binding forces operating between them and will correlate the behaviour of such solids with their structure.

Main types of crystalline solids are as follows:

(i) Ionic solids
(ii) Metallic solids
(iii) Covalent solids
(iv) Molecular solids

However, there are certain solid substances known to exist, which exhibit properties and characteristics of more than one of these types. This may be due either to the presence of two different types of bonds at the presence of bonds which are intermediate in character. Each type is described briefly in the following sections.

13.11.1 Ionic Solids

The constituent particles of ionic solids are oppositely charged ions. The strong electrostatic force of attractions between the ions is called an ionic bond. These crystals are formed between constituents,

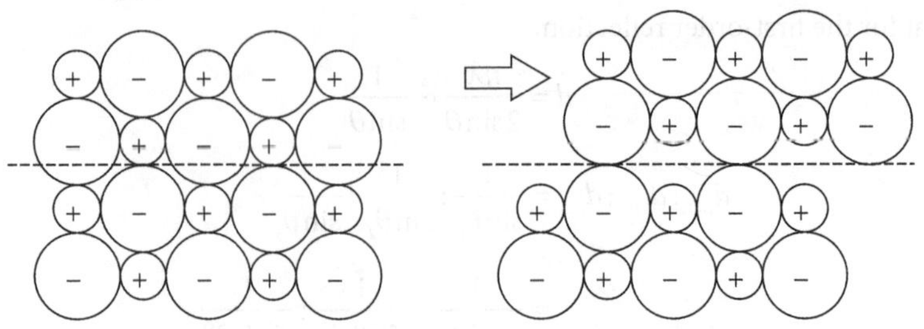

Figure 13.20 Movement of layers relative to each other

which are formed by complete transfer of electrons from one atom to another. NaCl, CsCl, CaO and ZnS are some common examples of ionic solids.

13.11.1.1 General Properties of Ionic Solids

The general properties of such solids are as follows:

(i) They do not conduct electricity in the solid state because their constituent ions are held in fixed positions. However; they conduct electricity in the fused or dissolved state.
(ii) They exhibit high melting points because the forces of attraction in these solids are very strong for the same reason they have high heats of fusion.
(iii) They are hard and brittle. This is because of the movement of layers of ions relative to each other brings ions of the same charge near each other. This causes strong repulsions which lead to the breakdown of the crystal as shown in Fig. 13.20.

13.11.2 Metallic Solids

The constituent units of metallic solids are positive ions which are held together by the free moving electron cloud due to the delocalized electrons. The electron cloud belongs to the entire crystal. Cu. Fe, Ag, Au, Na and K are well-known examples of metallic solids.

13.11.2.1 General Properties of Metallic Solids

(i) They are malleable and ductile. In other words, metals can be beaten into sheets and can be drawn into wires. This is because the layers of metallic crystals can be easily displaced with respect to each other. The movement goes not produce any repulsive effect due to the presence of delocalized electrons everywhere, which prevent repulsion between the positive ions.
(ii) They also exhibit metallic lustre, i.e., their surface appears shiny. Since the electrons are free and are not tied to any bond, they can absorb and emit light of all wavelengths.
(iii) The metals exhibit high electrical and thermal conductivities because electrons can move freely throughout the lattice.

13.11.3 Covalent Solids

The constituent unit in covalent solids is the atom. These atoms are bound to other atoms by shared electron pair bonds. These solids are formed when a large number of atoms are held by strong covalent bonds. This bonding extends throughout the crystal in three dimensions. For example, in diamond, each carbon atom is attached to four other carbon atoms by covalent bonds to form a huge crystal. We may also have two or more different elements linked together by these forces which result in building the entire lattice. In the example of quartz, the arrangement of silicon and oxygen atoms forms such a three-dimensional network. Silicon carbide (carborundum, SiC) has a structure which is very similar to that of diamond except that in this solid alternate carbon atom positions are occupied by silicon atoms.

13.11.3.1 General Properties of Covalent Solids

(i) Covalent solids usually have high melting points.
(ii) They are quite hard and are bad conductors of electricity.
(iii) They have high latent heat of fusion.

All these properties are essential due to the fact that every atom in a covalent solid is held rigidly in its position by strong covalent bonds.

13.11.4 Molecular Solids

The constituent units of molecular solids are molecules (either polar or non-polar) except in solidified noble gases where the units are atoms. The bonds present within the molecule are covalent bonds, whereas the forces which operate between different molecules of the crystal lattice are either *dipole-dipole interactions or van der Waals' forces.* Examples of such solids are iodine, sulphur, phosphorus (non-polar), ice, sugar (polar) etc.

Dipole–dipole interactions are the forces of attraction between positively charged end of one polar molecule and negatively charged end of another. van der Waals' forces are weak forces of attraction which operate between non-polar molecules.

13.11.4.1 General Properties of Molecular Solids

(i) The molecular solids are soft and vapourize very readily. This is due to the weak intermolecular forces.
(ii) For the same reason as above, the molecular solids have low melting points and low heats of fusion.
(iii) These solids do not conduct electricity. This is because their electrons are localized in the bonds in each molecule and are unable to move about from one molecule to another on applying electric field.

Some of the physical properties associated with different types of solids have been summarized in Table 13.4.

Table 13.4 Some physical properties associated with the four bond types.

Crystal Type	Constituent Units	Binding Forces	Physical Property: Hardness	Brittleness	Melting Points	Heats of Fusion	Thermal and Electrical Conductivities	Any Other Important Property	Examples
Ionic	Positive and negative ions	Strong electrostatic attractions between ions	Quite hard and brittle	Very high	Relatively high	Very high	Very low (high electrical conductivity in molten state)	—	NaCl, CaO
Metallic	Positive ions in a sea of electrons	Electrostatic attraction between +ve ions and sea of electrons	Variable	Very low	Variable	Moder-ate	Very high	Metallic lustre, malleable and ductile	Cu, Fe, Ag
Covalent	Atoms	Covalent bonds	Very hard	Medium	Very high	High	Very low	—	Diamond, SiC, SiO_2
Molecular	Molecules	van der Waals' forces	Very soft	Low	Very low	Low	Very low	—	CO_2, N_2, CCl_4

REVISION QUESTIONS

1. (a) How do crystalline substances differ from amorphous substances?
   (b) What do you understand by the term anisotropy and isotropy?
2. (a) Explain the terms: axis of symmetry, plane of symmetry and centre of symmetry. Give suitable examples.
   (b) Define the terms: axis of symmetry, mirror plane and inversion centre.
3. (a) Show diagrammatically the various axes of symmetry and mirror planes present in a simple tetragonal unit cell.
   (b) Show diagrammatically the various axes of symmetry and mirror planes present in a simple cube.
4. (a) Write note on 'symmetry elements of a crystal'.
   (b) A five-fold axis of symmetry cannot be present in any crystal. Why?
5. (a) What do you understand by interfacial angles in crystallography?
   (b) Briefly describe the 'law of rational indices'.
6. Write short note on
   (a) Miller indices (b) Types of crystals
   (c) Space lattice (c) Unit cell
7. The crystallographic axes of a unit cell are *a, b* and *c*. Draw a plane of which the Miller indices are (221). Can there be more than one plane with the same Miller indices?
8. (a) What are Miller indices? Illustrate 111 and 110 planes in a cubic crystal.
   (b) Show that for the given Miller indices in a crystal there is a whole set of equally spaced parallel planes.
   (c) Take a cubic system and draw the planes with Miller indices (112) and (2 10).
9. (a) Write note on 'crystal system'.
   (b) What are the crystallographic dimensions of their unit cells if some solids belong to the following crystal systems: (a) triclinic (b) tetragonal.
   (c) Write a short note on physical properties of crystals.
10. (a) Classify crystals, giving examples.
    (b) Identify the crystal system to which each of the following belongs from the given data:

    (i) $a = b = c = 1$ Å $\alpha = \beta = \gamma = 90°$
    (ii) $a \neq b \neq c$ $\alpha = \beta = \gamma = 90°$
    (iii) $a = b = c = 9$ Å $\alpha = \beta = \gamma \neq 90°$
    (iv) $a = b \neq c$ $\alpha = \beta = \gamma = 90°$
    (v) $a \neq b \neq c$ $\alpha = \beta = 90°, \gamma = 120°$
    (vi) $a \neq b \neq c$ $\alpha \neq \beta \neq \gamma \neq 90°$

    (c) Identify the crystal system to which some solids having the following crystallographic dimensions for their, unit cell belong:

    (i) $a \neq b \neq c$ $\alpha = \beta \neq \gamma \neq 90°$
    (ii) $a \neq b \neq c$ $\alpha = \gamma = 90° \neq \beta$

11. (a) Briefly explain the terms 'unit cell' and crystal lattice.
    (b) Write short note on 'Bravais lattices'.

12. (a) Name the different types of crystal systems. Give names and draw diagrams of the different types of spice lattices associated with the cubic system.
    (b) Show diagrammatically that a face-centred cubic lattice can represented as a rhombohedral lattice.
13. (a) Enumerate various elements of symmetry of a cubic type of unit cell.
    (b) What is the law of rational indices?
14. (a) Calculate the number of points per unit cell in:
    (i) a face-centred cubic lattice (ii) a body-centred orthorhmobic lattice
    (iii) an end-centred monoclinic lattice (iv) simple tetragonal lattice
15. (a) Explain the following with examples:
    (i) Primitive unit cell and non-primitive unit cell
    (ii) Plan of symmetry and axis of symmetry
    (b) Define and explain the law of constancy of interfacial angles and the law of rational indices
16. What do you understand by:
    (a) axis of four-fold symmetry and
    (b) axis of three-fold chemistry.

    How many such axes are present in cubic crystals?
17. Calculate Miller indices of a crystal plane which is cut through the crystal axes $2a$, $-3b$, $-c$.
18. A crystal plane has intercepts of 3, 4 and 2 units with x-, y- and z-axes, respectively. Calculate its Miller indices.
19. Derive Braggs equation for the diffraction of X-rays by crystal lattice. How is the equation used in elucidating the crystal structure?
20. Give different types of crystals. Explain crystal diffraction.
21. Explain what is meant by ionic crystal, molecular crystal and covalent crystal. Give examples.

MULTIPLE CHOICE QUESTIONS

1. A crystalline solid has
   (a) definite geometrical shape
   (b) flat faces
   (c) sharp edges
   (d) all of these
2. Amorphous substances are isotropic because
   (a) they have same value of any property in all directions
   (b) they have different values of physical properties in different directions
   (c) they have definite geometric shape
   (d) none of the above
3. The elements of symmetry are
   (a) plane of symmetry
   (b) axis of symmetry
   (c) centre of symmetry
   (d) all of these
4. The amorphous solid among the following is
   (a) table salt (c) plastic
   (b) diamond (d) graphite
5. A crystalline solid does not have one of the following properties. It is
   (a) anisotropy
   (b) sharp melting points
   (c) isotropy
   (d) definite and regular geometry
6. For tetragonal crystal system, which of the following is not true?
   (a) $a = b \neq c$ (c) $a \neq b \neq c$
   (b) $\alpha = \beta = \gamma = 90°$ (d) none of these

7. If there are four atoms in unit cell in a cubic system, it is an example of
   (a) simple cubic unit cell
   (b) body-centred cubic unit cell
   (c) face-centred cubic unit cell
   (d) none of these
8. Which is incorrect for a hexagonal crystal system?
   (a) $a = b = c$
   (b) $a = b \neq c$
   (c) $\alpha = \beta = 90°, \gamma = 120°$
   (d) none of these
9. The Bragg's equation for diffraction of X-rays is
   (a) $n\lambda = 2d^2 \sin \theta$ (c) $n\lambda = 2d \sin^2 \theta$
   (b) $n\lambda = 2d \sin \theta$ (d) $n\lambda = d \sin \theta$
10. In Bragg's equation $n\lambda = 2d \sin \theta$, '$n$' represents
    (a) the number of moles
    (b) the principal quantum number
    (c) the Avogadro's number
    (d) the order of reflection
11. The permitted co-ordination number in an ionic crystal is six, the arrangement of anions around the cation will be
    (a) plane triangular
    (b) tetrahedral
    (c) octahedral
    (d) body-centred cubic
12. NaCl is an example of
    (a) covalent solid (c) ionic solid
    (b) metallic solid (d) molecular solid
13. Particles of quartz are packed by
    (a) ionic bonds
    (b) van der Waals' forces
    (c) hydrogen bonds
    (d) covalent bonds
14. Crystal structure of NaCl is
    (a) fcc (c) Both (a) and (b)
    (b) bcc (d) None of these
15. The coordination number of Na^+ in cubic closest packing (ccp) structure of NaCl is
    (a) 4 (c) 8
    (b) 6 (d) 12
16. An fcc crystal contains how many atoms in each unit cell?
    (a) 4 (c) 6
    (b) 3 (d) 8
17. The number of atoms in a unit cell of fcc structure is
    (a) 1 (c) 2
    (b) 3 (d) 4
18. In a tetragonal crystal,
    (a) $a = b = c; a = \beta = 90° \neq \gamma$
    (b) $a = \beta = \gamma = 90°; a = b \neq c$
    (c) $a = \beta = \gamma = 90°; a \neq b \neq c$
    (d) $a = \beta = 90°; \gamma = 120°; a = b \neq c$
19. In a covalent solid, the lattice points are occupied by
    (a) molecules (c) atoms
    (b) positive ions (d) negative ions
20. Bragg's law is given by equation
    (a) $n\lambda = 2\theta \sin \theta$ (c) $2n\lambda = d \sin \theta$
    (b) $n\lambda = 2d \sin \theta$ (d) $n\lambda = ½\, d \sin \theta$

ANSWERS

1. (d)	5. (c)	9. (b)	13. (d)	17. (b)
2. (a)	6. (c)	10. (d)	14. (a)	18. (d)
3. (d)	7. (a)	11. (c)	15. (a)	19. (a)
4. (c)	8. (a)	12. (c)	16. (b)	20. (b)

Chemical Bonding 14

CHAPTER OBJECTIVES

14.1 Introduction

14.2 Valence Bond (VB) Theory

14.3 Molecular Orbital (MO) Theory

14.4 Relative Energies of Molecular Orbitals and Filling of Electrons

14.5 Stability of Molecules

14.6 Molecular Orbital Configurations

14.7 Comparison of Valence Bond (VB) Theory and Molecular Orbital (MO) Theory

One of the notable successes of quantum theory is its explanation of the forces that hold the atoms together. It may be said, without fear of contradiction that there can be neither qualitative nor quantitative understanding of the chemical bond if we restrict ourselves to the concepts of classical mechanics.

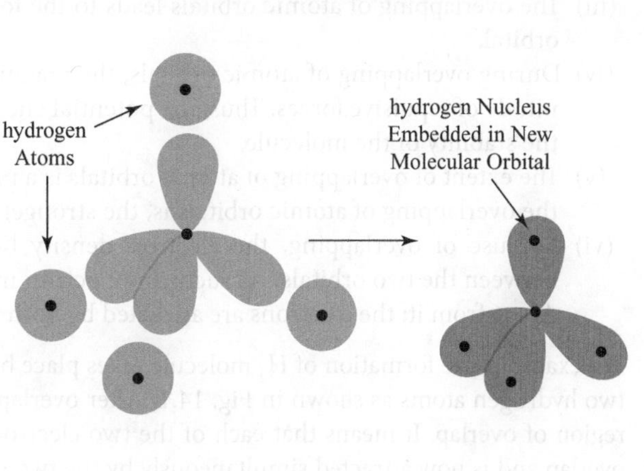

14.1 INTRODUCTION

We know that most of the substances exist in the form of aggregates of atoms called molecules. The attractive force which holds together the constituents (atoms or ions) in a molecule is called a chemical bond. We already know that bonding between two atoms can take place only if the process results in the lowering of energy and that bond formation can take place in different ways resulting into different kinds of bonds such as ionic, covalent, metallic and hydrogen bonds. The concept of covalent bond formation on sharing of electrons (Lewis concept) was purely qualitative approach. It could not answer a number of basic questions such as:

(i) Why a covalent bond formed at all?
(ii) How are electrons arranged in the molecule?
(iii) What are the forces of attractive interaction in a molecule?
(iv) What is the bond energy in a covalent bond?
(v) How does the sharing of electrons explain the shapes of resulting molecules?

The development of wave mechanics has provided answers to these questions and led ultimately to the development of more comprehensive theories of covalent chemical bonding. These are:

(i) Valence bond theory (VBT)
(ii) Molecular orbital theory (MT)

14.2 VALENCE BOND (VB) THEORY

According to valence bond theory of chemical bonding, developed by Linus Pauling (1930), a covalent bond is formed by *the overlapping of half-filled valence orbitals of two atoms with their electrons having spins in opposite directions*. It is assumed that the atoms interact to form the molecule, in which the atomic orbitals *retain* a large part of their *individual character*, i.e., electrons in the molecule are *localized* as if they are in isolated atoms. The main features of this theory are as follows:

(i) A covalent bond is formed when a half-filled valence orbital of one atom overlaps (partially merges) with a half-filled valence orbital of another atom of same or different element.
(ii) The two electrons of overlapping atomic orbitals must have their spins in opposite directions.
(iii) The overlapping of atomic orbitals leads to the formation of a new orbital known as molecular orbital.
(iv) During overlapping of atomic orbitals, the magnitude of attractive forces overpowers the magnitude of repulsive forces. Thus, the potential energy of the system decreases which accounts for the stability of the molecule.
(v) The extent of overlapping of atomic orbitals is a measure of the strength of covalent bond. More the overlapping of atomic orbitals is, the stronger is the covalent bond formed.
(vi) Because of overlapping, the electron density becomes maximum in the overlapping space between the two orbitals. As such, any electron may belong to anyone of the atoms at any time. Apart from it; the electrons are attracted by both the nuclei instead of one.

For example, the formation of H_2 molecule takes place by the overlap of half-filled 1*s* atomic orbitals of two hydrogen atoms as shown in Fig. 14.1. After overlapping, the electron density becomes high at the region of overlap. It means that each of the two electrons spends maximum time in the region of the overlap and is now attracted simultaneously by the two nuclei resulting in the formation of a new bond between them.

This theory *ignores* the fact that the electronic distribution of isolated atoms must be *affected* by the presence of other nuclei in the molecule. In spite of this serious limitation, the VB theory finds wide application in determining the *shapes* of molecules.

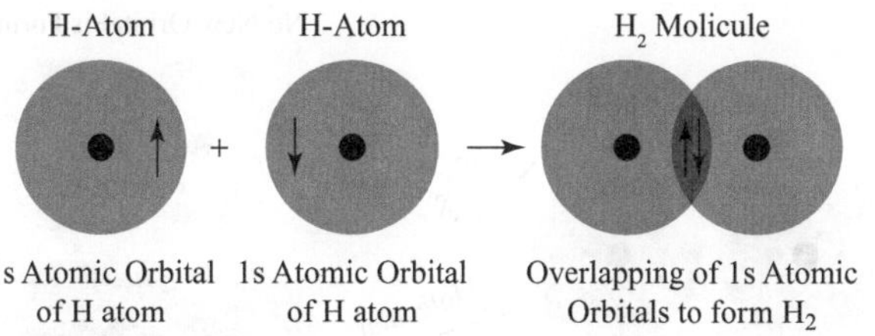

Figure 14.1 Formation of chemical bond by overlap of atomic orbitals

14.3 MOLECULAR ORBITAL (MO) THEORY

There is another approach to chemical bonding known as *molecular orbital (MO) theory*, developed mainly by Mulliken (1932), which explains the bonding characteristics in a better way.

The molecular orbital theory considers the entire molecule as a unit with all the electrons moving under the influence of all the nuclei present in the molecular. This approach recognizes that *each electron* belongs to the *molecule as a whole* and may move within the entire molecule.

14.3.1 Molecular Orbitals

When the atoms to be bonded come close together, the orbitals of the bonded atoms *lose their individual character* and fuse (overlap) to form larger orbitals called molecular orbitals. Thus, like atomic orbitals, there are molecular orbitals in a molecule. The only difference is that in atomic orbitals, electrons move under the influence of only *one nucleus* (i.e., AO's are *monocentric*), while in molecular orbitals, electrons move under the influence of *many nuclei* (i.e., MO's are *polycentric*).

Molecular orbital may, therefore, be defined as "the region in space associated with all the nuclei of the molecule where the probability of finding a particular electron is maximum". As in the case of atomic orbitals, each molecular orbital can accommodate at the most two electrons with opposite spins.

It may be noted that electrons in molecular orbitals are not confined to an individual atom; they belong to the *entire molecule* and are said to be *delocalized* with respect to the individual atoms.

Some important features of molecular orbital theory are as follows:

(i) Like an atomic orbital which is around the nucleus of an atom, there are molecular orbitals which are around the nuclei of a molecule.
(ii) The molecular orbitals are entirely different from the atomic orbitals from which they are formed.
(iii) The valence electrons of the constituent atoms are considered to be moving under the influence of nuclei of participating atoms in the molecular orbital.
(iv) The molecular orbitals possess different energy levels like atomic orbitals in an isolated atom.
(v) The shape of molecular orbitals is dependent upon the shapes of atomic orbitals from which they are formed.
(vi) Molecular orbitals are arranged in order of increasing energy just like atomic orbitals.
(vii) The number of molecular orbitals formed is equal to the number of atomic orbitals combining in bond formation.
(viii) Like atomic orbitals, the filling of electrons in molecular orbitals is governed by the three principles such Aufbau principle, Hund's rule and Pauli's exclusion principle.

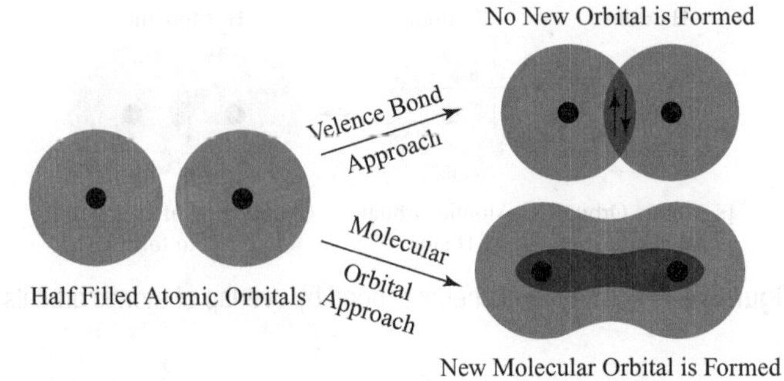

Figure 14.2 Comparison of valence bond approach and molecular orbital approach

The comparison of VBT and MOT in terms of the new molecular orbitals formed after bond formation is diagrammatically represented in Fig. 14.2. The shape of molecular orbital formed in case of valence bond approach depicts that the identity of atomic orbitals is not lost. Whereas in case of molecular orbital approach, the original identity of atomic orbitals is completely lost.

The main difference between VBT and MOT is that the molecular orbitals formed in a compound are quite different from the constituent atomic orbitals. When the atomic orbitals combine to molecular orbitals, they lose their identity and merge into each other completely.

14.3.2 Conditions for Atomic Orbitals to Form Molecular Orbitals

(i) For the atomic orbitals to combine and form molecular orbitals, the following conditions must be satisfied:

(a) The combining atomic orbitals should be of a comparable energy; for example, 1*s* atomic orbitals of two atoms can combine but 1*s* orbital of one atom cannot combine with 2*s* orbital of the other.

Similarly, 2*s* orbitals cannot combine with 2*p* orbital. Such combinations are possible only for the heteronuelear diatomic molecular of the type AB.

(b) The combining atomic orbitals must overlap to a large extent: greater the overlap is, stabler is the molecule formed. The combining atomic orbitals must have proper orientation, so that they are able to overlap to a considerable extent. For instance, overlapping is not possible if *s* and *p* orbitals with orientation shown in Fig. 14.3(a) approach each other.

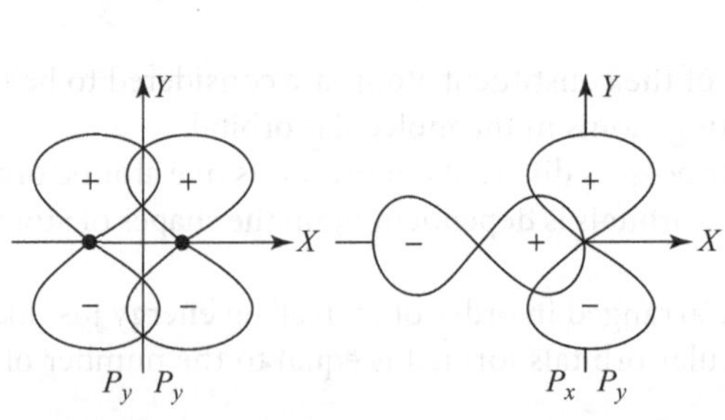

Figure 14.3 (a) p_y and p_y orbitals will combine as they have same symmetry about *x*–axis. (b) p_x and p_y orbitals will not combine as they do not have symmetry about *x*–axis

It may be mentioned here that orbitals which do not have the correct symmetry for combination (e.g., p_x and p_y) are called non-bonding, as there is no overall change in energy. Any stabilization which occurs from overlapping + with + is neutralized by an equal amount of overlap of + with –.

(ii) Thus, based on symmetry, the allowed and forbidden combination of various atomic orbitals is as follows, assuming x-axis as the internuclear axis.

First Orbital	Second Orbital (Allowed)	Second Orbital (Forbidden)	Type of Molecular Orbital Formed
s	s or p_x	p_y, p_z	σ, σ^*
p_x	s or p_x	p_y, p_z	σ, σ^*
p_z	p_z	s, p_x, p_y	π, π^*
p_y	p_y	s, p_x, p_z	π, π^*

First Orbital	Allowed Combinations	Forbidden Combinations	Type of Molecular Orbital Formed
s	s–s; s–p_z	s–p_x; s–p_z	σ, σ^*
p_x	$p_x - p_x$	$p_x - p_y$; $p_x - p_z$	π, π^*
p_z	$p_y - p_y$	$p_x - p_y$; $p_y - p_z$	π, π^*
p_y	s–p_z, $p_z - p_z$	$p_x - p_z$; $p_y - p_z$	σ, σ^*

14.3.3 Difference Between Atomic and Molecular Orbitals

The main differences have been shown in Table 14.1.

Table 14.1 Difference between atomic and molecular orbitals.

Atomic Orbitals	Molecular Orbitals
1. An electron in an atomic orbital is under the influence of only one nucleus.	An electron in a molecular orbital is under the influence of nuclei of two or more atoms of a molecule.
2. Their existence is because of inherent property of the atoms.	They are formed by the combination of atomic orbitals of comparable energies.
3. They are less stable than bonding molecular orbitals, which are more stable than anti-bonding molecular orbitals.	They are less or more stable than atomic orbitals.
4. They have simple shapes.	They have complex shapes.
5. They are represented by s, P, d, f etc.	They are represented by σ, σ^*, π, π^*. etc.

14.3.4 Formation of Bonding and Anti-bonding Molecular Orbitals (LCAO Method)

An electron in an atom is described by a wave function, ψ called an atomic orbital. Similarly, the behaviour of an electron in a molecule is described by a molecular wave function called molecular orbital. The most convenient way of working out the wave functions for molecular orbitals is to adopt the method of linear combination or atomic orbitals (LCAO).

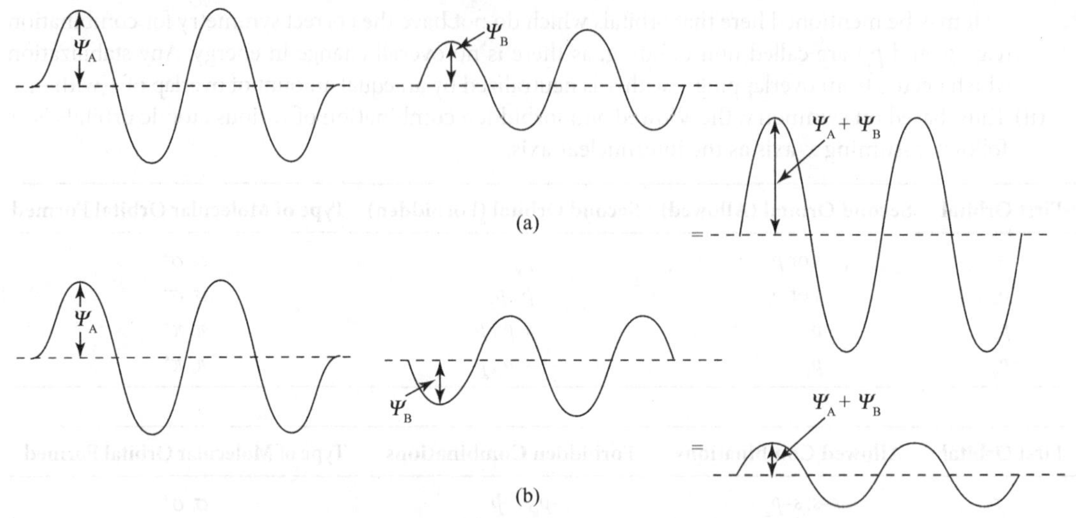

Figure 14.4 (a) Additive combination of waves (b) Subtractive combination of waves

Quantum mechanics shows that linear combination of two functions gives not one but two combinations and hence two molecular orbitals; a bonding orbital and an anti-bonding orbital.

(i) **Addition when two waves are in phase.** If ψ_A and ψ_B are the atomic wave functions of the combining atoms A and B, then the wave functions of orbitals can be obtained by the *addition* or *subtraction* of the wave functions of the two atomic orbitals. The additive effect of the electron waves because of the addition combination of atomic orbital wave functions when the two waves are in *phase* is shown in Fig. 14.4(a).

(ii) **Subtraction when two waves are not in phase.** Similarly, the subtractive effect of the electron waves because of the subtraction combination of atomic orbital wave functions when the two waves are *out of phase* is shown in Fig. 14.4(b).

Thus, the wave function of bonding orbital is given as

$$\psi_b = \psi_A + \psi_B \text{ (bonding orbital stabilizes molecule)} \tag{14.1}$$

The wave function of anti-bonding orbital is given as:

$$\psi_a = \psi_A - \psi_B \text{ (anti-bonding orbital destabilizes molecule)} \tag{14.2}$$

where ψ_b is called *bonding* molecular orbital and ψ_a is called *anti-bonding* molecular orbital.

Squaring Eq. (14.1), we get the expression for the electron density (probability) of the bonding molecular orbital.

$$\psi_b^2 = (\psi_A + \psi_B)^2 = \psi_A^2 + \psi_B^2 + 2\psi_A\psi_B \tag{14.3}$$

It is evident that probability ψ_b^2 is greater than the sum of the probabilities $\psi_A^2 + \psi_B^2$ by an amount $2\psi_A\psi_B$, i.e., in the molecular orbital, there *is greater electron density in the region between the nuclei* than the two isolated atoms. In this molecular orbital, *both the nuclei are shielded from each other and the attraction of*

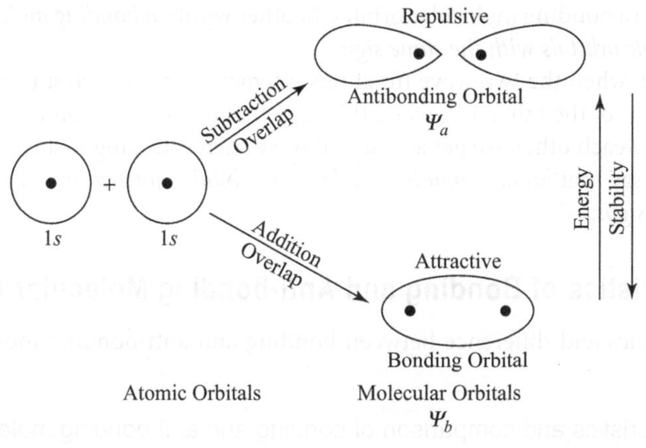

Figure 14.5 Molecular orbitals formed by the combinations of two 1s atomic orbitals

the nuclei for the electrons is increased. As a result, the energy of the molecule is *lowered* and this accounts for the stability of the bond A–B. The orbital ψ_b is, therefore, termed as bonding molecular orbital.

On the other hand, squaring Eq. (14.2) gives the expression for the probability of the anti-bonding molecular orbital.

$$\psi_a^2 = (\psi_A - \psi_B)^2 = \psi_A^2 + \psi_B^2 - 2\psi_A\psi_B \tag{14.4}$$

Evidently, probability ψ_a^2 is less than ψ_A^2 by an amount $2\psi_A\psi_B$, i.e., in this molecular orbital, there is *less electron density in the region between the nuclei* than the two isolated atoms. This results in *greater repulsion* between the nuclei and hence the orbital ψ_a represents the state of *higher energy* as compared to the energy of the individual atoms. Such an orbital, obviously cannot lead to the formation of a chemical bond and is, therefore, termed as an **anti-bonding molecular orbital.**

We have seen above that the combination of the two atomic orbitals forms two molecular orbitals; one bonding (ψ_b) and the other anti-bonding (ψ_a). The *bonding molecular orbital* has *lower energy* than that of atomic orbitals where as the *anti-bonding molecular orbital has higher* energy than that of atomic orbitals from which it is formed as shown in Fig. 14.5.

14.3.5 Bonding and Anti-bonding Molecular Orbitals in Terms of Wave Functions

We can interpret bonding and anti-bonding molecular orbitals by considering the interacting AD's as each being a stationary wave, having a crest (+ sign) and a trough (– sign) as shown in Fig. 14.6. *It may be mentioned here that signs* (+) *and* (–) *simply determine the symmetry of the wave functions and have nothing to do with the electrical charges.*

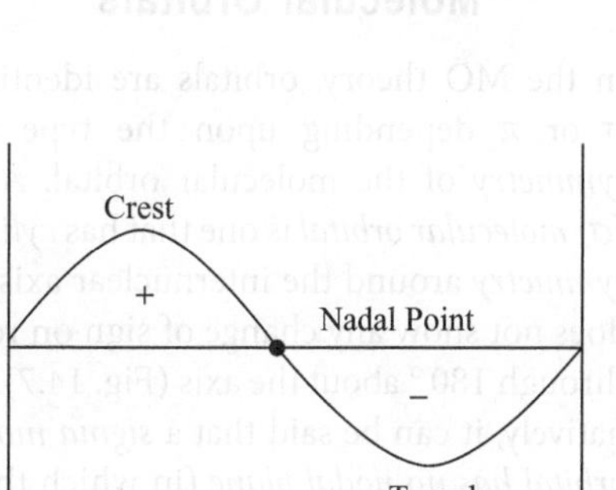

Figure 14.6 Stationary wave showing crest (+) and trough (–)

When the two functions interact in such a way that *crest* (+ sign) of one coincide with the *crest* (+ sign) of the other, i.e., when the wave functions of the combining atomic orbitals are in *phase* and *reinforce* each other, we get a resultant wave corresponding to a bonding molecular orbital. In other words, *a bonding molecular orbital is formed by the overlap of atomic orbitals with the same sign.*

On the other hand, when the two wave functions interact so that the *crest* (+ sign) of one coincides with the *trough* (– sign) of the other, i.e., when the wave functions of the combining atomic orbitals are *out of phase* and *cancel* each other, we get a resultant wave corresponding to an anti-bonding molecular orbital. Thus, we can say that an *anti-bonding molecular orbital is formed by the combination of atomic orbitals with opposite signs.*

14.3.6 Characteristics of Bonding and Anti-bonding Molecular Orbitals

The main characteristics and difference between bonding and anti-bonding molecular orbitals are as follows (Table 14.2).

Table 14.2 Characteristics and comparison of bonding and anti-bonding molecular orbitals.

Bonding Molecular Orbital	Anti-bonding Molecular Orbital
1. It is formed by the addition overlap of wave functions of atomic orbitals.	It is formed by the subtraction overlap of wave functions of atomic orbitals.
2. It has high electron density in the region between the two nuclei and this accounts for the stability of the bond.	The electron density is concentrated away from the inter-nuclear regions, i.e., the probability of finding the electron in between the nuclei is negligible.
3. It possesses lower energy than the atomic orbital from which it is formed.	It possesses higher energy than the atomic orbitals from which it is formed.
4. It is formed when the lobes of combining atomic orbitals have the same sign.	It is formed when the lobes of combining atomic orbitals have opposite sign.
5. Electron present in bonding M.D., contributes to attraction and bonding M.D. is represented by σ or π.	Electron present in anti-bonding M.D. contributes to repulsion and anti-bonding MO is represented by σ^* or π^*.

14.3.7 Combination of Atomic Orbitals—Sigma (σ) and Pi (π) Molecular Orbitals

In the MO theory, orbitals are identified as σ or π depending upon the type of the *symmetry* of the molecular orbital. *A sigma (σ) molecular orbital* is one that has *cylindrical symmetry* around the internuclear axis, i.e., it does not show any change of sign on rotation through 180° about the axis (Fig. 14.7). Alternatively, it can be said that a *sigma molecular orbital has no nodal plane* (in which the electron density is zero) *along the internuclear axis.* The bonding orbital is designated simply as σ orbital and anti-bonding as σ^*.

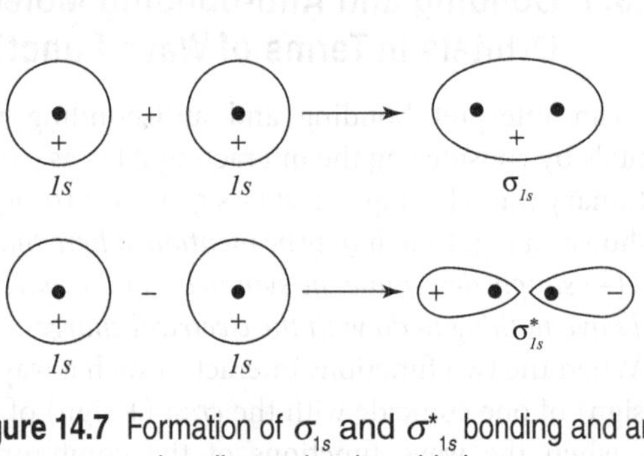

Figure 14.7 Formation of σ_{1s} and σ^*_{1s} bonding and anti-bonding molecular orbitals

Such a sigma (σ) orbital is also formed when any two p-atomic orbitals overlap in end-on (along their axes) position (Fig. 14.8).

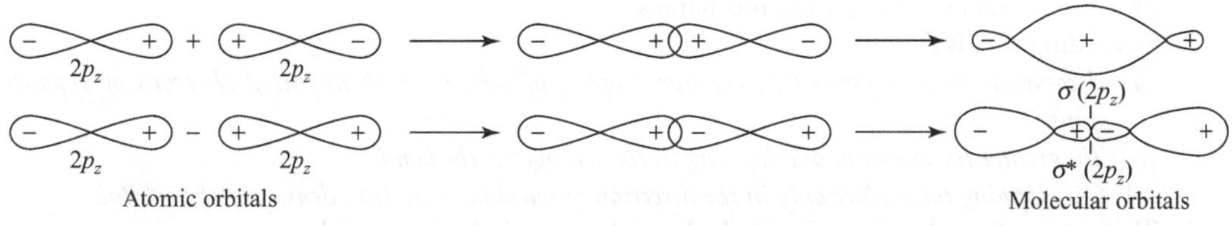

Figure 14.8 Formation of σ_{2p_z} and $\sigma^*_{2p_z}$ molecular bonding and anti–bonding molecular orbitals

When atomic orbitals overlap *laterally* (side-to-side), the resulting molecular orbital is called *a pi* (π) *orbital.* A π-type orbital *does not possess cylindrical symmetry* about the internuclear axis, i.e., if we rotate this molecular orbital through 180°, it will change sign. In contrast to σ-orbital, π-*MO* has a *nodal plane* (zero electron density) *along the internuclear axis* as shown in Fig. 14.9.

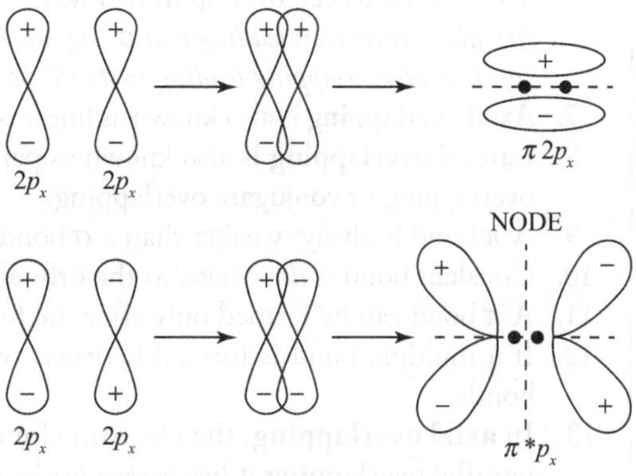

Figure 14.9 Bonding and anti-bonding molecular orbitals from $2p_x$ atomic orbitals

It is clear from the above discussion that the main difference between the two types of orbitals is in the number of nodal planes. The σ-orbital has a nodal plane while the π-orbital has one. Since the energy of a molecular orbital is directly related to the number of nodal planes, the π-orbital is *more energetic* than the σ-orbital. This explains why a π-bond is a *weaker* bond than a σ-bond.

Difference between σ and π Molecular Orbitals

Sigma (σ) MO	Pi (π) MO
1. It is formed from bonds by the end on end overlapping of s–s, s–p, pop orbitals.	It is formed from Jr bonds by the side-wise overlap of pop or pod orbitals.
2. Since the overlap is large, the bonds are quite strong.	Since the overlap is not very large, the bonds formed are weak.
3. The electron density is maximum along the internuclear axis.	The electron density is maximum both above and below the plane, which contains the internuclear axis.
4. σ-MO has no nodal plane while σ^* has only one.	π-MO has, one nodal plane while π^* has two.
5. Bond σ and σ^* are symmetric with respect to rotation about the internuclear axis.	Both π and π^* are anti-symmetric w.r.t. rotation about the internuclear axis.

MEMORY FOCUS

1. VBT proposed by Heitler and London and later developed by Pauling is based upon principal of wave mechanics or quantum mechanics.
2. According to **VBT**
   (a) A covalent bond is formed by the overlapping of orbitals with unpaired electrons of opposite spins.
   (b) The greater the extent of overlapping is, the stronger is the bond.
   (c) Overlapping takes place only in the direction of maximum electron density of the orbital.
3. The extent of overlapping of orbitals depends upon their shapes and size.
4. Relative strength of overlapping of orbitals in the order *p–p* > *s–p* > *s–s* overlapping present in H_2 is *s–s*; F_2 is *p–p*, H-F is *s–p*.
5. The *s–s* overlapping in H_2 is exceptionally stronger due to small size of the 1*s* orbital.
6. Two *p*-orbitals can overlap in two ways:
   (a) along their axes leading to a σ (sigma) bond formation.
   (b) lateral overlapping leading to π (Pi) bond formation.
7. **Axial overlapping** is also known as linear overlapping or end overlapping or heard on overlapping.
8. **Lateral overlapping** is also known as parallel overlapping or sidewise overlapping or sidewise overlapping or conjugate overlapping.
9. A π bond is always weaker than a σ bond due to smaller extent of overlapping.
10. Covalent bond is *directional* as the orbitals can overlap in a particular direction only.
11. A π bond can be formed only after the formation of a bond.
12. If a multiple bond is formed between two atoms, one is a σ bond and the remaining are π bonds.
13. In **axial overlapping**, the electron cloud lies *symmetrical* to the internuclear axis, while in **parallel overlapping** it lies *perpendicular* to the axis on either side.
14. In σ bond formation, one lobe of *p*-orbital gets involved in overlap, while in π bond, both the lobes get involved.
15. When *s*-orbitals are involved in the overlapping, the bond formed is always a sigma bond.
16. Hybrid orbitals never form a Pi bond.
17. VBT can explain (a) directional nature of the covalent bond, (b) bond angles to some extent, (c) shapes of simple molecules and (d) bond strengths.
18. *Paramagnetic nature of oxygen molecule cannot be explained by VBT.*

Numerical Problems

Example 1 Assuming *Z*-axis to be the internuclear axis label the molecular orbitals obtained by the combination of the following atomic orbitals

(i) 2 *s* + 2*s* (ii) $2p_z + 2p_z$ (iii) $2p_y - 2p_y$ (iv) $2p_x - p_x$

Solution: By the combinations of atomic orbitals, molecular orbitals are formed

(i) $2s + 2s \rightarrow \sigma_{25}$ (iii) $2p_y - 2p_y \rightarrow \pi^*2p_y$

(ii) $2p_z + 2p_z \rightarrow \sigma 2p_z$ (iv) $2p_x - 2p_x \rightarrow \pi^*2p_x$

Example 2 **Discuss the shape of molecular orbitals from the combination of following atomic orbitals**

(i) $2p_x + 2p_x$
(ii) $2p_z - 2p_z$

Solution: Combinations of atomic orbitals that give rise to molecular orbitals are given below:

(i) $2p_x + 2p_x \rightarrow \pi 2p_x$
(ii) $2p_z - 2p_z \rightarrow \sigma^* 2p_z$

14.4 RELATIVE ENERGIES OF MOLECULAR ORBITALS AND FILLING OF ELECTRONS

Two general criteria, which determine the energy of the molecular orbitals, are

(i) Initial energy of the atomic orbitals.
(ii) The extent of the overlap of the atomic orbitals.

It is obvious that molecular orbitals formed from lower energy atomic orbitals *have lower energy* than the molecular orbitals formed from higher energy atomic orbitals.

Furthermore, as the σ-overlap is *much more effective* than π-overlap, σp-molecular orbital is of lowest energy, even though originally all the three p-orbitals are of equal energy.

The relative energies of the molecular orbitals are obtained experimentally from *spectroscopic* data. The sequence in the *order of increasing energy* is given below:

$$\sigma\, 1s < \sigma^*\, 1s < \sigma\, 2s < \sigma^*\, 2s < \sigma\, 2p_z < \pi\, 2p_x = \pi\, 2p_y < \pi^*\, 2p_x = \pi^*\, 2p_y < \sigma^*\, 2p_z$$

It may be noted that $\pi 2P_y$ and $\pi 2p_x$ bonding molecular orbitals are *degenerate* (i.e., having equal energy but different orientations). Similarly, $\pi^* 2P_y$ and $\pi^* 2p_x$ anti-bonding molecular orbitals are also degenerate.

The sequence of increasing energy level of different molecular orbitals is represented graphically in Fig. 14.10.

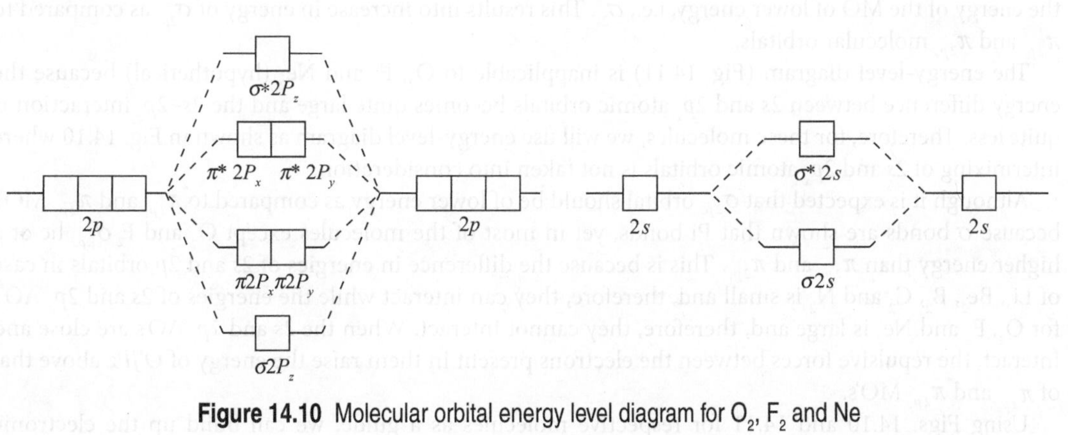

Figure 14.10 Molecular orbital energy level diagram for O_2, F_2 and Ne

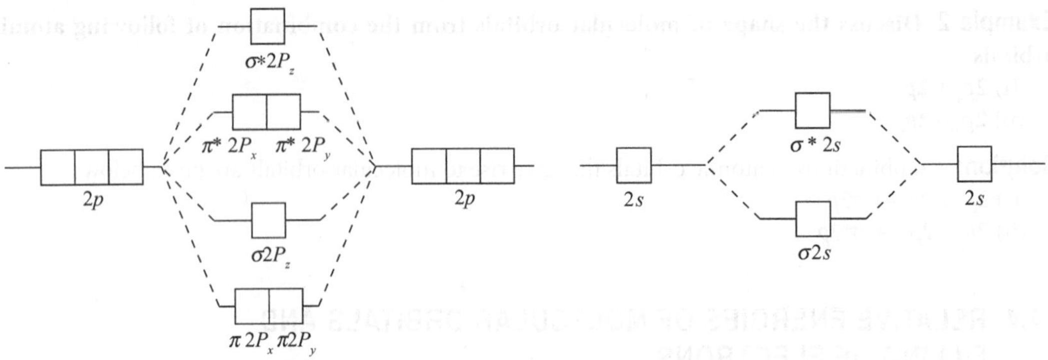

Figure 14.11 Molecular orbital energy diagram for Li_2, Be_2, B_2; and N_2 molecules

However, it has been found experimentally that in some of the diatomic molecules, $\sigma 2p_z$ MO is higher in energy than $\pi 2p_y$ and $\pi 2p_x$ MO's. Therefore, the order of increasing energy of these MO's changes to

$$\sigma_{1s} < \sigma^*_{1s} < \sigma_{2s} < \sigma^*_{2s} < \pi_{2p_x} = \pi_{2p_y} < \sigma_{2p_z} < \pi^*_{2p_x} = \pi^*_{2p_y} < \sigma^*_{2p_z}$$

The difference between the two types of sequences in energy levels is that for molecules O_2, F_2 and Ne (Hypothetical) the σ_{2p_z} molecular orbital is lower in energy than π_{2p_x} and π_{2p_y}, while in case of Li_2, Be_2 (Hypothetical) B_2, C_2 and N_2 molecules, the σ_{2p_z} MO's has higher energy than π_{2p_x} and π_{2p_y} MO's i.e., σ_{2s} and σ^*_{2s} orbitals get lowered in energy whereas and are raised in energy by an equivalent amount. The increase in energy is so much that it is higher in energy than even and MO's. This leads to a new molecular orbital energy-level diagram as shown in Fig. 14.11.

This new energy-level diagram is used for writing molecular orbital configurations of molecules such as H_2, He_2 (hypothetical), Li_2, Be_2 (hypothetical), B_2, C_2 and N_2.

In the atoms of these molecules, the 2*s* and $2p_x$ atomic orbitals lie fairly close together, thus they intermix to a large extent. After intermixing, the resulting molecular orbitals σ_{2s} and σ_{2p_z} have similar symmetry and tend to occupy same positions. These molecular orbitals thus experience maximum repulsions and their interaction naturally increases the energy of the MO of higher energy, i.e., σ_{2p_z} and decreases the energy of the MO of lower energy, i.e., σ_{2s}. This results into increase in energy of σ_{2p_z} as compared to π_{2p_y} and π_{2p_x} molecular orbitals.

The energy-level diagram (Fig. 14.11) is inapplicable to O_2, F_2 and Ne_2 (hypothetical) because the energy difference between 2*s* and $2p_x$ atomic orbitals becomes quite large and the $2s$–$2p_z$ interaction is quite less. Therefore, for these molecules, we will use energy-level diagram as shown in Fig. 14.10 where intermixing of 2*s* and $2p_x$ atomic orbitals is not taken into consideration.

Although it is expected that σ_{2p_z} orbital should be of lower energy as compared to π_{2p_x} and π_{2p_y} MO's because σ bonds are shown that Pi bonds, yet in most of the molecules except O_2 and F_2 σ_{2p_z} lie at a higher energy than π_{2p_x} and π_{2p_y}. This is because the difference in energies of 2*s* and 2*p* orbitals in case of Li_2, Be_2, B_2, C_2 and N_2 is small and, therefore, they can interact while the energies of 2*s* and $2p_z$ AO's for O_2, F_2 and Ne_2 is large and, therefore, they cannot interact. When the 2*s* and $2p_z$ AO's are close and interact, the repulsive forces between the electrons present in them raise the energy of *O']Pz* above that of π_{2p_x} and π_{2p_y} MO's.

Using Figs. 14.10 and 14.11 for respective molecules as a guide, we can build up the electronic structures of simple diatomic molecules. The *filling of electrons* in various molecular orbitals follows similar rules to those for filling atomic orbitals. Since molecular orbitals belong to the *entire* molecule, the *total number of electrons* are taken into consideration and accommodated in molecular orbitals according to the following rules:

(i) **Aufbau principle:** The molecular orbital of lowest energy is filled up first.
(ii) **Pauli exclusion principles:** Each molecular orbital can accommodate a maximum of two electrons having opposite spins.
(iii) **Hund's rule:** Whenever two equal energy molecular orbitals are available, electrons prefer to remain unpaired. This is particularly significant for π-orbitals because π-bonding and π-anti-bonding molecular orbitals have equal energies.

14.5 STABILITY OF MOLECULES

14.5.1 Stability of Molecules in Terms of Bonding and Anti-bonding Electrons

Since electrons in bonding orbitals (N_b) *increase* the stability of the molecule whereas those in anti-bonding orbitals (N_a) *decrease* its stability, the stability of the molecule, on the whole, depends on their relative numbers. Thus,

(i) If $N_b > N_a$, the molecule is stable.
(ii) If $N_b < N_a$, the molecule is unstable.
(iii) Even if $N_b = N_a$, the molecule is unstable. This is due to the fact that anti-bonding effect is somewhat stronger than the bonding effect.

It may be pointed out here that in the last two cases, i.e., (ii) and (iii), the molecules do not actually exist.

14.5.2 Stability of Molecules in Terms of Bond Order

The relative stability of a molecule is further evaluated by a quantity known as bond order. It is defined as the "number of covalent bonds formed in molecule and is given by one-half of the difference between the number of electrons in bonding orbitals and those in anti-bonding orbitals", i.e.,

$$\textbf{Bond order} = \frac{1}{2} \text{ [No. of electrons in bonding orbitals – No. of electrons in anti-bonding orbitals]}$$

$$= \frac{1}{2}[N_b - N_a] \tag{14.5}$$

The bond order of 1, 2 and 3 corresponds to single, double and triple bonds, respectively. It may be mentioned that according to MO theory, even a fractional bond order is possible. Thus:

(i) The stability of a molecule is *directly related* to its bond order. A negative or zero bond order means an unstable molecule whereas a positive bond order means a stable molecule. *Higher the bond order is, more stable is the molecule.*
(ii) **Dissociation energy:** The stability of a diatomic molecule can be expressed in terms of bond dissociation energy which is directly proportional to the bond order. For example, fluorine molecule having one bond order has lower dissociation energy (160 kJ mol^{-1}) than oxygen molecule (dissociation energy kJ mol^{-1}) having a bond order equal to 2 and so on.

(iii) Bond order, however, is *inversely related* to bond length. *Higher the bond order is, shorter is the bond length.* For example, nitrogen molecule having bond order 3 has shorter bond length (110 pm) than oxygen molecule (bond length 121 pm) having a bond order equal to 2.

(iv) **Diamagnetic and paramagnetic behaviours:** If all the electrons in a molecule are paired the substances is *diamagnetic*. On the other hand, if there are unpaired electrons in the molecule, the substance is paramagnetic. More the number of unpaired electrons in its molecule is, greater is the paramagnetism of the substance.

14.6 MOLECULAR ORBITAL CONFIGURATIONS

While writing electronic configurations of molecules or ions, the following points must be kept in mind:

(i) To begin with, the formula of the molecule or ion is written followed by the square bracket starting with the symbol.

(ii) For molecules of the second period such as Li_2, Be_2, B_2, N_2, O_2 and F2, the letters "K.K" are written for the first shell of both the combining atoms. The letter "K" denotes the filled inner K shell of each combining atom. These electrons do not take part in bonding. For H_2 and He_2 molecules, K.K letters are not used.

(iii) The number of electrons in any molecular orbital is written on the right-hand corner above a closed bracket (). For example, if σ_{1s} molecular orbital has one electron, it is represented as $(\sigma_{1s})^1$.

(iv) After writing the complete molecular orbital configuration, the square bracket is closed with the symbol].

Keeping in view above rules, the molecular orbital configuration for oxygen molecule can be written. As there are six valence electrons in each oxygen atom, the distribution of 12 electrons in molecular orbitals will be as shown below.

$$O_2\left[KK(\sigma_{2s})^2(\sigma^*_{2s})^2(\sigma_{2pz})^2\pi(2p_x)^2(\pi_{2py})^2(\pi^*_{2px})^1(\pi^*_{2py})^1\right]$$

14.6.1 Bonding in Some Homonuclear Diatomic Molecules and Ions – Electronic Configurations

Based on the above principles, we would proceed to learn how to draw the molecular orbital pictures of some diatomic homonuclear molecules and ions.

(i) Hydrogen molecular ion, Hi: This is the simplest molecule that can exist. Its presence can be detected *spectroscopically* by passing an electric discharge through hydrogen gas under low pressure.

Hydrogen molecular ion is formed by the combination of one hydrogen atom having one electron and a hydrogen positive ion having no electron.

In this ion, *only one electron is available to hold two protons together*. The *single available electron occupies the lowest energy σ1s bonding molecular orbital while the anti-bonding orbital remains vacant* (Fig. 14.12).

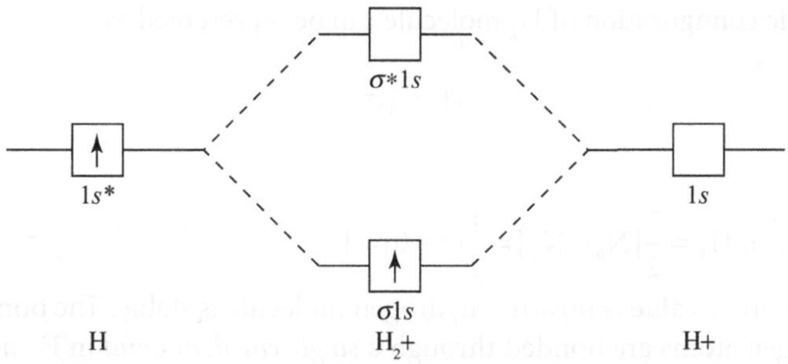

Figure 14.12 Molecular orbital energy level diagram for H_2^+ ion

Thus, the *electronic configuration* of H_2^+ molecule ion can be represented as

$$H_2^+ = (\sigma_{1s})^1$$

Thus,

1. **Bond order:** The bond order in $H_2^+ = \frac{1}{2}[N_b - N_a] = \frac{1}{2}[1-0] = \frac{1}{2}$.
2. **Stability:** *The positive value of the bond order* means that the hydrogen molecular ion has some stability.
3. **Magnetic character:** Due to the presence of an *unpaired* electron in the bonding orbital, the molecule ion should be paramagnetic. The H_2^+ molecule ion has actually been found to be stable and paramagnetic.

(ii) Hydrogen molecule, H_2: Hydrogen molecule is made up of two hydrogen atoms, each of which possesses one electron in 1*s* atomic orbital.

When the two atomic orbitals overlap, two available electrons in H_2 molecule occupy the lowest energy σ_{1s} bonding molecular orbital while the anti-bonding orbital remains unoccupied. In accordance with Pauli exclusion principle, the two electrons in σ_{1s} molecular orbital must have opposite spin (Fig. 14.13).

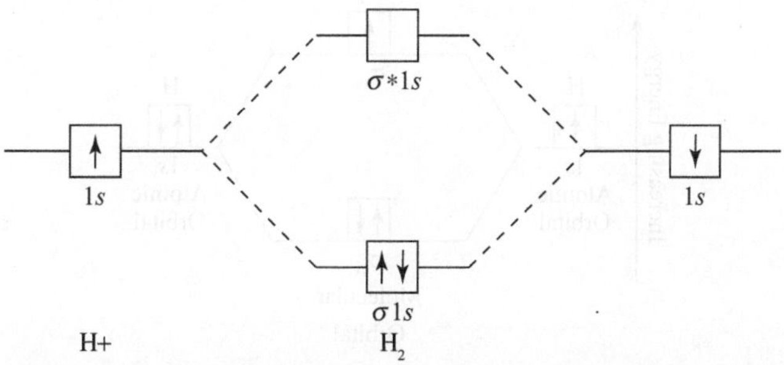

Figure 14.13 Molecular orbital energy level diagram for H_2 molecule

Thus, the electronic configuration of H_2 molecule can be represented as

$$H_2 = (\sigma_{1s})^2$$

Thus,

1. The bond order in $H_2 = \frac{1}{2}[N_b - N_a] = \frac{1}{2}[2-0] = 1$
2. Since the bond order value is positive, hydrogen molecule is stable. The bond order 1 shows that the two hydrogen atoms are bonded through a *single covalent bond* in H_2 molecule.
   Bond dissociation energy of H_2 = 433 kJ mol^{-1}; bond length = 74 pm
3. Since there is no *unpaired* electron in any of its molecular orbitals, H_2 molecule is *diamagnetic*.

(iii) Hydrogen molecular ion, H_2^-: This can be formed by the combination of one hydrogen atom having one electron and a hydrogen negative ion having two electrons in their 1*s* orbitals, respectively. Thus, the ion involves three electrons.

The two electrons occupy the lowest energy σ_{1s} bonding molecular orbital while the third electron goes to the higher energy anti-bonding σ^*_{1s} molecular orbital according to Aufbau and Pauli exclusion principles (Fig. 14.14).

The *electronic configuration* of H_2^- molecule ion can be represented as

$$H_2^- = \left[(\sigma_{1s})^2(\sigma^*_{1s})^1\right]$$

Thus,

1. The *Bond order* in $H_2^- = \frac{1}{2}[N_b - N_a] = \frac{1}{2}[2-1] = \frac{1}{2}$
2. **Stability:** The *positive value* of the bond order means that the hydrogen molecular ion H_2^- has some stability.
3. **Paramagnetic nature:** Since it has one unpaired electron in the anti-bonding molecular orbital, H_2^- ion is *paramagnetic*.

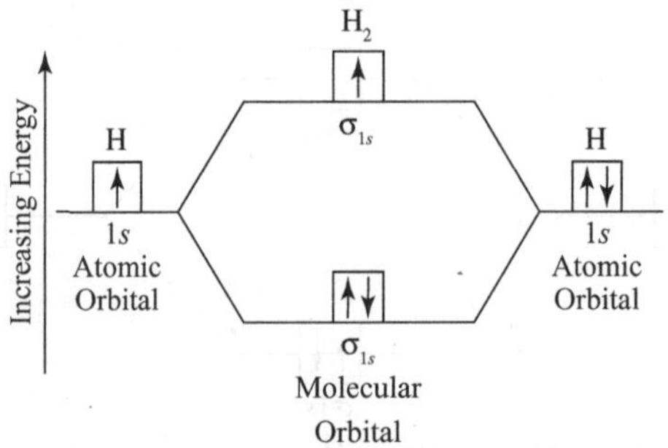

Figure 14.14 Molecular orbital energy level diagram for H_2^- ion

MEMORY FOCUS

Comparison of stabilities of H_2^+, H_2 and H_2^-

The molecular orbital configuration for all the hydrogen species, H_2^+, H_2 and H_2^-, are summarized as below:

Molecule/ion	Total Number of Electrons	Molecular Orbital Configuration	Bond Order
H_2^+	1	$(\sigma_{1s})^1$	$\frac{1}{2}$
H_2	2	$(\sigma_{1s})^2$	1
H_2^-	3	$(\sigma_{1s})^2(\sigma_{1s}^*)^1$	$\frac{1}{2}$

(1) Since bond order values are positive, all the hydrogen species, H_2^+, H_2 and H_2^-, are stable. Now, greater the bond order is, more stable is the molecule or the ion. Thus, *H_2 molecule is the most stable* and the *order of stability* for these molecules or ions will be: $H_2 > H_2^+ > H_2^-$.

(2) Although H_2^+ and H_2^- have the same bond order, i.e., 1/2 yet H_2^- is slightly less stable than H_2^+. It is because H_2^- contains one electron in the anti-bonding orbital which results in repulsion and decreases the stability. However, H_2^+ does not contain any electron in the anti-bonding orbital.

(3) It may be noted that H_2^+ and H_2^- molecule ions having an *unpaired electron* are *paramagnetic* while H_2 molecule having paired electrons is diamagnetic.

14.6.2 Helium ion, He_2^+

It has three available electrons. Two of these occupy a σls bonding orbital while the third occupy $\sigma^* ls$ anti-bonding orbital. The anti-bonding effect of one electron is greater than the bonding effect. The bond in the helium molecule ion is, therefore, weaker than the single electron bond in the hydrogen molecule ion.

The structure is written as

$$He_2^+ = [(\sigma 1s)^2 (\sigma * 1s)^1]$$

$$\text{Bond order} = \frac{1}{2}[N_b - N_a] = \frac{1}{2}(2-1) = \frac{1}{2}$$

Since $N_b > N_a$, the molecular ion is stable. Due to the presence of unpaired electron, it should be paramagnetic (Fig. 14.15).

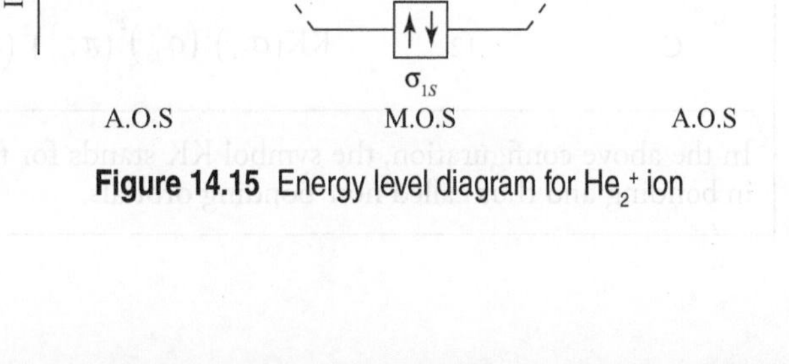

Figure 14.15 Energy level diagram for He_2^+ ion

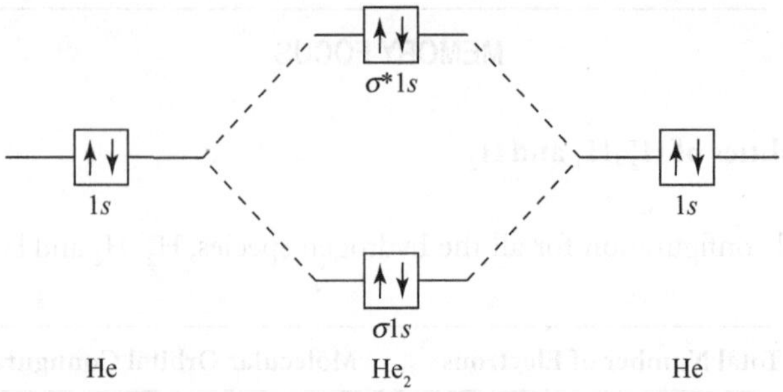

Figure 14.16 Molecular orbital energy level diagram for He_2 molecule hypothesis

Hypothetical helium molecule, He_2: The molecule He_2 does not exist because two of the four, available 1*s* electrons pass into a σ1s bonding orbital while the other two passes into σ^* 1*s* anti-bonding orbitals. Thus, these are arranged as $He_2 = (\sigma 1s)^2, (\sigma^* 1s)^2$. The filled anti-bonding orbital more than counteract the filled bonding orbital, so that no bond results (Fig. 14.16).

$$\text{Bond order} = \frac{1}{2}[N_b - N_a] = \frac{1}{2}(2-2) = 0$$

The molecule is unstable and does not exist since the bonding and anti-bonding electrons cancel each other.

MEMORY FOCUS

Li_2, Be_2, B_2 and C_2 molecules

The molecular orbital configurations with bond order and magnetic properties of these molecules are given below.

Molecule	Total Number of Electron	Molecular Orbital Configuration	Bond Order	Magnetic Nature
Li_2	6	$KK(\sigma_{2s})^2$	$\frac{1}{2}(2-0)=1$	Diamagnetic
Be_2	8	$KK(\sigma 2s)^2(\sigma^* 2s)^2$	$\frac{1}{2}(2-2)=0$	Does not exist
B_2	10	$KK(\sigma_{2s})^2(\sigma^*_{2s})^2(\pi_{2py})^1(\pi_{2px})^1$	$\frac{1}{2}(4-2)=1$	Paramagnetic
C_2	12	$KK(\sigma_{2s})^2(\sigma^*_{2s})^2(\pi_{2py})^2(\pi_{2px})^2$	$\frac{1}{2}(6-2)=0$	Diamagnetic

In the above configuration, the symbol KK stands for the closed or full K shells which are not used in bonding and thus called non-bonding orbitals.

14.6.3 Nitrogen Molecule, N_2

The nitrogen atom has electronic configuration 1s2, 2s2, 2p3. The N_2 molecule is well known. Its molecular orbital configuration (Fig. 14.17) may be written as $N_2\ [KK(\sigma_{2s})^2(\sigma^*_{2s})^2(\pi_2py)^2(\pi_2px)^2(\pi_2pz)^2]$. It has eight electrons in bonding and two electrons in anti-bonding orbitals. Since there are six electrons in excess in bonding orbitals, the bond order in N_2 is

$$\frac{1}{2}[N_b - N_a] = \frac{1}{2}(8-2) = 3$$

The molecule has 1σ and 2π bonds.

This is in conformity with its very high bond energy of 945 kJ/mol and very short bond length of 110 pm. The molecule is *diamagnetic*, as there are no unpaired electrons.

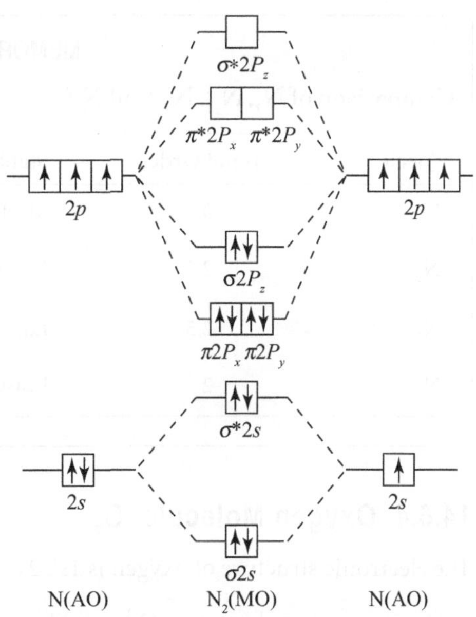

Figure 14.17 Molecular orbital energy level diagram for N_2 molecule

MEMORY FOCUS

Characteristics of N_2, N_2^+, N_2^- and N_2^{2-}

Species	Bond Order	Stability	Bond Length	Magnetic Character
N_2	14	$KK(\sigma_{2s})^2(\sigma^*_{2s})^2(\pi 2p_y)^2 (\pi 2p_x)^2(\sigma 2p_z)^2$	$\frac{1}{2}(8-2)=3$	Diamagnetic
N_2^+	13	$KK(\sigma_{2s})^2(\sigma^*_{2s})^2(\pi 2p_y)^2 (\pi 2p_x)^2(\sigma 2p_z)^2$	$\frac{1}{2}(7-2)=2\frac{1}{2}$	Paramagnetic as it has one unpaired electron
N_2^-	15	$KK(\sigma_{2s})^2(\sigma^*_{2s})^2(\pi 2p_y)^2 (\pi 2p_x)^2(\sigma 2p_z)^1$	$\frac{1}{2}(8-3)=2\frac{1}{2}$	Paramagnetic as it has one unpaired electron
N_2^{2-}	16	$KK(\sigma_{2s})^2(\sigma^*_{2s})^2(\pi 2p_y)^2 (\pi 2p_x)^1(\sigma 2p_z)^1$	$\frac{1}{2}(8-4)=2$	Paramagnetic as it has two unpaired electrons

(Continued)

MEMORY FOCUS (*Continued*)

Comparison of N_2, N_2^+, N_2^- and N_2^{2-}

Species	Bond Order	Stability	Bond Length	Magnetic Character
N_2	3	Most stable	Shortest	Diamagnetic
N_2^+	2.5	Less stable	Highest	Paramagnetic
N_2^-	2.5	Equality	Equal to above	Paramagnetic
N_2^{2-}	2	Least	Largest	Paramagnetic

14.6.4 Oxygen Molecule, O_2

The electronic structure of oxygen is $1s^2$, $2s^2$, $2p^4$. The structure of the O_2 molecule is written as

$$O_2\left[KK(\sigma 2s)^2(\sigma^* 2s)^2(\sigma 2p_z)^2(\pi 2p_y)^2(\pi 2p_x)^2(\sigma^* 2p_y)^1(\sigma^* 2p_x)^1\right]$$

The way in which this is built up from two oxygen atoms is shown in Fig. 13.18.

The lower molecular orbitals fill up quite normally but there are only two electrons available for the π^*p2x and π^*2py-orbital which, together, could hold four. If both the electrons were allotted to either π^*2px or the π^*2py orbitals, all electrons would be paired. The application of Hund's rule of maximum multiplicity shows, however, that the molecule contains two unpaired electrons (Fig. 14.18). These unpaired electrons account for the strong *paramagnetism* of oxygen and this explanation of it rather surprising fact was one of the *earliest successes of the application of molecular orbital theory.*

$$\text{Bond order} = \frac{1}{2}\ (8–4) = 2$$

The bond energy is 498 kJ mol^{-1}. The positive value of bond order means that oxygen molecule is stable. The bond order 2 shows that the two oxygen atoms in a molecule of oxygen are connected through *two covalent* bonds, i.e., a double bond. The value of obtained for bond dissociation energy (498 kJ mol^{-1}) supports the prediction. The bond length is 121 pm.

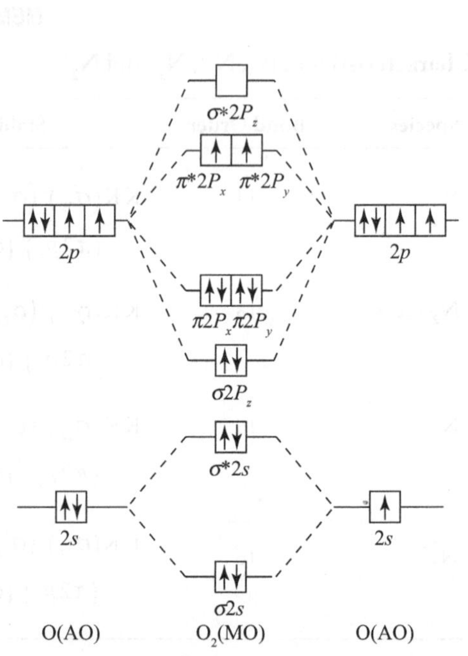

Figure 14.18 Molecular orbital energy level diagram for O_2 molecule

Thus, the MO theory accurately predicts the following characteristics of O_2 molecule.

(i) Oxygen is a *diradical** molecule (two unpaired electrons).
(ii) Oxygen is *a paramagnetic* substance (two unpaired electrons).
(iii) Oxygen molecule has a *double bond* (bond order = 2).
(iv) Oxygen molecule is *quite stable* and not so reactive at room temperature (bond dissociation energy = 498 kJ mol^{-1}).

Oxygen molecule ion, O_2^+: It has 15 electrons in the molecular orbitals, one electron less than O_2. These are arranged in the energy orbitals as

$$O_2^+\left[KK(\sigma 2s)^2(\sigma * 2s)^2(\sigma 2p_z)^2(\pi 2p_x)^2(\pi 2p_y)^2(\pi * 2p_x)^1\right]$$

The bond order is $\frac{1}{2}$ (8–3) = 2.5 which shows one *a* bond, and two π bonds with an unshared electron in the anti-bonding MO π^*2p_x. The ion O_2^+ will be paramagnetic.

MEMORY FOCUS

Comparison of stabilities of O_2^+, O_2, O_2^- and O_2^{2-}

The molecular orbital configuration for all the oxygen species O_2^+, O_2, O_2^- and O_2^{2-} are summarized as shown below.

Molecule/ion	Total Ion of electrons	Molecule Orbital Configuration	Bond Order	Magnetic Nature
O_2^+	15	$KK(\sigma_{2s})^2(\sigma_{2s}^*)^2(\sigma_{2pz})^2(\pi_{2px},\pi_{2py})^4(\pi_{2px}^*)^1$	21	Paramagnetic
O_2	16	$KK(\sigma_{2s})^2(\sigma_{2s}^*)^2(\sigma_{2pz})^2(\pi_{2px},\pi_{2py})^4(\pi_{2px}^*,\pi_{2py}^*)^2$	2	Paramagnetic
O_2^- (superoxide ion)	17	$KK(\sigma_{2s})^2(\sigma_{2s}^*)^2(\sigma_{2pz})^2(\pi_{2px},\pi_{2py})^4(\pi_{2px}^*,\pi_{2py}^*)^3$	11	Paramagnetic
O_2^{2-} (peroxide ion)	18	$KK(\sigma_{2s})^2(\sigma_{2s}^*)^2(\sigma_{2pz})^2(\pi_{2px},\pi_{2py})^4(\pi_{2px}^*,\pi_{2py}^*)^1$	1	Diamagnetic

(1) Since bond order values are positive, all the oxygen species O_2^+, O_2, O_2^- and O_2^{2-} are stable.

(2) As already stated, greater the bond order is, more stable will be the molecule or the ion. Thus, oxygen molecule cation O_2^+ is *most stable* and the order of *decreasing stability* for these molecules or ions will be:

$$O_2^+ > O_2 > O_2^- > O_2^{2-}$$

(3) It may be noted that O_2^+, O_2 and O_2^- having *unpaired electrons are paramagnetic*, out of these, O_2 is the most paramagnetic having two unpaired electrons while O_2^{2-} ion having *paired* electrons is *diamagnetic*.

14.6.5 The Fluorine Molecule, F_2

Fluorine atom has electronic configuration ls^2, $2s^2$, $2p^5$. There are 14 valence electrons (seven from each fluorine atom) to be filled in MOs of F_2 molecule. The molecular orbital configuration

$$F_2\left[KK(\sigma_{2s})^2(\sigma_{2s}^*)^2(\sigma_{2px})^2(\pi_{2py})^2(\pi_{2pz})^2(\pi_{2py}^*)^2(\pi_{2pz}^*)^2\right]$$

and its bond order is $\frac{1}{2}$ (8–6) = 1 σ bond. It is diamagnetic since it has no unpaired electrons. The bond dissociation energy is 158 kJ mol^{-1} and bond length is 142 pm.

14.6.6 Hypothetical Neon Molecule, Ne_2

The 20 electrons here can be distributed into five bonding and five anti-bonding orbitals. The B.O. thus becomes zero, showing thereby that Ne_2 cannot exist as a stable molecule just as He_2.

14.6.7 Molecular Orbital Electronic Configuration of Some Common Heteronuclear Molecules

The method of writing electronic configuration of simple heteronuclear molecules such as CO and NO is similar to the method followed in the case of homonuclear compounds. We have to add up the atomic numbers of the concerned atom to know the total number of electrons. These electrons are filled up in the molecular orbitals following Autbau principle, Pauli's principle and Hund's rule as usual. For example

(i) **Electronic configuration of CO:**
We know the total number of electrons = 6 + 8 = 14

$$\therefore \quad CO = KK(\sigma_{2s})^2\sigma_{2s}^*(\pi_{2px})^2(\pi_{2py})^2(\pi_{2pz})^2$$

$$\text{Bond order} = \frac{8-2}{2} = 3$$

(ii) **Electronic configuration of NO:**
Total number of electrons = 7 + 8 = 15

$$\therefore \quad NO = KK(\sigma_{2s})^2(\sigma_{2s}^*)^2(\pi_{2px})^2(\pi_{2py})^2(\pi_{2pz})^2(\pi_{2px}^*)^1$$

$$\text{Bond order} = \frac{8-3}{2} = 2\frac{1}{2}$$

(iii) **Electronic configuration of NO^+:**
Total number of electrons = 7 + 7 = 14

$$\therefore \quad NO^+ = KK(\sigma_{2s})^2(\sigma_{2s}^*)^2(\pi_{2px})^2(\pi_{2py})^2(\sigma_{2pz}^*)^2$$

$$\text{Bond order} = \frac{8-2}{2} = 3$$

(iv) **Electronic configuration of CN^-:**

Total number of electrons = 6 + 7 + 1 = 14

$$\therefore \quad CN^- = KK(\sigma_{2s})^2(\sigma_{2s}^*)^2(\pi_{2px})^2(\pi_{2py})^2(\sigma_{2pz})^2$$

$$\text{Bond order} = \frac{8-2}{2} = 3$$

14.7 COMPARISON OF VALENCE BOND (VB) THEORY AND MOLECULAR ORBITAL (MO) THEORY

14.7.1 Points of Similarly

(i) In both the theories, the resultant electron distributions for given molecules are similar (rules for filing of orbitals are similar in Aufbau, Pauli's and Hund's principles).
(ii) Both require that in the description of normal covalent bond, the essential interaction is the sharing of the bonding electron density by both nuclei and concentration of the electron density between the nuclei.
(iii) Both require that in bond formation, the combining orbitals must be of nearly the same energy. Orbitals of the bonded atoms overlap and have appropriate symmetry about the molecular axis in order to form a bond.
(iv) Both theories are approximations because of impossibility of describing molecules with an exact Schrödinger equation.
(v) Both theories account for the directional nature of bond.
(vi) Both predict the non-existence of helium molecule.

14.7.2 Points of Difference

VBT	MOT
1. Only the electrons present in the outermost (valence) shell take part in the formation of covalent bond.	All the electrons of all the atoms of the molecules are affected and take part in the bond formation.
2. The atoms retain their individual identity.	The atoms lose their identity.
3. It considers the interaction between atoms.	It starts with the nuclei of the constituent atoms.
4. Resonance phenomenon is a part and parcel of VBT.	It does not consider the resonance phenomenon at all.
5. It does not predict the presence of bonding and anti-bonding molecular orbitals.	It does consider bonding and anti-bonding molecular orbitals; the bonding molecular orbital causes attraction while the anti-bonding molecular orbital causes repulsion.
6. It cannot explain the paramagnetic nature of O_2.	It explains the paramagnetic nature of O_2.
7. It does not explain the bonding in electron-deficient compounds.	It does explain the bonding in electron-deficient compounds.
8. The properties such as bond order and electronic configuration in the excited state cannot be predicted.	It does predict the bond order and the electronic configuration in the excited state.
9. It cannot explain the existence of species such as H_2^+ and He_2^+.	It can explain their existence.
10. The wave functions include the ionic contribution.	It does not include the ionic contribution.
11. It is simple to apply.	It is difficult to apply

NOTEWORTHY POINTS

1. Valence bond theory fails to explain the paramagnetic behaviour of certain substances such as O_2 and geometry of non-linear molecules.
2. An atomic orbital is monocentric while a molecular orbital is polycentric.
3. During the formation of MOs from AOs, the energy of the anti-bonding MO is raised above the energy of the combining AOs by an amount more than that by which the energy of the bonding MO has been lowered. The total energy of two MOs, however, remains the same as that of the two original AOs.
4. In molecules such as B_2, C_2 and N_2, the energy of σ_{2pz} MO is higher than that of π_{2px} and π_{2py} MOs. This is due to the mixing of 2*s* and 2*p*-orbitals in case of B_2, C_2 and N_2.
5. Though bond order for H_2^- and H_2^- is same (i.e., 1/2), yet H_2^- is slightly less stable than H_2^+ because H_2^- has one electron in the anti-bonding orbital resulting in repulsion (decrease in stability).
6. Order of stability and bond dissociation energies are
   (a) $O_2^+ > O_2 > O_2^- > O_2^{2-}$ (b) $N_2 > N_2^+ = N_2^- > N_2^{2-}$
7. Order of bond lengths
   (a) $O_2^{2-} > O_2^- > O_2 > O_2^+$ (b) $N_2^{2-} > N_2^- = N_2^+ > N_2$

Numerical Problems

Example 3 With the help of MO theory, predict which of the following species is diamagnetic:

$$\mathbf{H_2^+, O_2 \text{ and } O_2^{2+}}$$

Solution The species in which all the electrons are paired will be diamagnetic.

$$H_2^+ : (\sigma_{1s})^1$$
$$O_2 : KK(\sigma_{2s})^2(\sigma_{2s}^*)^2(\pi_{2px})^2(\pi_{2py})^2(\pi_{2px}^*)^1(\pi_{2py}^*)^1$$
$$O_2^{2+} : KK(\sigma_{2s})^2(\sigma_{2s}^*)^2(\pi_{2pz})^2(\pi_{2px})^2(\pi_{2py})^2$$

Example 4 How is magnetic moment of a species related to the number of unpaired electrons?

Solution: Magnetic moment is directly related to the number of unpaired electron. Larger the number of unpaired electrons is, larger is the magnetic moment. The magnetic moment (μ_{so}) due to spin only is given by the relation

$$\mu_{so} = \sqrt{n(n+2)} \text{ (B)}O_2^- \text{ (C)}NO^+ \text{ (D)}CN^+$$

where *n* is the number of unpaired electrons and B.M. is the unit of magnetic moment called Bohr Magneton.

Example 5 Which of the following have identical bond orders?

(A) CN^- (B) O_2^- (C) NO^+ (D) CN^+

Solution:

(A) The electronic configuration of CN^- for (6 + 7 + 1 = 14) electrons is given below:

$$KK(\sigma_{2s})^2(\sigma_{2s}^*)^2(\pi_{2px})^2(\pi_{2py})^2(\pi_{2pz})^2$$

$$\text{Bond order} = \frac{8-2}{2} = 3$$

(B) The electronic configuration of O_2^{-1} for (8 + 8 + 1 = 17) electrons is given below.

$$KK(\sigma_{2s})^2(\sigma_{2s}^*)^2(\sigma_{2px})^2(\pi_{2px})^2(\pi_{2py})^2(\pi_{2px}^*)^2(\pi_{2py}^*)^1$$

$$\text{Bond order} = \frac{8-5}{2} = 1.5$$

(C) The electronic configuration of $N0^+$ for (7 + 8 −1 = 14) electrons will be same as for CN^- having 14 electrons.

Bond order would also be the same, i.e., 3.

(D) The electronic configuration of CN^+ for (6 + 7—1 = 12) electron is given below.

$$KK(\sigma_{2s})^2(\sigma_{2s}^*)^2(\pi_{2px})^2(\pi_{2py})^2$$

$$\text{Bond order} = \frac{6-2}{2} = 2$$

Thus, the bond order values for [A] and [C] are identical.

Example 6 **Which species out of 112, H are paramagnetic and why?**

Solution: The electron configuration is given below:

$$H_2 = (\sigma_{1s})^2, \quad H_2^- = (\sigma_{1s})^2(\sigma_{1s}^*)^1, \quad H_2^+ = (\sigma_{1s})^1$$

H_2^+ and H_2^1 have unpaired electrons in them, so they are paramagnetic.

REVISION QUESTIONS

1. Draw the *s*-bonding and anti-bonding molecular orbitals obtained from the "s" atomic orbitals.
2. What do the acronym L.C.A.O. stand for? What is meant by it?
3. What is mean by bond order? How does it determine the stability of a molecule?
4. What are the conditions for atomic orbital to form molecular orbital to form molecular orbitals? Can a bond be formed if electrons in the molecular bonding and anti-bonding orbitals are same?
5. Compare the bond order for H^+_2 is and H_2. Why bond length of H^+_2 is longer than that of H_2.

6. What information does MO theory provide for O_2, O^+_2, O^-_2 and 2 O_2^{2-} mole molecular species with regard to:
   (a) Stability (b) Bond length
7. With the help of energy-level diagram, explain the bond order and magnetic nature of oxygen.
8. Give difference between bonding and anti-bonding orbitals.
9. Explain why oxygen molecule is paramagnetic.
10. Give with example the composition of $sP^3 d$ or $sP^3 d^2$ hybrid orbitals.
11. What type of hybridization is present
    (a) PF_5 and (b) $[CO(NH_3)_6]^{3+}$?
12. Which one is more stable and why?
    H_2, H_2^+ and H_2^- – Which species is paramagnetic?
13. Explain the magnetic behaviour of N_2 molecule with help of an energy-level diagram.
14. Draw and discuss the shapes of molecular orbitals framed by the combination of *p*-orbitals.
15. With the help of energy-level diagram, explain the bond order and magnetic nature of oxygen.
16. Give energy sequence of molecular orbitals. Explain paramagnetic behaviour and bond order in O_2 molecule.
17. Calculate bond order of O_2^{-1} ion and explain its magnetic behaviour.
18. What are the conditions for the combination of atomic orbitals?
    Can a bond be formed if electrons in bonding molecular orbitals are as the electrons in anti-bonding molecular orbitals?
19. Draw and discuss the shapes of molecular orbitals framed by the combination of *p*-orbitals.
20. What is bond order? How is it related to the bond length and bond dissociation energy?
21. Explains the bond order of nitrogen molecule with the help of molecular orbital energy-level diagram.
22. Calculate the bond order of O_2^{-1} and O_2^{-2} and explain their magnetic behaviour.

MULTIPLE CHOICE QUESTIONS

1. Which of the following molecular orbital has the lowest energy?
   (a) $\sigma 2p_z$ (c) $\pi^* 2p_x$
   (b) $\sigma^* 2p_s$ (d) $\pi^* 2p_y$
2. For a homonuclear diatomic molecule, the energy of σ_{2s} orbital is
   (a) $> \sigma^*_{2s}$ orbital
   (b) $< \sigma^*_{2s}$ orbital
   (c) $> \sigma^*_{1s}$ orbital
   (d) both (b) and (c) are correct
3. Which sequence correctly describes a relative bond strength of oxygen molecule, superoxide ion and peroxide ion?
   (a) $O_2 < O_2^- < O_2^{2-}$ (c) $O_2 < O_2^- > O_2^{2-}$
   (b) $O_2 > O_2^- > O_2^{2-}$ (d) $O_2 > O_2^- < O_2^{2-}$
4. In the molecular orbital diagram for O_2^+ ion, the highest occupied orbital is
   (a) σMO orbital (c) π*MO orbital
   (b) πMO orbital (d) σ*MO orbital
5. In the formation of N_2 molecule, according to MOT, the outermost electron goes to
   (a) πMO orbital (c) σ MO orbital
   (b) *p* hybrid orbital (d) 2*p* orbital
6. Which of the following theory provides good explanation about the paramagnetic behaviour of oxygen?
   (a) Resonance theory
   (b) V.S.E.P.R. theory
   (c) Molecular orbital theory
   (d) Valence bond energy

7. Which of the following has unpaired electron in anti-bonding MO?
   (a) C_2 (c) O_2
   (b) N_2 (d) Both C_2 and N_2.
8. The axial overlap between two orbitals leads to the formation of a
   (a) multiple bond (c) Pi bond
   (b) sigma bond (d) none of these
9. Which is paramagnetic and has bond order?
   (a) O_2 (c) F_2
   (b) N_2 (d) H_2^+
10. The bond order of super oxide ion O_2^{2-} is
    (a) 2.5 (c) 2
    (b) 1.5 (d) 1.
11. For a stable molecule, the value of bond order should be
    (a) negative
    (b) positive
    (c) zero
    (d) no relationship of stability and both order
12. How many bonds (bond order) does B_2 have?
    (a) 0 (c) 2
    (b) 1 (d) 3
13. Which of the following is paramagnetic?
    (a) O_2 (c) O_2–2
    (b) N_2 (d) H_2
14. Which of the following is paramagnetic and also has a bond order equal to 0.5?
    (a) O_2 (c) He_2
    (b) N_2 (d) H_2^+
15. In a nitrogen molecule there are
    (a) one sigma and one pi bond
    (b) two sigma and one pi bond
    (c) one sigma and two pi bonds
    (d) three sigma bonds
16. Which one of the following has a zero bond order?
    (a) F_2 (c) O_2
    (b) N_2 (d) He_2
17. The molecular orbitals formed as a result of LCAO obey
    (a) Pauli's Exclusions principal
    (b) Hunds Rule
    (c) Aufbau principle
    (d) All of these
18. A molecule or ion is stable if
    (a) $N_b = N_a$ (c) $N_b < N_a$
    (b) $N_a < N_b$ (d) $N_b - N_a$ = +ve
19. Valence bond theory was proposed by
    (a) Rutherford (c) Heitler and London
    (b) Neil Bohr (d) Mund and mulliken

ANSWERS

1. (a)	5. (d)	9. (a)	13. (a)	17. (d)
2. (b)	6. (c)	10. (b)	14. (d)	18. (c)
3. (b)	7. (c)	11. (b)	15. (c)	19. (c)
4. (c)	8. (b)	12. (b)	16. (d)	

Phase Equilibria 15

CHAPTER OBJECTIVES

- 15.1 Introduction
- 15.2 Explanation of the Terms Involved
- 15.3 Mathematical Statement of Phase Rule
- 15.4 Phase Diagrams
- 15.5 One-Component Systems
- 15.6 The Water System
- 15.7 Sulphur System
- 15.8 Application of Phase Rule to Two-Component Systems (Liquid–Solid Phase Diagram)
- 15.9 Type A – Simple Eutectic System
- 15.10 Type B – System in which Two Components Form a Stable Compound (Zinc–Magnesium Alloy System)
- 15.11 Type C – The Two-Component Form: A Compound with Incongruent Melting Point
- 15.12 Thermal Analysis (Cooling Curve)

It cannot be said too often that science is not mathematics, but reasoning; not equipment, but inquiry.

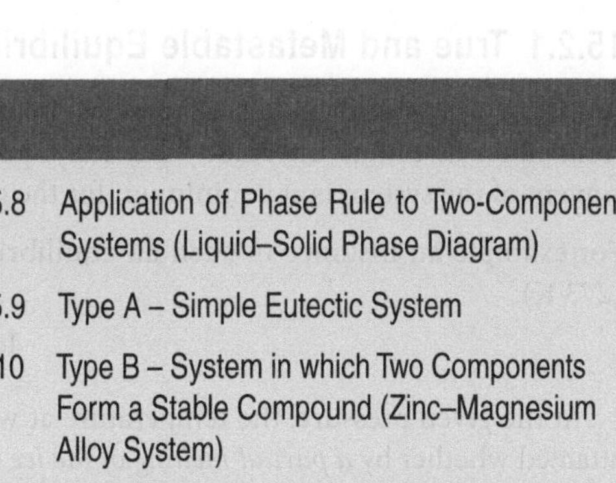

15.1 INTRODUCTION

J. Willard Gibbs, in 1875, as a result of his mathematical and thermodynamic studies, put forward a rule known as **phase rule.** This rule is applicable to all **heterogeneous systems** in equilibrium without any exception and with its help we can predict *qualitatively the effect of temperature, pressure and concentration on a heterogeneous system in equilibrium.*

It is assumed that the equilibrium is not influenced by gravity, electrical or magnetic forces or by surface action and is influenced only by temperature, pressure and concentration. The maximum number of degree of freedom is taken as three.

Mathematically, Gibbs phase rule may be stated as

$$F + P = C + 2 \quad \text{or} \quad F = C - P + 2$$

where F = number of degrees of freedom, C = number of components and P = number of phases.

15.2 EXPLANATION OF THE TERMS INVOLVED

Before proceeding to the statement of phase rule, it will be necessary to define and explain in detail certain terms which are employed frequently in this connection. These terms are (i) *true and metastable equilibrium*, (ii) *number of phases*, (iii) *number of components* and (iv) *degrees of freedom of a system.*

15.2.1 True and Metastable Equilibrium

A state of true equilibrium is said to exist in a system when the same state can be realized by approach from either direction. Thermodynamically speaking, true equilibrium is attained when the free energy content of the system is at a minimum for the given values of the variables.

For example an instance of such an equilibrium is ice and liquid water at 1 atm pressure and 0 °C (273 K).

$$\text{Ice} \rightleftharpoons \text{Water}$$

At the given pressure, the temperature at which the two phases are in equilibrium is the same; it is attained whether by *a partial melting of the ice* or by *a partial freezing of the water.*

On the other hand, water at −5 °C (268 K) can be obtained by careful cooling of the liquid, but not by fusion of ice.

Water at −5 °C is said to be in a state of metastable equilibrium. Such a state can be realized only by careful approach from one direction. It may be preserved provided the system is not subjected to a sudden "shock", "stirring" or "seeding" by the solid phase. As soon as a crystal of ice is introduced, solidification sets in rapidly, and the temperature rises to 0 °C.

15.2.2 Phase

A heterogeneous system consists of various homogeneous parts in contact with each other but with distinct boundaries. The various homogeneous parts which also have distinct physical and chemical properties, and which may (if necessary) be mechanically separated from each other are known as phases.

Hence, a phase may be defined as "any part of a system which is homogeneous in itself, physically distinct, and mechanically separable from other parts of the system".

The number of phases, (P), in a heterogeneous system is given by the number of such physically distinct and homogeneous parts constituting the system.

Essential Conditions for a Phase

(i) It should be homogeneous within itself. It may be homogeneous part of a heterogeneous system.
(ii) It should be *physically distinct.*

(iii) It is separated from other parts of system by *well-defined boundary surface.*

(iv) There must be a *dynamic equilibrium* between different phases of this system due to exchange of chemical species.

Examples

(i) Gaseous phases: *Gases are completely miscible in all proportions* and, therefore, the *mixtures of two or more gases* are homogenous and constitute only *one phase.* There is no boundary surface between the gases or vapours. Air, which is a mixture of nitrogen, oxygen, carbon dioxide, water vapour etc., is composed of one phase only and is thus homogenous.

(ii) Liquid phases: The number of liquid phases in a system is given by the number of liquids present depending on their mutual solubilities.

1. *Two or more completely miscible liquids* present in a system, such as water, alcohol and acetone, constitute only *one liquid phase* as there will be no surface of separation between them when mixed.
2. *Two immiscible liquids*, such as water and mercury, water and carbon disulphide and water and carbon tetrachloride, in contact form *two different phases.*
3. In case of *two partially miscible liquids such as water and ether, the number of liquid phases depends on the composition, i.e., the ratio in which the two liquids are present.* When the amount of one liquid is very large and the amount of the other is very small, the system is a homogeneous solution and hence only *one phase.*

   At *intermediate compositions*, the system is composed of two homogeneous solutions – one rich in ether and the other rich in water and, therefore, *two phases.*

   In such systems composed of partially miscible liquids, the number of phases depends on temperature as their mutual solubility varies with temperature.

(iii) Solid phases: A heterogeneous mixture of solid substances consists of *as many phases as there are substances present.* Different crystalline forms of a substance or allotropic forms of a substance constitute different phases and these are considered different solids for this purpose. However, a solid solution (or mixed crystals) is regarded as a single phase, as it is considered as one solid truly dissolved in another solid.

Thus, the number of liquid and solid phases depends on the mutual solubilities of different species concerned in a system and chemical interactions between them.

Remember:

(i) Miscible gases and miscible liquids constitute a *single* phase.

(ii) Solid + liquid, solid + gas, liquid + gas, two solids and two immiscible liquids constitute *two* phases.

(iii) Different solids (except in solutions) are regarded as different phases; howsoever, intimately, they may be mixed.

15.2.3 Components

In phase rule, the term component does not have its literal meaning, i.e., does not represent the number of constituents. The composition of a phase in a heterogeneous system is expressed quantitatively in terms of the molecular species constituting it. Some of these molecular species present in the system may be in chemical equilibrium and hence the concentration of these species is not independent of each other. By fixing the concentration of some of the species, the concentrations of other species which are in equilibrium with them are automatically fixed due to equilibrium between them.

The number of components (*C*) of a system at equilibrium is defined as 'the smallest number of independently variable constituents by means of which the composition of each phase present can be expressed either directly or in the form of a chemical equation'.

It may also be defined as 'the minimum number of molecular species in terms of which the compositions of all the phases can be quantitatively expressed'. It is important to note that the number of components of a system may or may not be the same as the actual number of substances present in the system, i.e., the constituents of the system. Only those constituents of an equilibrium mixture which can undergo *independent* variation are regarded as components.

Examples

(i) Water system consists of the equilibrium

$$\underset{\text{(Solid)}}{\text{Ice}} \rightleftharpoons \underset{\text{(Liquid)}}{\text{Water}} \rightleftharpoons \underset{\text{(Gaseous)}}{\text{Vapour}}$$

This system consists of one component only, i.e., it is a *one-component system* because the composition of each of the three phases present can be directly expressed as H_2O. It may be noted that hydrogen and oxygen are not the components of the system; first, because in H_2O, hydrogen and oxygen are combined in a fixed ratio of (1:8) by weight, i.e., they are not independently variables and they are also not present, as such, in the equilibrium mixture.

(ii) *Sulphur system* consisting of four phases – monoclinic(*s*), rhombic(*s*), liquid (*l*) and vapours (*g*). It is also *one-component system* because the composition of each phase can be expressed in terms of composition of one constituent – sulphur only.

(iii) *Solution of sugar in water*. The composition of solution of sugar in water is described by specifying that sugar and water are present in the solution phase. If the system has vapour as well as solid sugar in addition to solution, the composition of the three phases can be expressed again by sugar and H_2O, and in all two species are required to express the composition of the phases. The system has two chemical species and the number of components is also two.

(iv) *Decomposition of $CaCO_3$* by heat can be expressed according to the equilibrium

$$\underset{\text{(Solid)}}{CaCO_3} \rightleftharpoons \underset{\text{(Solid)}}{CaO} + \underset{\text{(Gas)}}{CO_2}$$

Thus, there are three different constituents forming three different phases; however, the composition of each phase can be expressed in terms of any two of the constituents. For example,

1. *If the components chosen are CaO and CO_2, then*

Phase		**Composition**
$CaCO_3$	→	$CaO + CO_2$
CaO	→	$CaO + 0\,CO_2$
CO_2	→	$0\,CaO + CO_2$

2. *If the components chosen are $CaCO_3$ and CO_2, then*

Phase		Composition
$CaCO_3$	→	$CaCO_3 + 0\ CO_2$
CaO	→	$CaCO_3 - CO_2$
CO_2	→	$0\ CaCO_3 + CO_2$

3. *If the components chosen are $CaCO_3$ and CaO, then*

Phase		Composition
$CaCO_3$	→	$CaCO_3 + 0\ CaO$
CaO	→	$0\ CaCO_3 + CaO$
CO_2	→	$CaCO_3 - CaO$

Thus, in all these cases, the smallest number of constituents which can fix the composition of the phases present at equilibrium is two; hence, "*dissociation of $CaCO_3$ by heat*" is a *two-component system.*

(v) *Dissociation of ammonium chloride in a closed vessel.* One-component system does not always mean that only one chemical species is present. For example, in a solid–vapour system of ammonium chloride,

(a) $NH_4Cl\ (s) \rightleftharpoons NH_3\ (g) + HCl\ (g)$

the solid phase is crystalline ammonium chloride and the vapour phase consists of mostly of hydrochloric acid gas and ammonia gas with small amounts of ammonium chloride. It is a one-component system though there are three molecular species in the vapour phase because all the three species are related by the following equations.

(b) $NH_4Cl\ (s) \rightleftharpoons NH_3\ (g) + HCl\ (g)$

In this system concentration of NH_3 and that of HCl in the solid phase as well as in the gaseous phase is same. In the NH_4Cl (s) one mole of NH_3 and 1 mole of HCl are present combined with each others whereas in the gaseous state they the same concentration (1 mole each).

It may be emphasized here that number of components need not be the same as the number of chemical species. As long as NH_3 and HCl are present in equivalent amounts, the composition of each phase may be represented as NH_4Cl. Hence, it is *one-component system.*

However, if an excess of NH_3 or HCl is introduced, the relation (b) will no longer be valid and the composition of the gaseous phase can no longer be the same and the composition of this phase can no longer be represented by NH_4Cl alone but one more component is required. Hence, it becomes *two-component system.*

(vi) *A system of NaCl, KCl and H_2O* is three-component system because the composition of all five phases—which are solid NaCl, solid KCl, solid H_2O, solution and vapour—can be given by three species—NaCl, KCl and H_2O.

(vii) *A system consisting of NaCl, KBr and H_2O* is a four-component system (note the difference in these last two examples—the first one has two salts with common chloride ion). This system can yield maximum seven phases under varying conditions, solid KBr, solid KC1, solid NaBr, solid NaCl, solid H_2O, solution and vapour. Obviously, the composition of NaBr and KC1 cannot be given by only NaCl, KBr and H_2O. If, however, one more species, say KCl, is included the composition of NaBr can be expressed as follows:

$$\text{NaBr:}\quad NaCl + KBr - KCl + 0H_2O$$

and thus the system is a four-component system.

(viii) A dilute solution of sulphuric acid in water may be represented as

$$H_2SO_4 + H_2O \rightleftharpoons H_3O^+ + HSO_4^-$$

$$HSO_4^- \rightleftharpoons H^+ + SO_4^{2-}$$

Here, the number of species is five, which are H_2SO_4, H_2O, H_3O^+, HSO_4^- and SO_4^{2-}. The number of relations is two.

Therefore, the number of components $C = S - (R + 1) = 5 - (2 + 1) = 2$. Hence, it is a two-component system.

15.2.4 Degrees of Freedom or Variance

The number of degrees of freedom or variance of a system (F) is "the smallest number of variable factors, i.e., temperature, pressure and concentration of the components which must be arbitrarily fixed in order that the condition of the system may be completely defined."

Examples

(i) *Ice-water–vapour system.* In this system, there are three phases of one component (H_2O). These three phases can co-exist in equilibrium only at one particular temperature and under one particular pressure. Any variation of these factors will lead to the ultimate disappearance of one or more of the phases and, therefore, to the destruction of the system. Hence, this system has no degrees of freedom or in other words it is *invariant*.

(ii) *Water-vapour system.* This system consists of two phases of one component. The vapour pressure of water is definite at a definite temperature independent of the amount of water present. It follows, therefore, that if the temperature is fixed, the vapour pressure is also fixed and conversely, fixing the pressure fixes the temperature. Thus, the system has only one degree of freedom; alternatively, in other words, it is univariant.

(iii) *For a gas*, the temperature, pressure and concentration (volume) are related according to the equation $PV = RT$. Thus, in order that the system may be completely defined, any two factors out of P, V and T must be fixed. Hence, it follows that a gas has two degrees of freedom or it is *bivariant*.

(iv) The system consisting of saturated sodium chloride solution in contact with solid salt and vapour is completely defined if we mention only temperature, since, the solubility of sodium chloride (hence the composition of solution) a vapour pressure has a definite value at a certain temperature. Hence, it is a system with one degree of freedom.

However, the system unsaturated solution of sodium chloride in water in equilibrium with its vapour would be a bivariant. In this case, besides temperature, the composition of the solution has also to be fixed to define the system completely.

The system may be described as *non-variant*, *univariant*, *bivariant*, and *trivariant* accordingly the number of degrees of freedom is *zero*, *one*, *two* or *three*, respectively.

15.3 MATHEMATICAL STATEMENT OF PHASE RULE

The phase rule gives relationship among the number of phases, components and degrees of freedom of a heterogeneous system.

The Gibbs phase rule may be stated as, 'In a heterogeneous system in equilibrium, the number of degrees of freedom plus the number of phases is equal to the number of components plus two.'

Mathematically,

$$F + P = C + 2 \quad \text{or} \quad F = C - P + 2$$

where F = number of degrees of freedom, C = number of components and P = number of phases.

It is assumed that the equilibrium is not influenced by gravity, electrical or magnetic forces or by surface action and is influenced only by temperature pressure and concentration. The maximum number of degree of freedom is taken as three.

The above equation is merely able to fix the number of variables involved, but it cannot give any quantitative relationship between these variables. For example, for pure liquid water in equilibrium with its vapour (two phases, one component), the number of degrees of freedom is one. Thus, either the pressure or the temperature, but never both, may vary independently, an arbitrary change in one of the intensive variables results in a certain, fixed change in the other. The quantitative relationship between these changes cannot be predicted from phase rule equation.

The phase rule is valid only under the following assumptions:

(i) The system is in thermal and mechanical equilibrium. Consequently, the pressure and temperature are the same in all the phases of system.

(ii) Surface contributions as well as contributions from any *electric or magnetic field* to the extensive properties of the system are negligible.

(iii) Interphase surfaces are deformable and permeable to components.

MEMORY FOCUS

1. A state of true equilibrium is said to exist in a system when the same state can be realized by approach from either directions.
2. Metastable equilibrium can be realized only by careful approach from one direction. Water at –5°C is said to be in a state of metastable equilibrium.
3. Phase may be defined "as any part of a system which is homogeneous in itself, physically distinct and mechanically separable from other parts of the system".
4. The minimum number of molecular species in terms of which the compositions of all the phases can be quantitatively expressed are called components (C) of that system.
5. The number of degrees of freedom or variance of a system (F) is "the smallest number of variable factors, i.e., temperature, pressure and concentration of the components which must be arbitrarily fixed in order that the condition of the system may be completely defined."
6. Phase rule may be defined as: "In a heterogeneous system in equilibrium, the number of degrees of freedom plus the number of phases is equal to the number of components plus two."

Numerical Problems

Example 1 **Calculate the number of components and degree of freedom in the following:**

(i) I_2 (s) in equilibrium with its vapour.

(ii) An aqueous solution of NaCl and Na_2SO_4.

Solution: (i) There are two phases of the system: solid and vapour.

The component is one, since composition of both the phases can be expressed by one species, i.e., I_2. The degree of freedom can be calculated from phase rule:

$$F = C - P + 2 = 1 - 2 + 2 = 1$$

(ii) There is only one phase, i.e., solution.

The components of the system are three, namely NaCl, Na_2SO_4 and water. The degree of freedom can be calculated from phase rule equation.

$$F = C - P + 2 = 3 - 1 + 2 = 4$$

Example 2 Calculate the degree of freedom of the following systems:

(i) Unsaturated solution of NaCl in equilibrium with its vapour.

(ii) Na_2SO_4 in water in a closed container at 32.4 °C.

Solution: (i) The system has two phases (solution and vapour) and it is a two-component system. The degree of freedom as calculated from phase rule equation.

$$F = C - P + 2 = 2 - 2 + 2 = 2$$

(ii) The system has again two phases (solution and vapour) and it is a two-component system. Since temperature is fixed at 32.4 °C, we will apply *condensed phase rule equation* to evaluate the degree of freedom.

$$F = C - P + 1 = 2 - 2 + 1 = 1$$

Therefore, the system is univariant.

15.4 PHASE DIAGRAMS

Phase diagram is the complete description of behaviour of phases in equilibrium. The number of phases that exist together when the system is in equilibrium depends upon the conditions of temperature and pressure or conditions of temperature and composition; pressure being kept constant. The conditions are determined experimentally and then substituted in temperature–pressure or temperature–composition diagram. The diagrams so obtained giving the conditions of equilibria between various phases of a substance are called **phase diagrams** or **equilibrium diagrams.**

Phase diagrams are thus maps which show the conditions under which the various phases are capable of stable existence and the condition under which one phase can be transformed into another. The diagram contain *a number of lines, areas and points of intersection*; the significance of each of which has been discussed below under the study of different systems. It is very easy to describe phase behaviour of a system by such diagrams and to know the conditions under which various phases will be present in the systems.

Let us consider one-component systems with different number of phases.

(i) **Single Phase:** When a pure substance is in single phase and is under consideration, it is evident from the phase rule equations that a maximum of two degree of freedoms are required to be fixed in order to define the system.

$$F = C - P + 2 \qquad [\because C = 1, P = 1]$$
$$= 1 - 1 + 2 = 2$$

All such systems can be completely described by stating two variables. The two variables are temperature and pressure, concentration being out of question as there is only one component. Such systems are referred to *bivariant systems.*

(ii) Two Phase: When we are considering a pure substance, i.e., one-component system in two phases in contact with one another, it is evident that only one degree of freedom is required to be fixed in order to define the system.

$$F = C - P + 2 \qquad [\because C = 1, P = 2]$$
$$= 1 - 2 + 2 = 1$$

All such systems can be completely described by stating only one variable – either temperature or pressure. If temperature is specified, the pressure of the system is automatically fixed and, vice versa. Such systems are referred to as *univariant systems.*

(iii) Three Phases: When there are three phases in equilibrium in one-component system, it is evident that the system has no degree of freedom.

$$F = C - P + 2 = 1 - 3 + 2 = 0$$

Such a system is referred to as non-variant or invariant system.

15.5 ONE-COMPONENT SYSTEMS

The general shape of the phase diagram for one-component system is shown below in Fig. 15.1(a) and that in case of water system in Fig. 15.1(b).

The inclination of curve OB representing solid liquid equilibrium in Fig. 15.1(a) in general is always away from the *Y*-axis, whereas for water, it is inclined towards *Y*-axis. This explains why the effect of pressure on melting point of ice is different from that of other solids in general. It is clear from the curves that melting point of ice decreases, while that of other solids increases with increase of temperature.

In phase diagrams of one-component system, bivariant systems are represented by 'areas', univariant systems by 'lines' or 'curves' and non-variant systems by 'points'.

From the phase rule equation, it is quite clear that when there is one component existing in one phase, the value for the degree of freedom (F) is two. Hence, all systems of one component can be completely described graphically, by stating two variable factors; pressure and temperature on appropriate axis. From the statement of phase rule, in the systems of one component, it will be observed that:

(i) When the number of phases is reduced to one, the systems will have two degrees of freedom and are known as *bivariant systems.* A specified area represents a single phase and to locate any point in any area we have to specify both the variables factors.

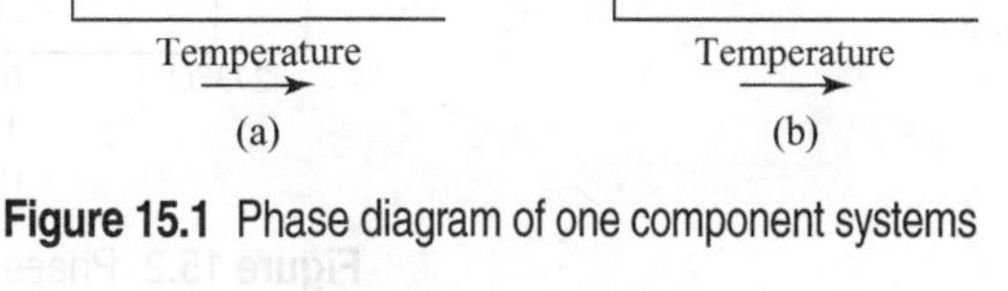

Figure 15.1 Phase diagram of one component systems

(ii) When the number of phases is two, the systems will possess one degree of freedom and are known as *univariant systems*. A line or curve represents two phases in equilibrium and to locate a point on any curve, we have to specify only one of the two variables factors.

(iii) When the three phases co-exist, systems will possess no degree of freedom and are known as *non-variant systems*. Three phases can coexist only at certain fixed points, which are self defined. We have not to specify any variable factor.

15.6 THE WATER SYSTEM

This is most typical example of the one-component system, in which the same chemical compound exists in the following three phases:

$$\underset{(s)}{\text{Ice}} \rightleftharpoons \underset{(l)}{\text{Water}} \rightleftharpoons \underset{(g)}{\text{Vapours}}$$

The conditions of temperature and pressure at which the various phases can exist have been determined experimentally and summed up in Fig. 15.2.

The phase diagram consists of:

(i) Three stable curves (OA, OB and OC) and a metastable curve (OA)′
(ii) Three areas
(iii) Definite point O

A. Curves: (1) OA: This is the **vapour–pressure curve of water**, representing the equilibrium between liquid water and vapour at different temperatures. At any point on the curve, the two phases are present. The curve starts from the point O, the freezing point of water (0.0098 °C at 4.578 mm.) and ends at A, the critical temperature (374 °C at 218 atms.) beyond which the two phases merge with each other. Consider any point on the curve. For fixed value of pressure, temperature has also a

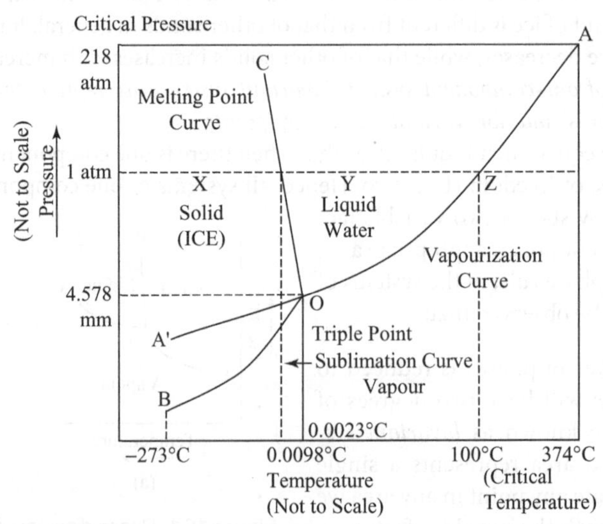

Figure 15.2 Phase diagram of water system

fixed value. Therefore, by keeping the pressure constant, an attempt is made to raise the temperature then the liquid phase will disappear and if the temperature is lowered, then vapour phase will disappear. Therefore, to define the system completely at any time, it is necessary to mention temperature or pressure. In other words, pressure being fixed, it is not necessary to mention temperature; it is automatically fixed up. Thus, water vapour system is univariant. It also follows from the phase rule equation:

$$F = C - P + 2 \quad H_2O_{(g)} \rightleftharpoons H_2O_{(g)}$$

$$F = 1 - 2 + 2 = 1$$

At 100 °C, the vapour pressure of water equals the atmosphere pressure (760 mm). This is, therefore, the boiling point of water. The curve OA extends up to the critical temperature of water (374 °C), since above this temperature liquid water cannot exist.

(2) OA : The dotted curve OA′, a continuation of OA curve represents the ***vapour–pressure of super-cooled water***. It is a metastable curve. On slight disturbance, the super-cooled phase at once changes to solid ice and the curve merges into OB.

$$H_2O_{(l)(\text{super-cooled})} \rightleftharpoons H_2O_{(S)}$$

The special features of vapour–pressure curve are:

(i) (a) *There is no break in the vapour–pressure curve as long as the solid phase does not separate.*

(ii) (b) *The vapour–pressure of the super-cooled water, which is an unstable phases is greater than that of solid (ice) which is the stable phase at that temperature.*

(3) OB: It is the **sublimation curve or the vapour pressure curve of ice.** It gives various values of temperature and pressure at which ice and vapours can exist together.

$$H_2O_{(\text{ice})} \rightleftharpoons H_2O_{(g)}$$

The curve starts from the point O, the freezing point of water and ends at B (−273 °C). As temperature decreases, the vapour–pressure of ice tends to become insignificant. Again we see that for a particular temperature, there can be one and only one pressure value and vice versa. In other words, ice–vapour system is univariant, i.e., $F = 1 + 2 + 2 = 1$.

(4) OC: It is the ***melting point curve*** **and** represents the equilibrium between ice and liquid water at various pressures.

In other words, it shows the effect of pressure on the melting point of ice. *It should be noted that the line OC is inclined towards the pressure axis.*

$$H_2O_{(s)} \rightleftharpoons H_2O_{(l)}$$

The slope of the curve OC clearly indicates that expansion takes place on freezing of water and that melting point of ice is lowered by increase of pressure.

The inclination of curve towards pressure axis can be explained based on Clapeyron–Clausius equation viz.

$$\frac{dP}{dT} = \frac{\Delta H_f}{T(V_l - V_s)} = -ive \qquad [V_l < V_s]$$

where ΔH_f = change in molar heat of fusion.

Since the density of ice is less than that of water, V(S) is greater than V(I), so the expression on the right-hand side of the above equation is negative. Hence, dP/dT should also have negative sign. This has been shown by inclination of curve OC towards pressure axis. This means that the increase of pressure must lower and decrease of pressure must increase the freezing point of water. The curve shows that freezing point of water is lowered by 0.0075 °C by 1 atm increase in pressure.

At any point on the curve OC, two phases (ice and liquid water) are present at equilibrium. Hence, the system is univariant, i.e., degree of freedom is one.

B. Areas: The areas, i.e., the region bounded by two lines give the conditions of temperature and pressure under which the single phase, i.e., ice or liquid water or vapour is capable of stable existence. Any point lying outside the curves falls within the areas AOB, BOC and COA. The area below AOB – water vapours, area BOC – liquid water and area COA represents ice. These areas are bivariant, as to locate any point between the boundaries; temperature as well as pressure must be fixed, thus indicating two degrees of freedom. This also follows from phase rule equation:

$$F = C - P + 2, \quad F = 1 - 1 + 2 = 2$$

C. Triple Point: The three curves OA, OB and OC meet at a point O known as triple point. At this point, the three phases (ice, water and vapours) are in equilibrium. It should be noted that it is only a point, indicating that there is only one set of condition under which it is possible for the three phases to exist together, as indicated by the phase rule. The point O is a self-defined point corresponding to definite temperature 0.0098 °C and definite pressure, *viz*, 4.58 mm and is different from the ordinary freezing point, i.e., 0 °C at 760 mm Hg If either pressure or temperature be changed, then one of the three phases will disappear. Since on changing either pressure or temperature, one of the phases disappears, therefore, the system at the point O has no degree of freedom, i.e., it is invariant. This also follows from phase rule equation:

$$H_2O_{(s)} \rightleftharpoons H_2O_{(l)} \rightleftharpoons H_2O_{(g)}$$

$$F = C - P + 2 = 1 - 3 + 2 = 0$$

MEMORY FOCUS

Water System

System as given in the Phase Diagram	Phases in Equilibrium	Degrees of Freedom
Curves		
OA	Water ⇌ Vapour	1
OB	Ice ⇌ Vapour	1
OC	Ice ⇌ Water	1
OA′ (metastable)	Water ⇌ Vapour	1
Areas		
AOB	Vapour	2
COB	Ice	2
AOC	Liquid water	2
Triple point		
O	Ice ⇌ Water ⇌ Vapour	0

15.7 SULPHUR SYSTEM

Sulphur shows two allotropic crystalline forms (rhombic and monoclinic) capable of existence under suitable conditions. These two forms can also exist in equilibrium with the liquid and the vapour phases. The system, therefore, consists of **four phases**, viz., (i) *rhombic sulphur*, (ii) *monoclinic sulphur*, (iii) *sulphur vapour* and (iv) *sulphur liquid*. This is a typical example of one-component system exhibiting polymorphism and solid transformation.

According to phase rule equation, all the four phases cannot co-exist at equilibrium. On substituting the values of component and phases in phase rule equation,

$$F = C - P + 2$$

i.e., we get

$$F = -1$$

$$F = 1 - 4 + 2 = -1,$$

This is, of course, impossible and so for a system consisting of single component; it is not possible to have four phases in equilibrium together. However, it is possible to have three or lesser number of phases to exist together. The system will be invariant, univariant or bivariant and accordingly three, two or single phases are present.

The pressure temperature diagram (Fig. 15.3) consists of *six stable curves* AB, BC, CD, BF, CF and FG and *four metastable curves*. The curves divide the diagram into *four areas* ABFG, ABCD, GFCD and BCF.

(A) Areas: The areas are bivariant to locate any point in the areas where both temperature and pressure are to be fixed up. Each area represents the conditions for the stable existence of a single phase. In areas, thus, there are two degrees of freedom which also follows from the phase rule equation

$F = C - P + 2$; putting the values $C = 1, P = 1$, we should get $F = 1 - 1 + 2 = 2$.

In the area ABFG, only S_R (rhombic sulphur) exists while below in the area ABCD, only s_V (sulphur vapour) exists; to the right of GFCD, S_L (liquid sulphur) is present and in the region enclosed by BCF, S_M (monoclinic sulphur) exists.

(B) Curves: (1) AB: ***It is the vapour pressure curve of rhombic sulphur***. The two phases present in equilibrium along this curve are S_R and S_V. The curve starts from the point B and ends at point A (60 °C). The point B corresponds to 95.6 °C which is a transition temperature. Rhombic sulphur is stable below this temperature where as monoclinic sulphur is stable above this temperature. At this temperature, both the forms exist in equilibrium.

If heating is done rapidly, then it is possible to heat rhombic sulphur above the transition point BE. BE *is the metastable curve* and on slight disturbance, it merges into curve BC, the vapour pressure curve of monoclinic sulphur.

(2) BC: ***It is the vapour pressure curve of monoclinic sulphur.*** The two phases present in equilibrium along this curve are S_M and S_V. The curve starts from the point B, the transition temperature of S_R and ends at C, the melting point of S_M (120 °C).

Except at the point B, where the three phases S_R, S_M and S_V exist and at any other point all along the curve, only two phases S_V and S_M are in equilibrium. For a fixed value of temperature, pressure is automatically fixed. The degree of freedom, therefore, is one.

If the temperature of monoclinic sulphur is lowered, then generally it would change into rhombic sulphur at 95.6 °C, but it is possible to prevent this conversion by carefully cooling it. Then, we will go along a curve BH which is a *metastable vapour pressure curve of monoclinic sulphur*.

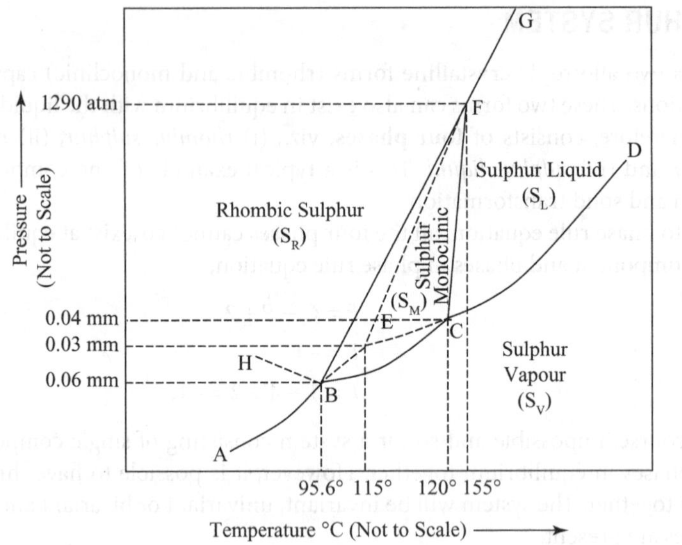

Figure 15.3 Pressure–temperature phase diagram of the sulphur system

(3) CD: ***It is the vapour pressure curve of liquid sulphur***. The two phases present in equilibrium along this curve are liquid sulphur, S_L, and vapour sulphur, S_V. The curve starts at C, the melting point of S_M, and ends at D, the critical point, beyond which out of the phases S_L and S_V, only S_V (vapour sulphur phase) can exist. Since for any point on the curve for a fixed value of pressure, the temperature has a fixed value; therefore, the system such as the above-discussed curves has one degree of freedom.

The dotted curve CE is the continuation of the curve DC and represents *metastable vaporization curve of liquid sulphur.*

(4) BF: ***It is the transition curve*** and represents the equilibrium between rhombic sulphur and monoclinic sulphur. The effect of pressure on the transition temperature of rhombic sulphur is obtained from this curve. The slope of curve BF is away from the vertical pressure axis, indicating that the transition point is raised by the rise of pressure. The curve ends at the point C, after which monoclinic sulphur phase disappears.

(5) CF: ***It is the melting point curve of monoclinic sulphur.*** It represents the equilibrium between monoclinic sulphur and liquid sulphur and is deduced from the effect of pressure on the melting point of monoclinic sulphur. The curve is inclined away from the pressure axis showing that the melting point is raised by the rise of pressure. The curve ends at F beyond which monoclinic sulphur ceases to exist in a stable condition. The system, such as the other curves, is univariant.

(6) EF: This line represents the equilibrium between rhombic sulphur and liquid sulphur and is the *metastable melting point curve of rhombic sulphur.* Since the curves BF and CF meet at F, for which conditions are 155 °C and 1290 atmospheres, it follows that if liquid sulphur were cooled at a pressure higher than 1290 atmospheres, then rhombic sulphur would crystallize at once, no monoclinic sulphur being formed. The curve passes through the point F and beyond this point it becomes a stable melting point curve FG of rhombic sulphur.

(C) Triple Points: There are three stable triple points, viz., B, C and F, in the phase diagram. At the **Point B,** *three phases in equilibrium are rhombic, monoclinic and vapour sulphurs.* If one of the variables is changed, one of the phases disappears. Similarly, C is also a triple point between *monoclinic, vapour and liquid sulphurs.* The point is again invariant. Further, lines BF and CF intersect at F to yield a three-phase equilibrium between *rhombic, monoclinic and liquid sulphurs.* Since on changing one of the variables, either pressure or temperature, one of the phases disappears; therefore, the system at the triple point has no degree of freedom. This also follows from phase rule equation:

$$F = C - P + 2 = 1 - 3 + 2 = 0$$

Part B is obtained at 95.6 °C and 0.006 mm Hg, while point C is obtained at 120 °C and 0.04 mm Hg and point F is obtained at 155 °C and 1290 atm pressure. Three phases in equilibrium at these points are:

$$B(S_R - S_M); \; C(S_M - S_L - S_V); \; \text{and} \; F(S_R - S_M - S_L)$$

(*System enclosed in brackets represents metastable curves*)

MEMORY FOCUS

Sulphur System

Name of System	Phases is Equilibrium	Degrees of Freedom
Areas		$F = C - P + 2$
		$F = 1 - 1 + 2 = 2$ (bivariant)
(i) To the left of ABFG	S_R	Two
(ii) Enclosed by BCF	S_M	"
(iii) To the right of GFCD	S_L	"
(iv) Below ABCD	S_V	"
Curves		$F = 1 - 2 + 2 = 1$ (univariant)
(i) AB (BE)	$S_R \rightleftharpoons S_V$	One
(ii) BC (BH)	$S_M \rightleftharpoons S_V$	"
(iii) CD (CE)	$S_L \rightleftharpoons S_V$	"
(iv) FG (FE)	$S_R \rightleftharpoons S_V$	"
(v) BF	$S_R \rightleftharpoons S_M$	"
(vi) CF	$S_M \rightleftharpoons S_L$	"
Triple points		$F = 1 - 3 + 2 = 0$ (invariant)
(i) B	$S_R \rightleftharpoons S_M \rightleftharpoons S_V$	Zero
(ii) C	$S_M \rightleftharpoons S_l \rightleftharpoons S_v$	"
(iii) F	$S_R \rightleftharpoons S_L \rightleftharpoons S_V$	"
(iv) E (metastable)	$S_R \rightleftharpoons S_L \rightleftharpoons S_V$	"

15.8 APPLICATION OF PHASE RULE TO TWO-COMPONENT SYSTEMS (LIQUID–SOLID PHASE DIAGRAM)

A two-component system can be defined as one in which it is possible to express the composition of all the phases, in terms of two substances. A two-component system may have as many as three degrees of freedom. This is because when $P = 1$, then

$$F = C - P + 2 = 2 - 1 + 2 = 3$$

Thus, the system will have, besides temperature and pressure variables, one concentration variable for each phase present at equilibrium. Naturally, for the graphic representation of all the possible conditions of equilibrium of such a system, a three-dimensional model is required in which pressure, temperature and concentration would be plotted along the three axes at right angles to one another (Fig. 15.4).

However, in actual practice, a simple plain diagram with two variables only is considered; the third variable is considered as constant. In this way, use is made of the following three types of two dimentional diagrams; because it is difficult to draw and interpret a three dimensional model.

(i) Pressure–temperature (p–t) diagrams
(ii) Temperature–composition (t–c) diagrams
(iii) Composition–pressure (c–p) diagrams

However, in most of the cases, it is usually convenient to keep pressure constant at the atmosphere value and thus a t–c diagram is often met with. A diagram drawn for a constant value of temperature is called *isothermal*, while one obtained by keeping composition constant is called *isoplethal* diagram.

In case of only solid–liquid phases, small changes of pressures have very little effect on the system. In this equilibrium, pressure will not have considerable effect because only gaseous systems, which involve large changes of volume, are affected by pressure. Such systems are called **condensed systems.** The degree of freedom in such a case is reduced by one as the vapour pressures of solids are very small and experiments are performed at atmospheric pressure. Thus, we may use the **reduced phase rule equation** in such cases.

$$F' = C - P + 1$$

where F' represents the number of degrees of freedom which the system can possess in addition to pressure, namely, temperature and concentration (or composition) of the constituent. Therefore, solid–liquid equilibria are represented on temperature composition diagram which are plotted by using thermal analysis procedure for different compositions of two-component systems.

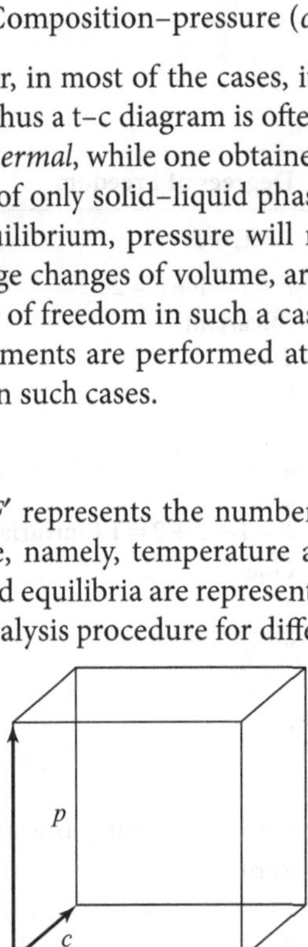

(a)

Figure 15.4 Space model for a two component system

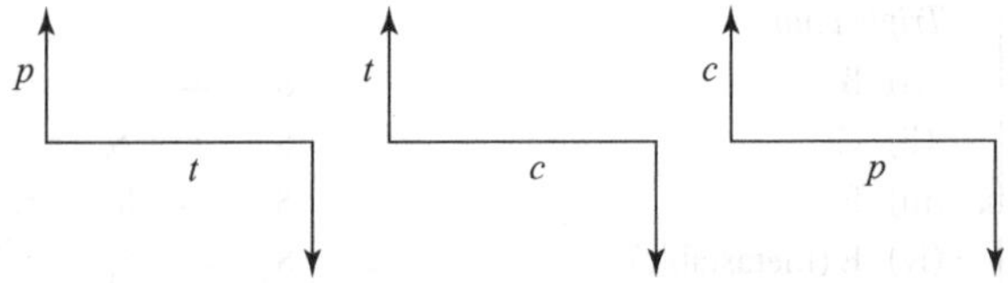

(b)

Figure 15.4 Two–dimensional split–up

For the two-component system among the various varieties, the following cases are discussed.

Type A: *The two components are not miscible in solid state and form a simple eutectic mixture.*
Type B: *The two components form a compound with congruent melting point.*
Type C: *The two components form a compound with incongruent melting point.*

15.9 TYPE A—SIMPLE EUTECTIC SYSTEM

A binary system consisting of two substances, which are miscible in all proportions in the liquid phase, but do not react chemically, is known as the *eutectic system*. The word eutectic means easy to melt. For example, a mixture of lead and silver and of potassium iodide and water are common eutectic systems.

Two solid substances capable of forming solid solution with each other have the property of lowering each other's freezing point; and the minimum freezing point attainable corresponding to the eutectic mixture is termed as the *eutectic point*. Systems giving rise to eutectic point are known as eutectic systems.

MEMORY FOCUS

Some Eutectic Systems

Components and their Melting Points	Eutectic Composition	Eutectic Temperature
Ag (961 °C); Cu (1083 °C)	(71.8% Ag + 29.2% Cu)	778 °C
Ag (961 °C); Pb (327 °C)	(2.6% Ag + 97.4% Pb)	303 °C
Zn (419 °C); Cd (323 °C)	(67% Zn + 33% Cd)	270 °C
Zn (419 °C); Al (658.7 °C)	(95.64% Zn + 4.36% Al)	380.5 °C
Bi (273 °C); Cd (323 °C)	(60% Bi + 40% Cd)	140 °C
KCl (773 °C); H_2O (O °C)	(20% KCl + 80% H_2O)	−11 °C
KI (682 °C); H_2O (0 °C)	(52% KI + 48% H_2O)	−22 °C

Out of the above-noted systems, only Pb–Ag systems have been discussed in this book.

15.9.1 Characteristics of Eutectic Point

(i) Eutectic point represents the lowest or limiting temperature at which a liquid phase can exist in the system or it is the maximum temperature up to which a solid phase can exist.
(ii) No other mixture containing the two components will have a melting point lower than the eutectic temperature.
(iii) Eutectic point has precise values of temperature and composition and it represents an invariant system.
(iv) If the liquid is cooled to just below the eutectic point, both the components of the eutectic simultaneously solidify without any change in the composition or temperature of the liquid phase.
(v) A eutectic system can maintain its temperature constant for long periods.
(vi) When the liquid is cooled below the eutectic point, the components solidify in the form of small crystals intimately mixed with each other which fill in the spaces between the larger crystals of the pure components which are already separated out.

(vii) Eutectic mixtures appreciably contribute towards the strength of the solid structure in case of alloys.

(viii) Eutectic are mixtures of the components but not their compounds.

15.9.2 Use of Eutectic Systems

(i) Eutectic mixtures are used in preparing "**solders**" used for joining two metal pieces together. Example: Pb–Sn solders.

(ii) They are used as "**safety fuses**" in various equipments such as automobiles, fire extinguishers and boilers. Example: Wood's metal (an alloy containing, Bi, Pb, Sn and Cd).

15.9.3 Lead–Silver System

It is the simplest case of two-component solid–liquid system involving eutectic formation. The two components are lead and silver. The four phases involved in the system are: (i) solid lead, (ii) solid silver, (iii) solution of lead and silver in the molten state and (iv) vapour.

In this system, pressure will not have considerable effect because the vapour pressure of solids and molten metals is very small and can be neglected. The phase diagram, therefore, is drawn at atmospheric pressure so that no gaseous phase is present.

According to the reduced phase rule equation, when one phase is present, $F = 2–1+1 = 2$. Since $F = 2$, only two variables (excluding pressure which is considered constant) must be specified in order to describe the systems and that the equilibrium can be represented in two dimensions by means of temperature–composition diagram (Fig. 15.5).

15.9.3.1 Curves and Points

(i) Point A: It represents the ***melting or freezing point of pure lead*** (327 °C) and **point B** represents the ***melting or freezing point of pure silver*** **(961 ϒC)**. Thus, at *point A*, liquid lead is in equilibrium with solid lead and at *point B* liquid silver is in equilibrium with solid silver.

(ii) Curve AO: It represents the ***freezing point of lead to*** which successive small amounts of silver are added. It is seen that the addition of silver to lead lowers the melting point of lead. All along

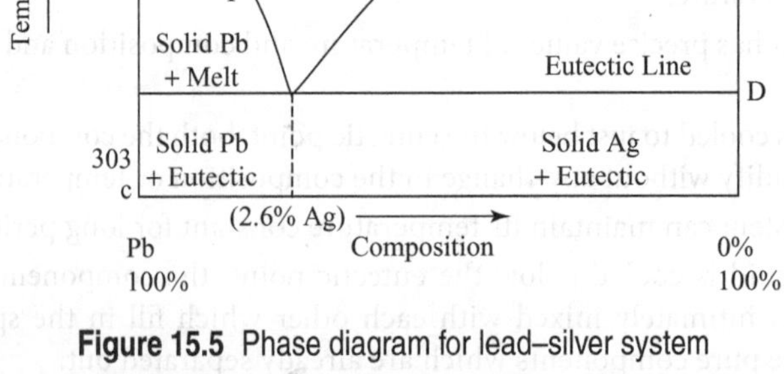

Figure 15.5 Phase diagram for lead–silver system

this curve, the silver, which is added, goes into the solution while the separation of solid lead takes place. *Thus, along this curve, solid lead is in equilibrium with liquid melt* (solution of silver in liquid lead) *of varying composition*. The curve AO represents the freezing point of lead from separate the melt containing varying amounts of silver. Along this curve, two phases (solid lead and liquid melt) are present in equilibrium. Therefore,

$$F = C - P + 1 = 2 - 2 + 1 = 1, \text{ i.e., the system is univariant.}$$

(iii) Curve BO: It represents the **freezing point of silver** to which successive small amounts of lead are added. It is observed that the melting point of silver is lowered, gradually by the addition of lead into it. All along this curve, the added lead goes into the solution and the separation of solid silver takes place. Thus, *along this curve, solid silver is in equilibrium with liquid melt* (solution of lead in liquid silver) *of varying composition*. The curve BO represents the freezing point of silver separated from the melt containing varying amounts of lead. Along the curve BO, two phases (solid silver and melt) are present and the system is univariant as shown below.

$$F = C - P + 1 = 2 - 2 + 1 = 1$$

At the point O in the phase diagram, lead and silver freeze out simultaneously. At the point, the composition of the melt is the same as that of the solid mixture of lead and silver. The liquid melt freezes out, as if it was a pure compound. The whole of the melt freezes out at constant temperature. This point O is called **eutectic point** (Greek – low melting) and the solid mixture is called ***eutectic*** which has a characteristic composition for each system.

The eutectic even though melts at a constant temperature, it is not a chemical compound of lead and silver as (i) *its composition does not correspond to any simple chemical formula* and (ii) *its composition depends on temperature.*

It may be made clear here that the phase diagram 15.5 will not be exactly similar at different pressures and the eutectic composition will also vary with pressure.

Below point O in Fig. 15.5, only solid phase exists and the cooling of this solid eutectic is represented by a dotted line.

Eutectic Point O is the point where two curves AO and BO meet. At this point, solid lead and solid silver are in equilibrium with the melt, i.e., three phases are present. Therefore, there is no degree of freedom $F = C - P + 1 = 2 - 3 + 1 = 0$.

The point O in Fig. 15.5 represents the lowest possible temperature (303 °C) below which a liquid phase cannot exist and beyond which the liquid phase cannot be enriched in either composition by freezing out the other component. *Such a liquid mixture of two components Pb and Ag, which has the lowest freezing point compared to all other liquid mixtures is called the* **eutectic mixture.** The mixture freezes out completely at a constant temperature and on freezing gives the two components as solid phase in the same proportion in which they are present in the liquid phase. The temperature corresponding to the eutectic point O is called the **eutectic temperature.** In the above system, we see that the alloy of silver 14.5% with lead 97.4% melts at 303 °C, a temperature lower than the melting point of silver (961 °C) and of lead (327 °C).

In the region or area above the curve AOB, only the liquid alloy (melt of silver and lead) exists, i.e., there is only one phase and it has two degrees of freedom, i.e., temperature and concentration besides pressure. This system is, therefore, bivariant.

In the area AOC: Solid lead and liquid melt exist. The composition of the liquid melt at any temperature can be obtained from the curve AO. At any point in this area, solid lead is in equilibrium with a liquid melt of definite composition.

In the area BOD: Solid silver and liquid melt exist. The composition of the liquid melt at any temperature can be obtained from the curve BO as in the case of area AOC.

In these areas AOC and AOD, there is one degree of freedom temperature or composition besides pressure.

Below line COD, there are two phases (solid lead and solid silver). In the region marked as solid Pb + eutectic, we find lead crystals and solid eutectic crystals whereas in the region marked solid Ag + eutectic, we find silver crystals and solid eutectic crystals. This can be examined under a microscope. Again this system has one degree of freedom besides pressure.

$$F = C - P + 1 = 2 - 2 + 1 = 1$$

The phase diagram provides very useful information about the system.

15.9.4 Pattinson's Process for Desilverization of Lead

Galena (PbS, an ore of lead) is usually associated with silver. When galena is worked out for the extraction of lead, some traces of silver pass into it because silver is soluble in lead to some extent. Such a lead is known as *argentiferous* lead. The desilverization of lead can be easily understood from the phase diagram of lead silver system (shown in Fig. 15.5).

The agentiferous lead consisting of very small percentage of silver is first of all heated to a temperature well above its melting point; therefore, the system consists only of liquid phase represented by the point x in the figure. It is then allowed to cool gradually along the line xy without any change in composition till the point y is reached. As soon as this point is attained, lead starts separating out and the solution will contain respectively increasing amount of silver. Further cooling will shift the system along the line YO. Lead continues to separate out and the melt continues to be richer and richer in silver, till the eutectic point O is reached where an alloy containing about 2.6% of silver is obtained.

The increase in the relative amount of Ag during the cooling process of molten argentiferous lead is shown in Fig. 15.6. The amount of Ag present in traces in the alloy increases up to 2.4% by gradually cooling the molten mass.

This alloy is then further refined by the process of cupellation. This above principle is used in Pattison's Process (i.e., the process of increasing the relative proportions of silver in the alloy) for the desilverization of lead.

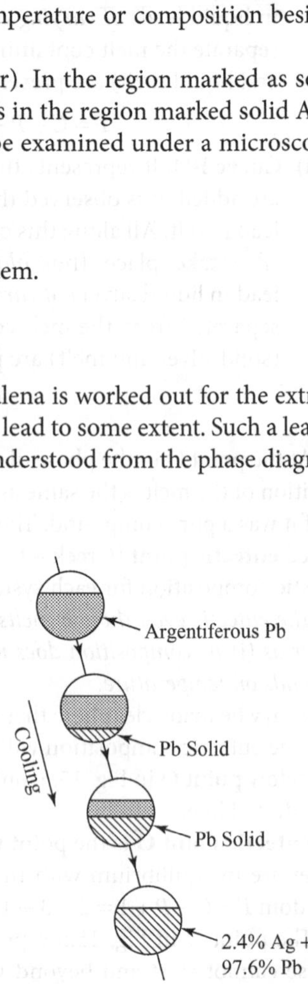

Figure 15.6 Cooling of argentiferous lead in patterson process

15.10 TYPE B—SYSTEM IN WHICH TWO COMPONENTS FORM A STABLE COMPOUND (ZINC–MAGNESIUM ALLOY SYSTEM)

In this type of binary alloys, the components form one or more compounds having definite proportions. These compounds melt into the liquid state, having the same composition as the solid. Therefore, such compounds are called *compounds with congruent melting points. Thus, a compound is said to possess* **congruent melting point,** *if it melts sharply at a constant temperature into a liquid, having the* **same composition** *as that of the solid.*

Let us illustrate such a system by Zn–Mg binary alloys (Fig. 15.7). Zinc and magnesium melt at 420 °C and 650 °C (points A and B), respectively, and they form a compound $MgZn_2$, i.e., in which the ratio of Mg and Zn in molecular proportions is 1:2. This compound melts at *B* (590 °C) which is congruent in point.

This phase diagrams may be considered as equivalent to two phase diagrams of Pb–Ag type (as discussed earlier) placed side-by-side. The one on the left consists of Zn and $MgZn_2$ system; while the other on the right is formed by the two components, $MgZn_2$ and Mg.

Curves: The curves AE_1 and BE_2 represent the compositions of liquids in equilibrium with solids Zn and solid Mg, respectively. The central portion of curve E_1BE_2 has a maxima at B. This gives the conditions of equilibrium of liquid with solid compound, $MgZn_2$. At B, the maxima, the liquid has the same composition as the solid in equilibrium. Thus, the temperature at B (590 °C) represents the melting point of the compound $MgZn_2$. As $MgZn_2$ can co-exist as solid and liquid of same composition, so $MgZn_2$ is said to possess *congruent melting point*. The composition corresponding to maxima (congruent melting point) is 33.3% Mg and 66.7% *Zn*, i.e., in the ratio of 1:2 as $MgZn_2$. The sharpness of maxima gives the information about the stability of the compound formed. A sharp maxima indicates that the compound does not dissociate appreciably; whereas a *flat maxima indicates* an *appreciable dissociation*, as in the present case.

All along these three curves, two phases are in equilibrium curve AE_1 is called freezing point curve of Zn to which Mg is being added gradually. The two phases in equilibrium are Zn(s) and melt. Similarly, curve BE_2 is called the freezing point curve of magnesium to which Zn is being added gradually. The two phases in equilibrium are Mg(s) and melt. Curve $E_1C_ME_2$ is called the freezing melting point curve

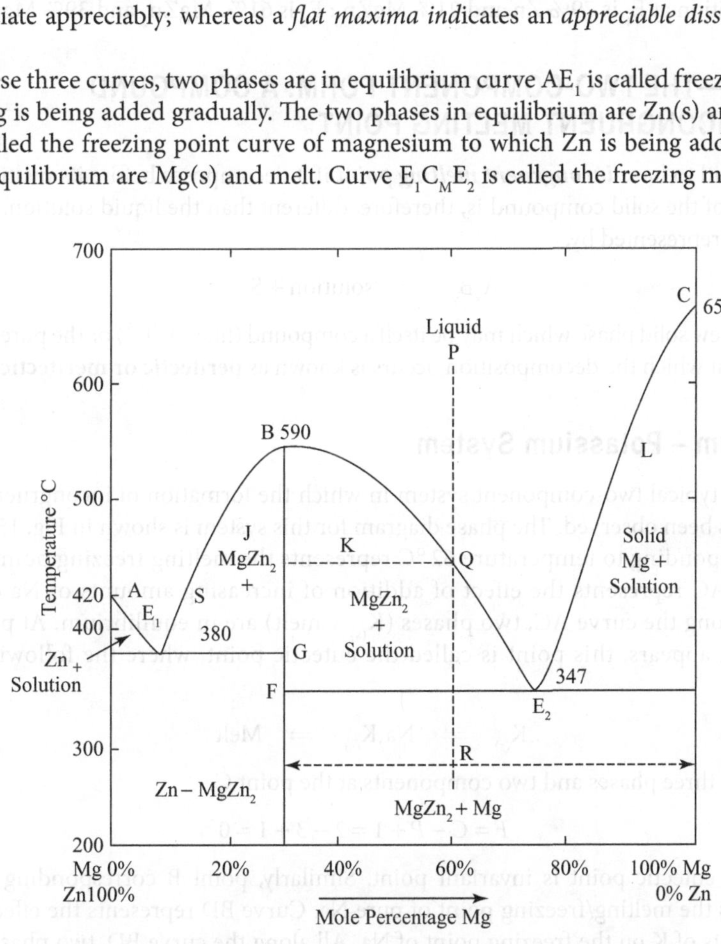

Figure 15.7 Temperature composition diagram for Zn–Mg system

of the compound $MgZn_2$, where the two phases in equilibrium are solid $MgZn_2$ and melt. According to phase rule,

$$F = C - P + 1 = 2 - 2 + 1 = 1$$

The degree of freedom along these curves is one, i.e., the curves are univariant. It means that to locate any point on any of these curves, either temperature or composition is to be fixed.

15.10.1 Eutectic Points and Congruent Melting Point

Points E_1 and E_2 represent eutectic points. At these points, three phases are in equilibrium. At these points, three phases are in equilibrium. At E_1, the three phases are in equilibrium.

At E_1, the three phases are Zn(s), melt and $MgZn_2$(s). At E_2, the three phases are Mg(s), melt and $MgZn_2$(s).

According to phase rule:

$$F = C - P + 1 = 2 - 3 + 1 = 0.$$

Molar composition at E_1 is 79% Zn and 21% $MgZn_2$: E_2 is 61%, $MgZn_2$ and 39% Mg.

15.11 TYPE C—THE TWO-COMPONENT FORM: A COMPOUND WITH INCONGRUENT MELTING POINT

A compound is said to have ***incongruent melting point*** if it decomposes before its melting temperature. The composition of the solid compound is, therefore, different than the liquid solution. The situation at this point may be represented by,

$$A_xB_y \rightleftharpoons \text{solution} + S_1$$

Here, S_1 is the new solid phase which may be itself a compound (like ($A_x'B_y'$) or the pure constant A or B. This temperature at which the decomposition occurs is known as **peritectic or meritectic temperature.**

15.11.1 Sodium – Potassium System

Na–K system is a typical two-component system in which the formation of incongruent melting compound (Na_2K) has been observed. The phase diagram for this system is shown in Fig. 15.8.

Point A corresponding to temperature 62 °C represents the melting freezing point of pure potassium and curve AC represents the effect of addition of increasing amounts of Na on the freezing point of K. All along the curve AC, two phases ($K_{(s)}$ + melt) are in equilibrium. At point C, another solid phase Na_2K appears, this point is called the eutectic point, where the following equilibrium exists

$$K_{(s)} \rightleftharpoons Na_2K_{(s)} \rightleftharpoons \text{Melt}$$

Thus, there are three phases and two components at the point C.

$$\therefore \qquad F = C - P + 1 = 2 - 3 + 1 = 0$$

Therefore, the eutectic point is invariant point. Similarly, point B corresponding to temperature 97.5 °C represents the melting/freezing point of pure Na. Curve BD represents the effect of addition of increasing amounts of K on the freezing point of Na. All along the curve BD, two phases ($Na_{(s)}$ + melt)

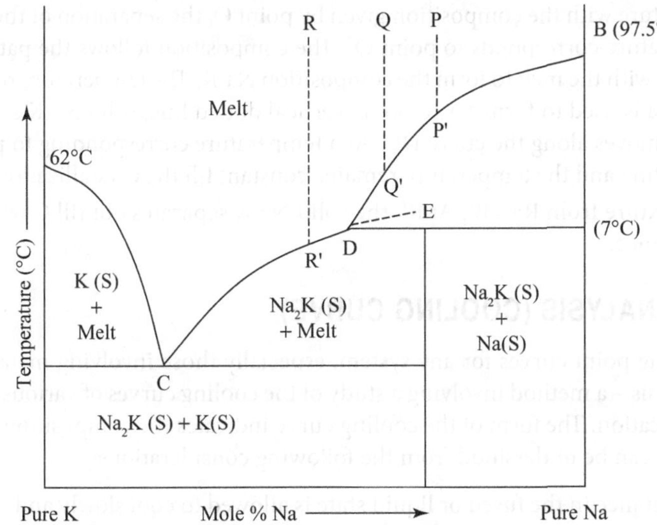

Figure 15.8 Phase diagram of Na–K system

are in equilibrium. At point D, the solid phase Na_2K being formed all along the curve decomposes, where the following equilibrium exists:

$$Na_2K_{(s)} \rightleftharpoons Na_{(s)} + \text{Melt}$$

The point D where the compound Na_2K decomposes yielding solid Na and a melt of composition corresponding to point D is called *peritectic or incongruent melting temperature* (7 °C)

Here, $$F = C - P + 1 = 2 - 3 + 1 = 0$$

The system is, therefore, invariant at incongruent point. The transition of solid Na into compound Na_2K will take place at constant temperature till the whole of the solid sodium disappears.

The dotted curve is the metastable state of the compound and the point E is the hypothetical melting point. The compound Na_2K formed undergoes dissociation at D below the melting point and the composition of the liquid is not the same as that of the solid compound. The point D is called **incongruent melting point.** This point represents the limit of the existence of the compound Na_2K as there are three phases present at point D.

15.11.1.1 Cooling of Mixtures along three Isopleths: PP′, QQ′ and RR′

(i) Mixture at point P contains less K than the compound Na_2K.

(ii) Mixture at point Q contains more K than Na_2K.

(iii) Mixture at R contains more K than the peritectic D but less than the eutectic point C.

(iv) On cooling a mixture of composition at P, pure Na separates at P′, the mother liquor becomes richer in K temperature and composition move along BD. D is the maximum temperature at which stable Na_2K can exist. Thus, D is a true transition point. At a temperature corresponding to point D, solid B reacts with the melt to form the compound Na_2K. The temperature remains constant till the solidification of Na and Na_2K is completed.

(v) On cooling a mixture with the composition given by point Q, the separation of the solid Na starts when the temperature corresponds to point Q′. The composition follows the path Q′D, at point D, solid Na reacts with the melt to form the composition Na_2K. The temperature remains constant till whole solid Na is used to form Na_2K. Some vertical dotted line, solid Na_2K separates out and the composition moves along the curve DC. At a temperature corresponding to point C, solid K also starts separating and the temperature remains constant till the crystallization is completed.

(vi) Cooling third mixture from R to R′. At R′, the solid Na_2K separates out till C when solid K also shows its appearance.

15.12 THERMAL ANALYSIS (COOLING CURVE)

The shape of the freezing point curves for any system, especially those involving metals, can be determined by thermal analysis – a method involving a study of the cooling curves of various compositions of a system during solidification. The form of the cooling curve indicates the composition of the solid. The principle of the method can be understood from the following considerations:

(i) When a pure substance in the fused or liquid state is allowed to cool slowly and the temperature is noted at definite times, the graphic representation of the rate of cooling will be continuous curve (Fig. 15.9). When the freezing point is reached and the solid makes its appearance, it is indicated by a break in the continuity of the cooling curve (point b) and the temperature will remain constant, until the liquid is completely solidified. Thereafter, the falling temperature will again be continuous along curve *cd*.

(ii) If a mixture of two solids in the fused state be cooled slowly and the cooling curve is obtained in a similar manner. We likewise obtain a continuous cooling curve, so long as the mixture (or solution) is in the liquid state. When a solid phase begins to form, the rate of cooling abruptly alters and the cooling curve exhibits a break. However, the temperature does not remain constant, as in the previous case of cooling of a pure substance. The temperature decreases continuously, but at a different rate and if the mixture forms a eutectic, the fall of temperature continues, till the eutectic point is reached. The system now becomes invariant from the viewpoint of the phase rule and the temperature remains constant; until solidification is complete. Thereafter, the fall of temperature becomes uniform, but the rate of fall is quite different from the previous one.

From the cooling curve for any mixture of a definite composition, it is possible to obtain its: (i) freezing point and (ii) the eutectic temperature.

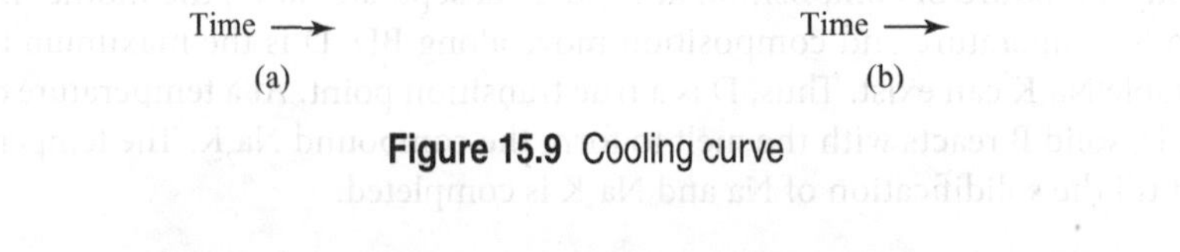

Figure 15.9 Cooling curve

(a) The freezing point varies with the composition of the system, but the eutectic point remains constant for a given system.

(b) The nearer the composition of the system to the eutectic is, the shorter is the portion *bc* and the more prolonged is the halt *cd*.

(c) If the mixture coincides with the eutectic composition, the curve shows no break corresponding to be, but the break appears only at the eutectic point (c).

(d) If the cooling curves of a series of alloys of known compositions are worked out and their freezing points are noted, by plotting freezing point against composition, T-C curve is obtained for the alloy system. However, in order to complete the diagram, it is necessary to know the freezing points of the pure components also.

(e) Now, the cooling curve of an alloy of the same metals, but of unknown composition is determined and its freezing point located in the T-C diagram. The composition corresponding to this freezing point yields file composition of file alloy.

REVISION QUESTIONS

1. State phase rule and explain the terms involved.
2. State and explain what is meant by the terms phase, component and degree of freedom? How are the related?
3. How many phases and components are present in the following systems?
   (a) A mixture of molten lead, tin and bismuth.
   (b) Two ice cubes floating on water in a closed container in the presence of water vapour.
4. Find out the number of components present in the following systems:
   (a) Silver (s) $\rightleftharpoons$ solution of silver and lead (*l*) + lead (s) + vapours of silver and lead
   (b) Water (*l*) $\rightleftharpoons$ water vapour (g)
5. Ammonium chloride is heated in a closed vessel. The system can be either a one-component system or a two-component system. Explain.
6. What is meant by (i) true equilibrium and (ii) metastable equilibrium? What is the thermodynamic criterion of phase equilibrium?
7. Draw a phase diagram for one-component water system. Label it and discuss the importance of various points, lines and areas at equilibrium.
8. What is meant by triple point of water? Why is it different from the normal melting point of ice?
9. Draw a labelled phase diagram of sulphur system and discuss its salient features.
10. Draw and explain a phase diagram for one-component system from comprising more than one solid–solid phase. Determine the degree of freedom of sulphur system at the transition point.
11. The melting point curve of water system has a negative slope whereas the melting point curve of sulphur has a positive slope. Explain.
12. Discuss the salient features of the phase diagram of lead–silver system.
13. Explain how argentiferous lead containing a very low percentage of silver can be concentrated by cooling. State the maximum limit of concentration thus obtained.
14. Sketch the cooling curves for the following alloys of Pb–Ag system and explain what happens as their melts are cooled:
    (a) An alloy of eutectic composition (b) An alloy of any other composition.

15. Explain, with suitable example, the following terms used in the phase rule study of heterogeneous equilibria:
    (a) Triple point
    (b) Transition point
    (c) Invariant system
    (d) Eutectic point
    (e) Eutectic
    (f) Metastable equilibrium
16. Explain why KCl–NaCl–H_2O should be regarded as a three-component system, while aqueous solution of KCl–NaBr should be as four-component system.
17. (a) What is meant by the number of components of a system?
    (b) What is condensed phase rule? When is it applied?
18. (a) Define and explain the term components of a system with suitable examples.
    (b) Draw a labelled phase diagram of water system and discuss the metastable curve and principle of freeze drying.
19. (a) Giving suitable examples, explain the terms: phase and degrees of freedom.
    (b) What do you understand by reduced phase rule?
    (c) Derive Gibb's phase rule equation.
20. (a) Giving examples, explain the term: components of a system.
    (b) Differentiate between true equilibrium and metastable equilibrium.
    (c) Justify the statement, "The eutectic is a mixture and not a compound."
21. (a) Define the various terms involved in phase rule. Explain the thermodynamic criterion of phase rule equilibrium.
    (b) Name the scientist who introduced the term eutectic.
    (c) Mention any two limitations of phase rule.
22. (a) Draw and compare the cooling curve of a pure substance and a mixture of two solids. Mention the striking difference between the curves.
    (b) Draw and describe the phase diagram of water system. Explain what is metastable equilibrium.
23. (a) What is eutectic? Is it a compound or not?
    (b) Describe cooling curve method for determination of melting point and eutectic point.
    (c) Explain the term metastable equilibrium.
24. (a) Draw a labelled phase diagram of Na-K system. Explain its silent features.
    (b) Explain the following:
        (i) Reduced phase-rule equation
        (ii) Congruent melting point
        (iii) Triple point
        (iv) Peritectic point
25 Draw and discuss the phase diagram of lead-silver system. Explain the eutectic point arid Pattison's process for the desilverization of lead.
26. Define the term with suitable example:
    (a) Phase
    (b) Component
    (c) Degree of freedom

27. Draw a phase diagram for one-component waters system. Label it and discuss the importance of various points, lines and areas at equilibrium.
28. (a) Define the following terms used in phase equilibria with examples: phase, component, degree of freedom, eutectic, freezing point.
    (b) With the help of neat, clean and labelled diagram discuss the phase diagram of sulphur system.

MULTIPLE CHOICE QUESTIONS

1. Mathematically, the phase rule can be expressed as
   (a) $F+P=C+2$
   (b) $F=C-P+2$
   (c) $P=C-F+2$
   (d) all of these
2. The phase rule was first discovered by
   (a) Nernst
   (b) Gibbs
   (c) Arrhenius
   (d) *Le* Chatelier
3. A mixture of three gases O2, N_2 and CO_2 is
   (a) one-phase system
   (b) two-phase system
   (c) three-phase system
   (d) four-phase system
4. A mixture of two immiscible liquids constitutes a system having the number of phase equal to
   (a) zero
   (b) one
   (c) two
   (d) three
5. A mixture of two liquids has the number of phases equal to
   (a) zero
   (b) one
   (c) two
   (d) three
6. A saturated solution of sodium chloride is a
   (a) one-phase system
   (b) two-phase system
   (c) three-phase system
   (d) none of these
7. A mixture of gases O_2 and N_2 constitutes one phase only. The number of components in the system is
   (a) zero
   (b) one
   (c) two
   (d) three
8. The decomposition of $CaCO_3$ in a closed vessel is represented by the equation

$$CaCO_3(s) \rightleftharpoons CaO(s) + CO_2(g)$$

   The number of phases and components, respectively, are
   (a) two and three
   (b) three and two
   (c) two and two
   (d) three and three
9. A saturated solution of sodium chloride in contact with solid has the phases and components equal to
   (a) two and two
   (b) two and three
   (c) three and two
   (d) three and three
10. The total number of variable factors which must be specified so that the remaining variables are fixed automatically and the system is completely defined. It is known as
   (a) a phase
   (b) a component
   (c) degrees of freedom
   (d) none of these
11. For a pure gas and mixture of gases, the degrees of freedom are
   (a) two and two
   (b) two and three
   (c) three and two
   (d) three and three
12. For one-component system the phase rule is
   (a) $F=3-P$
   (b) $F=2-P$
   (c) $F=1-P$
   (d) none of these
13. At a triple point
   (a) both the temperature and pressure are fixed
   (b) only the temperature is fixed
   (c) only the pressure is fixed
   (d) sometimes pressure and sometime temperature is fixed
14. The vapour pressure of the metastable phase is always ___ than that of the stable phase at the same temperature.
   (a) lower
   (b) higher
   (c) constant
   (d) none of these

15. The temperature at which a polymorphic substance changes from one form to another is called
(a) equilibrium temperature
(b) triple point
(c) Boyle's temperature
(d) transition temperature

16. When a single phase is present in a two-component system, the degree of freedom is
(a) zero (c) two
(b) one (d) three

17. The decomposition of NH_4Cl is represented by the equation

$$NH_4Cl(s) \rightleftharpoons NH_3(g) + HCl(g)$$

The number of components present in the system is
(a) zero (c) two
(b) one (d) three

18. A drop of water is placed in a stoppered bottle. How many phases are present in the system?
(a) zero (c) two
(b) one (d) three

19. A piece of molten ice is placed in a beaker covered with a water glass. How many phases are present in the system?
(a) zero (c) two
(b) one (d) three

20. The transition temperature of a substance is that temperature at which
(a) one enantiomer changes into another enantiomer
(b) one allotropic form changes to another
(c) all the three phases (solid, liquid and gas) can co-exist in equilibrium
(d) none of these

21. The reduced phase rule for a condensed system is
(a) $F = C - P + 2$ (c) $F' = C - P$
(b) $F' = C - P + 1$ (d) $F = C - P + 3$

22. The phase rule is applicable to
(a) homogenous systems
(b) reversible system
(c) irreversible system
(d) heterogeneous system whether physical or chemical

23. At a triple point,
(a) three phases co-exist in equilibrium
(b) the vapour pressure is equal to the atmospheric pressure
(c) there are three components in equilibrium
(d) there are three degrees of freedom

24. The sulphur system has four phases: rhombic, monoclinic, liquid and vapour sulphur. It is
(a) one-component system
(b) two-component system
(c) three-component system
(d) four-component system

25. Which one of the following is not a heterogenous system?
(a) $CaCO_3(s) \rightleftharpoons CaO(s) + CO_2(g)$
(b) Water vapour $\rightleftharpoons$ Water
(c) $NH_4Cl(s) \rightleftharpoons NH_3(g) + HCl(g)$
(d) $H_2 + I_2 \rightleftharpoons 2HI$

26. For a bivariant system, the degrees of freedom are
(a) one (c) three
(b) two (d) four

27. A one-component system has four phases. Can the four phases co-exist in equilibrium?
(a) no (c) sometimes
(b) yes (d) none of these

ANSWERS

1. (d)	6. (b)	11. (b)	16. (d)	21. (b)	26. (c)
2. (b)	7. (c)	12. (a)	17. (d)	22. (d)	27. (c)
3. (a)	8. (b)	13. (a)	18. (c)	23. (a)	
4. (c)	9. (a)	14. (b)	19. (c)	24. (a)	
5. (b)	10. (c)	15. (d)	20. (b)	25. (d)	

Pharmaceutical Physical Chemistry

Laboratory Manual on Pharmaceutical

Theory Guides

Practical Decides

This part covers 20 Experiments as per syllabus of Pharmaceutical Physical Chemistry

List of Experiments

Experiment 1 503

Find out the refractive index of the given liquid by means of Abbe refractometer.

Experiment 2 508

Find out the specific rotation and molecular rotation of cane sugar polarimetrically and also find the concentration of the unknown solution (concentration lies between 1% and 5%) Calculate the intrinsic rotation for cane sugar.

Experiment 3 512

Find out the order of reaction and rate constant for the inversion of cane sugar by acids, polarimetrically.

Experiment 4 514

Find out the rate constant of the hydrolysis of methyl acetate catalysed by an acid and also to find the half life period of the reaction.

Experiment 5 520

Find out the cell constant of the conductivity cell.

Experiment 6 524

Determine the dissociation constant of acetic acid and verify Ostwald's dilution law.

Experiment 7 528

Find out the strength of hydrochloric acid solution (approximate strength N/10) by titrating it against standard sodium hydroxide solution conductometrically.

Experiment 8 535

Find the surface tension of the given liquid by drop number method, at room temperature.

Experiment 9 540

Find out the partition coefficient of iodine between carbon tetrachloride and water.

Experiment 10 543

Find out the partition coefficient of benzoic acid between water and benzene at room temperature and the molecular state of benzoic acid in benzene as compared to its solution in water.

Experiment 11 **548**

Find the relative and absolute, viscosity of the given liquid at room temperature by Ostwald's viscometer.

Experiment 12 **552**

Determine the pH of a solution by the use of indicators.

Experiment 13 **555**

Determine the pH (varies between 6 and 9) of a given solution by buffer solution method.

Experiment 14 **558**

Determine the solubility of a given salt at room temperature and also to draw its solubility curve.

Experiment 15 **562**

Determine the molecular weight of a given substance by Rast method.

Experiment 16 **566**

Determine the molecular weight of a volatile liquid by Victor Meyer's method.

Experiment 17 **573**

Determination of R_f (retentions factor) value of amino acid by Thin Layer Chromatography (TLC) and identifications of the amino acid present.

Experiment 18 **577**

Separate Cu^{2+} and Cd^{2+} ions in a given mixture using paper chromatography and to determine the R_f values.

Experiment 19 **580**

Determine the heat of neutralisations of a strong acid (HCl) and a strong base (NaOH).

Experiment 20 **584**

Determine the heat of dissolutions of a solid in water at room temperature (Solid electrolyte may be taken, NH_4Cl, KCl, KNO_3 and $BaCl_2$, $2H_2O$).

Refractometry

REFRACTIVE INDEX

When a ray of light passes from a rarer medium (say air) into a denser medium (say a liquid), it is bent or refracted towards the normal. The ratio of the sine of the angle of incidence and that of refraction is constant and characteristic of that liquid (Snell's law). The constant ratio is called the refractive index of the liquid and may be represented as

$$n = \frac{\sin i}{\sin r}$$

According to the wave theory of light, the ratio of the sines of the angles of incidence and refraction is identical with the ratio of the velocities of light in the two media.

Thus, $$n = \frac{\sin i}{\sin r} = \frac{\text{velocity of light in air}}{\text{velocity of light in liquid}}$$

When a ray of light passes from a rarer medium to a denser medium, it can be shown from the law of refraction that

$$\frac{\sin i}{\sin r} = \frac{n_2}{n_1}$$

where n_1 is the index of refraction of the rarer medium and n_2 is the index of refraction of the denser medium. The angle of incident can never be evidently greater than 90° and when it is 90°, the above equation is reduced to

$$\sin r = \frac{n_2}{n_1}$$

or $$\sin 90 = 1$$

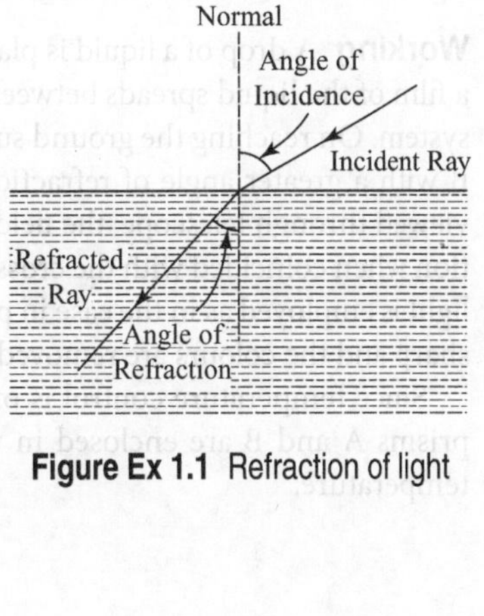

Figure Ex 1.1 Refraction of light

MEASUREMENT OF REFRACTIVE INDEX

For the convenient and rapid determination of the refractive index of liquids, Abbe refractometer is commonly employed for quick determination.

Abbe Refractometer The principle of this instrument is the same as used in the Pulfrich refractometer. It is principally designed for quick determination of the refractive index of small quantities of liquids.

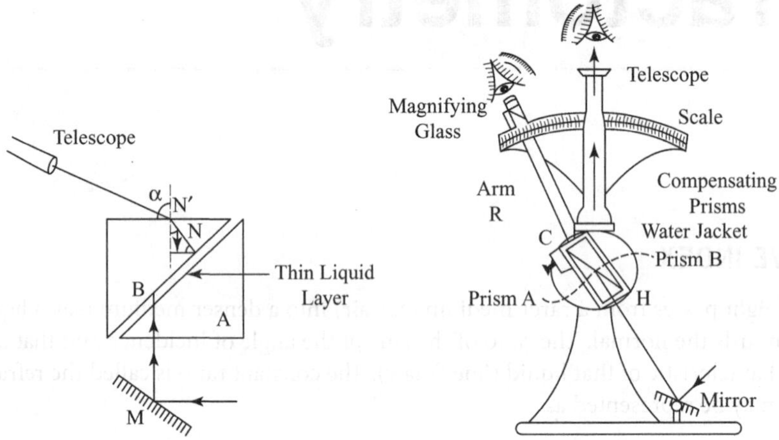

Figure Ex 1.2 (a) Optical system of Abbe Refractometer shows the course of light ray (b) Abbe Refractometer

Construction A general idea of the construction of this instrument is given in Fig. Ex 1.2(a) & (b). A and B are flint glass prisms, whose inclined surfaces are kept in close contact and a very small quantity of a liquid (just one drop is dropped) in between these surfaces. The hypotenuse surface of B is polished while that of A is finely ground. The two prisms are housed in metal casings hinged at H. The faces of two prisms can be held in contact with clamp C; A and B can be rotated about a horizontal axis immediately beneath a telescope (T). To the metal case carrying the prism, is attache d arc arm (R), which moves along a graduated scale S, the reading on which gives directly the refractive index.

Working A drop of a liquid is placed upon the surface of prism A. On clamping the two prisms A and B, a film of the liquid spreads between them. Light reflected by mirror M is then directed towards the prism system. On reaching the ground surface of A, it is scattered into the liquid film. No ray can, however, enter B with a greater angle of refraction than that of ray corresponding to 'grazing incidence'. Hence, when viewed through telescope, the field of view divided into bright and dark portions. The edge of bright portion when coincided with the cross wire of the telescope gives the refractive index on the scale. If the white light is employed as in the case in practice, a diffused coloured border is seen in the telescope. This is made sharp and the colours are removed by the adjustments of two prisms attached at the nose of the telescope.

Since temperature control is of great importance in determining the refractive index of liquids, the prisms A and B are enclosed in water enclosed in a water jacket (J) in order to maintain a constant temperature.

Experiment 1

OBJECT

To find out the refractive index of the given liquid by means of Abbe refractometer.

APPARATUS

Abbe refractometer, ordinary light lamp, thermometer, etc.

THEORY

Explained as above.

PROCEDURE

The method consists of determining refractive index of the liquid by means of Abbe refractometer as explained above.

OBSERVATIONS

Room temperature $= t$ °C.
Refractive index of the liquid

(a) ....
(b) ....
(c) ....

Mean value = ...

RESULT

The refractive index of the liquid is ....

PRECAUTIONS

(i) The polished surface of the prisms of Abbe refractometer should not be scratched.
(ii) Only 2–3 drops of the liquid are sufficient.

VIVA-VOCE

Q 1. Define refractive index.
Ans. Refractive index of a medium is defined as the ratio of velocity of light in vacuum to that in the medium.

Q 2. List the factors on which refractive index of a medium depends.
Ans. The refractive index of a medium depends upon the temperature at which it is determined and the wave length of light used during its determinations.

Q 3. Under what condition the standard refractive index is measured?
Ans. The D-line of sodium at 20 °C is used for the standard measurements of refractive index.

Q 4. Give the units of refractive index, if any.
Ans. There are no units associated with refractive index. It is a ratio.

Q 5. State Snell's law of refractions. How is it related to refractive index?
Ans. The refractive index of a liquid is given by Snell's law,

According to it,
$$n = \frac{\sin i}{\sin r}$$

where i is the angle of incidence and r is the angle of refraction.

Q 6. What do you mean by the term specific refraction?
Ans. The term specific refraction (R_s) is expressed as: $R_s = \frac{n^2 - 1}{n^2 + 2} \times \frac{1}{d}$

where d is the density and n is the refractive index of the liquid.

Q 7. Define the term molar refraction.
Ans. It is defined as the product of specific refraction and molecular mass. It is given as $R_M = \frac{n^2 - 1}{n^2 + 2} \times \frac{M}{d}$

Q 8. How does molar refraction is related to chemical constitution?
Ans. The value of molar refraction is characteristic of a substance and is temperature-independent. It is an additive as well as constitutive property and can be used in deciding the structure of liquid substances.

Q 9. What is a refractometer? On which principle it is based?
Ans. The instrument that is used for the measurement of refractive index of liquids is called a refractometer. It is based upon the measurement of the position of the critical ray, i.e., the critical angle by a suitable device is called refractometer. The principle of Abbes refractometer can be understood

from Fig. Ex 1.2(a). When a beam of light passing through a lower prism A, the surface of the prism being ground serves as a diffusing screen providing rays in every directions. The experimental liquid placed between the two prisms allows critical phenomenon because the refractive index of the prism A is smaller than that of the liquid. A particular ray going along the grazing incidence i.e. be at an angle only slightly less than 90° will pass through the prism B at the critical angle (α). According to critical angle phenomenoon, sum $\alpha = \frac{n}{N}$ where N is the refractive index of prism B and n that of the liquid. The critical ray coming out at an angle (α) is observed through the telescope T. The refractive and the liquid and the angle (α) are related by the equation:

$$n = \sin \text{angle ABC} \sqrt{(N^2 - \sin^2 \alpha)} - \cos \text{angle BAC} \sin \alpha$$

All rays entering the liquid at angles less than the grazing incidence will pass through the prism B at angle less than the critical angle. All other rays will suffer total internal reflection and so they will not pass through prism B at all. So, the band of light viewed through the telescope finishes sharply at the point where the critical angle out of upper face of prims B. Then, the angle (α) can be accurately measured.

Q 10. What is critical angle?

Ans. When the angle of incidence is made so large that it approximately equals to 90°, called grazing incidence, the angle of refraction achieves its maximum possible value. This angle is called the critical angle of refraction.

Q 11. How refractive index is used in qualitative analysis?

Ans. Refractive index is a characteristic property of a liquid just like its boiling point and so it is used in determining the identity and purity of a liquid.

Q 12. Give two examples illustrating the use of refractive index measurement in quantitative analysis.

Ans. The immersion refractometer is useful in determining the concentrations of aqueous and alcoholic solutions. Refractometers can be used for the analysis of binary mixtures, e.g., glycerine-water mixtures.

Polarimetry

A beam of light vibrates in all the planes perpendicular to its direction of propagation. When this light passes through a **Nicol prism** at the proper angle to the optic axis, only the plane-polarized light is transmitted. Only the plane-polarized light is being transmitted through all the angles to the optic axis. All the vibrations of light after polarization take place only in one plane which is perpendicular to the plane of polarization. The Nicol prism serves as the **polarizer.** If this emergent beam is allowed to fall upon another Nicol prism, the beam will pass through if it has parallel orientation and will be completely stopped if the second Nicol prism is at 90° to the first one. The second Nicol prism serves as **analyzer.** Two Nicol prisms, serving as a polarizer and as an analyser, along with a light source, a sample tube and an eyepiece form the basic parts of the **polarimeter.**

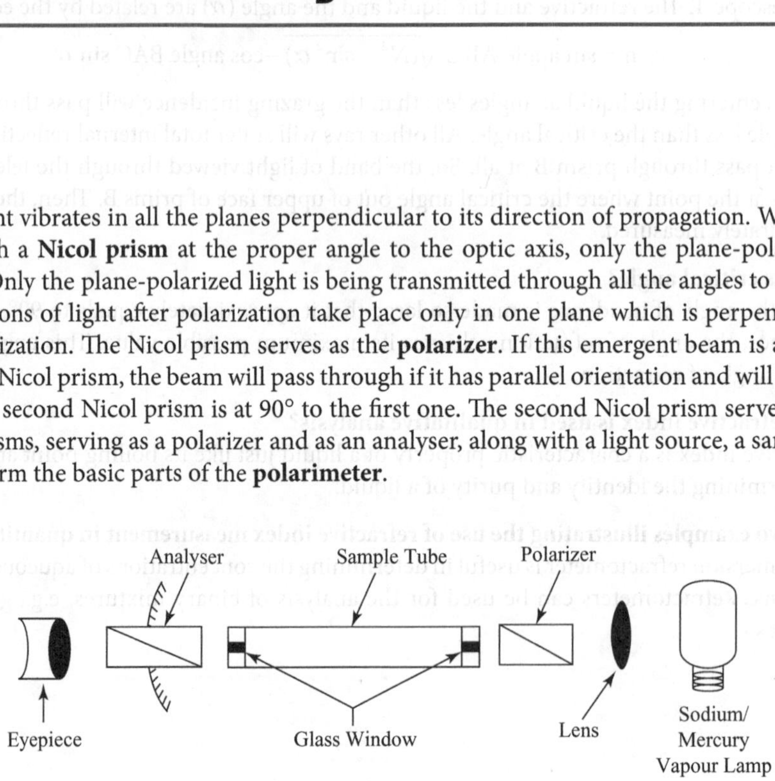

Figure Ex 2.1 The basic assembly of a polarimeter

OPTICAL ACTIVITY

Certain substances such as carbohydrates, proteins and steroids in solution produce a change in the direction of vibration of the linearly **polarized light**. Generally, optically active substances such as polypeptides, proteins and nucleic acids whose structures do not coincides with their mirror images, having helical structures and sugars containing asymmetric carbon atoms produce optical activity. A polarimeter measures the angle of rotation of linearly plane-polarized light when it passes through optically anisotropic liquids or solutions containing the above substances. Such substances are classified as *dextro-* or *laevo*-rotatory substance depending upon their power of rotation of the plane-polarized light in a clockwise or anticlockwise direction, respectively.

SPECIFIC ROTATION

Optical rotations are measured with the sodium D-yellow line (a doublet: 589, 589.6 nm) and for more precise work, with green mercury line (546.1). The specific rotation $[\alpha]^{t}_{\lambda}$ of a substance in solution depending on a particular wavelength of light λ, and a temperature of t °C, is given by

$$[\alpha]^{t}_{\lambda} = \frac{100R}{LC} \tag{2.1}$$

where R is the rotation of the plane-polarized light expressed in degrees, L is the path length in decimetres (length of sample tube) and C is the concentration of solute in grams per 100 cm^3 of solution. The molecular rotation of any substance in solution is expressed as

$$[M]^{t}_{\lambda} = \frac{MR}{LC} \tag{2.2}$$

where M is the molecular mass of the substance.

In particular, the angle of rotation R is measured. The sodium lamp, serving as the monochromatic radiation source, is switched on at least an hour before the measurement. The polarimeter is provided with a 0–360° graduated main scale and a vernier scale. Before taking the measurements with the sample solution, the zero correction is usually done with the pure solvent. The cross wire provided in the scale should coincide with the 0° mark of the scale. If it does not coincide, the corresponding positive or negative correction is made. The main scale reads the full numbers and the vernier scale reads the decimal numbers.

Experiment 2

OBJECT

To find out the specific rotation and molecular rotation of cane sugar polarimetrically.

APPARATUS

Polarimeter, thermometer, beakers, measuring flasks, sodium lamps.

THEORY

The specific rotation of cane sugar is obtained from the Eq. (2.1). The cane sugar concentration is obtained by interpolating its optical rotation in the calibration curve drawn between rotation and concentration with a series of solutions of known concentrations. The molecular rotation is obtained from Eq. (2.2), where M represents the molecular mass of cane sugar (342).

PROCEDURE

Switch on the sodium lamp half an hour before starting the experiment. Now, prepare 20% (w/v) of cane sugar solution in water by taking accurately 50 g of cane sugar in a 250 cm^3 volumetric flask. Prepare 15%, 10%, 7.5%, 5% and 2.5% (w/v) cane sugar solutions (25 cm^3 each) in standard volumetric flasks by dilution of the stock solution. The solution should be absolutely clear. Take the zero reading by filling the tube with water. No air bubbles should be left in the path of the polarized light. Clean the glass plates at both ends of the sample tube. Note down the reading. Now, fill the tube with 20% cane sugar solution avoiding air bubbles and then place it in the polarimeter. Rotate the knob of the main scale clockwise or anticlockwise until equal illumination on both sides of the cross wire is obtained. Now, take the reading by adjusting the vernier scale. Repeat the above procedure for other cane sugar solutions and the unknown cane sugar solution supplied. Care should be taken to rinse the polarimeter tube with the respective solutions each time.

OBSERVATIONS AND RESULTS

The results of the observation may be tabulated as follows:

Specific rotation of cane sugar in solutions

Temperature = ... K

Length of sample tube (l) = ...dm

Zero reading with distilled water* = ...°

Sr. No.	Concentration of Cane Sugar Solution (c) % w/v	Rotation Reading (r_1) Before Correction	Actual Rotation (R) = $(r_1 - r)°$
1.	1 %		
2.	2 %		
3.	3 %		
4.	4 %		
5.	5 %		
6.	Unknown solution		

On plotting R vs. c, a straight line passing through the origin will be obtained. The concentration of the unknown sugar solution may be obtained by interpolating its optical rotation in the above calibration curve and recorded.

Concentration of the unknown cane sugar solution = ...% w/v.

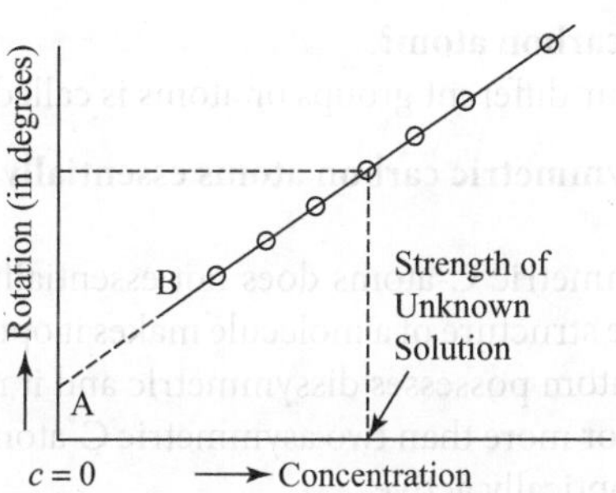

Figure Ex 2.2 Graphical determination of concentration

VIVA-VOCE

Q 1. What is a polarimeter?

Ans. It is an instrument used to measure the change in the rotation of a plane polarized light made by an optically active substance.

Q 2. What do you mean by plane polarized light?

Ans. An ordinary ray of light is considered to be vibrating in all the planes at right angle to its direction of propagation. However, plane polarized light has vibrations only in one plane at 90° to its direction of propagation.

Q 3. What is optical activity?

Ans. The property of the rotation of plane polarized light, i.e., unidirectional light through a certain angle is known as optical activity and the substances possessing this property are called optically active substances.

Q 4. Define the two forms of optically active substances.

Ans. The substances rotating the plane polarized light towards the right (clockwise) are called dextrorotatory (+) form and which rotate it towards the left (anti-clockwise) are called laevorotatory (–) form.

Q 5. State the factors controlling the magnitude of optical rotation.

Ans. The magnitude of rotation depends upon the following factors: (i) nature of the substance, (ii) length of the liquid column through which light passes, (iii) concentration of the solution, (iv) nature of the solvent, (v) temperature of the solution and (vi) wavelength of the liquid used.

Q 6. What do you mean by specific rotation? How is it expressed.

Ans. Specific rotation is defined as the angle of rotation of polarized light by a liquid, which in a volume of 1 ml contains 1 gram of the optically active substance when the length of the column through which the light passes is one decimetre. It is expressed as $[\alpha]_D^t = \frac{\alpha V}{lm} = \frac{\alpha V}{l\rho}$ where ρ is the density of the substance.

Q 7. What is the cause of optical activity?

Ans. Optical activity is caused due to the presence of asymmetry in the molecule. An asymmetric molecule must satisfy the following two conditions: (i) The object and its mirror image are not superimposable. (ii) It cannot be divided into two identical parts by a single plane.

Q 8. What is an asymmetric carbon atom?

Ans. A carbon atom having four different groups or atoms is called as asymmetric carbon atom.

Q 9. Does the presence of asymmetric carbon atoms essentially by making an organic compound optically active?

Ans. No, the presence of asymmetric C-atoms does not essentially make a molecule optically active. Overall dissymetric in the structure of a molecule makes it optically active. A molecule containing only one asymmetric C-atom possesses dissymmetric and it must be optically active. However, a molecule containing two or more than two asymmetric C-atoms may or may not be dissymmetric, so it may or may not be optically active.

Q 10. Give an example of optically active organic compound having one asymmetric carbon atom.

Ans. Lactic acid (α-hydroxy propionic) is an example of optically active compound having one asymmetric C-atom in it. Its structure is

$$H_3C-\overset{\displaystyle H}{\underset{\displaystyle OH}{\overset{|}{\underset{|}{C}}}}-COOH$$

and it exits as D (+)-lactic acid and L (–)-lactic acid.

Q 11. What is a racemic mixture?

Ans. It is a mixture of both the forms of an optically active compound in molecular proportions. It is, therefore, optically inactive form.

Q 12. Give examples of optically active organic compounds containing more than one asymmetric C-atom.

Ans. Tartaric acid having the structure CHOH — COOH | CHOH — COOH possesses two asymmetric C-atoms as shown below:

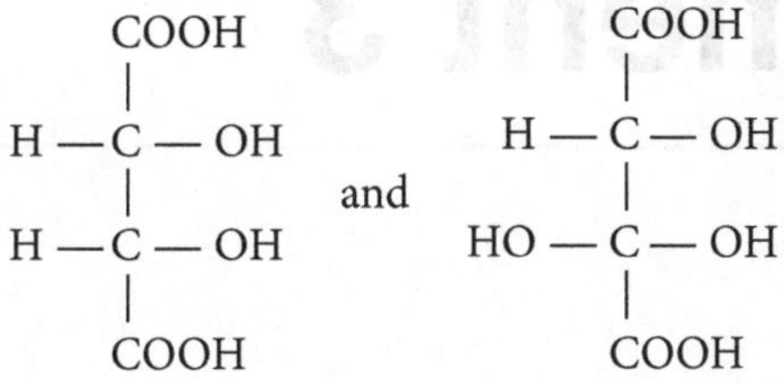

Glucose is another example of optically active compound having four asymmetric C-atoms

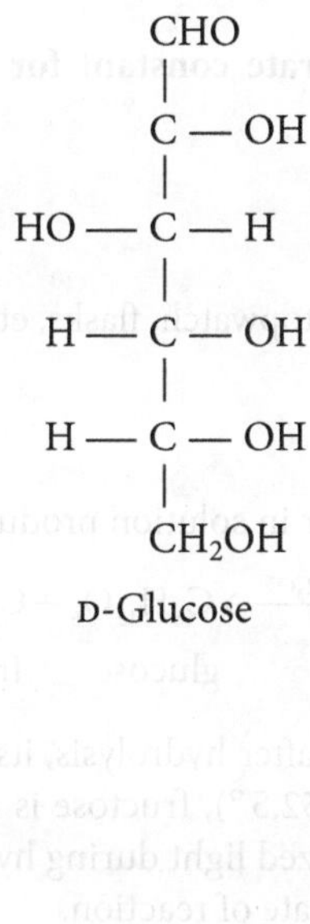

D-Glucose

Experiment 3

OBJECT

To find out the order of reaction and rate constant for the conversion of cane sugar by acids, polarimetrically.

APPARATUS

Polarimeter, sodium lamp, thermometer, stopwatch, flasks, etc.

THEORY

The acid-catalyzed hydrolysis of cane sugar in solution produces glucose and fructose.

$$\underset{\text{Cane sugar}}{C_{12}H_{22}O_{11}} \xrightarrow{H_3O^+} \underset{\text{glucose}}{C_6H_{12}O_6} + \underset{\text{fructose}}{C_6H_{12}O_6}$$

Cane sugar is *dextro*-rotator; however, after hydrolysis, its solution becomes *laevo*-rotatory, because although cane sugar is *dextro*-rotatory (+52.5 °), fructose is more *laevo*-rotatory (–92.0 °). The change in the angle of rotation of the plane-polarized light during hydrolysis of cane sugar can be conveniently followed by a polarimeter to measure the rate of reaction.

If R_0, R_t and R_∞ are the rotations at the beginning, at time t and at the completion of the reaction respectively $(R_0 - R_\infty)$ will be proportional to the initial concentration of cane sugar and $(R_t - R_\infty)$ will be proportional to the concentration of unhydrolyzed cane sugar at time t. Thus the rate expression will be

$$K = \frac{2.303}{t}\log\frac{a}{a-x} = \frac{2.303}{t}\log\frac{R_0 - R_\infty}{R_t - R_\infty}$$

PROCEDURE

Switch on the sodium lamp. Prepare 50 cm³ each of 0.1 N hydrochloric acid solution and 40% (w/v) cane sugar solution. Take the 'zero reading' with water in the polarimeter tube after half an hour of switching on the lamp. Mix the two solutions in a conical flask and shake them well. Start the stopwatch immediately after mixing. Load the polarimeter tube with the solution mixture avoiding air bubbles and clean its glass plates at both the ends.

Take polarimetric readings initially at intervals of 5 min and subsequently at intervals of 10 min. The angle of rotation of the solution is equal to the actual polarimetric reading minus the 'zero reading' taken with water. Take 8–10 readings in total. Warm the solution mixture contained in the stoppered conical flask for about one hour at 60 °C to complete the reaction. Measure angle of rotation to get the value of R_∞. The initial reading of R_0 may be obtained by mixing the sugar solution (40% w/v) with equal volume of distilled water, taking the polarimetric reading and making the necessary 'zero correction'.

Observations and Results The tabulation of experimental readings may be done in the following manner.

Temperature = ... K
Zero reading = ...°

Sr. No.	Time (s)	Polarimetric Reading (I1)	Angle of Rotation (*R*)	log $\frac{R_0 - R_\infty}{R_t - R_\infty}$
1.	0			
2.	5			
3.	10			
4.	20			
5.	30			
6.	40			
7.	50			
8.	∞			

CALCULATION

The value of initial concentration $(a) = R_0 - R_\infty$ and the amount of cane sugar hydrolyzed $(x) = R_0 - R_t$
Therefore, the value $(a - x) = R_0 - R_\infty - R_0 + R_t = R_t - R_\infty$.

Alternatively, log $R_0 - R_\infty$ may be plotted against t on a graph paper to get a straight line passing through the origin. The slope $R_t - R_\infty$ of the straight line gives the value of k.

RESULT

The value of rate constant (k) = ...

Since the value of rate constant (k) determined at different intervals of time comes out to be nearly constant by applying the equation of the first-order reaction, the order of the reaction under study is 1.

VIVA-VOCE

Please refer to Experiment 2 for VIVA-VOCE

Experiment 4

OBJECT

To determine the value of rate constant (K) of the hydrolysis of ethyl acetate catalyzed by hydrochloric acid.

THEORY

The hydrolysis of ethyl acetate catalyzed by dil. HCl is a reaction of first order, as the concentration of water is in large excess and does not change significantly during the progress of the reaction. It is thus a pseudo-unimolecular reaction.

$$\underset{\text{Ethyl acetate}}{CH_3-\underset{\underset{O}{\|}}{C}-OC_2H_5} + H_2O \xrightarrow{\text{dil. HCl}} \underset{\text{Acetic acid}}{CH_3-\underset{\underset{O}{\|}}{C}-OH} + \underset{\text{Ethyl alcohol}}{C_2H_5OH}$$

The value of velocity constant is calculated from the equation of first-order reaction.

$$K = \frac{2.303}{t} \log_{10} \frac{a}{a-x}$$

where a = initial concentration and $(a-x)$ = concentration after time t.

The values of a, $(a-x)$ and 't' are taken from the experimentally determined values. In an actual experiment, the solution of known concentration of ethyl acetate and hydrochloric acid are mixed together and kept in a water bath at a room temperature or as such at room temperature. Known volumes of the reaction mixture are withdrawn at intervals of 5–10 minutes, added to pounded ice or chilled water in a conical flask and titrated against standard NaOH solution using phenolphthalein as indicator. The volume of NaOH used corresponds to the total acid *i.e.*, HCl and acetic acid formed by hydrolysis in the reaction mixture. The final (infinite) reading is taken after keeping the remaining reaction mixture in water bath for 24 hrs or keeping it in hot water at 50–60 °C for an hour.

If V_0 = volume of NaOH used at time $t = 0$, V_t = volume of NaOH used after time 't', and V_∞ = volume of NaOH used for a sample of the reaction mixture kept in water bath at 50–60 °C. Then, $a = V_\infty - V_0$ and $(a-x) = V_\infty - V_t$.

$$K = \frac{2.303}{t} \log_{10} \frac{V_\infty - V_0}{V_\infty - V_t}$$

The value of 'K' is determined for the different sets of values corresponding to different intervals of time.

APPARATUS/REAGENTS REQUIRED

Six 50 ml and one 250 ml conical flasks with cork, burette, pipettes (5 ml and 10 ml), glass trough, iron stand, stop watch and NaOH (N/10) solution.

PROCEDURE

(i) Take 100 ml of 0.5 N. HCl in a clean dry 250 ml conical flask, cork it and set it in a water bath or trough containing water at a room temperature. In addition, take about half test tube of ethyl acetate in conical flask and set it into the same water bath or trough. Let the two solutions attain the same temperature of the bath.

(ii) In the mean time, fit the burette properly and fill it with 0.1 N NaOH solution. In addition, add about 15 g of the pounded ice in each of the six 50 ml conical flasks.

(iii) Pipette out 5 ml of the ester from the conical flask and add it to the flask containing 100 ml of 0.5 N HCl.

(iv) Shake the contents, pipette out 10 ml of the reaction mixture and transfer it at once to first conical flask containing pounded ice or chilled water. Titrate this against 0.1 N NaOH solution taken in the burette and adding a drop of phenolphthalein as indicator to the titration flask. The appearance of light pink colour confirms the end point. The volume of 0.1 N NaOH used against the withdrawn sample of the ethyl acetate and dil. HCl mixture is taken as V_0.

(v) In addition, note the time of draining out 10 ml of the mixture solution into the titration flask in the above case and if this is 10 sec, then in the subsequent estimations, the mixture solution should be added to the chilled water or pounded ice 5 sec earlier than the time interval of 10 min.

(vi) Pipette out 10 ml of mixture solution and add it to the conical flask containing pounded ice/chilled water after 10 min as suggested above. Titrate against 0.1 N NaOH. This gives V_t after 10 min.

(vii) Repeat the above procedure after every 10 min so as to get six readings. The time interval should be increased to 15 min after two or three readings if the readings (volume of 0.1 N NaOH used) do not differ much from each other.

(viii) Finally, remove the flask from the water bath or trough and place it in a separate water bath at a temperature of 50–60 °C for about 1 hr. However, in order to save time, this should be done after the first reading by withdrawing about 20 ml of the reaction mixture in a separate 50 ml conical flask and keeping it in water bath at 50–60 °C for about 1 hr. Pipette out 10 ml of the reaction-mixture and titrate against 0.1 N NaOH. This gives V_∞.

OBSERVATIONS

Temperature of the water bath = …. t°C.

S. No.	Time (minutes)	Burette Readings Initial	Final	Vol. of NaOH Solution Used	(Say)
1.	0				V_0
2.	10				V_{10}
3.	20				V_{20}
4.	30				V_{30}
5.	40				V_{40}

CALCULATIONS

As
$$K = \frac{2.303}{t} \log \frac{a}{a-x} = \frac{2.303}{t} \log \frac{V_\infty - V_0}{V_\infty - V_t} \text{ minute}^{-1}$$

When time = 10 min,
$$K = \frac{2.303}{10} \log \frac{V_\infty - V_0}{V_\infty - V_{10}} \text{ minute}^{-1}$$

Similarly, calculate the value of rate constant (K) at other intervals of time. The mean value of K obtained from the different sets of the readings gives the value of the rate constant, K.

RESULT

Since the values of K are constant, the hydrolysis of ethyl acetate catalyzed by acid is a first-order reaction.

PRECAUTIONS

(1) During the progress of the hydrolysis of ester, the conical flask should be kept stoppered after pipetting out the solution of the reaction mixture.
(2) Sufficient amount of pounded ice or chilled water should be taken in the flask.
(3) As the rate of reaction is very sensitive to temperature changes, the temperature of the bath should be maintained constant within ±0.5 °C throughout the course of the experiment.

GRAPHICAL METHOD

The rate equation for the first-order reaction

$$t = \frac{2.303}{K} \log \frac{a}{a-x}$$

$$= \frac{2.303}{K} \log a - \frac{2.303}{K} \log(a-x)$$

$$= -\frac{2.303}{K} \log(a-x) + \frac{2.303}{K} \log a,$$

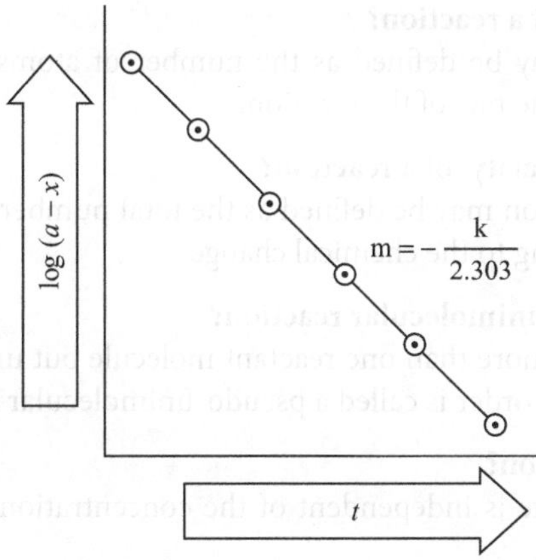

Figure Ex 4.1 Curve between *t* and log (*a* – *x*)

which indicates that a plot of *t* against log $(a - x)$.

or log $(V_\infty - V_t)$ will be a straight line whose slope is given by $-\dfrac{2.303}{K}$.

Thus, $$-\frac{2.303}{K} = \tan, = \frac{OA}{OB} \quad \text{or} \quad K = 2.303 \times \frac{OB}{OA}$$

VIVA-VOCE

Q 1. What is chemical kinetics?
Ans. It is a branch of chemistry, which is concerned with the study of rates of reaction and the mechanism by which the reactions change into products.

Q 2. What is meant by 'rate of a reaction'?
Ans. It is defined as the rate of change of concentration of the reactant or the product per unit time.

Q 3. What are the units of the rate of a chemical reaction?
Ans. If the concentration is expressed in moles per litre and time in seconds, the units of the rate of chemical reaction will be mole l^{-1} s^{-1}.

Q 4. Name the factors which influence the rate of a chemical reaction.
Ans. (i) Concentration of the reactants (ii) Nature of the reactants
(iii) Surface area of the reactants (iv) Temperature
(v) Catalyst (vi) Radiations.

Q 5. What is the effect of concentration on the rate of a reaction?
Ans. The rate of a reaction generally increases with the increase in the concentration of the reactants.

Q 6. What is the effect of temperature on the rate of a reaction?
Ans. In general, the rate of a reaction increases with the increases in temperature.

Q 7. What is meant by 'order' of a reaction?

Ans. The order of a reaction may be defined as the number of atoms, molecules, and ions whose concentrations determine the rate of the reaction.

Q 8. What is meant by 'molecularity' of a reaction?

Ans. The molecularity of a reaction may be defined as the total number of atoms of molecules which participate in the step leading to the chemical change.

Q 9. What is meant by pseudo-unimolecular reaction?

Ans. A reaction which involves more than one reactant molecule but under given conditions follows the rate equation of the first order is called a pseudo-unimolecular reaction.

Q 10. What is a zero-order reaction?

Ans. A reaction in which the rate is independent of the concentrations of the reactants involved is called a zero-order reaction.

Q 11. What do you mean by a second-order reaction?

Ans. A reaction whose rate depends upon the second power of the concentration of the reactant.

Q 12. What is the order of a reaction whose rate is determined by the variation of one concentration term only?

Ans. First-order.

Q 13. Is the order of a reaction always a whole number?

Ans. No, zero and fractional values of order are also possible.

Q 14. Are the order and molecularity of a reaction always the same?

Ans. No, but the two may be identical in case of simple reactions.

Q 15. Can the order of a reaction be deduced from its chemical equation?

Ans. No, it is always to be determined experimentally.

Q 16. Cite any one example of zero-order reactions.

Ans. Photochemical combination between hydrogen and chlorine on exposure of UV light; decomposition of ammonia on platinum surface at 800 °C.

Q 17. What are the units of rate constant of (i) zero-order, (ii) first-order, (iii) second-order, (iv) third-order, and (v) half-order reactions?

Ans. (i) L mol^{-1} s^{-1}, (ii) s^{-1}, (iii) mol L^{-1} s^{-1}, (iv) mol^2 L^{-2} s^{-1}, (v) $mol^{1/2}$ $L^{-1/2}$ s^{-1}

Q 18. How does half-life period of a first-order reaction vary with temperature?

Ans. Half-life *decreases* with the *rise* of temperature, because $t_{1/2}$ α $1/k$ and k increases with the rise of temperature.

Q 19. What type of graph is obtained when rate is plotted against time in a: (i) zero-order reaction and (ii) first-order reaction?

Ans. A straight line: (i) parallel to concentration axis and (ii) inclined to both rate and concentration axis.

Q 20. How is rate constant related to concentration of reactants?

Ans. $K = \dfrac{1}{c^{n-1} \cdot \text{time}}$, when n = order of reaction. ($n \neq 1$)

Q 21. The rate for the reaction: $2NO + O_2 \longrightarrow 2NO_2$ is: $= k[NO][O_2]$. What is the possible mechanism for this reaction?

Ans. One possibility is $NO + O_2 \xrightarrow{\text{Slow}} NO_3; NO_3 + NO \xrightarrow{\text{Fast}} 2NO_2$.

Q 22. Why is a reaction speeded up in the presence of a catalyst?

Ans. A catalyst *lowers* the activation energy; therefore, more molecules in the reactants possess sufficient energy to react.

Q 23. Why is a reaction speeded up with the increase in temperature?

Ans. An increase in temperature causes a greater fraction of the molecules to have an energy at least equal to activation energy.

Q24. Why is a reaction speeded up on increasing the concentrations of reactants?

Ans. An increase in concentration causes a greater number of collisions per second; therefore, a greater number of effective collisions per second.

Q 25. Define 'threshold energy'?

Ans. It may be defined as the minimum amount of energy which the reactant molecules must possess, so that their collisions result in an actual chemical change.

Q 26. What is activation energy?

Ans. The excess energy which the reactant molecules, having energy lower than threshold energy, must attain, so that their collisions may result in an actual chemical change is called activation energy.

Q 27. Why does the rate of a reaction increase with the increase in temperature?

Ans. The increase in temperature causes an increase in the number of molecules moving energy equal to or higher than threshold energy. Consequently, a much larger number of collisions result in an actual chemical change, i.e., the rate of reaction increases.

Q 28. Does activation energy vary with temperature?

Ans. No, but the reaction rate usually increases with the rise of temperature and vice-versa.

Experiment 5

OBJECT

Determination of the cell constant at different concentrations and infer the results obtained.

THEORY

Cell Constant

If R is the resistance of a solution measured by a cell with electrodes of cross-sectional area a cm^2 and l cm apart, the specific conductivity of the solution is given by,

$$K = \frac{1}{R} \times \frac{l}{a} = \frac{x}{R}$$

where $\frac{l}{a} = x$ is the cell constant.

The cell constant, *i.e.*, the ratio l/a, cannot be obtained from the geometrical dimensions of the cell, as both l and a, they are not accurately known. It is, therefore, necessary to calibrate the cell with a solution of known specific conductivity. Usually, 0.1 or 0.01 N KCl solution is used. Suppose K is specific conductivity of 0.1 N or 0.01 N KCl solution and K′ is the conductivity measured by the cell in question, then cell constant is given by,

$$x = \frac{K}{K'} = \frac{\text{Specific conducivity (known)}}{\text{Specific conducivity (observed)}}$$

If R′ is the resistance of a solution, of which the specific conductivity is to be determined, measured by the same cell, then specific conductivity of unknown solution $= \frac{x}{R'}$.

Once the cell constant is determined, great care must be taken not to change the position or shape of the electrodes during subsequent conductivity measurements.

Measurement of Cell Constant

The cell constant is determined by substituting the value of specific conductance of 0.02 N KCl solution at 25 °C as determined by Kohlrausch (0.002768) and the value of conductance as observed with given cell in relationship.

Specific conductance of KCl solution at various temperatures

Temperature (C°)	1.0 N	0.1 N	0.02 N	0.01 N
10	0.0832	0.00932	0.0011995	0.001019
15	0.0925	0.01048	0.002243	0.001147
20	0.1020	0.01167	0.032501	0.001278
25	0.1120	0.01289	0.002768	0.001412
30		0.01412	0.003036	0.001552

$$\text{Cell constant} = \frac{0.002768\ (\text{specific conductance})}{\text{observed conductance}}$$

For temperature other than 25 °C, the specific conductance of KCl solution may be taken from the table at temperature of the experiment.

APPARATUS REQUIRED

Conductometer, conductivity cell, thermostat, copper wires, beakers etc.

MATERIALS REQUIRED

0.02 N KCl solution, acetic acid, doubly distilled water etc.

PROCEDURE

(i) Set up a thermostat at 25 °C. [A water bath with tap water will serve the purpose for work at room temperature.]
(ii) Prepare 250 ml of exactly 0.02 N KCl solution by dissolving 0.3725 gm of KCl in 250 ml of distilled water.
(iii) Rinse the conductivity cell first with double distilled water and then with 0.02 N KCl solution.
(iv) Connect the lead of the conductivity bridge to the main.
(v) Transfer nearly 50 ml of 0.02 N KCl solution in a beaker and dip the conductivity cell in it. See that the electrodes of the cell are dipping into the solution. Support the conductivity cell in the thermostat at 25 °C.
(vi) Wait for about 15 min, so that the conductivity cell will acquire the temperature of thermostat. Connect the terminals marked *ext* or *earth* to a good electrical connection.
(vii) Now, switch on the instrument and wait for about 1 min. Maintain the sensitivity control at an almost-minimal position.
(viii) Connect the conductivity cell to the terminals marked *'cell'*.
(ix) By putting the *'multiplier'* switch in expected range in the conductance units, select the appropriate range by trial-and-error method. It is indicated by the shadow in the magic eye.

(x) Obtain the balance or null point on the dial which is indicated by the maximum width of the shadow on the magic eye. Then, increase the sensitivity of null detection and obtain a more precise balance.

(xi) The conductance of the unknown solution is given by the dial reading multiplied by the factor indicated by the multiplier.

(xii) Repeat the experiment a number times to get several determinations of the conductance of 0.02 N KCl solution. The mean of these values gives the observed conductance of the solution.

DETERMINATION AND CALCULATIONS

Temperature = ...t°C

S. No.	Strength of KCl Solution	Observed Conductance
1.	0.02 N	..... $ohms^{-1}$
2.	0.02 N	..... $ohms^{-1}$
3.	0.02 N	..... $ohms^{-1}$
4.	0.02 N	..... $ohms^{-1}$
5.	0.02 N	..... $ohms^{-1}$

$$\text{Mean value} = \frac{1+2+3+4+5}{5} = ...\text{ohms}^{-1}$$

$$\text{Cell constant} = \frac{\text{specific conductance}}{\text{observed conductance}}\,\text{cm}^{-1}$$

The value of specific conductance of 0.02 N KCl solution at 25 °C determined by Kohlrausch is 0.002768 ohm^{-1} cm^{-1}.

$$\therefore \quad \text{Cell constant } (x) = \frac{0.002761}{\text{observed conductance}}\,\text{cm}^{-1}$$

PRECAUTIONS

(i) As the temperature has a marked influence on the conductivity, all conductivity measurements should be made at controlled temperature.

(ii) The electrodes must be platinized to improve the sharpness of the sound.

(iii) All the solutions must be prepared in conductivity water.

(iv) Once the cell constant is determined, the distance between the electrodes and their shapes must remain uncharged throughout the experiment.

(v) The electrodes must always be washed with that solution which is to be filled in it.

(vi) A separate voltage stabilizer must be used to save the instrument from damage of voltage fluctuations of the mains.

VIVA-VOCE

Q 1. What are conductors?

Ans. The substances which allow the passage of electric current flow through them are called conductors.

Q 2. What are electrolytes?

Ans. The substances which allow the current to pass through them and in which chemical changes take place during the passage of current are called electrolytes, e.g., aq. NaCl, HCl etc.

Q 3. What is tile difference between metallic conductance and electrolytic conductance?

Ans. The conduction which occurs without producing a chemical change during the passage of electricity is called metallic conduction; however, the conductance which occurs with chemical change during the passage of current is called electrolytic conduction.

Q 4. What is Ohm's law?

Ans. According to Ohm's law, the strength of current (I) flowing through a conductor is directly proportional to the potential difference (E) applied across the conductor and is inversely proportional to the resistance (R) of the conductor.

Mathematically, $$I = \frac{E}{R}$$

Q 5. What is conductance? What are its units?

Ans. The conductance implies the ease with which the current flows through a conductor. It is defined as the reciprocal of resistance. It is represented by C.

$$C = \frac{1}{R}$$

Its units are ohm^{-1} or mho.

Q 6. What is specific conductance? What are its units?

Ans. Specific conductance is defined as the conductance of 1 cm^3 (1 cc) of the solution. The unit of specific conductance is ohm^{-1} cm^{-1}.

Q 7. What is equivalent conductance?

Ans. The product of specific conductance and volume V in millilitre of the solution containing 1 g equivalent of the electrolyte.

Q 8. What are the units of equivalent conductance?

Ans. ohm^{-1} cm^2 (g Eq.)$^{-1}$.

Q 9. What is molecular conductance?

Ans. The product of specific conductance and the volume V in millilitre of the solution containing 1 g mole of the electrolyte.

Q 10. What is the effect of dilution on conductance?

Ans. Conductance increases with dilution because the number of ions increases on dilution. Specific conductance decreases with dilution because the number of ions/cc. decreases on dilution. Equivalent conductance or molecular conductance increases with dilution.

Q 11. What is cell constant and what are its units?

Ans. $$\text{Cell constant} = \frac{\text{specific conductance}}{\text{observed conductance}}$$

The unit of cell constant is cm^{-1}.

Experiment 6

OBJECT

To determine the dissociation constant of acetic acid and verify Ostwald's dilution law.

APPARATUS AND REAGENTS

Conductometer, burette, pipette, beaker, volumetric flask (100 ml), acetic acid, conductivity water.

THEORY

Suppose a weak electrolyte AB dissociates in *V* litre of solution with initial concentration of one mole per litre. At equilibrium, its concentration will be as under:

Initial state:	1 mole				
	AB	$\rightleftharpoons$	A^+	+	B^-
Equilibrium state:	$(1-\alpha)$ mole		α mole		α mole

where *a* is degree of dissociation.
On applying the law of mass action, we get

$$K = \frac{[A^+][B^-]}{[AB]}$$

where K is the dissociation constant of the substance.
Putting the value of amount of $[A^+]$, $[B^-]$ and [AB] in the above equation, we get

$$K = \frac{a/v \times a/v}{1 - a/v}$$

$$K = \frac{a^2}{(1-a)}$$

Since concentration is inversely proportional to volume i.e. $c \propto 1/v$

Therefore, $$K = \frac{a^2c}{(1-a)}$$

and in case of weak electrolyte, $(1 - a)$ may be taken as 1. Since a is very small as compared to 1.

Therefore, $$K = a^2c, a = \sqrt{K/c}$$

Alternatively, $$a = \sqrt{Kv} \text{ or } a = \sqrt{v}$$

Hence, the degree of dissociation is directly proportional to square root of volume or inversely proportional to that of concentration. The value of a can also be calculated from conductivity

$$a = \lambda/\lambda_\infty$$

where λ = equivalent conductivity at particular concentration and λ_∞ = equivalent conductivity at infinite dilution. The value of λ_∞, according to Kohlrausch's law, is given by,

$$\lambda_\infty = \lambda a + \lambda_c$$

where λa and λ_c are the ionic conductance of anion and cation of the substance. These values are used for an ion and can be seen from the tables.

To verify Oswald dilution, we find out the volume of K at different dilution of the weak electrolyte. If the values come out to be constant, then the law is verified, otherwise not.

PROCEDURE

(i) Prepare exactly N/10 solution of acetic acid from 5N standard CH_3COOH solution. Take 50 ml of this solution in a beaker and note its conductance.

(ii) Now, take out 25 ml of acid solution with the help of pipette and add 25 ml of conductivity water to make the solution 50 ml again. Now, the strength of the solution is N/20. Mix the solution and note corresponding conductance.

Again, withdraw 25 ml of N/20 acid and add 25 ml water to make the strength of solution (N/40). Note the conductance again.

(iii) Proceeding in same manner and determine the conductance of N/80, N/160, N/320 and so on.

OBSERVATION

Cell constant = ………… Cm^{-1}

$a_\infty = \lambda_H + \lambda_{CH_3COO^-} = 350 + 40.8 = 390.8$

S.No.	Strength of CH_3COOH	Conductance (c) mhos	Specific Cond. K = cell const. × c	Equivalent Cond. $\lambda = K/c$	$a = \lambda/\lambda_\infty$	$K = a^2/v$ or a^2c
1.	N/10					
2.	N/20					
3.	N/40					
4.	N/80					
5.	N/160					
6.	N/320					
7.	N/640					

CALCULATION

The degree of ionization (a) of acetic acid at any dilution can be obtained from the relation

$$a = \frac{\lambda v}{\lambda \infty} = \frac{\lambda v}{\lambda a + \lambda c}$$

RESULT

The dissociation constant of acid at 25 °C is ....

Since the values of K (the dissociation constant) of acetic acid at different dilutions come out to be constant, the law is verified.

PRECAUTIONS

(i) Sufficient time should be given to solutions to attain the equilibrium temperature.
(ii) Solution should be prepared in conductivity water.
(iii) Conductivity cell should be completely dipped in solution.

VIVA-VOCE

Q 1. What are weak acids? Give examples.

Ans. Weak acids are substances which ionize partially to give H^+ ions in solutions, e.g. HCOOH, H_3C-COOH, HCN etc.

Q 2. What are strong acids? Give examples.

Ans. Strong acids are substances which ionize fully to give H^+ ions in solutions, e.g. HCl, HNO_3, H_2SO_4 etc.

Q 3. How can we compare the strength of weak acids?

Ans. The strength of weak acids is compared by comparing the value of their dissociation constants. The higher the value of dissociation constant of an acid, the stronger it would be comparatively.

Q 4. What do you understand by dissociation constant of a weak acid?

Ans. Let HA is considered to be a weak acid; its dissociation constant can be expressed as

$$HA \rightleftharpoons H^+ + A^-$$

$$\frac{[H^+][A^-]}{[HA]} = K_A,$$

where K_A is called of the weak acid. The greater the value of K_A, the stronger would be the acid. The strength of an acid depends upon the H^+ ions that it gives in solution and the values of K_A is directly related the amount of $[H^+]$ ions present in the solution.

Q 5. What is degree of ionization? How is it related to the strength of an acid?

Ans. The extent to which an acid ionizes in solution is referred to as its degree of ionizations. In other words, it is the fraction of the total ionization of the acid.

$$\text{Degree of ionization } (\infty) = \frac{\text{Number of ions available at a particular instant}}{\text{Total number of ions on complete ionization}}$$

Q 6. How can the value of degree of ionization be evaluated?

Ans. The degree of ionization (∞) of a weak acid at any dilution can be calculated as

$$\alpha = \frac{\lambda_v}{\lambda_\infty} = \frac{\lambda_v}{\lambda_a + \lambda_c}$$

Thus, α can be evaluated by measuring λ_v (the equivalent conductance at dilution V experimentally and λ_α can be known with the help of Kohlrausch Law ($\lambda_\alpha = \lambda_a + \lambda_c$), where λ_a and λ_c are the ionic conductances of anions and cations, respectively.

Q 7. State Kohlranusch law.

Ans. The law states that equivalent conductance of an electrolyte at infinite dilution or zero concentration is the sum of two values, one depending upon the anion and the other that of the cation.

Mathematically, $\lambda_\alpha = \lambda_a + \lambda_c$

Q 8. State Ostwald dilution law.

Ans. According to Ostwald's dilution law:

$$K = \frac{\alpha^2}{(1-\alpha)V}$$

For weak electrolyte, the value of α can be neglected in comparison to unity. Therefore,

$$K = \frac{\alpha^2}{V} \text{ or } \alpha = \sqrt{KV} = K\sqrt{V}$$

It means that the degree of dissociation is directly proportional to dilution.

Q 9. How do you verify Ostwald dilution formula?

Ans. The value of dissociation constant K of the weak acid (say CH_3COOH) can be calculated by applying Ostwald dilution formula provided the degree of dissociation of weak acid at particular dilution is known. The value of degree of dissociation (α) can be calculated from conductivity measurements by applying the formula, $\alpha = \frac{\lambda_\alpha}{\lambda_c + \lambda_a}$. If the value of dissociation constant (K) calculated at different dilutions comes out to constant, the law is verified, otherwise not.

Q 10. Can Ostwald dilution formula be applied to strong acids or not? Explain your answer.

Ans. Ostwald dilution formula gives satisfactory results only with weak electrolytes, whereas it does not hold good for strong electrolytes. The formula is based upon Arrhenius theory, which fails completely when applied to strong electrolytes.

Experiment 7

OBJECT

To find the strength of hydrochloric acid solution (approximately N/10) by titrating it against sodium hydroxide solution conductometrically.

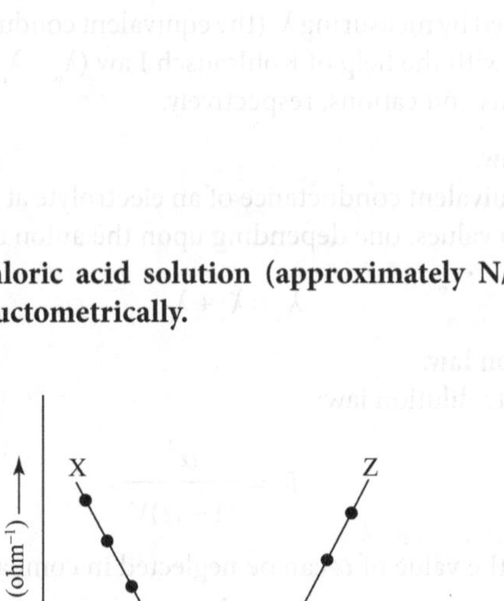

Figure Ex 7.1 Conductance curve of strong acid vs. strong base

THEORY

Conductometry can be used to detect the equivalence point (end point) of a titration. This method is based upon the measurement of conductance during the course of a titration. The Conductance varies differently before and after the equivalence point. This is due to the reason that electrical conductance of a solution depends upon the number of ions present and their ionic mobilities, i.e., speeds.

When conductance values are plotted against volume of titrant added, two straight lines are obtained; the point of intersection of the lines gives the end point. For studying HCl vs. NaOH titration, a known volume of HCl is taken in a beaker and NaOH solution in the burette. The conductance of acid solution is noted initially as well as after successive additions of small amounts of NaOH solution. Conductance of acid solution in the beginning is very high due to the presence of highly mobile H^+ ions. On adding

NaOH solution, the H^+ ions present in solution combine with OH^- ions released from NaOH to form water. Thus, the H^+ ions replaced by slow-moving Na^+ ions, decreasing the conductance of solution.

$$[H^+ + Cl^-] + [Na^+ + OH^-] \rightarrow Na^+ + Cl^- + H_2O$$

When neutralization is complete, further addition of NaOH will cause the conductance to increase due to excess of highly mobile OH^- ions. The conductance will thus be minimum at the equivalence point. Thus, if conductance values are plotted against the volume of NaOH added, a curve of the type *xyz* is obtained. The point of intersection corresponds to the end point.

REQUIREMENTS

Conductivity bridge, conductivity cell, burette, pipette 0.01N KCl solution, 0.1N NaOH solution and approximately 0.01N HCl.

PROCEDURE

The conductivity measurements are made by making use of conductivity bridges or conductometers, which are based upon the principle of Wheatstone bridge. The arrangement is shown in the following labelled diagram.

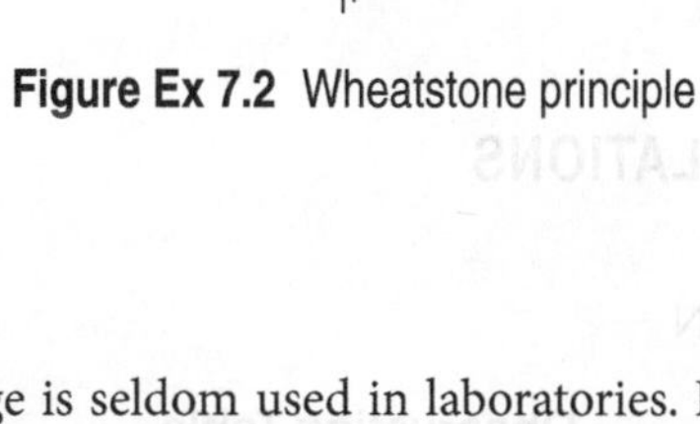

Figure Ex 7.2 Wheatstone principle

IMPORTANT NOTE

Nowadays, the Wheatstone bridge is seldom used in laboratories. In its place, automatic conductivity bridges are in vogue now. These conductivity bridges or conductometers give either observed conductivity or observed resistance. The whole of the system is inbuilt and these conductometers are A.C. mains operated.

CALIBRATION OF THE INSTRUMENT

1. Switch on the instrument and wait for half-a-minute.
2. Take the conductivity cell and connect it to the cell terminals on the instrument.
3. Take 50 ml of 0.01 N KCl solution in a 100 ml beaker and place it on the base plate of cell. Immerse the cell in the solution, so that the platinum plates of the cell are completely immersed in solution.

4. Select the proper conductance range by putting the multiplier switch in the proper range.
5. Push the cal./meas. switch to cal. position.
6. Set temperature control at 25°C and adjust calibration control to get 1000 on the screen position of decimal point.
7. Measure the temperature of the standard solution and set the temperature control to the value of temperature.
8. Now, push the cal./meas. switch to meas. position.
9. Adjust calibration control switch to get the desired value of conductivity of 0.01 KCl on the screen.
10. After the calibration of the instrument, remove the cell from the KCl solution and wash with distilled water.

TITRATION OF HCl vs. NaOH SOLUTION

1. Pipette out 50 ml of given HCl solution in a 100 ml beaker.
2. Immerse the conductivity cell in the solution, so that the electrodes completely dip in the solution.
3. Select the proper conductance range and put the cal./meas. switch to meas. position.
4. Measure the temperature of the sample solution and set the temperature control to the values of measured temperature.
5. Note down conductance of the solution.
6. From the burette, add 0.1 N NaOH solution in 0.5 ml lots. Mix the solution with the help of a glass rod carefully and note the conductance of mixed solution.
7. Keep on adding NaOH solution in 0.5 ml lots and note the conductance value till 10–12 ml of solution is added.
8. Plot a graph between observed conductance values along y-axis against the volume of the alkali added along x-axis. The point of intersection gives the amount of alkali required for neutralization of acid.

OBSERVATION AND CALCULATIONS

Volume of HCl taken = 50 ml
Normality of NaOH solution = 0.1 N

Observation Table

S.No.	Volume of NaOH Added (ml)	Observed Conductance (ohm^{-1})
1.	0.5	
2.	1.0	
3.	1.5	
4.	2.0	
5.	2.5	
and so on		

Suppose that the volume of NaOH used at equivalence point = A ml

Applying normality equation

$$N_1V_1 = N_2V_2$$
$$\text{(HCl)} \quad \text{(NaOH)}$$
$$N_1 \times 50 = 0.1 \times \text{A}$$
$$N_1 = \frac{0.1 \times \text{A}}{50}\text{N}$$
$$= \text{XN.}$$

RESULT

We know that strength = normality × eq. wt.

Therefore, the strength = $X \times$ eq. wt. of HCl

$$= X \times 36.5 = \text{z g/litre}$$

NOTE

HCl being monobasic acid its eq. wt. = mol. wt. = 36.5

Strength of given HCl solution = z g/litre

PRECAUTIONS

1. The solution taken in the burette should be about 10 times stronger than that taken in the beaker, so that the volume change of latter solution is negligible on the addition of the format solution.
2. After every addition of NaOH solution, the solution must be stirred thoroughly.

VIVA-VOCE

Q 1. What is conductance and what are its units?

Ans. The reciprocal of resistance is called conductance. Its units are ohm^{-1} or Ω^{-1} or mho or siemen (S).

Q 2. What is cell constant and what are its units?

Ans. $\text{Cell constant} = \dfrac{\text{distance between parallel electrodes } (l)}{\text{area of cross section of electrode } (A)}$

Its unit is cm^{-1}.

Q 3. What is specific conductance and what are its units?

Ans. Specific conductance is defined as the conductance of 1 cm^3 (one) of the solution. It is denoted by K (kappa).

$$\text{Specific conductance} = \frac{\text{cell constant}}{\text{observed resistance}}$$

Q 4. What is equivalent conductance?

Ans. Equivalent conductance (Δ eq.) is the conductance of all the ions produced by dissociation of 1 gm equivalent of a solute in solution.

$$\lambda \text{ eq.} = \frac{\text{sq. conductance (K)} \times 1000}{N}$$

where N = normality of solution in gm equivalent per litre

Q 5. What are the units of equivalent conductance?

Ans. The units of equivalence conductance are ohm^{-1} cm^2 $(gm\ eq.)^{-1}$.

Q 6. What is molecular conductance?

Ans. Molecular conductance (λ_m) is the conductance of all the ions given by dissociation of one mole of a solute in solution.

$$\lambda_m = \frac{\text{sq. conductance (K)} \times 1000}{M}$$

where M = molarity of solution in mol per litre.

Q 7. What are units of molecular conductance?

Ans. Its units are ohm^{-1} cm^2 mol^{-1}.

Q 8. What is the effect of dilution on conductance?

Ans. Conductance increases with dilution because on dilution, dissociation increases and hence the number of ions increases.

Q 9. What is the effect of dilution on specific conductance?

Ans. Specific conductance decreases with dilution because the number if ion per c.c. of solution decreases on dilution, although the total number of ions in the given volume increases due to increase in dissociation.

Q 10. What is a conductivity cell?

Ans. The vessel in which the measurement of conductivity of the solution is to be made is known as conductivity cell.

Q 11. What is cell constant? How is it determined?

Ans. The observed conductivity must be multiplied by a factor $1/a$ say that x in order to get the value of specific conductance and is known as cell constant.

Specific conductivity = observed conductivity × cell constant.

Cell constant is determined by taking a solution of known specific conductivity and the observed conductivity measured; usually, we take N/50 KCl solution in a conductivity cell, at the room temperature 25°C.

$$\text{Cell constant} = \frac{\text{specific conductivity}}{\text{observed conductivity}} = \frac{0.002768}{\text{observed conductivity}}$$

Q 12. Why ordinary water is unsuitable for conductivity measurements? What is conductivity water?

Ans. Ordinary water is unsuitable for conductivity measurements because it possesses large conductance due to the materials dissolved from the container and due to CO_2 and NH_3 dissolved from air. Therefore, water is specially purified by distilling it a number of times after addition of a little $KMnO_4$. Such water is known as conductivity water and should have a conductivity not more than $2–3 \times 10^{-8}$ ohm^{-1}.

Q 13. How does the conductance change when a weak acid such as acetic acid is titrated against a strong base such as sodium hydroxide?

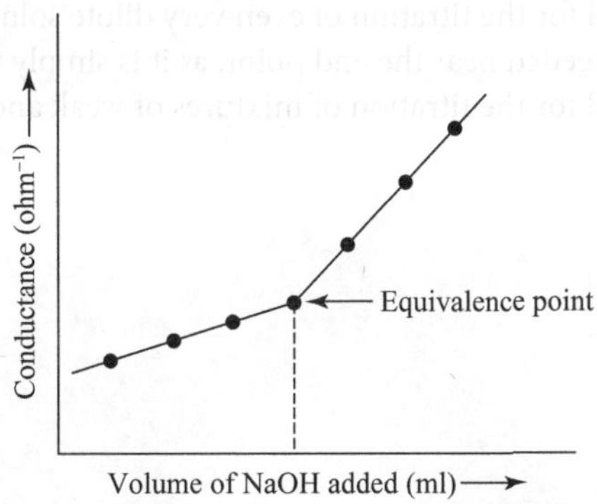

Figure Ex 7.3 Conductometric titration curve of CH_3COOH vs. NaOH

Ans. Acetic acid is a weak acid; therefore, its conductance is low. As NaOH is added, sodium acetate is produced which is highly ionized and, therefore, the conductance begins to increase. After end point, NaOH added introduces fast-moving OH^- ions and hence the conductance begins to increase sharply.

$$CH_3COOH + [Na^+ + OH^-] \rightarrow [CH_3COO^- + Na^+] + H_2O$$

Q 14. How does the conductance vary when a strong acid is titrated against a weak base?

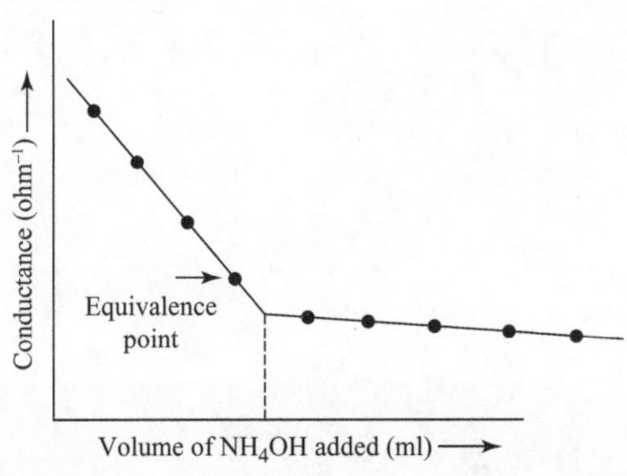

Figure Ex 7.4 Conductometric titration curve of HCl vs. NH_4OH

Ans. Let us consider the titration of HCl against NH_4OH solution.

$$[H^+ + Cl] + NH_4OH \rightarrow [NH_4^+ + Cl^-] + H_2O$$

The conductance of HCl solution is initially high because of the presence of fast-moving H^+ ions. As NH_4OH solution is added, the fast-moving H^+ ions are replaced by slower NH_4^+ ions and hence the conductance decreases. When the end point is passes, the further addition of NH_4OH does not cause much change in the conductance because NH_4OH is weakly ionized.

Q 15. What are the advantages of conductometric titrations?

Ans. (a) These titrations can be used for coloured solutions where ordinary indicators fail to give the end point.

(b) These can be used for the titration of even very dilute solutions of the order of 10^{-4} M.

(c) No extra care is needed near the end point, as it is simply the intersection of two lines.

(d) These can be used for the titration of mixtures of weak and strong acids.

Experiment 8

OBJECT

To determine the surface tension of the given liquid at room temperature by drop count method using a stalagmometer.

THEORY

The surface tension of the given liquid is determined relative to water at the room temperature by using stalagmometer. The number of drops for the same volume of water and the given liquid is counted and let these be as n_1 and n_2 respectively. Now, if d_1 and d_2 are the densities of water and the given liquid at the room temperature as determined separately by using the specific gravity bottle or pyknometer, then the surface tension γ_2 of the given liquid can be calculated by using the simplified relationship as

$$\frac{\gamma_1}{\gamma_2} = \frac{n_2}{n_1} \times \frac{d_1}{d_2}$$

MATERIALS

Stalagmometer, distilled water, unknown liquid, beaker, rubber tube with screw-pinch cock, clamp stand and the specific gravity bottle or pyknometer.

DESCRIPTION OF THE APPARATUS

The stalagmometer consists of a bubbled dropping tube with a capillary. The lower end of capillary is well ground, polished and flattened in order to provide larger dropping surface. There are two marks C and D above and below the bulb B. The upper end of the apparatus is connected with a rubber tube fitted with a screw-pinch cock in order to control the rate of flow so that the spherical drop may be formed.

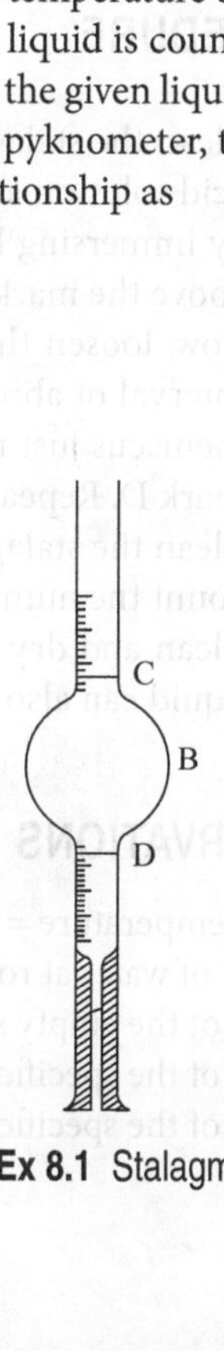

Figure Ex 8.1 Stalagmometer

Hence, by simply counting the number of drops for the given liquid and water and knowing their density, we can find the relative surface tension of the liquid. The advantages of the drop-pipette method are that it is very convenient and quick. It can also be employed for determinations at different temperatures by placing the whole apparatus in a thermostat.

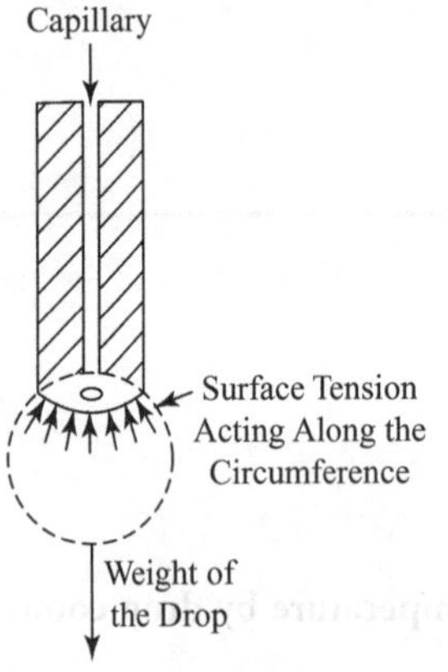

Figure Ex 8.2 Formation of a drop at the capillary end

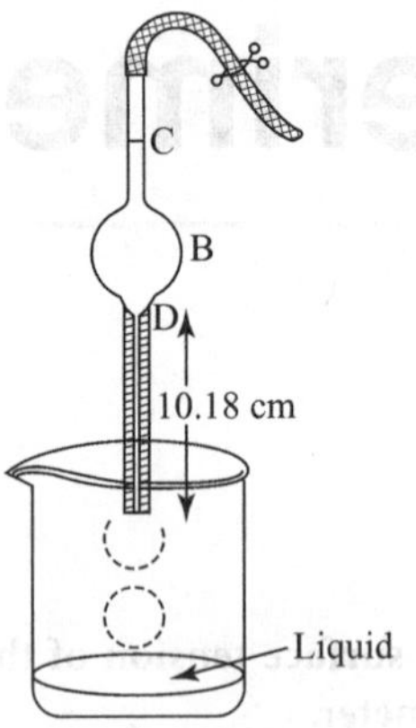

Figure Ex 8.3 Determination of surface tension by drop count method

PROCEDURE

1. Clean the stalagmometer, a specific gravity bottle or pyknometer thoroughly first with chromic acid solution and wash finally with distilled water and then dry.
2. By immersing lower end in a beaker containing distilled water, suck up the water until it rises above the mark C and tighten the screw of the screw-pinch.
3. Now, loosen the screw of the screw-pinch carefully so that the liquid drops start falling at an interval of about 2–3 sec in successive drops. The counting of drops is started when the water meniscus just reaches the upper mark C and stopped when the meniscus just passes the lower mark D. Repeat to get three readings and take the mean value.
4. Clean the stalagmometer and dry it. Fill it with liquid until it rises above the upper mark C and count the number of drops as before.
5. Clean and dry the specific gravity bottle and measure the density of liquid. The density of the liquid can also be determined by the pyknometer.

OBSERVATIONS AND CALCULATIONS

Room temperature = ……….. °C

Density of water at room temperature = d_1

Weight of the empty specific gravity bottle or pyknometer = w_1 g

Weight of the specific gravity bottle or pyknometer with water = w_2 g

Weight of the specific gravity bottle or pyknometer with liquid = w_3 g

$$\therefore \quad \text{Weight of water} = (w_2 - w_1) = m_1 \text{ g(say)}$$

and $\qquad$ Weight of the liquid $= (w_3 - w_1) = m_2\,g$

Surface tension of water at room temperature = say (16.8)

Liquid	Number of Drops (1)	(2)	(3)	Mean Value
Water				n_1
Given liquid				n_2

CALCULATIONS

Density of the given liquid relative to water at room temperature.

$$\frac{m_2}{m_1} = d_2$$

Also
$$\frac{\gamma_1}{\gamma_2} = \frac{n_2}{n_1} \times \frac{d_1}{d_2}$$

$$\gamma_2 = \gamma_1 \times \frac{n_2}{n_1} \times \frac{d_2}{d_1}$$

where γ_1 = the surface tension of water at room temperature.

Substituting the values of γ_1, n_1, n_2, d_1 and d_2, the surface tension of the given liquid at room temperature thus becomes known.

RESULT

The relative surface tension of the liquid with respect to water at room temperature is ....

PRECAUTIONS

1. The stalagmometer and the specific gravity bottle or pyknometer should be cleaned properly and dried before use.
2. Fit the stalagmometer vertically.
3. The rate of the fall of drops should be adjusted so that the drop interval is at least 2–3 sec in successive drops. The number of drops per minute should be in between 15 and 20.
4. The drops should fall from the tip of the stalagmometer under their own weight rather than pushing them by force.
5. Wash and dry the stalagmometer after use.

VIVA-VOCE

Q 1. How do you define the term surface tension?

Ans. The surface tension of a liquid is the force acting per centimetre along the surface of the liquid at right angle to any line on the surface of the liquid in any direction.

Q 2. Give the CGS and the SI units of surface tension?

Ans. The CGS units are dynes per cm and the SI units are Newton per metre. One SI unit is 1000 times the CGS units.

Q 3. What do you understand by the surface energy of liquid?

Ans. The surface energy of a liquid is the work required to extend the surface of the liquid by one unit area.

Q 4. What are the units of surface energy?

Ans. Its units are ergs/cm^2 in the CGS system and J/m^2 in the SI units.

Q 5. How the surface tension and the surface energy are related to each other?

Ans. The value of the surface energy for any liquid surface is numerically equal to the value of the surface tension of liquid in either set of units.

Thus, if for a liquid, the surface tension equals γ dynes/cm or $\gamma \times 10^3$ N/m. Surface energy will equal γ ergs/cm or $\gamma \times 1000$ J/m.

Q 6. What method do you use in the laboratory for the determination of surface tension?

Ans. Surface tension can be measured by using stalagmometer or a drop-pipette method.

Q 7. What is the shape of liquid drop falling from capillary?

Ans. Liquid drops are spherical in shape.

Q 8. Why liquid drops are spherical in shape?

Ans. Due to surface tension, the surface of liquid tends to become minimum. The shape of sphere has minimum surface area.

Q 9. Small insects are seen moving over the surface of water tanks. Why?

Ans. The surface of a liquid behaves like a stretched membrane due to the property of surface tension. The molecules on the surface are experiencing a new downward pull due to molecules present in the hulk of the liquid.

Q 10. What instrument is used for measuring the density of a liquid?

Ans. The density of a liquid can be measured by using the specific gravity bottle or a pyknometer.

Q 11. What formulas are used for calculating the surface tension of a liquid?

Ans. $$\gamma_1 = \frac{n_w d_1}{n_1 d_w} \times \gamma_w$$

where γw represents the surface tension of the water at same temperature, nw and nl represents the number of drops from the same volume of the water and the liquid respectively. dw and dl represent the density of water and the liquid.

Q 12. What do you understand by the interpolation of graph?

Ans. The graph is plotted by taking two physical quantities on the two axis of the graph. Knowing the value of either of the two values that of the other can be found out by interpolation. Knowing the numerical value of the property on Y-axis, its corresponding value on X-axis can be known by interpolation.

Q 13. What do you understand by extrapolation of graph?

Ans. By extrapolation, we mean the extension of the curve obtained on the graph to either of two axes. The curve is extended towards the particular axis on which that physical quantity to be determined has been taken. The corresponding point at which the plot touches the axis gives the required value.

Q 14. Why liquids rise in a capillary tube?

Ans. The rise of liquid is due to the inward pull of surface tension acting on the surface which pushes the liquid into the capillary tube.

Q 15. Why mercury level falls and that of other liquids rise in capillary tube?

Ans. Liquids like mercury which do not wet the glass, their level inside fall below the level outside. Moreover, the upper surface of liquids that wet the glass is concave and that of mercury is convex. Surface tension becomes operational and pulls up the liquids that wet the glass. The liquid rises in the capillary and is pulled up due to surface tension acting all along the circumference of the tube along with the wall of the tube.

Q 16. What is the effect of the rise in temperature on the surface tension of the liquids?

Ans. The surface tension of the liquids decreases with the increase in temperature and becomes zero at all temperature where the difference between the liquid and vapour disappears.

Q 17. Why surface tension decreases with rise in temperature?

Ans. The decrease in the surface tension of liquid with increase of temperature is due to the fact that with rise in temperature, the kinetic energy of the molecules (and hence the motion of molecules) increases and therefore the intramolecular forces of attraction decrease.

Q 18. When and why does the surface tension become zero?

Ans. The surface tension of a liquid becomes zero at its critical temperature since, at this temperature, the surface of separation between the liquid and its vapour disappears.

Q 19. What are surface active agents?

Ans. The substances which decrease the surface tension of water when dissolved in it are called surface active agents, e.g. soap, detergents, alcohol etc.

Q 20. What is interfacial tension?

Ans. Interfacial tension is defined as the force acting per unit length along the surface of separation of the immiscible or partially miscible liquids in contact with each other.

Q 21. Why is the surface tension of water much higher than those of organic liquids?

Ans. It is because the intermolecular forces in water are much stronger than those in organic liquids.

Experiment 9

OBJECT

To determine the partition co-efficient of iodine between water and carbon tetrachloride.

THEORY

Iodine distributes itself in two immiscible solvents as water and carbon tetrachloride and has the same molecular state of I_2 in both of these solvents. If C_1 and C_2 are the concentrations of iodine in water and carbon tetrachloride layers in contact with each other and forming the heterogeneous equilibrium, then the ratio of these concentrations, i.e., $\frac{C_1}{C_2}$ would be constant at a fixed temperature.

Mathematically, $$K = \frac{C_1}{C_2} = \frac{\text{Concentration of } I_2 \text{ in the water layer}}{\text{Concentration of } I_2 \text{ in the } CCl_4 \text{ layer}}$$

where K is distribution or partition coefficient

APPARATUS/REAGENTS REQUIRED

Four glass-stoppered reagent bottles of about 250 ml capacity, separating funnel, burette, pipette (5 and 10 ml), conical flask, glazed tile, burette stand, saturated solution of iodine in carbon tetrachloride, $\frac{N}{10}$ and $\frac{N}{100}$ hypo solution and starch indicator.

PROCEDURE

(i) Take four glass-stoppered bottles of about 250 ml capacity. Wash these bottles thoroughly first with a suitable reagent and then distilled water. Dry these bottles and label them as 1, 2, 3 and 4.

(ii) Using measuring cylinder take out 10, 20, 30 and 40 ml of this solution in different glass bottles.

(iii) Add 40, 30, 20 and 10 ml of CCl_4 in each bottle respectively. Then add 50 ml of distilled water in each bottle.

(iv) Stopper these bottles and shake each bottle vigorously for about 20 min mechanically or otherwise by using an electric shaker. Sufficient shaking of each bottle is necessary for accurate results.

(v) Keep these bottles undisturbed for some time so that two separated layers are formed distinctly. In case some droplets are found sticking to the walls of any bottle, then that bottle is swirled gently to bring down those droplets.

(vi) Pour the contents of each bottle separately into a separating funnel and collect carbon tetrachloride and water layers of each bottle in separate beakers or vessels.

(vii) Pipette out 10 ml of carbon tetrachloride layer from the separated layers of bottle no. 1 into a titration flask. Alternatively, pipette out 10 ml of carbon tetrachloride layer directly from bottle no. 1 by dipping the nozzle of the pipette into the CCl_4 layer.
Add about 1 g of potassium iodide (or 10 ml of 10% KI solution), five drops of starch solution and titrate against $\frac{N}{10}$ hypo solution. Disappearance of the blue colour marks the end point.
Repeat to get two concordant readings. Repeat this process with CCl_4 layer of other bottles in the same way.

(viii) Pipette out 10 ml of aqueous layer of bottle no. 1 into a titration flask and titrate against $\frac{N}{100}$ hypo solution as given in step (vii).

(ix) Carry out similar titration for upper and lower layers respectively in other bottles.

OBSERVATIONS

Room temperature = ...°C
Volume of water layer used in each titration = ... ml
Volume of CCl_4 layer used in each titration = ... ml

S. No.	Percentage of Iodine (V/V)	Titration with Water Layer			Normality of Hypo Solution	Titration with CCl_4 Layer			Normality of Hypo Solution
		Initial Reading	Final Reading	Volume of Hypo used V_1 ml		Initial Reading	Final Reading	Volume of Hypo used V_2 ml	
1.	10%				N/100				N/10
2.	20%				N/100				N/10
3.	30%				N/100				N/10
4.	40%				N/100				N/10

CALCULATIONS

$$K = \frac{\text{Concentration in the } H_2O \text{ layer}}{\text{Concentration in the } CCl_4 \text{ layer}}$$

$$\text{Concentration in the water layer} = \frac{\text{Normality of hypo used with the water layer} \times V_1}{\text{Volume of the water layer used}} \text{ g equi/L}$$

$$\text{Concentration of the } CCl_4 \text{ layer} = \frac{\text{Normality of hypo used with the water layer} \times V_2}{\text{Volume of the } CCl_4 \text{ layer used}} \text{ g equi/L}$$

CALCULATE VALUES OF *K* FOR EACH SET

1. 10% solution   *K*..............
2. 20% solution   *K*..............
3. 30% solution   *K*..............
4. 40% solution   *K*..............

RESULTS

1. The value of *K* in different sets comes fairly constant; hence molecular state of iodine in water and CCl_4 is the same.
2. The distribution coefficient of iodine between water and CCl_4 is ..... (at °C).

PRECAUTIONS

1. Do not use rubber stoppers as iodine solution in CCl_4 can react with it.
2. The reagent bottles should be thoroughly cleaned and dried.
3. The stoppers are kept properly pressed during shaking of the bottles.
4. Loosen the stopper after shaking for releasing the pressure, if any, developed during the process of shaking the contents in the bottle.
5. Use only the fresh starch solution as indicator in hypo titrations.

VIVA-VOCE

Please refer to VIVA-VOCE under Experiment 10.

Experiment 10

OBJECT

To determine the partition coefficient of benzoic acid between water and benzene at room temperature and the molecular state of benzoic acid in benzene as compared to its solution in water.

THEORY

If the solute associates in the solvents, the distribution law expressed is

$$\frac{C_1}{\sqrt[n]{C_2}} = K$$

where n is the number of solute molecules which associate in solvent 2.

Benzoic acid exists as normal molecules in water and as dimer in benzene. Therefore, $\frac{C_1}{\sqrt{C_2}}$ should be constant. This can be verified experimentally.

APPARATUS REAGENTS REQUIRED

Five reagent bottles (250 ml) with stoppers
Conical flasks, pipette (10 ml), burette, beakers, measuring cylinder
0.1 N NaOH benzene, benzoic acid, phenolphthalein

METHOD

1. Prepare 1% solution of benzoic acid in water.
2. Take four reagent bottles and label them 1, 2, 3 and 4. To each bottle, add 50 ml of benzene. Then add from the measuring cylinder the following quantities of 1% benzoic acid solution and distilled water.

Stopper and shake the bottles vigorously. Leave them for a considerable time (50–60 min) during which shake the bottles vigorously at frequent intervals. Then swirl each bottle gently to bring down the droplets sticking to its walls. Let the contents of the bottle stand quietly till the two layers separate. The upper layer is of benzene and the lower layer is of water.

Pipette out carefully 10 ml of benzene layer from one bottle and deliver it to the titration flask. Add about 20 ml of distilled water and shake. Add one to two drops of phenolphthalein indicator and titrate

Bottle Number	Volume of 1% Benzoic Acid Solution (ml)	Volume of Distilled Water (ml)
1.	50	....
2.	40	10
3.	30	20
4.	20	30

by releasing 0.1 N NaOH solution from the burette. Stop when pink colour is obtained. Note the volume of alkali used. Similarly do one more titration and take the mean of the two titre values. Similarly repeat the titration of the benzene layer from the remaining three bottles also.

Close the upper end of pipette tightly with the finger and push its nozzle into the water layer (lower layer) in one bottle. The benzene will not enter into it. Then suck up the aqueous layer and fill the pipette up to the mark. Deliver the liquid to the titration flask and titrate it with 0.1 N NaOH solution using phenolphthalein as indicator. Similarly, repeat the titration and take the mean of the two titre values. Similarly repeat the titration of the benzene layer from the remaining three bottles also.

OBSERVATIONS

Room temperature = ...
Volume of benzene layer used in each titration = ... ml
Volume of water layer used in each titration = ... ml

BURETTE READINGS

	Water Layer C_1			Benzene Layer C_2		
Bottle No.	Initial Reading	Final Reading	Volume of NaOH used (V_1 ml)	Initial Reading	Final Reading	Volume of NaOH used (V_2 ml)

CALCULATIONS

$$K = \frac{C_1}{\sqrt{C_2}}$$

where C_1 = Concentration in the water layer and C_2 = Concentration in the benzene layer.

$$C_1 = \frac{\text{Volume of NaOH used with the water layer} \times \text{Normality of NaOH}}{\text{Volume of the water layer used}} \text{g equi/L}$$

$$C_2 = \frac{\text{Volume of NaOH used with the benezene layer} \times \text{Normality of NaOH}}{\text{Volume of the benezene layer used}} \text{g equi/L}$$

It is seen that the values of C_1/C_2 are not constant, but those of $\frac{C_1}{\sqrt{C_2}}$ are found to be constant. Hence, the complexity of molecules of benzoic acid in benzene is 2.

Plot a graph between C_1 and $\sqrt{C_2}$. A straight line is obtained which verifies that the complexity of molecules of benzoic acid in benzene is 2.

RESULT

1. The value of partition coefficient between water and benzene is ..... at .... °C.
2. The molecular state of benzoic acid in benzene is dimer.

PRECAUTIONS

1. The reagent bottles should be thoroughly cleaned and dried.
2. The stoppers are kept properly pressed during shaking of the bottles.
3. Preferably glass stopper and non-rubber stopper bottles should be used as organic reagent can react with the rubber.
4. Loosen the stopper after shaking in order to release the pressure, if any, developed during the process of shaking of the contents in the bottles.

VIVA-VOCE

Q 1. Define distribution.

Ans. Whenever a solute is dissolved in two immiscible solvents, there exists a mathematically constant ratio of the concentrations of the distributing solute in the two phases at a fixed temperature and this ratio is independent of the amount of the solute and the immiscible liquids provided the solute remains in the same molecular state in both the solvents.

Q 2. What is partition coefficient?

Ans. Nernst distribution law can be mathematically stated as

$$\frac{\text{Concentration of x in A}}{\text{Concentration of x in B}} = \frac{C_A}{C_B} = K$$

where K is called the partition coefficient.

Q 3. What is meant by the same molecular state of the distributing substance?

Ans. The solute is said to exist in the same molecular state if it does not dissociate or associate or enter into chemical combination with either of the immiscible solvents in contact with each other.

Q 4. What do you mean by association of distributing substance?

Ans. By the association of the solute, we mean that two or more molecules of the same solute may combine with one another to form dimers or trimers in a particular solvent.

Q 5. How does benzoic acid exist in benzene and water phases?

Ans. In benzene, benzoic act as a dimer, i.e., two molecules of benzoic acid combine to form a bigger molecule and as such its molecular concentrations becomes one half of its concentration in water. However, in water, it remains in its normal state.

Q 6. How is iodine estimated in CCl_4 and water layers?

Ans. Iodine can be estimated iodimetrically, i.e., titration with standard hypo solution. The partition coefficient of iodine in favour of CCl_4 is higher than that of water. Therefore, iodine is CCl_4 layer is titrated with N/10 hypo solution and is aqueous layer with N/100 hypo solution.

Q 7. Name the indicator and describe the end point in hypo-titrations.

Ans. Freshly prepared starch solution is used as indicator in hypo-titrations. Disappearance of bluish colour marks the end point in these titrations. Hypo solution is taken in the burette.

Q 8. Write the chemical reaction involved in iodine titration.

Ans. $$\underset{\text{Sodium thiosulphate}}{2Na_2S_2O_3} + I_2 \longrightarrow \underset{\text{Sodium tetra-thionate}}{Na_2S_4O_6} + 2NaI$$

Ionic equations can be written as

$$2S_2O_3^{2-} + I_2 \rightarrow S_4O_6^{2-} + 2I^-$$

Q 9. Can iodine titration be regarded as redox titration? Explain your answer.

Ans. Yes, the reaction of hypo with iodine is a redox reaction, in which I_2 acts a mild oxidizing agent. It oxidizes $S_2O_3^{2-}$ ions to $S_4O_6^{2-}$ ions and I_2 is reduced to iodine (I−) ions

Q 10. Which will be lower layer when I_2 is distributed between water and carbon tetrachloride?

Ans. Carbon tetrachloride will form the lower layer because its density is more than water and is heavier than water.

Q 11. Why do we use different concentrations of the hypo solution in estimating iodine in water and CCl_4 layers?

Ans. It is because I_2 has greater solubility in CCl_4 as compared to water. The concentration of I_2 in CCl_4 layer would be higher and so it is titrated with N/10 hypo solution, whereas concentration of I_2 in aqueous layer being lower, it is titration with N/100 hypo solution.

Q 12. Why we use a modified distribution law formula in case of distribution of benzoic acid between benzene and water?

Ans. We use the modified formula $\left(\dfrac{C_1}{\sqrt{C_2}} = K\right)$ in this case because benzoic acid exist as dimer in the benzene layer. It undergoes association in benzene layer.

Q 13. Represent the physical state of Benzoic acid in water and benzene layers.

Ans. In water layer, benzoic acid exists in its normal molecular state as C_6H_5COOH. However, in benzene layer, it exists as a dimer, i.e., (C_6H_5COOH). The structure of the dimer is represented as

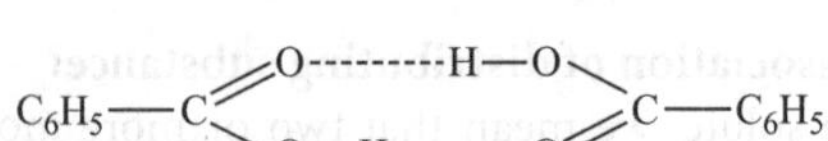

Q 14. Write the modified formula for distribution law in case there is dissociation of the solute in second layer.

Ans. The modified formula would be

$$\frac{C_1}{C_2(1-\alpha)} = K,$$

where α is the degree of dissociation of the solute in the second layer solvent.

Q 15. How is benzoic acid estimated in the water and benzene layers.

Ans. Benzoic acid in the two layers can be estimated by titration with a standard solution of strong base (say N/10 NaOH) using phenolphthalein as indicator.

Experiment 11

OBJECT

Determine the relative viscosity of the liquid with respect to water at room temperature by Ostwald's viscometer.

APPARATUS

Distilled water, stop watch, viscometer (Ostwald's), unknown liquid, pyknometer or R.D. bottle, pipette etc.

DESCRIPTION

Ostwald's viscometer is a U-shaped glass tube as shown in the Fig. Ex 11.1. In one arm, the bulb A is connected with a fine capillary. The lower end of the capillary is connected with a U tube provided with a bulb B in the second arm. The bulbs are necessary to maintain the hydrostatic pressure during the flow of liquid. Through the capillary tube, the liquid flows with measurable speed. There are two marks C and D above and below the bulb A. The upper end of the bulb A is attached with a rubber tube. The liquid flows under its own weight. The density of liquid is determined either by the pyknometer or by the specific gravity bottle.

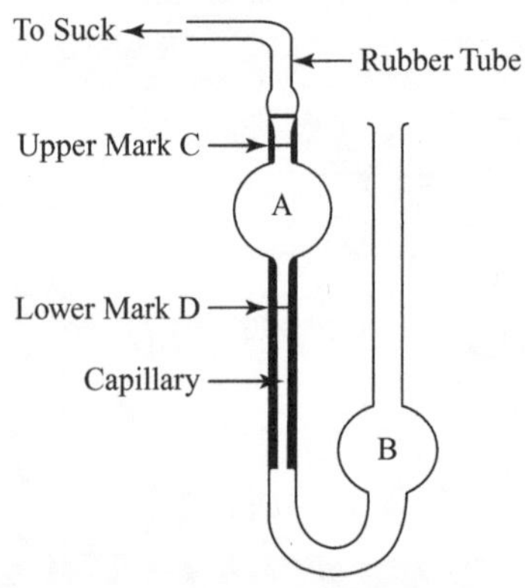

Figure Ex 11.1 Ostwald's viscometer

THEORY

The measurement of viscosity by viscometer is based on the following equation:

$$\frac{\eta_1}{\eta_2} = \frac{d_1 t_1}{d_2 t_2}$$

where d_1 and d_2 are the densities of two liquids and t_1 and t_2 are times required to flow for equal volumes of two liquids through the same length of the capillary tube.

η_1 = the viscosity of the liquid

η_2 = the viscosity of water

Then, $\frac{\eta_1}{\eta_2}$ = relative viscosity of liquid with respect to water

Relative viscosity has no unit.

PROCEDURE

Clean the viscometer with chromic acid and then wash thoroughly with distilled water. It is finally washed with alcohol or acetone and then dried. A sufficient volume of distilled water is introduced by pipette in bulb B so that the bend portion of the tube and half or a little more than half of bulb B are filled up. Clamp the viscometer in quite vertical position. Through the rubber tube attached to upper arm of a bulb A, suck up water until it rises above the upper mark C and allow it to flow under its own weight. The time of flow of water from C to D is counted by starting the stop watch as the meniscus just reaches upper mark C and stopping the watch as the meniscus just passes the lower mark D. Take at least three or four readings with water and then take the mean of them.

Now, remove the water from viscometer and dry it. Introduce in bulb B the same volume of liquid and measure the time of flow of liquid as before. Take at least three or four readings and then take the mean value.

Now, wash and dry the pyknometer (or specific gravity bottle) and then weight it empty. Then, fill it with distilled water and weigh. Remove water and dry. Then, fill it with experimental liquid and weigh. Note the room temperature.

OBSERVATIONS

(1) Room temperature ....
(2) Time of flow between C and D:

	Water		Liquid under Experiment	
S.No.	Time of Flow	Mean (t_2)	Time of Flow	Mean (t_1)
1.	....			....
2.	....			....
3.	....			....
4.	....			....

(3) Density of liquid (d_1) and of water (d_2)

Mass of empty of the specific gravity bottle (or pyknometer) = ... g.
Mass of the specific gravity bottle (or pyknometer) + water = ... g.
∴ Weight of water = ... g.
Mass of the specific gravity bottle (or pyknometer) + liquid = ... g.
∴ Mass of liquid = ... g.

CALCULATIONS

$$\frac{\text{Density of liquid } (d_1)}{\text{Density of water } (d_2)} = \frac{\text{Weight of liquid}}{\text{Weight of water}}$$

$$\text{Relative viscosity of liquid} = \frac{\eta_1}{\eta_2} = \frac{d_1 t_1}{d_2 t_2}$$

By inserting the respective values, the relative viscosity of liquid is calculated.

RESULT

The relative viscosity of this liquid with respect to water at room temperature is ....

PRECAUTIONS

(1) The viscometer should be thoroughly cleaned before use.
(2) Viscometer must be strictly kept in vertical position.
(3) Observe accurately when the meniscus of liquid and water are just passing the upper and lower marks.
(4) Same volumes of water and liquid should be taken in viscometer.

VIVA-VOCE

Q 1. Define viscosity.
Ans. It may be defined as the property of a liquid by the virtue of which it offers resistance to flow.

Q 2. What is meant by coefficient of viscosity?
Ans. It is defined as the force per unit area required to maintain velocity difference of 1 cm/s between two parallel layers 1 cm apart.

Q 3. What are the units of coefficient of viscosity?
Ans. It is expressed in units of dynes-sec-cm^{-2} or poise. Lower values of viscosity are, however, generally expressed in centipoises (= 0.01 poise) or millipose (= 0.001).

Q 4. What are the dimensions of viscosity coefficient?
Ans. It has dimensions of $ML^{-1} T^{-1}$.

Q 5. What is fluidity?
Ans. The reciprocal of the coefficient of viscosity is called fluidity. It is the measure of the ease with which a liquid flows.

Q 6. What is the effect of temperature on viscosity?
Ans. Viscosity of a liquid decreases rapidly with the rise in temperature.

Q 7. How does viscosity change when a solid melts or a liquid vaporizes?
Ans. It decreases in both the cases.

Q 8. What is the effect of hydrogen bonding on viscosity?
Ans. The presence of hydrogen bonds results in an increase in viscosity.

Q 9. Glass is a super cooled liquid. What do you think its viscosity should?
Ans. It should have high viscosity.

Q 10. How is viscosity related to molecular weight?
Ans. For similar substances, it increases with the increase in molecular weight.

Q 11. How can you increase the viscosity of a solution?
Ans. By adding lyophilic colloids and substances such as sugar and gum.

Q 12. What is meant by molecular viscosity?
Ans. It is equal to the product of viscosity and the molecular surface, i.e., molecular viscosity $\eta \times \left(\frac{M}{d}\right)^{2/3}$.

Q 13. What do you understand by the terms 'streamlined' and 'turbulent' liquid flows?
Ans. If various layers of a liquid flow smoothly over each other without intermixing, the flow is said to be streamlined. It is said to be turbulent if intermixing of layers takes place.

Q 14. Define rheochor.
Ans. It may be defined as the molar volume of the liquid at the temperature at which its viscosity is unity.

Experiment 12

OBJECT

To determine the pH of given solution by the use of indicators.

APPARATUS

Beakers, test tubes, test tube stand, indicators (methyl violet, methyl orange, methyl red, bromothymol blue, phenol red and phenolphthalein).

THEORY

Electrical conductivity of water is small and even the purest water dissociates into hydrogen and hydroxyl ions. The dissociation of water can be represented as

$$2H_2O \rightleftharpoons H^+_3O + OH^-$$

In simple form, $H_2O \rightleftharpoons H^+ + OH^-$

The equilibrium constant for this reversible reaction is given by,

$$K = \frac{[H^+][OH^-]}{[H_2O]}$$

The active mass of the undissociated water is taken as unity at a given temperature. Thus, the equilibrium constant equation may be written in a simplified form as $[H^+][OH^-] = [H_2O]\ K = K\omega$, where $K\omega$ is called the ionic product or dissociation constant of water.

Measurements have shown that hydrogen ion concentration in water at 25 °C is 1×10^{-7} M. The value of $K\omega$ at 25 °C is accordingly $(1 \times 10^{-7})(1 \times 10^{-7}) = 10^{-14}$ moles^2/litre^2.

Since water dissociates into H^+ and OH^- ions, all aqueous solutions (acidic, basic or neutral) contain both hydrogen and hydroxyl ions. Since hydrogen ion concentrations commonly met with vary from 1 to 10^{-14} gm ions per litre, a logarithmic scale has been found more convenient for expressing the concentrations of H^+ and OH^- ions.

Sorensen (1909) suggested the use of pH for expressing the $[H^+]$ or $[OH^-]$ of any solution with the help of a logarithmic scale. Since the concentrations of hydrogen ions of most solutions are less than 1 gm ion per litre, the logarithm of these concentrations would be negative. Therefore, the pH of solution is defined by the equation.

$$pH = \log 10 \frac{1}{[H^+]} = -\log_{10}[H^+]$$

or

$$H^+ = 10^{-pH}$$

Thus, the pH value of solution is the logarithm of the reciprocal of the concentration of hydrogen ion.

INDICATORS

Indicators are either weak organic acids or bases and they also dissociate into ions. For example, an indicator which is a weak acid when added to water will dissociate as

$$H(ind) \rightleftharpoons H^+ + (ind^-)$$

Acid form, H(ind), has a different colour from the dissociated form (ind^-).

Let us take the example of phenolphthalein, which may be represented as HPh. It is considered as weak acid and it dissociates as in the following equation:

$$\underset{\text{(colourless)}}{HPh} \rightleftharpoons H^+ + \underset{\text{(pink)}}{Ph^-}$$

In the presence of acid, the above equilibrium will shift towards left due to common ion effect of H^+ ions from the acid and the solution will be colourless. In the presence of a base (e.g. NaOH), the H^+ ions will combine with OH^- ions from the base to form water and the equilibrium will shift towards right and the solution will be pink due to the presence of the Ph^- ions.

$$HPh \rightleftharpoons H^+ + Ph^-$$
$$NaOH \rightleftharpoons OH^- + Na^+$$
$$H^+ + OH^- \rightleftharpoons H_2O$$

Similarly, methyl orange may be considered as a weak base (MeOH) and it will dissociate as under:

$$\underset{\text{(yellow)}}{MeOH} \rightleftharpoons \underset{\text{(red)}}{Me^+} + OH^-$$

In the presence of base, the equilibrium will shift towards the left due to common ion effect of OH^- ions and the solution will be yellow. However, in presence of acid (e.g. HCl), the equilibrium will shift towards right and the colour will be red.

The particular shade of the indicator produced will depend upon the pH of the solution. For methyl orange, the pH range is between 3 and 4. Below pH 3, the colour of the solution is red and above pH 4, the colour is yellow. The pH range of an indicator depends on the dissociation constant of the indicator acid or base. Suitable indicators are available to cover range from 0 to 14.

The pH ranges of indicators which we shall use in this experiment are:

Acidic solution basic solution

Indicator	pH Range	Colour Change
Methyl violet	0.3–3.2	Yellow to violet
Methyl orange	3.1–4.4	Red to yellow
Methyl red	4.2–6.3	Red to yellow
Bromothymol blue	6.0–7.6	Yellow to blue
Phenol red	6.8–8.4	Yellow to red
Phenolphthalein	8.2–10.0	Colourless to pink

pH of solutions

PROCEDURE

1. Take six test tubes and add 5 ml of 0.1 M sodium carbonate solution to each of them.
2. To each solution, add two drops of an indicator solution, starting at the top of the above given list. (Add two drops of methyl violet to the first test tube, two drops of methyl orange to the second, two drops of methyl red to the third and so on.)
3. Record the colours and estimate the approximate pH of the sodium carbonate solution.

OBSERVATIONS

Solutions	Methyl Violet	Methyl Orange	Methyl Red	Bromothymol Blue	Phenol Red	Phenol pH Approx. Phthalein-p.pt
Na_2CO_3 (0.1 M)						
Acetic acid (0.1M)						
Na_2HPO_4 (0.1 M)						

Repeat the above procedure to determine the pH 0.1 M acetic acid and 0.1 M sodium phosphate solutions. Record the colours and the approximate pH values in the form of a table.

VIVA-VOCE

Please refer to VIVA-VOCE under Experiment 13.

Experiment 13

OBJECT

To determine the pH value (varies between 6 and 9) of a given solution by buffer solution method.

APPARATUS

Test tubes, test tube holders, B.D.H. universal indicator etc.

THEORY

We know that indicators change their colours with change in pH value of the solution in which they are put. Consult the table given in the previous experiment to compare the colour with the colour chart provided there. The solution under examination is mixed with a suitable indicator and the colour that appears is compared with a series of mixture of buffer solutions of known pH and of the indicator selected already. The buffer solution which gives the identical colour as the test solution will have the same pH. For accurate pH determination, indicators with short ranges are employed to get distinct colour changes. By mixing a series of indicators suitably, the colour change is made to vary from pH 1 to 12. Such indicators are known as universal indicators, which give only approximate value of pH.

PROCEDURE AND OBSERVATIONS

1. Take 10 ml of given solution, add 0.5 ml of the universal indicator and note the colour change.
2. By comparing the colour with the colour chart, determine the approximate pH of the solution within 1 to 2 units. Suppose that the value lies between 7 and 8.6.
3. Now, select an indicator whose pH range lies within these values. Phenol red indicator is found to be suitable within this range, as pH range is 6.8–8.4. It has yellow colour at pH 6.8 and red at pH 8.4.
4. Prepare a series of buffer solutions differing successively in pH by about 0.2 unit, covering the pH range of the solutions under test. For this, mix equal volume (50 ml) of KH_2PO_4 and 0.2 M NaOH to prepare the buffer.

5. Take test tubes of exactly the same dimensions and add 1 ml of buffer solution of definite pH value in each test tube, so that the pH value in the tubes is 7.0, 7.2, 7.4, 7.6, 7.8, 8.0, 8.2, 8.4 and 8.6. Now, mark the pH value on each test tube.
6. Now, add 0.5 ml of phenol red indicator in each tube. The tube with pH 7.0 would be yellow in colour and that with pH 8.4 and 8.6 would be red in colour. The other tubes would have colours in between yellow and red.
7. Now, take 10 ml of the solution under test in a tube of the same dimension and add 0.5 ml of phenol red into it.
8. Find out a complete match of colour with that of the buffer solution. The pH of the solution under test will be the same as that of the buffer solution.
9. In case, if complete matching is not observed, i.e., the colour lies between those two successive standards, then, the pH of the solution will be between those of two standards.

RESULT

The pH of the given solution is approximately ....

VIVA-VOCE

Q 1. Define the term pH.

Ans. pH value of a solution is defined as the negative power to which $[H^+]$ concentration of a solution is raised to express the H^+ concentration of the solution.

$$[H^+] = 10^{-pH} \text{ (for dilute solution)}$$

Taking logarithms, we get

$$-\log [H^+] = pH$$

pH value of a solution may also be defined as the negative logarithm of $[H^+]$ or reciprocal of logarithm of $[H^+]$ of the solution.

Q 2. What is the effect of temperature on pH?

Ans. The pH value of the solution increases with the increase of temperature.

Q 3. What is the possible pH of pure water?

Ans. Less than 7, since the concentration of H^+ in water at room temperature is more than 10^{-7} mol L^{-1}.

Q 4. What is the effect of dilution on pH of an acidic solution?

Ans. pH increases towards 7.

Q 5. What is the pOH of pure water at 25°C?

Ans. pOH $= 14 - pH = 14 - 7 = 7$.

Q 6. What are buffers?

Ans. Substances which resist the change in pH value of a solution on keeping or on diluting or on addition or small amounts of an acid or a base into the solution are called buffers.

Q 7. Define two types of buffers.

Ans. **1. Acidic buffer:** It is a mixture of a weak acid and its salt with a strong base. For example, a mixture of CH_3COOH and CH_3COONa.

2. Basic buffer: It is a mixture of a weak base and its salt with the strong acid. For example, a mixture of NH_4OH and NH_4Cl.

Q 8. What chemicals would you use to make a buffer of pH (i) 5 and (ii) 10?

Ans. (i) $H_3CCOOH + H_3CCOONa$ and (ii) $NH_4OH + NH_4Cl$.

Q 9. Explain the buffer action of acidic buffer.

Ans. Let us take the following example:

$$H_3CCOOH \rightleftharpoons H_3CCOO^- + H^+$$

$$H_3CCOONa \rightarrow H_3CCOO^- + Na^+$$

Due to common ion effect of H_3CCOO^- ion, the dissociation of the weak acetic acid is further suppressed. The small amount of H^+ ions from any source will combine with H_3CCOO^- ion to form H_3CCOOH whose dissociation has already been suppressed.

$$H_3CCOO^- + H^+ \rightarrow H_3CCOOH$$

The small amount of OH^- from any source will react with H_3CCOOH as follows:

$$H_3CCOOH + OH^- \rightarrow H_3C\quad COO^- + H_2O$$

Thus, the pH value remains constant.

Q 10. Explain the buffer action of a basic buffer.

Ans. Let us take the following example:

$$NH_4OH \rightleftharpoons NH_4^+ + OH^-$$

$$NH_4Cl \rightarrow NH_4^+ + Cl^-$$

Due to common ion effect of NH_4^+ ions, the dissociation of weak base NH_4OH is further suppressed. The OH^- ions from any source will combine with NH_4^+ ions to form NH_4OH whose dissociation is already suppressed.

The H^+ ions from any source will react with NH_4OH as follows:

$$NH_4OH + H^+ \rightarrow NH_4^+ + H_2O$$

The pH value of the solution remains constant.

Experiment 14

OBJECT

To determine the solubility of a given salt at room temperature and also to draw its solubility curve.

APPARATUS

Thermostat, beakers, stirrer, thermometer etc.

THEORY

The solubility of the salt is determined by preparing its saturated solution at a number of temperatures, and then a curve is plotted between the solubility as ordinate and temperature as abscissa.

PROCEDURE

Take about 50 ml of distilled water in a beaker and add to it some amount of the given salt, say KCl and stir the contents. Now, heat the solution to about 85–90 °C and go on adding a large amount of the salt. Please note that the salt should be in excess at 85–90 °C. Now, cool the solution and as the temperature falls to say 80 °C, pipette out about 4–5 ml of the supernatant liquid almost quickly and transfer it to a previously weighed beaker or watch glass. Then, again pipette out about 4–5 ml of the supernatant liquid, when the temperature falls to 70 °C and again transfer it to a previously weighed another beaker or watch glass. Repeat the above process at temperatures 60 °C, 50 °C, 40 °C, 30 °C and room temperature. (The solution should be cooled in a larger beaker containing water.)

Now, weigh each beaker or watch glass. Dry the solution in an electric oven at 120 °C or on a water bath till whole of water evaporates. When only dry residue of the solid is left behind, weigh all the beakers or watch glasses.

S. No.	Temperature (°C)	Weight of Watch Glass (g)	Weight of Watch Glass + Solution (g)	Weight of Watch Glass + Residue (g)	Weight of Residue (g)	Weight of Water (g)
1	80	a_1	b_1	c_1	$c_1 - a_1$	$c_1 - b_1$
2	70	a_2	b_2	c_2	$c_2 - a_2$	$c_2 - b_2$
3	60	a_3	b_3	c_3	$c_3 - a_3$	$c_3 - b_3$
4	50	a_4	b_4	c_4	$c_4 - a_4$	$c_4 - b_4$
5	40	a_5	b_5	c_5	$c_5 - a_5$	$c_5 - b_5$
6	30	a_6	b_6	c_6	$c_6 - a_6$	$c_6 - b_6$
7	Room temperature	a_7	b_7	c_7	$c_7 - a_7$	$c_7 - b_7$

CALCULATIONS

Solubility of the given salt in 100 g of water at

$$80°C = \frac{c_1 - a_1}{c_1 - b_1} \times 100$$

$$70°C = \frac{c_2 - a_2}{c_2 - b_2} \times 100 \quad \text{and so on.}$$

A curve is plotted with solubility as ordinate and temperature as abscissa and on joining the points, we get the solubility curve.

RESULT

Solubility of the given salt at room temperature = …… %.

PRECAUTIONS

(i) While pipetting out, only the supernatant liquid should be taken out.

(ii) To ensure complete evaporation of the solvent, repeated weighing and drying are made till the weight is constant.

(iii) While pipetting hot solutions, usually small crystals separate on the surface of the liquid, which get into the pipette. To avoid this, use a pipette to the tip of which is tied a piece of dry filter paper.

(iv) The pipette should be warmed by sucking hot water repeatedly at that temperature at which the measurement is to be made, otherwise the solute will deposit at the cold wall of the pipette as soon as it comes in contact with the pipette.

VIVA-VOCE

Q 1. Define solubility of a solid in a solvent.

Ans. The amount of the solute dissolved in 100 g of the solvent at a definite temperature to form a saturated solution is called its solubility.

Q 2. What is a saturated solution?

Ans. Whenever a solid solute is dissolved in a particular solvent (say water) at a definite temperature, a stage is ultimately reached when no more solid goes into the solution. The solution at this stage is called a saturated solution. The solubility of any solid is, therefore, fixed at a particular temperature.

Q 3. What is the effect of temperature on the solubility of a solid in a liquid?

Ans. The process of dissolution of a solid in a liquid is usually accompanied by the adsorption of heat, i.e., it is an endothermic process. So, according to Le-Chateleir principle, such a process is favoured by the increase of temperature. So with rise of temperature, the solubility of a solid in a liquid increases. In some cases, dissolution of lime in water is accompanied by evolution of heat. In such cases, the solubility would decrease with rise of temperature.

Q 4. What is the nature of solubility equilibrium? Explain.

Ans. Solubility equilibrium is dynamic in nature. It means that when the equilibrium is established, the process does not stop. It only appears to have stopped. At this stage, the rate of dissolution of solid in the liquid becomes equal to the rate of precipitation. The process is taking place in both the directions with equal rates.

Q 5. How can you determine the solubility of a salt at a particular temperature?

Ans. The solubility of a salt is determined by preparing its saturated solution at a number of temperatures and then a curve between the solubility and the temperature is plotted from the curve. The solubility at a particular temperature can be calculated from the graph.

Q 6. Define the enthalpy of a solution.

Ans. The heat of solution of a substance in a particular solution is defined as the heat change (i.e., the amount of heat evolved or absorbed) when 1 mole of the substance is dissolved in such a large volume of the solvent that further addition of the solvent does not produce any more heat range.

Q 7. Define the heat of hydration.

Ans. The amount of heat change (i.e., the heat evolved or absorbed) when one mole of the anhydrous salt combines with the required number of moles of water so as to change into the hydrated salt is called the heat of hydration. The may be represented as

$$CuSo_{4(s)} + 5H_2O \rightarrow CuSo_4\ 5H_2O_{(s)} \quad \Delta H = -78.2 \text{ kJ}$$

Q 8. How does the heat of hydration of an anhydrous salt can be calculated?

Ans. It can be calculated by applying Hess's Law. The process of dissolution of anhydrous salt takes place in two steps. The anhydrous salt will first get hydrated, and the hydrated salt will then get dissolved to form the solution.

So, ΔH (hydration) can be calculated as

ΔH (hydration of anhydrous salt) = ΔH (solution) anhydrous salt – ΔH (solution) hydrated salt

Q 9. State Hess's Law of constant heat summation.

Ans. The total amount of heat evolved or absorbed in a reaction in the same whether the reaction takes place in one step or in a number of steps. In other words, the total amount of heat change in a reaction depends only upon the nature of the initial reactants and the nature of the final products and is independent of the path or the manner by which this change is brought about.

Q 10. What is the main consequence of Hess's Law which make it very important for indirectly calculating the heat of reactions of some reactions?

Ans. All the calculations of Hess's Law are based upon the following consequence of the law:

The thermochemical equations can be treated as algebraic equations, which can be added subtracted, multiplied or divided by any number.

Q 11. Why we usually study enthalpy change and not internal energy change?

Ans. It is because most of the processes are carried out in open vessels, i.e., at constant pressure (say atmospheric) and not in a closed vessel, i.e., at constant volume.

Q 12. What is the sign of ~ H for endothermic reactions any why?

Ans. In the case of endothermic reaction ΔH is always positive. It is because

$$\Delta H = H_P - H_R \quad \text{and} \quad H_R < H_P$$

Experiment 15

OBJECT

To determine the molecular weight of the given substance by Rast method.

THEORY

The molecular weight of a non-volatile substance is determined by Rast method employing the principle of depression of freezing point. The following equation is applicable.

$$m = \frac{1000\,K_f w}{W \times \Delta T_f}$$

where m = mol. wt. of the solute, K_f = molar depression constant of the solvent, w = weight of the solute in gram, W = weight of the solvent in gram and ΔT_f = depression in freezing point of the solvent.

APPARATUS AND MATERIALS

1. A small Pyrex or Corning tube.
2. Sulphuric acid or liquid paraffin or glycerine bath.
3. Thin glass rod.
4. Thiele's tube.
5. Thermometer with the calibration of 0.1°C.
6. Camphor.
7. Experimental solute whose mol. wt. is to be determined.

PROCEDURE

1. Purify camphor by the process of sublimation.
2. Take a cork. Make a groove into it such that the tube can fit into the groove and stand erect.

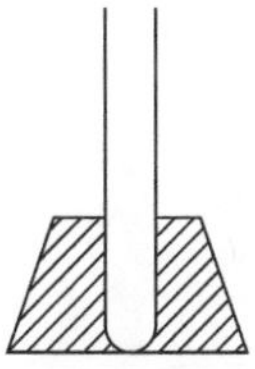

Figure Ex 15.1 Cork supporting a tube

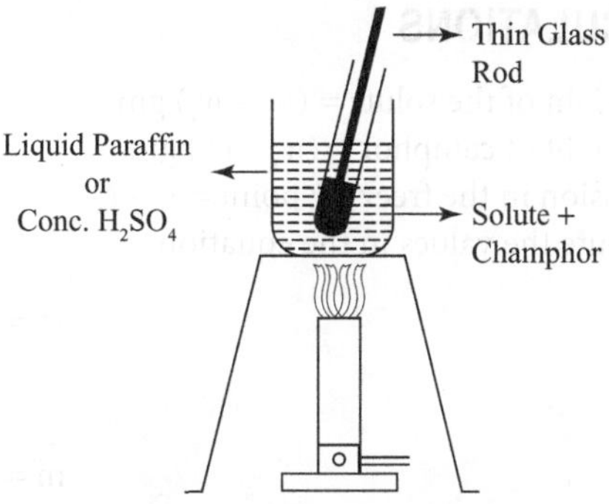

Figure Ex 15.2 Preparing a uniform mixture of the solute and the solvent

3. Clean the test tube. Dry it by placing in the oven for 30 min at 120°C. Fix it in the groove of the cork and place it along with the cork in the pan of the weighing balance to weigh it.
4. Take approximately 50 mg of the solute in the tube and note down the exact weight. Again, take approximately 500 mg of camphor in the tube and note down the exact weight.
5. Dip the tube in a small beaker containing conc. H_2SO_4 or liquid paraffin or glycerine. Heat the beaker on a low flame until the constituents melt. Stir the constituents of the tube with a thin glass rod to mix them uniformly. Remove the tube from the beaker and allow it to cool. The mixture will solidify.
6. Take out the mixture from the tube with the help of a thin spatula on a watch glass. Fill this mixture in a capillary tube and determine the melting point in a Thiele's tube. In addition, determine the melting point of pure camphor.

OBSERVATIONS

Let,

The weight of cork + empty tube = w_1 gm

The weight of cork + tube + solute = w_2 gm

The weight of cork + tube + solute + camphor = w_3 gm

The melting point of pure camphor

1st observation = °C

2nd observation = °C

3rd observation = °C

The mean value = t_1°C (say)

The melting point of the solute + camphor

1st observation = °C

2nd observation = °C

3rd observation = °C

The mean value = t_2°C (say)

Molal depression constant of camphor = 39.7°C

CALCULATIONS

The weight of the solute = $(w_2 - w_1)$ gm
The weight of camphor = $(w_3 - w_2)$ gm
Depression in the freezing point = $t_1 - t_2$
Substitute the values in the equation

$$m = \frac{1000 \times K_f \times w}{W \times \Delta T_f}$$

or

$$m = \frac{1000 \times 39.7 \times (w_2 - w_1)}{(w_3 - w_2) \times (t_1 - t_2)}$$

PRECAUTIONS

1. Camphor to be used for the experiment should be purified by sublimation.
2. The concentration of the solute in the solvent should not be more than 10% (50 mg in 500 mg), as the colligative law is applicable to dilute solutions.
3. It is necessary to mix homogenously the solute and the solvent before finding out the melting point of the mixture.
4. Slow heating should be done while finding out the melting point of pure camphor and the mixture.

VIVA-VOCE

Q 1. What is the principle of Rast method of determination of molecular weight?
Ans. It is based on the principle of depression of freezing point.

Q 2. Define freezing point.
Ans. Freezing point of a liquid is the temperature at which both the liquid and solid form of the substance co-exist and vapour pressure of both the phases becomes equal.

Q 3. What do you understand by depression in freezing point?
Ans. Freezing point of any solution of a non-volatile solid in a liquid is always lower than that of the pure solvent. The difference between the freezing point of a solvent and that of its solution is called depression in freezing point.

Q 4. Why does depression in freezing point take place?
Ans. The vapour pressure on the surface of a solid is always less than that of its liquid form at a particular temperature. At the freezing point, the vapour pressure of the two forms becomes equal. The vapour pressure on the pure liquid solvent being higher, the temperature has to be lowered in case of solution, so that the vapour pressure on the two phases becomes equal.

Q 5. Define molal depression constant.
Ans. It is the depression in freezing point produced by dissolving 1 gm mole of a solute in 1000 gm of the solvent.

Q 6. Write the equation between the mol. wt. and the weights of the solute and the solvent.

Ans.

$$m = \frac{1000 \times K_f \times w}{W \times \Delta T_f}$$

Q 7. Why is the camphor taken as solvent?

Ans. It is because camphor has a high value of molal depression constant of 39.7. Thus, the addition of a small quantity of the solute produces a big depression in freezing point, which can be observed comfortably.

Q 8. What are the limitations of the Rast method?

Ans. Rast method is applicable only to those compounds which dissolve in camphor to give a uniform mixture.

Q 9. What is the principle of cooling mixture of ice and salt?

Ans. It works on the principle of depression of freezing point.

Experiment 16

OBJECT

To determine the molecular weight of a volatile liquid by Victor Meyer's method.

THEORY

A known weight of the volatile liquid is heated and made to displace its own volume of air which is then collected and measured under known conditions of temperature and pressure. The volume of the air displaced is then converted to the volume at NTP by applying the equation

$$\frac{P_1V_1}{T_1} = \frac{P_2V_2}{T_2}.$$

It is assumed that vapours of the liquid behave ideally. The molecular weight of the liquid can be derived from the fact that one mole of gas at NTP occupies a volume of 22.4 litres (22,400 ml).

APPARATUS AND MATERIALS REQUIRED

(i) Victor Meyer's apparatus consists of:
   (a) *A hard glass tube*, called 'Victor Meyer's tube'. This tube has a cylindrical bulb at the lower end and a side tube near the upper end. The upper end is funnel shaped and is fitted with a cork.
   (b) *An outer jacket of copper* or corning glass containing a suitable liquid.
   (c) A small stoppered bottle is known as 'Hofmann's bottle'. The compound is weighed in this bottle and introduced into the Victor Meyer's tube.
   (d) *A graduated tube* for collecting the displaced air, i.e., one eudiometer tube.

(ii) Two iron stands with two clamps and two rings.
(iii) One Bunsen burner and a wire gauze.
(iv) 100 ml-measuring cylinder.
(v) One levelling glass jar.
(vi) One large glass disc.
(vii) Glass wool or asbestos.
(viii) Sample bottle containing the experimental liquid.
(ix) Analytical balance.

(x) A very reliable set of fractional weights.
(xi) Benzene, chloroform, dichloroethylene and carbon tetrachloride may be used as experimental liquids.

PROCEDURE

(i) The Victor Meyer's tube is cleaned and dried and a little glass wool or an asbestos pad or a 0.5-cm thick layer of sand is placed at its bottom. It is then tightly corked and introduced into the outer copper jacket.
(ii) Fill the bulb of the jacket with a liquid, the boiling point of which is 30–40°C higher than the boiling point of the liquid under test. In this case, place 250–300 ml of water in the bulb of the outer jacket. This is enough to boil for half an hour. It is then placed on a tripod stand and clamped on a heavy iron stand with two clamps.
(iii) Fix the Victor Meyer's tube in the outer jacket with the help of a split-hole rubber cork (which has been cut into two halves) fitted with a small bent glass tube in one side for steam to go out. *The bottom of the Victor Meyer's tube should be about an inch higher than the water level in the outer jacket, i.e., it should not dip in water.*

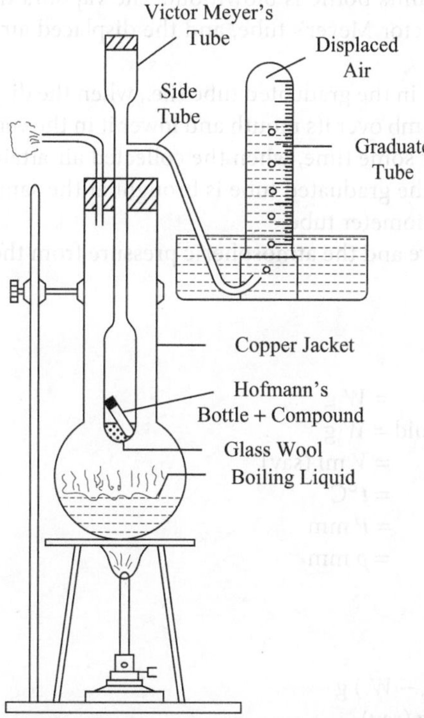

Figure Ex 16.1 Victor Meyer's apparatus with a graduated tube

A few pieces of broken porcelain are added carefully to the liquid to prevent *its bumping during boiling. The side tube of the Victor Meyer's tube is dipped in water contained in a trough. The tip of the side tube should not be more than 1 cm below the water surface.* The water above the tip obstructs the flow out of the displaced air.

(iv) Weigh accurately the dry Hoffmann's bottle and then weight into it about 0.1–0.2 g of the liquid. *The bottle should not be filled more than one-fourth.*

(v) Heat the bulb of the jacket on a low flame and as soon as the water begins to boil, adjust the flame so that the steam just escapes through the bent glass tube. Continue the boiling of water. The vapours of this liquid heat the Victor Meyer's tube, and as a result the air inside the tube expands. Now, some of it escapes through the side tube and air bubbles and thus appear in the trough of water. When the air stops expanding, no more air bubbles appear in the trough of water. When the temperature of the inner tube becomes uniform, no more air bubbles escape from the apparatus. The graduated tube or the eudiometer tube full of water is then placed over the end (tip) of the side tube in the trough.

(vi) The cork of the Victor Meyer's tube is now momentarily removed and the loosely stoppered Hoffmann's bottle containing the weighed quantity of the liquid under examination is introduced quickly into the Victor Meyer's tube. The mouth of the Victor Meyer's tube is then at once closed with a rubber stopper. The stopper should not be forced too far, otherwise some air may be displaced from the apparatus.

Due to high temperature inside, vaporization of the compound takes place immediately and the stopper of the Hoffmann's bottle is blown out. The vapours thus formed displace an equal volume of air from the Victor Meyer's tube, and the displaced air is collected in the graduated (eudiometer) tube.

(vii) When no more air collects in the graduated tube (i.e., when the displacement of air is complete), take it out by putting a thumb over its mouth and lower it in the vertical position into a tall levelling jar full of water. After some time, when the collected air attains the room temperature, the water inside and outside the graduated tube is brought at the same level by raising or lowering the graduated tube or eudiometer tube.

(viii) Note the room temperature and the atmospheric pressure from the barometer.

OBSERVATIONS

Weight of empty Hoffmann tube $= W_1$ g
Weight of Hoffmann bottle + liquid $= W_2$ g
Volume of air collected $= V$ ml (say)
Room temperature $= t$°C
Atmospheric pressure $= P$ mm
Aqueous tension of water at t°C $= p$ mm

CALCULATIONS

Weight of the liquid taken $= (W_2 - W_1)$ g
$= W$ g (say)

Pressure of dry air $= (P - p)$ mm

From gaseous equation, we have

$$\underset{\text{(Experimental)}}{\frac{P_1V_1}{T_1}} = \underset{\text{(NTP)}}{\frac{P_0V_0}{T_0}}$$

∴ Volume (V_0) of dry gas at NTP is given by,

$$V_0 = \frac{P_1V_1}{T_1} \times \frac{T_0}{P_0}$$
$$= \frac{(P-p)V}{(t+273)} \times \frac{273}{760} \text{ ml} = v \text{ ml (say)}$$

∴ W g of the liquid in the vaporous state occupies a volume of v ml at NTP.

∴ Weight of the substance (or liquid) that would occupy a volume of 22,400 ml at NTP $= \frac{W \times 22400}{v}$

∴ Molecular weight of the liquid $= \frac{22400 \times W}{v}$

RESULT

The molecular weight of the given liquid = ...

Correction of aqueous tension: This correction is necessary, as the air displaced from inside the Victor Meyer's tube A due to evaporation of the liquid is the same as that of the atmospheric air. However, when the displaced air is collected over water, it gets saturated with water vapours. If the atmospheric pressure is P mm of Hg and aqueous tension is p mm of Hg, then the true pressure of displaced air, if it was to occupy the same volume as that of the moist air and at the same temperature, would have been equal to $(P - p)$ mm of Hg. However, it is an impossible situation that the atmospheric air should be absolutely dry. If the percentage of humidity in the atmosphere is x%, then the correct pressure of air would be equal to,

$$\left[P - P\left(1 - \frac{x}{10}\right)\right] \text{ mm of Hg}$$

PRECAUTIONS

The vaporization of liquid should take place rapidly. If it takes place slowly, diffusion and condensation on the upper and lower parts of the tube may occur. The volume of air expelled will then be too small. For the same reason, the volume of air expelled should be read as soon as it becomes constant.

(i) Sometimes, the stopper of the Hoffmann's bottle becomes fixed and the liquid is prevented from vaporizing or it vaporizes very slowly. *So the stopper of the Hoffmann's bottle should be loosened (or removed altogether if the liquid is not too volatile) before it is dropped into the Victor Meyer's tube,* so that as soon as it is inserted in the Victor Meyer's tube, it should open spontaneously.

(ii) *Care must be taken that the weighing bottle is not filled so full that the liquid wets the stopper and thus some of it gets lost.*

(iii) *The fractional weights used for weights must be reliable as error of a few milligrams will cause a high percentage of error in the result.* The error in the determination of molecular weight by this method should not exceed 5%.
(iv) The apparatus should be air tight.
(v) The volume should be noted after *careful levelling.*
(vi) The liquid taken in the outer jacket must be such that its boiling point is about 30–40°C higher than the boiling point of the substance so that it may cause quick vaporization of its experimental liquid.
(vii) The Hoffmann's bottle should not be filled more than three-fourthfull, otherwise if it is taken in a large quantity, the vapours of the liquid will displace all air from the Victor Meyer's tube and start passing through water where they will condense.
(viii) Heating should be strong and uniform, otherwise water will pass from the Victor Meyer's tube.
(ix) While introducing the Hofmann's bottle, the cork of the Victor Meyer's tube should be slightly opened and that too for a short time.

VIVA-VOCE

Q 1. Define molecular mass of a substance.
Ans. Molecular weight of any substance is the relative mass of one molecule of that substance as compared with an atom of C-12 taken as 12 a.m.u.

Q 2. Is the molecular mass the actual mass of the molecule of the substance? Comment on your answer.
Ans. No, the actual masses of the molecules are very small; we, therefore, express molecular masses as the relative masses. To begin with, the weight of an H-atom being the smallest atom was taken as 1 a.m.u. and the weight of other atoms and molecules were compared with it. At present, C-12 taken as 12 a.m.u. is the accepted standard for comparison.

Q 3. Can we determine the molecular mass of any liquid by Victor Meyer's method? Comment.
Ans. No, the molecular masses of only volatile liquids can be determined.

Q 4. What to you understand by molar volume of any gas?
Ans. Molar volume is volume occupied by one mole (i.e., gram molecular mass) of any substance at normal temperature and pressure (NTP). Its numerical value is the same for all gases, i.e., 22.4 l at NTP.

Q 5. Why the liquid taken in the outer jacket of Victor Meyer's apparatus should have boiling point 30–40 °C higher than that of the liquid under trial?
Ans. The liquid under trial will get volatilized only if the temperature of air inside the Victor Meyer's tube is 30–40 °C higher than the boiling point of the liquid to be vaporized.

Q 6. Why the Hoffmann's bottle is not filled more than one fourth?
Ans. When the volume of liquid taken in bottle is more, its vapours may not only displace the air and the vapours themselves, but it may pass through water and would be condensed. The volume of the vapours formed will not be measured correctly.

Q 7. Why the volume of the air displaced is measured after careful levelling in a levelling jar?

Ans. It is because we wish to measure the volume of air displaced at room temperature and atmospheric pressure. When the level of water inside the graduated tube and outside in the levelling jar is the same, then it is being measured at atmospheric pressure. Atmospheric pressure is acting on the water outside, i.e., in the levelling jar and the air enclosed in the graduated tube exerts its pressure on water level inside the tube.

Q 8. Can you suggest an alternative method for calculating the molecular mass of the volatile liquid? Explain the method.

Ans. The alternative method can be the determination of vapour density. Vapour density of a gas or vapour is defined as *the number of times a certain volume of it is heavier than the weight of an equal volume of hydrogen at the same temperature and pressure.*
If V ml of a gas or vapour at N.T.P. weighs Wg, then
Weight of 1 ml of vapour = W/V g

$$\text{Vapour density (V.D.)} = \frac{\text{Weight of certain volume of gas}}{\text{Weight of the same volume of hydrogen}}$$

$$= W/V \times 0.00009 = \frac{W}{0.00009 \times V}$$

($\therefore$ 1 ml of H_2 at N.T.P. weighs 0.00009 g)

$$\text{Molecular weight} = 2 \times \text{V.D}$$

$$= 2 \times \frac{W}{0.00009 \times V}$$

Q 9. At what stage the Hoffmann bottle carrying the liquid should be dropped into the Victor Meyer inner tube?

Ans. When the air inside the Victor Meyer tube has attained uniform temperature, which will be indicated when no more air bubble escapes from the water contained in the trough.

Q 10. Define vapour density. How is it related to molecular mass?

Ans. Vapour density of a gas or vapour is defined as the number of times a certain volume of it is heavier than the weight of an equal volume of hydrogen at the same temperature and pressure.

Molecular mass = 2 × Vapour Density.

Q 11. How will you calculate the molecular mass from vapour density measurements?

Ans. $$\text{Vapour density (V.D.)} = \frac{\text{Weight of certain volume of gas}}{\text{Weight of the same volume of } H_2 \text{ gas}}$$

$$\text{V.D.} = \frac{\text{Weight of V ml of gas}}{\text{Weight of V ml of } H_2 \text{ gas}}$$

If V ml of a gas or vapour at N.T.P. weighs Wg, then

$$\text{Weight of 1ml of vapour} = \frac{W}{V} g$$

Weight of 1 ml of H_2 at N.T.P = 0.00009g

$$\text{Therefore V.D} = \frac{W}{V} \bigg/ 0.00009 = \frac{W}{0.00009 \times V}$$

$$\text{Molecular mass} = 2 \times \text{V.D} = 2 \times \frac{W}{0.00009}$$

Q 12. State Avogadro's law.

Ans. Equal volumes of gases under similar conditions of temperature and pressure contain equal number of molecules.

Q 13. How will you calculate molecular mass from Avogadro's law?

Ans. We know that one mole of every gas at N.T.P. occupies the same volume, i.e., 22400 ml. If Wg of a volatile substance given V ml of dry vapour at N.T.P., then according to Avogadro's law.

$$\text{Molecular mass (M)} = \frac{W}{V} \times 22400$$

Experiment 17

OBJECT

Determination of R_f (retention factor) value of amino acid by thin-layer chromatography (TLC) and identification of the amino acid present.

THEORY

TLC is a special type adsorption chromatography in which a thin layer of the adsorbent on a flat surface is utilized instead of column of adsorbent. Elution and development of the chromatogram is accomplished by capillary movement of the solvent up the thin layer of the adsorbent.

In this technique, the material to be identified or separated is dissolved in a suitable solvent. It is spotted near one end of the plate covered with a thin layer of silica gel or alumina containing some binding material such as calcium sulphate. The spot is then developed by evaporating the solvent, and the spotted material is retained on the plate. The spot is then developed into a chromatogram by placing the coated plate into a solvent or a solvent mixture. The solvent rises through the adsorbent layer by capillary action and components of the mixture applied ascent at different rates, depending upon their different affinities to the adsorbent. This results in the separation of the components spotted from one another. When the solvent front reaches near the upper edge, the plate is taken out from the developing solvent. It is then dried, visualized and examined.

The substance can be identified by determining its R_f value. It is the ratio of distances moved through by the solute and solvent on a chromatogram.

$$R_f = \frac{\text{Distance moved by the solute}}{\text{Distance moved by the solute front}}$$

It should be noted that the value of R_f is always less than unity, and its value for various substances is entirely different. The R_f value of the sample is compared with that of the pure compound. If the R_f value in both cases is the same, the sample is pure.

MATERIALS REQUIRED

(i) The amino acids such as L-lysine, L-leucine and DL-alanine.
(ii) Silica gel for TLC.

(iii) Microscopic slides (glass plates).
(iv) Butan-1-ol, propanol, ninhydrin reagent.
(v) Small jars with lids.
(vi) Wide-mouth stoppered bottles.
(vii) *Preparation of silica gel slurry:* Take 35 g of silica gel in 100 ml of chloroform or 100 ml of chloroform–methanol mixture (2:1). Shake and stir the slurry for about 3 min. The slurry may be stored for sometime in a tightly sealed container such as wide-mouth stoppered bottle and must be shaken again before use.
(vii) *Preparation of the solution of amino acids:* Prepare a solution of DL-alanine, L-leucine and L-lysine hydrochloride by dissolving 5 mg of each separately in 0.5 ml of distilled water, measured with a graduated 1 ml pipette. The solution may have to be warmed. Mix one drop of each solution to provide a mixture of the three amino acids. Dilute the remainder of each solution to 1 ml to give the solution of the respective amino acids.

PROCEDURE

TLC involves the following steps:

1. **Preparation of a thin-layer plate**

   Dip two clean and dry microscopic slides held together back to back by crucible tongs into the slurry. Dip in the slurry slowly in a continuous movement. Remove the slides slowly and allow them to drain on the edge of the container. Do not coat 1 cm of the slides from the top. The thickness of the film can be controlled by adjusting the ratio of the absorbent to solvent in the slurry. Separate the slides by handling only the top edges. Place them on a sheet of filter paper with the films facing upwards and let them dry for 5 min. Remove the excess adsorbent from the edges.
2. **Applications of the sample (spotting the plates)**

   The applicator is made by pulling the centre of an open melting point capillary in a flame to a very fine capillary and breaking it at the centre. Apply two sample drops at least 1 cm apart and about 1 cm above the lower end of the plate. One of the drop is that of any of the pure amino acid and the other drop is that of the mixture of the amino acids. Allow the drops to dry in the air for some time.
3. **Development of the chromatogram plate**

   Prepare the developing solvent by making 70 ml of propan-1-ol with 30 ml of concentrated aqueous ammonia soak a filter paper strip and stick it to the interior of the jar, so that the jar remains saturated with the vapours placed into the jar. Carefully introduce the developing solvent by means of a pipette so that the lower edge of absorbent layer is immersed in the solvent. Cover the mouth of the jar and allow the chromatogram to develop.
4. **Visualization (location of components)**

   Remove the chromatogram from the developing jar and dry it at 100°C for 10 min. Spray with ninhydrin reagent prepared by dissolving 0.2 g of the reagent in 100 ml of butan-ol. Heat at 110°C for 5–10 min. In case the spots are not visible, the plate may be placed in a jar containing a few crystals of iodine. The iodine vapours are absorbed into the areas of the plate containing organic compounds. Brown spots appear on the white background.
5. **Checking the purity of the sample**

   Three spots from the mixture of the three amino acids, one above the other, are seen in a line and one spot from the pure amino acid. Their R_f values can be determined.

$$R_f = \frac{\text{Displacement of the compounds } (d_1)}{\text{Displacement of the solvent front } (d_2)}$$

where d_1 = the distance between centres of the initial spot and the located spot and d_2 = the distance between the centre of the initial spot and the solvent front

PRECAUTIONS

(i) Bottles containing the slurry must be properly stoppered, since the solvents in which the slurry is prepared are highly volatile.
(ii) The glass plates must be absolutely clean.
(iii) The chromatographic plate must be handled with great care.
(iv) Before spotting, the chromatic plate should be heated to about 100–200°C for activation.
(v) There should be a thin or uniform film of slurry on the glass plates.
(vi) The slurry may be made thinner by adding more solvent if it is very thick. In case the slurry is too thin, it can be made thicker by the addition of more adsorbent.

VIVA-VOCE

Q 1. Define the term chromatography?
Ans. Chromatography is an analytical technique of identification of compounds and separation of mixture of solutes, brought about by the differential movement of the individual solutes through a porous medium under the influence of a solvent.

Q 2. What is the basis of chromatographic technique?
Ans. This technique is based on preferential adsorption of solutes from a mobile phase onto the stationary solid phase. The extent of adsorption of a single component depends on the polarity of the molecule, the activity of the adsorbent and the polarity of the mobile liquid phase.

Q 3. What is adsorption chromatography?
Ans. The use of solid as a stationary phase with a liquid mobile phase is known as the adsorption chromatography. It is further divided into (i) column chromatography, (ii) paper chromatography and (ii) TLC.

Q 4. What is TLC?
Ans. In this technique, a thin layer of suitable adsorbent in a suitable solvent is pasted on a glass plate, which acts as a stationary phase. The substance to be identified or the mixture to be separated is dissolved into a suitable solvent (mobile phase) to move up the stationary phase of adsorbent on the plate. The separation of the components takes place due to their different affinities for the adsorbent. The developed spots are located by visualization, followed by the identification of the compounds.

Q 5. List some common adsorbents in the order of their increasing adsorption activity.
Ans. Some common adsorbents in the order of their increasing adsorption power are as given below: cellulose, starch, sugars, magnesium sulphate, calcium sulphate, silicic acid (silica gel), magnesium oxide, aluminium oxide (alumina) and activated charcoal.

Q 6. What are eluents?

Ans. The solvents employed for removing the adsorbed components from the surface of adsorbent in column chromatography are called eluents. The eluting power of a solvent depends upon its polarity. The solvents used for developing the chromatogram in TLC are also called eluents. Petroleum ether, cyclohexane, carbon tetrachloride and benzene are examples of eluents with low polarity whereas ethanol water and acetic acid are eluents with high polarity.

Q 7. On what factors the choice of an eluent depends?

Ans. The choice of a suitable solvent for elution depends on the type of the adsorbent and the components to be separated or identified. The eluting power of a solvent depends on its polarity. The type of solvent polarity depends upon the similarity of the components to be separated. A proper selection of eluent is made by trial by mixing the solvents of the desired polarity.

Q 8. Give the advantages of TLC over paper chromatography.

Ans. TLC is applied in preference to the paper chromatography on account of the following advantages:

(i) It is far more rapid than paper chromatography and gives quick and reliable results.
(ii) Sharp spots are obtained as compared to paper chromatography where the spots are diffused.
(iii) Acidic or alkaline solution can be used for the location of spots, which is not possible in paper chromatography.
(iv) TLC chromatoplates can be heated, if required. This is not possible in paper chromatography.
(v) The spots can be scrapped out for further analysis.

Q 9. Enlist the steps involved in TLC technique.

Ans. TLC involves the following steps:

(i) Preparation of a thin-layer plate.
(ii) Application of the material to be separated.
(iii) Development of chromatogram plate in a solvent.
(iv) Location of the components, i.e., visualization.
(v) Calculation of R_f value.

Q 10. What do you understand by the term retention factor?

Ans. The movement of any substance relative to the solvent front in a given chromatographic system is constant and characteristic of a substance. This constant is expressed as R_f value and is identified as

$$R_f = \frac{\text{Distance moved by the solute}}{\text{Distance moved by the solute front}}$$

Experiment 18

OBJECT

To separate Cu^{2+} and Cd^{2+} ions present in a given mixture using paper chromatography and to determine the R_f values.

THEORY

Paper chromatography is a special case of partition chromatography. In this case, the stationary phase is the paper, which contains about 22% water absorbed on 78% of the cellulose. Thus, the stationary phase is water adsorbed on cellulose which provides an inert support. The solvent used for developing is called the irrigent and is the mobile phase. The irrigent is usually solvent or a mixture of solvents.

The separation of the mixtures into its components takes place by partitioning of the components between stationary phase containing water and the mobile phase (irrigent). The mobile phase travels through the filter paper by capillary action. The paper chromatographic technique can be carried out in three ways— (i) ascending, (ii) descending and (iii) circular—depending on the way the solvent travels on the paper. We shall be using ascending paper chromatography.

MATERIALS REQUIRED

(i) Whatmann filter paper strip (20 × 2 cm size).
(ii) A glass cylinder (25 × 7 cm size) with a rubber cock containing a hook at the centre.
(iii) 1% aqueous solution of copper sulphate and cadmium sulphate or cadmium chloride.
(iv) The developing solution (irrigent).
   (a) Ethanol (90 ml) and 5-N HCl (10 ml), or
   (b) *n*-butanol saturated with 3-N HCl
(v) *Visualizing reagents:* Colourless solution of $(NH_4)_2S$ prepared by bubbling H_2S gas in dilute ammonium hydroxide solution.

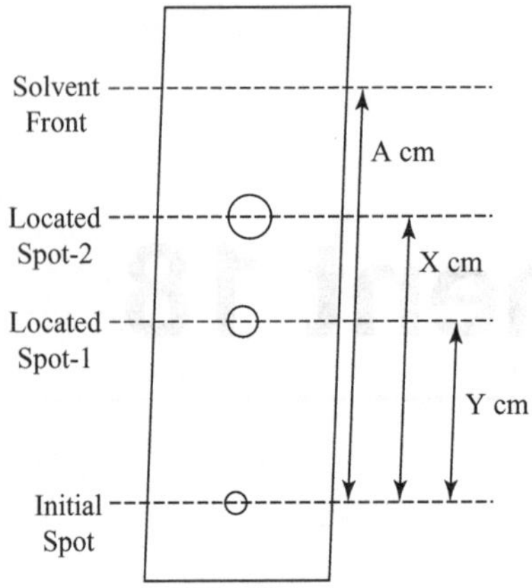

Chromatogram

PROCEDURE

A Whatmann filter paper strip (20 × 2 cm size) is taken, and a line is drawn with pencil approximately 4 cm from one end. A 2% solution of a mixture of Cu^{2+} and Cd^{2+} ion is prepared in distilled water in a test tube. With the help of a fine capillary tube, the filter paper is spotted with the given mixture solution at the centre of the horizontal line drawn on it. The process is repeated two to three times to make sure that adequate amount of the mixture is left on the paper. The applicator may be drawn from the capillary tube used for determining the melting point.

The developing solvent (20–25 ml) is put in glass cylinder water fitted with rubber cork having a hook at the centre. The spotted filter paper is then hanged straight through the hook. The initial line should be well above the developing solvent, so that around 1 cm of the filter paper is dipped into the solvent. The solvent slowly rises in the paper and carries the components of the spotted mixture up to different heights on the paper. When the solvent has risen approximately 15 cm, the filter paper strip is taken out and the solvent front is marked with pencil. The filter paper is dried in air.

The dried filter is sprayed with $(NH_4)_2S$ solution and dried again first in air and then in an electric oven at 100–110 °C. Two spots, one black (due to copper) and the other yellow (due to cadmium), appear on the chromatogram. There R_f values are calculated.

OBSERVATIONS

Temperature ....°C

Irrigent used ....

Colour of the Spot	Distance Between the Centre of the Spot and the Original and the Original Line	Distance Between the Solvent Front and the Original Line	R_f Value
Black (Cu^{2+})	X cm	A cm	X/A
Yellow (Cd^{2+})	Y cm	A cm	Y/A

RESULT

R_f value of Cu^{2+} = ...
R_f value of Cd^{2+} = ...

VIVA-VOCE

Q 1. What is chromatography?
Ans. It is a technique for purifying organic compounds and for separating components of a mixture.

Q 2. What is the basic principle of chromatography?
Ans. It is based on the difference in the rate of the migration of individual components of a mixture through a stationary phase under the influence of a moving phase.

Q 3. Name some of the different chromatographic techniques used.
Ans. Column chromatography, thin-layer chromatography, gas chromatography and paper chromatography.

Q 4. What is a paper chromatography?
Ans. It is basically a partition chromatography. In this case, water adsorbed on the paper which is made up of cellulose serves as the stationary phase and organic solvent as the moving phase.

Q 5. What is meant by R_f value?
Ans. R_f also known as retention factor of a substance is the ratio of the distance travelled by the solute from the original line and the distance travelled by the solvent from the original line.

Q 6. What is chromatogram?
Ans. In paper chromatography, the mixture is applied on the strip of the filter paper. It is then placed in a suitable solvent. After some time, when the solvent front is moved appreciably, paper is taken out and dried in the air. This is called known as chromatogram.

Q 7. How does the liquid rise through the paper?
Ans. Due to capillary action.

Q 8. How many types of paper chromatography are used?
Ans. Ascending, descending and circular paper chromatography. In ascending paper chromatography, the solvent (the mobile phase) moves in the upward direction. On the other hand, in descending paper chromatography the solvent moves downwards. However, in circular paper chromatography, the solvent moves horizontally.

Q 9. What are the factors on which the R_f depends?
Ans. Nature of the compound, solvent and the temperature.

Q 10. Is it possible to have R_f value greater than one? Explain your answer.
Ans. No, it is always less than one. The distance moved by the solvent front is greater as compared to that over any compound or ion carried by it.

Q 11. What is the use of value R_f?
Ans. It is constant for a compound with a particular solvent at a particular temperature, so it is also used for the identification of compounds.

Experiment 19

OBJECT

To determine the heat of neutralization of a strong acid, (HCl) and a strong base (NaOH).

THEORY

Heat is given out during neutralization of an acid with an alkali. Known quantities of the standard solutions of the acid and alkali are mixed in a calorimeter and the change in temperature is noted. Thus, the heat of neutralization may be defined *as the amount of heat given out or evolved when 1 gm equivalent of the acid is completely neutralized by 1 gm equivalent of the base in the dilute solutions.*

Heat of neutralization of a strong acid or strong base is always found constant, i.e., 57.3 kJ. This is due to the fact that strong acids (HCl) and strong base (NaOH) are completely ionized in dilute solutions. Thus, the heat of neutralization is regarded as the combination of H^+ and OH^- ions to form the undissociated molecule of water.

$$NaOH + HCl \rightarrow NaCl + H_2O + Q$$

or

$$H^+ + OH^- \rightarrow H_2O + 57.3\ kJ$$

APPARATUS AND OTHER MATERIALS REQUIRED

(i) 250 ml and wide-mouth polythene bottle or thermos flask fitted with a rubber stopper having two holes, one for the thermometer and the other for the glass stirrer.
(ii) Thermometer (graduated to 0.1 °C or 0.2 °C).
(iii) Beakers.
(iv) 0.5 N sodium hydroxide solution.
(v) 0.5 N hydrochloride acid solution.
(vi) Magnifying glass.

PROCEDURE

(a) Determination of water equivalent of 250-ml polythene bottle or thermos flask.

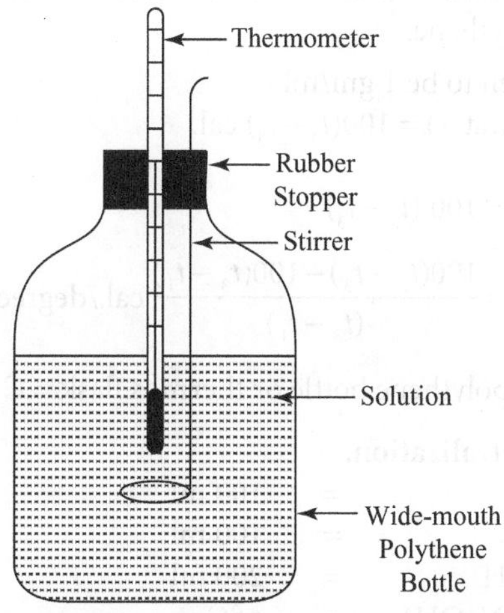

(i) Take the wide-mouth polythene bottle as mentioned above.

(ii) Place a thermometer (graduated to 0.1 °C) in the polythene bottle or thermos flask, so that its bulb nearly touches the bottom of the beaker.

(iii) Pipette out 100 ml of distilled water into the 250 ml polythene bottle and record the temperature to the nearest 0.1 °C.

(iv) Heat some water in a separate beaker to a temperature about 10–20 °C higher than the room temperature.

(v) Measure out 100 ml of the warm water with a pipette into another beaker and record its temperature.

(vi) Add the warm water quickly to the water in the 250 ml polythene bottle or thermos flask and stir the contents well.

(vii) Read the highest temperature (to 0.1 °C) attained immediately after mixing.

Calculate the water equivalent of 250 ml polythene bottle.

(b) Determination of heat of neutralization.

(i) Measure accurately 100 ml of 0.5 N hydrochloric acid into polythene bottle or thermos flask and record its temperature.

(ii) Take 0.5 N NaOH solution in a separate beaker and note its temperature. It should be the same as that of the acid taken in the polythene bottle. Pipette out 100 ml of 0.5 N NaOH solution and transfer it into bottle or thermos flask quickly.

(iii) Stir well and record the temperature (when it becomes steady).

OBSERVATIONS AND CALCULATIONS

(a) Determination of the water equivalent of polythene bottle or thermos flask.

Volume of water taken = 100 ml
Initial temperature of water = t_1 °C
Volume of hot water = 100 ml

Temperature of the hot water = t_2°C
Temperature of the mixed water = t_3°C
Heat gained by the beaker and water = $(MS + 100)(t_3 - t_1)$ cal., where M = weight of polythene bottle and S = specific heat of the polythene.

Specific gravity of water is taken to be 1 gm/ml
Heat lost (or given out by hot water) = $100(t_2 - t_3)$ cal.
Now, heat gained = heat lost
Therefore, $(MS + 100)(t_3 - t_1) = 100(t_2 - t_3)$

Therefore,
$$MS = \frac{100(t_2 - t_3) - 100(t_3 - t_1)}{(t_3 - t_1)} \text{ cal./degree} = C$$

Therefore, water equivalent of the polythene bottle or thermos flask = C cal./degree

(b) Determination of heat of neutralization.

Volume of the 0.5 N HCl = 100 ml
Volume of the 0.5 N NaOH = 100 ml
Total volume of HCl and NaOH = 200 ml
Initial temperature of HCl (or NaOH) = t_4°C
Final temperature of HCl + NaOH = t_5°C
Rise in temperature = $(t_5° - t_1°)$
Heat given out by the solution = $(200 + C)(t_5 - t_4)$ cal.
= x cal.

The quantity x is evolved by the neutralization of 100 ml of 0.5 N HCl.
Hence, the heat of neutralization for 1000 ml of 1 N HCl (or 1 gm equivalent)

$$= \frac{x}{100} \times 1000 \times \frac{1}{0.5}$$
$$= 20x \text{ cal.}$$

VIVA-VOCE

Q 1. What is neutralization?

Ans. It is the process of interaction between the hydroxyl ions (OH^-) of a base and hydrogen ions (H^+) of an acid. In other words, it will nothing but het of formation of water when both the acid and the base are strong.

$$OH^- + H^+ \rightarrow H_2O; \quad \Delta H = -57.3 \text{ kJ}$$

Q 2. Is the heat of neutralization exothermic or endothermic?

Ans. The process of neutralization is exothermic.

Q 3. Define heat of neutralization of an acid.

Ans. It is defined as the change in heat content of the system when 1 gm equivalent of an acid is neutralized by 1 gm equivalent of a base in dilute solutions.

Q 4. Define heat of neutralization of a base.

Ans. It is defined as the heat evolved when 1 gm equivalent of the base is completely neutralized by a strong acid using dilute solutions.

Q 5. Why heat of neutralization of strong acids and bases is constant?

Ans. The heat of neutralization in case of strong acids and bases is always the same in respective of the acid and the base taken. It is because strong acids and bases are fully ionized in aqua solution. In other words, the heat of neutralization in such cases is merely the heat of formation of water by the combination of hydrogen and hydroxyl ions.

Q 6. Why the heat of neutralization of a strong acid with a weak base and vice versa are different in different cases?

Ans. Weak acids and weak bases are not fully ionized. In such cases, the heat of neutralization will involve not only the heat of combination of hydrogen ions and hydroxyl ions but also the heat of ionization of a weak acid or the weak base. The heat of neutralization will be different in such cases depending upon the degree of ionization of the weak acid or the base.

Q 7. Which of the following reactions would have greater value of heat of neutralization and why?

(i) HCl (aq) + NaOH (aq) = NaCl (aq) + H_2O; $\Delta H = -57.3$ kJ

(ii) CH_3COOH (aq) + NaOH (aq) = CH_3COONa (aq) + H_2O; $\Delta H = -55.2$ kJ

Ans. The value of the heat of neutralization in the first case would be numerically greater than that of second case. It is because in the first case, both the acids and the base are strong; however, in the second case, the acid (CH_3COOH) is weak. A part of the heat is utilized in bringing about the complete neutralization of the weak acid.

Q 8. Which of the following reactions would have greater value of heat of neutralization and why?

(i) CH_3COOH (aq) + NaOH (aq) = CH_3COONa (aq) + H_2O; $\Delta H = -55.2$ kJ.

(ii) HCl (aq) + NH_4OH (aq) = NH_4Cl (aq) + H_2O; $\Delta H = -51.5$ kJ.

Ans. The numerical value of the heat of neutralization in the first is greater than that of second. It is because in the first case, the acid (CH_3COOH) is weak and in the second case, the base (NH_4OH) is weak. The heat of the ionization of acetic acid is different from that of the weak base. The numerical value of the heat of ionization of acetic acid is lesser than that of ammonium hydroxide. The part of the heat required for the complete dissociation of acetic acid is lesser in comparison to that of ammonium hydroxide.

Experiment 20

OBJECT

To determine the heat of dissolution of a solid in water at room temperature (solid electrolyte may be taken NH_4Cl, KCl, KNO_3 and $BaCl_2$ $2H_2O$).

THEORY

When a solid crystal is dissolved in water, it involves two processes.

(1) Crystal lattice is broken, this consumes *lattice energy* and thus it is an endothermic process.
(2) Water molecules attach to the solute particles, i.e., hydration takes place. As a result, energy is liberated. This is called *hydration energy* and thus is an exothermic process.

The difference between the lattice energy and hydration energy gives the heat of the solution.

Therefore, the heat of solution = lattice energy – hydration energy

Thus, from the above, it is quite clear if lattice energy < hydration energy, then heat is evolved and the heat of solution is negative and if lattice energy > hydration energy, then the heat of the solution is positive and heat is absorbed. Thus, the heat of solution is defined as

The amount of heat evolved or absorbed when 1 gm mole of solute is dissolved in a large excess of the solvent, so that further addition of the solvent produces no heat change.

The heat of solution of a substance varies with the number of gram molecules of water taken for dissolving 1 gm molecule of the substance.

APPARATUS AND OTHER MATERIALS REQUIRED

250 ml wide-mouth polythene bottle, a glass test tube, rubber cork, stirrer thermometer (graduated to 0.2°C), solid electrolyte (may be KNO_3, NaCl, NH_4Cl or $BaCl_2{\cdot}2H_2O$) about 6.0 gm and magnifying lens.

PROCEDURE

(a) Determination of water equivalent of the flask.

(i) Take a 250 ml wide-mouth polythene bottle with a rubber stopper having two holes, one for the thermometer and the other for the stirrer.
(ii) Place a thermometer (graduated to 0.2°C) in the polythene bottle, so that its bulb nearly touches the bottom of the beaker.
(iii) Pipette out 100 ml of distilled water into the 250 ml polythene bottle and record the temperature to the nearest 0.2°C.
(iv) Heat some water in a separate beaker to a temperature about 10–20°C higher than the room temperature.
(v) Measure out 100 ml of the warm water with a pipette into another beaker and record its temperature.
(vi) Add the warm water quickly to the water in the 250 ml polythene bottle and stir the contents well.
(vii) Read the highest temperature (to 0.2°C) attained immediately after mixing.

Calculate the water equivalent of the 250 ml polythene bottle.

(b) Determination of heat of dissolution of the solid electrolyte.

(i) Powder about 6.0 gm of the experimental solid electrolyte (NaCl) and place it in a test tube. Cork the tube with a rubber cork and weigh it accurately.
(ii) Pipette out 200 ml of distilled water in to the 250 ml polythene bottle (whose water equivalent has been determined) and record its temperature to the nearest 0.2°C.
(iii) Take some water in a separate beaker and dip the weighed test tube containing the experimental solid in water for some time. When the salt has taken the temperature of the water (say after 10–15 min), remove the tube and roughly dry it from outside.
(iv) Pour the contents of the test tube quickly into the 250 ml polythene bottle containing 200 ml distilled water. Stir the salt briskly till it is dissolved. Record its temperature to the nearest 0.2°C.
(v) Weigh the test tube with a proper cork after pouring out its content into 250 ml polythene bottle.

OBSERVATION AND CALCULATIONS

(a) Determination of water equivalent of the 250 ml polythene bottle.
Please see the previous experiment.

(b) Determination of heat of dissolution of the solid.
Weight of the tube + substance = m_1 gm.
Weight of the test tube after pouring out the substance = m_2 gm
Therefore, the weight of the substance dissolved = $(m_1 - m_2) = m$ gm.
Temperature of the distilled water before adding the substance = t_4°C.
Final temperature of solution after adding substance and dissolving completely = t_5°C.
Water equivalent of 250 ml polythene bottle = C (as calculated earlier).

Heat evolved or absorbed for dissolution of m gm of the substance per 200 gm of water.

$$= (200 + C + m)\ (t_5 - t_4)\ \text{cal} = Q\ \text{cals.}$$

Therefore, the heat evolved or absorbed for $\dfrac{m}{\text{Mol. wt.}}$ moles = Q cals.

$$\left(\text{No. of moles} = \frac{\text{wt. in gm}}{\text{Mol. wt.}}\right)$$

Hence, evolved or absorbed for one mole of electrolyte

$$= \frac{Q \times \text{Mol. wt. cal.}}{m}$$

VIVA-VOCE

Q 1. Define the heat of the solution (dissolution).

Ans. The heat of solution is the change in enthalpy when 1.0 gm mole of a solid is dissolved in excess of water.

Q 2. Why is the heat of solution determined in a dilute dilution?

Ans. It is because heat changes involve change with dilution.

Q 3. When is the heat absorbed during the process of dissolution?

Ans. During the process of dissolution, the heat is absorbed when crystal lattice is broken (lattice energy).

Q 4. Which type of electrolytes dissolves with evolution of heat?

Ans. Those which form hydrates.

Q 5. What is meant by integral heat of dissolution?

Ans. It is the change in heat content when 1 mole of a substance is dissolves in a specified quantity of the solvent.

Q 6. If dilution of solution cannot be infinite, what is done to prescribe the quantity of solvent?

Ans. It is fixed as 1 mole of the solvent.

Q 7. Give an example in which heat of solution is exothermic.

Ans. H_2SO_4 or NaOH.

Index

A

Abbe Nollet, 115
Abbe refractometer, 502, 504
absolute reaction rate theory, 265
absolute scale, 3, 36, 126–127, 148, 154
absolute value, 130, 136, 387
absolute zero, 3, 16, 42, 127, 145
absorbance, 196–198, 277, 282
absorption coefficient, 196–197, 199–200, 216
absortivity, 196–197
acetaldehyde, 55, 94, 228, 256
acetic acid, 115, 160, 240, 302, 305, 307, 309, 311, 321, 323, 329, 337, 365, 380, 514, 526
acetone, 47, 50, 57, 67, 92, 303
acid base catalysts, 182
activated complex, 185–186, 231, 257, 265, 268
adhesion forces, 51
adiabatic process, 123, 125, 133
adsorbate, 160, 162, 173
adsorption, 160–162
    theory of, 168
    types of, 161
        langmuir adsorption, 161, 165, 167, 188–189
        van der Waals' adsorption, 161
adsorption isotherm, 159, 163–165, 167, 169, 174, 188, 190
aggregation, 2
    states of, 2
        solid, 2, 4, 7, 489, 490, 492, 575, 585
        liquid, 2, 4, 7, 30, 32, 36, 44, 46, 89
        gas, 44, 86, 109, 116, 122, 124, 127, 138, 140
Albert Einstein, 301
alkaline solutions, 256
ammonia, 76, 160, 179, 188, 194, 231, 313, 574
ammonia molecule, 76
ammonium chloride, 473, 493
ammonium thiocyanate, 372
amorphous solid, 412, 418, 433, 438
analyzer, 506
angle of rotation, 70, 73, 506, 510, 512
anions, 183, 303, 305, 350, 365, 368, 371, 380, 392, 527
anisotropic, 412, 414, 506
anode, 303, 350, 352, 367, 371, 374, 377, 380, 387, 397
anode mud, 353
Anon, 43
anti-bonding molecular orbital, 445, 447–448, 451, 456, 465
antiseptic action, 51
appreciable dissociation, 489
argentiferous, 488, 493
argentoferous lead, 342
arrhenius, 60, 259–260, 265302, 306, 375
arrhenius equation, 259–260, 265
Arrhenius theory, 302, 306, 375, 302, 305, 324, 349, 375, 527
    limitation of, 306
asymmetric carbon atom, 71–73, 506, 510
atmospheric pressure, 2, 7, 36, 42, 44, 81, 99, 136, 267, 388, 484, 486, 496, 568, 571
atomic orbital, 273, 276
atomic orbital, 291–292, 294, 296, 298, 442, 444, 447, 450, 453, 464, 465
atomic parachors, 53–54
attraction correction, 22
Avogadro's hypothesis, 11
Avogadro's number, 178
axial overlapping, 450
azeotropic mixture, 91–92, 94

B

Babo's law, 112
basicity, 88, 315
bauxite, 353
behaviour of real gas, 24, 36, 127
benzene, 46, 51, 57, 64, 69, 91, 103, 112, 332, 334, 339, 344, 346, 499, 543, 546, 567, 576
binary alloy, 488–489
bivariant, 474, 481
blackbody radiations, 277, 279
bond moments, 78–79
Boyle point, 19
Boyle's temperature, 19–20, 37, 39, 496
Bragg's equation, 430, 432, 439
    application of, 432
Bragg's method of crystal analysis, 430
bravais lattice, 411, 428, 432, 437
bronsted acid, 183, 309, 311, 313, 324, 326
bronsted-lowry, 301, 308, 311, 324
    limitation of, 311
buffer action, 322–323, 557
buffer solution method, 555
buffer solutions, 321–323, 555

applications of, 323
characterestics of, 321
types of, 321
bulk liquid, 46, 52
bulk phenomenon, 162

C

cadmium iodide, 370
camphor, 562–563
capillary action, 52, 573, 577, 579
capillary tube, 47, 50, 56, 106, 211, 539, 548, 563, 578
carbon tetrachloride, 51, 58, 238, 251, 267, 327, 330, 339, 345, 471, 540, 546, 567, 576
carbon dioxide, 19, 26, 28, 144, 307, 471
carbon disulphide, 51, 60, 86, 94, 304, 328, 338, 471
catalyatic promoter, 182
Catalyst, 156, 159, 179, 182, 187, 191, 219, 226, 242, 258, 270, 312
positive catalyses, 179
negative catalyses, 179
cathode, 212, 350, 353, 367, 369, 371, 377, 388, 397, 408,
cations, 302, 365, 368, 378, 381
cell constant, 360, 379, 382, 520, 522, 532
cell potential, 383, 389, 396, 402, 404, 409
Centigrade scale, 7, 127
centipoise, 61, 550
characteristics of bonding molecular orbitals, 448
characteristics of catalytic reactions, 181–182
characteristics of liquids, 44–45
Charcoal, 160, 162, 168, 190, 367, 575
Charles law, 3, 17
chemical thermodynamics, 119–120
chemical constitution, 52, 62, 66, 81
chemical kinetics, 222, 261
chemical potential, 169, 172, 329
chemisorptions, 161, 163, 165, 189
chloroform, 51, 92, 94, 179, 339, 567, 574
chromatogram, 573–574, 576, 578
chromatographic technique, 577
chromic acid, 549
classical mechanics, 274–275
classification of Photochemical Reactions, 205
classification of physical properties, 43, 46
additive property, 46
colligative property, 46
constitutive property, 46
Clausius inequality, 146
co-efficient of attraction, 22
coefficient of viscosity, 58–59, 62, 80, 550, 58
cohesion, 51
colligative properties, 85, 95, 112, 114, 116, 376
condensed systems, 484
conductivity, 350, 355, 357, 360, 362, 364, 366, 378, 415
conductivity cells, 359
conductometry, 528
configurations of molecules, 452, 454
congruent melting point, 469, 485, 488–490, 494
conjugate acid, 309–311, 313, 325
consecutive reactions, 256–258, 269
contact catalysis, 179, 187
co-ordination number, 418, 439
correspondence principle, 283
co-volume, 20
critical angle, 65, 504
critical angle phenomenon, 65
critical pressure, 31, 34, 42
critical temperature, 30–32, 36, 40, 48, 479
Cryoscopy, 103
crystal, 107, 274, 411
crystal lattice, 188, 416, 435, 437
crystalline solid, 413, 415, 418, 433, 438
crystallographic designations, 411, 423
Crystallography, 412, 416, 424, 426, 437
laws of, 424
cyclic process, 132, 134, 142, 144
cyclohexane, 92, 576

D

Dalton's law of partial pressure, 6, 11, 37, 89
de Broglie, 273, 275, 282, 288, 291, 298
Debye–Huckel–Onsagar, 376–377
Debye–Huckel–Onsagar equation, 377
decinormal calomel electrode, 391
definite composition, 93, 487, 492
degree of hydrolysis, 343
degree of ionization, 303, 313, 324, 375, 526, 583
democratic properties, 95
denominator, 96–97, 244, 249, 282, 342, 370
derivation of gas laws, 1, 9
detergent, 47, 51, 539
deviations from boyle's law, 18–19
compressibility factor, 18
effect of pressure, 18
effect of temperature, 19
dextrorotatory, 70, 72, 239
diamagnetic, 454, 457, 459, 461, 464
diatomic molecule, 76, 80, 203, 452, 454, 466
difference between atomic and molecular orbitals, 445
diffusion, 2, 6, 11, 17, 21, 37, 44, 105, 569
diffusion in solutions, 105
dilute dilution, 586
dipole moment, 43, 75–79, 81, 418
application of, 76
unit of, 75
dissociation, 88, 113, 115, 203, 205, 266, 303, 334

distilled water, 49, 100, 245, 327, 509, 513, 521, 536, 540, 543, 548, 558, 574, 578, 581, 585
distribution coefficient, 328, 331, 340, 342, 344, 346, 542
Dunstan's rule, 62
dynamic equilibrium, 45, 122, 140, 165, 267, 302, 305, 375, 471,
dynamo, 383

E

ebulisscopic constant, 100
ebullioscopy, 100
effective volume, 20, 23, 36
Eigen functions, 296
electric battery, 351
electric or magnetic field, 475
electrical conductance, 356–357, 359, 528
electrochemical cell, 350, 384, 389, 394, 396, 407, 409
electrochemical change, 349, 384
electrochemical equivalent, 351, 353, 378
Electrochemistry, 349–350, 357, 384
electrode potential, 384–385, 387–389, 396, 399, 401, 406
electrodes, 306, 350, 358, 360, 364, 367, 372, 374, 375, 379, 382, 388, 393, 407, 520, 522, 530
electrolysis, 131, 303, 324, 350–353, 360, 367, 372, 374, 378, 381, 384, 395, 408,
  application of, 352
electrolytes, 113, 115, 173, 175, 301–302, 306, 349–350, 353, 355, 358, 364, 366, 375–376
  strong electrolytes, 376, 381, 527
  weak electrolytes, 301–303, 305, 355, 364, 366, 375, 527
electrolytic cell, 350, 378, 384, 406, 408
electrolytic conduction, 350, 523
electrolytic solution pressure, 386
electromagnetic radiation, 278, 298
electromotive chemistry, 350
electronegative atom, 75, 77
elevation in boiling point, 46, 85, 99–100, 108, 112
emitter, 277, 282
emulsion, 51
enantiomers, 73–74
enantiomorphs, 72
endothermic, 122, 136, 144, 155, 190, 194, 208, 257, 266, 268, 271, 317, 408, 560, 582
endothermic reactions, 144, 194, 257, 268, 561
enthalpy, 190, 194, 266, 271, 560
enthalpy function, 136
Entropy, 120
  units of, 147
  physical significance of, 147
enzyme catalysis, 183, 189, 192, 270
eosin, 211, 219
Eotvas, 48
equilibrium diagrams, 476
equivalent conductance, 305, 358, 361, 363–364, 366, 376, 379, 523, 527, 532
Erwin Schrödinge, 273
ethanol, 92, 182, 184, 574, 576,
ethylene, 54, 66, 112
eutectic point, 482, 485, 487, 490, 492, 494
  characteristics of, 482
  use of, 486
eutectic system, 469, 485–486
eutectic mixture, 485–486
eutectic temperature, 485, 487, 492
excitance, 279
excluded volume, 20, 37
experimental liquid, 49–50, 65, 505, 549, 566, 570

F

factors affecting electrolytic conduction, 355
Fahrenheit scale, 7
Faraday's first law of electrolysis, 351
Faraday's second law of electrolysis, 351, 367
feasible reaction, 143
Felex Franks, 43
fluorescein, 211
fluorescence, 193, 201, 211–212
fluorine molecule, 453, 462
fractional distillation, 93–94, 116
freezing point, 116, 118, 126, 303, 478, 480, 485, 487, 490, 492, 495, 562, 564
Freundlich, 164–166, 168, 188, 190
Freundlich adsorption isotherm, 164, 168, 188
fusion, 116, 136, 147–148

G

Galena, 416, 488
gas laws, 2
  avogadro's law, 4
  boyle's law, 2
  charles law, 3
gaseous phase, 44, 231, 471, 473, 486
gaseous solution, 85–86
gaseous state, 2, 30, 44, 145, 189, 238, 473
Gay-Lussac, 3, 40
Gibbs' adsorption equation, 159, 169, 173–174, 176
glass electrode, 390–393, 406
  advantages of, 392
  limitations of, 393
Glycerol, 61, 212
Goldschmidt, 423
Graham's law of diffusion, 6, 11, 37
gravitation, 21, 128
Grazyna Stochel, 193
Grotthus–Drapper, 201

H

heat content, 119–120, 136, 138, 582, 586

heat of adsorption, 161, 163, 177, 190
Heisenberg, 273, 275, 282, 290, 298
helium gas, 41
helium molecule ion, 457
heterogeneous catalysis, 159, 179, 181, 185–186, 188, 192
 mechanism of, 184
heterogeneous system, 469–470, 474, 496
hexatriene, 67, 69
high pressure, 2, 17, 20, 22, 25, 27, 36, 38, 40, 124, 143, 165, 168, 188, 190, 231
high quantum yield, 206
Hittorf, 367, 369, 371–372, 380
Hittorf's number, 369
Hittorff's method, 371
Hofmann's bottle, 566, 570
homogeneous catalysis, 179, 184, 188
 mechanism of, 184
homologous series, 53, 62, 67
homonuclear molecules, 454
horizontal platinum ring, 50
Hund's rule, 443, 453, 460, 462
hydration energy, 584
hydrocarbons, 53, 59, 86, 173
hydrochloric acid, 75, 118, 194, 207, 240, 305, 310, 365, 370, 391, 473, 512, 514, 528, 581
hydrogen bonding, 27, 61, 92, 112, 551
hydrogen chloride, 231, 350,
hydrogen electrode, 387–388, 392
hydrogen gas, 19–20, 131, 387, 398, 454
hydrogen molecular ion, 454–456
hydronium ion concentration, 315
hypertonic solution, 108, 115
hypotenuse surface, 502
hypothetical, 2, 17, 147, 325, 452, 458, 462
hypothetical helium molecule, 458
hypothetical Neon Molecule, 462
hypotonic solution, 108

I

ideal gas, 2, 5, 12, 15, 17–18, 20, 25, 27, 36, 38, 40, 42, 118, 122, 135, 140, 150
ideal gas equation, 5, 12, 20, 27, 36
ideal gas temperature scal, 127
ideal or perfect gases, 17
 characteristics of, 17
ideal pressure, 22, 24, 38
ideal solution, 85, 90–91, 93, 112, 116
ideal-crystal, 417
immiscible solvents, 328, 336, 344, 540, 545
improper rotation, 421
incongruent melting point, 469, 485, 490
indices, 64, 66, 412, 423–424, 426, 437
inherent energy, 130
inhibitors, 188
instantaneous rate, 223, 226
instantaneous reactions, 222
insulator, 354
intermolecular attraction, 22, 39, 48, 51, 59, 92,
intermolecular distances, 44
intermolecular forces, 27, 44, 52, 61, 82, 90, 92, 130, 411, 539
internal energy, 119–120, 125, 130, 132, 134, 136, 145, 157, 561
intrinsic energy, 130, 134
inversion temperature, 38, 40, 139
ionic equilibrium, 302
ionic product of water, 313–314, 316, 324
ions, 302, 305, 309, 314
irreversible process, 123–124, 142, 149, 155, 161
irrigent, 577
isobaric process, 123, 125, 151
isochoric process, 123, 125, 133, 151
isomeric compounds, 67
isothermal change, 132
isotherms, 28–29, 34, 163, 167, 189, 191
isotonic solutions, 107–108, 116, 118
isotropic, 418, 438, 506

J

J. Willard Gibbs, 469
J.D. van der Waals, 23
James Joule, 129
Joule–Thomson effect, 38, 119, 138, 140

K

Kekule structure, 57
Kelvin scale, 3, 377
Ketones, 67
kinetic theory of gases, 2, 6, 8, 12, 37, 39
Kirchoff's law, 277–278
Kohlrausch, 350, 364–366, 380, 522, 525, 527
Kohlrausch law, 364
 application of, 365
Konowaloff's law, 112

L

Laevorotatory, 70, 72, 239
laevo-rotatory, 506, 512
Lambert's law, 195–197, 199, 217
Lambert–Beer's law, 197
 limitations of, 198
Langmuir, 161, 164–165, 168, 176, 188, 191
Langmuir adsorption isotherm, 165, 167, 188
Langmuir's theory, 167, 176, 188
lateral overlapping, 450
lattice, 188, 416–417, 428, 430, 432, 434, 438, 584
lattice energy, 584
law of corresponding sttes, 34–35
law of symmetry, 425–426
laws of photochemistry, 193, 201
lead–silver system, 486, 493
Lewis acid, 311, 313, 324, 326

Lewis concept, 301, 311
limitation of, 312
liquefaction commences, 29
liquid amonia bottle, 39
liquid paraffin, 562
liquid solution, 64, 86, 116, 490
liquid state, 2, 30, 38, 43, 412, 418, 488, 492
liquification of gas, 36
Lorentz, 66
low quantum yield, 206, 217
Lummer, 279

M

mass fraction, 89
mass percentage, 87
Maxwell's electromagnetic wave theory, 274, 282
measurement of cell constant, 520
measurement of refractive index, 65–66, 502, 504
measurement of surface tension, 48, 50
stalagmometer or drop pipette method, 48
capillary rise method, 50
Du Nouy tensiometer method, 50
mechanism of heterogeneous catalysis, 185
meniscus, 47, 50, 80, 536, 549
Mesoform, 73
method of electrochemical reactions, 384
direct redox reaction, 384
indirect redox reaction, 384
migration of Ions, 349, 367, 380
Miller, 412, 423–425, 437
millipoise, 60
mobile phase, 575, 577, 579
modern theory of strong electrolytes, 350, 376
molal elevation constant, 100, 102, 115
molality, 88, 98, 100, 104, 116, 118
molar absorption coefficient, 197–198, 200
molar heat, 46, 137, 147, 152, 162, 479
molar heat of vaporization, 46, 148, 153
molar volume, 39, 52, 56, 62, 551, 570
molarity, 87–88, 116, 118, 270, 359, 379
mole fraction, 66, 88, 90, 95, 98, 116, 330
molecular forces, 2, 27, 30, 36, 44, 61, 82, 90, 92, 130, 411, 539
molecular formula, 53, 72, 80, 331, 335
molecular orbital, 443, 445, 447, 450, 454, 458, 462–463
molecular orbital theory, 443, 460, 466
molecular Velocities, 12
molecules, 443, 445, 453–454
momentum, 8, 276, 283, 287, 297
monocentric, 443, 464
monochromatic light, 69
Mulliken, 443

N

Naumam, 423
negative catalyst, 179, 181
negative or under perfect deviation, 22, 25
Nernst equation, 383, 399–401, 404, 406
applications of, 400
Nernst's distribution law, 328, 344, 346
Newtonian mechanics, 274, 297
Nicol prism, 69, 506
nitrogen gas, 19, 237
nitrogen molecule, 14, 454, 459, 466
non-aqueous solution, 86, 115
non-conductor, 354
non-electrolyte, 101, 107, 117, 173, 355
non-polar covalent bond, 74–75
non-radiative transitions, 215
non-variant systems, 478
non-volatile solute, 90, 96, 99, 102, 112, 116
normal calomel electrode, 391
normal temperature and pressure, 5, 570
numerator, 244, 249, 342, 370

O

occlusion, 162, 190
operand, 285
operator, 285
types of, 286
optical activity, 43, 46, 70–72, 80, 506, 510
optical exaltation, 67
optical isomers, 72
optically active substances, 73, 506, 510
orbital wave functions, 446
osmotic pressure, 105–107, 112, 114, 116, 118, 303
Ostwald viscometer, 64
Ostwald's dilution law, 304–305
oxidation, 351, 378, 381, 384, 386, 388
oxidation potential, 386–387, 397, 405
oxygen atoms, 435, 460
oxygen molecule ion, 461

P

parachor, 52
atomic, 53
structural, 54
paraldehyde structure, 55
paramagnetic, 450, 454, 456, 460, 463
Parke's process, 342, 347
partition coefficient, 328, 336, 338, 340, 342, 344, 540, 543, 545
partition law, 328, 346,
parts per million, 89
peritectic or incongruent melting temperature, 491

pH scale, 301, 315–316
phase diagrams, 469, 476, 489,
phase equilibrium, 45, 86, 483, 493
phase rule, 120, 469–470
phases of sulphur system, 472
    liquid, 472
    monoclinic, 472
    rhombic, 472
    vapour, 472
phosphorescence, 193, 201, 211–212, 215
photo polymerization, 194
photochemistry, 194, 201
photohalogenation, 194
pinch cock, 49, 353
Planck's constant, 40, 207, 276, 281, 283, 289, 299
Planck's distribution law, 281
Planck's radiation law, 273, 281–282
plasmolysis, 115
poiseuille equation, 59
polar molecules, 74, 76, 81, 303, 435
polarimeter, 239, 506, 508, 512
polarity of bonds, 43, 74,
    types of, 74
        non-polar covalent, 74
        polar covalent, 74
polarized light, 69, 72, 506, 508, 510, 512
polarizer, 69, 506
polycentric, 443, 464
polyvinyl alcohol, 59
positive or over perfect deviation, 25
potassium iodide, 338, 485, 541
pressure axis, 17, 479–480, 482
primary photochemical process, 203–204
principle of continuity of state, 30
principle of corresponding states, 35
Pringsheim, 279
probability factor, 264, 268
promoters, 188
propionic acid, 63
pseudounimolecular reaction, 230
Pulfrich refractometer, 502
pyknometer, 60, 535, 538, 548, 550

Q

quantization of energy, 281
quantum efficiency, 193, 205, 207, 211, 216, 218
quantum mechanics, 213, 274–275
quinine, 54–55

R

racemic mixtures, 72
radial probability, 273, 292–295, 297
radiation transitions, 216
radiationless, 214
Ramsay and Shields, 48,
Raoult's Law, 86, 89–90, 93, 96, 98, 112, 115, 117, 174, 346,
rast method, 103, 562, 564
rate constant, 225–226, 228, 232, 235
rate equation, 221, 225, 516, 518
Rayleigh–Jeans law, 282, 296
real gas, 17–18, 20, 23, 26, 36, 38, 40, 138
redox reactions, 383
reduced phase rule equation, 484, 486, 494
reduction, 388–389, 392, 394, 396
reduction electrode potential, 388, 393, 401
reduction potential, 386, 388
reference electrodes, 389–390, 406
refractive index, 122, 124, 198, 414, 501
    measurement of, 502
resistivity, 357
resolution, 73
reverse osmosis, 115
    exo osmosis, 115
    endo osmosis, 115
reversible process, 123–124, 132, 134, 140, 142, 148–149, 155, 161, 304
    characteristics of rheochor, 551
rhombic sulphur, 427, 481–483
Robert Boyle, 2
root mean square, 8–9, 13, 16, 40, 42
rubber tube, 49, 535, 548

S

safety fuses, 486
saturated calomel electrode, 391
Schrödinger wave equation, 274, 288–289, 291, 296, 299
secondary reactions, 204, 206, 218
second-order reaction, 228, 242–243
    characteristics of, 243
second-order reflection, 431
Semipermeable membrane, 105–107, 115
simply as onsagar equation, 377
singlet and triplet states, 213
singlet ground state, 213
Snelle's law, 69
sodium – potassium system, 490
sodium acetate, 333, 365, 533
sodium chloride, 116, 301, 353, 365, 370, 426, 474, 495
solders, 486
solid angle, 415–416, 420
solid Solution, 85–86, 471, 485
solid state, 44, 102, 376, 411, 417, 419, 421, 423, 427, 429, 434
solubility curve, 558
solute, 66, 87, 90, 96, 331
solution, 86
    introduction, 86
solvent, 418, 507, 510, 540, 543, 545, 560
space lattice, 416–417
specific conductance, 357–359, 361, 364, 379
specific reaction rate, 225, 228
specific resistance, 357
specific rotation, 70, 73, 80, 507, 510
spherical molecules, 58
spin multiplicity, 213
spontaneous processes, 124,143, 149, 152
stability of molecules, 453

stalagmometer, 48–49, 80, 535, 537
standard temperature and pressure, 5
standard electrode potential, 387–388, 401, 406
    measurement of, 388
standing wave, 288, 290
stannous chloride, 250
Stefan–Boltzmann, 273, 279, 281
steric factor, 264–265
successive members, 53
succinic acid, 329, 331, 333, 347
Sugden, 56
sulphur system, 469, 472, 481–482, 493, 495
super-cooled liquids, 413
surface active agents, 47, 539
surface energy, 46, 48, 81, 83, 171, 538
surface tension, 46, 48, 51, 535, 537, 539
    definition of, 46
symmetrical structure, 76

T

temperature coefficient, 258, 260, 329, 345
tetrahedral chiral centres, 73
theories of reaction rates, 221, 264
theory of electrolytic dissociation, 303
thermal or dark reactions, 198
thermochemical reactions, 194
thermodynamic equilibrium, 122, 140, 267, 268
thermodynamic process, 122, 154
thermodynamic properties, 124
    extensive property, 124
    intensive properties, 124
thermodynamic system, 119, 121
    objectives of, 120
    limitations of, 120
thermodynamics, 119–120, 122, 125, 130
    first law of, 131
    second law of, 142–144
thermometer, 126, 503, 508, 512, 558, 562, 580, 584
thin-layer chromatography, 573, 579
third-order reaction, 221, 228, 247, 249, 254, 269
    characteristics of, 248
Thomas Hobbes, 1
transference number, 369, 379
transmittance, 196, 198, 200
transpiration, 51
transport number, 350, 369–370, 372, 375, 379
triplet excited state, 213, 215
types of crystalline solids, 433
    covalent solid, 435
    Ionic solid, 433
    metallic solid, 434
    molecular solid, 435
types of unit cells, 411, 417, 428

U

unity, 56, 62
univariant systems, 477

V

Valence bond theory, 442, 464, 467
van der Waals, 25, 27, 31, 33, 35, 37, 161, 168, 185, 435, 439
Van der Waals' equation, 20, 24, 33, 36
    attraction correction, 22–23
    volume correction, 20–21
Vant Hoff, 72, 107, 109, 113, 115, 117
Vant' Hoff and Le bel, 72
velocity, 7, 13, 16, 39, 41, 58, 64, 69, 201, 207, 226, 237, 256, 258, 269, 367
velocity constant, 225, 237, 256, 258, 269
vessel, 2, 6, 11, 20, 22, 28, 44, 106, 121, 135, 271, 395, 398, 473, 493, 495, 532, 541
vice versa, 30, 57, 133, 357, 477, 479, 519, 583
Victor Meyer's method, 566, 570
viscosity, 57–61
    coefficient of, 58
    effect of temperature on, 60
    factors affecting, 61
    measurement of, 59
viscous flow, 60
volume correction, 20, 23, 26
volume percentage, 87

W

W. Ostwald, 85
W.C. McCLewis, 221
W.H. Bragg, 430
wave functions, 273, 290, 296, 445, 447
wave mechanics, 275, 282, 287, 442
Weiss, 412, 423
Weiss indices, 423, 425
Wheatstone bridge method, 359
WIien's displacement law, 279, 281

Z

zero-order reaction, 227, 229, 231–232, 234, 247, 254, 269, 271, 518
    characteristics of, 232
zeroth law of thermodynamics, 119, 126
zone axis, 416